Innovations in Sustainable Agriculture

Muhammad Farooq • Michele Pisante
Editors

Innovations in Sustainable Agriculture

Springer

Editors
Muhammad Farooq
Department of Crop Sciences,
College of Agricultural and
Marine Sciences
Sultan Qaboos University
Al-Khoud, Oman

Michele Pisante
Agronomy and Crop Sciences Research
and Education Center
University of Teramo
Teramo, Italy

ISBN 978-3-030-23168-2 ISBN 978-3-030-23169-9 (eBook)
https://doi.org/10.1007/978-3-030-23169-9

This Springer imprint is published by the registered company Springer Nature Switzerland AG.
The registered company address is: Gewerbestrasse 11, 6330 Cham, Switzerland

Foreword

The term "sustainable agriculture" implies multidimensional approach to managing agroecosystems encompassing (i) environmental, (ii) economic, and (iii) social components of sustainability. This book edited by Prof. Muhammad Farooq and Prof. Michele Pisante is a comprehensive treaty on this timely theme of global significance. In addition, this book also addresses several related and pertinent issues such as resource-use efficiency, water and nutrient management, pest management, genetic resources and biodiversity, and climate change. It is indeed a comprehensive treatise of the subject, based on synthesis of the existing knowledge by authors who are extremely about this important topic.

The "sustainability" issue has been a popular theme ever since publication of the 1987 report by the World Commission on Environment and Development chaired by Gro Harlem Brundtland (former Prime Minister of Norway): *Our Common Future*. This report was also the basis of the Agenda 21 and of the "Rio Declaration on Environment and Development." The thematic focus has attracted attention of the scientists, policy makers, civic societies, and general public toward an attempt to addressing the emerging global issues. It is in this context that the theme of "Sustainable Agriculture" of the present volume is timely and crucial. It is also pertinent to addressing the Agenda 2030 or the "Sustainable Development Goals" of the UN adopted in 2015.

Major questions related to agricultural sustainability have been raised ever since the dawn of settled agriculture about ten millennia ago when the world population was merely 10–20 million and even more strongly in 1798 by Thomas Malthus who wrote the "Principle of Population: As it Affects The Future Improvement of Society." Whereas the questions (i.e., resource availability, use efficiency, food and nutritional security, environment quality) have more or less remained the same over millennia, the answers and strategies to address these questions have changed with every generation depending on the population size, lifestyle, and the technical knowledge available at the specific time.

The human population of 7.7 billion in 2019 is destined to reach 9.8 billion by 2050 and 11.2 billion by 2100. Furthermore, the growing affluence of the expanding middle class in emerging economics is increasing demands on the finite and nonrenewable resources that are also prone to degradation and pollution because of misuse and mismanagement. Therefore, the need to meet the food and other demands of the growing and richer population must be reconciled with the absolute necessity of improving and restoring the environment. These two must go hand in hand.

Therefore, the concept of multidimensional sustainability deliberated in this book (i.e., environmental, economic, and social dimensions) must also encompass the focus on "institutional sustainability.". It is the weak, poor, and unsustainable institutions throughout the developing world that have led to the widespread problems of soil degradation and desertification, water pollution and scarcity, air pollution and gaseous emissions, malnourishment and undernourishment, extinction of biodiversity and weakening of ecosystem services, etc. Strong institutional sustainability is also essential to translating "science into action." A related but somewhat different issue is that of "political will," which is also essential to implementation of pertinent programs which protect, restore, and use the finite resources so that demands of the present society are met without jeopardizing those of the future generations.

The Ohio University,
Columbus, OH, USA

Rattan Lal

Preface

Over the millennia, agriculture has evolved over time between the growing demand for food and the progressive decline of natural resources to meet the needs of a society that do not stop, increase rapidly and with new lifestyles, as never first registered. And new challenges are facing the transition from agriculture to production, from migration from rural to urban areas, and from the need to recognize the value and conservation of various aspects such as economic, social, and productive efficiency and soil quality, water quality, greenhouse gas emissions, and biodiversity.

There is no future without agriculture, there cannot be agriculture without innovation, and there can be no innovation without the knowledge that can improve our lives, thanks to the ever-present alliance between plants and people who, through agriculture, use them. This demands pragmetic solutions to manage the agroecosystems rationally and to guarantee, with the current limits of the biosphere, sufficient food for the world population that will exceed 9 billion people by 2050.

Based on these considerations in this book, we have brought together researchers specializing in different disciplines and working in different regions of the world, united by the rigor method of scientific approach to tackle together many of the current and emerging aspects of the sustainability of agricultural production. For an ordered reading and for the accurate thematic analysis, the book is divided into 7 sections and 20 chapters as detailed below.

Part I Introduction

- Chapter "Sustainable Agriculture and Food Security" describes the basic and evolved concepts of sustainable agriculture and food security.

Part II Ecological Sustainability

- Chapter "Integrating Conservation into Agriculture" elaborates and integrates conservation with agriculture for sustainable agriculture.
- Chapter "Microbial Applications for Sustainable Agriculture" describes the new scientific evidence of microbial applications for sustainable agriculture.

- Chapter "Innovation System Approach for Urban Agriculture: Case Study of Mexico City" discusses the innovation system approach for urban agriculture with focus on Mexico City.

Part III Resources Use Efficiency for Sustainable Agriculture

- Chapter "Sustainable Soil Management" illustrates the agronomic principles and practices of sustainable soil management.
- Chapter "Sustainable Water Management" aims to answer five questions on sustainable water management.
- Chapter "Sustainable Nutrient Management" covers the principles and applications of sustainable nutrient management.
- Chapter "Alternative Fertilizers and Sustainable Agriculture" analyses the experiences and issues and proposes options for the alternative fertilizers and sustainable agriculture.

Part IV Sustainable Pest Management

- Chapter "Sustainable Weed Management" describes sustainable weed management between climate change and agronomic and environmental issues.
- Chapter "Sustainable Management of Insect-Pests" discusses sustainable management of insect pests.
- Chapter "Sustainable Management of Plant Diseases" analyzes management strategies as an important contribution to the sustainable management of pathogens and diseases.

Part V Genetic Resources and Crop Improvement for Sustainable Agriculture

- Chapter "Conservation of Biodiversity and Genetic Resources for Sustainable Agriculture" covers the challenges on conservation of biodiversity and genetic resources.
- Chapter "New Breeding Techniques for Sustainable Agriculture" describes the innovative breeding techniques for sustainable agriculture.

Part VI Agricultural Sustainability in Changing Climate

- Chapter "Sustainable Agriculture and Climate Change" illustrates the climate change challenges for the sustainability of agriculture.
- Chapter "Carbon Sequestration for Sustainable Agriculture" describes the experiences on carbon sequestration and sustainable agriculture.
- Chapter "Use of Biochar in Sustainable Agriculture" covers the use of biochar for sustainable agriculture.
- Chapter "Managing Drylands for Sustainable Agriculture" highlights strategies for the management of dry lands for sustainable agriculture.
- Chapter "Crop-Livestock Interaction for Sustainable Agriculture" discusses for crop-livestock interaction for sustainable agriculture.

Part VII Use of IT Tools and Modeling for Sustainable Agriculture

- Chapter "Information Technology for Sustainable Agriculture" provides an overview of potential applications of information technology tools in sustainable crop production systems.
- Chapter "Spatializing Crop Models for Sustainable Agriculture" introduces the application and case studies of spatializing crop models for sustainable agriculture.

Our heartfelt thanks is addressed to all the authors who with infinite generosity of their time, proposals, and active participation have contributed to the realization of this book, the result of a harmonious teamwork. A special thanks to the reviewers, for the importance of the contribution offered from the scientific point of view, in the validation of concepts, objectives, interpretation of results, and evidence on the different topics covered and finally, detail not negligible, for having read and reread the different drafts of the manuscript. Of course, if there were still errors, they would be only ours.

We are grateful to Professor Rattan Lal for the clarity of the foreword, full of experience and illuminating wisdom, and for his trust. We also thank Ms Melanie van Overbeek, Assistant Editor, Agronomy, Springer Dordrecht, the Netherlands, for her patience and trust in us during this book project.

We thank the Sultan Qaboos University, Muscat, Oman, and the University of Teramo, Italy, for their continued support.

Muscat, Oman
Teramo, Italy

Muhammad Farooq
Michele Pisante

Contents

Contributors

Tanveer Abbas Institute of Soil and Environmental Sciences, University of Agriculture, Faisalabad, Pakistan

Aftab Afzal Department of Botany, Hazara University, Mansehra, Pakistan

Tahseen Afzal Institute of Soil and Environmental Sciences, University of Agriculture, Faisalabad, Pakistan

J. Álvaro-Fuentes Soil and Water Department, Estación Experimental de Aula Dei (EEAD), Spanish National Research Council (CSIC), Zaragoza, Spain

J. L. Arrúe Soil and Water Department, Estación Experimental de Aula Dei (EEAD), Spanish National Research Council (CSIC), Zaragoza, Spain

Saeed A. Asad Centre for Climate Research and Development, COMSATS University, Islamabad, Pakistan

Muhammad Zahir Aziz Institute of Soil and Environmental Sciences, University of Agriculture, Faisalabad, Pakistan

Paul Kayode Baiyeri Department of Crop Science, University of Nigeria, Nsukka, Nigeria

Sofia Bajocco Research Centre for Agriculture and Environment, Council for Agricultural Research and Economics, Roma, Italy

Ali Ahsan Bajwa School of Agriculture and Food Sciences, The University of Queensland, Gatton, QLD, Australia

The Centre for Crop Science, Queensland Alliance for Agriculture and Food Innovation (QAAFI), The University of Queensland, Gatton, QLD, Australia

Prasanta Chitta Bhowmik Stockbridge School of Agriculture, University of Massachusetts, Amherst, MA, USA

Simone Bregaglio Research Centre for Agriculture and Environment, Council for Agricultural Research and Economics, Bologna, Italy

C. Cantero-Martínez Crop and Forest Sciences Department, EEAD-CSIC Associated Unit, Agrotecnio, University of Lleida (UdL), Lleida, Spain

Giovanni Cappelli Research Centre for Agriculture and Environment, Council for Agricultural Research and Economics, Bologna, Italy

Bhagirath Singh Chauhan The Centre for Crop Science, Queensland Alliance for Agriculture and Food Innovation (QAAFI), The University of Queensland, Gatton, QLD, Australia

Muhammad Jehanzeb Masud Cheema Center for Advanced Studies in Agriculture and Food Security, University of Agriculture, Faisalabad, Pakistan

Faculty of Agricultural Engineering and Technology, University of Agriculture, Faisalabad, Pakistan

Sara D'Egidio Agronomy and Crop Sciences Research and Education Center, University of Teramo, Teramo, Italy

Hans Dieleman College of Sciences and Humanities, Autonomous University of Mexico City, Mexico City, Mexico

Muhammad Farooq Department of Crop Sciences, College of Agricultural and Marine Sciences, Sultan Qaboos University, Al-Khoud, Oman

Department of Agronomy, Faculty of Agriculture, University of Agriculture, Faisalabad, Pakistan

The UWA Institute of Agriculture, The University of Western Australia, Crawley, WA, Australia

Harriet Ndofor Foleng Department of Animal Science, University of Nigeria, Nsukka, Nigeria

Angelica Galieni Research Centre for Vegetable and Ornamental Crops, Council for Agricultural Research and Economics, Monsampolo del Tronto, Italy

Ankit Garg Department of Civil and Environmental Engineering, Shantou University, Shantou, China

Alessandra Gentile Hunan Agricultural University, Horticulture and Landscape College, Changsha, China

Department of Agriculture, Food and Environment, University of Catania, Catania, Italy

Fabrizio Ginaldi Research Centre for Agriculture and Environment, Council for Agricultural Research and Economics, Bologna, Italy

Muhammad Dildar Gogi Integrated Pest Management Laboratory, Department of Entomology, University of Agriculture, Faisalabad, Pakistan

Nirmali Gogoi Department of Environmental Science, Tezpur University, Tezpur, Assam, India

Maria Lodovica Gullino Center of Competence AGROINNOVA, University of Torino, Grugliasco, Italy

DISAFA, University of Torino, Grugliasco, Italy

Ayse Gul Ince Vocational School of Technical Sciences, Akdeniz University, Antalya, Turkey

Muhammad Wajid Javed Integrated Genomics Cellular Developmental and Biotechnology Laboratory, Department of Entomology, University of Agriculture, Faisalabad, Pakistan

Mehmet Karaca Department of Field Crops, Akdeniz University, Antalya, Turkey

Amir Kassam University of Reading, Reading, UK

Rupam Kataki Department of Energy, Tezpur University, Tezpur, Assam, India

Muhammad Azeem Khan Center for Advanced Studies in Agriculture and Food Security, University of Agriculture, Faisalabad, Pakistan

Faculty of Agricultural Engineering and Technology, University of Agriculture, Faisalabad, Pakistan

Muhammad Jamal Khan Faculty of Veterinary and Agricultural Sciences, The University of Melbourne, Dookie, VIC, Australia

Ndubuisi Samuel Machebe Department of Animal Science, University of Nigeria, Nsukka, Nigeria

Stefano La Malfa Department of Agriculture, Food and Environment, University of Catania, Catania, Italy

Marcello Mastrorilli Research Centre for Agriculture and Environment, Council for Agricultural Research and Economics, Bari, Italy

Subham C. Mondal Department of Environmental Science, Tezpur University, Tezpur, Assam, India

Faisal Nadeem Department of Agronomy, Faculty of Agriculture, University of Agriculture, Faisalabad, Pakistan

Muhammad Naveed Institute of Soil and Environmental Sciences, University of Agriculture, Faisalabad, Pakistan

Ahmad Nawaz Integrated Pest Management Laboratory, Department of Entomology, University of Agriculture, Faisalabad, Pakistan

Cynthia Ebere Nwobodo Department of Agricultural Extension, University of Nigeria, Nsukka, Nigeria

Giancarlo Pagnani Agronomy and Crop Sciences Research and Education Center, University of Teramo, Teramo, Italy

Michele Pisante Agronomy and Crop Sciences Research and Education Center, University of Teramo, Teramo, Italy

D. Plaza-Bonilla Soil and Water Department, Estación Experimental de Aula Dei (EEAD), Spanish National Research Council (CSIC), Zaragoza, Spain

Abdul Rehman Department of Crop Sciences and Biotechnology, Dankook University, Cheonan-si, South Korea

Muhammad Sanaullah Institute of Soil and Environmental Sciences, University of Agriculture, Faisalabad, Pakistan

Banashree Sarma Nature Learning Center- Assam, Assam State Biodiversity Board, Guwahati, Assam, India

Tanvir Shahzad Department of Environmental Sciences and Engineering, Government College University Faisalabad, Faisalabad, Pakistan

Sulman Siddique Institute of Soil and Environmental Sciences, University of Agriculture, Faisalabad, Pakistan

Davide Spadaro Center of Competence AGROINNOVA, University of Torino, Grugliasco, Italy

DISAFA, University of Torino, Grugliasco, Italy

Fabio Stagnari Agronomy and Crop Sciences Research and Education Center, University of Teramo, Teramo, Italy

Muhammad Sufyan Biological Control Laboratory, Department of Entomology, University of Agriculture, Faisalabad, Pakistan

Aman Ullah Department of Agronomy, Faculty of Agriculture, University of Agriculture, Faisalabad, Pakistan

D. Villegas Sustainable Field Crops Program, Institute of Agriculture and Food Research and Technology (IRTA), Lleida, Spain

Abdul Wakeel Institute of Soil and Environmental Sciences, University of Agriculture, Faisalabad, Pakistan

Michael Walsh The University of Sydney, Narrabri, NSW, Australia

Liyong Xie College of Agronomy, Shenyang Agricultural University, Shenyang, People's Republic of China

Muhammad Yaseen Institute of Soil and Environmental Sciences, University of Agriculture, Faisalabad, Pakistan

Hongliang Zhao College of Agronomy, Shenyang Agricultural University, Shenyang, People's Republic of China

Raffaella Zucaro Research Centre for Policies and Bioeconomy, Council for Agricultural Research and Economics, Roma, Italy

Part I
Introduction

Sustainable Agriculture and Food Security

Muhammad Farooq, Abdul Rehman, and Michele Pisante

Abstract Demand of increase in food production to feed the rapidly growing global population has pose serious threat to the agricultural sustainability. Climate change also offers serious challenge to global food security situation as it will negatively affect agricultural yields, particularly in low income countries. The increased agricultural intensification to produce more food from the existing cropland has put the environmental sustainability at stake due to increased emissions, loss of biodiversity, soil health due to increase use of chemical fertilizers and pesticides. Therefore, there is need to develop a multidimensional approach for agriculture sustainability without damaging social, economic and environmental integrity. This chapter highlights the role of social, economic and environmental sustainability in agricultural sustainability and food security. It further describes current food security status and proposes how sustainable intensification can help reduce the adverse effect of intensification on social, environment and economical components of agriculture. It also highlights the mitigation strategies to achieve the food security and safety in changing climate.

Keywords Food security · Agriculture sustainability · Crop intensification · Climate mitigation · Malnutrition

---

M. Farooq (✉)
Department of Crop Sciences, College of Agricultural and Marine Sciences, Sultan Qaboos University, Al-Khoud, Oman

Department of Agronomy, Faculty of Agriculture, University of Agriculture, Faisalabad, Pakistan

The UWA Institute of Agriculture, The University of Western Australia, Crawley, WA, Australia
e-mail: farooqcp@squ.edu.om

A. Rehman
Department of Crop Sciences and Biotechnology, Dankook University, Cheonan-si, South Korea

M. Pisante
Agronomy and Crop Sciences Research and Education Center, University of Teramo, Teramo, Italy

M. Farooq, M. Pisante (eds.), *Innovations in Sustainable Agriculture*,
https://doi.org/10.1007/978-3-030-23169-9_1

1 Introduction

The world population according to UN organization estimates will be expected between 8.3 and 109 billion by 2050; however, the existing trend point to the latter figure. This population growth rate will need increase food supply from 50% to 75% depending upon the region (Prosekov and Ivanova 2018). Furthermore, the global climate change will also affect the food production in many parts of the world. Global climate change during the last century lead to heat waves due to rising temperature, atmospheric CO_2 level, frequent spells of drought in some places while higher precipitation on others (OECD 2016). Climate changes and anthropogenic activities have created serious challenges to agriculture sustainability due to natural resource depletion. Furthermore, this situation will lead to lower agricultural yields (Waggoner 1983; Adams et al. 1998; Nelson et al. 2009; Gliessman 2015), threat to the food security and food and feed safety (Miraglia et al. 2009). Food security is physical, social and economic accessibility of healthy, safe and nutritious food to the people at all times, which can fulfill their dietary requirements and food preference for healthy and active life (Fig. 1; FAO 2001). However, to achieve this situation agricultural sustainability is very important as first component of food security is food supply which is linked to agriculture system to meet the food demand which depends on agro-climatic condition of crop and pasture production (Tubiello et al. 2007).

In order to fulfill the increasing food demand, agriculture extension and intensification has helped in increasing food production to achieve the food security (FAO 2009; Tilman et al. 2011; Stevenson et al. 2013). Agriculture intensification without sustainability will result at major environmental costs, as many scientists are expecting stagnant or yield reduction owing to low land expansion and degradation

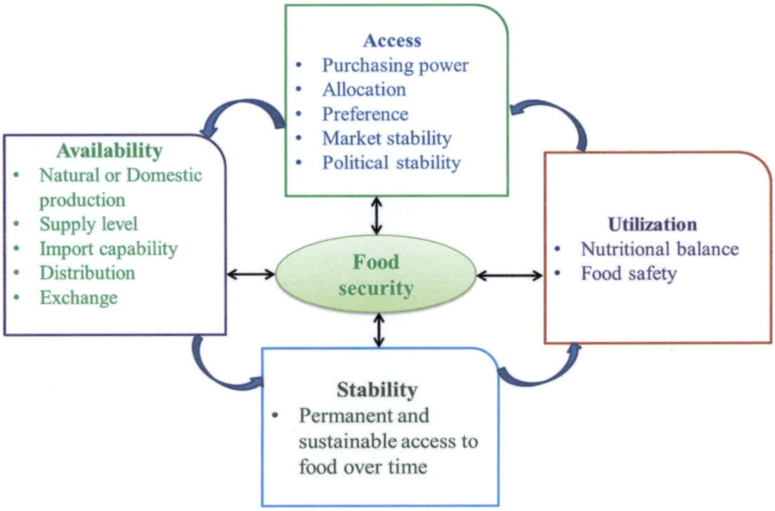

Fig. 1 Components of food security

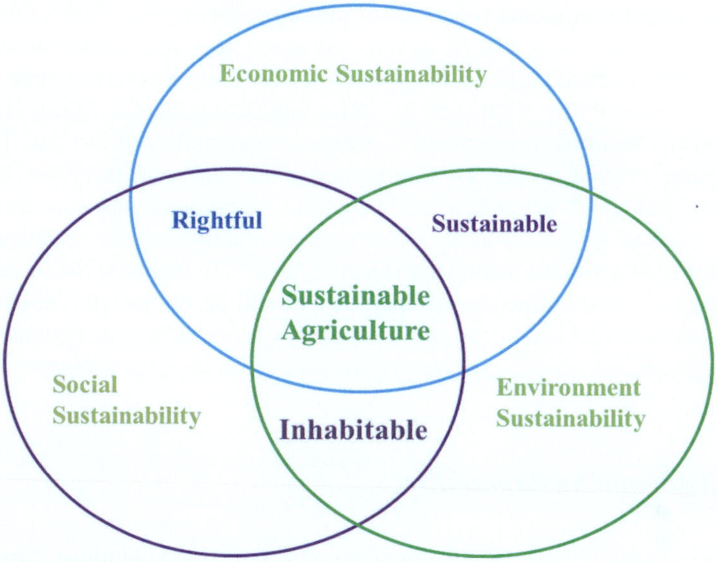

Fig. 2 Components of sustainable agriculture

of land and natural resources due to climate change (FAO 2009; Ray et al. 2012; Stevenson et al. 2013; Eitelberg et al. 2015). Sustainable agriculture holds the key to global food security in view of the oscillating climate, social and market situation. Sustainable agriculture uses integrated animal and plant production system practices that are aimed to improve the farmer income with environmental integrity and social sustainability (Fig. 2). In recent years, the sustainability of agriculture has got ground among academics, policy makers and practitioners due to growing awareness about social and environmental concerns among stakeholders.

This chapter encompasses the work done on agricultural sustainability including social, economic and environmental sustainability, global food security situation and role of sustainable intensification in agriculture sustainability and food security. Furthermore, how climate change mitigation can help in achieving food security will also be discussed.

2 Sustainability in Agriculture

Sustainability is a broader term and pillars of sustainability approach which represent individual indicators linked with a metrics to represent different domains of sustainability (Smith et al. 2017). Sustainability in Agriculture is measured as economic stability, social stability and ecological/environmental sustainability (Smith et al. 2017). In the last several decades, many developing countries have been facing robust population growth, reduced arable land, urban expansion, change in food habits and oscillating global food market pressure (Huang et al. 2009; Qiang et al. 2013; Huang and Yang 2017), which influence the agriculture sustainability (Table 2). The enhanced agriculture production has resulted a conflict between environmental sustainability, food security and safety (Miraglia et al. 2009). The green revolution has resulted in improved economic sustainability of farmers. However, the increased cost of production due to high prices of seed, fertilizer, labor and machinery and yield stagnation has also led to economic unsustainability. The increasing risk of agro-environment from grain production in future can threaten the sustainability of agricultural land use (Qi et al. 2018). To maintain the sustainability of agriculture, a multidimensional approach should be optimized considering the social, environmental and economic sustainability approaches for agricultural production system.

2.1 Economic Sustainability

Economic sustainability is vital component of sustainable agriculture. Green revolution in Asia has substantially enhanced the agriculture production through improved provision and use of inputs (new cultivars, chemical fertilizers and pesticides) (Hazell 2009). The advancement of agriculture technology boosted the grain yield in Asia by 3.57% annually during 1965–1982 (Rosegrant and Hazell 2000). The yield of rice and wheat has increased from 150% to 250% respectively in subcontinent and Indonesia (FAO 2014). This rise in agricultural crop yields lead to many fold increase in per capita income of farmers. For instance, in two provinces of India (Haryana and Punjab) the poverty declined from 35.2% to 8.1% and 28.1–8.4% in 2004/2005. In Pakistan, post green revolution has improved the food security situation as per capita caloric intake has increased from 2462 in 2000 to 1748 in early 1960s (Evenson 2005). Moreover, population density, agriculture credit facilities, market situation, food habits are also key determinants of economic sustainability (Table 2).

The green revolution has resulted in improved economic sustainability of agriculture production system in South Asia. During past few decades, the increase cost of production due to high prices of seed, fertilizer, labor and machinery and yield stagnation has led to economic unsustainability. Recently, Zulfiqar and Thapa (2017) studied the economic, environmental and social sustainability of Agriculture of Pakistan. To study the economic sustainability, they used overall crop production and stability of crop production as indicators, while environmental indicators were crop diversification, soil salinization, fertilizer use (organic and inorganic) and pesticide usage. Moreover, food security and employment of rural labor force were social sustainability indicators. They found regional differences for agriculture sustainability across Pakistan. Their findings revealed the tendency of unsustainable production in all provinces (Punjab, Sindh, Balouchistan and KPK) as the farmers in Sindh and Punjab are using more chemical fertilizers, pesticide and pumping ground water for irrigation purposes. The lack of sustainability in Balochistan and KPK was due to low use of fertilizer and pesticides and even no application in some areas. Furthermore, groundwater pumping for irrigation in coastal areas further increases the unsustainability of agriculture.

Arable landholding and soil fertility also influence the economic sustainability of agriculture. In a study, Wasąg and Parafiniuk (2015) assessed the ecological and social sustainability in Roztocze Region of Poland. They found that bigger holdings 70 ha UAA (utilized agricultural area) have ecological stability due to high organic matter in soil. Although these holdings have high organic matter decomposition but the balance remained positive due to increased addition of manure as these holdings have large scale of animal production. Similarly, social sustainability was also observed in holding >30 ha UAA due to increased mechanization which reduced the workload i.e. <100 man hours per ha UAA (Table 1). Niemmanee et al. (2015) studied the existing agricultural systems cover on the environmental, economic andsocialcondition of Samut Sakhonn Province, Thailand and suggested suitable pattern for sustainable agriculture. The major characteristics of this sustainable agriculture system were use of mixed cropping systems with string bean (most supplementary plants) and chili, less use of pesticides and chemical fertilizers and more use of manures, crop residues and application of knowledge gained through training and its dissemination.

In conclusion; the higher use of chemical fertilizers, pesticides and other inputs along with fluctuating agricultural markets has resulted in increased agricultural production over the years however, there is also indication of economic unsustainability due to increased use of inputs and its high input costs.

2.2 *Environmental Sustainability*

The global population will rise increase to 9.7 billion by mid of this century, which will result in increased food demand and environmental degradation due to agriculture intensification. The biggest challenge in this situation is achieving food security with agriculture sustainability worldwide. Food security ensures the healthy and continuous food availability over time; whereas, sustainable agriculture ensures the maintenance of agro ecosystem resilience (Table 2; Skaf et al. 2019). Increase in food demand associated with agriculture intensification to achieve food security is posing threat to environmental sustainability. The environmental integrity is at stake due to intensive agriculture which resulted in fertilizer driven eutrophication (Scherer and Pfister 2015), acidification (Tian and Niu 2015) and water scarcity due high use of water for irrigation (Scherer and Pfister 2016).

The rise in food demand has increased the competition for water, land and inputs for food production. Moreover, this competition has resulted in the use of high

Table 1 Effect of sustainable intensification on environmental, economic and social sustainability

Country	Cropping system	Fertilizer use/ crop residue	Description	Reference
India	CA in maize-wheat-mungbean	29.8 t ha^{-1} residue incorporation	↓soil bulk density ↑organic matter (72%),↑MBC (213), ↑MBN (293), ↑DHA (210%),↑ APA (49%), ↑bacteria (28%), ↑fungi (68%), ↑actinomycetes (98%),↑ system yield (39%)	Choudhary et al. (2018)
India	CA in rice-wheat-mungbean	30.95 t ha^{-1} residue incorporation	↑organic matter (83%), ↑MBC (117%), ↑MBN (171%), ↑DHA (140), ↑APA (42), bacteria (26%), fungi (61%), actinomycetes (92%), ↑ system yield (39%)	Choudhary et al. (2018)
India	Rice-wheat-legume crop rotation with CA		Improved carbon sequestration, soil health, with higher grain yield and food security	Samal et al. (2017)
Malawi	Sole maize	No fertilizer	One food crop per year with 12% food sufficiency and profit of USD 188	Snapp et al. (2018)
Malawi	Sole maize	69 kg N/ha 9 kg P/ha	One food crop per year with 96% food sufficiency and profit of USD 935	Snapp et al. (2018)
Malawi	Pigeonpea-maize intercrop	34.5 kg N/ha 4.5 kg P/ha	Two food crop per year with 92% food sufficiency and profit of USD 1054	Snapp et al. (2018)
Malawi	Doubled up legume/Maize rotation	17.3 kg N/ha 2.3 kg P/ha	Three food crop per year with 100% food sufficiency and profit of USD 637	Snapp et al. (2018)
Pakistan	Mungbean-chickpea		Cost effective and sustainable intensification crop rotation for dryland areas of Pakistan	Hassan et al. (2015)
China	Monocropped maize systems		Monocropped maize is effective alternative to wheat-maize system for economic and environmental sustainability in North China plain	Cui et al. (2018)
Europe	Multiple cropping		Multiple cropping by replacing luxury crops with food crops and using zero tillage and water deficit irrigation can help in utilization of 16 M ha land with conservation of 11 m ton soil and 17 billion m^3 H_2O and will ensure the food security of 229 M more people	Scherer et al. (2018)
Spain	Olive oil cropping system		SI increased the olive production from 36.8% to 64.4% with a 40–60% increase in revenue/unit of energy invested with maintained energy consumption	Sánchez-Escobar et al. (2018)

↑ = increase; ↓ = decrease; *CA* conservation agriculture, *MBC* microbial biomass carbon, *MBN* microbial biomass nitrogen, *DHA* Dehydrogenase activity, *APA* alkaline phosphatase activity

Table 2 Factors affecting socio-economic and environmental sustainability

Country	Approaches	Description	Reference
Kenya	Population density	The areas with population density below 600 persons/km^2 showed improved organic matter, better crop residue management with better nutrient use efficiency and availability (N and P), increased crop yield and food security	Willy et al. (2019)
Kenya	Population density below 600 persons/km^2	The areas with population density above 600 persons/km^2 showed high fertilizer and manure use, increased labor and capital cost, soil fertility and low nutrient use efficiencies	Willy et al. (2019)
Finland	Meat replacement with bean	Plant proteins replace with meat due to socio-economic and cultural aspects. Moreover, high bean production and consumptions ensures social and economic sustainability with better food security	Jallinoja et al. (2016)
Ecuador	Improving the ecosystem water management	Improved water management through changing in crop pattern, reduction in virtual water use with high crop water productivity will ensure water and food security	Salmoral et al. (2018)
	Sustainable branding	Creating awareness about food products e.g. organic production, ethical features will encourage consumer to pay for organic produce which will improve the environmental and economic sustainability	Franco and Cicatiello (2019)
Poland	Land holding	Farmers with more arable land (>70 ha UAA) better economic and social sustainability due to increased animal production, mechanization and better soil fertility owing to increase animal manure, relative to farmers with small land holdings (<30 ha UAA)	Wasąg and Parafiniuk (2015)
Mexico	Urban agriculture	Restoration of historic agriculture practices (floating gardens, Aztecs etc.). Improved food production with sound and balanced social, ecological and environmental dimensions	Dieleman (2017)
Brazil	Botanical pesticides	Nano formulations (18%) of essential oil of *Lippia sidoides* (thymol –68.5%) effectively control the population of *S. zeamais* populations	Oliveira et al. (2017)
Pakistan	Allelopathic plant water extracts	Application of crop water extracts of sorghum, mustard, rice and brassica in combination with low doses of Pendimethalin help in weed management with higher yield and environment safety	Jabran et al. (2008), (2010)
England and Wales	Integrated farm management	Use of traditional and modern farming method in integrated manner improves the productivity, with better social and environmental sustainability	Rose et al. (2019)
Lebanon	Cropping system selection	Olive production system is eco-friendly due to energy and agricultural input requirement, while citrus is harmful for economic and environment sustainability due high fuel, energy, water and fertilizer cost	Skaf et al. (2019)

UAA utilized agricultural area

inputs and some farming practices which are damaging the environment and major contributor of anthropogenic GHG emissions (Reilly et al. 1996). Post green revolution intensification of agriculture has resulted in soil degradation in the form of compaction, erosion, loss of organic matter, pesticide contamination, low biodiversity and increased soil salinization and waterlogging etc. (Kibblewhite et al. 2008; Schiefer et al. 2016; Turpin et al. 2017; Shah et al. 2017; Shahzad et al. 2018). Therefore, sustainable agriculture should be able to restore soil quality by use of non-chemical fertilizer and pesticides (organic fertilizers, bio-fertilizers and biopesticides), crop rotation with increased diversity to meet the global food production with sustainable environment and soil health (Verma et al. 2015; Zhang et al. 2016). Skaf et al. (2019) studied the environmental sustainability in nine different cropping systems of Lebanon by following the environmental accounting methods i.e. material flow accounting (MFA) (emission and energy accounting and their impact) and gross energy requirement (GER). They found that at farm level, citrus had the highest environmental cost due to high inputs use (fertilizers, water and diesel), whereas, olive production resulted in lowest MFA and labor resulted in less harmful for environment and use of mass is lowest environmental cost (Table 2).

Increasing use of chemical fertilizers and intensive use of arable land has substantially increased the grain production but has seriously deteriorated the upstream of environment of agriculture, farmland and downstream communities (Vitousek et al. 2009; Gomiero et al. 2011; Schreinemachers and Tipraqsa 2012). The use of pesticide has resulted substantial increase in yield globally to meet the food demand of human population growing at rapid pace. Nevertheless, this has resulted in accumulation of harmful substances in food and decline in environmental health with development of pest resistance. For sustainable food security, it is necessary to produce more food which is safe and nutritious, in the era of decreasing arable land and limiting water resources. However, use of alternative substance to synthetic chemicals has got attention due to rapid deterioration and harmful effects of pesticides on human health with increase in food safety. Use of allele chemicals from plant orientation has been effective in controlling the pest in agriculture (Campos et al. 2018; Rehman et al. 2018). Most of the botanical pesticides are developed from the essential oils produced by plants, these oils play many vital role in plant biology as they help in defense against biotic and abiotic stresses, attract pollinators and seed dissemination (Table 2: Isman 2000, 2006; Regnault-Roger et al. 2012; Pavela and Benelli 2016). These plant based compounds can be used as insecticides, fungicides or replants (Isman 2006). Use of traditional and modern farming method in integrated manner improves the productivity, with better social and environmental sustainability (Table 2: Rose et al. 2019)

In conclusion: agriculture intensification associated with high use of chemical inputs is deteriorating environment, soil health, microbial diversity in rhizosphere. However, use of organic fertilizers, botanical pesticides and natural predators can help in environmental sustainability with sustainable agricultural yields.

3 Is Global Food Security Situation Sustainable?

Agriculture faces some serious challenges as it has to address the issue of ~1 billion people who sleep hungry daily, with around an expected addition of ~2 billion people in world population by mid of this century (Rosenzweig and Parry 1994). Globally around 821 million people are suffering from malnutrition in 2017, which make around 10.9% of global population (Fig. 3; FAO 2018). In spite of all efforts the number of hungry and malnourished people is rising. In a recent survey of World Bank around 83 million people in 45 countries were starving. The undernourished population is below 5% in developed countries while it goes upto 13% and 20% in Asia and Africa respectively (Prosekov and Ivanova 2018). The number of unnourished population increased from 218.7 in 2015 to 243.2 million in Africa, while it is 519.6 million (19.7%) in 2016 compared to 508.3 in 2015 (18.3%) (Fig. 3; Prosekov and Ivanova 2018). This number is further escalated in 2017 as 20.4% of the African population is undernourished (Figs. 3 and 4). In 2013, around 19.8% of the African population is undernourished with the rate is alarmingly high in Eastern (31.9%) and Central Africa (40.9%); whereas, in Asia around 12.4% population is undernourished with highest number of undernourished population lives in Southern Asia (15.9%). Likewise, 6.1% population of Latin America and Caribbean is undernourished (FAO 2018). Moreover, the food consumption pattern is changing as the people with high income eat more and nutritious food while the situation is opposite in the developing countries.

According to FAO (2009), with stable population growth, the possibility of hunger eradication by mid of this century is questionable. The major cause of malnutrition and hunger are natural calamities, wars, poverty and high population growth rate. In a recent study, Prosekov and Ivanova (2018) selected the five countries with a decrease or increase in food security index by the end of 2017. They found an

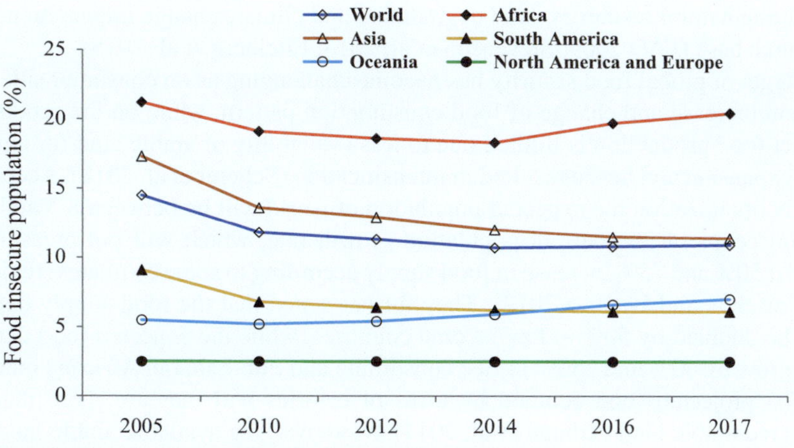

Fig. 3 Global food security situation. (Source: FAO 2018)

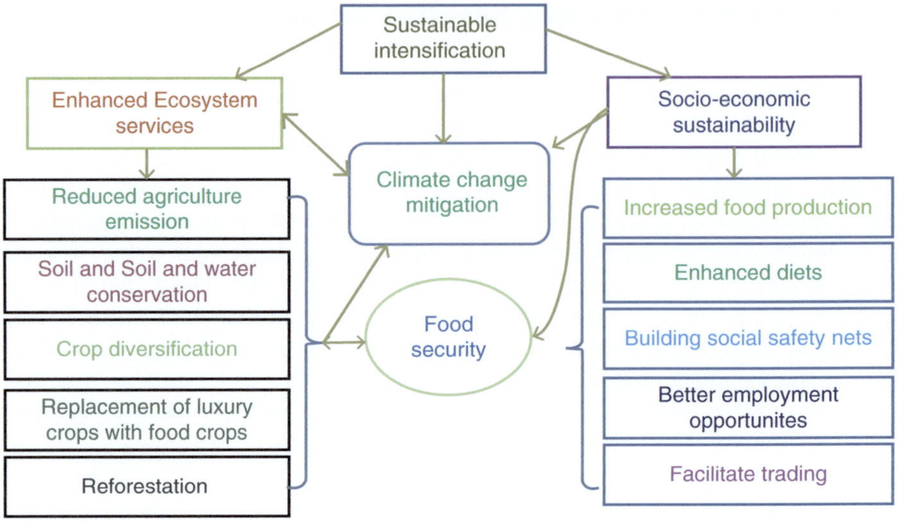

Fig. 4 Role of Sustainable intensification in climate change mitigation and food security

improvement in food security index in Nicaragua/Bangladesh (+1.3), Ecuador (+1.4), Paraguay (+2.0) and Sierra Leone (+2.6); whereas a decrease in food security index was observed in Venezuela (−7.1), Qatar (−6.0), Madagascar (−4.7), Congo (Dem. Rep.) (−3.8) and Yemen (−3.4). The decline in food supply in war affected countries is obvious but the decline in some peaceful countries is due to financial crisis globally.

In order to feed the global population, food production has kept pace by increased agricultural intensification and expansion (FAO 2009; Stevenson et al. 2013). However, in future, some estimates an increase in food production (Ewert et al. 2005), while, others expect decrease or stagnant agricultural yields due to rapidly depleting natural resources, land degradation and climate change impact on natural resource base (FAO 2009; Stevenson et al. 2013; Eitelberg et al. 2015).

Target of global food security has become challenging as on consumer side population increase and change of food consumption pattern, while on the production aspect food production is limited due to less availability of arable land for agriculture expansion and hence resulted in intensification (Scherer et al. 2018). According to UN organization the expected population of world will be between 8.3 and 10.9 billion people at the current population growth rate, which will put pressure for almost 50% and 75% increase in food supply according to some estimates (reviewed by Prosekov and Ivanova 2018). They further stated that the food supply demand will be doubled by 50% in low income countries, while the expected food demand will grow by 60% and 250% in rice consuming and Sub-Saharan African countries.

The projected food demand in terms of calories will increase more than the expected arable land (Tilman et al. 2011). Moreover, the available arable land also provides feed, fuel, fiber, timber, helping in regulating ecosystem through flood control, water purification, carbon sequestration and providing habitat to fauna and flora (Scherer et al. 2018). The potential arable land is expected to be less productive in future compared to current agricultural land as the recent enhancement in food production was attained at the cost of intensification rather than expansion (Foley et al. 2011). There are still some yield gaps in many parts of globe in spite of agriculture intensification (crop cycle) (Mueller et al. 2012) and harvest gaps (cropping frequency) (Ray and Foley 2013; Yu et al. 2017) that could be narrowed.

4 The Sustainable Intensification of Agriculture

The green revolution in 1960s has led to humongous increase in yield of staple crops due to development of input responsive crop cultivars which have come at cost of environmental integrity. In order to minimize the effect of agriculture practices on the environment, many alternative approaches have been put (e.g. organic agriculture, conservation agriculture, agroecology, ecological intensification, sustainable farming systems and sustainable intensification) (Petersen and Snapp 2015). Among these approaches sustainable Intensification (SI) is a relatively new addition and was proposed by Jules Pretty (Pretty 1997). The earlier on SI focused on approaches which can help in improving agricultural yields to meet rising food demand and also ensures environmental integrity (Pretty 1997, 1999).

Sustainable intensification is needed to handle the problem of global food security and environmental change. Local climatic conditions define the potential and need for sustainability of agricultural practices. Nevertheless, the application of these practices depends on social and economic factors, as farmers need to adopt new farming practices while consumer demand affect the economic viability of these adaptation (Scherer et al. 2018). The SI approach deems imperative in low income countries to meet the increasing global food demand (Table 1; Tilman et al. 2011). The SI can increase food and economic sustainability, especially in areas which have more and fertile agricultural land but have lower yields (Drechsel et al. 2001; Pisante et al. 2012; Vanlauwe et al. 2014).

A large number of scholars conceptualize ‘sustainable intensification’ as approached aiming at increasing the food production from the existing cultivated land in such a way with lower environmental impact and sustainable food production in future (Garnett et al. 2013). SI approach is complementary to climate smart agriculture (CSA) as SI it focuses on adaptability to climate change with lower emissions per unit of output. The CSA emphasizes the improvement of risk management, information flows and local organizations/institutes to support adaptation (Campbell et al. 2014). Moreover, CSA serves as basis for encouraging and enabling intensification. However, for adaptation instead of narrow intensification, there is need to include diverse cropping systems along with local planning for adaptation, development of efficient governess system with more asset diversity. Both SI and CSA agriculture are critical for global food security as they are part of multi-deft approach aimed at lowering waste and consumption, development and facilitating social safety net and trade and improving availability of healthy and nutritious diet

(Campbell et al. 2014). Schut et al. 2016 studied the sustainable intensification in African highlands and found that sustainable intensification faces constraints of economic and institutional nature. They further reported that institutional constraints include poor functioning or absence of markets, policies, low capabilities and finance resources and interaction between stakeholders.

In developing countries agriculture intensification is aimed at producing more food and income from existing agricultural resources. In order to attain this target, agriculture intensification can help in improving the sustainable agriculture yields with profit and social sustainability. However, SI requires better and improved agriculture technology and inputs (crop management practices, improved seed and fertilizer etc.), natural resource management (soil fertility, erosion control, reforestation, increase biodiversity etc.) and institutional reforms and innovation (policy, social infrastructure, easy access to finance, inputs, market and services) (Pretty et al. 2011; Vanlauwe et al. 2014). The integrated emergence of these innovations make smart and efficient use of existing agro-ecological, financial and human resource use across levels of different systems in a specific context (Robinson et al. 2015).

Snapp et al. (2018) conducted a participatory action research using four technologies i.e. sole maize with no and recommended fertilizer, pigeon pea –maize intercrop, doubled up legume rotation (pigeoneer intercropping in groundnut) followed by maize with half of recommended fertilizer and visualized the SI performance and tradeoffs using radar charts. They found that pigeon pea-based technologies have more environmental gains than sole maize plantation due to more biomass production, nitrogen fixation and cover duration. The domain for human and social capacity building were better for legume intergradation particularly due to diversity in diet, farmer preference (especially females) and food security over sole maize (Table 1). Furthermore, legume system was more beneficial on marginal soils due to less risk of crop failure than unfertilized maize. Niemmanee et al. (2015) studied the impact of existing agricultural systems cover on the social, economic and environmental condition of Thailand and suggested mixed cropping system with chilli as secondary and string bean as supplementary crop. They further suggested increasing use of manures and crop residues as fertilizer source and reducing the use of synthetic fertilizers and pesticides. Moreover, application of knowledge gained through trainings to the production system management and by sharing it with the farming community can help in sustainable agriculture system.

The crop yield of cereals and oil crops saw a humongous increase of 135% between during last five decades while an increase of only 27% was reported in arable land (arable land expansion varies across region) (Burney et al. 2010). Intensification without sustainability has led to numerous problems globally (Bennett et al. 2014). Furthermore, SI is a pervasive reorganizing of food systems to not only limit the environmental impact but also increase the human nutrition, animal welfare and rural economies with sustainable development (Garnett et al. 2013). In conclusion, food demand should be met through existing farmland as cultivation of new lands will have major environmental effects and costs. Therefore, intensification combined with prices and policies will have positive impact on land sparing.

5 Mitigating Impact of Climate Change in Agriculture for Food Security

Global average temperature has increased by 0.13 °C per decade since 1950s, and a faster increase (0.2 °C) is expected for next two to three decades which will have larger impact on cultivated land area (IPCC 2007). Agriculture production system contributes substantially towards global warming and accounts for 19–29% of total global greenhouse gases (GHGs) emissions, and most of these are directly coming from agricultural activities directly in the form of CH_4 and N_2O and indirectly through agricultural driven change in soil cover (Vermeulen et al. 2012). For instance, Yue et al. (2017) studied the GHGs emission from 26 crops and 6 livestock products in China and found that meat had the highest carbon footprints (CF), while lowest CFs value were observed for vegetables. Furthermore, methane emission from fertilizer and paddy were the major contributor of CFs from crop production. Climate change will have more and in general negative impact on agriculture in areas with lower latitude (Vermeulen et al. 2012; IPCC 2013). It is expected that climatic variation in future will increase the intensity and frequency of droughts, floods and will increase the risk for livestock and crop producers (Thornton and Gerber 2010). Furthermore, climate change will limit the food access to both urban and rural population due to low income, high risk and disrupted markets (Vermeulen 2014). Climatic variation is major contributor towards land degradation and change in soil cover particularly in drylands causing rapid soil deterioration.

Climate change will significantly affect the crop yield and future food availability (Table 2). For instance, Lobell et al. (2011) studied the climate trends with global crop production over three decades and found that climate change has reduced the yield of wheat and maize by 5.5% and 3.8% respectively, whereas the yield gains or losses for rice and soybean balanced out due to losses in some countries while gain in others. Al-Amin and Ahmed (2016) studied the effect of climate change on food security of Malaysia and potential climate change adaptations over 50 year time span. They predicted a 30–35% food sustainability gap below the national baseline in 2015 and the gap widens over time due to climatic change influence on agricultural yields. Nevertheless, application of certain adaptation strategies can narrow the food security gap from 5% to 20% over time. Recently, Agovino et al. (2018) constructed index of sustainable agriculture (ISA) over the period of 2005–2014 according to 16 variables and studied the climate change impact on agricultural production in 28 European Union countries. They ISA provide the ranking of EU countries based on social, economic and environmental sustainability. They found (a) negative bidirectional relationship between agricultural yield and climate change (b) negative bidirectional relationship between SA and climate change and lastly (3) conventional agriculture have negative impact on SA. A decade ago, Ravi et al. (2010) reported that extreme climatic eve-n would increase the incidence of wind and water erosion, which will shift the soil cover at faster rate in dryland areas. Farmers with poor resources, small landholdings are more vulnerable to climate change. Nevertheless, the negative effect of climate change on agriculture production and food security can be ameliorated through sustainable agriculture approaches through minor modifications in crop and livestock production practices, changing the cropping and food systems to ensure socio-economic and environmental sustainability (Table 3).

Agriculture is one of the principal factors affecting climate change by directly contributing GHGs emission through anthropogenic activities (14%) and land use (17%). Moreover, it is expected that low and middle-income countries will be the major contributors of agriculture emission in the future (Smith et al. 2007). Although the industrialized countries have substantially reduced the GHGs emissions, nevertheless developing countries face the problem of high carbon emissions. Climate smart agriculture is a pragmatic option to improve the food security with better climate change adaptations and mitigations. Furthermore, in developing countries, climate change mitigation is a co benefit as priority remains with adaptation and food security (Campbell et al. 2014). Parihar et al. (2018) in a 5 year study observed the diversified crop rotations and conservation agriculture (CA) impact on soil health, GHGs emission and food security in north-western Indo-Gangetic plains and found that CA practices in maize based cropping systems can help in reducing the GHGs emission and reduced the soil degradation. They further reported that CA in maize based cropping systems (maize-wheat-mungbean, maize-chickpea-sesbania and maize-maize-sesbania) improved the carbon sequestration, soil mineral N, with reduced N_2O emission and soil degradation.

In dryland areas, rainwater harvesting can help in mitigating the issue of climate change induced soil degradation in agro-ecosystem (Lal 2001). Maintaining the soil fertility can help in mitigating the problem of food security through increased food production under climatic variations (Wagstaff and Harty 2010). Adoption of eco-friendly sustainable agricultural approach can help in maintaining soil fertility and limit land degradation (Lovo 2016). Several practices such as use of cover crops, intercropping, crop diversification, and agroforestry can help in maintaining the agricultural production and soil conservation (Mensah 2015). Crop diversification can help in mitigating the climate change by providing options of increased diversity of marketable produce, development of innovative approaches and better functioning of agricultural system (McCord et al. 2015). Intercropping with legumes and trees can help in root proliferation which will help in improving water and nutrient uptake (Lithourgidis et al. 2011). Soil fertility holds the key for sustainable management of agricultural systems for improved biodiversity and agricultural production (Ponisio et al. 2015; Garbach et al. 2016). Furthermore, balance crop rotations can also help in improving the soil health, soil organic matter buildup and carbon sequestration (Omonode et al. 2007). Crop rotation also helps in breaking the pest cycle, improve disease resistance and crop yield (Katsvairo and Cox 2000; Krupinsky et al. 2006). Furthermore, integrated livestock and crop production systems help in reducing the land degradation with improved soil health, fertility and better land utilization helping in increased economic benefit and thus can contribute towards social and environmental sustainability by lowering poverty and reduced use of chemical fertilizer and pesticides (Gupta et al. 2012). For instance, Devendra and Thomas (2002) reported that nutrient transfer from pasture to crop land through

Table 3 Climate change mitigation strategies

Region	Strategy	Description	Reference
Andean regions	Biodiversity Quinoa	It is highly nutritious and can grow on diverse climatic conditions and marginal lands with high economic returns	Ruiz et al. (2014)
Tanzania	Soil conservation (Mulching, bund making and terracing) and agroforestry	A large number of farmers use terracing (28%), bund making (46%) and mulching (57%) to enhance soil productivity and replacing coffee with agroforestry to mitigate global warming effect	Mulangu and Kraybill (2013) and Kajembe et al. (2016)
Malaysia	Irrigation scheduling, crop diversification, IPM, Conservation agriculture, better weather and climate information system	These adaptation strategies can help in reducing the negative effective of climate change on agriculture yield and will help in ensuring sustainable food production	Al-Amin and Ahmed (2016)
Ecuador	Conservation payments to international C price	Reduce deforestation and GHG emissions	Ortega-Pacheco et al. (2019)
	Sustainable intensification	Increased yield with reduction in emission of 161 GtC annually	Burney et al. (2010)
India	Sustainable intensification (Rice-wheat-legume) and CA	Increased carbon sequestration, better soil health with higher food production	Samal et al. (2017)
India	Diversified maize rotations	The higher SOC and mineral-N, with lower N_2O fluxes and lower global warming potential with high food security and soil health were found in maize–wheat–mungbean and maize–chickpea–Sesbania than in maize–maize–Sesbania cropping system	Parihar et al. (2018)
Italy	Precision Agriculture (PA) and conservation tillage (CT)	Minimum tillage and no-tillage reduced the soil carbon losses by 17% and 63% respectively than conventional tillage with reduced carbon emission. In addition, PA practices optimized the fertilizer and fossil fuel consumption. Adoption of PA and CT reduced the CO_2 emission by 56%	Cillis et al. (2018)
China	Change of crop cultivar (13%), crop type (9%), soil management (16%) and planting dates (5%)	Crop diversification, crop variety, planting date and soil management were best suited adaption to climate change for better crop production and economic return	Kibue et al. (2015)
India	Efficient fertilizer use	Adoption of these cost effective strategies can reduce 50% of GHG emission	Sapkota et al. (2019)
	Zero tillage		
	Rice water management		
Canada	Organic waste managment	Application of organic waste of food industry to soil can help in reduction the cost of N fertilizer application by soil nitrate recycling and also reducte the GHGs emission	Rashid et al. (2010)

manure substantially helps in maintaining soil fertility and crop yield. Livestock provides cheapest labor and efficient route to intensification through nutrient cycling.

Extensive work should be developed on the topic of climate change and potential implications for food safety, which include developing models (on the basis of the available information and on the generation of reliable new data) in order to obtain more information on the spatial distribution of risk determinants for food systems under different scenarios of climate change (Miraglia et al. 2009).

In conclusion, climatic variation can largely affect agricultural yields globally. However, understanding the previous impact of climate change and devising of new policies, introduction of new crops, crop rotations, crop and livestock integration can help in mitigating the adversities of climate change through reduced GHGs emissions, better soil fertility and agricultural yield.

6 Conclusion

The rise in food demand has led to intensification. The agricultural yields have increased to many folds during last 60 years with a little increase in agriculture land. This intensification has helped in meeting the food demand but it is deteriorating environmental integrity and also poses threat to social and economic sustainability as the agricultural yields will decline in future from the same crop land due to intensification. Sustainable intensification can help in maintaining the agricultural productivity without decreasing agricultural yield and ensuring food security along with social and economic sustainability. It can also help in reducing the GHG emissions from agriculture. Climate change negatively affects food security with increase in GHG emission. However, sustainable crop and livestock intensification, crop diversification, intercropping, carbon sequestration can help in reducing the emissions from agriculture and improving the food security situation.

References

Adams MR, Hurd BH, Lenhart S, Leary N (1998) Effects of global climate change on agriculture: an interpretative review. Clim Res 11:19–30

Agovino M, Casaccia M, Ciommi M, Ferrara M, Marchesano K (2018) Agriculture, climate change and sustainability: the case of EU-28. Ecol Indic

Al-Amin AQ, Ahmed F (2016) Food security challenge of climate change: an analysis for policy selection. Futures 83:50–63

Bennett E, Carpenter SR, Gordon LJ, Ramankutty N, Balvanera P, Campbell B, Cramer W, Foley J, Folke C, Karlberg L, Liu L, Lotze-Campen H, Mueller ND, Peterson GD, Polasky S, Rockström J, Scholes RJ, Spierenburg M (2014) Toward a more resilient agriculture. Solutions J 5:65–75

Burney JA, Davis SJ, Lobell DB (2010) Greenhouse gas mitigation by agricultural intensification. PNAS 107:12052–12057

Campbell BR, Thornton P, Zougmoré R, van Asten P, Lipper L (2014) Sustainable intensification: what is its role in climate smart agriculture? Curr Opin Environ Sustain 8:39–43

Campos EV, Proença PL, Oliveira JL, Bakshi M, Abhilash PC, Fraceto LF (2018) Use of botanical insecticides for sustainable agriculture: future perspectives. Ecol Indic. https://doi.org/10.1016/j.ecolind.2018.04.038

Choudhary M, Jat HS, Datta A, Yadav AK, Sapkota TB, Mondal S, Meena RP, Sharma PC, Jat ML (2018) Sustainable intensification influences soil quality, biota, and productivity in cereal-based agroecosystems. Appl Soil Ecol 126:189–198

Cillis D, Maestrini B, Pezzuolo A, Marinello F, Sartori L (2018) Modeling soil organic carbon and carbon dioxide emissions in different tillage systems supported by precision agriculture technologies under current climatic conditions. Soil Tillage Res 183:51–59

Cui J, Yan P, Wang X, Yang J, Li Z, Yang X, Sui P, Chen Y (2018) Integrated assessment of economic and environmental consequences of shifting cropping system from wheat-maize to monocropped maize in the North China Plain. J Clean Prod. https://doi.org/10.1016/j.jclepro.2018.05.104

Devendra C, Thomas D (2002) Crop–animal interactions in mixed farming systems in Asia. Agric Syst 71:27–40

Dieleman H (2017) Urban agriculture in Mexico City; balancing between ecological, economic, social and symbolic value. J Clean Prod 163:S156–S163

Drechsel P, Gyiele L, Kunze D, Cofie O (2001) Population density, soil nutrient depletion, and economic growth in sub-Saharan Africa. Ecol Econ 38:251–258

Eitelberg DA, Vliet J, Verburg PH (2015) A review of global potentially available cropland estimates and their consequences for model-based assessments. Glob Change Biol 21:1236–1248

Evenson RE (2005) The green revolution and the gene revolution in Pakistan: policy implications. Pak Dev Rev 44:359–386

Ewert F, Rounsevell MDA, Reginster I, Metzger MJ, Leemans R (2005) Future scenarios of European agricultural land use: I. Estimating changes in crop productivity. Agric Ecosyst Environ 107:101–116

Foley JA, Ramankutty N, Brauman KA, Cassidy ES, Gerber JS, Johnston M, Mueller ND, O'Connell C, Ray DK, West PC, Balzer C, Bennett EM, Carpenter SR, Hill J, Monfreda C, Polasky S, Rockstrom J, Sheehan J, Siebert S, Tilman D, Zaks DPM (2011) Solutions for a cultivated planet. Nature 478:337–342

Food Agriculture Organization (FAO) (2001) The state of food insecurity in the world. Food and Agriculture Organization, Rome

Food Agriculture Organization (FAO) (2009) High Level Expert Forum – how to feed the world in 2050. Food and Agriculture Organisation of the United Nations (FAO), Rome

Food Agriculture Organization (FAO) (2014) AGROSTAT. Food and Agirculture Organization (FAO), Rome

Food Agriculture Organization (FAO) (2018) The state of food security and nutrition around the world. Food and Agriculture Organisation of the United Nations (FAO), Rome

Franco S, Cicatiello C (2019) The role of food marketing in increasing awareness of food security and sustainability: food sustainability branding. Encyclop Food Secur Sustain 3:27–31

Garbach K, Milder JC, DeClerck FA, de Wit MM, Driscoll L, Gemmill-Herren B (2016) Examining multi-functionality for crop yield and ecosystem services in five systems of agroecological intensification. Int J Agric Sustain. https://doi.org/10.1080/14735903.2016.1174810

Garnett T, Appleby MC, Balmford A, Bateman IJ, Benton TG, Bloomer P, Burlingame B, Dawkins M, Dolan L, Fraser D (2013) Sustainable intensification in agriculture: premises and policies. Science 341:33–34

Gliessman SR (2015) Agroecology: the ecology of sustainable food systems, 3rd edn. Taylor & Francis Group, LLC, Boca Raton

Gomiero T, Pimentel D, Paoletti MG (2011) Is there a need for a more sustainable agriculture? Crit Rev Plant Sci 30:6–23

Gupta V, Rai PK, Risam KS (2012) Integrated crop-livestock farming systems: a strategy for resource conservation and environmental sustainability. Indian Res J Ext Educ 2:49–54

Hassan A, Ijaz SS, Lal R, Barker D, Ansar M, Ali S, Jiang S (2015) Tillage effect on partial budget analysis of cropping intensification under dryland farming in Punjab, Pakistan. Arch Agron Soil Sci 62:151–162. https://doi.org/10.1080/03650340.2015.1043527

Hazell PBR (2009) The Asian green revolution. A paper prepared for the project on millions fed: proven successes in agricultural development. IFPRI discussion paper 00911.

Huang J, Yang G (2017) Understanding recent challenges and new food policy in China. Glob Food Sec 12:119–126

Huang J, Liu Y, Martin W, Rozelle S (2009) Changes in trade and domestic distortions affecting China's agriculture. Food Policy 34:407–416

IPCC (Intergovernmental Panel on Climate Change IPCC) (2013) Summary for policymakers. In: Stocker TF, Qin D, Plattner G-K, Tignor MMB, Allen SK, Boschung J, Nauels A, Xia Y, Bex V, Midgley PM (eds) Climate Change the physical science basis. contribution of working group I to the fifth assessment report of the Intergovernmental Panel on Climate Change. Cambridge University Press, Cambridge

IPCC (Intergovernmental Panel on Climate Change) (2007) In: Solomon S et al (eds) The physical science basis. Contribution of working group I to the fourth assessment report of the Intergovernmental Panel on Climate Change. Cambridge University Press, Cambridge. www.ipcc.ch/publications_ and_data/ar4/wg1/en/spm.html

Isman MB (2000) Plant essential oils for pest and disease management. In: Crop Prot XIVth International plant protection congress, pp. 603–608

Isman MB (2006) Botanical insecticides, deterrents, and repellents in modern agriculture and an increasingly regulated world. Annu Rev Entomol 51:45–66

Jabran K, Cheema ZA, Farooq M, Basra SMA, Hussain M, Rehman H (2008) Tank mixing of allelopathic crop water extracts with pendimethalin helps in the management of weeds in canola (*Brassica napus*) field. Int J Agric Biol 10:293–296

Jabran K, Cheema ZA, Farooq M, Hussain M (2010) Lower doses of pendimethalin mixed with allelopathic crop water extracts for weed management in canola (*Brassica napus*). Int J Agric Biol 12:335–340

Jallinoja P, Nivaa M, Latvala T (2016) Future of sustainable eating? Examining the potential for expanding bean eating in a meat-eating culture. Futures 83:4–14

Kajembe JI, Lupala W, Mugasha FN (2016) The role of selected agroforestry trees in temperature adaptation on coffea arabica: a case study of the Moshi district, Tanzania. In: Lal R (ed) Climate change and multi-dimensional sustainability in African agriculture. Springer, Heidelberg

Katsvairo T, Cox WJ (2000) Economics of cropping systems featuring different rotations, tillage, and management. Agron J 92:485–493

Kibblewhite M, Ritz K, Swift M (2008) Soil health in agricultural systems. Philos Trans R Soc B: Biol Sci 363:685–701

Kibue GW, Pan G, Zheng J, Zhengdong L, Mao L (2015) Assessment of climate change awareness and agronomic practices in an agricultural region of Henan Province, China. Environ Dev Sustain 17:379–391

Krupinsky JM, Tanaka DL, Merrill SD, Liebig MA, Hanson JD (2006) Crop sequence effects of 10 crops in the northern Great Plains. Agric Syst 88:227–254

Lal R (2001) Soil degradation by erosion. Land Degrad Dev 12:519–539

Lithourgidis AS, Vlachostergios DN, Dordas CA, Damalas CA (2011) Dry matter yield, nitrogen content, and competition in pea-cereal intercropping systems. Eur J Agron 34:287–294

Lobell DB, Schlenker W, Costa-Roberts J (2011) Climate trends and global crop production since 1980. Science:1204531

Lovo S (2016) Tenure insecurity and investment in soil conservation. Evidence from Malawi. World Dev 78:219–229

McCord PF, Cox M, Schmitt-Harsh M, Evans T (2015) Crop diversification as a smallholder livelihood strategy within semi-arid agricultural systems near Mount Kenya. Land Use Policy 42:738–750

Mensah EJ (2015) Land tenure regimes and land conservation in the African drylands: the case of northern Ghana. J Land Use Sci 10:129–149

Miraglia M, Marvin HJP, Kleter GA, Battilani P, Brera C, Coni E, Cubadda F, Croci L, De Santis B, Dekkers S, Filippi L, Hutjes RA, Noordam M, Pisante M, Piva G, Prandini A, Toti L, Van Den Born G, Vespermann A (2009) Climate change and food safety: an emerging issue. Food Chem Toxicol 47:1009–1021

Mueller ND, Gerber JS, Johnston M, Ray DK, Ramankutty N, Foley JA (2012) Closing yield gaps through nutrient and water management. Nature 490:254–257

Mulangu F, Kraybill D (2013) Climate Change and the future of mountain farming on Mt. Kilimanjaro. In: Mann S (ed) The future of mountain agriculture. Springer, Heidelberg

Nelson G, Rosegrant M, Koo J, Robertson R, Sulser T, Zhu T, Ringler C, Msangi S, Palazzo A, Batka M, Magalhaes M, Valmonte-Santos R, Ewing M, Lee D (2009) Climate change: impact on agriculture and costs of adaptation. IFPRI, Washington, DC

Niemmanee T, Kaveeta R, Potchanasin C (2015) Assessing the economic, social, and environmental condition for the sustainable agricultural system planning in Ban Phaeo District, Samut Sakhonn Province, Thailand. Proc Soc Behav Sci 197:2554–2560

OECD (2016) Agriculture and climate change: towards sustainable, productive and climate-friendly agricultural systems. OECD Meeting of Agriculture Ministers, Background note. Available from: https://www.oecd.org/tad/sustainableagriculture/4_background_note.pdf

Oliveira AP, Santana AS, Santana EDR, Lima APS, Faro RRN, Nunes RS, Lima AD, Blank AF, Araújo APA, Cristaldo PF, Bacci L (2017) Nanoformulation prototype of the essential oil of *Lippia sidoides* and thymol to population management of *Sitophilus zeamais* (Coleoptera: Curculionidae). Ind Crops Prod 107:198–205

Omonode RA, Vyn TJ, Smith DR, Hegymegi P, Ga'l A (2007) Soil carbon dioxide and methane fluxes from long-term tillage systems in continuous corn and corn–soybean rotations. Soil Tillage Res 95:182–195

Ortega-Pacheco DV, Keeler AG, Jiang S (2019) Climate change mitigation policy in Ecuador: Effects of land-use competition and transaction costs. Land Use Policy 81:302–310

Parihar CM, Parihar MD, Sapkota TB, Nanwal RK, Singh AK, Jat SL, Nayak HS, Mahala DM, Singh LK, Kakraliya SK, Stirling CM, Jat ML (2018) Long-term impact of conservation agriculture and diversified maize rotations on carbon pools and stocks, mineral nitrogen fractions and nitrous oxide fluxes in inceptisol of India. Sci Total Environ 640–641:1382–1392

Pavela R, Benelli G (2016) Essential oils as ecofriendly biopesticides? Challenges and constraints. Trends Plant Sci 21:1000–1007

Petersen B, Snapp S (2015) What is sustainable intensification? Views from experts. Land Use Policy 46:1–10

Pisante M, Stagnari F, Grant CA (2012) Agricultural innovations for sustainable crop production intensification. Italian J Agron 7(4):300–311

Ponisio LC, M'Gonigle LK, Mace KC, Palomino J, De Valpine P, Kremen C (2015) Diversification practices reduce organic to conventional yield gap. Proc R Soc London 282:20141396

Pretty J (1997) The sustainable intensification of agriculture. Nat Resour Forum 21:247–256

Pretty J (1999) Can sustainable agriculture feed Africa? New evidence on progress, processes and impacts. Environ Dev Sustain 1:253–274

Pretty J, Toulmin C, Williams S (2011) Sustainable intensification in African agriculture. Int J Agric Sustain 9:5–24

Prosekov AY, Ivanova SA (2018) Food security: the challenge of the present. Geoforum 91:73–77

Qi X, Fu Y, Wang RY, Ng CN, Dang H, He Y (2018) Improving the sustainability of agricultural land use: an integrated framework for the conflict between food security and environmental deterioration. Appl Geogr 90:214–223

Qiang W, Liu A, Cheng S, Kastner T, Xie G (2013) Agricultural trade and virtual land use: The case of China's crop trade. Land Use Policy 33:141–150
Rashid MT, Voroney RP, Khalid M (2010) Application of food industry waste to agricultural soils mitigates greenhouse gas emissions. Biores Technol 101:485–490
Ravi S, Breshears DD, Huxman TE, D'Odorico P (2010) Land degradation in drylands: Interactions among hydrologic–aeolian erosion and vegetation dynamics. Geomorphology 116:236–245
Ray DK, Foley JA (2013) Increasing global crop harvest frequency: recent trends and future directions. Environ Res Lett 8:44041
Ray DK, Ramankutty N, Mueller ND, West PC, Foley JA (2012) Recent patterns of crop yield growth and stagnation. Nat Commun 3:1293
Regnault-Roger C, Vincent C, Arnason JT (2012) Essential oils in insect control: low risk products in a high-stakes world. Annu Rev Entomol 57:405–424
Rehman S, Shahzad B, Bajwa AA, Hussain S, Rehman A, Cheema SA, Abbas T, Ali A, Shah L, Adkins S, Li P (2018) Utilizing the allelopathic potential of Brassica species for sustainable crop production: a review. J Plant Growth Regul:1–14
Reilly J, Baethgen W, Chege FE, Van De Geijn SC, Erda L, Iglesias A, Kenny G, Patterson D, Rogasik J, Rötter R, Rosenzweig C (1996) Agriculture in a changing climate: impacts and adaptation. In: Watson RT, Zinyowera MC, Moss RH (eds) Climate change 1995; impacts, adaptations and mitigation of climate change: scientific-technical analyses, Contribution of working group II to the second assessment report of the Intergovernmental Panel on Climate Change. Cambridge University Press, Cambridge, UK, pp 427–467
Robinson LW, Ericksen PJ, Chesterman S, Worden JS (2015) Sustainable intensification in drylands: what resilience and vulnerability can tell us. Agric Syst 135:133–140
Rose DC, Sutherland WJ, Barnes AP, Borthwick F, Ffoulkes C, Hall C, Moorby JM, Nicholas-Davies P, Twining S, Dicks LV (2019) Integrated farm management for sustainable agriculture: lessons for knowledge exchange and policy. Land Use Policy 81:834–842
Rosegrant MW, Hazell PBR (2000) Transforming the rural Asian economy: the unfinished revolution. Accessed on 24 Dec 2018. Available online at https://www.adb.org/publications/transforming-rural-asian-economy-unfinished-revolution
Rosenzweig C, Parry ML (1994) Potential impact of climate change on world food supply. Nature 367:133–138
Ruiz KB, Biondi S, Oses R, Acuna-Rodriguez IS, Antognoni F, Martinez-Mosqueira EA, Coluibaly A (2014) Quinoa biodiversity and sustainability for food security under climate change. A review. Agron Sustain Dev 34:349–359
Salmoral G, Khatun K, Llive F, Lopez CM (2018) Agricultural development in Ecuador: a compromise between water and food security? J Clean Prod 202:779–791
Samal SK, Rao KK, Poonia SP, Kumar R, Mishra JS, Prakash V, Mondal S, Dwivedi SK, Bhatt BP, Naik SK, Choubey AK (2017) Evaluation of long-term conservation agriculture and crop intensification in rice-wheat rotation of Indo-Gangetic Plains of South Asia: Carbon dynamics and productivity. Eur J Agron 90:198–208
Sánchez-Escobar F, Coq-Huelva D, Sanz-Cañada J (2018) Measurement of sustainable intensification by the integrated analysis of energy and economic flows: case study of the olive-oil agricultural system of Estepa, Spain. J Clean Prod 201:463–470
Sapkota TB, Vetter SH, Jat ML, Sirohi S, Shirsath PB, Singh R, Jat HS, Smith P, Hillier J, Stirling CM (2019) Cost-effective opportunities for climate change mitigation in Indian agriculture. Sci Total Environ 655:1342–1354
Scherer L, Pfister S (2015) Modelling spatially explicit impacts from phosphorus emissions in agriculture. Int J Life Cycle Assess 20:785–795
Scherer L, Pfister S (2016) Dealing with uncertainty in water scarcity footprints. Environ Res Lett 11:54008
Scherer LA, Verburg PH, Schulp CJE (2018) Opportunities for sustainable intensification in European agriculture. Glob Environ Change 48:43–55

Schiefer J, Lair GJ, Blum WEH (2016) Potential and limits of land and soil for sustainable intensification of European agriculture. Agric Ecosyst Environ 230:283–293

Schreinemachers P, Tipraqsa P (2012) Agricultural pesticides and land use intensification in high, middle and low income countries. Food Policy 37:616–626

Schut M, van Asten P, Okafor C, Hicintuka C, Mapatano S, Nabahungu NL, Kagabo D, Muchunguzi P, Njukwe E, Dontsop-Nguezet PM, Sartas M (2016) Sustainable intensification of agricultural systems in the Central African highlands: the need for institutional innovation. Agric Syst 145:165–176

Shah AN, Tanveer M, Shahzad B, Yang G, Fahad S, Ali S, Bukhari MA, Tung SA, Hafeez A, Souliyanonh B (2017) Soil compaction effects on soil health and crop productivity: an overview. Environ Sci Pollut Res 24:10056–10067

Shahzad B, Tanveer M, Che Z, Rehman A, Cheema SA, Sharma A, Song H, ur Rehman S, Zhaorong D (2018) Role of 24-epibrassinolide (EBL) in mediating heavy metal and pesticide induced oxidative stress in plants: a review. Ecotoxicol Environ Saf 147:935–944

Skaf L, Buonocore E, Dumontet S, Capone R, Franzese PP (2019) Food security and sustainable agriculture in Lebanon: an environmental accounting framework. J Clean Prod 209:1025–1032

Smith P, Martino D, Cai Z, Gwary D, Janzen H, Kumar P, McCarl B, Ogle S, O'Mara F, Rice C, Scholes B (2007) Greenhouse gas mitigation in agriculture. Philo Trans R Soc B Biol Sci

Smith A, Snapp S, Chikowo R, Thorne P, Bekunda M, Glover J (2017) Measuring sustainable intensification in smallholder agroecosystems: a review. Global Food Secur 12:127–138

Snapp SS, Grabowski P, Chikowo R, Smith A, Anders E, Sirrine D, Chimonyo V, Bekunda M (2018) Maize yield and profitability tradeoffs with social, human and environmental performance: is sustainable intensification feasible? Agric Syst 162:77–88

Stevenson JR, Villoria N, Byerlee D, Kelley T, Maredia M (2013) Green revolution research saved an estimated 18 to 27 million hectares from being brought into agricultural production. Proc Natl Acad Sci 110:8363–8368

Thornton PK, Gerber P (2010) Climate change and the growth of the livestock sector in developing countries. Mitig Adapt Strat Glob Chang 15:169–184

Tian D, Niu S (2015) A global analysis of soil acidification caused by nitrogen addition. Environ Res Lett 10:24019

Tilman D, Balzer C, Hill J, Befort BL (2011) Global food demand and the sustainable intensification of agriculture. Proc Natl Acad Sci U S A 108:20260–20264

Tubiello FN, Soussana JF, Howden SM (2007) Crop and pasture response to climate change. Proc Natl Acad Sci 104:19686–19690

Turpin N, Ten Berge H, Grignani C, Guzmán G, Vanderlinden K, Steinmann HH, Siebielec G, Spiegel A, Perret E, Ruysschaert G, Laguna A, Giráldez JV, Werner M, Raschke I, Zavattaro L, Costamagna C, Schlatter N, Berthold H, Sandàn T, Baumgarten A (2017) An assessment of policies affecting sustainable soil management in Europe and selected member states. Land Use Policy 66:241–249

Vanlauwe B, Coyne D, Gockowski J, Hauser S, Huising J, Masso C, Nziguheba G, Schut M, Van Asten P (2014) Sustainable intensification and the African smallholder farmer. Curr Opin Environ Sustain 8:15–22

Verma JP, Jaiswal DK, Meena VS, Kumar A, Meena RS (2015) Issues and challenges about sustainable agriculture production for management of natural resources to sustain soil fertility and health. J Clean Prod 107:793–794

Vermeulen SJ (2014) Climate change, food security and small-scale producers. CCAFS info brief. CGIAR research program on Climate Change, Agriculture and Food Security (CCAFS)

Vermeulen SJ, Campbell BM, Ingram JSI (2012) Climate change and food systems. Annu Rev Environ Resour 37:195

Vitousek PM, Naylor R, Crews T, David MB, Drinkwate LE, Holland E (2009) Agriculture. Nutrient imbalances in agricultural development. Science 324:1519–1520

Waggoner PE (1983) Agriculture and a climate changed by more carbon dioxide. Changing climate. National Academy Press, Washington, DC, pp 383–418

Wagstaff P, Harty M (2010) The impact of conservation agriculture on food security in three low veldt districts of Zimbabwe. Trócaire Dev Rev:67–84

Wasąg Z, Parafiniuk S (2015) Ecological and social sustainability of agricultural production in the Roztocze Region. Agric Agric Sci Proc 7:287–291

Willy DK, Muyanga M, Jayne T (2019) Can economic and environmental benefits associated with agricultural intensification be sustained at high population densities? A farm level empirical analysis. Land Use Policy 81:100–110

Yu Q, Wu W, You L, Zhu T, van Vliet J, Verburg PH, Liu Z, Li Z, Yang P, Zhou Q, Tang H (2017) Assessing the harvested area gap in China. Agric Syst 153:212–220

Yue Q, Xu X, Hillier J, Cheng K, Pan G (2017) Mitigating greenhouse gas emissions in agriculture: from farm production to food consumption. J Clean Prod 149:1011–1019

Zhang D, Yan M, Niu Y, Liu X, van Zwieten L, Chen D, Bian R, Cheng K, Li L, Joseph S, Zheng J, Zhang X, Zheng J, Crowley D, Filley TR, Pan G (2016) Is current biochar research addressing global soil constraints for sustainable agriculture? Agric Ecosyst Environ 226:25–32

Zulfiqar F, Thapa GB (2017) Agricultural sustainability assessment at provincial level in Pakistan. Land Use Policy 68:492–502

Part II
Ecological Sustainability

Integrating Conservation Into Agriculture

Amir Kassam

Abstract The intensification of tillage agriculture has been aimed to increase crop yields. However, this has also been causing degradation of agricultural lands and natural resources. In this regard, Conservation Agriculture (CA) is an alternate paradigm, which integrate the conservation into agriculture and make that regenerative, conserving and resilient than the conventional tillage agriculture. CA is, thus, an ecosystem approach to improve and sustain productivity, and increase profits and resource base. In this chapter, some aspects of conservation that are integrated into agriculture when practicing CA are described. The potential of CA in improving productivity, economic, social and environmental benefits to farmers and the society at large have been discussed.

Keywords Biomass soil mulch cover · Conservation Agriculture · Land degradation · Tillage agriculture

1 Introduction

We have been aware of the degradation of agricultural lands and natural resources caused by conventional tillage agriculture since the days of the 'Dust Bowles' in the American mid-west since the 1930s and 1940s. Despite this, tillage agriculture has continued to become more and more intensified over the years, and any attention to soil health management has been mainly related replenishing plant nutrients in the soil with mineral fertilizers. The history of agricultural development tells us that the intensification of tillage agriculture has been driven mainly by the need to increase crop yields along certain lines of thinking, often referred to as the Green Revolution mind-set, namely: to increase crop yields, inputs, particularly of agrochemicals to feed and protect the crop, must be increased, and the genetic make-up of the seeds or crops should be such that they are able to respond to increased production inputs and

---

A. Kassam (✉)
University of Reading, Reading, UK
e-mail: amirkassam786@gmail.com

M. Farooq, M. Pisante (eds.), *Innovations in Sustainable Agriculture*,
https://doi.org/10.1007/978-3-030-23169-9_2

produce a high yield. Little concern has been directed to the fact that mechanical tillage at any level of agricultural development and farm power availability leads over time to destruction of soil health and function because of the damage to soil biology, soil organic matter and soil structure and porosity. Similarly, little concern has been paid to the destruction of the natural enemies of pests in modern agriculture through the application of heavy doses of biocides. Over time, and particularly since the WWII, modern farming has become a destructive force towards nature and towards ecosystem functions and services. In general, it would be true to say that modern tillage farming has led to the degradation and loss of the natural resource base, including biodiversity in the soil, in the cropping system and above the ground, and in the landscape (Pretty 2002). It has also led to high levels of soil erosion, water pollution and loss of ecosystem societal services such as clean water, carbon sequestration, carbon, nutrient and water cycling, regulatory processes, habitats for wildlife and natural enemies of pests, pollination services etc. (Kassam et al. 2013). Estimates vary but as a result of soil tillage, agricultural lands have been lost or abandoned at annual rates of around 7–12 M ha per year over the past 70 years (Montgomery 2007; Gibbs and Salmon 2015). A recent study puts the annual global cost of land degradation due to land use and cover change at 300 billion USD (Khonya et al. 2016).

Thus, overall, it would seem that modern conventional agriculture has done little to explicitly integrate conservation into agriculture as the main focus has been on increasing crop and livestock production, in terms of both yields and factor productivity in rainfed and irrigated agriculture based on the narrow Green Revolution approach of genetic improvement and agrochemical inputs to feed and protect crops and livestock. However, while the multi-lateral and bilateral donors and multinational seed and agrochemical companies have been promoting the Green Revolution paradigm, farmers in different parts of the world have also been reacting to address and overcome the inherent degradation consequences. They along with extension agronomists and machine companies began to replace intensive tillage with no-tillage and introduce soil and water conservation practices that eventually led to an alternate paradigm now generally referred to as Conservation Agriculture (CA).

This chapter elaborates on some of the aspects of conservation that are integrated into agriculture when practicing CA. Being a new paradigm, its potential in this area remains to be more fully unlocked through scientific research and farmer practice. However, enough is known about CA from global scientific and empirical evidence to be able to describe how conservation is integrated into agriculture in CA systems, and how this leads to productivity, economic, social and environmental benefits to farmers and to society.

2 What Is Conservation Agriculture?

FAO defines CA as an approach to managing agroecosystems for improved and sustained productivity, increased profits and food security while preserving and enhancing the resource base and the environment (Kassam et al. 2009; FAO 2011).

CA is characterised by the practical application of three linked principles, along with other context-specific complementary good agricultural practices of crop and production management, namely:

1. Continuous no or minimal mechanical soil disturbance (implemented by the practice of no-till seeding or broadcasting of crop seeds, and direct placing of planting material into untilled soil; and causing minimum soil disturbance from any cultural operation, harvest operation or farm traffic);
2. Maintenance of a permanent biomass soil mulch cover on the ground surface (implemented by retaining crop biomass, root stocks and stubbles and cover crops and other sources of ex situ biomass); and.
3. Diversification of crop species (implemented by adopting a cropping system with crops in rotations, and/or sequences and/or associations involving annuals and perennial crops, including a balanced mix of legume and non-legume crops).

CA systems are now in existence in all continents in all land-based agriculture, supporting the notion that CA principles are universally applicable to all agricultural landscapes and land uses with locally formulated and adapted practices. The three individual principles when applied concomitantly constitute the ecological foundation of CA production systems. If the three principles are applied separately, they do not constitute a CA system. For example, use of no-till practice on its own does not qualify the production system to be CA based, unless it is linked to the application of the other two practices of soil mulch cover and diversified cropping. Generally, CA practices provide a sustainable foundation upon which to overlay other complementary practices that can strengthen the system further in terms of integrated crop, soil, nutrient, water, pest, labour, energy and farm power management. In some instances, controlled traffic can add further resilience and stability to the production and land management system.

In 2015/2016, CA covered some 180 M ha of global cropland, corresponding to 12.5% of the global cropland area (Kassam et al. 2018). Since 2008/2009, the cropland under CA has been increasing at an annual rate of some 10 M ha. While much of the spread is concentrated in North and South America and Australia, CA has also been spreading in Europe, Asia and Africa as more attention is directed to understanding its relevance in sustainable agriculture and land management. CA is now practiced in all agro-climatic regions, by small-scale and larger-scale farmers, and has shown to be beneficial with any farm power, manual, animal traction or motorized. About half of the CA area is in the low income regions and half in the industrialized regions.

Soil erosion and land degradation is a common consequence of tillage agriculture regardless of the tillage implement used. Loss of top soil, loss of structure, loss of soil health and function and soil compaction can be caused by regular soil tilling using a hand hoe, or animal drawn or motorised implements such as simple ards and tines to inversion ploughs as well as non-inversion tillage implements such as disc harrows and rippers. Motorized rotovators and roto-tillers pulverise the top soil to the depth of 10–15 cm across 100% of the crop area. The resulting loss of productivity was the main driver that made farmers in North and South America in the 1960s and 1970s to move away from tillage and adopt the practice of no-till with stubble mulching to establish crops. Over time, retaining more ground mulch cover and diversified cropping were found to be important and became popular because of their role in increasing soil organic matter and improving soil health and function. In many geographical areas such as in Brazil, USA and Australia, this way of agricultural land management also led to regenerating or rehabilitating degraded or abandoned agricultural lands, including extending farming into marginal drylands that were considered too risky (Kassam et al. 2013).

There have been other drivers that have led the farmers in all continents to adopt CA. These have included: (1) increasing cost of agrochemical, energy and labour inputs along with decreasing productivity (use and output efficiency) and pollution of the environment; (2) biodiversity loss and land-mediated ecosystem societal services such as clean water, carbon sequestration, pollination services; (3) increased vulnerability to climate change and poor ability to contribute to climate change mitigation; and (4) poor ability to serve the needs of resource poor smallholders.

Unlike the conventional tillage agriculture with its bare and unprotected soil surfaces, CA offers to all farmers a way to sustainably manage agricultural lands and production, with minimum negative externalities. In the industrialised nations, conventional tillage agriculture involves a strong tendency towards mono-cropping and heavy use of agrochemicals causing pollution, leading to sub-optimal yields and low return to investments, and to poor climate change adaptability and mitigation. Thus, CA appeals to a wide range of farmers because it is regenerative and much more self-protecting, with greater yields and yield stability as well as higher factor productivity, thus better and more reliable income. For smallholders' livelihoods, these benefits are extremely important for sustaining a decent quality of life.

3 What Is 'Conserved' When Practicing Conservation Agriculture?

In light of the above, it would not be an exaggeration if one were to conclude that with conventional tillage agriculture, there is no explicit attempt to integrate conservation into agriculture. The resulting agriculture production systems at any level of development are all generally unsustainable (Montgomery 2007; Kassam et al. 2013, 2017). The end points with any form of tillage-based agricultural systems are: (1) soil erosion, land degradation and abandonment of agricultural land; (2) loss of land productivity and land potentials; (3) loss of biodiversity and environmental quality; and (4) dysfunctional agroecosystems and loss of ecosystem societal services.

The opposite is true with CA systems because CA pays attention to establishing a dynamic ecological foundation of production systems such that the natural resource base and its agricultural potential and ecosystem functions are conserved, enhanced and maintained at the optimum level. This means that all three CA practices work together along with complementary good agricultural practices to offer an optimal performance of maximum efficiency, adaptability and resilience (and profitability and stability) and maximum output with minimum input (FAO 2011, 2016; Kassam et al. 2013).

CA has often been described as being a resource-conserving sustainable production paradigm (FAO 2011). The following sections elaborate four aspects of CA production systems and land management that reflect the integration of this resource-conserving feature in CA systems. These are: (1) soil health and biology; (2) relationship between healthy soil and crop response; (3) biodiversity above, below and at the ground surface; and (4) ecosystem societal services.

3.1 Soil Health and Biology in CA Systems

Soil health has been defined as the capacity of soils to function as a living system within ecosystem and land use boundaries, to sustain plant and animal productivity, maintain and enhance water and air quality, and promote plant and animal health. Soil health emphasises a unique property of biological systems, since inert components cannot be sick or healthy. Management of soil health means management of the living portion of the soil to maintain essential functions of soil to sustain plant and animal productivity and health (FAO 2008).

Conservation Agriculture makes a special effort to enhance and sustain soil health because it encourages soil life to flourish and soil biology to function in a manner that leads to the creation of the soil as a living biological system. Such a system is inhabited by all kinds of microorganisms including protozoa, fungi, bacteria, viruses, and mesofauna such as earthworms and arthropods including insects such as termites, ants, spiders, beetles and bumble bees. In conventional tilled soils with low organic matter, soil life is minimal.

CA is particularly beneficial to below ground soil biota and biodiversity which is believed to improve four main aggregate ecosystem functions (Swift et al. 2008): (i) decomposition and humification of organic matter and root exudates brought about by the enzymatic activity of bacteria and fungi, and facilitated by soil animals such as mites, millipedes, earthworms and termites; (ii) nutrient cycling which is closely associated with organic decomposition, with transformations mediated through microorganisms; (iii) soil structure maintenance through the activities of plant roots, earthworms, termites, ants and some other soil mesofauna in the soil that form channels, pores, aggregates and mounds, and moving particles from one horizon to another; and (iv) disease and pest control through for example the regulations of activities of pathogens by the microbiovore and micropredator portions of the soil biota that feed on microbial and animal pests respectively.

Fungi such as mycorrhiza and earthworms and termites find it almost impossible to survive, and optimal relationships between various microorganisms are never established. For example, glomalin, sugar-protein, serves as a ‘biological cement’ that is essential in the formation of micro and macro aggregates and in aggregate stability and for soil structure and porosity. Glomalin plays an important role in carbon sequestration and in the creation of soils that behave like water sponges, holding maximum amount of water while maintaining high infiltration and drainage rates. Glomalin is produced by mycorrhiza in healthy undisturbed soils. Similarly, earthworms are essential in incorporating surface crop biomass into the soil, creating soil aggregates and adding biological nitrogen to soils. They also create a network of stable biopores in the soil which contributes to drainage, aeration and growth of roots. Thus, there appears to be a large list of functions carried out by soil microorganisms and many of these play beneficial roles in agricultural production and in soil and water conservation and in nutrient retention and carbon sequestration.

In terms of its effect on soils, CA adds up to 1 mm soil per year; organic matter and soil biota increases at about 0.1–0.2% per year until reaching a saturation, after which organic matter is still required to maintain the plateau level as losses continue to occur through microbial activity; a diversity of rooting systems from a diversified crop rotations and associations provide for more efficient use of soil nutrients and biologically fixed nitrogen; soil structure through the workings of a diverse set of microorganisms is more stable and porous with lower bulk density and higher cation exchange capacity, and soil erosion and degradation is stopped or reversed (see Fig. 1).

The effect of CA on water include: fuller recharge of aquifer (permanent soil pore structure due to soil biodiversity, plant roots and root exudates maintaining the soil micro and macro pores); improved water quality (less leaching and erosion due to surface protection with plant organic matter cover from a diverse source of crops); more soil available water to crops due to higher soil organic matter (SOM) (1% SOM = 150 m^3/ha); reduced surface water losses (decrease in soil evaporation); and better water efficiency (crop water requirements decrease by some 30%) and increased water productivity (more crop per drop). Additionally, there is a significant reduction in flood risk under CA. This is because water infiltration rates under

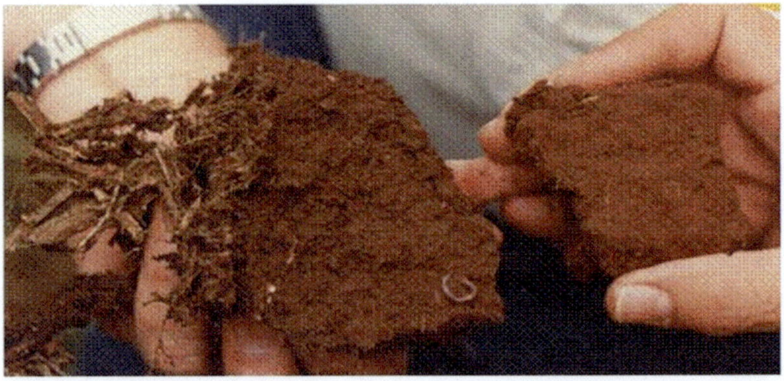

Fig. 1 Soil comparison in a farmer's trial – Clods of topsoil from adjacent plots, Parana, Brazil. (Photo credit: Francis Shaxson)

CA are more than 120 mm/h compared to 20 or 30 mm/h for soils under tillage-based farming. Similarly, there are beneficial impacts of the diversified CA cropping system on weeds, pathogens and insect-pests, reducing overall use of pesticides; and CA-based agrobiodiversity above and below the ground surface contributes to climate change adaptation and resilience of the production system, and to climate change mitigation due to greater carbon sequestration and lower greenhouse gas emission.

3.2 Biodiversity Above the Ground Surface in CA Systems

All the three defining interlinked CA practices promote biological control of pests (weeds, insects, pathogens), and lend themselves to additional complementary integrated pest management practice. CA systems promote the enhancement of biodiversity at and above the ground surface as well as below the ground level. The biodiversity in the diversified cropping system, the continuous ground cover with biomass mulch and the minimum mechanical soil disturbance all contribute to setting up food webs above the ground surface as well as below the ground surface.

Weeds are suppressed through weed seeds rotting away in undisturbed soils and cover crops and biomass mulch as well as crop rotations and associations suppress weed infestation and growth. For example, CA-based push-pull pest control (Fig. 2) provide good control of maize stem borer which are pushed by Desmodium under sown cover crop from the maize field to Napier grass or Brachiaria grass planted at the field boundary which serves to attract the stem borer to lay eggs inside their stems. When the larvae are hatched, they are killed by the exudate from the host

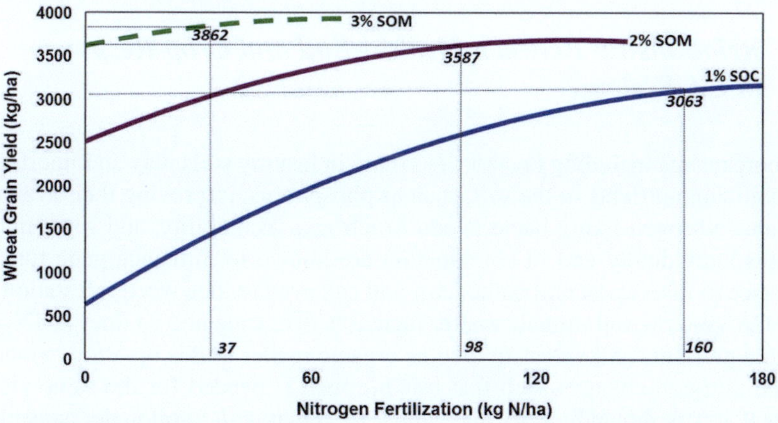

Fig. 2 Yield response to nitrogen fertilization in wheat in Portugal. Values in italic represent the economically optimal N-fertilization rate and the respective yield for the different levels of SOM. The dashed green line is the modelled yield response for a SOM level of 3%. (Source: Carvalho et al. 2012)

plants. The Desmodium cover crop also wipes out the parasitic weed Striga, and pumps into the soil biologically fixed nitrogen, as well as improves soil structure and its water retention capacity (Khan et al. 2017).

In other CA systems, natural enemies of pest are nurtured through the existence of biomass soil mulch cover which promotes tiny neutral arthropods upon which natural enemies of pest feed. Such habitats within CA cropping system serve to retain natural enemies of pest in readiness to protect the crop should there be a pest attack.

In the case of weeds, a crop intercropping association involving of multi-purpose cover crop such as Dolichos or Mucuna or pumpkin can offer effective protection against weed infestation (Owenya et al. 2011). Equally, crop rotations have been shown to be effective in reducing the number of weed species and their density (Anderson 2015). There are reports of alellopathic control of weeds infestation too through the use of appropriate crop combination and including allelopathic crops in the associations and rotations (Farooq and Siddique 2015).

Farmers have also shown that through planting green, which involves crimper rolling a cover crop, it is possible to seed the next crop without the use of herbicides. This approach has been shown to work for smallholder and for large-scale farmers (Gullickson 2018; Duiker 2017).

The main benefit from the above features of CA systems is greater natural self-protection and resilience, deceased cost of pesticides and reduced pollution of the environment. Additionally, due to minimum or no mechanical disturbance of the soil at ground level and the availability of biomass and stubble protection, wildlife including ground nesting birds and hares also benefit. So not only agricultural biodiversity is enhanced and conserved in CA systems, but also wild and natural biodiversity is enhanced and conserved too.

3.3 Relationship Between Healthy Soil and Crop Response in CA Systems

Microorganisms including bacteria and fungi in healthy soils play an important role in mobilizing nutrients in the soil, such as phosphorus, improving their availability to plants. Nitrogen fixing bacteria add to nitrogen availability, and some microorganisms individually and in combination are known to influence gene functions, resistance to insect pest and pathogens, and can even reduce weed infestation.

In CA systems soil organic matter increases over time and so does soil's ability to retain nutrients and water. In fact, as organic matter builds up, crop response to applied nitrogen changes such that less nitrogen is needed for the same yield. In Portugal, it was shown by Carvalho et al. (2012) that with a soil under conventional tillage containing 1% soil organic matter it took 160 kg N ha^{-1} to produce 3 t ha^{-1}

of wheat grain (Fig. 2). However, when the same soil built up 2% soil organic matter through continuous CA, 3 t ha^{-1} yield was obtained with 37 kg N ha^{-1}. A crop modelling exercise showed that at 3% soil organic matter content, a yield of 3.5 t ha^{-1} would be achieved without any application of N.

Transforming a tillage-based production system to a CA-based system is a time-related biological process. When implemented correctly, CA systems offer a range of benefits in terms crop and cropping system response that correspond to the mobilization of greater crop and land potentials, and actual crop and land performance. The intensity and range of benefits generally increase over time as new and healthier soil productivity and resource conservation equilibrium is established, including:

Higher and stable yields, factor productivity and profit:

Increased yields, factor productivity, farm production and profit, depending on the level of initial degradation and yield, and the agroecological potential of the location (Basch et al. 2012; Soane et al. 2012; Jat et al. 2014; Li et al. 2016; Kassam et al. 2013, 2017).

Higher nutrient productivity:

A 50% or more decrease in fertilizer required if already applying higher rates, and greater nutrient productivity with increased soil organic matter level. In cases where mineral fertilizers are not available, integrated nutrient management can provide the required nutrition from local sources (Sims and Kassam 2015; Lalani et al. 2016, 2017; Kassam et al. 2017).

Lower or no use of pesticides (herbicides, insecticides, fungicides):

A 20–50% decrease in pesticides if already applying higher rates, and greater output per unit of pesticide applied. Where pesticides are not used or available, integrated pest management built within CA cropping systems can achieve adequate control with less labour and cost (Lindwall and Sonntag 2010; Lalani et al. 2016, 2017; Khan et al. 2017).

Lower use of fossil fuel, labour, time and machinery:

Up to 70% less machinery, energy and labour cost. In manual production systems there can be a 50% reduction in labour requirement as there is much less labour required for seedbed preparation and weeding (Sims and Kassam 2015; Freixial and Carvalho 2010) (Table 1).

Decreased erosion and runoff and improved soil water balance:

Decreased soil erosion and water runoff, increase water infiltration, water retention and up to 40% reduced water requirement and increased water productivity in rainfed and irrigated conditions (Derpsch 2003; Basch et al. 2012; Jat et al. 2014; Nkonya et al. 2016; Vlek et al. 2017; Fig. 3).

Table 1 Summary of annual expenses for maintenance and repair of tractors and of tillage/drilling implements, for fuel and labour for a farm near Évora, South Portugal

Summary of annual expenses			
	Conventional tillage (year 2000)	No-till (year 2003)	Reduction (%)
Maintenance and repair of tractors	10.450,47 €	1.507,15 €	85
Maintenance and repair of tillage/drilling implements	8.158,41 €	1.840,40 €	77,5
Fuel	17.460 €	7.110 €	60
Labour	25.000 €	15.000 €	40
Total annual	61.068,88 €	18.347,55 €	70

Source: Freixial and Carvalho (2010)
Farm power – 4 tractors with 384 HP under tillage and 2 tractors with 143 HP under no-till

Fig. 3 Erosion and runoff on conventionally tilled bare soil near Cordoba, Spain. (Photo credit: Francis Shaxson)

Increased biomass and greater livestock carrying capacity:

More biomass (along with greater yields) becoming available for livestock with time as soil health improves, thus decreasing the initial 'conflict situation' and opening up the possibility of increased livestock carrying capacity and stocking rates (Landers 2007; FAO 2009, 2012, 2013; Owenya et al. 2011). However, production of livestock is not a necessary component of sustainable agriculture nor can it be assumed that is it a core element of responsible production and consumption.

3.4 Ecosystem Societal Services from CA Systems

Conventional tillage agriculture systems are known for their destruction of the soil and most of the soil- and landscape-mediated ecosystem services which farming communities and rural and urban societies require and rely upon for their livelihoods and as part of their life support system. Ecosystem functions and services are provided by nature to societies as well as to the living world as a whole. Four types of services are recognized – supporting services such as soil formation, water, nutrient and carbon cycling, natural vegetation, general circulation of atmospheric, etc.; regulatory services such as groundwater flows, aquifer recharge, stream flows etc.; provisioning services such as tangible 'goods' that result from supporting and regulatory services such as clean water resources and supplies, carbon sequestration, biological nitrogen fixation, control of soil erosion and degradation, pollination services etc., and cultural services such as sacred natural sites and phenomena, recreational areas, conservation areas for nature and wildlife etc. These ecosystem services function at the farm level in the individual fields (in-situ) as well at the landscape level (ex-situ) covering large landscapes and watersheds, and even provinces Examples of in-situ and ex-situ ecosystem services resulting from CA-based land use are provided by Kassam et al. (2013).

Conventional tillage agriculture through soil and landscape erosion and degradation leads to dysfunctional provisioning, regulatory and supporting ecosystem services. In the industrialized and low income countries, tillage agriculture has led to large scale wind and water erosion, soil and land degradation, pollution of ground water and water bodies including dead zones in the oceans, inland water bodies and in underground aquifers, destruction of agricultural birds, chemical contamination of domestic and irrigation water supplies, increased risks of flooding, increased sediment load in streams and rivers. Agriculture is also regarding as a major source of greenhouse gasses as a result of the use of fossil fuel for agricultural operations, emission of CO_2, CH_4 and N_2O particularly from agricultural soils, and from agriculture driven deforestation.

On the other hand, CA systems have the ability of minimizing the negative impact on ecosystem services because with CA it is possible to farm more closely with nature than against nature. A greater range of soil functions and soil-mediated ecosystem services include:

Adaptability to climate change:

Greater adaptability to climate change in terms of more stable yields, and lower impact of climate variability (Thierfelder et al. 2015; Kassam et al. 2017; Gonzalez-Sanchez et al. 2017a, b).

Climate change mitigation:

Increased contribution to climate change mitigation from enhanced soil carbon sequestration, reduced greenhouse gas emissions, and decreased use of fossil fuel (Haugen-Kozyra and Goddard (2009). Additionally, lower carbon and environmental footprint due to reduced use of manufactured inputs such as agrochemicals and machinery (Basch et al. 2012; Corsi et al. 2012; Gonzalez-Sanchez et al. 2012, 2017a, b).

Lower environmental cost to society:

Lower environmental cost to society from decreased levels of water pollution, and damage to infrastructure such as roads, bridges and riverbanks as well as water bodies due to reduced erosion and floods (ITAIPU 2011; Mello and van Raij 2006; Basch et al. 2012; Nkonya et al. 2016; Vlek et al. 2017).

Rehabilitation of degraded lands and ecosystem services:

Rehabilitation of degraded lands and ecosystem services from all agricultural land under use, as well as from abandoned agricultural land in which the eroded topsoil and the soil profile can be rebuilt (Kassam et al. 2013).

Opportunity for establishing large scale ecosystem service programmes:

Greater opportunity for establishing large scale, community-based, cross-sectorial ecosystem service programmes such as the watershed services programme in the Parana Basin in Brazil (ANA 2011; ITAIPU 2011; Mello and van Raij 2006; Kassam et al. 2013).

4 Concluding Comments

CA represents a new way of thinking – an alternate paradigm – about integrating conservation into agriculture that is regenerative, conserving and much more resilient that conventional tillage agriculture. CA is based on an ecosystem approach to sustainable agriculture production and land management and by definition integrates conservation into agriculture by emulating nature as much as possible which results in greater output, productivity and profit as well as the harnessing of ecosystem societal services. This is because CA systems aim at optimization and try to maximize output with minimum inputs. This is not the case with conventional tillage agriculture which has not been able to move away from high input-dependent agriculture that has led to top soil loss and agricultural land degradation globally. Conventional tillage systems are vulnerable to climate change and are not fully able to adapt to climate change nor can they mitigate climate change as they cannot convert the soil into a sink for carbon. The opposite is true for CA systems.

CA principles are applicable to all land-based production systems, including rainfed and irrigated, annual and perennial systems, mixed crop-livestock systems, orchards and plantation, organic systems, rice-based systems, agroforestry and pasture and rangeland systems. In all these systems CA provides an ecological foundation for sustainable production, making the systems resource conserving and regenerative. Some of the benefits of CA systems to farmers and society are not available from conventional tillage systems, and the erosion and degradation cannot be stopped unless farmers move away from tillage agriculture in to productive and climate smart CA systems.

While much needs to be discovered about CA systems, enough is known to show that conventional tillage systems have run their course and their productivity and profitability are no longer optimal. Sustainable production intensification in the future must be based on CA systems as they are adapted to smallholder as well as large-scale farmer, rich and poor farmer, women and men farmers, and offer ways to rehabilitate degraded and abandoned agricultural lands.

References

ANA (2011) Programa Produtor de Água (Programme for water producers). http://www.ana.gov.br/produagua/

Anderson RL (2015) Integrating a complex rotation with no-till improves weed management in organic farming. A review. Agron Sustain Dev 35:967–974

Basch G, Kassam A, Friedrich T, Santos FL, Gubiani PI, Calegari A, Reichert JM, dos Santos DR (2012) Sustainable soil water management systems. In: Lal R, Stewart BA (eds) Soil water and agronomic productivity. Advances in Soil Science. CRC Press, Boca Raton, pp 229–289

Carvalho M, Basch G, Calado JMG, Barros JFC (2012) Long term effect of tillage system and crop residue management on soil carbon content of a Luvisol under rainfed Mediterranean conditions. Agrociencia (Special Issue):183–187

Corsi S, Friedrich T, Kassam A, Pisante M, de Sà JM (2012) Soil organic carbon accumulation and greenhouse gas emission reductions from conservation agriculture: a literature review. Integrative Crop Management 16. FAO, Rome

Derpsch R (2003) Published with slight changes as a special publication by no-till on the Plains Inc., Underst Process Water Infiltr 2(3):1–4

Duiker SW (2017) Planting green – a new cover crop management technique. Field Crop News. https://extension.psu.edu/planting-green-a-new-cover-crop-management-technique

FAO (2008) Investing in sustainable crop intensification: the case for improving soil health. Integrated crop management, vol 6. FAO, Rome

FAO (2009) Enhancing crop-livestock systems in conservation agriculture for sustainable production intensification: a farmer discovery process going to scale in Burkina Faso. Integrated crop management, vol 17. FAO, Rome, p 42

FAO (2011) Save and grow: a policymaker's guide to the sustainable intensification of smallholder crop production. FAO, Rome, 102p

FAO (2012) Conservation agriculture and sustainable crop intensification in Karatu district, Tanzania. Integrated crop management, vol 15. FAO, Rome

FAO (2013) Forest management and conservation agriculture experiences of smallholder farmers in the eastern region of Paraguay. Integrated crop management, vol 18. FAO, Rome

FAO (2016) Save and grow in practice: maize, rice and wheat. FAO, Rome

Farooq M, Siddique KHM (2015) Conservation agriculture. Springer, Basel

Freixial R, Carvalho M (2010) Aspectos prácticos fundamentales en la implantación de la agricultura de conservación/siembra directa en el sur de Portugal. In: Proceedings of the European Congress on conservation agriculture: towards agro-environmental climate and energetic sustainability. Madrid, pp 361–369

Gibbs HK, Salmon JM (2015) Mapping the world's degraded lands. Appl Geogr 57:12–21

González-Sánchez EJ, Ordoñez-Fernández R, Carbonell-Bojollo R, Veroz-Gonzalez O, Gil-Ribes J (2012) Meta-analysis on atmospheric carbon capture in Spain through the use of conservation agriculture. Soil Tillage Res 122:52–60

Gonzalez-Sanchez EJ, Mkomwa S, Conway G, Kassam A, Ordonez-Fernandez R, Moreno-Garcia M, de Repullo-Ruiberriz TM, Gil-Ribes J, Basch G, Veroz-Gozalez O, Trivino-Tradas P, Holgado-Cabrera A, Miranda-Fuentes A, Carbonell-Bojollo R (2017a) Making climate change mitigation and adaptability real in africa with conservation agriculture. European Conservation Agriculture Federation (ECAF), 143p

Gonzalez-Sanchez EJ, Moreno-Garcia M, Kassam A, Holgado-Cabrera A, Trivino-Tradas P, Carbonell-Bojollo R, Pisante M Veroz-Gozalez O, Basch G (2017b) Conservation agriculture: making climate change mitigation and adaptability real in Europe. European Conservation Agriculture Federation (ECAF), 154p. https://bitly.com/

Gullickson G (2018) Planting into green cover crops. Successful farming. https://www.agriculture.com/crops/soybeans/planting-into-green-cover-crops

Haugen-Kozyra K, Goddard TM (2009) Conservation agriculture protocols for green house gas offsets in a working carbon markets. In: Paper presented at the IV World Congress on Conservation Agriculture, 3–7 February 2009, New Delhi, India

ITAIPU (2011) Cultivando Agua Boa (Growing good water). https://www.itaipu.gov.br/

Jat RA, Sahrawat KL, Kassam AH (eds) (2014) Conservation agriculture: global prospects and challenges. CABI, Wallingford, 393p

Kassam AH, Friedrich T, Shaxson TF, Pretty JN (2009) The spread of conservation agriculture: justification, sustainability and uptake. Int J Agric Sustain 7(4):292–320

Kassam A, Basch G, Friedrich T, Shaxson F, Goddard T, Amado T, Crabtree B, Hongwen L, Mello I, Pisante M, Mkomwa S (2013) Sustainable soil management is more than what and how crops are grown. In: Lal R, Stewart RA (eds) Principles of soil management in agro-ecosystems. Advances in soil science. CRC Press, Boca Raton

Kassam A, Mkomwa S, Friedrich T (eds) (2017) Conservation agriculture for Africa: building resilient farming systems in a changing climate. CABI, Wallingford

Kassam AH, Friedrich T, Derpsch R (2018) Global spread of conservation agriculture. Int J Environ Stud. Published online at: https://doi.org/10.1080/00207233.2018.1494927

Khan ZR, Midega CAO, Pittachar JO, Murage A, Pickett J (2017) Climate-smart push-pull: a conservation agriculture technology for food security and environmental sustainability in Africa. In: Kassam AH, Mkomwa S, Friedrich T (eds) Conservation agriculture for Africa: building resilient farming systems in a changing climate. BABI, Wallingford, pp 151–166

Lalani B, Dorward P, Holloway G, Wauters E (2016) Smallholder farmers' motivations for using conservation agriculture and the roles of yield, labour and soil fertility in decision making. Agric Syst 146:80–90

Lalani B, Dorward P, Holloway G (2017) Farm level economic analysis – is conservation agriculture helping the poor? Ecol Econ 141:144–153

Landers J (2007) Tropical crop-livestock system in conservation agriculture: the Brazilian experience. Integrated crop management, vol 5. FAO, Rome

Li H, He J, Bharucha ZP, Lal R, Pretty J (2016) Improving China's food and environmental security with conservation agriculture. Int J Agric Sustain. https://doi.org/10.1080/14735903.2016.1170330

Lindwall CW, Sonntag B (eds) (2010) Landscape transformed: the history of conservation tillage and direct seeding. Knowledge impact in society. University of Saskatchewan, Saskatoon

Mello I van Raij B (2006) No-till for sustainable agriculture in Brazil. In: Proceedings of world association of soil and water conservation, P1, pp 49–57

Montgomery DR (2007) Dirt: the erosion of civilizations. University of California Press, Berkeley/Los Angeles

Nkonya E, Mirzabaev A, von Braun J (eds) (2016) Economics of land degradation and improvement – a global assessment for sustainable development. IFPRI and ZEF. Springer, Cham

Owenya MZ, Mariki LW, Kienzle J, Friedrich T, Kassam A (2011) Conservation Agriculture (CA) in Tanzania: the case of the Mwangaza B CA farmer field school (FFS), Rhotia Village, Karatu District, Arusha. Int J Agric Sustain 9(1):145–152

Pretty J (2002) Agri-culture: reconnecting people, land and nature. Earthscan

Sims B, Kassam A (2015) East Anglia branch visit to Tony Reynolds at Thurlby Grange farm: no-till farming in the Lincolnshire fens. TAA Agric Dev 26:51–52

Soane BD, Ball BC, Arvidsson J, Basch G, Roger-Estrade FMJ (2012) No-till in northern, western and south-western Europe: a review of problems and opportunities for crop production and the environment. Soil Tillage Res 118:66–87

Swift MJ, Bignell DE, Moreira FMS, Huising EJ (2008) The inventory of soil biological diversity: concepts and general guidelines. In: Moreira FMS, Huising EJ, Bignell DE (eds) A handbook of tropical soil biology: sampling and characterization of below-ground biodiversity. Earthscan, London, pp 1–16

Thierfelder C, Bunderson WT, Jere ZD, Mutenje M (2015) Development of conservation agriculture (CA) systems in Malawi: lessons learned from 2005 to 2014. Exp Agric 52(4):579–604. https://doi.org/10.1017/S0014479715000265

Vlek PLG, Khamzina A, Tamene L (eds) (2017) Land degradation and the sustainable development goals: threats and potential remedies. CIAT publication no. 440. International Center for Tropical Agriculture (CIAT), Nairobi, Kenya, 67p. Available at: http://bit.ly/2rrkxiQ

Microbial Applications for Sustainable Agriculture

Aftab Afzal and Saeed A. Asad

Abstract Agriculture in the current era is highly dependent on chemical fertilizers, pesticides and weedicides. Excessive applications of these chemicals on crop plants has increased the production cost, jeopardized the environment and has depleted the non-renewable resources. Potential threats to non-renewable resources and soil, water, air environments have led to seek alternative approaches for sustainable crop production and clean environment. To lessen these adversaries, not only scientific community, but industry and farmers are also continuously involved in research, development and adoption of new sustainable technologies. The tiny organisms in rhizosphere have shown their potential to play ubiquitous role in sustainable agricultural development and have been in continuous use since over the last century. In this chapter, different aspects of microbial applications for sustainable agriculture are elaborated. Applications of bacteria-containing biofertilizers, their types and benefits to crops have been discussed. Reports on plant growth promotion through phytohormones, siderophores and enzymes production by rhizobacteria are also detailed. Moreover, sustainable control of plant diseases through biocontrol and amelioration of abiotic stresses including; drought, salinity, climate change and heavy metals by using rhizobacteria are also encompassed in this chapter.

Keywords Applied microbiology · Biofertilizer · Biocontrol · Phytohormone · Siderophore · Abiotic stresses · Drought · Salinity · Climate change

---

A. Afzal
Department of Botany, Hazara University, Mansehra, Pakistan

S. A. Asad (✉)
Centre for Climate Research and Development, COMSATS University, Islamabad, Pakistan
e-mail: saeed.asad@comsats.edu.pk

M. Farooq, M. Pisante (eds.), *Innovations in Sustainable Agriculture*,
https://doi.org/10.1007/978-3-030-23169-9_3

1 Introduction

Agriculture is the mainstay of economies around the world. The livelihood of people does substantially rely on agricultural products in such economies. Because of intensive nature of today's agricultural practices, it happened to be a high input agriculture and costly as well. This high input agriculture has threatened the whole biogeochemical cycle. The imbalanced use of chemical fertilizers, pesticides, weedicides, fungicides has led to the development of an alarming situation by polluting soil, edibles and atmosphere. The future of agriculture, especially in developing countries is under threat because of a speedy decline in natural resources particularly the reserves of rock phosphate and fossil fuel (Clair and Lynch 2010). As a result, agriculture at time is not sustainable and is leading towards unwise and unjustified use of non-renewable resources. To cope with this alarming situation a sustainable agricultural approach is a need of the hour.

Soil bacteria known as Plant Growth Promoting Rhizobacteria (PGPR) have shown potential for sustainable agriculture. Plant growth-promoting rhizobacteria (PGPR) were first defined by Kloepper and Schroth (1978) as the bacteria inhabiting the rhizosphere, colonizing the roots when they are inoculated on the seeds and have the ability to improve plant growth (Aziz et al. 2012). Recently these eco-friendly microbes have also been recognized as a tool for combating abiotic stresses in crops (Jha and Subramanian 2018). Weller and Thomashow, (1994) reported that rhizosphere is rich source of nutrients and root exudates as a result number and diversity of bacteria is also rich in this zone generally 10–100 times as compare to bulk soil. Bacteria, fungi, actinomycetes, protozoa, and algae generally colonize in surrounding of the roots. However, bacteria are the most predominant microbes existing in the rhizosphere (Kaymak 2010). On the basis of occupancy, PGPR could be categorized in to (i) ectorhizospheric (ii) rhizoplanic or (iii) endo-rhizospheric (Gray and Smith 2005). Later on, PGPR were classified into extracellular plant growth promoting rhizobacteria (ePGPR) and intracellular plant growth promoting rhizobacteria (iPGPR) (Viveros et al. 2010). The ePGPRs may exist in the rhizosphere, on the rhizoplane or in the spaces between the cells of root cortex while iPGPRs locate generally inside the specialized nodular structures of root cells. Out of total rhizospheric bacteria, a small fraction (2–5%) may be plant growth promoters (Antoun and Prevost 2005). The PGPRs have diversified bacterial species, however the predominant are species of *Bacillus* and *Pseudomonas* (Podile and Kishore 2006). These microbial populations, when inoculated, enhance plant growth which is a proven fact (Nehra 2011; Bhattacharyya and Jha 2012).

The success of inoculation depends on (i) survival of inoculated bacteria on seed, (ii) ability to reproduce in the spermosphere (region around seed), (iii) ability to attach to the phyllo sphere and (iv) the ability of inoculated bacteria to colonize the extending root system (Kloepper 1993) and of course the inoculation method. Most of the time, PGPR fail in the field due to incapability to survive and colonize plant roots (Bloemberg and Lugtenberg 2001). Basically this colonization process is controlled by a variety of bacterial traits and specific genes. These traits include; motility, chemotaxis in response to seed and root secretions, production of pili or fimbriae, production of cell surface components, protein secretions and quorum sensing. Now the mutants are being generated to study the expression of these traits in order to comprehend the involvement of these traits in colonization process (Lugtenberg et al. 2001). To detect gene expression during colonization, reporter transposons and in vitro expression technology (IVET) are being employed (Roberts et al. 1998; Rainey 1999). The location of individual rhizobacteria and its metabolic activity in the rhizosphere can be monitored by using molecular markers such as green fluorescent protein, gfp, rfp, lux, gus or fluorescent antibodies and by using confocal laser scanning microscopy (Bloemberg et al. 2000). By combining these techniques with an rRNA-targeting probe it was revealed that bacteria colonized at the root tip were most active (Lübeck et al. 2000).

The PGPR can increase the plant growth either directly or indirectly (Glick 1995; Akhtar and Siddiqui 2009) through various mechanisms. The direct modes of action include; nitrogen fixation, solubilization of phosphorous and various other minerals (e.g. K, Zn), phytohormone production and reducing the level of ethylene by producing ACC- deaminase (Vessey 2003; Ahemad and Kibret 2014). The rhizobacteria with these direct mechanisms act as biofertilizers or Phytostimulators in the absence of plant pathogens (Lugtenberg and Kamilova 2009). On the other hand, some PGPR improve plant growth indirectly by suppressing plant pathogens (especially soilborne plant pathogens) using different mechanisms (Labuschagne et al. 2010). So, the beneficial rhizobacteria can promote plant growth as well as plant health through environment friendly way (Calvo et al. 2014). For decades, a large number of PGPRs have been investigated and some of them have been marketed as biofertilizers/bio pesticides, including the Genera: *Pseudomonas*, *Bacillus*, *Azosprillum*, *Azobacter*, *Enterobacter*, *Klebsiella*, *Variovorax* and *Serratia* (Glick 2012).

The interactions of PGPR with plants have been used commercially (Podile and Kishore 2006) and is very applicable for sustainable agriculture. These bacteria have been reported to interact with a variety of crop plants including maize, wheat, oat, barley, peas, canola, soy, potatoes, tomatoes, lentils, radicchio and cucumber (Gray and Smith 2005). There is a strong growing market for microbial inoculants worldwide with an annual growth rate of approximately 10% (Berg 2009). So it's a scientifically and technologically proven fact that PGPR can be applied for sustainable crop production and for environment friendly agricultural practices. To highlight the applied aspects of microbes and their potential as biological agents for fertilization and disease control a thorough review has been done. In current review, five different applications of microbes (bacteria) for sustainable agriculture have been discussed. These applications included: (i) biofertilizers, when they are used to enhance nutrient availability (ii) Phytostimulators, helping in plant growth via plant growth regulators (iii) bio pesticides, while protecting plants from phytopathogens (iv) as Bioremediators, their application for cleaning the soil polluted with heavy metals and (v) abiotic stress ameliorators, when rhizobacteria are being employed for reducing the risk of abiotic stress conditions.

2 Historical Perspectives

The involvement of rhizobacteria in plant growth promotion is historical and inoculation of plants with useful bacteria is not a new idea, it goes back to many centuries. The benefits of growing legumes before non-legumes were well known to farmers by experience. In late nineteenth century, the rhizobium inoculants were used in USA for the first time as a biological fertilizer named as 'Nitragin' and subsequently this practice was used with legumes in many countries (Bagnasco et al. 1998). Kloepper and Schroth (1978) for the first time used the term "plant growth promoting rhizobacteria (PGPR)" for these beneficial bacteria.

The practice of mixing "naturally inoculated" soil with seeds became a recommended method of legume inoculation in the USA by the end of the nineteenth century (Smith 1992). A decade later, the *Rhizobium* sp. was registered for plant inoculation, as the first patent ("Nitragin") (Nobbe and Hiltner 1896). Ultimately, rhizobia inoculation to legumes became common practice. Many small companies are producing *Rhizobium* inoculants in different countries since centuries. In Brazil for example nitrogen fertilization is done through Rhizobia (Döbereiner et al. 1994). Similarly, rhizobial inoculants were contributing significantly to legume production in Australia, New Zealand, Egypt, South Africa, North America, Eastern Europe and somewhat in Southeast Asia. In the USA, Brazil, and Argentina however inoculation of soybean made a major agricultural impact. Inoculant technology has not been much successful in Asia, Africa, and Central and South America due to poor quality of inoculants (Eaglesham 1989). In the 1930s and 1940s *Azotobacter* inoculation was done on a large scale in Russia. But this practice could not bring noticeable results, so it was abandoned at that time (Rubenchik 1963). In 1930s *Bacillus megaterium* was used for phosphate solubilization on large scale in Eastern Europe (Macdonald 1989). Two major advancements in biofertilizer technology occurred in the late 1970s (1) Plant growth and yield of non-legumes was improved significantly with inoculation of *Azospirillum* (Döbereiner and Day 1976), due to its direct effect on plant metabolism (Bashan and Holguin 1997), and (2) *Pseudomonas fluorescens* and *P. putida* were largely investigated and proven to be effective biocontrol agents (Défago et al. 1992; Kloepper and Schroth 1981; Glick 1995; Glick and Bashan 1997). At the end of twentieth century other bacteria like *Bacillus*, *Flavobacterium*, *Acetobacter*, and several other microorganisms were also investigated to be potential PGPR (Kloepper 1994; Tang 1994; Tang and Yang 1997). The first commercial inoculant of PGPR (Free living or associative rhizobacteria) was only possible at the end of last century (Fages 1992; Tang 1994; Tang and Yang 1997).

3 Application of Microbes as Nutrient Mobilizers (Biofertilizers)

Biofertilizer is a substance which contains living microorganisms and promotes growth by increasing the supply or availability of primary nutrients to the host plant, when applied to seeds, plant surfaces, or soil (Vessey 2003). Bio-fertilizers provide "eco-friendly"organic agro-input. Biofertilizer technology has shown promise for integrated nutrient management through biological N fixation (BNF) and improving P availability to crops. Bio-fertilizers, containing *Rhizobium*, *Azotobacter*, *Azospirillum* and blue green algae (BGA) have been in use since a long time. Rhizobium inoculant is used for leguminous crops. One of the rhizobacteria '*Azotobacter*' is used for the inoculation of crops like wheat, maize, mustard, cotton, potato and other vegetable crops. Another very promising rhizobacteria *Azospirillum* is generally recommended for inoculation onto sorghum, millets, maize, sugarcane and wheat. Nitrogen fixing cyanobacterial genus, *Nostoc* or *Anabaena* or *Tolypothrix* or *Aulosira* (Blue green algae) are used as inoculations for paddy crop. One of the blue green algae '*Anabaena*' can fix N up to 60 kg/ha/season in association with water fern Azolla. Biofertilizers can be classified on the basis of basic nutrient they provide to the crops. Three major classes are described in the paragraphs below.

3.1 Nitrogen Biofertilizers

Nitrogen (N) is the most essential and primary macronutrient for plants. Nitrogen-fixing microorganisms can transform atmospheric nitrogen into available nitrogen (inorganic compounds usable by plants) through conversion of N_2 into NH_3. These microorganisms play a pivotal role in N cycle, because more than 90% of N fixation is carried out by these organisms (Encyclopedia Britannica 2018). Generally, there are two kinds of N fixing microorganisms (Bacteria). The first kind, symbiotic (mutualistic) bacteria, includes *Rhizobium* (symbiotic with leguminous plants) and *Frankia* (symbiotic with actinorhizal plants). The second type, non-symbiotic bacteria may be free-living (e.g., *Azotobacter*, *Beijerinckia*, *Clostridium*, cyanobacteria (*Anabaena* and *Nostoc*) *Gluconacetobacter diazotrophicus* and *Azocarus*) or associative/endophytic (e.g., *Azospirillum*) (Bhattacharyya and Jha 2012).

The first bacterium of this type was isolated by Beijerinck from the nodules of legumes in 1888 and named as *Bacillus radiocicola*. However, Frank (1889) renamed it as *Rhizobium leguminosarum* (Fred et al. 1932), which was retained in *Bergey's Manual of Determinative Bacteriology* (Holt et al. 1994). Salvagiotti et al. (2008) while analyzing the data of publications from 1966 to 2006, derived from 108 field studies in 17 countries mostly related to soybean N fixation and fertilization, concluded that biological N fixation has a major contribution (50–60%) in soybean N fertilization; however, increasing N fertilizer rates badly affect N fixation.

Rhizobium inoculation on other legumes is also beneficial and helps in profound nodulation and increasing yield of important legumes.

3.1.1 Non-symbiotic N-Fixers

3.1.1.1 Free-Living Nitrogen Fixers

Free living bacteria (rhizospheric bacteria) have the capability to inhabit soil and biologically fix N, without any host. *Azotobacter*, *Beijerinckia*, *Clostridium*, *Cyanobacteria* (*Anabaena* and *Nostoc*) *Gluconacetobacter diazotrophicus* and *Azocarus* are examples of free living N fixers. Vadakattu and Paterson (2006) reported that free-living N fixers contributed 20 kilograms N per hectare per year in an intensive wheat rotation farming system in Australia (30–50% of the total needs). Non-symbiotic N_2 fixation (by free-living bacteria in soils or associated with the rhizosphere) is important in providing some amount of N particularly in low input cropping systems worldwide. Due to use of indirect methods of measurement of N fixed, non-symbiotic N fixers could not get good name, however isotope-based direct methods indicate agronomically significant amounts of N_2 fixation both in annual crop and perennial grass systems. New molecular technologies should be employed to determine the potential of free living N fixers. This knowledge should assist the development of new plant-diazotrophic combinations for specific environments and more sustainable exploitation of N_2-fixing bacteria as inoculants for agriculture (Roper and Gupta 2016).

3.1.1.2 Associative Nitrogen Fixers

Some bacterial species live in close association with host plant, either on the surface of roots or inside the root (in intercellular spaces). Species of *Azospirillum* are peculiar example of such species which form association with important cereal crops such as; rice, wheat, corn, oats, and barley. These bacteria are able to fix atmospheric Nitrogen which is useful for the plants. It is reported that such bacteria can fix upto 52 mg N_2 g^{-1} malate (Stephan 1979). One of the most used plant growth promoting bacteria (PGPB) is *Azospirillum brasilense.* This bacterial specie has been used in Brazil, Argentina, Mexico, India and Europe. Statistically significant increases in yield varying from 5% to 30% have been achieved as a result of inoculation of *A. brasilense*. Analyses of field experiments have shown that 60–70% of inoculation with Azospirillum was successful (Yaacov and Robin 1995). Associative N fixation (ANF) proved to be an important source of N to unfertilized switchgrass and to temperate grasslands. This was concluded by Roley et al. (2018) when he used to measure N fixation potential of associative bacteria.

Fig. 1 Extensive nodulation of a peanut root after inoculation with *Bradyrhizobium* strain 32H1. (Source: Wagner 2011)

3.1.2 Symbiotic Nitrogen Fixers

Symbiotic N_2-fixing bacteria include members of the family rhizobiaceae (*Rhizobium, Sinorhizobium, Bradyrhizobium, Mesorhizobium* and *Azorhizobium*, collectively termed as 'rhizobia' which forms symbiosis with leguminous plants (Zahran 2001). Alfalfa, beans, chickpea, clover, cowpeas, lupines, peanut, soybean and vetches are important legumes used in agricultural systems. About 50% of the global area devoted to legumes is under the cultivation of soybean and represent 68% of the total global legume production (Vance 2001). The symbiotic nitrogen-fixing bacteria (Rhizobia) enter the root hairs of host plants, where they multiply and stimulate formation of root nodules. A typical example of nodule formation in peanut is shown in Fig. 1. The nodules are the sites of N fixation where it is reduced to ammonia in presence of a complex enzyme system, the 'Nitrogenase'. The ammoniac N is available for plant nutrition. This natural process is expedited by application of inocula of rhizobial strains to seeds of legume crops for abundant nodule formation and maximum plant growth (Encyclopedia Britannica 2018; Laranjo et al. 2014).

A water fern *Azolla* also form symbiotic relationship with a cyanobacterium *Anabaena Azolla*. *Azolla* fronds allow the *Anabaena* to colonize at the cavities formed at its base. After colonization the cyanobacteria fix a plenty of N in its specialized cells called heterocyst. This symbiotic relationship is being employed as biofertilizer for at least 1000 years in wetland paddies in Southeast Asia. During the growing season, up to 600 kg N ha^{-1} $year^{-1}$ is fixed by *Azolla* "blooms" in rice paddies (Postgate 1982; Fattah 2005).

Actinomycetes, *Frankia alder* (Alnus sp. actinorhizal plants) is another example of symbiotic association (Benson and Silvester 1993). The tree genera such as the temperate-region *Alnus* and *Myrica*, the arid-region *Acacia*, and the tropical-region

Casuarina and *Ceanothus* are typical examples of actinorhizal plants making symbiotic association with actinomycetes (*Frankia*). In the latter region, efforts are being made to develop a crop-rotation system with agro-forestry that utilizes leguminous trees (e.g., *Leucaena leucocephala*) able to incorporate significant amounts of nitrogen into the soil for the subsequent benefit of crop production. This type of association needs to be investigated and exploited further for its potential use in sustainable agriculture system.

3.2 Phosphate Solubilizing Biofertilizers

Phosphorus (P) is the second most essential macronutrient for plants after nitrogen (N), and is applied to soil in the form of phosphate fertilizers. However, most of the P applied to soil or native soil P becomes unavailable to plants because it forms chemical bonding with the metal ions present in the soil (Ca^{++}, Fe^{++} or Al^{+}), thus forming insoluble compounds (Malboobi et al. 2009). Phosphate solubilizing bacteria (PSB) are beneficial bacteria capable of solubilizing inorganic phosphorus from insoluble compounds (Chen et al. 2006). This is one of the most important traits of the rhizospheric bacteria and plays a significant role in P nutrition of crop plants. It is generally accepted that these bacteria solubilize P by producing low molecular weight organic acids which can chelate the cations chemically bonded to P thus releasing P from these insoluble inorganic compounds. Such bacteria (PSB) are being used for preparing biofertilizers for increasing P availability to plants. The major issue is to optimize the P fertilizer rates without compromising the yield and to minimize P loss from soil. These bacteria have attracted the attention of agriculturists for sustainable crop production (Zandi and Chalaras 2014). About 50% of the crop requirement of phosphatic fertilizer can be saved by using PSB with rock phosphate (Saleem et al. 2013). Accordingly, it is reported that P fertilizer application can be reduced to 50% by co-inoculating the phosphate-solubilizing bacteria (PSB) with PGPR without compromising crop yields (Jilani et al. 2007; Yazdani et al. 2009). Inoculation of seeds with PSB can reduce the use of P fertilizers upto 50% (equivalent to 30 kg P_2O_5 ha^{-1}). Alternatively, fertigation or hydroponic methods can also be employed to inoculate PSB to fields. *Bacillus megaterium, Pseudomonas putida* (P13), *Pantoea agglomerans* (P5), *Microbacterium laevaniformans* (P7) strains are highly effective for insoluble phosphate solubilization. A consortium of bacteria is more effective and solubilizes phosphate at faster rate than single strain inoculum. Peter et al. (2016) used a consortium of four PGPR (marketed as Mammoth P) and reported that it solubilizes phosphate at much faster rate as compared to any strain inoculated alone. Romano et al. (2017) reported recently that bacteria inhabited under phosphorus deficient conditions produce iron-chelating molecules (Siderophores). It was suggested by the author that some bacteria can interact with both of these elements (Phosphorus and iron) and can improve the availability of these essential and limiting plant nutrients.

4 Biofertilizers Impacts on Crop Yield and Nutrient Uptake

According to a meta-analysis conducted recently (Schutz et al. 2018), Arbuscular Mycorrhizal Fungi (AMF), and other biofertilizers with N fixing and P solubilizing capability are most effective in increasing crop yield and nutrient uptake. Co-inoculation of bacteria with both traits (N fixation and P solubilization) is more beneficial for improving crop yield as compare to single inoculation (Fig. 2). Similarly, across all crop categories (Table 1), an average yield increase of 16.2 ± 1.0% was recorded by inoculation with biofertilizers as compared to non-inoculated controls (Fig. 3a). It was also noted that legumes showed greater response upon inoculation and response of root crops was relatively poor. Phosphorus use efficiency (PUE) and nitrogen use efficiency (NUE) was also improved as a result of biofertilization (7.5 ± 0.8 kg yield per kg P and 5.8 ± 0.6 kg yield per kg N fertilizer). The nutrient use efficiency was most profound in legumes as compared to other crops (Fig. 3b, c).

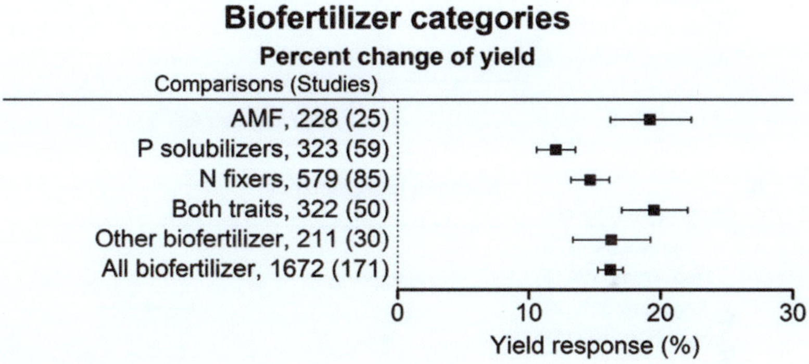

Fig. 2 Percentage change of yields in response to the application of various categories of biofertilizers. Mean values and 95% confidence intervals of the back-transformed response ratios are shown. AMF and N-fixers in combination with P solubilizers showing more pronounced effect. (Adapted from Schutz et al. 2018)

Table 1 Crops included in meta-analysis

Category	Crops included
Cereals	Barley, durum wheat, rice, spring wheat, winter wheat, pearl millet, maize, sorghum, kamut, silage maize, ryegrass, finger millet
Legumes	Blackgram, chickpea, peanut, horsegram, kidney bean, mung bean, fenugreek, lentil, snap bean, soybean, runner bean, pigeon pea
Root crops	Garlic, potato, turmeric, sugar beet, cassava
Vegetables	Eggplant, tomato, cabbage, watermelon, pepper, okra, cucumber, melon
Other crops	Dill, anise, rapeseed, cotton, sesame, fennel, coriander, sunflower, mustard, sugarcane

Source: Schutz et al. (2018)

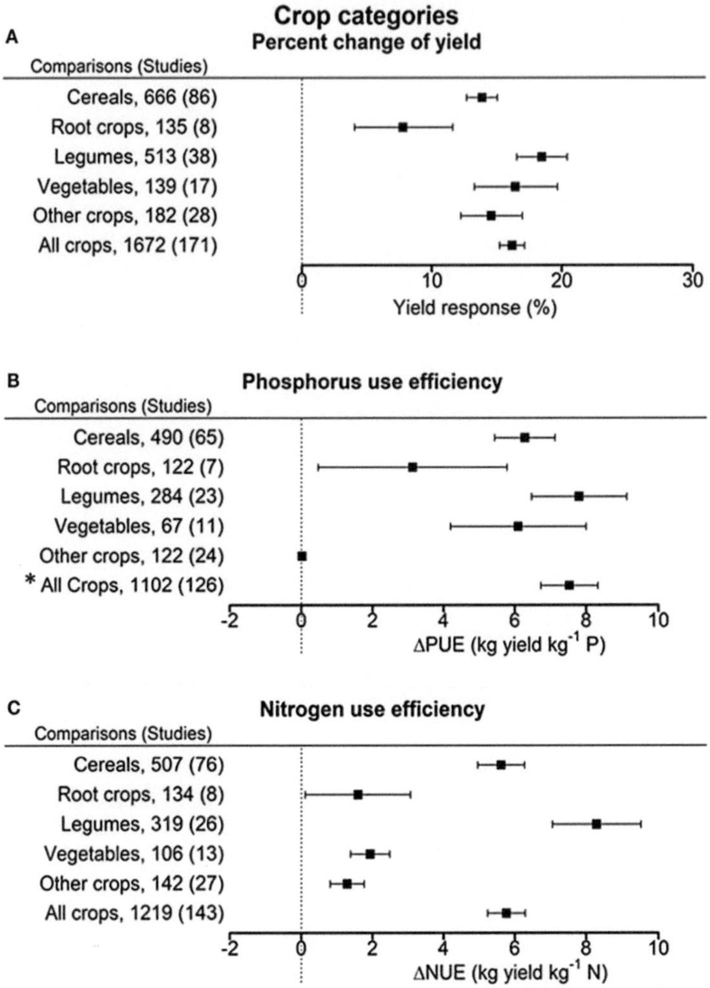

Fig. 3 Percentage change of yields (**a**), change in phosphorus use efficiency (PUE) (**b**), nitrogen use efficiency (NUE) (**c**), in response to biofertilizer application. ∗The high value for all crops is caused by the outlier calculation that resulted in different pairs being excluded for the full sample and the sub-samples. (Adapted from Schutz et al. 2018)

5 Microbial Applications as Plant Growth Promoters (Phytostimulators)

Soil microorganisms are ubiquitous to impart myriads of benefits for plant growth and health leading to successful survival of flora. These tiny creatures exhibit unique characteristics which directly or indirectly regulate some core functions of plants (Berg 2009). Diazotrophs including; *Rhizobium* and *Azospirillum* significantly improve plant growth while producing phytohormones, nitrogen fixation, phosphate solubilization. While several bacterial (e.g. *Pseudomonas*) and fungal (*Trichoderma* and *Coniothyrium*) genera are well studied for their involvement in improving plant health and barring different plant diseases. Plant beneficial microorganisms including plant growth promoting rhizobacteria, mycorrhizal fungi and antagonists are the special creatures for partial substitution or possible replacement of artificial chemicals. These microbes or microbial products can be successfully employed to meet the increasing food demand without any toxicity or environmental concerns (Glick 2012).

Mycorrhizal fungi living in symbiotic association with plants are of two types; ectomycorrhizae and arbuscular mycorrhizae (AM) where the later are most abundant in soil environments. The AM form symbiotic relationship with plants (Willis et al. 2013). These fungi play pivotal role in enhancing the productivity of several field crops by penetrating deeply in to the soil, for more nutrients and water especially under nutrient and water limiting environments (Guo et al. 2010) than non-mycorrhizal plants. Mycorrhizae promote plant growth via secretion of metabolites including; amino acids, phytohormone, vitamins and/or through speeding up the mineralization processes. They also enhance the supply of phosphate, where around 80% of phosphorous taken up by mycorrhizal plant is supplied by AM fungi (Marschner and Dell 1994). Moreover, these fungi also help plant to take up other macro and micro nutrients including N, Zn, K, Cu and Mg especially when these nutrients are in less soluble forms.

Microorganisms influence plant growth through; secretion of metabolites, enzymes, inducing systemic resistance, protecting from pathogens and diseases and most importantly from environmental stresses (Shameer and Prasad 2018). Regardless of their modes of actions, these microorganisms could involve anyone or multiple of the following metabolites for enhancing plant growth and deterring pathogenicity. Paragraphs below detail the involvement and mechanisms through which these microorganisms alleviate the plants against stresses.

5.1 Phytohormone Producers

Among many of the physiological attributes, phytohormone production/metabolization ability of these tiny creatures is well recognized (Okon and Labandera-González 1994). Benefits imparted by microorganisms are actually an outcome of multiple physiological activities occurring at the same time. This series of activities idea originates from the "additive hypothesis" proposed in the last decade of previous century (Bashan and Levanony 1990; Bashan and de Bashan 2010). Phytohormone production by microorganisms is one of the mechanisms to explain this hypothesis. Almost all of the microbial genera involved in plant growth are capable to produce or metabolize phytohormones. All bio inoculants or biofertilizers including; *Bacillus, Azospirillum, Pseudomonas, Enterobacter, Erwinia, Azotobacter, Serratia, Klebsiella, Alcaligenes, Flavobacterium, Arthrobacter, Burkholderia, Bacillus,*

Acinetobacter, Azotobacterium, Xanthomonas and *Rhizobium* (Bhattacharyya and Jha 2012) have demonstrated phytohormone production under controlled and natural settings.

Phytohormones produced by different microbial genera and physiological functions attributed to them are detailed in Table 2. Auxins, gibberellins, cytokinins, ethylene and abscisic acid are the most studied microbial hormones involved in the plant growth and development one way or the other. *Auxins* or naturally occurring auxin molecule, indole-3-acetic acid (IAA) are known to induce cell elongation in the subapical regions of the stem. The major attributes of these hormones vary from lateral and adventitious roots initiation, root/shoot elongation, photo and gravitropism and cell division (Teale et al. 2006). Infact, *Azospirillum* has proved to be a model species for elaborating and understanding the role of auxins in plant growth and even in maintaining plant and rhizobial interactions (Berg 2009). Despite a wide range of physiological activities attributed to auxins produced by *Azospirillum*, only few commercial bio inoculants have been formulated containing *Azospirillum* sp. (Cassán et al. 2014) Majority of the plant growth promoters and bio fungicides respectively contain *Bacillus* and *Trichoderma* as is evident from data in Table 3. As described in previous paragraphs, direct promotion of plant growth by microbial inoculants is through secretion of phytohormones. *Acetobacter diazotrophicus*, *Azospirillum* sp. *Azospirillum lipoferum* and *Azospirillum brasilense* all have been well studied for producing indole3-acetic acid (IAA), ethylene, gibberellic acid (GA3) and abscisic acid (ABA) respectively (Bastian et al. 1998; Strzelczyk et al. 1994). Auxin production by *Azospirillum* in the root zone is considered as the major factor for enhancing plant growth and development of root system in Gramineae plants. Moreover, these auxins also regulate other rhizosphere bacteria such as nodule formation and improve the symbiotic relationships between rhizobia and legumes. Therefore, any alteration in the concentration of auxin could severely impact nodule formation (Mathesius et al. 1997). A study conducted by Burdman et al. (1996) revealed that *Phaseolus vulgaris* seedlings inoculated with *A. brasilense* exhibited increased root flavonoids and enhanced expression of nod gene in *Rhizobium* compared with control. Auxins facilitate adventitious roots penetration thereby providing more nutrients for bacterial and plant growth. These characteristics make auxins key metabolite to regulate plant-microbe interactions in terms of Phyto stabilization and pathogenicity (Ahemad and Khan 2012a).

Another important phytohormone, Gibberellic acid (GA) alleviate the drought stress and play crucial role in the initiation of flowering and hypocotyls elongation (Yamaguchi 2008; Vandenbussche et al. 2005). Major physiological development in plants from seed germination to photosynthetic activity, light interception, nutrient use efficiency, fruit growth and delayed dormancy in major plant genera are attributed to the presence of gibberellic acid. This hormone actively relives the plant against abiotic stresses and maintains the continued growth and development of stressed plant organs (Iqbal et al. 2011). GAs produced by *Bacillus* and *Azospirillum* inoculants resulted in increased uptake of N in wheat plants thereby alleviating the plants against drought and salinity stresses (Shaddad et al. 2013). Other plant bacteria, *Pseudomonas*, *Bacillus* and *Azotobacter* and *actinomycetes* were reported to

Table 2 Phytohormones and other plant beneficial metabolites produced by different microorganisms and their effects on plant growth/health. Data presented below is research data collected from in vitro experimentation

Microorganisms/ biofertilizers	Phytohormones/ metabolites produced by beneficial microbes	Influence on plant growth/ health	References
Bradyrhizobium japonicum	Auxins (IAA), siderophores, antibiotics, cell wall degrading enzymes	Phosphate solubilization, improved germination and increased biomass of plant	Chandra and Pareek (2007) and Afzal and Bano (2008)
Mycobacterium	Auxins (IAA)	Increased plant resistance against pathogens	Egamberdiyeva (2007)
Pseudomonas	Siderophores	Inhibited fungal growth on plant roots	Nowak et al. (1994)
Fluorescens			
Bacillus sp.	Auxin (IAA) and spore formation	Increased shoot length by up to 40% and increased the number and length of adventitious roots	Ahmed and Hasnain (2010)
Burkholderia	Auxin (IAA), reduced acetylene to ethylene	Improved germinations percentage and increased rice yield up to 23%	Govindarajan et al. (2008)
Bradyrhizobium sp.	HCN, Auxins (IAA) and siderophores.	Phosphate solubilization, significantly increased plant biomass and wheat yield	Afzal and Bano (2008)
Sphingomonas	Gibberellins (GAs)	Enhanced the plants competitive ability for space and nutrients	Innerebner et al. (2011) and Khan et al. (2014)
Enterobacter cloacae	Auxins (IAA)	Phosphate solubilization	Bhattacharyya and Jha (2012)
Serratia mercescens	Auxin, HCN and siderophore production	Significantly improved plant biomass	Selvakumar et al. (2008)
Acinetobacter sp.	Auxins, ACC deaminase and producing antifungal metabolites	Phosphate solubilization	Indiragandhi et al. (2008)
Actinomycetes	Cell wall degrading enzymes (e.g. cellulases)	Induced resistance against soil born pathogen, *R. solani*	Schmidt et al. (2001)
Enterobacter asburiae	Auxins (IAA), HCN, exopolysaccharides, and siderophores	Phosphate solubilization	Ahemad and Khan (2012b)
Streptomyces	Auxins (IAA) and siderophores	Resistance against soil borne pathogens	Verma et al. (2011)
Rhizobium leguminosarum	Cytokinins, antibiotics and cell wall degrading enzymes	Enhanced minerals and P solubilization for plant uptake	Zahir et al. (2010)
Azotobacter chroococcum	Gibberellins, kinetin, IAA	Phosphate solubilization	Ahemad and Kibret (2014)

Table 3 List of commercially available biocontrol products

Microorganism's type	Biocontrol Agent	Commercial mame/ company	Target pathogen
Bacteria	*Agrobacterium radiobacter*	Galltrol/AgBioChem Inc. USA	Crown gall disease caused by *Agrobacterium tumefaciens*
		www.agbiochem.com	
		Nogall/Bio Care Technology, Australia	Crown gall disease caused by *Agrobacterium tumefaciens*
		http://bio-caretechnology.com/	
	Bacillus sp.	Companion/Growth Products Ltd. NY, USA/	*Rhizoctonia*, *Pythium*, *Fusarium*, and *Phytophthora*
		http://www.growthproducts.com	
		HiStick N/T/Helena Agri-Enterprises, USA	*Fusarium, Rhizoctonia, Aspergillus*
		https://helenaagri.com	*Rhizoctonia solani, Fusarium spp.,*
		Kodiak/ Bayer crop Science, USA	*Alternaria spp., and Aspergillus spp.* that attack roots powdery mildew, downy mildew,
		https://www.bayer.com	
		Serenade/Bayer crop Science, USA	Cercospora leaf spot, early blight, late blight, brown rot, fire blight
		https://www.cropscience.bayer.us	
		YieldShield/Bayer crop Science, USA	Soil borne fungal pathogens causing root diseases
		https://www.cropscience.bayer.us	
		Rhizo-Plus/Disha Chemicals, India	*R. solani*, *Fusarium spp. Alternaria spp.*, *Sclerotinia* and *Verticillium*
		http://www.theagrihub.com	
	Pseudomonas sp.	BioJet Spot-Less/Eco Soils Systems, Inc., San Diego, Ca	Dollar spot, Anthracnose, *Pythium aphanidermatum, Michrochium patch* (pink snow mold)
		https://www.nasdaq.com	
		Bio-save/Jet Harvest Solutions, Florida, USA	*Botrytis cinerea*, *Penicillium spp.*, *Mucor pyroformis, Geotrichum candidum*
		https://jetharvest.com	
		BlightBan/Nufarm Americas Inc. USA	*Erwinia amylovora*, and russet inducing bacteria
		http://www.nufarm.com	
		Cedomon/Nutrilita, Lithuania	Leaf stripe, net blotch, *Fusarium* sp., spot blotch, leaf spot
		http://www.nutrilita.lt	

(continued)

Table 3 (continued)

Microorganism's type	Biocontrol Agent	Commercial mame/ company	Target pathogen
		Conquer/Mauri Foods, Australia	*Pseudomonas tolassii*
		http://www.maurianz.com	
		Victus/Sylvan Spawn Laboratories, USA	*Pseudomonas tolassii*
		https://www.manta.com	
Fungi	*Ampelomyces quisqualis*	AQ10/Bioguard, CBC Group, Europe	Powdery mildew
		http://www.biogard.it	
	Candida oleophila	Aspire/Ecogen Inc. USA	*Botrytis* spp., *Penicillium* spp.
		https://www.bloomberg.com	
	Coniothyrium minitans	Contans WG/Intercept WG/Bayer Crop Science, South Africa	*Sclerotina sclerotiorum* and *S. minor*
		https://www.cropscience.bayer.co.za	
	Myrothecium verrucaria (killed)	DiTera/Valent, North America	Parasitic nematodes
		https://www.valent.com	
	Trichoderma sp./*Gliocladium* sp.	Plantshield/Rootshield/T-22 Planter box Soilgard Primastop/Bioworks, NY, USA	Pythium spp., Rhizoctonia solani, Fusarium spp.
		https://www.bioworksinc.com	

Adopted and modified from Gardener and Fravel (2002)
These biocontrol products are registered with the environment protection agency (EPA) of USA

produce GAs which significantly influenced nutrient uptake and growth improvement of inoculated plants of wheat as compared to control (Shaddad et al. 2013).

Cytokinins play key role in cell division, primary root growth and senescence. In fact, many of the cytokinins genes are expressed in roots highlighting their involvement in root development. Wide range of microbial species produce cytokinins in plant roots enhancing their growth and development thereby resulting in more nutrients uptake in plants. Cytokinins produced by *Bacillus megaterium* had a significant role in promoting plant growth as noted by Ortíz-Castro et al. (2008). These bacterial cytokinins influenced the root architecture, increased root hair length and lateral root formation. Endophytic bacteria, *Bacillus* isolated from *Arabidopsis thaliana* exhibited the potential to increase the root/shoot growth as compared to control plants (Wang et al. 2015).

The plant hormone, ethylene is recognized as the regulator of plant growth and development. In response to environmental stresses, plants up regulate the production of ethylene to initiate the defense mechanisms, but increased production of ethylene will induce a range of abnormalities including growth inhibition and delayed flowering. This increased level of ethylene can be easily reduced by using chemicals such as; cobalt ion (Co^{2+}) and silver ion (Ag^{+}), but because of their toxicity and higher price make them last choice for farmers. Hence keeping the balance in ethylene production is of paramount importance for agricultural crops productivity and microbes can potentially modulate the ethylene production. For example, microorganisms decrease the level of ethylene through ACC-deaminase enzyme production which is widely reported in fungi, bacteria and stramenopiles (Nascimento et al. 2014). Worth noting that ethylene reduction by microorganisms is not always in the favor of plant. For example, under saline environment, ethylene reduces root growth to avoid salt pollution. Under such environment, ethylene reduction by microbes may increase root growth but may also result in disastrous effects on overall growth of plant and food chain toxicity (Desbrosses et al. 2009).

Like ethylene, abscisic acid is also called stress hormone, synthesized in plants in response to environmental stresses and expressing the stress resistance genes (Sah et al. 2016). ABA ameliorates the salinity stress by regulating the photosynthetic apparatus and is also important hormone for mediating the plant-microbial interactions as many plant growth promoting bacteria such as; *P. fluorescens, A. brasilense, Variovorax paradoxus* and *B. licheniformis* produce ABA (Dodd et al. 2010; Cohen et al. 2015). A study by Cohen et al. (2015) revealed that plant inoculated with abscisic acid producing PGPR, *P. fluorescens* enhanced the ABA hormone thereby increasing their ability to withstand better under drought conditions as compared to uninoculated controls. Moreover, inoculation with PGPR decrease the hormone accumulation in roots thereby regulating shoot/root and root/shoot hormonal signaling and resulting changes in ABA may reduce the plant sensitivity to water deficiency. Qin et al. (2016) reported that tomato plants inoculated with halotolerant PGPR exhibited enhanced growth. Role of PGPR to influence ABA concentrations makes it an ideal choice for inducing resistance against abiotic stresses in plants and withstand harsh environments without jeopardizing the yield potential.

5.2 Siderophore Producers

Siderophores are low molecular weight iron chelating metabolites having great affinity for iron. Out of approximately 500 known siderophores, chemical formulae of >200 have been worked yet (Shameer and Prasad 2018). These water soluble compounds can be grouped in to extracellular and intracellular ones (Hider and Kong 2010). In fact, siderophores are the key instrument to release unavailable iron and make it available to the living biota (Rajkumar et al. 2010). Iron mostly exists in Fe^{3+} form which remains insoluble and hence unavailable for plant uptake. Siderophores released by microorganisms' act as iron solubilizing agents especially under iron-limiting conditions (Ahemad and Khan 2012a).

Microbes used in the formulations of biofertilizers are both gram positive and gram negative and interestingly both forms of bacteria are equipped with the ability to reduce Fe^{3+} to Fe^{2+} in their membranes. This reduced form of iron is released by siderophores in to the cell making it available to plants via gating channels connecting outer and inner membranes of the cell (Mahanty et al. 2017). Various mechanisms through which plants take up iron liberated by bacterial siderophores include either direct uptake of Fe-siderophore complexes, or through chelating/releasing iron and ligand exchange reaction (Thomine and Lanquar 2011). Model plant, *Arabidopsis thaliana* accumulated an increased level of Fe synthesized by *Pseudomonas fluorescens* from Fe-pyoverdin complex which significantly improved plant growth compared with control plants (Parray et al. 2016). *Pseudomonas* produce a mixture of Fe-pyoverdin which has key role for iron uptake by *A. thaliana.* Fe availability is of paramount importance especially for plants under stressed environments, where these siderophores alleviate the plants against heavy metal stresses (Rajkumar et al. 2010). Several studies investigating the benefits of siderophores revealed that plants were able to take up the iron once inoculated with siderophore producing *Pseudomonas* bacteria (Hider and Kong 2010). Mung bean (*Phaseolus vulgaris*) plants inoculated with these bacteria in iron deficient soils, showed less chlorotic symptoms than control plants. Iron supply is imperative for plants exposed to heavy metal stress, where siderophores produced by microorganisms alleviate heavy metal stresses to plants. This siderophore triggered uptake of iron help plants to survive under Fe-limiting conditions (Guerinot and Ying 1994).

5.3 Enzymes Production

Plant diseases/pathogens have deleterious effects on agricultural productivity and pose a multiplying challenge for ensuring the food security. Amongst those pathogens, soil borne pathogens are the most devastating agents hampering the agricultural productivity (Newbery et al. 2016; Kashyap et al. 2017). According to Savary et al. (2012), direct yield losses because of diseases and weeds are approximately 40% of agricultural produce. For controlling these plant diseases, use of pesticides has rewarded in terms of yield increase but compromising on the quality as well as heralding challenges for sustainable production. To substitute or lessen the use of chemicals, plant beneficial microorganisms have imparted marvelous benefits in terms of biological control of pathogens. This is perhaps due to the fact that onset of green revolution and indiscriminate uses of herbicides, pesticides and chemical fertilizers has posed several adverse effects to the environment (Tilman 1998). Many of these chemicals have been reported to be carcinogenic (Damalas and Eleftherohorinos 2011). To lessen the adversaries triggered by these toxic means of controlling pathogens and diseases, biological control is employed for controlling agricultural pests mainly for economic and sustainability. Various microorganisms exhibit hyperparasitic action to hydrolyze pathogen cell wall through extracellular enzymes (Chemin and Chet 2002). For instance, chitinase produced by *Serratia plymuthica* significantly reduce spore germination of *Botrytis cinerea* (Gaffney et al. 1994). Soil bacteria perform excellently to control soil borne plant pathogens. *Bacillus* controls various fungal diseases through secretion of various lytic enzymes which inhibit mycelial growth of various fungal species (Yu et al. 2002). Overall three mechanisms including: the secretion of antibiotics, competing for nutrients and space and mycoparasitism by microorganisms suppress the pathogen growth. Interestingly, many of the *Bacillus* strains exhibit mycoparasitic characteristic because of their tendency towards physical interactions (Abdullah et al. 2008). Many plant growth promoting rhizobacteria such as, *Pseudomonas, Staphylococcus, Burkholderia, Ochrobactrum, Enterobacter* and *Stenotrophomonas* exhibit antagonistic potential (Tariq et al. 2017). This antagonistic potential is evident from the fact that many plant beneficial microorganisms secrete lytic enzyme to hydrolyze compounds like, hemicellulose, chitin and protein to hamper the activities of pathogens including the lysis of fungal cell wall (Neeraja et al. 2010). *Serratia marcescens* reduce the mycelial growth of soil borne pathogen, *Sclerotium rolfsii* through overexpression of chitinases (Ordentlich et al. 1988). Moreover, during this cell wall degradation dead organic matter and plant residues are also decomposed for carbon supplies. Similarly, Lysobacter controls *Pythium* and *Bipolaris* fungal species through glucanase and these enzymes also reduce the plant biotic stresses by directly parasitizing the phytopathogens (Palumbo et al. 2005; Haran et al. 1996). Apart from these cell wall degrading enzymes, certain PGPRs strains (e.g. *Enterobacter cloacae, Azospirillum brasilense, Bacillus, Rhizobium, Pseudomonas etc.*) contain 1-aminocyclopropane-1-carboxylate (ACC) deaminase enzyme which regulates the production of gaseous hormone, ethylene in plants. In fact, this enzyme hydrolyzes the ethylene precursor, ACC in to ammonia and ketobutyrate thereby inhibiting the ethylene production under stressed environments including flooding, drought, salinity heavy metals. ACC-deaminase containing PGPRs relieve the plants against such stresses and improve plant growth and development (Saleem et al. 2007).

6 Application of Microbes as Bio Pesticides/Bio Control Agents

Agricultural productivity remains under threat due to biotic factors such as plant pathogens. Currently, these plant pathogens are being controlled through chemical method i.e. pesticides/fungicides application. Although this method is effective and convenient but it has proven to be a potential threat to environment and all life forms on earth. Hence, the use of biological method i.e. microbial inoculants is environment friendly as well as sustainable approach for profitable agricultural productivity (Shafi et al. 2017). *Bacillus* and *Pseudomonas* spp. are two PGPR that have been reported to be effective bio-control agents (Gong et al. 2006; Leonardo et al. 2006).

Among these bacterial species, *Bacillus subtilis, Bacillus amyloliquefaciens*, and *Bacillus cereus* are the most effective species at controlling plant diseases through various mechanisms (Francis et al. 2010). *Bacillus* spp. have the ability to form spores which allows these PGPR to survive in a wide range of environmental conditions, thus facilitating the effective formulation of biofertilizer (Perez-Garcia et al. 2011). Bacillus-based biocontrol agents are playing a significant role in biopesticide industry. Shafi et al. (2017) also reported that most of the *Bacillus* sp. are very effective against multiple types of plant pathogens. These biocontrol agents have the ability to combat disease causing soil borne pathogens by using a variety of mechanisms. Production of antimicrobial compounds (lipopeptides, antibiotics), competition for nutrients and space and induction of host resistance (induced systemic resistance) are the major mode of actions employed by these bacteria.

One of the most effective modes of action of rhizobacteria in suppression of soil borne pathogens is *Antibiosis* (Handelsman and Stab 1996). Fungal plant pathogens are inhibited by several groups of antibiotics produced by biocontrol agents inoculated to most of the crops (Haas and Defago 2005). Soilborne infections of cereal crops like wheat, rice, maize, chickpea, and barley are suppressed by antibiotics produced by these biocontrol agents (Raaijmakers et al. 1999).

Pseudomonas fluorescens, P. putida, P. aeruginosa, Bacillus subtilis and other *Bacillus* spp. are most effective PGPR with market potential as Bio pesticides. The PGPR isolates are prepared by using different inert carrier materials and fermented in solid or liquid forms and marketed in packets or bottles (Fig. 4). The method of application of bacterial inoculants may be seed treatment, bio-priming, seedling dip, soil application, foliar spray, fruit spray, hive insert, sucker treatment and sett treatment. Application of PGPR inoculants in a consortium is more effective as compare to single strain inoculants for inhibition or suppression of soilborne plant pathogens and for better plant growth (Ji et al. 2006). Efficacy of antagonists can be

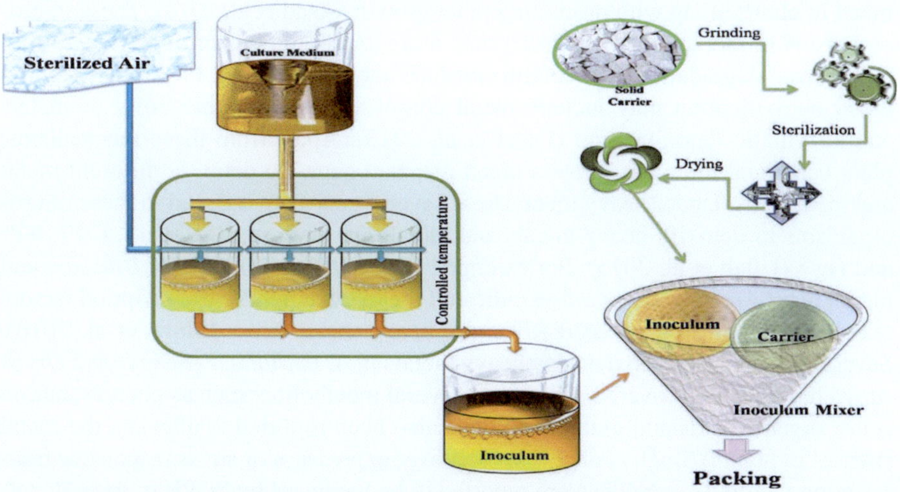

Fig. 4 A generalized sketch of the biofertilizer/bio-pesticide preparation by industry where PGPRs are preserved in an appropriate carrier molecule and packaged for commercial application at farmer's end. (Adapted from Tabassum et al. 2017)

improved by supplementation of chitin in the formulation. These inocula are being commercialized in many countries including; India, China, Japan, Germany, Australia, USA. In North America for example, more than 33 products of beneficial rhizobacteria are commercially available for their application in field or greenhouse. It is important to mention that some PGPRs are potential threats to human beings for example *Pseudomonas aeruginosa, P. cepacia* and *Bacillus cereus*. Hence, careful measures must be taken before their large-scale application for pest and disease management (Nakkeeran et al. 2005). The commercialized bio pesticides available in international market are listed in Table 3.

7 Application of Microbes as Bio Remediators of Contaminated Soils

Many microorganisms impart synergistic effects on plants through improving plant growth, accumulating heavy metals, reducing the toxicities of heavy metals and mitigating the effects of other environmental and edaphic factors such as; drought, over wetting, temperature extremes, climate change stresses and salinity. For example, plant growth promoting rhizobacteria used as biofertilizer also intensify the phytoremediation process (Sobariu et al. 2017).

Over the last many decades, heavy metals (HMs) have posed serious threats to both plants and animals. Moreover, HMs have devastatingly compromised the food safety and security via food chain contamination, soil degradation, stunted plant growth and hampering microbial community (Ashraf et al. 2017). Empirical evidences indicate that certain bacterial species enhance the accumulation of heavy metal in plants along with promoting plant growth (Asad et al. 2018). These microorganisms in fact are capable to degrade inorganic pollutants through transformation, rhizo- degradation and volatilization (Ullah et al. 2015). Physiology behind metal detoxification may include metal complexation, impermeability of metals and enzymatic detoxification (Pavel et al. 2013). Apart from these mechanisms, plant beneficial microbes possess metal resistant genes to detoxify different metal and metalloids. Under heavy metal stress several genes are induced in these microorganisms to detoxify heavy metals and metalloids such as; Zn^{+2}, Cu^{+2}, Cd^{+2}, Ni^{+2} and Hg^{+2} (Ullah et al. 2015). For example, transcriptome analysis of *Brassica* and model plant *Arabidopsis thaliana* indicated the involvement of transcription factors (TFs), bZIP, bHLH and AP2/ERF under heavy metal stress (Singh et al. 2016). Several target proteins to detoxify heavy metals in *A. thaliana*, *Zea mays* and *Oryza sativa* have been discovered. Moreover, several metabolites such as phenols, amino acids, organic acids and glutathione have also been reported to alleviate the metal stresses in plants (Singh et al. 2016). The over expression of stress responsive transcription factors (e.g. bZIP) were reported to be mediated by PGPR in *Arabidopsis* and Chickpea (Tiwari et al. 2017). Similarly, phytohormones (SA, ABA, ethylene and JA) released by plant beneficial microorganisms have also been reported to be involved to alleviate the heavy metal stressed plants. Perhaps induction of stress signaling genes in the presence of plant beneficial microorganisms elucidates a complex interaction between microorganism, plant and HMs in stress response and tolerance which warrants further investigations to understand this complex network of interactions between plants and microbes under metal stress (Tiwari and Lata 2018).

The microbial populations in heavy metal contaminated environment mostly belong to notable genera, *pseudomonas*, *Arthrobacter*, *Bacillus* and *Rhizobia* (Pires et al. 2017). Many of the plant growth promoting attributes i.e. nitrogen fixation and nitrogenase activities are very sensitive to heavy metals stresses but resistant strains of these microorganisms have been noted to carry out these activities successfully at contaminated sites. According to Checcucci et al. (2017) symbiotic relationship between rhizobia and legume are well researched for heavy metal detoxification and improving quality of contaminated sites. Amongst fungal genera, *Basidiomycota, Ascomycota* and *arbuscular mycorrhiza* have been reported to reduce heavy metal toxicity and improve the degraded soil quality (Narendrula-Kotha and Nkongolo 2017). In fact, these functions are primarily accomplished by binding of heavy metal ions on the cell surface or transporting in to the cell and changing the metal toxicity and deterioration in soil (Gadd 2010). However, metal-microbe interactions are very complex and success rate very much depends on physico-chemical properties of soil, concentration of HM in soil and microbial composition.

8 Microbial Applications as Abiotic Stress Ameliorators

Microbial inoculants are being investigated for their potential as ameliorators of following abiotic stresses.

8.1 Drought Stress

Drought is one of the major limitations toward reduced agricultural productivity in both arid and semi-arid habitats. Drought affects nitrogen fixation and major constraint for reduced legumes production (Serraj 2009). In legumes, drought is equally detrimental for nitrogen fixation during pre and post nodule formation; during post nodule formation drought causes reduced root development. The water content of rhizosphere is a potent factor determining the nutrients and oxygen supplies to plants and microorganisms (Gestel et al. 1993). These interactions among microorganisms, water and plant roots in fact formulate the soil structure which is a key determinant of soil health and hence crop productivity. For example, soil moisture levels administer the production and consumption of protein and polysaccharides by the bacteria thereby influencing the soil structure (Roberson and Firestone 1992). Similarly, exopolysaccharides released by microbes bind soil particles forming macro and micro aggregates having greater or less than 250 μm diameter respectively (Oades 1993) and hence helping plant roots to creep through these aggregates. Moisture stress may alter the biological and physico-chemical properties of soil rendering it unfit for soil biodiversity and agricultural productivity. Bacterial species such as *Pseudomonas* and *Azospirillum* commonly used as biofertilizers successfully survive under stressed environments because of exopolysaccharides (EPS) which enhance water retention and regulate carbon sources. Therefore, it becomes imperative to manage the moisture stressed or drought affected soils for meeting the food demands and use of PGPRs could provide a sustainable option for managing such soils. EPS producing microorganisms based applications may bridge this gap thereby alleviating the stressed plants and enhance productivity. A study conducted by Sandhya et al. (2009) on sunflower inoculated with *Pseudomonas putida* strain GAP-P45 revealed that almost one third of the microbial isolates used in the study could tolerate drought stress up to a level of −0.73 Mpa. The most exciting part of the investigation was that EPS production of studied strains was more prominent under stressed conditions and it continued to increase with increasing stress. Moreover, these strains expressed growth promoting properties through production of several metabolites including; HCN, phosphate solubilization, ammonia, IAA and GA production under water limiting conditions which is pre-requisite for sustainable agricultural productivity under limited water availability. Arbuscular mycorrhiza (AM) fungi have been reported to rescue plants from drought stress. Under water limiting conditions, AM increase the nitrogen availability (Bowles et al. 2018) This is perhaps because; under drought conditions these fungi absorb water more efficiently due to alterations in root architecture or most probably due to regulation of abscisic acid concentration under drought conditions (Khalvati et al. 2005; Jahromi et al. 2008). Hyphae of AM fungi penetrate deeply in to the soil in the thirst of acquiring more water and nutrients and also improve soil structure, the key factor for enhanced productivity of crops. Enhanced growth and yield of several important fruit crops (Peach, apple, citrus) is observed through AM fungi colonization (Nunes et al. 2010)

8.2 Salinity Stress

Salinity disturbs the uptake of mineral nutrients and their distribution within plant body. Moreover, it negatively influences the plant metabolism consequently reduce the quality and quantity of agricultural productivity (Silveira et al. 2003). Reduced water content of plants and creating drought like conditions, significant decline in uptake of essential nutrients, decreased photosynthetic rates, reduced biomass of plant is all attributed to the salinity stress in plants (Ben-Asher et al. 2006). Salinity has also been reported to cause abnormalities in the soil biodiversity and many genera of plant beneficial microbes have disappeared from the saline environments (Andronov et al. 2012). However, many microbial species still survive in such toxic ecologies. For example, several microbial species such as *Rhizobia*, *Azospirillum* and *Bacillus* are able to survive under such environments but exhibit varying abilities to tolerate salinity (Lloret et al. 1995). All these bacterial species are well researched for their plant growth promotion, N_2 fixation, disease suppression and plant growth hormones production characteristics (Naz et al. 2009). Ethylene production is aggravated in response to salinity resulting in stunted root growth. Madhaiyan et al. (2007) while studying the effects of exogenous application of ACC (1-aminocyclopropane-1-carboxylic acid) in *Brassica campestris* observed enhanced level of ethylene and stunted root growth in treated plants compared with control. Plant growth promoting rhizobacteria (PGPR) enhance plant growth under salinity stress conditions, which is possibly because of production of ACC-deaminase to hydrolyze ethylene precursor. So perhaps use of PGPRs is one of the plausible options to reduce salinity induced ethylene production (Mayak et al. 1999). Ethylene production in plants is enhanced in response to biotic and abiotic stresses, however PGPRs applications inhibited ethylene production significantly under these stresses, indicating the involvement of PGPRs in plant management under stressed environments (Ahmad et al. 2011). PGPR exhibit the characteristic to maintain an equilibrium in ionic concentration thereby increasing the growth and yield of crops due to reduced ethylene production (Nadeem et al. 2009). Different bacterial strains have varying potential for propagating the ACC-deaminase activity, most probably because other growth promoting activities including phosphate solubilization, production of lytic enzyme, chitinase activity, N_2 fixation are also continued along with ACC-deaminase activity (Nadeem et al. 2009; Ahmad et al. 2011). Under saline environments, a higher K^+/Na^+ ratio is very important which increases the plant tolerance against salinity (Hamdia et al. 2004). Many plant growth promoting bacteria are capable to help plants tolerate exceeded level of Na^+ by secreting exopolysaccharides (EPS) which reduce the level of Na^+ uptake (Nadeem et al. 2014) through biofilm formation (Qurashi and Sabri 2012). Interestingly, these exopolysaccharides also help plants to withstand water limiting environments and protect the microorganisms against drought stress (Sandhya et al. 2009).

8.3 Climate Change Stress

Agriculture and the linked food security are largely dependent on the natural environment, hence facing critical threats from climate change. Combination of abiotic and biotic stresses have multiplied the risks for sustainable crop production particularly in the sub-tropical regions around the world. Extreme weather events including drought, floods, torrential rains, increasing temperatures has certainly jeopardized the regional as well as global food security with considerable shifts in cropping pattern and the associated reduced yields of major agricultural crops. For example, South Asia being the hotspot of climate change has witnessed significant reductions in paddy and wheat yields because of increasing temperature, less rain followed by increasing water stresses influenced by climate change. Abiotic stresses including temperature, salinity, drought cause approximately 50% yield losses in agricultural productivity (Kaur et al. 2018). Evolving and adapting cost effective and sustainable technologies for sustainable crop production under extreme environments has always been a challenge. A plethora of literature exists detailing the technologies for adapting to the changing climate scenarios whereby developing drought resistant varieties and resource management among others have proved to be very effective in combating climate change stresses (Venkateswarlu and Shanker 2009). Use of microorganisms for promoting plant growth, controlling pathogens/diseases and nutrient management under climate change stresses has attracted plausible attention of research community (Grover et al. 2011). These microorganisms reside in the plant rhizosphere and thereof transmit several direct and indirect benefits to the plant (Saxena et al. 2005). Major microbial genera capable to induce tolerance in plants exposed to climate change stresses include; *Bacillus*, *Pseudomonas*, *Rhizobium*, *Paenibacillus*, *Azospirillum*, *Burkholderia*, *Achromobacter*, *Enterobacter*, *Methylobacterium* and *Microbacterium*. These microorganisms protect the plants against frost, higher temperature, over wetting, drought and other climate change stresses (Grover et al. 2011). Therefore, using these microorganisms to alleviate the agricultural crops could be an effective technology for enhancing the agricultural productivity on sustainable basis.

Although, mechanisms for alleviating the plants against climate change abnormalities are under researched and warrants further investigation. However, production of auxins, gibberellins and root exudates to increase the growth and surface area of roots to quench and uptake more nutrients by microorganisms is considered as one possible explanation of helping plants withstand the abiotic stresses including climate change (Egamberdieva and Kucharova 2009). For example, PGPR have been found to be involved in rescuing the vegetable and oil seed crops against many abiotic stresses (Barassi et al. 2006). Similarly, inoculation of *Paenibacillus* and *Azospirillum brasilense* in *Arabidopsis* and wheat respectively, relieved the plants against drought stress which helped the plants to maintain better water status and minerals Ca, Mg and K (Timmusk and Wagner 1999). A study conducted by Mayak et al. (2004) involving inoculation of tomato with *Achromobacter piechaudii* induced systemic resistance against drought in inoculated plants as compared to control.

The role of stress hormone 'ethylene' is widely known to reduce root/shoot growth under stress conditions. ACC-deaminase producing bacteria including *Rhizobia*, *Pseudomonas* degrade the plant ACC and enhance nitrogen and energy supply to plants. Glick (2007) noted that these bacterial strains lessen the abnormalities caused by ethylene. Hence, role of ACC-deaminase producing PGPR is pivotal for agriculture management under stressed environments as these microbes induce longer roots enhancing water uptake efficiency of plants under water limiting conditions (Zahir et al. 2008). Arbuscular mycorrhiza (AM) fungi have also been reported for rescuing the stressed plants. This AM fungi induced stress resistance is mediated by several enzymes such as peroxidase (POD), superoxide dismutase (SOD), catalase (CAT) and abscisic acid (ABA). During abiotic stress, activities of these enzymes were enhanced by AM fungi which improved the osmotic

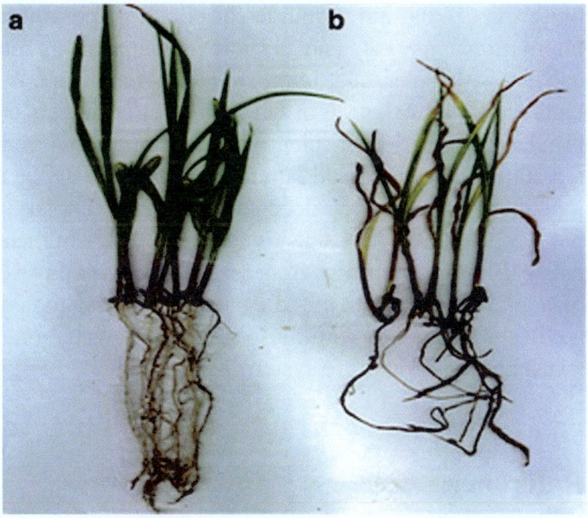

Fig. 5 Example of suppression of *Pythium ultimum* root rot in 4-week-old sorghum seedlings by bacterial strains isolated from the rhizosphere of wild grasses in South Africa. (**a**) Plants inoculated with *P. ultimum* and treated with rhizobacterial isolates. (**b**) Control plants that were treated only with *P. ultimum* developed visible root rot and necrotic leaves. (Adapted from Idris et al. 2007)

adjustment thereby increasing the drought resistance in citrus seedlings (Wu et al. 2005). Similarly, lavender plants inoculated with mycorrhizae possessed better water, N and K contents and exhibited greater biomass than uninoculated plants under drought stress (Marulanda et al. 2007). Growth hormone, ABA is also suggested behind AM fungi reduced drought resistance in plants (Aroca et al. 2008). Extreme flood events have deleterious effects on crops causing irreparable losses to the peasants and altering the socioeconomic balance in the affected areas. In wetlands, AM fungi are well established and alleviate the submerged crops after flooding events. *Glomus intraradices* colonization of *Pterocarpus officinalis* seedlings substantially improved resistance of inoculated plants through improved P uptake (Fougnies et al. 2007). Resistance against flooding is also mediated by enhanced proline content and osmotic adjustment in submerged plants as observed by Neto et al. (2006). Abiotic and biotic stresses triggered by rapidly changing climate scenarios, environmental and edaphic factors pose serious challenges for sustainable agriculture and use of beneficial plant microbes such as; PGPRs and AM fungi could prove to be an effective, environment friendly option for sustainable and enhanced agricultural productivity under compromised environments (Fig. 5).

9 Conclusions and Future Research

Scientific literature and research on presence/survival of beneficial rhizobacteria in rhizosphere, illustrations of specified mode of action, characterization of plant beneficial traits and evaluation of inoculants in pot and field trials has revealed that PGPR has proved to be potential candidates for sustainable agricultural development. Applications of beneficial rhizobacteria (microbial inoculants) as biofertilizers and bio pesticides has been successful to make a reasonable space in international market. Bio pesticide/Biocontrol agents (BCAs) have shown more consistency in market as compared to biofertilizers. Application of microbial inoculants for bioremediation of heavy metals/pollutants and abiotic stress amelioration is catching attention of scientific community however practical application of these techniques is limited and is a matter of future scope of this technology. Microbial inoculants have limited acceptance perhaps because of the complications during field application, their sensitivity toward environmental changes and most importantly lack of farmer's awareness. To cope with these challenges there is a dire need of comprehensive integrated agricultural management policy. Integration of these renewable resources with those of non-renewable ones in a wise way could lead to a sustainable agriculture system.

References

Abdullah TM, Ali YN, Suleman P (2008) Biological control of *Sclerotinia sclerotiorum* with *Trichoderma harzianum* and *Bacillus amyloliquefaciens*. Crop Prot 27:1354–1359

Afzal A, Bano A (2008) Rhizobium and phosphate solubilizing bacteria improve the yield and phosphorus uptake in wheat (*Triticum aestivum* L.). Int J Agric Biol 10:85–88

Ahemad M, Khan MS (2012a) Effects of pesticides on plant growth promoting traits of *Mesorhizobium* strain MRC4. J Saudi Soc Agric Sci 11:63–71

Ahemad M, Khan MS (2012b) Effect of fungicides on plant growth promoting activities of phosphate solubilizing *Pseudomonas putida* isolated from mustard (Brassica compestris) rhizosphere. Chemosphere 86:945–950

Ahemad M, Kibret M (2014) Mechanisms and applications of plant growth promoting rhizobacteria: current perspective. J King Saud Uni Sci 26:1–20

Ahmad M, Zahir ZA, Asghar HN, Asghar M (2011) Inducing salt tolerance in mung bean through co-inoculation with rhizobia and plant-growth promoting rhizobacteria containing 1-aminocyclopropane-1-carboxylate- deaminase. Can J Microbiol 57:578–589

Ahmed A, Hasnain S (2010) Auxin-producing Bacillus spp: auxin quantification and effect on the growth of *Solanum tuberosum*. Pure Appl Chem 82:313–319

Akhtar MS, Siddiqui ZA (2009) Use of plant growth promoting rhizobacteria for the biocontrol of root-rot disease complex of chickpea. Aus Plant Pathol 38:44–50

Andronov EE, Petrova SN, Pinaev AG, Pershina EV, Rakhimgalieva SZ, Akhmedenov KM, Sergaliev NK (2012) Analysis of the structure of microbial community in soils with different degrees of salinization using T-RFLP and real-time PCR techniques. Eurasian Soilless Sci 45:147–156

Antoun H, Prévost D (2005) Ecology of plant growth promoting rhizobacteria. In: Siddiqui ZA (ed) PGPR: biocontrol and biofertilization. Springer, Dordrecht, pp 1–38

Aroca R, Vernieri P, Ruiz-Lozano JM (2008) Mycorrhizal and nonmycorrhizal *Lactuca sativa* plants exhibit contrasting responses to exogenous ABA during drought stress and recovery. J Exp Bot 59:2029–2041

Asad SA, Rehman M, Ahmad R, Umer M (2018) Differential uptake of cadmium and chromium in *Brassica oleraceae* in response to application of plant growth promoting rhizobacteria. Int J Agric Biol 20(7):1613–1622. https://doi.org/10.17957/IJAB/15.0682

Ashraf MA, Hussain I, Rasheed R, Iqbal M, Riaz R, Arif MS (2017) Advances in microbe-assisted reclamation of heavy metal contaminated soils over the last decade: a review. J Environ Manag 198:132–143

Aziz ZFA, Saud HM, Rahim KA, Ahmed OH (2012) Variable responses on early development of shallot (*Allium ascalonicum*) and mustard (*Brassica juncea*) plants to *Bacillus cereus* inoculation. Malaysian J Micro 8:47–50

Bagnasco P, De La Fuente L, Gualtieri G, Noya F, Arias A (1998) Fluorescent p*seudomonas* spp. as biocontrol agents against forage legume root pathogenic fungi. Soil Biol Biochem 30:1317–1322

Barassi CA, Ayrault G, Creus CM, Sueldo RJ, Sobero MT (2006) Seed inoculation with *Azospirillum* mitigates NaCl effects on lettuce. Sci Hortic 109:8–14

Bashan Y, de Bashan L (2010) How the plant growth-promoting bacterium *Azospirillum* promotes plant growth: a critical assessment. Adv Agron 108:77–136

Bashan Y, Holguin G (1997) *Azospirillum*-plant relationships: environmental and physiological advances (1990-1996). Can J Microbiol 43:103–121

Bashan Y, Levanony H (1990) Current status of *Azospirillum* inoculation technology: *Azospirillum* as a challenge for agriculture. Can J Microbiol 36:591–608

Bastian F, Cohen A, Piccoli P, Bottini R, Luna MV, Baraldi R (1998) Production of IAA and gibberellins A1 and A3 by *Acetobacter diazotrophicus* and *Herbaspirillum seropedicae* in chemically-defined culture media. Plant Growth Regul 24:7–11

Ben-Asher J, Tsuyuki I, Bravdo BA, Sagih M (2006) Irrigation of grape vines with saline water, I. leaf area index, stomatal conductance, transpiration and photosynthesis. Agric Water Manag 83:13–21

Benson DR, Silvester WB (1993) Biology of Frankia strains, actinomycete symbionts of actinorhizal plants. Microbiol Rev 57:293–319

Berg G (2009) Plant-microbe interactions promoting plant growth and health: perspectives for controlled use of microorganisms in agriculture. Appl Microbiol Biotechnol 84:11–18

Bhattacharyya PN, Jha DK (2012) Plant growth-promoting rhizobacteria (PGPR): emergence in agriculture. World J Microbiol Biotechnol 28:1327–1350

Bloemberg GV, Lugtenberg BJ (2001) Molecular basis of plant growth promotion and biocontrol by rhizobacteria. Curr Opin Plant Biol 4:343–350

Bloemberg GV, Wijfjes AH, Lamers GE, Stuurman N, Lugtenberg BJ (2000) Simultaneous imaging of *Pseudomonas fluorescens* WCS365 populations expressing three different auto fluorescent proteins in the rhizosphere: new perspectives for studying microbial communities. Mol Plant-Microbe Interact 13:1170–1176

Bowles TM, Jackson LE, Cavagnaro TR (2018) Mycorrhizal fungi enhance plant nutrient acquisition and modulate nitrogen loss with variable water regimes. Glob Chang Biol 24:171–182

Burdman S, Volpin H, Kigel J, Kapulnik Y, Okon Y (1996) Promotion of nod gene inducers and nodulation in common bean (*Phaseolus vulgaris*) roots inoculated with *Azospirillum brasilense* Cd. Appl Environ Microbiol 62:3030–3033

Calvo P, Nelson LM, Kloepper JW (2014) Agricultural uses of plant bio stimulants. Plant Soil 383:3–41

Cassán F, Vanderleyden J, Spaepen S (2014) Physiological and agronomical aspects of phytohormone production by model plant-growth-promoting rhizobacteria (PGPR) belonging to the genus *Azospirillum*. J Plant Growth Regul 33:440–459

Chandra R, Pareek RP (2007) Effect of rhizobacteria in urdbean and lentil. Ind J Pulses Res 15:152–155

Checcucci A, Bazzicalupo M, Mengoni A (2017) Exploiting nitrogen fixing rhizobial symbionts genetic resources for improving phytoremediation of contaminated soils. In: Anjum NA, Gill SS, Tuteja N (eds) Enhancing cleanup of environmental pollutants. Biological approaches. Springer, Cham, pp 275–288

Chemin L, Chet I (2002) Microbial enzymes in biocontrol of plant pathogens and pests. In: Burns RG, Dick RP (eds) Enzymes in the environment: activity, ecology and applications. Marcel Dekker, New York, pp 171–225

Chen YP, Rekha PD, Arun AB, Shen FT, Lai W-A, Young CC (2006) Phosphate solubilizing bacteria from subtropical soil and their tricalcium phosphate solubilizing abilities. Appl Soil Ecol 34:33–41

Clair SB, Lynch JP (2010) The opening of Pandora's Box: climate change impacts on soil fertility and crop nutrition in developing countries. Plant Soil 335:101–115

Cohen AC, Bottini R, Pontin M, Berli FJ, Moreno D, Boccanlandro H (2015) *Azospirillum brasilense* ameliorates the response of *Arabidopsis thaliana* to drought mainly via enhancement of ABA levels. Physiol Plant 153:79–90

Damalas CA, Eleftherohorinos IG (2011) Pesticide exposure, safety issues, and risk assessment indicators. Int J Environ Res Public Health 8:1402–1419

Défago G, Berling CH, Burger U, Haas D, Kahr G, Keel C, Voisard C, Wirthner P, Wüthrich B (1992) Suppression of black root rot of tobacco and other root diseases by strains of *Pseudomonas fluorescens*: potential applications and mechanisms. In: Hornby D (ed) Biological control of soil-borne plant pathogens. International Wallingford, Oxon UK, pp 93–108

Desbrosses G, Contesto C, Varoquaux F, Galland M, Touraine B (2009) PGPR- *Arabidopsis* interactions is a useful system to study signaling pathways involved in plant developmental control. Plant Signal Behav 4:319–321

Döbereiner J, Day JM (1976) Associative symbioses in tropical grasses: characterization of microorganisms and dinitrogen-fixing sites. In: Newton WE, Nyman CJ (eds) Proceedings of the 1st international symposium on nitrogen fixation. Washington State University Press, Pullman, pp 518–538

Döbereiner J, Baldani VLD, Olivares FL, Reis VM (1994) Endophytic diazotrophs: the key to BNF in gramineous plants. In: Hegasi NA, Fayez M, Monib M (eds) Nitrogen fixation with non-legumes. The American University in Cairo Press, Egypt, pp 395–408

Dodd IC, Zinovkina NY, Safronova VI, Belimov AA (2010) Rhizobacterial mediation of plant hormone status. Ann Appl Biol 157:361–379

Eaglesham ARJ (1989) Global importance of rhizobium as an inoculant. In: Campbell R, Macdonald RM (eds) Microbial inoculation of crop plants. IRL Press, Oxford/New York, pp 29–48

Egamberdieva D, Kucharova Z (2009) Selection for root colonizing bacteria stimulating wheat growth in saline soils. Biol Fertil Soils 45(6):563–571. https://doi.org/10.1007/s00374-009-0366-y

Egamberdiyeva D (2007) The effect of plant growth promoting bacteria on growth and nutrient uptake of maize in two different soils. Appl Soil Ecol 36:184–189

Fages J (1992) An industrial view of *Azospirillum* inoculants: formulation and application technology. Symbiosis 13:15–26

Fattah QA (2005) Plant resources for human development. In: 3rd international botanical conference 2005. Bangladesh Botanical Society, Dhaka, Bangladesh

Fougnies L, Renciot S, Muller F, Plenchette C, Prin Y, de Faria SM, Bouvet JM, Sylla SND, Dreyfus B, Ba AM (2007) Arbuscular mycorrhizal colonization and nodulation improve tolerance in Pterocarpus officinalis Jacq. seedlings. Mycorrhiza 17:159–166

Francis I, Holsters M, Vereecke D (2010) The gram-positive side of plant-microbe interaction. Environ Microbial 12:1–12

Frank B (1889) Ueber die Pilzsymbiose der Leguminosen. Ber Dtsch Bot Ges 7:332–346

Fred EB, Baldwin I, McCoy M (1932) Root nodule bacteria and leguminous plants. Uni Wisconsin Studies in Sci

Gadd GM (2010) Metals, minerals and microbes: geomicrobiology and bioremediation. Microbiologica 156:609–643

Gaffney TD, Lam ST, Ligon J, Gates K, Frazelle A, Di Maio J, Hill S, Goodwin S, Torkewitz N, Allshouse AM (1994) Global regulation of expression of antifungal factors by a *Pseudomonas fluorescens* biological control strain. Mol Plant-Microbe Interact 7:455–463

Gardener BBM, Fravel DR (2002) Biological control of plant pathogens: research, commercialization, and application in the USA. Online. Plant Health Progress https://doi.org/10.1094/PHP-2002-0510-01-RV

Gestel VM, Merckx R, Vlassak K (1993) Microbial biomass responses to soil drying and wetting: the fate of fast- and slow-growing microorganisms in soils from different climates. Soil Biol Biochem 25:109–123

Glick BR (1995) The enhancement of plant-growth by free-living bacteria. Can J Microbiol 41:109–117

Glick BR (2007) Promotion of plant growth by bacterial ACC deaminase. Crit Rev Plant Sci 26:227–242

Glick BR (2012) Plant growth-promoting bacteria: mechanisms and applications. Scientifica:1–15

Glick BR, Bashan Y (1997) Genetic manipulation of plant growth-promoting bacteria to enhance biocontrol of phytopathogens. Biotechnol Adv 15:353–378

Gong M, Wang JD, Zhang J, Yang H (2006) Study of the antifungal ability of *Bacillus subtilis* strain PY-1 in-vitro and identification of its antifungal substance (Iturin A). Acta Biochim Biophys Sin 38:233–240

Govindarajan M, Balandreau J, Kwon SW (2008) Effects of the inoculation of *Burkholderia vietnamiensis* and related endophytic diazotrophic bacteria on grain yield of rice. Microb Ecol 55:21–37

Gray EJ, Smith DL (2005) Intracellular and extracellular PGPR: commonalities and distinctions in the plant-bacterium signaling processes. Soil Biol Biochem 37:395–412

Grover M, Ali SKZ, Sandhya V, Rasul A, Venkateswarlu B (2011) Role of microorganisms in adaptation of agriculture crops to abiotic stresses. World J Microbiol Biotechnol 27:1231–1240

Guerinot ML, Ying Y (1994) Iron: nutritious, noxious, and not readily available. Plant Physiol 104:815–820

Guo Y, Ni Y, Huang J (2010) Effects of rhizobium, arbuscular mycorrhiza and lime on nodulation, growth and nutrient uptake of lucerne in acid purplish soil in China. Trop Grasslands 44:109–114

Haas D, Defago G (2005) Biological control of soil-borne pathogens by fluorescent pseudomonads. Nat Rev Microbiol 3:307–319

Hamdia MA, Shaddad MAK, Doaa MM (2004) Mechanism of salt tolerance and interactive effect of *Azospirillum bransilense* inoculation on maize cultivars grown under salt stress conditions. Plant Growth Regul 44:165–174

Handelsman J, Stab EV (1996) Biocontrol of soil borne plant pathogens. Plant Cell 8:1855–1869

Haran S, Schickler H, Chet I (1996) Molecular mechanisms of lytic enzymes involved in the biocontrol activity of *Trichoderma harzianum*. Microbiologica 142:2321–2331

Hider RC, Kong X (2010) Chemistry and biology of siderophores. Nat Prod Rep 27:637–657

Holt JG, Krieg NR, Senath PHA, Staley JT, Williams ST (1994) Bergey's manual of determinative bacteriology, 9th edn. Williams and Wilkins, Baltimore

Idris AH, Labuschagne N, Korsten L (2007) Screening rhizobacteria for biological control of Fusarium root and crown rot of sorghum in Ethiopia. Biol Control 40:97–106

Indiragandhi P, Anandham R, Madhaiyan M, Sa TM (2008) Characterization of plant growth-promoting traits of bacteria isolated from larval guts of diamondback moth, *Plutella xylostella* (Lepidoptera: Plutellidae). Curr Microbiol 56:327–333

Innerebner G, Knief C, Vorholt JA (2011) Protection of *Arabidopsis thaliana* against leaf pathogenic *Pseudomonas syringae* by Sphingomonas strains in a controlled model system. Appl Environ Microbiol 77:3202–3210

Iqbal N, Nazar R, Iqbal MRK, Masood A, Nafees AK (2011) Role of gibberellins in regulation of source sink relations under optimal and limiting environmental conditions. Curr Sci 100:998–1007

Jahromi F, Aroca R, Porcel R, Ruiz-Lozano JM (2008) Influence of salinity on the in-vitro development of *Glomus intraradices* and on the in-vivo physiological and molecular responses of mycorrhizal lettuce plants. Microb Ecol 55:45–53

Jha Y, Subramanian RB (2018) Effect of root-associated bacteria on soluble sugar metabolism in plant under environmental stress. In: Ahmad P et al (eds) Plant metabolites and regulation under environmental stress. Academic, pp 231–240

Ji P, Campbell HL, Kloepper JW, Jones JB, Suslow TV, Wilson M (2006) Integrated biological control of bacterial speck and spot of tomato under field conditions using foliar biological control agents and plant growth-promoting rhizobacteria. Soil Biol Biochem 103:117–130

Jilani G, Akram AR, Ali M, Hafeez FY, Shamsi IH, Chaudhry AN, Chaudhry AG (2007) Enhancing crop growth, nutrients availability, economics and beneficial rhizosphere microflora through organic and biofertilizers. Ann Microbiol 57:177–183

Kashyap PL, Rai P, Srivastava AK, Kumar S (2017) *Trichoderma* for climate resilient agriculture. World J Microbiol Biotechnol 33:155

Kaur J, Pandove G, Gangwar M (2018) Mitigating the impact of climate change by use of microbial inoculants. Pharma Innov J 7:279–288

Kaymak DC (2010) Potential of PGPR in agricultural innovations. In: Maheshwari DK (ed) Plant growth and health promoting bacteria. Springer, Germany, p 16

Khalvati MA, Hu Y, Mozafar A, Schmidhalter U (2005) Quantification of water uptake by arbuscular mycorrhizal hyphae and its significance for leaf growth, water relations, and gas exchange of barley subjected to drought stress. Plant Biol 7:706–712

Khan AL, Waqas M, Kang SM (2014) Bacterial endophytes Sphingomonas spp. LK11 produces gibberellins and IAA and promotes tomato plant growth. J Microbiol 52:689–695

Kloepper JW (1993) Plant growth-promoting rhizobacteria as biological control agents. In: Meting FB Jr (ed) Soil microbial ecology: applications in agricultural and environmental management. Marcel Dekker Inc, New York, pp 255–274

Kloepper JW (1994) Plant growth promoting rhizobacteria: other systems. In: Okon Y (ed) Azospirillum/plant associations. CRC Press, Boca Raton, pp 137–166

Kloepper JW, Schroth MN (1978) Plant growth-promoting rhizobacteria on radishes. In: Station de Pathologie, Proceedings of the 4th international conference on plant pathogenic bacteria, France, pp 879–882

Kloepper JW, Schroth MN (1981) Plant growth-promoting rhizobacteria and plant growth under gnotobiotic conditions. Phytopathology 71:642–644

Labuschagne N, Pretorius T, Idris AH (2010) Plant growth promoting rhizobacteria as biocontrol agents against soil-borne plant diseases. In: Maheshwari DK (ed) Plant growth and health promoting bacteria, microbiology monographs. Springer, Berlin, pp 211–230

Laranjo M, Alexandre A, Oliveira S (2014) Legume growth-promoting rhizobia: an overview on the *Mesorhizobium* genus. Microbiol Res 169:2–17

Leonardo D, Blanca LF, Landa B, Weller DM (2006) Host crop affects rhizosphere colonization and competitiveness of 2, 4-diacetylphloroglucinol-producing *Pseudomonas fluoresens*. Phytopathology 96:751–762

Lloret J, Bolaños L, Lucas MM, Peart JM, Brewin NJ, Bonilla I, Rivilla R (1995) Ionic stress and osmotic pressure induce different alterations in the lipopolysaccharide of a *Rhizobium meliloti* strain. Appl Environ Microbiol 61:3701–3704

Lübeck PS, Hansen M, Sørensen J (2000) Simultaneous detection of the establishment of seed-inoculated Pseudomonas fluorescens strain DR54 and native soil bacteria on sugar beet root surfaces using fluorescence antibody and in situ hybridization techniques. FEMS Microbiol Ecol **33**(1):11–19. https://doi.org/10.1111/j.1574-6941.2000.tb00721.x

Lugtenberg B, Kamilova F (2009) Plant-growth-promoting rhizobacteria. Annu Rev Microbiol 63:541–556

Lugtenberg BJJ, Dekkers L, Bloemberg GV (2001) Molecular determinants of rhizosphere colonization by pseudomonas. Annu Rev Phytopathol 39:461–490

Macdonald RM (1989) An overview of crop inoculation. In: Campbell R, Macdonald RM (eds) Microbial inoculation of crop plants. IRL Press, Oxford/New York, pp 1–9

Madhaiyan M, Poonguzhali S, Sa TM (2007) Characterization of 1- Aminocyclopropane-1-carboxylate deaminase (ACC) deaminase containing *Methylobacterium oryzae* and interactions with auxins and ACC regulation of ethylene in canola (*Brassica campestris*). Planta 226:867–876

Mahanty T, Bhattacharjee S, Goswami M, Bhattacharyya P, Das B, Ghosh A, Tribedi P (2017) Biofertilizers: a potential approach for sustainable agriculture development. Environ Sci Pollut Res 24:3315–3335

Malboobi MA, Parviz O, Mandana B, Elaheh S, Sara M, Bagher Y, Ali D, Kambiz MH (2009) Solubilization of organic and inorganic phosphates by three highly efficient soil bacterial isolates. World J Microbiol Biotechnol 25:1471–1477

Marschner H, Dell B (1994) Nutrient uptake in mycorrhizal symbiosis. Plant Soil 159:89–102

Marulanda A, Porcel R, Barea JM, Azcon R (2007) Drought tolerance and antioxidant activities in lavender plants colonized by native drought tolerant or drought sensitive Glomus species. Microb Ecol 54:543–552

Mathesius U, Shalaman H, Meijer D, Lugtenberg B, Spaink H, Weinman J, Rodam L, Sautter C, Rolfe B, Djordjevic M (1997) New tools for investigating nodule initiation and ontogeny: spot inoculation and micro targeting of transgenic withe clover roots shows auxin involvement and suggest a role for flavonoids. In: Stacey G, Mullin B, Gresshoff P (eds) Advances in molecular genetics of plant–microbe interactions. Kluwer Academic, Dordrecht

Mayak S, Tirosh T, Glick BR (1999) Effect of wild-type and mutant plant growth promoting rhizobacteria on the rooting of mung bean cuttings. Plant Growth Regul 18:49–53

Mayak S, Tirosh T, Glick BR (2004) Plant growth promoting bacteria that confer resistance to water stress in tomatoes and peppers. Plant Sci 166:525–530

Nadeem SM, Zahir ZA, Naveed M, Arshad M (2009) Rhizobacteria containing ACC deaminase confer salt tolerance in maize grown on salt affected soils. Can J Microbiol 55:1302–1309

Nadeem SM, Ahmad M, Zahir ZA, Javaid A, Ashraf M (2014) The role of mycorrhizae and plant growth promoting rhizobacteria (PGPR) in improving crop productivity under stressful environments. Biotechnol Adv 32:429–448

Nakkeeran S, Fernando WGD, Zaki A, Siddiqui ZA (2005) Plant growth promoting rhizobacteria formulations and its scope in commercialization for the management of pests and diseases. In: Siddiqui ZA (ed) PGPR: biocontrol and biofertilization. Springer, Dordrecht, pp 257–296

Narendrula-Kotha R, Nkongolo KK (2017) Microbial response to soil liming of damaged ecosystems revealed by pyrosequencing and phospholipid fatty acid analyses. PLoS One 12. https://doi.org/10.1371/journal.pone.0168497

Nascimento FX, Rossi MJ, Soares CRFS, McConkey BJ, Glick BR (2014) New insights into 1-aminocyclopropane-1-carboxylate (ACC) deaminase phylogeny. Evolution and ecological significance. PLoS One 9. https://doi.org/10.1371/journal.pone.0099168

Naz I, Bano A, Hassan T (2009) Morphological, biochemical and molecular characterization of rhizobia from halophytes of Khewra Salt range and Attock. Pak J Bot 41:3159–3168

Neeraja C, Anil K, Purushotham P, Suma K, Sarma P, Moerschbacher BM, Podile AR (2010) Biotechnological approaches to develop bacterial chitinases as a bioshield against fungal diseases of plants. Crit Rev Biotechnol 30:231–241

Nehra BSV (2011) Plant growth promoting rhizobacteria: a critical review. Life Sci Med Res:21–29

Neto D, Carvalho LM, Cruz C, Martin-Loucao MA (2006) How do mycorrhizas affect C and N relationships in flooded *Aster Trifolium* plants? Plant Soil 279:51–63

Newbery F, Qi A, Fitt BD (2016) Modelling impacts of climate change on arable crop diseases: progress, challenges and applications. Curr Opin Plant Biol 32:101–109

Nobbe F, Hiltner L (1896) Inoculation of the soil for cultivating leguminous plants. USA Patent No. 570 813

Nowak TB, Gould SJ, Kraus J, Loper JE (1994) Production of 2,4- diacetylphloroglucinol by the biocontrol agent *Pseudomonas fluorescens* Pf-5. Can J Microbiol 40:1064–1066

Nunes JLD, de Souza PVD, Marodin GAB, Fachinello JC (2010) Effect of arbuscular mycorrhizal fungi and indole butyric acid interaction on vegetative growth of 'Aldrighi' peach rootstock seedlings. Cienc Agrotec 34:80–86

Oades JM (1993) The role of biology in the formation, stabilization and degradation of soil structure. Geoderma 56:182–186

Okon Y, Labandera-Gonza'lez C (1994) Agronomic applications of *Azospirillum*: an evaluation of 20 years' worldwide field inoculation. Soil Biol Biochem 26:1591–1601

Ordentlich A, Elad Y, Chet I (1988) The role of chitinase of *Serratia marcescens* in biocontrol of *Sclerotium rolfsii*. Phytopathology 78:84–88

Ortíz-Castro R, Valencia-Cantero E, López-Bucio J (2008) Plant growth promotion by *Bacillus megaterium* involves cytokinin signaling. Plant Signal Behav 3:263–265

Palumbo JD, Yuen GY, Jochum CC, Tatum K, Kobayashi DY (2005) Mutagenesis of beta-1,3-glucanase genes in *Lysobacter Enzymogen* strain C3 results in reduced biological control activity toward bipolaris leaf spot of tall Fescue and Pythium damping-off of sugar beet. Phytopathology 95:701–707

Parray JA, Jan S, Kamili AN, Qadri RA, Egamberdieva D, Ahmad P (2016) Current perspectives on plant growth promoting rhizobacteria. Plant Growth Regul 35:877–902

Pavel V, Sobariu D, Fertu I, Statescu F (2013) Symbiosis in the environment bio management of soils contaminated with heavy metals. Eur J Sci Theol 9:211–224

Perez-Garcia A, Romero D, de Vicente A (2011) Plant protection and growth simulation by micro-organism: biotechnological applications of *Bacillus* in agriculture. Curr Opin Biotechnol 22:187–193

Peter B, Bell C, Mancini LM, Lee MN, Conant RT, Wallenstein MD (2016) Phosphorus mobilizing consortium Mammoth P (TM) enhances plant growth. Peer J. https://doi.org/10.7717/peerj.2121

Pires C, Franco AR, Pereira SIA, Henriques I, Correia A, Magan N (2017) Metal(loid)-contaminated soils as a source of culturable heterotrophic aerobic bacteria for remediation applications. Geomicrobiol J:1–9

Podile AR, Kishore GK (2006) Plant growth-promoting rhizobacteria. In: Gnanamanickam SS (ed) Plant-associated bacteria. Springer, Dordrecht, pp 195–230

Postgate JR (1982) The fundamentals of nitrogen fixation. Cambridge University Press, New York

Qin Y, Druzhinina IS, Pan X, Yuan Z (2016) Microbially mediated plant salt tolerance and microbiome-based solutions for saline agriculture. Biotechnol Adv 34:1245–1259

Qurashi AW, Sabri AN (2012) Bacterial exopolysaccharide and biofilm formation stimulate chick-pea growth and soil aggregation under salt stress. Braz J Microbiol 43:1183–1191

Rainey PB (1999) Adaptation of *Pseudomonas fluorescens* to the plant rhizosphere. Environ Microbiol 1:243–257

Raaijmakers JM, Bonsall RF, Weller DM (1999) Effect of population density of on production of 2,4-diacetylphloroglucinol in the rhizosphere of wheat. Phytopathology 89:470–475

Rajkumar M, Ae N, Prasad MN, Freitas H (2010) Potential of siderophore-producing bacteria for improving heavy metal phytoextraction. Trends Biotechnol 28:142–149

Roberson EB, Firestone MK (1992) Relationship between desiccation and exopolysaccharide production in soil *Pseudomonas* spp. Appl Environ Microbiol 58:1284–1291

Roberts DP, Yucel I, Larkin RP (1998) Genetic approaches for analysis and manipulation of rhizosphere colonization by bacterial biocontrol agents. In: Boland GJ, Kendall LD (eds) Plant-microbe interactions and biological control in soils, plants and the environment. Marcel Dekker Inc, New York, pp 415–431

Roley SS, Duncan DS, Liang D, Garoutte A, Jackson RD, Tiedje JM, Robertson GP (2018) Associative nitrogen fixation (ANF) in switchgrass (*Panicum virgatum*) across a nitrogen input gradient. Plos One. https://doi.org/10.1371/journal.pone.0197320

Romano S, Bondarev V, Kölling M, Dittmar T, Schulz-Vogt HN (2017) Phosphate limitation triggers the dissolution of precipitated iron by the marine bacterium Pseudo vibrio sp. FO-BEG 1. Front Microbiol 8:364

Roper MM, Gupta VVSR (2016) Enhancing non-symbiotic N2 fixation in agriculture. Open Agric J 10:7–27

Rubenchik LI (1963) *Azotobacter* and its use in agriculture. Soil Sci 4:280

Sah SK, Reddy KR, Li J (2016) Abscisic acid and abiotic stress tolerance in crop plants. Front Plant Sci 7:571

Saleem M, Arshad M, Hussain S, Bhatti A (2007) Perspective of plant growth promoting rhizobacteria (PGPR) containing ACC-deaminase in stress agriculture. J Ind Microbiol Biotechnol 34:635–648

Saleem MM, Arshad M, Yaseen M (2013) Effectiveness of various approaches to use rock phosphate as a potential source of plant available P for sustainable wheat production. Int J Agric Biol 15:223–230

Salvagiotti F, Kenneth GC, James ES, Daniel TW, Weiss A, Dobermann AR (2008) Nitrogen uptake, fixation and response to fertilizer N in soybeans: a review. Field Crop Res 108:1–13
Sandhya V, Ali SKZ, Grover M, Reddy G, Venkateswarlu B (2009) Alleviation of drought stress effects in sunflower seedlings by the exopolysaccharides producing *Pseudomonas putida* strain GAP-P45. Biol Fertil Soils 46:17–26
Savary S, Ficke A, Aubertot JN, Hollier C (2012) Crop losses due to diseases and their implications for global food production losses and food security. Food Sec 4:519–537
Saxena AK, Lata SR, Pandey AK (2005) Culturing of plant growth promoting rhizobacteria. In: Gopi KP, Varma A (eds) Basic research applications of mycorrhizae. I K International Pvt Ltd, New Delhi, pp 453–474
Schmidt CS, Lorenz D, Wolf GA (2001) Biological control of the grapevine dieback fungus Eutypa lata I: screening of bacterial antagonists. J Phytopathol 149:427–435
Schütz L, Gattinger A, Meier M, Müller A, Boller T, Mäder P, Mathimaran N (2018) Improving crop yield and nutrient use efficiency via Biofertilization, a global meta-analysis. Front Plant Sci 8:2204
Selvakumar G, Mohan M, Kundu S, Gupta AD, Joshi P, Nazim S, Gupta HS (2008) Cold tolerance and plant growth promotion potential of *Serratia marcescens* strain SRM (MTCC 8708) isolated from flowers of summer squash (Cucurbita pepo). Lett Appl Microbiol 46:171–175
Serraj R (2009) Effects of drought stress on legume symbiotic nitrogen fixation: physiological mechanisms. Ann Bot 104:1263–1280
Shaddad MAK, Abd El-Samad HM, Mostafa D (2013) Role of gibberellic acid (GA3) in improving salt stress tolerance of two wheat cultivars. Int J Plant Physiol Biochem 5:50–57
Shafi J, Tian H, Ji M (2017) Bacillus species as versatile weapons for plant pathogens: a review. Biotechnol Biotechnol Equip 31:446–459
Shameer S, Prasad TNVKV (2018) Plant growth promoting rhizobacteria for sustainable agricultural practices with special reference to biotic and abiotic stresses. Plant Growth Regul 84:603–615
Silveira JA, Viegas Rde A, da Rocha IM, Moreira AC, Moreira Rde A, Oliveira JT (2003) Proline accumulation and glutamine synthase activity are increased by salt induced proteolysis in cashew leaves. J Plant Physiol 160:115–123
Singh S, Parihar P, Singh R, Singh VP, Prasad SM (2016) Heavy metal tolerance in plants: role of transcriptomics, proteomics, metabolomics, and ionomics. Front Plant Sci 6:1143
Smith RS (1992) Legume inoculant formulation and application. Can J Microbiol 38:485–492
Sobariu DL, Fertu DIT, Diaconu M, Pavel LV, Hlihor RM, Drăgoi EN, Curteanu S, Lenz M, Corvini PFX, Gavrilescu M (2017) Rhizobacteria and plant symbiosis in heavy metal uptake and its implications for soil bioremediation. New Biotechnol 39:125–134
Stephan MP (1979) Physiological studies with azospirillum spp. In: Vose PB, Ruschel AP (eds) Associative N_2-fixation, vol I. CRC Press, Boca Raton, pp 7–14
Strzelczyk E, Kamper M, Li C (1994) Cytokinin-like-substances and ethylene production by *Azospirillum* in media with different carbon sources. Microbiol Res 149:55–60
Tabassum B, Khan A, Tariq M, Ramzan M, Khan MSI, Shahid N, Aaliya K (2017) Bottlenecks in commercialization and future prospects of PGPR. Appl Soil Ecol 121:102–117
Tang WH (1994) Yield-increasing bacteria and biocontrol of sheath blight of rice. In: Bowen GD, Ryder MH, Stephens PM (eds) Improving plant productivity with rhizosphere bacteria. CSIRO, Adelaide, pp 267–273
Tang WH, Yang H (1997) Research and application of biocontrol of plant diseases and PGPR in China. In: Ogoshi A, Kobayashi K, Homma Y, Kodama F, Kondo N, Akino S (eds) Plant growth-promoting rhizobacteria-present status and future prospects. Hokkaido University, Sapporo, pp 4–9
Tariq M, Noman M, Ahmed T, Hameed A, Manzoor N, Zafar M (2017) Antagonistic features displayed by plant growth promoting rhizobacteria (PGPR): a Review. J Plant Sci Phytopathol 1:38–43
Teale WD, Paponov LA, Palme K (2006) Auxin in action: signalling, transport and the control of plant growth and development. Nat Rev Mol Cell Biol 7:847–859

The editors of Encyclopædia Britannica (2018) Nitrogen-fixing bacteria. Encyclopædia Britannica, inc. https://www.britannica.com/science/nitrogen-fixing-bacteria

Thomine S, Lanquar V (2011) Iron transport and signaling in plants. In: Transporters and pumps in plant signaling. Springer, Berlin, pp 99–131

Tilman D (1998) The greening of the green revolution. Nature 396:211–212

Timmusk S, Wagner EGH (1999) The plant growth-promoting rhizobacterium *Paenibacillus polymyxa* induces changes in *Arabidopsis thaliana* gene expression: a possible connection between biotic and abiotic stress responses. Mol Plant-Microbe Interact 12:951–959

Tiwari S, Lata C (2018) Heavy metal stress, signaling, and tolerance due to plant-associated microbes: an overview. Front Plant Sci 9

Tiwari S, Prasad V, Chauhan PS, Lata C (2017) *Bacillus amyloliquefaciens* confers tolerance to various abiotic stresses and modulates plant response to phytohormones through osmoprotection and gene expression regulation in rice. Front Plant Sci 8:1510

Ullah A, Mushtaq H, Ali H, Munis MFH, Javed MT, Chaudhary HJ (2015) Diazotrophs-assisted phytoremediation of heavy metals: a novel approach. Environ Sci Pollut Res 22:2505–2514

Vadakattu G, Paterson J (2006) Free-living bacteria lift soil nitrogen supply. Farm Ahead 169:40

Vance C (2001) Symbiotic nitrogen fixation and phosphorus acquisition. Plant nutrition in a world of declining renewable resources. Plant Physiol 127:391–397

Vandenbussche F, Pierik R, Millenaar FF, Voesenek LACJ, Van Der Straeten D (2005) Reaching out of the shade. Curr Opin Plant Biol 8:462–468

Venkateswarlu B, Shanker AK (2009) Climate change and agriculture: adaptation and mitigation strategies. Indian J Agron 54:226–230

Verma VC, Singh SK, Prakash S (2011) Bio-control and plant growth promotion potential of siderophore producing endophytic Streptomyces from Azadirachta indica A. Juss. J Basic Microbiol 51:550–556

Vessey JK (2003) Plant growth promoting rhizobacteria as biofertilizers. Plant Soil 255:571–586

Viveros OM, Jorquera MA, Crowley DE, Gajardo G, Mora ML (2010) Mechanisms and practical considerations involved in plant growth promotion by rhizobacteria. J Soil Sci Plant Nutr 10:293–319

Wagner SC (2011) Biological nitrogen fixation. Nat Educ Knowl 3:15

Wang JF, Zhang YQ, Li Y, Wang XM, Nan WB, Hu YF, Zhang H, Zhao CZ, Wang F, Li P, Shi HY, Bi YR (2015) Endophytic microbes Bacillus sp. LZR216-regulated root development is dependent on polar auxin transport in *Arabidopsis* seedlings. Plant Cell Rep 34:1075–1087

Willis A, Rodrigues BF, Harris PJC (2013) The ecology of arbuscular mycorrhizal fungi. Crit Rev Plant Sci 32:1–20

Weller DM, Thomashow LS (1994) Current challenges in introducing beneficial microorganisms into the rhizosphere. In: O'Gara F, Dowling DN, Boesten B (eds) Molecular ecology of rhizosphere microorganisms: biotechnology and release of GMOs. VCH, New York, pp 1–18

Wu QS, Xia RX, Ying Y, Shen T, Xue B (2005) Effects of AM fungi on drought tolerance of citrus grafting seedling trifoliate orange/cara cara. J Appl Ecol 16:865–869

Yaacov O, Robin I (1995) The development of *Azospirillum* as a commercial inoculant for improving crop yields. Biotechnol Adv 13:415–424

Yamaguchi S (2008) Gibberellin metabolism and its regulation. Annu Rev Plant Physiol 59:225–251

Yazdani M, Bahmanyar MA, Pirdashti H, Esmaili MA (2009) Effect of phosphate solubilization microorganisms (PSM) and plant growth promoting rhizobacteria (PGPR) on yield and yield components of Corn (Zea mays L.). Proc World Acad Sci Eng Technol 37:90–92

Yu GY, Sinclair JB, Hartman GL, Bertagnolli BL (2002) Production of iturin A by *Bacillus amyloliquefaciens* suppressing *Rhizoctonia solani*. Soil Biol Biochem 34:955–963

Zahir ZA, Munir A, Asghar HN, Arshad M, Shaharoona B (2008) Effectiveness of rhizobacteria containing ACC-deaminase for growth promotion of peas (*Pisum sativum*) under drought conditions. J Microbiol Biotechnol 18:958–963

Zahir ZA, Shah MK, Naveed M, Akhter MJ (2010) Substrate dependent auxin production by *Rhizobium phaseoli* improves the growth and yield of *Vigna radiata* L. under salt stress conditions. J Microbiol Biotechnol 20:1288–1294

Zahran HH (2001) Rhizobia from wild legumes diversity, taxonomy, ecology, nitrogen fixation and biotechnology. J Biotechnol 91:143–153

Zandi P, Chalaras SK (2014) The importance of plant growth promoting thizobacteria (PGPR). In: Health and productivity of Agro-ecosystems, 5th international scientific agricultural symposium 'Argosym 2014' Jahorina, Bosnia and Herzegovina. https://doi.org/10.13140/2.1.4217.3442

Innovation System Approach for Urban Agriculture: Case Study of Mexico City

Hans Dieleman

Abstract This chapter presents an innovation system approach for urban agriculture. It argues that urban agriculture is a systemic concept – agriculture intertwined with urban dynamic – but that a systemic approach is often missing. Such an approach allows identifying strengths and weaknesses of urban agriculture for a particular city, region or country, in a comprehensive way. Based on these insights, more precise and targeted policies can be designed to stimulate urban agriculture and innovations needed in its context. The chapter illustrates this through the presentation of urban agriculture in Mexico City, presented in a number of elements of an innovation system, such as system boundaries, dynamics, institutions, knowledge, and learning cultures. Cultural dimensions are as yet only rarely recognized. The chapter describes how the cultural dimensions of urban agriculture are very important in understanding the case of Mexico City, and probably in much more cities.

Keywords Urban agriculture · Mexico city · Innovation systems · Learning cultures

1 Introduction

Urban agriculture is rapidly establishing itself as a new practice in many cities worldwide (WinklerPrins 2017). In Asia, Vietnam is a country with a long tradition of urban agriculture, and in Hanoi today, 80% of fresh vegetables and 40% of eggs are produced by urban and peri-urban agriculture (Kohlbacher 2015). In Africa as well, there are various countries with extended experience in urban agriculture. In Ghana's capital Accra, around 90% of all the fresh vegetables consumed comes from production within the city (Corbould 2013). In Latin America and the Caribbean, urban agriculture equally is already widespread in 23 countries in that region. It is practiced by 40% of

H. Dieleman (✉)
College of Sciences and Humanities, Autonomous University of Mexico City, Mexico City, Mexico
e-mail: hansdieleman@gmail.com

M. Farooq, M. Pisante (eds.), *Innovations in Sustainable Agriculture*,
https://doi.org/10.1007/978-3-030-23169-9_4

households in Cuba, and by 20% of the households in Guatemala while in 16 of the 23 countries surveyed, people earned some income from this activity (FAO 2014). In the North, urban agriculture has a long history, going back to the economic depression of the late nineteenth century. Community gardens were developed in many cities in both Northern America as well as Europe. Today, urban agriculture is seen as part of the urban ecological infrastructure, needed to meet demands of sustainability and urban resilience. Roof and vertical gardens contribute to the greening of cities as they curb air pollution, increase humidity, lower urban temperatures and reduce energy consumption as well as extreme temperature fluctuations within buildings (Dieleman 2016).

An impressive body of literature has seen the light in the past 20 years, ranging from professional reports, instructional guides and leaflets to academic articles and books. This is a positive development but, as Stefan Reyburn argues, there is a certain mismatch or imbalance in the literature. A large part of it is primarily case-based, founded on the use of mere personal observations and experiences gathered in fieldwork. Moreover, it has a rather technical orientation, focusing on one or more operational aspects of urban agriculture. As a result, Reuborn is of the view that urban agriculture has been taken in a conceptual way, even though its essence has a conceptual construct (Reyburn 2012).

The essence of urban agriculture is its location within a city, more than its set of mere agricultural activities. It is a practice taking place in the midst of, and interrelated with, a variety of urban dynamics: economic, geographical, sociological, cultural, anthropological infrastructural, and more. This is often reduced to specific problems like for instance the question of how to handle contaminated or poor soils, or the lack of knowledge among practitioners of urban agriculture, and the need of knowledge transfer. Cultural, sociological or anthropological factors, however, are always present on a fundamental and more invisible level, and co-shape the present and the future of urban agriculture.

In this chapter, urban agriculture has been discussed in a conceptual way, using the idea of an innovation system as its key concept. This aims at linking technical and operational aspects with economics and policies, as it aims at linking those with sociological and anthropological insights. Using this concept allows for really integrating the *urban* character of urban agriculture, not seeing the city as solely a context for agricultural activities, but to integrate the economic, social and cultural dynamics of a city (that what makes a city "urban") as an integral part of the agricultural activities. In developing this approach, the chapter presents urban agriculture in Mexico City as a case study, to illustrate to way the concept of innovation system is applied.

2 Towards an Innovation System Approach for Urban Agriculture

Innovation can simply be described as the invention and application of a new idea, device, product, service or method (cf. Frankelius 2009). It is a buzzword in sustainability, which makes sense as sustainability is about creating a new world with innovative ideas, practices and systems. With respect to urban agriculture, there equally exists a considerable amount of literature touching upon the phenomenon of innovation (Cunk et al. 2017; Driscoll 2017; Pfeiffer et al. 2013; Prain and De Zeeuw 2007). The published work points at various elements of which many have been introduced in this chapter: actors, interaction of actors, institutions, industries, governmental policies, learning cultures, technological development, and more. Frequently however, these aspects are not mutually linked and seen in a systemic perspective. To understand the potentials of innovation for a new practice as urban agriculture, or for a city, region or country, it is important to see all of these factors in their mutual interrelationship. This leads us to the concept of innovation systems, a concept introduced in the beginning of the ninety eighties, by Christopher Freeman and Bengt-Åke Lundvall (Freeman 1982; Lundvall 1985).

Innovation systems initially had a rather economic perspective, focusing on the production, distribution and consumption of new products, devices or technologies. The perspective gradually broadened, and the mere economic factors became part of wider innovation system (Lundvall 1988). With variables like legislation, education, knowledge transfer, entrepreneurial cultures and more (Lundvall 2007). These all, as a system, are supposed to make us understand better the dynamics of innovation in a particular field, country or region. There is a trend now to put ever more relative emphasize on sociological, political and – little by little – cultural factors, indicating that these are supposed to carry more weight than the mere economic factors, in explaining innovation processes (Lundvall 2016). Markatoua and Alexandroub (2015) argued that urban innovation systems should include the whole spectrum of societal challenges, as these form the unique societal aspects of cities (Markatoua and Alexandroub 2015).

A next step is to include cultural aspects and variables. As Tabellini (2010) convincingly showed, cultural characteristics of an economy are crucial in understanding the way the economy moves, functions and performs. They are equally crucial in determining the potential for change and transformation into the direction of, in our case, urban agriculture. Lundvall (2016) emphasized the importance of culture and language because culture make us "interpret identical signals in different ways", which is a starting point in creation and design. Alon-Mozes and Amdur (2010) gave an interesting example of culture in the field of urban agriculture, analyzing how the meaning of urban agriculture in Israel changed. It changed form a collective Zionist project into a personal project for people involved, allowing them to do physical exercise and stay in good physical shape through working with the soil. Redefining the meaning of urban agriculture ensured an ongoing interest, after its original meaning lost relevance. In innovation system literature, culture is very often seen as entrepreneurial cultures, and their degree of openness for innovation and change (Pohlmann et al. 2005). The example from Israel shows us that culture can also be relevant in the form of a certain national or group belief.

Coenen and Díaz López (2009) conceptualized an innovation system constructed around a number of key variables that all are non-economic by nature, such as System Boundaries, Activities, Actors and Networks, Institutions, Dynamics, drivers and barriers, Knowledge Transfer, Learning Cultures. Inside of these variables, economic processes certainly play a role, but they are not seen as the constituting elements of the system as such.

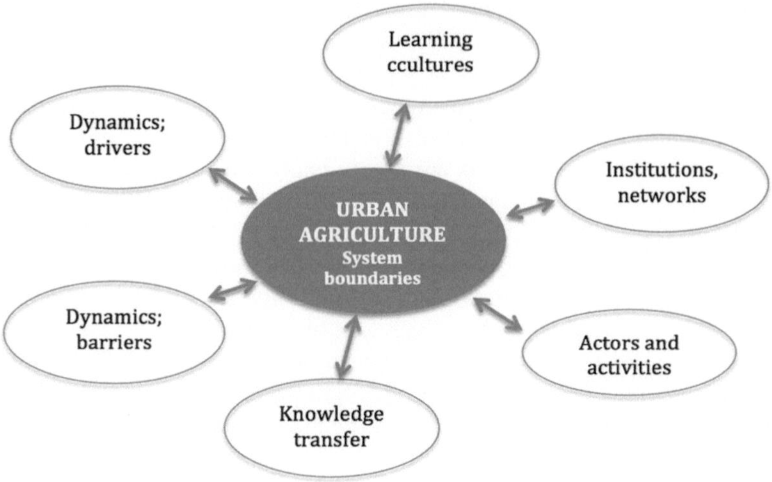

Fig. 1 Schematic representation of the key variables of the innovation system of urban agriculture

This chapter is structured with the elements mentioned by Coenen and Diáz López in mind (see Fig. 1). First it presents what urban agriculture actually is in Mexico City (system boundaries). In this description, the relevant actors, networks and institutions in urban agriculture in Mexico City are introduced. Then, it will focus on the drivers within the system (Dynamics), and will do so from a point of view of particularly social and cultural dynamics. This is inspired by an earlier work carried out by Dieleman, in which he interrelated ecological, economic, social and symbolic aspects of urban agriculture in Mexico City (Dieleman 2016). This work highlighted the relevance of cultural values. Both positive dynamics (drivers) are presented, as negative dynamics (barriers). Finally, the chapter interrelates all of those in an attempt to show how the innovation system approach sheds new light an urban agriculture, in the case of Mexico City.

3 Urban Agriculture in Mexico City; System Boundaries, Actors and Institutions

Mexico City is located in the Valley of Mexico, in the midst of the Mexican highlands. The city area is approximately 1479 km^2, with an average altitude of 2238 m above sea level, surrounded by mountains of up to 3880 m (Torres-Lima Pablo et al. 2000). The main soil types are litosoles, andosoles, feozem, regosoles and solonchak, and the climate is moderate, with a dry winter and a wed summer season. The mean temperatures range from 18 to 24 °C, and average annual rainfall is between 100 and 1400 mm (CETENAL 1977).

Urban agriculture in the Valley of Mexico is, seemingly paradoxical, older than the city itself. Before the colonization process, the center of what currently is Mexico City was the Aztec city called Tenochtitlán, build on an island located in a big lake that stretched throughout most of the valley. The Aztecs developed floating gardens around their city (the so-called *Chinampas*) to cultivate food for the inhabitants. Near the edges, around the lake, many communities existed with each having its own agricultural production. The cultural, symbolic meaning of these facts plays a significant role in the way urban agriculture is perceived today, something that will be analyzed in detail in this chapter. The Spanish conquistadores dried out the lake to create Mexico City, converting freshwater into land and later into urban space. This process reached its almost total completion long after the colonization period was finished, in the last three decades of the previous century. Near the end of the last century, the valley almost completely changed into urban area, due to mass migration from rural Mexico to the capital. Agricultural activities stayed and transformed into urban agricultural activities.

The response to the historical process of urbanization has been unplanned and rather chaotic, with little governmental organization or guidance, and with frequent violations of the few regulations that existed (Torres Lima et al. 2000). As a result, urban agriculture in Mexico City is far from a homogeneous or well-structured activity. The city usually is seen as composed of three zones with distinct features: a peri-urban, suburban and urban zone. Underneath this zone-division however, a diverse mix of activities take place, while the distinction of the peri- and suburban zone is far less clear than the difference of these two with the urban zone. In the description of the zones in this chapter, the peri-urban and suburban zones are combined.

3.1 The Peri- and Suburban Zone

Both the peri-urban and the suburban zones are in the south of the city (Fig. 2). The peri-urban zone has a total of 300 km^2 of farmland, which is divided in small plots that range in size from 1 to 3 ha (Torres-Lima and Rodríguez-Sánchez 2008). The zone consists of various small, traditional and indigenous communities such as Milpa Alta, San Mateo Xalpa, San Salvador Cuauhtenco, Magdalena Contreras and Cuajimalpa, and parts of Tláhuac. The suburban zone is found in lowland areas, particularly in the neighborhoods of Xochimilco and parts of and Tláhuac. These communities, as their names indicate, equally are pre-Hispanic and still maintain much of their traditional ways of working and living. The suburban zone used to be peri-urban but has been locked in by the ongoing urban sprawl, making them now part of the suburban zone of the city.

The communities of the peri-urban zone cultivate a variety of crops, including nopal, oats, potatoes, broccoli, carrots, lettuce, maize, tuna (fruit) and amaranth. To give an image of its magnitude, the 2012 harvest was valued at more than US$100 million and included 336,000 tons of nopal, 147,000 tons of forage oats, 12,500 tons

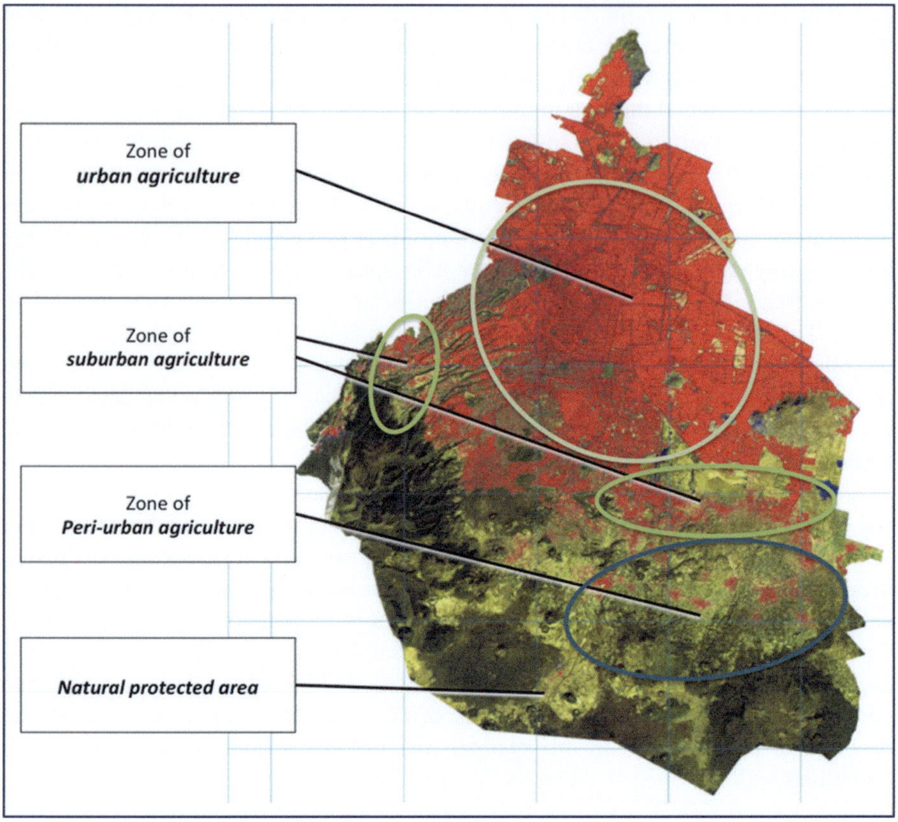

Fig. 2 Map of Mexico City with different urban agricultural zones

of potatoes and 15,000 tons of broccoli, carrots, lettuce, maize, tuna (fruit) and amaranth. Farms also raise livestock such as sheep, rabbits, pigs, horses and poultry. The animal population is estimated at some 6.650 head of cattle, 30.000 pigs, 10.000 sheep and 220.000 chickens (FAO 2014). Spaces inside the villages are used for milk and meat production in stables and to keep animals for work and transport (mules, donkeys and horses). The backyards are used for hens, turkeys, ducks, rabbits, pigs and birds of prey while the family orchard is used for the production of vegetables, fruit trees, and medicinal, ritual as well as ornamental plants. The space immediately around the village is dedicated to the intensive production of nopal, surrounded by a circle designated for the cultivation of maize, chile and beans (Losada et al. 2011).

For the communities and farmers in especially the peri-urban zone, it is not easy to generate sufficient income. Only 49% of the farmers in Milpa Alta and 25% of the farmers in Tlalpan can make a fulltime living from their agriculture (Torres-Lima and Rodríguez-Sánchez 2008). The others commute to the city center to find additional means of income, often in the informal economy. They tried to increase

Fig. 3 Impression of Aztec Chinampas, and an image of a modern Chinampa

and intensify production, but mainly with adverse effects. As a result, they now face, especially in the cultivation of nopal, an increasing amount of plagues, a new problem related to a more modern and intensified way of cultivation. Even though the use of agrochemicals to fight the plagues is legally prohibited, the enforcement of this law is very weak (FAO 2014). Moreover, the city government promotes a style of agriculture that is difficult to practice without the use of such chemicals. As a result, it remains very difficult for most of the farmers to generate a stable income. Later in this chapter, we will come back to that (Fig. 3).

The dominant production in the suburban zone is horticulture and floriculture, with some maize, using treated water for irrigation. On a yearly base 17,600 tons of flowers and 3,635,000 potted plants are cultivated. Sheep, rabbits, birds, horses and pigs are still raised in backyards and in some small (dairy) farms (Losada et al. 2011). Xochimilco is the municipality where the cultivation in the traditional *Chinampas* or floating gardens is still present. However, farmers are little by little changing towards the use of greenhouses. In the last two decades, the use of plastic greenhouses for flower and horticultural production has increased considerably, causing the abandonment and transformation of *Chinampas*. In 2006, the area of *Chinampas* in production in Xochimilco was estimated at 262 ha, with an annual loss rate of 31 ha, compared with 244 ha of greenhouses, with an annual growth rate of 14 ha (Merlín-Uribe et al. 2012). This has considerable environmental effects. It involves a loss of the lacustrine environment, which integrates water, trees and wetlands. This in turn has a negative effect on the water quality in the area. Furthermore, *Chinampa*s use less agrochemicals than greenhouses. Merlín-Uribe et al. show that 94% of the greenhouse farmers use chemical pesticides, while this is practiced by only 68% of the *Chinampa* owners. Twenty-six percent of them use purely organic measures, and 6% combines the two (Merlín-Uribe et al. 2012). Here as well various challenges present themselves that we will come back at later in this chapter (Fig. 4).

Some 20% of the food consumption of Mexico City is produced in the peri- and suburban zone of the city combined. The farmers transport every day 30.000 tons of products to the central market (the 'Central de Abasto'), that extends 328 ha of surface), while the food is distributed from thereon towards approximately 312 smaller markets in all parts of the city, the so-called "*Tianguis*" or mobile markets, that

Fig. 4 Huerto Romita en the downtown area of La Rome, Mexico City

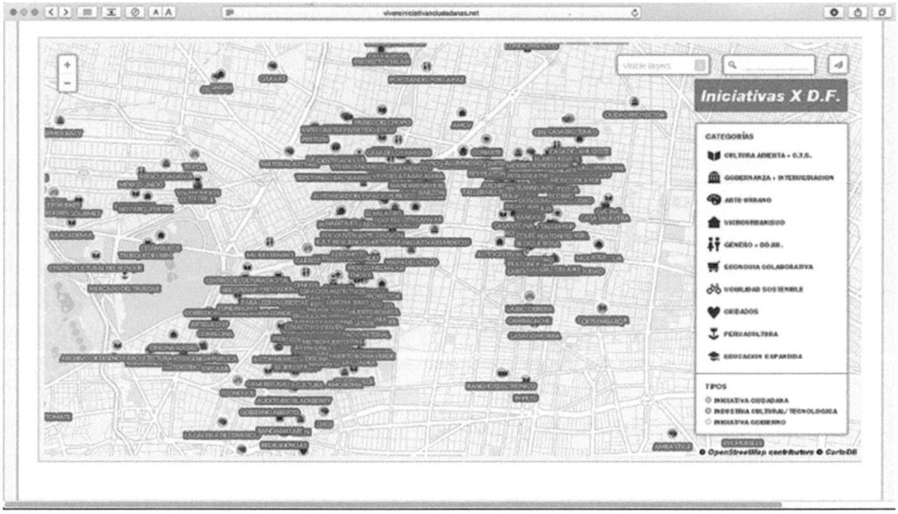

Fig. 5 Google map of citizen initiatives in Mexico City

equally have their roots in pre-Hispanic times (Soriano Robles 2005). In terms of employment and income generation however, urban agriculture is far less important. About 30.366 inhabitants of the city are involved in urban agriculture, in a total of 450 rural localities, in both the suburban (17.006 inhabitants) and the peri-urban zone (13.360 inhabitants). This corresponds with not more than 0.3% of the city's population (Fig. 5).

3.2 The Urban Zone

The center of Mexico City is densely populated, with an urban infrastructure comprised of main avenues, smaller roads and streets and parks with an organic mix of domestic, commercial and institutional functions within the neighborhoods. The areas and city municipalities surrounding the downtown city center, such as the municipalities of Iztapalapa, Iztacalco or Gustavo A. Madero, were largely constructed in the last decades of the previous century. They emerged in rather unplanned ways, with hardly any space for parks or recreational areas. They are real urban jungles of endless small homogeneous little houses of usually two floors with a flat roof, constructed without any architecture or feel for aesthetics. These urban areas have rather distinct forms of urban agriculture, concentrated in specially designed public gardens, rooftops and yards.

Agricultural activities inside of the urban zone are divers as well, showing a rich patchwork of projects initiated by governments, by NGO, private firms, NGO's and households. Between 2007 and 2012, local government invested 6 million US Dollar in 2800 projects, among other gardens in houses, collective housing units, schools and governmental buildings, and reached directly 15.700 inhabitants with these projects. Some 3000 families, especially the poorer, received support from the Government of Mexico City to create gardens on their rooftops, some with simple greenhouses to protect their crops from nightly mountain chill and occasional hail (Gaceta Oficial Del Distrito Federal 2012). On top of that, the city created, between 2010 and 2014, 22.000 m^2 of green roofs/gardens on public schools, hospitals, governmental buildings and some metro stations (Gaceta Oficial Del Distrito Federal 2012). In 2015, the construction of 10.000 more m^2 of green roofs was initiated (Dieleman 2016).

The central city government encourages the city municipalities to establish a department responsible for the creation urban gardens and the stimulation of domestic small-scale urban agriculture. As a result, initiatives now come from both the central city level and the level of the municipalities. The trickling down effect from the central city level to the municipalities however is rather slow. An additional number of 140 projects were realized in 16 of the city's municipalities, in vacant lots, backyards and roofs of private and public buildings. This includes the creation of public or semi-public urban gardens in 14 of the municipalities. These gardens have educational purposes and allow for the cultivation of plants and flowers for private consumption. The distribution of the crops thus realized is however not systematically planned or organized. Volunteers take what they want, and sometimes crops are not harvested at all.[1]

[1]This is, as far as I know not well documented, the statement is base don personal observations realized in various visits.

In 2015, a new public urban garden of 700 m^2 was opened in the rather poor municipality of Iztapalapa, within a public secondary school. The garden received the explicit label of being a "productive urban garden", where the students can work and take their share of the crops collectively produced to their homes. The garden focuses on the cultivation of cilantro, parsley, chile, chamomile and lavender, which all form part of the regular Mexican cuisine. The Cuauhtémoc municipality, the most central and downtown area of the city, is one of the most active in urban agriculture (Gaceta Oficial de la Ciudad de Mexico 2016). Since 2009, it is training citizens in becoming certified small-scale sustainable urban farmers. By now, some 500 citizens were certified. This municipality uses its public urban garden to give the training for the small-scale sustainable urban agriculture program. The municipalities of Miguel Hidalgo and Coyoacán initiated various projects for migrant families and single mothers, integrating in these projects the concept of microcredit in the form of small grants (1000–3000 Mexican pesos corresponding with or 50–150 US dollar) that enable them to invest in equipment to grow vegetables, to compost and to capture and use rainwater (Gaceta Oficial Del Distrito Federal 2012).

Besides the initiatives and policies generated by the government, Mexico City must be characterized by its large number of private initiatives, coming from NGO's, private industries, start-ups of young, recently graduated academics, and rather small-scale spontaneous bottom-up initiatives in neighborhoods and communities. One of the well-known communal gardens "*Huerto Romita*" (Romita garden) located in the fancy neighborhood of 'La Roma' (part of the Cuauhtémoc municipality).[2] It exists since 2007 and even though it is very small in surface (it has a 56 m^2 gardening center), it is very active in giving training in organic community vegetable production and in teaching permaculture techniques. It also helps in starting up school gardens and installs home and community gardens for city residents. *Huerto Romita* is well-known, and just one of many similar initiatives in the city.

The private sector as well shows various initiatives. Well-known is the urban planners group, "*Efecto Verde*",[3] who's imaginative and bold proposal is to cover 40% of the city's urban surface by 2030, with low-maintenance vegetation. The group is engaging in many projects that all partially contribute to their big objective, but a comprehensive plan to realize the 40% still has not been accepted by the city government. Besides this well-known company, many small startups enter the 'market' of urban agriculture as well. A nice example is the startup "*Solution Culture*", a company of three recently graduated industrial designers, who design green roofs, gardens and walls, primarily for companies located in Mexico City.[4]

Overlooking the entire spectrum of urban agricultural activities in Mexico City, one can only conclude that it is extremely divers, with many actors and institutions involved, and with a diverse range of activities. The indigenous farming communities are trying to maintain century-old traditions, agro-industries selling (especially

---

[2] http://www.huertoromita.com/centro-romita

[3] https://connectamericas.com/company/simbiosis-urbana-efecto-verde

[4] http://www.solutionculture.mx/nosotros

flowers), startups of young university graduates involved in the design of fashionable green roofs and green walls, NGO's, governmental agencies, many citizen initiatives and thousands of individual households and/or persons, in one way or other active in urban agriculture. This diversity reflects the complexity of the Mexican society, and is both a positive driver for urban agriculture, as it is also a potential barrier, as will be shown in the following sections.

4 Positive Drivers for Urban Agriculture in Mexico City

The diversity of urban agriculture in Mexico City reflects the complexity of the Mexican society in general. Yet, urban agriculture also has the capacity to unite the otherwise very divided country. This is an important driver for urban agriculture in Mexico City.

Mexico is a very complex country with on the one hand people living in modernity and postmodernity, and others still living in pre-modern conditions and cultures. Moreover, there is a deep sense of distrust among the various groups, resulting in a desire to minimize contact with those coming from other cultures and social-economic classes. There is in general a fear for "otherness", for people with different customs and lifestyles (Yépez 2010; Dieleman 2010). Mexico-City is not different; the situation may be even more extreme there. The city hosts many globally oriented well-educated individuals, living in the fashionable downtown area, choosing their own urban lifestyle, tailored to fit their specific wishes and desires. It hosts a modern national oriented middle class, living in often gated communities, or otherwise rather well secured neighborhoods. They live lifestyles that enroll around fixed jobs, family life, holidays and activities for children (Philip et al. coordinators (2015:2). Finally, it hosts many pre-modern oriented inhabitants, who arrived in the 1970s and 1980s, the period of mass migration towards the city. In numbers they are more than half of the city's population, living in the huge suburbs that make up almost 80% of the urban area. They still have rather regional and rural orientation, longing to continue – or return to – the way of living they lost because of their migration to the city (Dieleman 2010).

An interesting characteristic of urban agriculture is that it cuts right across all of the groups mentioned. More importantly, it has the potential to unite the Mexican society that is otherwise so very divided, as it touches upon, and mobilizes, two very distinctive features of the Mexican culture and society:

- The desire to reconnect with its largely lost pre-colonial past, and
- The longing for freedom and independence from societal institutions.

These two are key drivers for urban agriculture in Mexico City, drivers of a mainly cultural nature.

4.1 The Symbolic Meaning of Urban Agriculture

The previously mentioned *Chinampas* play a significant role as drivers for urban agriculture, as they are symbols of an almost lost past. It is important to give a short description of them in historic perspective. Around the year 1350, the Aztecs created the city of Tenochtitlán on one of the islands of what was at that time the lake of the Valley of Mexico. The city grew steadily and became the biggest urban settlement in pre-Hispanic Latin America, with at its peak around the year 1500 a population of approximately 250.000 inhabitants (Aguilar-Moreno 2007). To feed the ever-increasing population and to overcome land shortage, the Aztecs created their so-called '*Chinampas*'or floating gardens. The *Chinampas* increased the land area available for cultivation and were a model for numerous other cities in Mexico at that time (Aguilar-Moreno 2007). The *Chinampas* were constructed by staking out rectangular enclosures, ranging in size from 100 to 850 m^2, filled with mud and decaying vegetation and used for cultivation of mainly vegetables and aromatic flowers. On average 10–15 persons worked on one *Chinampa*. Cultivation was accomplished by the effective use of seedbeds, thus allowing for continuous planting and harvesting of crops (Evans 2013).

Soon after the Spanish 'conquistadores' took control of the Aztec land, however, in between 1519 and 1523, they started drying out the lake, creating land that later served as the foundation for contemporary Mexico City. Only in the suburban community of Xochimilco, the pre-colonial canals and *Chinampas* remained and still exist. In 1987, UNESCO declared them to be part of the UN World Heritage, underscoring their cultural importance, while taking a stand against their ongoing deterioration (Torres-Lima and Rodríguez-Sánchez 2008).

This history of Mexico City, combined with the previously mentioned desire to reconnect with the pre-colonial past, gives urban agriculture a positive connotation for most Mexicans today. This is not a small thing, on the contrary. Mexico is, as many other postcolonial countries, still suffering from its traumatic colonial history, resulting in a huge problem of uniting the divers and mixed population, and of creating one nation as a social whole, with a shared identity (Brushwood 1966; Hoy 1982; Yépez 2010; Dieleman 2010). Urban agriculture offers the city's inhabitants an opportunity to re-experience their past, in a symbolic way, and to be Mexican in an identity-full way, while they can at the same time be part of a global emerging movement of sustainability and food security. Urban agriculture stands for a tradition, an identity, as well as for contemporary values of sustainability and care for future generations. The *Chinampas* play a crucial role in this, as icons of a time largely gone by (Torres Lima et al. 1992).

For the indigenous farmers and communities like those of Milpa Alta or Cuajimalpa, the symbolic meaning of urban agriculture helps them in their struggle to continue living their traditional lifestyles. For the migrants of the 1970s and 1980s, as well as their second or third generation offspring, urban agriculture opens opportunities to reestablish parts of their rural lifestyle. Finally, for the more affluent population in the urban zone of the city looks, the symbolic meaning of urban agriculture helps them integrating their postmodern lifestyle with their Mexican identity.

While the city government is interested in urban agriculture for reasons of food security in the context of climate change, and private industries are interested in combatting contamination to change the city into an attractive workplace that appeals to a foreign and well-educated workforce, they find the citizens of Mexico City on their side, though for different reasons (Dieleman 2016). This mix of mutually reinforcing drivers creates a huge potential for urban agriculture in Mexico City.

4.2 *Citizen Bottom-Up Initiatives*

A second potential for urban agriculture in Mexico City is the longing for freedom and independence, a desire that has a particular meaning in the Mexican context. This context is partly, again, historic and rooted in the colonial past. But it is also contemporary, and rooted in the malfunctioning of Mexican democracy, government and the juridical system. These all are all seriously plagued by corruption, brutal inefficiency and clientelism (Philip et al. 2015). Mexicans have suffered throughout history from rulers that never were really interested in the wellbeing of their citizens. This shaped a particularly deep longing for freedom, respect and independence, and resulted in an active civil society that is remarkably active today (Vargas Hernández 2010).

This civil society is relatively quiet young, and its development accelerated in the second half of the ninety eighties of the previous century. The earthquake that struck Mexico City on September 19, 1985 is seen as an important catalyst in this development. This quake destroyed a considerable part of the city and resulted in the death of approximately 10.000 persons, with 250.000 people losing their homes (Quarantelli 1992). In the days after the quake, governmental responses were very inadequate, and citizens were obliged to organize themselves. They spontaneously took up the tasks of rescuing people, distributing food and providing shelter. Without this spontaneous civil response, the effects of the earthquake would have been much more detrimental (Quarantelli 1992). Yet something else happened. The aftermath of the earthquake awakened, in the words of the Mexican poet Homero Aridjis, a social earthquake that is still roiling in Mexico City and the entire country.[5] It is the social earthquake of the awareness that the Mexicans can take the course of life in their own hands, despite of a malfunctioning government. Vargas Hernández talks about the Mexican civil society as an emergent property of a failing political and institutional system, that accelerated in 1985 and never disappeared since (Vargas Hernández 2010).

[5] https://www.huffingtonpost.com/homero-aridjis/mexicos-1985-earthquake_b_8170324.html

Houtzager and Acharya conducted an exhaustive comparable study on citizenship in and Mexico City and Sao Paolo in Brazil. They arrived at the conclusion that in Mexico City the participation in associations for self-provisioning is particularly strong. Twenty-five percent of the total population participates, or participated, in associations, initiatives or actions organized by the civil society. This participation was realized by people coming from all types of education, from lower levels up to people with higher education (Houtzager and Acharya 2010). These data reveal an image of Mexico City as a vibrant city, full of bottom-up initiatives that together constantly create and recreate the city. This is very relevant for the future of urban agriculture in the city.

In 2013, the VIC, the "*Vivero de Iniciativas Ciudadanas*" or in English the 'Nursery of Citizens Initiatives', an NGO based in Spain, started recording and mapping citizen initiatives in Mexico City, in collaboration with the Spanish Cultural Center in Mexico.[6] It registered and mapped a total of 369 citizen initiatives, in various categories as 'Care and lifestyle', 'Collaborative economy', 'Micro-urbanism', 'Permaculture' and more. Many of them are directly or indirectly involved in urban agriculture, even though they are categorized under labels as micro-urbanism, permaculture or collaborative economy. The initiatives registered by VIC only form the top of the iceberg, as only those were included who have their own website and can be found on internet. The reason for this is that VIC created a Google Map style map of the initiatives, which allows visiting each of them online.

Research carried out by VIC indicates that 91% of the initiatives consider themselves as "bottom-up" initiative without any connection to an established institution, while 87% responded that their explicit goal was to practice alternative ways of living, with keywords as ethics, social responsibility, equity and sustainability. Ninety-one percent indicated that their objective is to contribute to those values through concrete actions, instead of using political action. The combination of the historically prompted interest in urban agriculture, and the active participation of many Mexicans in bottom-up initiatives that aim at creating a better – sustainable, ethical, equal – society, creates a very fertile cultural soil for urban agriculture in Mexico City.

5 Barriers for Urban Agriculture in Mexico City

A fertile soil is not enough to make agriculture flourish, and the same is true for a fertile cultural soil. Other conditions need to be fulfilled as well, such as the availability of knowledge and capacities, certain cultural outlooks, technologies, favorable policies and structural tendencies that help urban agriculture to develop. In this section, some of those, and places them in a wider context, of modernization and some key features of the Mexican culture in general have been described.

[6] http://viveroiniciativasciudadanas.net/2015/04/20/iniciativas-x-d-f/

5.1 *Structural Tendencies*

Several structural tendencies form serious threats for the further development of urban agriculture, especially in the peri-urban and suburban zones. One of the problems the farmers in these zones are facing is a loss of agricultural land. This is mainly due to the urban sprawl that continues to demand more land for housing, industrial as well as recreational activities. In relation to this, a second problem is the overexploitation of aquifers, because of the increasing water demand of the city. This has led to a serious decline in water supply, water quality and to ground subsidence (SEDEREC 2017). Despite of that however, as was mentioned in paragraph 3.1, there is an ongoing trend to substitute *Chinampas* for greenhouses, even though the last are considerably less sustainable in terms of the conservation of water quantity and maintaining water quality. In the peri-urban zone, the significant increase in the cultivation of nopal – in itself a response to changing market demands – has led to an enormous increase in plagues, for which the cure until now is the use of agrochemicals, something that is strongly rejected by the farmers themselves, and goes against the objective to create sustainable urban agriculture.

The supply of seeds is a third serious challenge, for the horticulture and floriculture in the suburban zone and for the cultivation of especially nopal, maize and broccoli in the peri-urban zone. Government stopped seed production in the 1980s and as a result, seed supply is now largely in the hands of large corporations. Many of them are foreign with just a few – large Mexican companies active in this field. As the cost of certifying seeds are very high, the farmers are increasingly dependent on those private corporations (FAO 2014).

Even though farmers, especially in the peri-urban zone, are encouraged to produce for local and national markets, their access to the wholesale market is limited, a fourth critical barrier. A vision of how to integrate urban agricultural production within mainstream markets however is missing. This involves designing new producer-consumer networks and structures beyond the incidental organization of fairs for indigenous products produced in the urban context, and beyond the sale of organic products for the middle and upper classes, willing to pay higher prices than low-income groups can afford to do (FAO 2014).

5.2 *Modern Thinking Regarding Politics and the Definition of Agriculture*

The policies developed in Mexico City target some of the challenges mentioned, but their effect is partial and, in some cases, potentially averse. This has to do with another cultural dimension, that of modernity and modern thinking. It may sound strange to mention modern thinking as a barrier for urban agriculture, but in Mexico City this is certainly the case. In general, modern thinking tends to divide the complex reality in different parts, to then analyze those parts separately and develop policies for each part in relative isolation (Dieleman et al. 2017). This obscures a systemic view on the whole. Moreover, modern thinking in general places strong emphasize on rationalization and individualization. In agriculture this manifests itself in an agro-industrial and an agro-entrepreneurial approach. In Mexico City this creates various concrete problems.

In territorial planning the focus on dividing and separation is very clearly present (Ruiz et al. 2014). Several years ago, the territorial planning of Mexico City declared the forests in the very south of the city to be protected natural area, and from that moment on, agricultural activities are strongly discouraged, and the use agrochemicals in this area is prohibited. It is a typical modern politics of protection by separation. The farmers in the peri-urban zone of Mexico City used these areas for centuries, respecting natural cycles without destroying nature in a structural way. The idea that agriculture stands in opposition to nature was never a reality for them, but this thinking is now imposed on them. This is even more problematic as they are pushed to increase their production, and the use of fertilizers and pesticides is stimulated, even though research programs for sustainable alternatives are put in motion.

Governmental planning in Mexico City, as virtually everywhere else, is compartmentalized in separate domains as economic policy, social policy, infrastructure, education, etc. Separate policies per sector often only very partially integrate with other policy domains. This is also true for the policy to stimulate urban agriculture in Mexico City. Economics, infrastructure, social programs, market etc. are mentioned, but largely remain a context that the urban agriculture programs don't try to influence. They are merely mentioned. There is no analysis of the structural challenges just mentioned, with the exception of the problematic of the seeds. Subsequently, there are no policies developed to curb those trends to favor urban agriculture (SEDEREC 2017). The policies mentioned are largely limited to *the subsystem of agriculture* within the city. The question is what the farmers can do, not how government can curb the trends affecting the future of urban agriculture. With respect to the problematic of the seeds, this is different. The city's Secretariat for the Environment created a system for the certification of organic production, the so-called Green Seal, and has set standards for organic agriculture in the conservation zone. Subsidies are provided to the farmers of Milpa Alta who preserve local maize varieties under traditional production systems with low environmental impact.

A third aspect to urban agriculture and modernization is the proposed change towards an agro-entrepreneurial approach. Many of the farmers in the peri-urban zone seek to increase their income, but for them this is never a separate objective. In their traditional way of living – and even worldview – work, family, economics, nature, agriculture all are related, within a spiritual explanation of how all fits together (hence: worldview). Being a farmer is indeed a way of *being*, far beyond a way of merely *doing* or a profession. On a yearly basis, the community of Milpa Alta, the largest in the peri-urban zone, has 43 religious celebrations and 16 pilgrimages in which the relationship between the land, the community, fertility, water and mother earth are celebrated (Losada 2005).

As Torres-Lima and Rodríguez-Sánchez (2008) rightfully observe, this risks losing the social cohesion in the communities. The challenge is to find ways to maintain the traditional culture and stimulate urban agriculture at the same time. This is as yet not fully recognized in the politics of urban agriculture. The modern agro-industrial and agro-entrepreneurial approach also favors the use of greenhouses over *Chinampas* in the suburban zone, while Chinampas are remarkably more efficient and sustainable, in terms of both the use and contamination of water.

Secondly, the policies do not really find an answer to the question how sustainability, traditions, modern techniques and practices can all go together, or be blended together in a convincing way. This is problematic, as the city government pretends to stimulate sustainability and respect indigenous practices, but sees agriculture through the standard lens of mainstream modern thinking. These various objectives do not organically go together however.

5.3 Organizational Learning Cultures

A second major barrier for an ongoing development of urban agriculture in Mexico City is the lack of appropriate organizational learning cultures among people, and especially groups, involved in urban agriculture. Organizational learning cultures do not address specific technical knowledge or skills, but focus on the way that teams, networks and organization function and collaborate. Learning cultures are oriented towards both individual capacities and group dynamics. Individual capacities are the levels of self-knowledge and capacity of self-reflection and self-discipline, and the capacities to develop personal professional trajectories, with specific ideas and visions for the long and the short term. Relevant aspects of group dynamics are capacities of effective professional communication, effectively dividing of tasks, meeting deadlines, giving feedback and collectively evaluating results.

In Mexico in general, this is a big challenge. Gordon (2010) analyzed Mexican business cultures using the dimensions that Hofstede developed in his famous cross-cultural analysis of 50 countries (Hofstede 1980). These dimensions are individualism versus collectivism, power distance, uncertainty avoidance and masculinity versus femininity. Especially the dimension of power distance is relevant in the context of urban agriculture in Mexico City. Mexico is one of the countries with the worlds' highest index for power distance, which means that people is reluctant to express a different opinion than their boss, and tend to conform to what their boss says, even when they disagree and/or know that their boss is wrong (Kesseli 2017). The Mexican macho culture makes this worse, as the macho tends to impose his ideas on others, disregarded of how the other thinks, feels or desires (Dieleman 2010). This prohibits creating more equal and open working relationships and genuine partnerships, which are essential for good teamwork. It equally discourages learning and teaching skills for teamwork, such as professional communication and collaboration skills, as they are not perceived to be that relevant.

A research project carried out with students of the undergraduate program in Environmental Sciences and Climate Change of the Autonomous University of Mexico City (Dieleman and Martinez-Rodriquez 2017), shows the relevance of these cultural phenomena for urban agriculture. The project investigated the cultures of 20 citizens initiatives included in the list of VIC. These 20 initiatives all focused on urban agriculture, as either their main activity, or as part of a broader range of activities. The research showed that organizational cultures indeed have a negative effect on the functioning of these initiatives, both in terms of their internal organization as well as in terms of their participation in networks. Many, if not most of these initiatives were founded rather charismatic persons with clear visions on how to accomplish their goals. And even though most of them favor horizontal and open working relationships, their coworkers reported that they were frequently not open for dialogues and remain closed to ideas of collaborators. This negatively impacted the motivation of various coworkers and diminishes the potentials of working with all of the energy and creativity available in a group or team. Various initiatives stayed on the level of where they started, and didn't show real development in their ways of working, and the services and products offered (Dieleman and Martinez-Rodriquez 2017).

This also affected the initiatives' participation in networks. The founders/directors often are convinced that they know very well how to organize their work, and do not expect much from the possible collaboration with others. Moreover, almost all of them expressed mistrust in governmental institutions, and avoided working with the private sector as well, to avoid entering market dynamics of profit making. This demonstrated – and is nourished by – the distrust mentioned before, and the fear for others and "otherness. It is however a serious barrier for the future development of urban agriculture in Mexico City. It is widely recognized that citizen initiatives need support from the party of local governments (Ostrom 1996; Sirianni 2009; Bakker et al. 2012; Pestof et al. 2012). Sirianni's main recommendation is that both the citizen initiatives as the government need to acquire skills and abilities to collaborate in networks, such as facilitating and moderating skills. On a personal level, this requires openness to the ideas and experiences of others and involves training in organizational learning skills to learn to co-work, co-produce and co-create.

It equally requires another view on the city, as a complex system full of bottom-up activities, action, reactions and emerging properties (Dieleman and Hernández Vázquez 2018; Dieleman 2012). Kagan et al. call this a city as a 'space of possibilities', that asks for proper stimulation, moderation and facilitation, to growth and fulfill its potentials (Kagan et al. 2018). This characteristic however is not recognized by the government of Mexico City, neither in the literature on urban agriculture. It is however a potentially very strong driver for changing the city.

6 Knowledge Transfer

In 2007, the city created a new secretary to stimulate small-scale urban agriculture, the so-called SEDEREC, the Secretary for Rural Development and Community Equality. The aim of SEDEREC is to stimulate urban agriculture, do research and development of sustainable practices and technologies and engage in knowledge transfer. The program aims at improving production planning, training, technology development, agro-processing and marketing. Through this program, the city together with Mexico's Federal Government, invested between 2007 and 2012 some US$24.6 million in horticulture, floriculture and crop and livestock production, US$37 million in the conservation and sustainable use of natural resources in primary production (SEDEREC 2017).

For the farmers in the peri-urban and suburban zone, it organizes trade fairs and exhibitions, helping them to promote their traditional food in local, national and international markets. In 2013, it signed an agreement with the Havana's Institute of Fundamental Research in Tropical Agriculture in Cuba, to establish a Program for Technology Transfer of Small Scale Sustainable Agriculture. This program is first targeted at the farmers in especially the peri-urban zone, yet also focuses on the public gardens in the urban zone and through them, on single groups and households in the urban zone. In that zone, the program helps developing training programs focusing on how to compost, use rainwater and cultivate native plants and crops in urban spaces such as rooftops and small yards. In addition, the programs developed teaching material focusing on market orientation and some basic administration. In the suburban zone, water quantity and quality is a major issue. The program focuses on increasing the capacity for rainwater harvesting and storage and treatment of wastewater (SEDEREC 2017).

For the peri-urban zone, knowledge transfer focuses on the introduction of improved technologies for processing particularly on nopal and maize. As mentioned however, the program promotes a modern agro-industrial and agro-entrepreneurial approach, which has various adverse effects on the way many farmers in the peri-urban and suburban zone envision how urban agriculture in Mexico City needs to develop itself. It contradicts the cultural orientation and wishes of many, possibly because this is hardly recognized by the city government, certainly as a driver for change and innovation. What is needed is a more comprehensive interdisciplinary or transdisciplinary program, which combines a mere agricultural approach with a cultural and historical approach. Even the FAO is recognizing this omission in the knowledge transfer activities of Mexico City (FAO 2014: 25).

This need also presents itself when reading SEDERC's 4th very comprehensive annual report (SEDEREC 2017). The document addresses virtually any aspect relevant for urban agriculture, including many of the topics presented in this chapter: technology transfer, creation of new markets, citizen initiatives, the positive attitude towards urban agriculture in the city, the link with tradition and the past, and many more. However, all topics are presented in a rather separate way, and an overall vision behind them on the future of urban agriculture for the city, is missing.

As already touched upon, two questions remain unsettled. The first is how the city envisions integrating an increase in urban agricultural food production with maintaining traditions. How can horticulture and floriculture grow in sustainable ways with *Chinampas*, instead of replacing them by greenhouses? And how can the cultivation of nopal, maize and beans grow within the century old tradition of communal agricultural practices, instead of through the concept of agro-entrepreneurship? The importance of all of them is mentioned (growth, tradition ecology, sustainability), but an analysis of how they may reinforce each other, or conflict with each other, is absent. Secondly, a vision is missing on the question how urban agriculture may be integrated in mainstream economic activities in the city. It is sympathetic to organize fairs for indigenous organic products, a few days per year in the city center. But more importantly in the long run is, how these products can find their way to supermarkets and the dinner tables of those Mexicans that are less conscious and less critical in their purchases. This long-term integrated perspective still is missing.

7 Conclusion: The Importance of Seeing Urban Agriculture as a Systemic Activity

This chapter presents urban agriculture as a systemic endeavor, meaning that its success and growth depend on a fairly large number of divering variables. The innovation system approach applied in this chapter focused on System boundaries, Actors and Activities, Institutions and Networks, Dynamics, Knowledge transfer and Learning Cultures. The identification of these variables is not an exact science; it depends on the perspective one uses and the specific context one works in. The context of Mexico City, the case study presented to illustrate how a systemic approach may look like, highlights in particular various cultural aspects as drivers and barriers for urban agriculture in Mexico City. These cultural aspects do not stand alone, but have their effects on public policies, knowledge transfer and the learning cultures analyzed. The approach presented here, aims at complementing a more widely applied approach of focusing primarily on case studies from a perspective of operational aspects, practical knowhow, techniques and technologies, and rather targeted public policies to stimulate urban agriculture.

The question to answer now is, if the model applied can give us some insights into probable future developments of urban agriculture as an innovation, a new idea, device, product, service or method, in Mexico City. The answer is as follows. The chapter shows a widely present interest in urban agriculture among virtually all the different actors and groups present in Mexico City. There is an enormous potential for its further growth and development, and this is above all culturally induced. The interest is not in the first place coming from a felt need to increase food production or improve the environmental quality of the city. Key words rather are: history, identity, independence, living a meaningful traditional and both contemporary Mexican life. For the farmers in the peri-urban and suburban zones of the city, economics do play a vital role, but many of them do not isolate mere economic concerns from cultural, social and even spiritual concerns.

The cultural relevance is recognized in academic circles and in the academic literature presented in this chapter, and it is often mentioned while talking with people in the field. Most of them however, see it as an interesting feature of urban agriculture, but not as a driver or dynamic force for the future development of urban agriculture in Mexico City. The concept of innovation changes this, and turns it into a driver, making us look at urban agriculture as a process, which can be stimulated or hampered. There is an immense potential for urban agriculture, once it is seen as a decisive characteristic of a future sustainable Mexico City. As an outstanding feature of the city that unites the past and has a promise of creating a sustainable future, while it potentially unites the divers, often antagonistic groups within the city. As a symbol, which is on the one hand based on a century old history and opens on the other hand a door to a sustainable future.

The city government is yet not recognizing urban agriculture in this way. It, somewhat implicitly, thinks of urban agriculture in terms of its food production in the peri-urban and suburban zones, and in terms of its environmental benefits and environmental education for the urban zone. It recognizes the cultural dimension, but as a mere contextual variable, not as a driver for innovation. The focus on increasing food production, even though the importance of organic production methods is recognized, tends to favor greenhouses over *Chinampas* in the suburban zone, and agro-industrial and agro-entrepreneurial practices over communal indigenous practices in the peri-urban zone. By contrast, it still underestimates the potentials of food production in the urban zone. That is why knowledge transfer was, paradoxically, characterized as a barrier for urban agriculture instead of as a driver, which it is supposed to be.

A second identified driver is the desire for freedom and independence, motivating many individuals to start producing parts of their own food on rooftops or in little yards. This also includes NGO's like Huerto Romita and others, and many private initiatives such as *Efecto Verde*, startups like *Solution Culture* and many spontaneous citizen initiatives. All of these initiatives together show an image of Mexico City as a vibrant system, and a real 'space of possibilities'. As mentioned, this characteristic again is not really recognized as such, and less as a driver for urban agriculture as an innovation process. It is however a potentially very strong driver for changing the city. Here as well however, we need to signal an important barrier. The knowledge transfer does not include training in organizational learning capacities, even though these are certainly needed to promote urban agriculture in Mexico City.

References

Aguilar-Moreno M (2007) Handbook to life in the Aztec world. Oxford University Press, Oxford/New York

Alon-Mozes T, Amdur L (2010) Urban agriculture in Israel: between civic agriculture and personal empowerment. In: Gianquinto GP, Orsini F (eds) Proceedings of the 2nd international conference of landscape and urban horticulture, Bologna 2010

Bakker J et al (2012) Citizens' initiatives: how local governments fill their facilitative role. Local Gov Stud 38(4):395–414

Brushwood JS (1966) Mexico in its novel: a nation's search for identity. University of Texas Press, Austin

CETENAL (1977) Carta edafológica E-14 A-49 (Ciudad de México). Escala: 1:50,000

Coenen L, Díaz López FJ (2009) Comparing systemic approaches to innovation for sustainability and competitiveness. Paper presented in International Conference "Innovation, Sustainability and Policy", 11-13 September 2008. GREThA, University Montesquieu Bordeaux IV, France

Corbould C (2013) Feeding the cities: is urban agriculture the future of food security? Future Directions International Pty Ltd. Available online: http://www.futuredirections.org.au/files/sap/Urban_Agriculture-_Feeding_the_Cities_1Nov.pdf

Cunk K (2017) In: Straus M, Zamfira R (eds) Approaching urban agriculture as a social innovation: guidelines for the development and implementation of an action plan editors. Association for Culture and Education PiNA, Koper

Dieleman H (2010) Maskers, Macho's, Mañana, over de Mexicaanse realiteit die altijd anders is. Scriptum Publishers, Schiedam

Dieleman H (2012) Organizational learning for resilient cities, through realizing ecocultural innovations. J Clean Prod 50(2013):171–180

Dieleman H (2016) Urban agriculture in Mexico city; balancing between ecological, economic, social and symbolic value. J Clean Prod, JCLP 6670, https://doi.org/10.1016/j.jclepro.2016.01.082

Dieleman H, Hernández Vázquez J (2018) Respuestas Urbanas al Cambio Climático; desde los Paradigmas de la Simplicidad y la Complejidad. In: Astrolabio, Revista de Ciencias y Humanidades del Colegio de Ciencias y Humanidades, UACM

Dieleman LJ, Martinez Rodríguez C (2017) Las iniciativas ciudadanas sustentables en la Ciudad de México; fuerzas, debilidades y potencial para el largo plazo. In: Tercero congreso "Instituciones, Gobierno y Sociedad", 27–28 de abril de 2017, Hermosillo, Sonora

Dieleman H, Nicolescu B, Ertas A (eds) (2017) Transdisciplinary and interdisciplinary education and research. The Atlas 2017. ISBN: 978-0-9998733-0-4

Driscoll L (2017) Urban farms: bringing innovations in agriculture and food security to the city. In: Policy brief, Berkeley Food Institute, August, 2017

Evans ST (2013) Ancient Mexico and Central America: archaeology and culture history, 3rd edn. Thames and Hudson, London

FAO (2014) Growing greener cities in Latin America and the Caribbean. Food and Agricultural Organization of the United Nations, Rome

Frankelius P (2009) Questioning two myths in innovation literature. J High Technol Managem Res 20(1):40–51

Freeman C (1982) Technological infrastructure and international competitiveness. Draft paper submitted to the OECD ad hoc group on science, technology and competitiveness, August 1982

Gaceta Oficial De la Ciudad de Mexico (2016) Administración Púbica de la Ciudad de Mexico, nr. 147. Available online: http://data.caprepol.cdmx.gob.mx/archivos/2016/14/XXVII/lpn_05.pdf

Gaceta Oficial Del Distrito Federal (2012) Programa de agricultura sustentable a pequenva escala de la Ciudad de Mexico. Available online: http://www.sederec.df.gob.mx/sites/default/files/prog_agricultura_0.pdf

Gordon G (2010) Managing in Mexico – an ethnographic comparison to theory and previous research. J Int Bus Cult Stud 2:74–84

Hofstede G (1980) Consequences: international differences in work-related values. Sage, Beverly Hills

Houtzager P , Acharya AK (2010) Associations, active citizenship, and the quality of democracy in Brazil and Mexico. Springer

Hoy T (1982) Octavio Paz: the search for Mexican identity. Rev Polit 44(3):370–385

Kagan S, Hauerwaas A, Holz V, Wedler P (2018) Culture in sustainable urban development: practices and policies for spaces of possibility and institutional innovations. City Cult Soc 13:32–45

Kesseli T (2017) The business environment and culture of Mexico from the perspective of Finnish companies. Thesis Spring 2017, School of Business and Culture, Seinäjoki University of Applied Sciences

Kohlbacher B (2015) Urban agriculture in Da Nang and the utilization of by- products from waste water and solid waste processing for urban agriculture. Draft Report. GIZ, Bangkok, Thailand

Losada T (2005) La vigencia de la tradición cultural mesoamericana en Milpa Alta, pueblo antiguo de la ciudad de México. In: Revista Mexicana de Ciencias Políticas y Sociales, vol. XLVII, núm. 195, septiembre-diciembre, 2005

Losada H, Rivera J, Cortes J, Vieyra J (2011) Urban agriculture in the metropolitan area of Mexico City. J Fields Actions 5. http://journals.openedition.org/factsreports/781

Lundvall B-Å (1985) Product innovation and user–producer interaction. Aalborg University Press, Aalborg

Lundvall B-Å (1988) Innovation as an interactive process: from user–producer interaction to the national innovation systems. In: Dosi G, Freeman C, Nelson RR, Silverberg G, Soete L (eds) Technical change and economic theory. Pinter, London

Lundvall B (2007) National innovation systems-analytical concept and development tool. Aalborg University and Tsinghua University. Conference in Copenhagen, Denmark

Lundvall B-Å (2016) The learning economy and the economics of hope. Anthem Press, London/New York

Markatoua M, Alexandroub E (2015) Urban system of innovation: main agents and main factors of success. Procedia Soc Behav Sci 195(2015):240–250. https://doi.org/10.1016/j.sbspro.2015.06.355

Merlín-Uribe Y, González-Esquivel CE, Contreras-Hernández A, Zambrano L, Moreno-Casasola P, Astier M (2012) Environmental and socio-economic sustainability of chinampas (raised beds) in Xochimilco, Mexico City. Int J Agric Sust. https://doi.org/10.1080/14735903.2012.726128

Ostrom E (1996) Crossing the great divide: coproduction, synergy, and development. World Dev 24(6):1073–1087

Pestoff V, Brandsen T, Verschuere B (Coord.) (2012) New public governance, the third sector, and co-production. Routledge Publishers

Pfeiffer A, Silva E, Colquhoun J (2013) Innovation in urban agricultural practices: responding to diverse production environments. Renewable Agric Food Syst 1–13. https://doi.org/10.1017/S1742170513000537

Philip G, Faust J, Thunert M (2015) Sustainable governance indicators, Mexico report. Bertelsmann Stiftung

Pohlmann M, Gebhardt C, Etzkowitz H (2005) The development of innovation systems and the art of innovation management—strategy, control and the culture of innovation. Tech Anal Strat Manag 17(1):1–7. https://doi.org/10.1080/09537320500044206

Prain G, de Zeeuw H (2007) Enhancing technical, organisational and institutional innovation in urban agriculture. In: Urban Agriculture Magazine 19, The RUAF Foundation, December 2007

Quarantelli EL (1992) Organizational response to the Mexico City earthquake of 1985: characteristics and implications. University of Delaware Disaster Research Center Preliminary Paper #I87

Reyburn S (2012) Les défis et les perspectives de l'agriculture urbaine. Environnement Urbain/Urban Environment 6. Available online: http://journals.openedition.org/eue/449

Ruiz N, Suárez M y Delgado J (2014) Urban segregation and local retail environments. Evidence from Mexico City. In: Habitat International. https://doi.org/10.1016/j.habitatint.2014.09.001. ISSN 0197–3975

SECEREC (2017) 4° Informe de Gobierno, Ciudad de Mexico. Available online: http://www.sideso.cdmx.gob.mx/documentos/2017/informes_gobierno/sederec/4TO%20INFORME.pdf

Sirianni C (2009) Investing in democracy: engaging citizens in collaborative governance. Brookings Institution Press, Washington, D.C.

Soriano Robles R (2005) Agricultura Urbana en la Ciudad de México y su Área Conurbada: Situación y Perspectivas. Revista Agraria -Nueva Epoca- Año II 1(1)·Enero–Abril 2005

Tabellini G (2010) Culture and institutions: economic development in the regions of Europe. J Eur Econ Assoc 8:677–716. https://doi.org/10.1111/j.1542-4774.2010.tb00537.x

Torres Lima P, Canabal Cristianii AB Gilberto Burela R (1992) Sustentabilidad, viabilidad económica y estrategias sociales del modelo chinampero. Universidad Autónoma Metropolitana-Unidad XochimHco. México, enero 1992

Torres-Lima P, Rodríguez-Sánchez L (2008) Farming dynamics and social capital: a case study in the urban fringe of Mexico City. Environ Dev Sustain 10:193–208. Springer

Torres-Lima P, Sánchez LMR, García BI (2000) Mexico City: the integration of urban agriculture to contain urban sprawl. In: Bakker N, Dubbeling M, Guendel S, Sabel Koschella U, de Zeeuw H (eds). (2000) Growing cities, growing food, urban agriculture on the policy agenda. DSE, Feldafing

Vargas Hernández JG (2010) A short study of indigenouse social movments and the political ecology in Mexico and Latin America. J Manage Res 2:1–12

WinklerPrins A (2017) Global urban agriculture. Johns Hopkins University, USA

Yépez H (2010) La increíble hazaña de ser mexicano. Editor: Temas de Hoy Méxic

Part III
Resources Use Efficiency for Sustainable Agriculture

Sustainable Soil Management

Fabio Stagnari, Angelica Galieni, Sara D'Egidio, Giancarlo Pagnani, and Michele Pisante

Abstract Soil is threatened by the increase of human population, intensive management, urbanization and degradation. Sustainable Soil Management (SSM) is one of the main key factors both for significant crop production and for environment conservation. Conservation tillage techniques, especially applied together with the permanent maintenance of mulch cover on the soil surface as well as the diversification of cropping system (Conservation Agriculture (CA) system), induce positive changes to soil properties and characteristics. Additional C is sequestered from the atmosphere reducing climate change and increasing net C accumulation in long periods. In term of physical aspects, soil porosity, thanks to increased density of storage pores and elongated transmission pores, increases, saturated and unsaturated hydraulic conductivity improves, more stable aggregates are found, and, as consequence, water holding capacity and water use efficiency ameliorate while soil erosion reduces. The soil pH, CEC, exchangeable cations, and soil principal macronutrients availability, especially at surface layer level, are found to improve as well as soil biota including both invertebrates and microorganisms thanks to an increased densities and diversity.

Thanks to the numerous positive effects on soil health, sustainable soil management in turn affect positively crop yield, thanks to better root growth development, higher water and nutrient availability, profitable interaction with microorganisms.

Keywords Sustainable soil management · Conservation practices · Soil organic carbon · Soil biodiversity · Soil structure

F. Stagnari · S. D'Egidio · G. Pagnani · M. Pisante (✉)
Agronomy and Crop Sciences Research and Education Center,
University of Teramo, Teramo, Italy
e-mail: mpisante@unite.it

A. Galieni
Research Centre for Vegetable and Ornamental Crops,
Council for Agricultural Research and Economics, Monsampolo del Tronto, Italy

M. Farooq, M. Pisante (eds.), *Innovations in Sustainable Agriculture*,
https://doi.org/10.1007/978-3-030-23169-9_5

1 Introduction

Sustainable Soil Management (SSM) is one of the main key factors for agricultural crop production, providing a range of regulating, supporting and cultural ecosystem services in addition to food, fodder, fuel, and fiber. While nature gaps are not necessarily related to the intensity of cultivation, agricultural intensification has in the past been associated with the creation of significant nature gaps. A challenge will be to create biodiverse farming systems that are productive, resilient and enablers of 'intensification without simplification' (Frison et al. 2011). Agricultural systems emphasizing these principles tend to display a number of broad features that distinguish them from the process and outcomes of conventional systems. These systems tend to be multifunctional across landscapes and economies (Dobbs and Pretty 2004; MEA 2005; IAASTD 2009). SSM is addressed for resource saving agricultural crop production aimed to achieve acceptable profits together with high and sustained production levels (Stagnari and Pisante 2010; Calzarano et al. 2018), while simultaneously preserving the environment. Interventions such as mechanical soil tillage are reduced to an absolute minimum and external inputs, such as agrochemicals and nutrients of mineral or organic origins, are applied at the optimum level and in a way and quantity that does not interfere with, or disrupt, the biological processes (FAO 2012). Benefits of SSM include improved moisture conservation and water infiltration, reduced run-off of pesticides and fertilizers, reduced consumption of fuel, improved organic matter content with associated carbon sequestration, improved diversity of soil, flora, and fauna, better wildlife habitat, better soil structure, reduced wind and water erosion, less labor and less investment in equipment (Cook et al. 2006; Huggins and Reganold 2008; Stagnari et al. 2009) (Fig. 1). Furthermore, in many environments soil erosion is reduced to a level below the soil regeneration one or it is avoided altogether, and water resources are restored in quality and quantity to levels recorded before the land was put under intensive agriculture (Montgomery 2007; FAO 2011). However, intensified production is still possible under SSM regime with benefits including lower capital costs, reduced inputs, flexibility in terms of adaptation, aggrandized ecosystem efficiency, and environmental protection. In this regard, SSM which greatly improve soil porosity and reduce soil temperature are advantageous practices to mitigate these releases into the environment.

For productive and remunerative agriculture, which at the same time preserves and enhances the natural resource base and environment, and positively contributes to harnessing environmental services, SSM represents a task for Sustainable Crop Production Intensification (SCPI) to not only reduce the impact of climate change on crop production but also mitigate the factors that cause climate change by reducing emissions and by contributing to carbon sequestration in soils (Pisante et al. 2012). Hence, it adapts to and mitigates climate change and leads to a more efficient use of inputs to reduce production costs. Intensification should also enhance biodi-

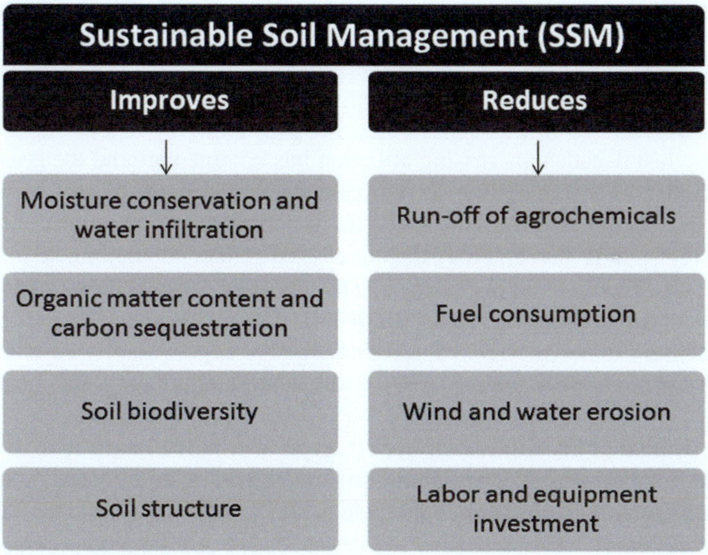

Fig. 1 Benefits of sustainable soil management (SSM). (Acquired from Cook et al. 2006; Huggins and Reganold 2008; Stagnari et al. 2009)

versity – above and below ground levels – in crop production systems to improve ecosystem services for better productivity and a healthier environment (Pisante et al. 2012). Investments in knowledge – especially in the form of science and technology – have featured prominently and consistently in most strategies to promote sustainable and equitable agricultural development worldwide.

Thus, the topic of SSM has a wide and complex scope as reflected in the list of ten tenets proposed by Lal (2009).

This chapter illustrates the agronomic principles of SSM in agricultural crop production and proposal for integrate sustainable soil management into sustainable farming and landscape management and what solutions are being implemented in different conditions. Section 2 describes Soil health and quality deepening some of the soil quality attributes, such as the organic carbon content as well as the physical and biological soil properties. Section 3 illustrates the contributions of SSM within agricultural practices and land management in line with the advances of contemporary agriculture with explanation of indicators for sustainability assessment. This is followed, in Section 4, by future perspectives for sustainable soil management, including how sustainable soil management has been able to restore degraded soils in different agricultural environments. Section 4 also offers some concluding remarks regarding the current trend toward sustainable soil management.

2 Soil Health and Quality

2.1 Definition and Attributes

During the last decades, increasing attention has emerged around the issues of soil preservation and sustainability, recognizing the unique role of soil in producing food, fiber, energy and for the adequate functioning of the global ecosystem. Nevertheless, soil is threatened by the increase in human population, intensive management, urbanization and degradation. The tendency towards preservation of soils goes beyond the agricultural scope (Blum 2003), so starting from the second half of the past century many interrelated terms emerged to cope with the "sustainability of soil" concept, like "soil quality" (SQ), "soil functionality" and "soil health" (SH) (Lal 2016).

SQ is defined as the "fitness for use" (Larson and Pierce 1991), and "capacity of the soil to function within ecosystems and land-use boundaries to sustain biological productivity, maintain environmental quality, and promote plant and animal health" (Doran and Parkin 1994; Karlen et al. 1997). It emerges the complexity and site-specificity of soil ecosystems (Bünemann et al. 2018) as well as the implicit several linkages between SQ and soil functions also termed as ecosystem services provided by soil resources (Daily 1997) including "sustaining plant and animal productivity (agricultural land), forest productivity (silvicultural land), air and water quality in relation to human health, contamination with heavy metals (minelands and urban lands) etc." (Lal 2016). High SQ promotes high crop yields (Fuentes et al. 2009; Turmel et al. 2015).

Differently to SQ, SH presents the soil as "a finite and dynamic living soil resource, directly related to plant health" (Lal 2016). Indeed, the term SH originates from the fact the SQ influences, through the quality of crops, animals, and human's health (Bünemann et al. 2018). SH would "capture the ecological attributes of the soil which have implications beyond its quality or capacity to produce a particular crop. These attributes are chiefly those associated with the soil biota, its biodiversity, its food web structure, its activity and the range of functions its performs" (Pankhurst et al. 1997). Doran and Zeiss (2000) and FAO (2011) gave specular definition of SH: "the capacity of soil to function as a vital living system to sustain biological productivity, maintain environment quality and promote plant, animal and human health". It is emphasized the presence of organisms, involved in several activities from nutrient cycling, symbiotic relationships with plant roots, pest, weed and disease control, to soil aggregate formation and aeration (Turmel et al. 2015). Within a context of sustainable crop production, SQ is obviously related to SH. Soils rich in organic matter content produce high biodiversity, showing high reserve of soil nutrients and moisture (Turmel et al. 2015) confirming that a healthy soil is also a high-quality soil.

In the individuation of suitable indicators to measure soil properties and processes (Kibblewhite et al. 2008) it should be taken into account that the concepts of SQ and SH should include also the capacity for emergent system properties, i.e. the self-organization of soils – feedbacks between soil organisms and soil structure (Lavelle et al. 2006) – and the adaptability to changing conditions (Bünemann et al. 2018). In the case of SQ quantification several indicators, which accurately summarize soil functions, exist (Bastida et al. 2006) and relate to measurable properties or characteristics of soils which can be distinguished on physical, chemical, and biological with the formers being the mainly applied. However, soil biological and biochemical properties may respond more rapidly to management activities and perturbations (Gianfreda and Ruggiero 2006; Paz-Ferreiro et al. 2007, 2013; Benintende et al. 2015; García-Orenes et al. 2012), characteristics equivalent to reliability and robustness (Arshad and Martin 2002); besides, further requirements include ease of sampling and measurement, comparability to data and easy interpretation. Bünemann et al. (2018) reviewed some biological or biochemical, additional or novel, SQ indicators based on: nematodes (Stone et al. 2016), (micro) arthropods (Rüdisser et al. 2015), soil suppressiveness (Janvier et al. 2007; Wu et al. 2015), genotypic and phenotypic community diversity (Hartmann et al. 2015; Kumari et al. 2017; Ritz et al. 2009), metabolomics (Vestergaard et al. 2017) and metaproteomics (Simon and Daniel 2011).

Additional simple and sophisticate methods, such as visual soil structure information (Shepherd et al. 2008; Mueller et al. 2013) and aggregate stability measurements by laser granulometry with sonication (Rawlins et al. 2013) have been developed to monitor soil physical quality. Spectroscopic techniques, e.g., near-infrared spectroscopy and remote sensing, offer the opportunity to measure various soil chemical, physical and biological parameters in a fast and inexpensive way (Kinoshita et al. 2012; Paz-Kagan et al. 2014; Gandariasbeitia et al. 2017; Ramirez-Lopez et al. 2018; Tziolas et al. 2019; Ostovari et al. 2018), also in combination with electrical conductivity and penetration resistance measurements (Veum et al. 2017).

A selection, due to the increase in both collinearity and complexity of the relationships between indicators and costs of measurements (O'Sullivan et al. 2017), is required in order to obtain a minimum dataset (MDS) for SQ assessment (Zornoza et al. 2015). The first MDS were proposed by Larson and Pierce (1991) and Doran and Parkin (1994), and successively adapted, modified or extended in studies based on statistical data reduction by multivariate techniques (Lima et al. 2013; Schipper and Sparling 2000; Shukla et al. 2006) and multiple regression (Kosmas et al. 2013), or by resorting to participatory approaches (Ritz et al. 2009; Bünemann et al. 2018). MDS is based on a number of SQ indicators (6–8) belonging to physical, chemical, and biological classes, to achieve a holistic image of SQ (Nannipieri et al.1990). Methodological transparency is imperative to allow wide application of MDS selection (Bünemann et al. 2018).

To summarize the information conveyed by each SQ indicator, the soil quality index (SQI) was proposed, which can help determining SQ trends (Karlen et al. 2001; Zobeck et al. 2015; Zornoza et al. 2015). SQI integrates the most relevant soil indicators into a single numerical measurement (Velasquez et al. 2007; Bastida et al. 2008; Calero et al. 2018) and is calculated using linear/non-linear scoring functions and additive/weighted additive methods (Askari and Holden 2015). For example, Calero et al. (2018) proposed the Field Soil Quality Index (FSQI) that integrates 18 morphological soil indicators; Puglisi et al. (2006) a complex SQI (Alteration index, AI1) based on soil enzymes, while Mijangos et al. (2010) developed an SQI based on several biochemical properties (dehydrogenase, urease, β-glucosidase, acid phosphatase and arylsulphatase) similar to others (Paz-Ferreiro and Fu 2016). In addition, the visual soil assessments (VSA) approaches, which target mainly soil physical and chemical attributes in relation to productivity, are often summarized in an overall soil quality rating (Shepherd et al. 2008; Guimares et al. 2011; Mueller et al. 2014; McKenzie et al. 2015).

2.1.1 Soil Organic Carbon

Soil C is by far the largest terrestrial C pool, storing more than double the quantity of C in vegetation or in atmosphere (Jia et al. 2017; Wu et al. 2009); it is composed by two distinct and related components, SOC and Soil Inorganic C (SIC). C sequestration, a potential for climate change mitigation, implies transferring atmospheric CO_2 into SOC and SIC stocks, that is storing it securely into long-lived pools (i.e. increasing in the soil C storage capacity; Lal 2004a).

SIC pool is approximately two to ten times larger than SOC in arid and semiarid regions (rainfall <500 mm/year) (Mi and Huang 2008), it comprises carbonates (CO^{-2}; solid pool) and bicarbonates (HCO^{-}; dissolved pool) of Ca^{+2}, Mg^{+2}, K^{+} and Na^{+} and the solid SIC pool fraction can further be classified as lithogenic inorganic C (primary carbonates) and pedogenic inorganic C (secondary carbonates). The former comes from the parent material of the soil, with no change in soil inorganic C content (Wu et al. 2009). The latter is formed through the reaction of HCO_3^{-} and CO_3^{-2} (in solution by dissolution of CO_2) with Ca^{+2} or Mg^{+2}, which leads to sequestration of atmospheric CO_2 (Lal 2016) – 2 moles of atmospheric CO_2 for every mole released during the precipitation of pedogenic carbonate (Schlesinger 1982). Weathering of Ca/Mg-bearing silicates represents another way of pedogenic inorganic C formation (Goddard et al. 2007). SIC sequestration, through formation of secondary carbonates, could reach 0.12–0.38 Mg C ha year^{-1} to 160 cm depth by irrigation and fertilization (Bughio et al. 2016). Besides, leaching of bicarbonates into the subsoil or shallow water table and its reprecipitation can also be high (Monger et al. 2015). SIC storage and dynamics have been mainly investigated at local or regional level, consequently a quantitative assessment of soil C storage as a baseline to estimate the overall C budget at a national or global scale and the identification of the areas where C sequestration should be concentrated (Wu et al. 2009) are needed.

SOC represent the majority of C in soils, with an average content of 2400 Pg to a soil depth of 2 m (Han et al. 2015). In arid climates the SOC pool to 1-m depth is about 30 tons/ha, increasing to 800 tons/ha in organic soils from cold regions (Lal 2004b). It is part of SOM (45–60%) and results in a heterogeneous mixture of organic materials including fresh litter, carbohydrates, and simple sugars, complex organic compounds, some inert materials, and pyrogenic compounds (Lal 2016). The SOC stock can be divided into labile or actively cycling pool, a slow pool, and a stable or passive, recalcitrant pool with varying residence times (Majumder et al.

2008). The rate of its turnover depends on the degree of protection mechanisms within the soil matrix (Dungait et al. 2012; Lal 2016), which can be distinguished in: physical, i.e. encapsulation within stable microaggregates (Six et al. 2000, 2002), formation of organo-mineral complexes (SOC storing for millennia) and transfer deep into the subsoil; chemical, i.e. formation of some recalcitrant compounds (von Lützow and Kögel-Knaber 2009); biological, i.e. microbial exudates that repel other organisms, transfer of SOC into biologically non-preferred soil spaces (Dungait et al. 2012); ecological, i.e. coupled cycling of C with H_2O, N, O, S and microelements, erosion control, deep translocation.

The labile C pool, which shows the most rapid turnover rates, deeply influence nutrient cycling (Majumder et al. 2007), crop production and it is sensitive to agronomic practices. Conversely, the highly recalcitrant or passive pool, being involved by microbial activities in a very slow times, it is not a suitable indicator for monitoring soil quality and productivity (Majumder et al. 2007).

High SOC levels in agroecosystems – threshold value in the rootzone of 1.5–2.0% – are necessary to ensure high physical, chemical and biological soil attributes, which in turn affect crop production. Indeed, SOC: (1) promotes better soil structure and particle aggregation; (2) guarantees water retention, thus increasing the soil water availability to crop, and ameliorates water use efficiency; (3) is associated to improved nutrient retention and use efficiency; (4) promotes microbial activity in the rhizosphere; and (5) regulates gaseous emissions (e.g., CO_2, CH_4, N_2O) (Fig. 2). SOC pool is also related to several ecosystem services, such as mitigation of the climate change and advance food security (Lal 2016) and can be considered

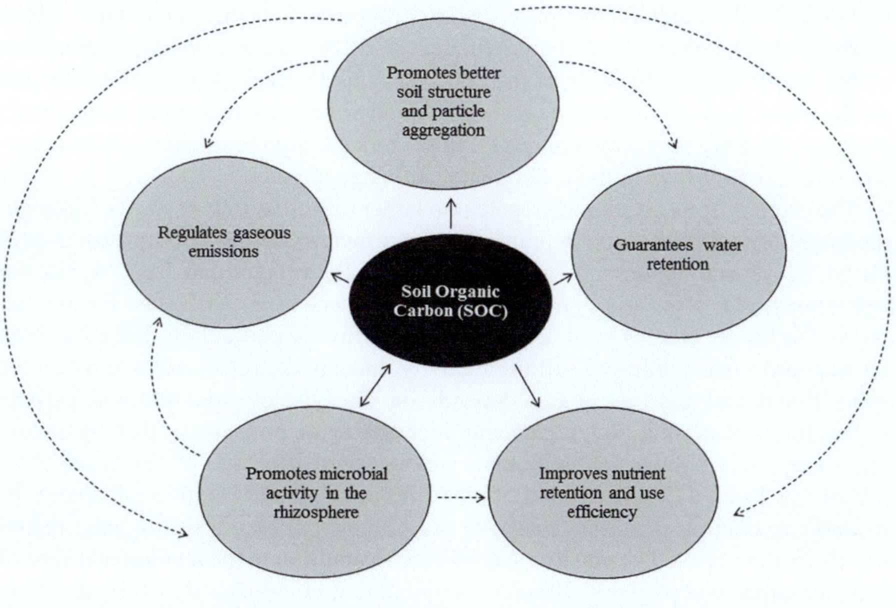

Fig. 2 The effects of soil organic carbon (SOC) in agroecosystems. (Acquired from Lal 2016)

an indicator of soil health (Winowiecki 2015). Soil C sequestration – i.e. increasing of both SOC and SIC stocks – as well as the maintenance of SOC pools are recognized as important strategies for building efficiency and resilience of the system (Corsi et al. 2012; Victoria et al. 2012) and are achieved through soil management practices which will be further exposed.

2.1.2 Soil Physical Properties

Physical properties have significant influence on the behavior of soil for agricultural uses and can be assessed, in addition to the laboratory analyses approach, visually or by touch, and measurable with appropriate scales. The principal ones are texture and structure, which are strictly interrelated to each other. Other characteristics such as porosity, softness, adhesion, plasticity, color, aeration and temperature depend on these main properties and have a direct impact on plant growth. Indeed, physical properties of the soil determine the supporting capability, movement, retention and availability of water and nutrients, penetration of the root system, flow of heat and air as well as they have strong influences on chemical and biological properties.

Soil texture, which refers to the prominent size range of mineral particles, i.e. relative proportion of its sand, silt, and clay contents, defined both qualitatively and quantitatively, is a static property affecting almost all the other soil properties. Land use capability and soil management practices largely depend on the texture (Whisler et al. 2016). Depending on the diameter of the soil fraction, the degree of water retention, aeration and workability can be determined. In general, sandy soils have poor water retention capacity, lower organic carbon content and biological activity, higher hydraulic conductivity, poor sealing properties for ponds and dams, higher leaching of nutrients and pollutants (Singh et al. 2016); the low organic carbon content in sandy loam soils is one of the major reasons for declining in soil health and productivity. Besides, the clayey soils have high water and nutrient holding capacity, poor aeration, very slow drainage unless cracked, high to medium organic matter content, medium to high swelling and shrinkage characteristics.

The primary soil particles aggregate into larger units in different ways of arrangement and organization such defining the soil structure. The soil disturbances produced by several agronomic practices, principally tillage, can have significant deleterious effects on soil aggregation (Mangalassery et al. 2013; Dal Ferro et al. 2014; Tuzzin de Moraes et al. 2016) since structure is particularly linked to both fertility and sustainability of soil productivity. Indeed, the relationship between the solid, liquid and gaseous phases depends on the type of structure: soil particle aggregation creates intra-aggregate and inter-aggregate pore space, thereby changing water, gases, solutes, and pollutants which in turn affect life of plants and other organisms. From a chemical point of view, minor or greater aeration influences the oxidation/reduction processes that take place, hence affecting soil organic matter transformation as well as nutrient availability. In addition to the amount and type of clay and other soil particles, structure development is influenced also by the presence of exchangeable ions (especially Ca; Rengasamy and Marchuk 2011), amount and type of organic matter (it provides food for fungi, bacteria and larger organism whose secretion act as cementing agents; Six et al. 2004), presence of iron and aluminum oxides (cementing agents), binding between organic and inorganic compounds (aluminium oxides, cations, clays; Clarholm et al. 2015) and vegetation (roots act as holding soil together; Afzalinia and Zabihi 2014). Soil structure deterioration induces soil compaction, increase bulk density thus reducing gaseous exchange between soil and atmosphere, water infiltration rate, water storage (enhancement of runoff and soil loss) as well as restricting root development (Batey 2009; Nawaz et al. 2013). It is indeed well acknowledged that when a good structure occurs, the roots have a greater capability to explore the soil, higher probability to uptake water and nutrients and to have more oxygen for their metabolic processes (Rabot et al. 2018). In addition, the number of organisms increases, with a better control of those useful, and the processes of transfer of nutrients from the organic matter as well. In addition, a good aggregation decreases also the detachability and transportability of soil particles by water or wind thus reducing runoff and soil erosion.

To maintain a high fertility of the soils, the natural resource base of food production and security, is hence essential to preserve soil physical aspects (in particular structure since texture is a static property; Gao et al. 2017), through appropriate agricultural practices in particular soil management which has been shown to affect dramatically soil quality (Singh et al. 2009; Van Wie et al. 2013).

2.1.3 Soil Biological Biodiversity and Activity

Soil fertility and biodiversity lie in the combination of activities of living organisms in the soil, which interact with each other, with plants and small animals within a network of biological activity. In such complex structure, minerals and organic components determine the habitat conditions and the availability of food resources. Larger organisms, such as earthworms, burrow through the soil, producing large pores that are important for water flow and retention, aeration, and root development, and help mixing organic materials into the soil favoring aggregate formation. Microorganisms, as well as microfauna, mesofauna, and macrofauna play essential roles in nutrient cycling and organic matter decomposition in the soil. Interactions among different organisms can have either beneficial or harmful effects on crops. It has been estimated that 1 g of soil contains up to 1 billion bacteria cells consisting of tens of thousands of taxa, up to 200 m fungal hyphae, and a wide range of mites, nematodes, earthworms, and arthropods (Bardgett 2005; Roesch et al. 2007). This wide and hidden diversity contributes to the total terrestrial biomass and it is intimately linked to above-ground biodiversity (Fierer et al. 2009; Wardle et al. 2004). It is due to this important fraction, the soil remains in good health and manages to perform important ecosystem functions, such as:

- the stability of the agroecosystem, subsequently a rich and structured pedological biodiversity guarantees regular functioning even under conditions of environmental stress;
- flow regulation and infiltration, water purification;
- promotion and maintenance of natural or agricultural productivity.

The heterogeneity of soil biota can be characterized in various ways. The most common method is based on the size of the organism, and divides soil biota in soil microflora (bacteria, fungi, green algae, etc.) and the soil fauna. Soil fauna is generally divided into three groups based on the average body size and life adaptation to the pore space, filled with water or in the air-filled pores (Cochran et al. 1994; Lavelle 2000). The three most commonly used categories consist in:

(i) microfauna, comprehending organisms less than 0.2 mm, such as Protozoa, Nematodes, Rotifers and Tardigrade. Many of these organisms live and move in the most humid areas of the ground, moving in the water held between the soil particles; (ii) mesofauna, fauna ranging from 0.2 to 2 mm in size, they live in the pore space full of soil air and inside the litter box. This category includes micro-arthropods (e.g. acaridae, springtails) and enchyroid worms; (iii) macrofauna which includes animals of sizes between 2 and 80 mm as *Anbridae, Lumbricidae*, Gastropods, Opilions, Spiders, Insect larvae, Ants, Coleoptera and others.

Another approach to the study of soil biota is considering organisms living in agricultural systems as part of larger food grids that act as food reserves for animals belonging to higher orders in the food web. The soil food web is a community of organisms, which live all or part of their life in the soil. It describes a complex living system in the soil and how it interacts with the environment, plants and animals (Moore 1994).

Soil biota can also be described through method used to classify soil organisms based on organisms' functions. The functions carried out by soil organisms belonging to the macrofauna category depend largely on the efficiency of their digestive systems and on the occurrence and abundance of the biological structures that they produce in the soil.

Lavelle (1997) suggested to classify invertebrates into three functional groups based on the role of soil fauna and their relationship with microflora: micropredators, litter transformers, and ecosystem engineers. The first group contains the smallest invertebrates, protozoa, and nematodes, the micro food networks that links microorganisms to their predators and their principal effect is to stimulate the mineralization of soil organic matter. Moreover, they do not produce organo-mineral structures.

In the litter-transformer group recognized by Lavelle (1997) we find large arthropods and mesofauna. They do not ingest mineral soil or dig the ground but ingest purely organic material. The "ecosystem engineers" or "ecological engineers" (Jones et al. 1994) are assisted by a series of other organisms (ants, termites, moles, etc..) that chop up and crumble the dead organic substance, making a first reduction of animal and vegetable residues. This macrofauna contributes to develop mutualistic relationships with the microflora (within their gut) and is able to excavate soil and produce a wide variety of organo-mineral structures, such as excretions, caverns, mounds, macropores, nests, and, galleries.

Several studies have shown that agricultural intensification and land use change, reduce microbial and faunal abundance and the overall diversity of soil organisms (Helgason et al. 1998; Mäder et al. 2002; de Vries et al. 2013; Stagnari et al. 2014b). This has triggered increasing concern that reduced biodiversity in soils may impair numerous ecosystem functions, such as nutrient acquisition by plants and the cycling of resources between above- and below-ground communities (Wall et al. 2010; van der Heijden et al. 2008; de Vries et al. 2013). Tillage, in particular, induces significant biophysical and biochemical changes, modifies the relationships between soil organisms within the soil ecosystem, and the diversity of communities.

3 Soil Management

3.1 Sustainable Agricultural Practices

Despite a precise and absolute definition of sustainable agriculture is impossible, due to the complex and contested nature of its notion, the concept of sustainable agriculture is continuously growing. This founds the main reasons in the need of providing food and other resources to an increasing world population, within a context characterized by growing issues that threaten such ability of agriculture. They include climate change, loss of biodiversity, land degradation through soil erosion, compaction, salinization and pollution, depletion and pollution of water resources, rising production costs as well as a decreasing number of farms and rural population.

Among the practices that more influence soil sustainability those pointed to soil management play a major role. Conventional tillage, defined as the mechanical manipulation of the soil for crop production, affects noticeability soil characteristic from water conservation, infiltration and evapotranspiration processes to soil temperature, biological and chemical traits with significant impact on environment as well as requires high energy costs for the mechanical operations (Bhatt and Khera 2006). As results it may adversely affect long-term soil productivity.

A sustainable soil management approach should be based on "conservation tillage" practices, although the term is often unclear and very variable among different countries and even among regions of a single country. In any case the basic assumptions are fixed in: reduced or no soil disturbance, direct drilling, crop rotation, and permanent soil cover. The most accepted definition is from the US Conservation Technology Information Center (CTIC 2000): any tillage and planting system that covers 30% or more of the soil surface with crop residue, after planting, in order to reduce soil erosion by water. Back to the assumptions, it emerges that tillage remains the main core while crop residues are considered as a by-product of tillage although scientific evidences indicate several benefits of leaving crop residues on the soil surface.

Conservation tillage techniques include *no-tillage* (otherwise called direct seeding, slot planting, zero-till) which involves soil is left undisturbed from harvest to planting which causes less than 25% of row width disturbance by planting equipment (e.g., coulters, disk openers, in-row chisels, roto-tillers); *mulch tillage*, where the soil is prepared or tilled in such a way that the plant residues or other materials are left to cover the surface to a maximum extent (>30%); *ridge-tillage*, consisting in left undisturbed soil from harvest to planting, except that planting is completed in a seedbed prepared on ridges with sweeps, disk openers, coulters or row cleaners and residue is left on the surface between ridges (>30%).

Besides, any seedbed preparation system that leaves 15–30% residue cover after planting is considered a reduced tillage, practically between conservation and conventional method. When the continuous no or minimum mechanical soil disturbance (no-till seeding/planting and weeding, and minimum soil disturbance with all other farm operations including harvesting) is applied together with the following two principles, (i) permanent maintenance of mulch cover on the soil surface (crop biomass, root stocks, stubble, cover crops and other ex situ biomass) and (ii) diversification of cropping system, the wider concept of Conservation Agriculture (CA) system is performed.

Such sustainable soil management induces positive changes to soil properties and characteristics. Soil organic matter is the key of soil fertility and it is mainly composed by C. Nevertheless, carbon SOC represents both a threat and an opportunity in the context of the global C cycle, climate change and soil productivity (Pisante et al. 2015). The opportunity is to manage soils in such ways as to sequester additional C from the atmosphere so enhancing soil fertility while reducing climate change. It is noteworthy to indicate that soil organic carbon level is a balance between the C organic C inputs and outputs of (e.g. Johnston et al. 2009; Lützow et al. 2006). Plant residues (above and below ground) root exudates and organic fertilization represent the main C inputs while outputs are the decomposition of organic matter by soil microorganisms and fauna leading to evolution of CO_2 to the atmosphere, leaching of soluble organic C compounds and particulate losses through erosion. The impact of conservation practices is noticeable (Palm et al. 2014): no-tillage and residue retention significantly increase SOM content in the top soil layers (0–5 cm or 0–10 cm) (Bissett et al. 2013), especially its particulate and more labile fractions (particulate organic matter, POM – >53 mm in size) (Álvaro-Fuentes et al. 2008; Martín-Lammerding et al. 2015; Blanco-Moure et al. 2016). Of course, the magnitude is affected by several aspects, from amount and type of crop residues to soil texture (Martín-Lammerding et al. 2015; Blanco-Moure et al. 2016). Under annual no-tillage, Rasmussen (1999) and During et al. (2002) indicated the important role of plant residues left on the soil surface in increasing the organic matter in the topsoil. Besides, other authors (Ismail et al. 1994; Lal 1997) reported a significantly higher SOC in soil with NT compared to tilled soil considering only the soil mechanical management so delivering C sequestration through decreased SOC decomposition. With these regard, it would seem that in fine textured soils, where clay- and silt-sized particles may chemically stabilize SOM, the enhancement in SOM content is primarily attributable to the reduction of soil mechanical disturbance (lower SOM decomposition), while in coarse textured soils, where tillage determines the greater SOM losses, due to lack of its physical protection, more emphasis is placed on the management of C inputs (residues retention) Chivenge et al. (2007).

However, it has to be highlighted that some works (Baker et al. 2007; Angers and Eriksen-Hamel 2008) indicates that the net accumulation of C under conservation tillage, whilst measurable in the long term, is less than reported in other papers and that SOC concentrates nearer the soil surface with respect to conventionally tilled soils. Since organic C is normally higher concentrated in topsoil than subsoils, the latter offer greater potential for increased storage thanks also to some evidence that organic C in subsoil is stabilized to a greater degree (Jenkinson and Coleman 2008), though the mechanisms involved are still poorly understood and debated (Fontaine et al. 2007; Salomé et al. 2010). Consequently, as plant roots represents a significant means of delivering organic C into subsoil, the role of cover crop management as well as crop rotation as "sustainable soil management" should play a major role (see the exploitation of rooting depths or exudation characteristics between among cover crops and arable crops) Carter and Gregorich (2010).

Positive influence of the conservation tillage on physical properties varies and are dependent on the particular system chosen although are limited to the upper few soil centimeters (Anikwe and Ubochi 2007). Some researchers have found that no-mechanical or very limited soil disturbance induce improvements of saturated and unsaturated hydraulic conductivity (Benjamin 1993); higher soil porosity has been observed under minimum tillage systems thank to the increased storage pores (0.5–50 mm) and to the density of elongated transmission pores (50–500 mm) McVay et al. (2006), Pagliai et al. (2004); moreover, more stable aggregates in the upper surface of soil have been associated with no-till soils (Lal 2007). As consequence water conservation (Stagnari et al. 2009, 2014a; Pisante et al. 2010) as well as water use efficiency (McVay et al. 2006; Su et al. 2007) are improved thanks to higher water holding capacity or moisture content in the topsoil (0–10 cm) which, in no-till soils, can reach values up to 25%. Nevertheless, some works report high infiltration rates under conventional tillage than no-tillage or minimum tillage: this is because in some circumstances CT create fast draining macro-pores, that could facilitate infiltration momentarily just after tillage, which has been demonstrated to dramatically reduced with time (Martınez et al. 2008), leading to a lower infiltration rate under CT than ZT overtime. In general, higher infiltration rates under NT are found (Aase and Pikul 1995; Shukla et al. 2003) thanks to the protection of the soil surface and effect of SOC but also due to increased activity of surface-feeding earthworms, leaving the root channels undisturbed, which in turn leads to the presence of numerous surface- connected macro-pores (Kemper et al. 1987; Lal and Shukla 2004). Besides, the higher water content in the topsoil, observed under conservation tillage, finds reasons not only in an increased water infiltration, but also in declined evaporation rates thanks to the more plant residues on soil surface which have been correlated with lower soil temperatures (Rasmussen 1999). These has been definitively demonstrated through the stable isotope technique (Busari et al.

2013) which registered more enriched soil water stable isotopes (δ18O and δD) near the soil surface under CT compared with ZT, indicating higher evaporation rates.

With and increased water infiltration rates conservation tillage practices as consequence protect the soil from surface runoff and erosion. Soil erosion presents a threat to agricultural productivity, especially in circumstances with low agronomic inputs, poor soil cover, in not resilient soils and where intense rainfall sometimes occurs (Govers et al. 2017). The amount of human-induced agricultural erosion has been estimated at 25,000–40,000 Tg year^{-1} for water erosion, ca. 5000 Tg year^{-1} for tillage erosion and 2000–3000 Tg year^{-1} for wind erosion (Govers et al. 2014). The mechanisms involved in soil erosion are reasonably well understood (see for example the text books, Morgan 2005; Kirkby and Morgan 1980) as well as its ecological effects on aquatic environments (Stagnari et al. 2016) due to the soil-eroded particles (clay, silt and organic matter) transportation of mineral elements to surface waters (Quinton et al. 2010) such as heavy metals, often contained in agrochemicals i.e. copper (Cu), zinc (Zn), lead (Pb), nickel (Ni), cadmium (Cd), chromium (Cr), fluorine (F), mercury (Hg) and S, resulting in pollution of drinking water resources (Stagnari et al. 2016). Although soil type, topography and weather conditions are significant variables in the erosion phenomena, it has been demonstrated that NT and surface residues retention reduce erosion by an order of magnitude in comparison with conventional systems (Prasuhn 2012). Conservation soil management practices also reduce surface and ground water pollution both by decreasing agrochemical runoff (Palm et al. 2014) and thanks to the higher activity of soil microorganisms and quantity of organic matter which allow faster agrochemical degradation (Busari et al. 2015), and binding of agrochemicals (Alletto et al. 2012). When all CA principles are adopted, over a range of different circumstances a strong reduction of soil erosion is registered with respect to conventional approach (Palm et al. 2014): from US, with values of decreasing runoff of about 15–89% (Lemke et al. 2011), to India, where Bhatt and Khera (2006) observed soil values of 22% lower. It is worthy to remember that that when no-till (or conservation tillage) is practiced without soil mulch cover, the effects can be disastrous with rapid surface sealing resulting in increased run-off and accelerated soil erosion (Giller et al. 2015).

A soil sustainable approach has been demonstrated to have significant influence also on soil chemical properties, such as pH, CEC, exchangeable cations, and soil principal macronutrients, especially at surface layer level (Lal 1997). Although it has been observed that tillage approach does not often have effect on soil pH (Rasmussen 1999), in some circumstances soil pH values has been reported to be lower under conservation systems compared to CT (Rahman et al. 2008). In such cases, the lower pH observed under ZT and residues management soils was attributed to the accumulation of organic matter up to soil surface (Rhoton 2000) which probably causes increases in the concentration of electrolytes which in turn reduces pH values (Rahman et al. 2008). In any case results are sometimes contradictory since Cookson et al. (2008) found a decreasing of pH values under increasing tillage as well as Lal (1997) observed a significantly higher soil pH in NT plots than conventionally managed soils. Such controversial data suggests that tillage may not directly affect soil pH, but several factors are involved (i.e. climatic condition, soil type etc.). Although tillage induced changes in cation exchange capacity and base saturation are usually small, conversion to a conservation tillage system is able to increases both as reported by several studies (Gallaher and Ferrer 1987; Lal 1976; Lal et al. 1990). The increasing in cation exchange capacity occur exclusively in the first 10–15 cm of the profile, and it is attributed to the increases content in organic matter.

Nutrient availability and concentrations could be affected by tillage practices (Lavado et al. 1999) which, although it is reported to influence the depth distribution of macro and micronutrients. concentrate their effects principally in soil surface (Wright et al. 2007). The influence of soil management (especially tillage) on N cycle has mainly been approached with net mineralization rates assessments (Gòmez-Rey et al. 2012). Indeed, crop residues contain N principally in organic form and during decomposition of crop residues, this organically bound N is made available for crop or microbial growth through N mineralization (Lupwayi et al. 2006). Conservation tillage enhance N availability to plants in the long-term (Rice et al. 1986; Galieni et al. 2016) thanks to an increased soil N retention and labile N pool (McCarty et al. 1995) in the first soil layers. In some circumstances has emerged that no-till and ridge-till promote significantly ($p < 0.05$) similar concentrations of soil organic N at the very soil surface thus restricting N mineralization in the short periods (Zibilske et al. 2002).

With regards to phosphorus availability numerous studies report greater amounts of extractable phosphorus at the surface of conservation tilled soils than conventionally tilled ones (Dick et al. 1991). In some cases, some researchers have found opposite results (Ismail et al. 1994; Karlen et al. 1991; Lal et al. 1990). Nevertheless, the majority of these studies showed higher surface concentrations of available phosphorus in conservation-tilled systems (Hargrove 1985; Ekeberg and Riley 1996). It has indeed observed that crops in dry areas suffer often from phosphorus deficiency, indicating that the mineral is very moisture sensitive. Since conservation practices determine soil moisture environments a consequently increased diffusion of phosphorus to plant roots has been recognized under no-till systems (Thomas and Frye 1984; Thomas 1986; Dick et al. 1991). Besides, another probable reason why plants do not suffer from poor nutrient uptake under no tillage is the greater activity of roots found in such circumstances (Hargrove 1985).

The influence of sustainable tillage techniques on the behavior of soil potassium is not as significant as for the other macronutrients (Thomas and Frye 1984; Thomas 1986) as numerous authors have found out (Ekeberg and Riley 1996). This because it is held tightly except in those soils with low cation exchange capacity. The major concern, linked to the lack of soil mixing under conservation tillage, regards the stratification of exchangeable potassium which highly concentrates in the surface layers when compared to conventional tillage systems (Blevins et al. 1983; Franzluebbers and Hons 1996; Hargrove 1985; Ismail et al. 1994; Lal et al. 1990).

Among the aspects of soil quality, soil biology is recognized as playing a major role and unlike physical and chemical soil properties which change slowly, biological soil properties are particularly affected by soil management (Bastida et al. 2008;

Yao et al. 2013). Soil biota, for definition, include both invertebrates (e.g. nematodes, earthworms,) and microorganisms (e.g. protists, bacteria, fungi). Earthworms – the major component of the soil macrofauna, which play a significant role in increasing aggregate stability, infiltration rates, macroporosity, saturated hydraulic conductivity and nutrient cycling – are more abundant in soils in terms of number and biomass under conservation practices (Nieminen et al. 2011; Stagnari et al. unpublished data; van Capelle et al. 2012; Bertrand et al. 2015). Such response implicates a set of factors ranging from reduced injuries to increased availability of organic matter at the soil surface. In the case of microbial biomass, the significant increasing values observed in multiyear experiments, under sustainable soil management practices must be brought back principally to the higher organic matter levels in the topsoil when reduced (Diacono and Montemurro 2012; Spiegel et al. 2015; Valckx et al. 2009; Briones and Schmidt 2017). van Capelle et al. (2012) reported an average increase in microbial biomass in the 0–10 cm soil layer of 63% when shifting from conventional tillage to conservation tillage highlighting a significant tillage effect on the vertical distribution of microbial biomass in the soil.

Thanks to the positive effects on soil health, sustainable management of soils (see conservation practices) in turn affect positively crop yield, thanks to better root growth development (Lal 1989; Boone and Veen 1994; Martınez et al. 2008), higher water and nutrient use efficiencies (Davis 1994; Lal 1993). Malhi and Lemke (2007) reported a 22% increase in root mass under NT compared with CT, attributed principally to higher number of worm channels as well as of biopores (Francis and Knight 1993) although it takes time before observing significant responses. This also emphasized the more resilient nature of soil maintained under a no-tillage system (Lal 1993) compared with other tillage systems.

4 Conclusions and Future Perspectives

Within a situation of threatening food production and security, directly through increasing losses and degradation of soils and due to climate change, new approaches in soil and water conservation in agriculture becomes unavoidable. Although the application of soil agronomic sustainable practices is not diffused under intensive agricultural systems, several studies have demonstrated, that whether applied within a context of conservative approach, many benefits arise, from the preservation of soil fertility and quality to the obtaining of higher crop yields and environmental benefits.

Although CA practices are profitable, at least when applied for long periods, in some countries they are confined to few circumstances probably due to the lack of technology transfer, mental barriers, scarce technical preparation and professional experience.

It has well acknowledged that the application of sustainable management practices results in the amelioration of agronomic, ecological, economic, and social aspects. Nevertheless, some different questions need to be investigated further in order to achieve better results. On this basis the matter which need clarification regard a wider understanding of soil ecology, which could lead to more precise management of soil organisms for beneficial purposes in agriculture. Furthermore, there is a growing interest in studying and understanding the role that precision agriculture could have in soil sustainable management, the applications of the tools of precision agriculture to better manage the variability of soil in terms of physical, chemical and biological aspects. Besides, the building up as and diffusion of easier to manage and apply soil quality indexes and indicators which are able to synthetize all the aspects of soil quality, are needed.

References

Aase JK, Pikul JL (1995) Crop and soil response to long-term tillage practices in the northern Great Plains. Agron J 87:652–656

Afzalinia S, Zabihi J (2014) Soil compaction variation during corn growing season under conservation tillage. Soil Tillage Res 137:1–6. https://doi.org/10.1016/j.still.2013.11.003

Alletto L, Coquet Y, Bergheaud V, Benoit P (2012) Water pressure head and temperature impact on isoxaflutole degradation in crop residues and loamy surface soil under conventional and conservation tillage management. Chemosphere 88:1043–1050. https://doi.org/10.1016/j.chemosphere.2012.05.021

Álvaro-Fuentes J, López MV, Cantero-Martínez C, Arrúe JL (2008) Tillage Effects on soil organic carbon fractions in mediterranean dryland agroecosystems. Soil Sci Soc Am J 72:541–547. https://doi.org/10.2136/sssaj2007.0164

Angers DA, Eriksen-Hamel NS (2008) Full-Inversion tillage and organic carbon distribution in soil profiles: a meta-analysis. Soil Sci Soc Am J 72:1370–1374. https://doi.org/10.2136/sssaj2007.0342

Anikwe MAN, Ubochi JN (2007) Short-term changes in soil properties under tillage systems and their effect on sweet potato (*Ipomea batatas* L.) growth and yield in an Ultisol in south-eastern Nigeria. Aust J Soil Res 45:351–358. https://doi.org/10.1071/SR07035

Arshad MA, Martin S (2002) Identifying critical limits for soil quality indicators in agroecosystems. Agric Ecosyst Environ 88(2):153–160

Askari MS, Holden NM (2015) Quantitative soil quality indexing of temperate arable management systems. Soil Tillage Res 150:57–67. https://doi.org/10.1016/j.still.2015.01.010

Baker JM, Ochsner TE, Venterea RT, Griffis TJ (2007) Tillage and soil carbon sequestration-What do we really know? Agric Ecosyst Environ 118:1–5. https://doi.org/10.1016/j.agee.2006.05.014

Bardgett R (2005) The biology of soil: a community and ecosystem approach. Biology of habitat series. Oxford University Press, Oxford. https://doi.org/10.1093/acprof:oso/9780198525035.001.0001

Bastida F, Moreno JL, Hernandez T, García C (2006) Microbiological degradation index of soils in a semiarid climate. Soil Biol Biochem 38:3463–3473. https://doi.org/10.1016/j.soilbio.2006.06.001

Bastida F, Zsolnay A, Hernández T, García C (2008) Past, present and future of soil quality indices: a biological perspective. Geoderma 147:159–171. https://doi.org/10.1016/j.geoderma.2008.08.007

Batey T (2009) Soil compaction and soil management – A review. Soil Use Manag 25:335–345. https://doi.org/10.1111/j.1475-2743.2009.00236.x

Benintende S, Benintende M, Sterren M et al (2015) Biological variables as soil quality indicators: effect of sampling time and ability to classify soils by their suitability. Ecol Indic 52:147–152. https://doi.org/10.1016/j.ecolind.2014.11.030

Benjamin JG (1993) Tillage effects on near-surface soil hydraulic properties. Soil Tillage Res 26(4):277–288

Bertrand M, Barot S, Blouin M, Whalen J, De Oliveira T, Roger-Estrade J (2015) Earthworm services for cropping systems. A review. Agron Sustain Dev 35:553–567. https://doi.org/10.1007/s13593-014-0269-7

Bhatt R, Khera KL (2006) Effect of tillage and mode of straw mulch application on soil erosion in the submontaneous tract of Punjab, India. Soil Tillage Res 88:107–115. https://doi.org/10.1016/j.still.2005.05.004

Bissett A, Richardson AE, Baker G, Kirkegaard J, Thrall PH (2013) Bacterial community response to tillage and nutrient additions in a long-term wheat cropping experiment. Soil Biol Biochem 58:281–292

Blanco-Moure N, Gracia R, Bielsa AC, López MV (2016) Soil organic matter fractions as affected by tillage and soil texture under semiarid Mediterranean conditions. Soil Tillage Res 155:381–389. https://doi.org/10.1016/j.still.2015.08.011

Blevins RL, Thomas GW, Smith MS et al (1983) Changes in soil properties after 10 years continuous non-tilled and conventionally tilled corn. Soil Tillage Res 3:135–146

Blum WEH (2003) European soil protection strategy. J Soils Sediments 3:242. https://doi.org/10.1007/BF02988670

Boone FR, Veen BW (1994) Mechanisms of crop responses to soil compaction. In: Developments in agricultural engineering. Elsevier, Amsterdam, pp 237–264

Briones MJI, Schmidt O (2017) Conventional tillage decreases the abundance and biomass of earthworms and alters their community structure in a global meta-analysis. Glob Chang Biol 23(10):4396–4419. https://doi.org/10.1111/ijlh.12426

Bughio MA, Wang P, Meng F et al (2016) Neoformation of pedogenic carbonates by irrigation and fertilization and their contribution to carbon sequestration in soil. Geoderma 262:12–19. https://doi.org/10.1016/j.geoderma.2015.08.003

Bünemann EK, Bongiorno G, Bai Z et al (2018) Soil quality – a critical review. Soil Biol Biochem 120:105–125. https://doi.org/10.1016/j.soilbio.2018.01.030

Busari MA, Salako FK, Tuniz C et al (2013) Estimation of soil water evaporative loss after tillage operation using the stable isotope technique. Int Agrophys 27:257–264. https://doi.org/10.2478/v10247-012-0093-8

Busari MA, Kukal SS, Kaur A et al (2015) Conservation tillage impacts on soil, crop and the environment. Int Soil Water Conserv Res 3:119–129. https://doi.org/10.1016/j.iswcr.2015.05.002

Calero J, Aranda V, Montejo-Ráez A, Martín-García JM (2018) A new soil quality index based on morpho-pedological indicators as a site-specific web service applied to olive groves in the Province of Jaen (South Spain). Comput Electron Agric 146:66–76. https://doi.org/10.1016/j.compag.2018.01.016

Calzarano F, Stagnari F, D'egidio S et al (2018) Durum wheat quality, yield and sanitary status under conservation agriculture. Agriculture 8:140. https://doi.org/10.3390/agriculture8090140

Carter MR, Gregorich EG (2010) Carbon and nitrogen storage by deep-rooted tall fescue (*Lolium arundinaceum*) in the surface and subsurface soil of a fine sandy loam in eastern Canada. Agric Ecosyst Environ 136:125–132. https://doi.org/10.1016/j.agee.2009.12.005

Chivenge PP, Murwira HK, Giller KE et al (2007) Long-term impact of reduced tillage and residue management on soil carbon stabilization: implications for conservation agriculture on contrasting soils. Soil Tillage Res 94:328–337. https://doi.org/10.1016/j.still.2006.08.006

Clarholm M, Skyllberg U, Rosling A (2015) Organic acid induced release of nutrients from metal-stabilized soil organic matter – The unbutton model. Soil Biol Biochem 84:168–176. https://doi.org/10.1016/j.soilbio.2015.02.019

Cochran VL, Sparrow SD, Sparrow EB (1994) Residue effects on soil micro-and macroorganisms. Manag Agric Res:163–184

Conservation Technology Information Center (CTIC) (2000) National crop residue management survey: 2000 survey results. 1220 Potter Dr., West Lafayeme, IN

Cook HF, Valdes GSB, Lee HC (2006) Mulch effects on rainfall interception, soil physical characteristics and temperature under *Zea mays* L. Soil Tillage Res 91:227–235. https://doi.org/10.1016/j.still.2005.12.007

Cookson WR, Murphy DV, Roper MM (2008) Characterizing the relationships between soil organic matter components and microbial function and composition along a tillage disturbance gradient. Soil Biol Biochem 40:763–777. https://doi.org/10.1016/j.soilbio.2007.10.011

Corsi S, Friedrich T, Kassam A, Pisante M, De Moraes Sà J (2012) Soil organic carbon accumulation and greenhouse gas emission reductions from conservation agriculture: a literature review. Integrated crop management, vol 16. FAO, Roma, IX+89 pp. ISBN 978-92-5-107187-8

Daily GC (1997) Nature's services. Island Press 1718 Connecticut avenue, N.W., Suite 300 Whashington, DC 20009. ISBN 1-55963-475-8

Dal Ferro N, Sartori L, Simonetti G et al (2014) Soil macro- and microstructure as affected by different tillage systems and their effects on maize root growth. Soil Tillage Res 140:55–65. https://doi.org/10.1016/j.still.2014.02.003

Davis JG (1994) Managing plant nutrients for optimum water use efficiency and water conservation. Adv Agron 53:85–121

de Vries FT et al (2013) Soil food web properties explain ecosystem services across European land use systems. Proc Natl Acad Sci U S A 110:14296–14301

Diacono M, Montemurro F (2012) Long-term effects of organic amendments on soil fertility: a review. Agron Sustain Dev 30:401–422

Dick WA, McCoy EL, Edwards WM, Lal R (1991) Continuous application of No-Tillage to Ohio soils. Agron J 83:65–73. https://doi.org/10.2134/agronj1991.00021962008300010017x

Dobbs TL, Pretty JN (2004) Agri-environmental stewardship schemes and "multifunctionality". Rev Agric Econ 26:220–237

Doran JW, Parking TB (1994) Defining and assessing soil quality. In: Doran JW, Coleman DF, Bezdicek DF, Stewart BA (eds) Defining soil quality for a sustainable environment. Special publication, vol 35. Soil Science Society of America, Madison, pp 3–22

Doran JW, Zeiss MR (2000) Soil health and sustainability: managing the biotic component of soil quality. Appl Soil Ecol 15:3–11. https://doi.org/10.1016/S0024-3205(98)00082-4

Dungait JA, Hopkins DW, Gregory AS, Whitmore AP (2012) Soil organic matter turnover is governed by accessibility not recalcitrance. Glob Chang Biol 18:1781–1796. https://doi.org/10.1111/j.1365-2486.2012.02665.x

Düring RA, Hoß T, Gäth S (2002) Depth distribution and bioavailability of pollutants in long-term differently tilled soils. Soil Tillage Res 66:183–195. https://doi.org/10.1016/S0167-1987(02)00026-0

Ekeberg E, Riley HCF (1996) Effects of mouldboard ploughing and direct planting on yield and nutrient uptake of potatoes in Norway. Soil Tillage Res 39:131–142. https://doi.org/10.1016/S0167-1987(96)01065-3

FAO / Food and Agriculture Organization of the United Nations (2012) What is conservation agriculture? Available on line at: http://www.fao.org/conservation-agriculture/overview/what-is-conservation-agriculture/en/

FAO/Food and Agriculture Organization of the United Nations (2011) Save and grow: a policy-maker's guide to the sustainable intensification of smallholder crop production, pp 1–98

Fierer N, Strickland MS, Liptzin D et al (2009) Global patterns in belowground communities. Ecol Lett 12:1238–1249. https://doi.org/10.1111/j.1461-0248.2009.01360.x

Fontaine S, Barot S, Barré P et al (2007) Stability of organic carbon in deep soil layers controlled by fresh carbon supply. Nature 450:277–280. https://doi.org/10.1038/nature06275

Francis GS, Knight TL (1993) Long-term effects of conventional and no-tillage on selected soil properties and crop yields in Canterbury. Soil Tillage Res 26:193–210

Franzluebbers AJ, Hons FM (1996) Soil-profile distribution of primary and secondary plant-available nutrients under conventional and no tillage. Soil Tillage Res 39:229–239. https://doi.org/10.1016/S0167-1987(96)01056-2

Frison EA, Cherfas J, Hodgkin T (2011) Agricultural biodiversity is essential for a sustainable improvement in food and nutrition security. Sustainability 3:238–253. https://doi.org/10.3390/su3010238

Fuentes M, Govaerts B, De León F, Hidalgo C, Dendooven L, Sayre KD, Etchevers J (2009) Fourteen years of applying zero and conventional tillage, crop rotation and residue management systems and its effect on physical and chemical soil quality. Eur J Agron 30:228–237

Galieni A, Stagnari F, Visioli G et al (2016) Nitrogen fertilisation of durum wheat: a case of study in mediterranean area during transition to conservation agriculture. Ital J Agron 11:12–23. https://doi.org/10.4081/ija.2016.662

Gallaher RN, Ferrer MB (1987) Effect of no-tillage vs. conventional tillage on soil organic matter and nitrogen contents. Commun Soil Sci Plant Anal 18:1061–1076. https://doi.org/10.1080/00103628709367883

Gandariasbeitia M, Besga G, Albizu I et al (2017) Prediction of chemical and biological variables of soil in grazing areas with visible- and near-infrared spectroscopy. Geoderma 305:228–235. https://doi.org/10.1016/j.geoderma.2017.05.045

Gao L, Becker E, Liang G et al (2017) Effect of different tillage systems on aggregate structure and inner distribution of organic carbon. Geoderma 288:97–104. https://doi.org/10.1016/j.geoderma.2016.11.005

García-Orenes F, Roldán A, Mataix-Solera J et al (2012) Soil structural stability and erosion rates influenced by agricultural management practices in a semi-arid Mediterranean agro-ecosystem. Soil Use Manag 28:571–579. https://doi.org/10.1111/j.1475-2743.2012.00451.x

Gianfreda L, Ruggiero P (2006) Enzyme activities in soil. In: Nucleic acids and proteins in soil. Springer, Cham/Heidelberg, pp 257–311

Giller KE, Andersson JA, Corbeels M et al (2015) Beyond conservation agriculture. Front Plant Sci 6:870. https://doi.org/10.3389/fpls.2015.00870

Goddard MA, Mikhailova EA, Christopher JP, Schlautman MA (2007) Atmospheric Mg^{2+} wet deposition within the continental United States and implications for soil inorganic carbon sequestration. Tellus Ser B Chem Phys Meteorol 59:50–56

Gómez-Rey MX, Couto-Vázquez A, González-Prieto SJ (2012) Nitrogen transformation rates and nutrient availability under conventional plough and conservation tillage. Soil Tillage Res 124:144–152. https://doi.org/10.1016/j.still.2012.05.010

Govers G, Van Oost K, Wang Z (2014) Scratching the critical zone: the global footprint of agricultural soil erosion. Procedia Earth Planet Sci 10:313–318. https://doi.org/10.1016/j.proeps.2014.08.023

Govers G, Merckx R, Van Wesemael B, Van Oost K (2017) Soil conservation in the 21st century: why we need smart agricultural intensification. Soil 3:45–59. https://doi.org/10.5194/soil-3-45-2017

Guimarães RML, Ball BC, Tormena CA (2011) Improvements in the visual evaluation of soil structure. Soil Use Manag 27:395–403. https://doi.org/10.1111/j.1475-2743.2011.00354.x

Han E, Kautz T, Perkons U et al (2015) Root growth dynamics inside and outside of soil biopores as affected by crop sequence determined with the profile wall method. Biol Fertil Soils 51:847–856. https://doi.org/10.1007/s00374-015-1032-1

Hargrove WL (1985) Influence of tillage on nutrient uptake and yield of corn. Agron J 77:763–768

Hartmann M, Frey B, Mayer J et al (2015) Distinct soil microbial diversity under long-term organic and conventional farming. ISME J 9:1177–1194. https://doi.org/10.1038/ismej.2014.210

Helgason T, Daniell TJ, Husband R, Fitter AH, Young JPW (1998) Ploughing up the wood-wide web? Nature 394:431

Huggins DR, Reganold JP (2008) No-till: the quiet revolution. Sci Am 299:70–77

IAASTD (2009) Agriculture at a crossroads: global report. International Assessment of Agricultural Knowledge, Science and Technology for Development. Global report

Ismail I, Blevins RL, Frye WW (1994) Long-term no-tillage effects on soil properties and continuous corn yields. Soil Sci Soc Am J 58:193. https://doi.org/10.2136/sssaj1994.03615995005800010028x

Janvier C, Villeneuve F, Alabouvette C et al (2007) Soil health through soil disease suppression: which strategy from descriptors to indicators? Soil Biol Biochem 39:1–23. https://doi.org/10.1016/j.soilbio.2006.07.001
Jenkinson DS, Coleman K (2008) The turnover of organic carbon in subsoils. Part 2. Modelling carbon turnover. Eur J Soil Sci 59:400–413. https://doi.org/10.1111/j.1365-2389.2008.01026.x
Jia Z, Kuzyakov Y, Myrold D, Tiedje J (2017) Soil organic carbon in a changing world. Pedosphere 27:789–791. https://doi.org/10.1016/S1002-0160(17)60489-2
Johnston SG, Keene AF, Bush RT et al (2009) Contemporary pedogenesis of severely degraded tropical acid sulfate soils after introduction of regular tidal inundation. Geoderma 149:335–346. https://doi.org/10.1016/j.geoderma.2008.12.013
Jones CG, Lawton JH, Shachak M (1994) Organisms as ecosystem engineers. In: Ecosystem management. Springer, New York, pp 130–147
Karlen DL, Berry EC, Colvin TS, Kanwar RS (1991) Twelve-year tillage and crop rotation effects on yields and soil chemical properties in northeast Iowa. Commun Soil Sci Plant Anal 22:1985–2003
Karlen DL, Mausbach MJ, Doran JW et al (1997) Soil quality: a concept, definition, and framework for evaluation (a guest editorial). Soil Sci Soc Am J 61:4–10
Karlen DL, Andrews SS, Doran JW (2001) Soil quality: current concepts and applications. Adv Agron 74:1–40
Kemper WD, Rosenau RC, Dexter AR (1987) Cohesion development in disrupted soils as affected by clay and organic matter content and temperature. Soil Sci Soc Am J 51:860–867. https://doi.org/10.2136/sssaj1987.03615995005100040004x
Kibblewhite MG, Jones RJA, Montanarella L et al (2008) Environmental assessment of soil for monitoring volume VI: soil monitoring system for Europe. Office for Official Publications of the European Communities, Luxembourg, p 72
Kinoshita R, Moebius-Clune BN, van Es HM et al (2012) Strategies for soil quality assessment using visible and near-infrared reflectance spectroscopy in a Western Kenya chronosequence. Soil Sci Soc Am J 76:1776–1788. https://doi.org/10.2136/sssaj2011.0307
Kirkby MJ, Morgan RPC (1980) Soil erosion. Wiley, Chichester, pp. 375
Kosmas C, Kairis O, Karavitis C et al (2013) Evaluation and selection of indicators for land degradation and desertification monitoring: methodological approach. Environ Manag 54:951–970. https://doi.org/10.1007/s00267-013-0109-6
Kumari KGID, Moldrup P, Paradelo M et al (2017) Effects of biochar on dispersibility of colloids in agricultural soils. J Environ Qual 46:143–152. https://doi.org/10.2134/jeq2016.08.0290
Lal R (1976) No-tillage effects on soil properties under different crops in western Nigeria. Soil Sci Soc Am J 40:762–768. https://doi.org/10.2136/sssaj1976.03615995004000050039x
Lal R (1989) Conservation tillage for sustainable agriculture: tropics versus temperate environments. Adv Agron 42:85–197
Lal R (1990) Soil erosion and land degradation: the global risks. In: Advances in soil science. Springer, New York, pp 129–172
Lal R (1993) Tillage effects on soil degradation, soil resilience, soil quality, and sustainability. Soil Tillage Res 27:1–8. https://doi.org/10.1016/0167-1987(93)90059-X
Lal R (1997) Residue management, conservation tillage and soil restoration for mitigating greenhouse effect by CO2-enrichment. Soil Tillage Res 43(1–2):81–107
Lal R (2004a) Carbon sequestration. Philos Trans R Soc B Biol Sci 363:815–830. https://doi.org/10.1007/978-3-642-37048-9_12
Lal R (2004b) Soil carbon sequestration impacts on global climate change and food secruity. Science 304(80):1623–1627
Lal R (2007) Carbon management in agricultural soils. Mitig Adapt Strateg Glob Chang 12(2):303–322
Lal R (2009) Laws of sustainable soil management. In: Sustainable agriculture. Springer, Dordrecht, pp 9–12

Lal R (2016) Soil health and carbon management. Food Energy Secur 5:212–222. https://doi.org/10.1002/fes3.96
Lal R (2017) Residue management, conservation tillage and soil restoration for mitigating greenhouse effect by CO_2-enrichment. Soil Tillage Res 43:81–107
Lal R, Shukla MK (2004) Principles of soil physics. Marcel Dekker Inc, New York, p 716. ISBN 0-8247-5324-0
Lal R, Logan TJ, Fausey NR (1990) Long-term tillage effects on a Mollic Ochraqualf in North-West Ohio. Soil Tillage Res 15:371–382
Larson WE, Pierce FJ (1991) Conservation and enhancement of soil quality. In: Evaluation for sustainable land management in the developing world: proceedings of the international workshop on evaluation for sustainable land management in the developing world, Chiang Rai, Thailand, Bangkok, September 1991, pp 15–21
Lavado RS, Porcelli CA, Alvarez R (1999) Concentration and distribution of extractable elements in a soil as affected by tillage systems and fertilization. Sci Total Environ 232:185–191
Lavelle P (1997) Faunal activities and soil processes: adaptive strategies that determine ecosystem function. In: Advances in ecological research. Academic, San Diego, pp 93–132
Lavelle P (2000) Ecological challenges for soil science. Soil Sci 165:73–86
Lavelle P, Decaëns T, Aubert M et al (2006) Soil invertebrates and ecosystem services. Eur J Soil Biol 42:S3–S15. https://doi.org/10.1016/j.ejsobi.2006.10.002
Lemke AM, Kirkham KG, Lindenbaum TT et al (2011) Evaluating agricultural best management practices in tile-drained subwatersheds of the Mackinaw River, Illinois. J Environ Qual 40:1215–1288. https://doi.org/10.2134/jeq2010.0119
Lima ACR, Brussaard L, Totola MR et al (2013) A functional evaluation of three indicator sets for assessing soil quality. Appl Soil Ecol 64:194–200. https://doi.org/10.1016/j.apsoil.2012.12.009
Lupwayi NZ, Clayton GW, O'Donovan JT et al (2006) Decomposition of crop residues under conventional and zero tillage. Can J Soil Sci 84:403–410. https://doi.org/10.4141/S03-082
Lützow MV, Kögel-Knabner I, Ekschmitt K et al (2006) Stabilization of organic matter in temperate soils: mechanisms and their relevance under different soil conditions – A review. Eur J Soil Sci 57:426–445. https://doi.org/10.1111/j.1365-2389.2006.00809.x
Mäder P et al (2002) Soil fertility and biodiversity in organic farming. Science 296(80):1694–1697
Majumder B, Mandal B, Bandyopadhyay PK, Chaudhury J (2007) Soil organic carbon pools and productivity relationships for a 34 year old rice-wheat-jute agroecosystem under different fertilizer treatments. Plant Soil 297:53–67. https://doi.org/10.1007/s11104-007-9319-0
Majumder M, Gangopadhyay TK, Chakraborty AK et al (2008) Fibre Bragg gratings in structural health monitoring-Present status and applications. Sensors Actuators A Phys 147:150–164. https://doi.org/10.1016/j.sna.2008.04.008
Malhi SS, Lemke R (2007) Tillage, crop residue and N fertilizer effects on crop yield, nutrient uptake, soil quality and nitrous oxide gas emissions in a second 4-yr rotation cycle. Soil Tillage Res 96:269–283. https://doi.org/10.1016/j.still.2007.06.011
Mangalassery S, Sjögersten S, Sparkes DL et al (2013) The effect of soil aggregate size on pore structure and its consequence on emission of greenhouse gases. Soil Tillage Res 132:39–46. https://doi.org/10.1016/j.still.2013.05.003
Martínez E, Fuentes JP, Silva P et al (2008) Soil physical properties and wheat root growth as affected by no-tillage and conventional tillage systems in a Mediterranean environment of Chile. Soil Tillage Res 99:232–244. https://doi.org/10.1016/j.still.2008.02.001
Martín-Lammerding D, Navas M, del Mar Albarrán M et al (2015) LONG term management systems under semiarid conditions: Influence on labile organic matter, β-glucosidase activity and microbial efficiency. Appl Soil Ecol 96:296–305. https://doi.org/10.1016/j.apsoil.2015.08.021
McCarty GW, Meisinger JJ, Jenniskens FMM (1995) Relationships between total-N, biomass-N and active-N in soil under different tillage and N fertilizer treatments. Soil Biol Biochem 27:1245–1250. https://doi.org/10.1016/0038-0717(95)00060-R
McKenzie DC, Moncada MP, Ball BC, Munkholm L (2015) Reduction of yield gaps and improvement of ecological function through local-to-global applications of visual soil assessment.

In: Visual soil evaluation: Realising potential crop production with minimum environmental impact. CABI, Wallingford, pp 31–48

McVay KA, Budde JA, Fabrizzi K et al (2006) Management effects on soil physical properties in long-term tillage studies in Kansas. Soil Sci Soc Am J 70:434–438. https://doi.org/10.2136/sssaj2005.0249

Mi H, Huang L (2008) Forest-based translation rule extraction. In: Proceedings of the conference on empirical methods in natural language processing. Association for Computational Linguistics, pp 206–214

Mijangos I, Albizu I, Epelde L et al (2010) Effects of liming on soil properties and plant performance of temperate mountainous grasslands. J Environ Manag 91:2066–2074. https://doi.org/10.1016/j.jenvman.2010.05.011

Millennium Ecosystem Assessment (2005) Ecosystems and human well-being: synthesis. Island Press, Washington, DC

Monger C, Sala OE, Duniway MC et al (2015) Legacy effects in linked ecological-soil-geomorphic systems of drylands. Front Ecol Environ 13:13–19. https://doi.org/10.1890/140269

Montgomery DR (2007) Soil erosion and agricultural sustainability. Proc Natl Acad Sci 104:13268–13272

Moore JC (1994) Impact of agricultural practices on soil food web structure: theory and application. Agric Ecosyst Environ 51:239–247. https://doi.org/10.1016/0167-8809(94)90047-7

Morgan RPC (2005) Soil erosion and conservation, 3rd edn. Blackwell Publishing Ltd, Oxford

Mueller L, Shepherd G, Schindler U et al (2013) Evaluation of soil structure in the framework of an overall soil quality rating. Soil Tillage Res 127:74–84. https://doi.org/10.1016/j.still.2012.03.002

Mueller L, Schindler U, Shepherd TG, Ball BC, Smolentseva E, Pachikin K, Eulenstein F (2014) The Muencheberg soil quality rating for assessing the quality of global farmland. In: Novel measurement and assessment tools for monitoring and management of land and water resources in agricultural landscapes of central Asia. Springer, Cham, pp 235–248

Nannipieri P, Grego S, Ceccanti B, Bollag JM, Stotzky G (1990) Ecological significance of the biological activity in soil. Soil Biochem 6:293–355 (Bollag JM, Stotzky G, eds). Marcel Dekker, New York

Nawaz MF, Bourrié G, Trolard F (2013) Soil compaction impact and modelling. A review. Agron Sustain Dev 33:291–309. https://doi.org/10.1007/s13593-011-0071-8

Nieminen TM et al (2011) Soil solution collection and analysis. In: Manual on methods and criteria for harmonized sampling, assessment, monitoring and analysis of the effects of air pollution on forests. United Nations Economic Commission for Europe convention on long-range transboundary air pollution. ICP Forests, Hamburg

O'Sullivan OS, Holt AR, Warren PH, Evans KL (2017) Optimising UK urban road verge contributions to biodiversity and ecosystem services with cost-effective management. J Environ Manag 191:162–171. https://doi.org/10.1016/j.jenvman.2016.12.062

Ostovari Y, Ghorbani-Dashtaki S, Bahrami HA et al (2018) Towards prediction of soil erodibility, SOM and $CaCO_3$ using laboratory Vis-NIR spectra: a case study in a semi-arid region of Iran. Geoderma 314:102–112. https://doi.org/10.1016/j.geoderma.2017.11.014

Pagliai M, Vignozzi N, Pellegrini S (2004) Soil structure and the effect of management practices. Soil Tillage Res 79:131–143. https://doi.org/10.1016/j.still.2004.07.002

Palm HW, Bissa K, Knaus U (2014) Significant factors affecting the economic sustainability of closed aquaponic systems. Part II: fish and plant growth. Aquac Aquar Conserv Legis 7:162–175

Pankhurst CE, Doube B, Gupta VVSR (1997) Biological indicators of soil health. CAB International, Oxford/New York

Paz-Ferreiro J, Fu S (2016) Biological indices for soil quality evaluation: perspectives and limitations. L Degrad Dev 27:14–25. https://doi.org/10.1002/ldr.2262

Paz-Ferreiro J, Trasar-Cepeda C, Leirós MC et al (2007) Biochemical properties of acid soils under native grassland in a temperate humid zone. New Zeal J Agric Res 50:537–548. https://doi.org/10.1080/00288230709510321

Paz-Ferreiro J, Marinho MDA, da Silva LFS, Motoshima ST, da Silva Dias R (2013) The effects of bulk density and water potential on multifractal characteristics of soil penetration resistance microprofiles measured on disturbed soil samples. Vadose Zo J:12

Paz-Kagan T, Shachak M, Zaady E, Karnieli A (2014) A spectral soil quality index (SSQI) for characterizing soil function in areas of changed land use. Geoderma 230:171–184. https://doi.org/10.1016/j.geoderma.2014.04.003

Pisante M, Corsi S, Kassam A, Friedrich T (2010) The challenge of agricultural sustainability for Asia and Europe. Transit Stud Rev 17:662–667. https://doi.org/10.1007/s11300-010-0181-z

Pisante M, Stagnari F, Grant CA (2012) Agricultural innovations for sustainable crop production intensification. Ital J Agron 7:300–311. https://doi.org/10.4081/ija.2012.e40

Pisante M, Stagnari F, Acutis M, Bindi M, Brilli L, Di Stefano V, Carozzi M (2015) Conservation agriculture and climate change. In: Conservation agriculture. Springer, Cham, pp 579–620

Prasuhn V (2012) On-farm effects of tillage and crops on soil erosion measured over 10 years in Switzerland. Soil Tillage Res 120:137–146. https://doi.org/10.1016/j.still.2012.01.002

Puglisi E, Del Re AAM, Rao MA, Gianfreda L (2006) Development and validation of numerical indexes integrating enzyme activities of soils. Soil Biol Biochem 38:1673–1681. https://doi.org/10.1016/j.soilbio.2005.11.021

Quinton JN, Govers G, Van Oost K, Bardgett RD (2010) The impact of agricultural soil erosion on biogeochemical cycling. Nat Geosci 3:311

Rabot E, Wiesmeier M, Schlüter S, Vogel HJ (2018) Soil structure as an indicator of soil functions: a review. Geoderma 314:122–137

Rahman MM, Moran MS, Thoma DP et al (2008) Mapping surface roughness and soil moisture using multi-angle radar imagery without ancillary data. Remote Sens Environ 112:391–402. https://doi.org/10.1016/j.rse.2006.10.026

Ramirez-Lopez L, Wadoux AMJC, Franceschini MHD et al (2018) Robust soil mapping at the farm scale with vis-NIR spectroscopy. Eur J Soil Sci:1–16. https://doi.org/10.1111/ejss.12752

Rasmussen KJ (1999) Impact of ploughless soil tillage on yield and soil quality: a Scandinavian review. Soil Tillage Res 53:3–14. https://doi.org/10.1016/S0167-1987(99)00072-0

Rawlins BG, Wragg J, Lark RM (2013) Application of a novel method for soil aggregate stability measurement by laser granulometry with sonication. Eur J Soil Sci 64:92–103. https://doi.org/10.1111/ejss.12017

Rengasamy P, Marchuk A (2011) Cation ratio of soil structural stability (CROSS). Soil Res 49(3), 280–285

Rhoton FE (2000) Influence of time on soil response to no-till practices. Soil Sci Soc Am J 64:700–709. https://doi.org/10.2136/sssaj2000.642700x

Rice CW, Smith MS, Blevins RL (1986) Soil Nitrogen availability after long-term continuous no-tillage and conventional tillage corn production. Soil Sci Soc Am J 50:1206–1210. https://doi.org/10.2136/sssaj1986.03615995005000050023x

Ritz K, Black HIJ, Campbell CD et al (2009) Selecting biological indicators for monitoring soils: a framework for balancing scientific and technical opinion to assist policy development. Ecol Indic 9:1212–1221. https://doi.org/10.1016/j.ecolind.2009.02.009

Roesch LF, Fulthorpe RR, Riva A, Casella G, Hadwin AK, Kent AD et al (2007) Pyrosequencing enumerates and contrasts soil microbial diversity. ISME J 1:283–290

Rüdisser J, Tasser E, Peham T et al (2015) The dark side of biodiversity: spatial application of the biological soil quality indicator (BSQ). Ecol Indic 53:240–246. https://doi.org/10.1016/j.ecolind.2015.02.006

Salomé C, Nunan N, Pouteau V et al (2010) Carbon dynamics in topsoil and in subsoil may be controlled by different regulatory mechanisms. Glob Chang Biol 16:416–426. https://doi.org/10.1111/j.1365-2486.2009.01884.x

Schipper LA, Sparling GP (2000) Performance of soil condition indicators across taxonomic groups and land uses. Soil Sci Soc Am J 64:300–311. https://doi.org/10.2136/sssaj2000.641300x

Schlesinger WH (1982) Carbon storage in the caliche of arid soils: a case study from Arizona. Soil Sci 133(4):247–255

Shepherd G, Stagnari F, Pisante M, Benites J (2008) Visual soil assessment–field guide for wheat. Food and Agriculture Organization of the United Nations, Roma. ISBN-978-92-5-105941-8. 50 pp

Shukla MK, Lal R, Owens LB, Unkefer P (2003) Land use and management impacts on structure and infiltration characteristics of soils in the North Appalachian Region of Ohio. Soil Sci 168:167–177. https://doi.org/10.1097/00010694-200303000-00003

Shukla MK, Lal R, Ebinger M (2006) Determining soil quality indicators by factor analysis. Soil Tillage Res 87:194–204. https://doi.org/10.1016/j.still.2005.03.011

Simon C, Daniel R (2011) Metagenomic analyses: past and future trends. Appl Environ Microbiol 77:1153–1161. https://doi.org/10.1128/AEM.02345-10

Singh P, Wu JQ, McCool DK et al (2009) Winter hydrologic and erosion processes in the U.S. Palouse region: field experimentation and WEPP simulation. Vadose Zo J 8:426–436. https://doi.org/10.2136/vzj2008.0061

Singh VK, Dwivedi BS, Singh SK et al (2016) Soil physical properties, yield trends and economics after five years of conservation agriculture based rice-maize system in north-western India. Soil Tillage Res 155:133–148. https://doi.org/10.1016/j.still.2015.08.001

Six J, Elliott ET, Paustian K (2000) Soil macroaggregate turnover and microaggregate formation: a mechanism for C sequestration under no-tillage agriculture. Soil Biol Biochem 32:2099–2103. https://doi.org/10.1016/S0038-0717(00)00179-6

Six J, Conant RT, Paul EA, Paustian K (2002) Six Stabilization mechanisms of SOM implications for C saturation of soils.pdf. Plant Soil 241:155–176. https://doi.org/10.1023/A:1016125726789

Six J, Bossuyt H, Degryze S, Denef K (2004) A history of research on the link between (micro) aggregates, soil biota, and soil organic matter dynamics. Soil Tillage Res 79:7–31. https://doi.org/10.1016/j.still.2004.03.008

Spiegel H, Zavattaro L, Guzmán Díaz MG, D'Hose T, Pecio A, Lehtinen T ... Ruysschaert G (2015) Impacts of soil management practices on crop productivity, on indicators for climate change mitigation, and on the chemical, physical and biological quality of soil. Catch-C project, p 41 Available online at: http://www.catch-c.eu/

Stagnari F, Pisante M (2010) Managing faba bean residues to enhance the fruit quality of the melon (Cucumis melo L.) crop. Sci Hortic (Amsterdam) 126:317–323

Stagnari F, Ramazzotti S, Pisante M (2009) Conservation agriculture: a different approach for crop production through sustainable soil and water management: a review. In: Organic farming, pest control and remediation of soil pollutants. Springer, Dordrecht, pp 55–83

Stagnari F, Galieni A, Speca S et al (2014a) Effects of straw mulch on growth and yield of durum wheat during transition to conservation agriculture in mediterranean environment. F Crop Res 167:51–63. https://doi.org/10.1016/j.fcr.2014.07.008

Stagnari F, Perpetuini G, Tofalo R et al (2014b) Long-term impact of farm management and crops on soil microorganisms assessed by combined DGGE and PLFA analyses. Front Microbiol 5:1–8. https://doi.org/10.3389/fmicb.2014.00644

Stagnari F, Jan S, Galieni A, Pisante M (2016) Sustainable agricultural practices for water quality protection. Water Stress Crop Plants A Sustain Approach 1:75–85

Stone D, Ritz K, Griffiths BG et al (2016) Selection of biological indicators appropriate for European soil monitoring. Appl Soil Ecol 97:12–22. https://doi.org/10.1016/j.apsoil.2015.08.005

Su Z, Zhang J, Wu W et al (2007) Effects of conservation tillage practices on winter wheat water-use efficiency and crop yield on the Loess Plateau, China. Agric Water Manag 87:307–314. https://doi.org/10.1016/j.agwat.2006.08.005

Thomas GW (1986) Mineral nutrition and fertilizer placement. In: Sprague MA, Triplett GB (eds) No-tillage and surface tillage agriculture: the tillage revolution. Wiley, New York, pp 93–116

Thomas GW, Frye WW (1984) Fertilization and liming. In: Phillips RE, Phillips SH (eds) No-tillage agriculture principles and practices. Van Nostrand Reinhold Company Inc, New York

Turmel MS, Speratti A, Baudron F et al (2015) Crop residue management and soil health: a systems analysis. Agric Syst 134:6–16. https://doi.org/10.1016/j.agsy.2014.05.009

Tuzzin de Moraes M, Debiasi H, Carlesso R et al (2016) Soil physical quality on tillage and cropping systems after two decades in the subtropical region of Brazil. Soil Tillage Res 155:351–362. https://doi.org/10.1016/j.still.2015.07.015

Tziolas N, Tsakiridis N, Ben-Dor E, Theocharis J, Zalidis G (2019) A memory-based learning approach utilizing combined spectral sources and geographical proximity for improved VIS-NIR-SWIR soil properties estimation. Geoderma 340:11–24

Valckx J, Govers G, Hermy M, Muys B (2009) ECOWORM–Erosiecontrole in akkerland door het beheer van regenwormgemeenschappen. In: Departement Aard-en omgevingswetenschappen. Leuven 44. IWT landbouwkundig onderzoek – eindrapport ecoworm (PROJECT 040681)

van Capelle C, Schrader S, Brunotte J (2012) Tillage-induced changes in the functional diversity of soil biota – A review with a focus on German data. Eur J Soil Biol 50:165–181. https://doi.org/10.1016/j.ejsobi.2012.02.005

van der Heijden MG, Bardgett RD, Van Straalen NM (2008) The unseen majority: soil microbes as drivers of plant diversity and productivity in terrestrial ecosystems. Ecol Lett 11:296–310

Van Wie JB, Adam JC, Ullman JL (2013) Conservation tillage in dryland agriculture impacts watershed hydrology. J Hydrol 483:26–38. https://doi.org/10.1016/j.jhydrol.2012.12.030

Velasquez E, Lavelle P, Andrade M (2007) GISQ, a multifunctional indicator of soil quality. Soil Biol Biochem 39:3066–3080. https://doi.org/10.1016/j.soilbio.2007.06.013

Vestergaard G, Schulz S, Schöler A, Schloter M (2017) Making big data smart – how to use metagenomics to understand soil quality. Biol Fertil Soils 53:479–484. https://doi.org/10.1007/s00374-017-1191-3

Veum KS, Sudduth KA, Kremer RJ, Kitchen NR (2017) Sensor data fusion for soil health assessment. Geoderma 305:53–61. https://doi.org/10.1016/j.geoderma.2017.05.031

Victoria R, Banwart S, Black H et al (2012) The Benefits of Soil Carbon: managing soils for multiple economic, societal and environmental benefits. UNEP Year B Emerg Issues Our Glob Environ 2012:19–33. https://doi.org/10.1038/nclimate1692

von Lützow M, Kögel-Knabner I (2009) Temperature sensitivity of soil organic matter decomposition-what do we know? Biol Fertil Soils 46:1–15. https://doi.org/10.1007/s00374-009-0413-8

Wall DH, Bardgett RD, Kelly E (2010) Biodiversity in the dark. Nat Geosci 3:297–298

Wardle DA, Bardgett RD, Klironomos JN et al (2004) Ecological linkages between aboveground and belowground Biota. Science 304(80):1629–1633

Whisler KM, Rowe HI, Dukes JS (2016) Relationships among land use, soil texture, species richness, and soil carbon in Midwestern tallgrass prairie, CRP and crop lands. Agric Ecosyst Environ 216:237–246. https://doi.org/10.1016/j.agee.2015.09.041

Winowiecki LA (2015) Landscape-scale assessments of soil health: local determinants of soil organic carbon in Ethiopia. Available online at: https://cgspace.cgiar.org

Wright AL, Hons FM, Lemon RG et al (2007) Stratification of nutrients in soil for different tillage regimes and cotton rotations. Soil Tillage Res 96:19–27. https://doi.org/10.1016/j.still.2007.02.005

Wu T, Ayres E, Li G et al (2009) Molecular profiling of soil animal diversity in natural ecosystems: incongruence of molecular and morphological results. Soil Biol Biochem 41:849–857. https://doi.org/10.1016/j.soilbio.2009.02.003

Wu T, Milner H, Díaz-Pérez JC, Ji P (2015) Effects of soil management practices on soil microbial communities and development of southern blight in vegetable production. Appl Soil Ecol 91:58–62. https://doi.org/10.1016/j.apsoil.2015.02.011

Yao R, Yang J, Gao P et al (2013) Determining minimum data set for soil quality assessment of typical salt-affected farmland in the coastal reclamation area. Soil Tillage Res 128:137–148. https://doi.org/10.1016/j.still.2012.11.007

Zibilske LM, Bradford JM, Smart JR (2002) Conservation tillage induced changes in organic carbon, total nitrogen and available phosphorus in a semi-arid alkaline subtropical soil. Soil Tillage Res 66:153–163. https://doi.org/10.1016/S0167-1987(02)00023-5

Zobeck TM, Steiner JL, Stott DE, Duke SE, Starks PJ, Moriasi DN, Karlen DL (2015) Soil quality index comparisons using fort cobb, Oklahoma, watershed-scale land management data. Soil Sci Soc Am J 79(1):224–238

Zornoza R, Acosta JA, Bastida F et al (2015) Identification of sensitive indicators to assess the interrelationship between soil quality, management practices and human health. Soil 1:173–185. https://doi.org/10.5194/soil-1-173-2015

Sustainable Water Management

Marcello Mastrorilli and Raffaella Zucaro

Abstract In agronomy, using water in a sustainable way means producing without waste and without having a negative impact on the environment. A sustainability indicator is the crop water use efficiency. From an economic point of view sustainability is the ability to continue extracting net positive social returns from a resource for an indefinite period of time. This chapter describes the ways to measure efficiency of crops in using water for growing and producing, and discusses the cost of final yield in terms of water requirements. Water use efficiency of the cultivated species, the mechanisms that influence the water-use efficiency at the farm scale and strategies for improving water efficiency have also been discussed. Economic efficiency and sustainability, and the policy instruments to reach an efficient and sustainable water management have also been discussed in this chapter.

Keywords Water-use efficiency · Water losses · Irrigation · Water policies · Water management · Economic efficiency of irrigation · Sustainability · Water policy instruments

1 Introduction

Despite being the most valuable natural resource and the factor that limits economic and social development (Chenoweth et al. 2014), the ecological consequences of the freshwater misuse are underestimated. Conversely, the sustainable management of water in agriculture does not aim at the immediate exploitation, but at the conservation of the whole agro-ecosystem (Mariolakos 2007).

M. Mastrorilli (✉)
Research Centre for Agriculture and Environment, Council for Agricultural Research and Economics, Bari, Italy
e-mail: marcello.mastrorilli@crea.gov.it

R. Zucaro
Research Centre for Policies and Bioeconomy, Council for Agricultural Research and Economics, Roma, Italy
e-mail: raffaella.zucaro@crea.gov.it

M. Farooq, M. Pisante (eds.), *Innovations in Sustainable Agriculture*,
https://doi.org/10.1007/978-3-030-23169-9_6

In developing countries agriculture consumes up to 90% of available water, but this percentage can hardly be reduced because the irrigated cropping systems produce 40% of food for a world population which continuously grows.

Increasing the crop productivity resulting from irrigation has an important role in the global health of an agro-eco-system. If more is yielded per unit of surface, no additional land will be cultivated at the expense of forests or hilly areas. If not strictly associated to the sound agro-techniques, cultivation systems in hilly areas do not reconcile with soil conservation, rather they facilitate soil erosion. Even the soil erosion is one of the causes that reduces the formation of water resources (Pisante et al. 2015; Stagnari et al. 2016). In fact, higher surface run-off, the less water accumulates in the soil profile. Water run-off, in turns, triggers soil erosion and sediment downstream transport (Ramazzotti et al. 2008). This causes the deposition of eroded soil particle into the water reservoirs and, as a consequence, the amount of stored water of lakes or surface channels (lower depths) and rivers (the distance between the banks decreases).

Over the last 50 years, water consumption has multiplied by 4 and threatens the consistency of water resources. As a reaction, many countries introduced measures to protect the fresh water and have given economic value to environmental services that promote water resources and preserve their quality. Among these services should be mentioned the conservation of eco-systems and habitats (De Groot and Hermans 2009; Beltrán-Przekurat et al. 2012).

Not always, however, irrigation has positive effects on the environment and food safety (Mastrorilli and Zucaro 2016). In semi-arid regions, where most of the cropping systems cannot produce without irrigation, in the absence of surface water, water is taken from the groundwater tables and brought to surface for supplying crops. If withdrawal exceed the ability of aquifers to restore ground water-table, the water use becomes unsustainable. Consequently, food safety, which would be achieved by bringing ground water to surface and increasing crop productivity, does not last long (Turral et al. 2010; Molden et al. 2013; Kaune et al. 2017; Rockström et al. 2017).

At global level, the amount of fresh water cannot be defined as a poor resource: water is present but is poorly distributed. In Italy every year, it rains average 300 km^3 of water, which is a quite abundant volume. What is worrying is the distribution of rain: it rains too much in certain seasons and in certain places, little at other periods of the year and in other areas (Pendergrass and Gerber 2016). This results in two opposite agronomic problems: remove excess water from the soil surface in the rainy season and bring water back to the soil in the drought periods.

In addition to geographic and seasonal distribution, water availability depends on the economic capacity of accessing water. If there were a free water market, industry, tourism and home hygiene would purchase water at higher prices than agriculture could pay for (Rubino et al. 2013).

'More crop per drop' is the slogan coined by FAO to promote a series of actions to increase the efficiency of water in agriculture (Luquet et al. 2005). To cope with demographic growth, FAO expects that by 2030 the availability of food will be necessary increased by 60%, despite the reduction of the areas to be cultivated and of soil fertility. Irrigation systems should reach this goal without resorting to additional water resources than those currently used. For dry cropping systems, the current production standards must be maintained or increased in response to an evapotranspiration demand that, due to climate change, is expected to be higher than the current one. In the Mediterranean area the rainfall regime tends to reduce effective rainfall, since each rain event becomes higher in intensity and the dry spells are longer and longer. This is called the 'tropicalization' of the rain regime.

Today, agriculture utilizes on average 70% of the freshwater available and in perspective this percentage cannot be exceeded. In the near future, farming systems will have to produce more despite the decline in water resources, lower water quality, and increased water access costs. Agronomy science and agriculture politics propose a series of sustainable solutions to increase the efficiency with which farming systems use water (Perry 2014). Operative strategies (ready to be transferred from the research centres) and normative instruments (regulating the environmental politics) are reported in the following paragraphs.

2 Sustainable Water Management in Farming Systems

The new water management models in agriculture are based on two different types of strategies, but that complement each other. The first one is to reduce water waste through the exact determination of the water crop requirements and the use of biological and physical criteria to schedule irrigation. The second strategy consists in improving the water distribution techniques through the increase in the performance of water distribution systems in the farm and the adoption of precision irrigation techniques taking into account the intrinsic variability of the soil characteristics.

2.1 Determination of the Crop Water Requirements

The development of a crop is related to the water displacement from the soil, through the plants to the atmosphere in the form of water vapour. A crop loses 99% of the water delivered in the form of irrigation and/or rain, this loss is called actual evapotranspiration (ET) and can be considered as the sum of soil evaporation and transpiration from plants. Determining accurately the ET is a necessary step to cover the crop water needs with irrigation, or to provide the right amount of water. It means neither too much for not wasting water it nor little for not inducing deleterious water stress. Evapotranspiration can be measured with sensors or estimated by models: research in this area has made remarkable progress over the last 50 years, although it is still a major topic of international agenda research, especially when it comes to arid and semi-arid environments (Rana and Katerji 2000). The ET measurement of a crop is objectively a complex activity because complex is the soil-plant-atmosphere continuum to investigate. The measurement methods considered more reliable, which do not disturb the system, are the so-called indirect micro-meteorological (Katerji and Rana 2006). Such techniques are based on the consideration that water vapour transport above a natural surface occurs mainly due to atmospheric turbulence. These methods (Bowen ratio, aerodynamic method, eddy covariance) predict the measurement of thermodynamic variables and atmospheric turbulence above a crop. The biophysics underlying these methods imposes some limitations on their operative applicability in the field: the parcels must be homogeneous and sufficiently extended; the sensors must be accurate and have a very rapid response time. In fact, the micro-meteorology is a technique still confined to avant-garde research centres. The only direct method to measure ET the weighing lysimeter, in which a portion of soil (for a depth equal to the area of influence of the root system and an area of few m^2) with the crop, is literally 'weighed' by a balance. In this way, the water lost from the system is directly measured. This method requires a complex infrastructure, a bunker below soil surface. It is very expensive and can be subject to precision problems, so it is once again a prerogative of specialized research centres (Lovelli et al. 2005). Given the extreme difficulty of the measurement, for applications, it is preferred to estimate ET through models. Currently, two are the most used models for determining the ET from well-watered crops, the first, and most popular, described in the FAO56 (from the 'Irrigation and Drainage' series), it is called the 'two-step' model. ET is given by the following equation:

$$ET = K_c ET_{ref}$$

This approach requires the knowledge of the crop coefficient (Kc) reported for each species and phenological stage. It is also necessary to estimate the so-called reference evapotranspiration (ETref) for a well-watered meadow with particular characteristics. In this case, the formula used, always provided by the FAO56 handbook, is the Penman-Monteith model, in which the resistance of the grass meadow is assumed constant over time and for any environment: it is 50 s m^{-1} and 70 s m^{-1}, for hourly or daily scale, respectively. This method has the advantage of using climatic variables measured by a standard agro-meteorological station and nowadays ETref calculation can be made easily by the local meteorological networks. It should be noted, however, that a widespread literature shows a marked variability of Kc in relation to the local climate, agronomic techniques, irrigation method, irrigation management and varieties, imposing precautions on the transferability of the coefficient values from a site to another. The second method of estimating the ET is based on the so-called one-step model, which directly calculates the ET of a crop, without calculating the reference ET. This type of model also applies formulas similar to those of Penman-Monteith. In this case, however, the crop resistance is specific to each species and is not constant but variable depending on the site climate and on the aerodynamic characteristics of the crop. The one-step model requires less intermediate calculations and, respect to the two-step model, it is less susceptible to errors. Therefore, the one-step model provides more accurate estimates, as it has been widely tested on many species. Its weakness is to require parameterization of crop resistance and specific micro-meteorological measurements above the crop.

Recently, a team of Italian-French researchers have developed a one-step operational model, which also uses only routine measurements from standard weather stations, overcoming the problems associated with its low applicability in the field (Katerji and Rana 2014). The Penman-Monteith model, cited above, is based on two physical principles: the energetic balance and the convective diffusion of sensitive and latent heat above a natural surface. The basic hypotheses are two: (1) the regime must be permanent, in natural conditions it is equivalent to accept the validity of this principle at a short time scale (from a few minutes to an hour) and (2) the fluxes must be conservative between the evapo-transpiring surface and the atmosphere reference layer, which in practice happens when these relationships are used on sufficiently large surfaces.

2.2 *Precision Irrigation*

Precision irrigation increases WUE and produces a return on the economy by optimizing the use of water in space and time, reducing waste, and lowering management costs (Sadler et al. 2005; Castrignanò et al. 2008; Pisante et al. 2012). Its use would therefore be particularly advantageous in those areas where water is the main constraining factor in the production process. The physical and chemical characteristics of the soil are not always identical, even within a cultivated plot (Castrignanò et al. 2006). If there is spatial variability, the agronomic crop management must also be site-specific (Ballesteros et al. 2014). Irrigation is the agro-technique that more than other ones are apt to be applied at variable rate. In semi-arid environments farmers tend to irrigate with 'precision', but with the intent of delivering 'exactly' the same amount of water to each parcel and the lack of uniformity in irrigation is considered as negative. This kind of irrigation assumes that the water requirement of each plant is exactly the same and ignores the differences due to the spatial variability present in any cropped field. The precision agriculture, which considers also the precision irrigation, takes into account the variability that exists within each parcel. The variability of canopies and soil properties is handled in two ways: (1) Automatic, when the irrigation follows immediately the measurements (on plant or soil); (2) Delayed, when water supplying is deferred respect to the measurement time. Whatever the mode of irrigation application is, four are the steps required by precision irrigation: data acquisition; interpretation; control with variable intensity application and evaluation. Precision agriculture in the recent years has focused on the application of deferred operations and the use of 'management zones', defined as the field portions that provide the same yield levels. Since water in the soil varies more in time than in space, precision irrigation must be carried out with criteria other than 'management zones'. Precision irrigation requires that the volume of water to be applied varies within the same parcel. Spatially variable irrigation can be practiced with irrigation systems capable of delivering variable water rates in combination with on-the-go sensors, which in real time measure the water content of the soil. Currently there are several sensors (ER- electrical resistivity, EMIelectromagnetic induction, GPR- ground penetrating radar, gamma ray meters, fluorimeters, multi- and hyperspectral spectrophotometers) that, when connected to a GPS – Global Positioning System – receiver, measure the texture, humidity and nutrient concentrations in the soil. Given the complexity of agricultural systems, only one sensor is of little use. For a more integrated representation, the latest approach is based on the data fusion from different sensors (Stellacci et al. 2012; Casa et al. 2015; Rinaldi et al. 2015).

3 Crop Water Use Efficiency

Any sustainable water management strategy at the farm scale aims at increasing efficiency of water used by the cultivated species and varieties. To achieve this goal, the biological, environmental and crop parameters which influence the crop efficiency in using water should be necessary identified directly in conditions of effective cultivation.

A series of strategies for an efficient use of the water resources are listed in the agronomic literature.

3.1 Measuring the Crop Water Use Efficiency

To determine the water use efficiency (WUE), two methods are used: eco-physiological and agronomic (Fereres et al. 2014). The eco-physiological approach focuses on the instantaneous relationship between photosynthesis and transpiration, per leaf area unit. The reference scale ranges from leaf to whole crop and, in some cases, has been extended to the territorial level (Chen and Coughenour 2004). With this approach, two goals are achieved: (1) describing the physiological processes that, according to theoretical models, determine the WUE; (2) measuring, at leaf level, photosynthesis and transpiration of the same species subjected to various water treatments and then to analyze the consequences on WUE.

The eco-physiological method is a complement to the results obtained from the agronomic approach, but alone it does not provide applicative conclusions. The photosynthesis and leaf transpiration data cannot be extrapolated to estimate the final yield of a crop or water consumption. Cultivation depends not only on photosynthesis rate, but also on the interaction of many factors, such as respiration, leaf expansion, distribution of assimilates, flowering, and fruit setting (Steduto et al. 1997).

The agronomic approach is based on two terms: seasonal evapotranspiration (ETc) and production (WUE = Yield/ETc). The time scale includes all phases of the crop cycle.

Yield and ETc are the key data to handle irrigated or dry cropping systems and to identify the best strategies to increase yields and reduce water consumption (Blum 2009). But this methodology alone does not provide the elements to explain all the obtained results.

Defining WUE in terms of commercial production (in place of final dry matter) is preferable for two reasons. The first, because yield represents a percentage of total biomass that varies according to genetic potential and agro-techniques. The second reason, because it is an indispensable economic parameter for evaluating the irrigation cost. Unfortunately, the water content in the products is not always known, so it becomes difficult to compare WUE values on the basis of the fresh weight.

Approaches combining the two methods of investigation (eco-physiological and agronomic) are the most interesting because they provide the elements to explain both potential productivity and yield reductions when sub-optimal water conditions or particular agro-technique measures are involved.

The two methods for determining WUE are correlated, as schematized by Hsiao et al. (2007), and fall into the 'efficiency chain'. Each chain ring is measured in terms of water (W) or biomass (M). The chain starts from the supply of water and ends in the agricultural product, involving competences ranging from hydraulic engineering to agronomy to plant physiology. The overall efficiency referring to the entire water transmission chain (E_{all}) is due to the ratio between the water transferred to the final agricultural product (M_{yld}) and the water taken from the source.

$$\underbrace{\frac{W_{fg}}{W_{vo}}\times\frac{W_{fd}}{W_{fg}}\times\frac{W_{rz}}{W_{fd}}\times\frac{W_{et}}{W_{rz}}\times\frac{W_{tr}}{W_{et}}\times\frac{M_{as}}{W_{tr}}}_{\text{Quantities (in volume)}}\times\underbrace{\frac{M_{bm}}{M_{as}}\times\frac{M_{yld}}{M_{bm}}}_{\text{Quantities (in weight)}}=\frac{M_{yld}}{W_{vo}}=E_{all}$$

water (W)	vo	withdrawal from the source (surface basin or water table)
	fg	delivered to the farm
	fd	distributed in the farm
	rz	held in the root zone
	et	evapo-transpired
	tr	transpired

organic matter (M)	as	assimilated
	bm	accumulated in the biomass
	yld	translocated in the final yield

3.2 *The WUE Values: Review from the Literature*

The specialized literature reports values of WUE, however they are affected by a great variability. To analyse and compare WUE data with a scientific method, a criterion must be adopted. A criterion is selecting from the bibliography WUE data determined by the agronomic method and referring to a field plot, as a spatial scale, and to the entire crop cycle, as a time scale. In addition, limiting the comparison to those WUE data referring to commercial production and measured evapotranspiration (ET). The dispersion of observed data reduces if these criteria are adopted. From a review by Katerji et al. (2008) some figures are reported here as an example. Cereals are the most studied annual species, although data on a species such as rice that uses high volumes of water is lacking (Darzi-Naftchali et al. 2017). Cereal data show a large dispersion of WUE values, even for those referring to the same species (Table 1). Differences exist not only for WUE data measured in different countries, but also for those referring to the same site.

The WUE values of those species whose commercial value depends on their fresh weight are obviously larger than those of species marketed in the form of grain.

For the latter, there are great differences: C4 species, such as corn, are characterized by higher WUE values than sunflower, soybean or legumes. These differences are explained not only by the photosynthetic mechanism but also by the chemical composition of the seeds. Corn seeds contain essentially starch, whereas sunflower seeds contain 50% oil, while leguminous seeds prevail in protein content. For the plant the biosynthesis of lipids and proteins is much more expensive in terms of energy than the synthesis of sugars.

Very few are the published WUE data referring to the tree species, although these are the crop systems that are normally irrigated. It should be underlined here that, from a methodological point of view, it is particularly complex to measure the ET of multi-annual crops, such as fruit trees (Losciale et al. 2010).

Despite all the precautions that can be taken in selecting the data in the literature, the WUE values show a wide range of variation for the same species. For wheat, the values of WUE are from 0.5 to 2.5 kg m^{-3}, for maize from 0.2 to 2.2 kg m^{-3}. The variability of WUE values applies to both autumn or spring sown crops, as well as rain-fed and irrigated crops.

Table 1 Water use efficiencies (WUE) data of cereal species in dry and irrigated cropping systems in the Mediterranean area

WUE (kg m^{-3})			
Rain-fed		Watered	
Wheat		Corn	
Syria	0.5–2.5	Turkey	1.6–2.1
Morocco	0.1–1.2	Turkey	0.2–1.2
Morocco	0.3–1.1	Italy	1.3–1.8
Israel	0.6–1.6	Italy	0.8–1.2
Italy	1.0–1.2	Lebanon	1.4–1.9
Italy	1.1–1.6	France	1.6
Turkey	1.3–1.5	Spain	1.5–2.2
Barley		Sorghum	
Italy	1.5–2.8	Italy	0.7–1.6

Source: Katerji et al. (2008)

To manage water in a sustainable way, it is very useful to understand the causes that determine this variability to develop the best agronomic strategies to rationalize water consumption.

3.3 Factors Influencing the Water Use Efficiency at the Farm Scale

To explain why the values of WUE vary, a number of factors of a different nature have to be analysed. Excluding experimental errors, the factors that affect WUE values can be traced back to three main causes (Katerji et al. 2008): the plant (differences between species, varieties, sensitivity of the stages), the soil water content (water resources, quality of irrigation water) and the environment (soil and climate, as well as atmospheric pollution and climate change).

The different causes act simultaneously and independently of each other. For example, a sunny day simultaneously causes soil dehydration, increases evaporative demand from the atmosphere, and promotes the causes of pollutant concentration in the air. It should be added that there are other causes that may alter the efficiency of water use. This is the case, for example, of biotic stress, pathogenic and insect attacks, and disease (Bethenod et al. 2005).

In the irrigated cropping systems, as well as in the dry ones, every agro-technique, together with the knowledge of the potentials and limits that depend on plant biology and the pedo-climatic environment, determines the efficiency with which crops transform the evapo-transpired water in commercial products. In addition, a number of agro-techniques are valid strategies for improving the efficient use of water by crops. This is, for example, the choice of the date of sowing of autumn-winter crops, crop density (Ritchie and Basso 2008), cut intervals of forage or biomass (Mastrorilli et al. 2002) crops, mulch (Deng et al. 2006).

3.3.1 Soil Water Content

The values of WUE for wheat ranges from a minimum of 0.5 kg m^{-3}, if the wheat is grown in rain-fed conditions, up to 2.5 kg m^{-3}, when supplemental irrigation is applied. Normally, the WUE values for dry wheat under full-field conditions vary from 0.43 kg m^{-3} in the Middle East and North Africa (de Fraiture and Wichelns 2007) to 0.76 kg m^{-3} in other areas of the Mediterranean climate (Sadras and Angus 2006).

Experimental evidence conducted in the Middle East is an example demonstrating how WUE does not depend on the amount of water received from the soil, in the form of rain or irrigation (Fig. 1). In fact, in the case of dry cultivation, if quantity and distribution of rainfall are favourable, and the good agronomic practices are applied, WUE values can reach up to 1 kg m^{-3}. Instead, although regularly irrigated

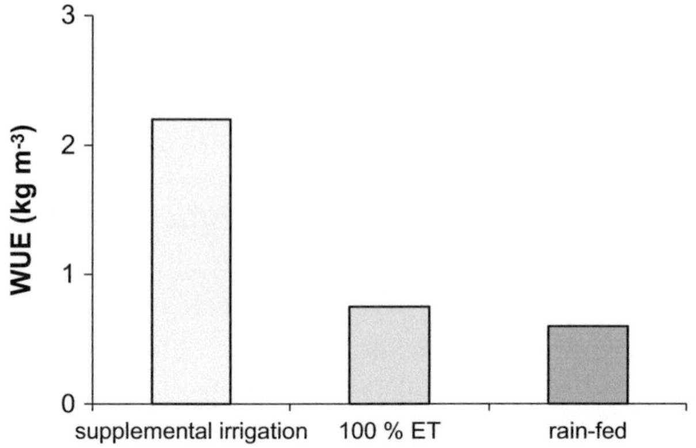

Fig. 1 WUE (kg m^{-3}) of winter wheat grown in Syria under three water regimes: full irrigation (100% ET), rain-fed, and supplemental irrigation. (Source: Oweis (1997)

(100% of evapotranspiration), the wheat WUE values do not exceed 0.75 kg m^{-3}. Instead, in case of supplemental irrigation (to supply water in certain phenological phases, when in the soil profile the water is not available for the crop), water is used much more efficiently (WUE = 2.5 kg m^{-3}) than in full-irrigated wheat. With supplemental irrigation, water is applied to the soil to relieve water stress during the phenological stages which are most susceptible to water scarcity. Although supplied in reduced volumes, the supplemental irrigation has a positive impact on final production.

The example described also shows that under the definition 'irrigated' crops are actually reported extremely different situations of plant water status and which are independent of the amount of water the crop has received.

The soil water status (measured through water content, moisture, water potential, water stored in the soil profile) is an imperfect parameter to characterize the actual plant water status. It could not be easy to derive the crop water status from the soil moisture. The main difficulty is to determine the depth of the soil layer explored by the root system. The deepening of the roots in the soil varies with the evolution of the crop cycle, the nature of the soil, the soil tillage, any biotic and abiotic stresses. To these variables must be added the uncertainty that comes from correctly quantifying the contribution to the crop water supply from the deep soil layers (capillary rise). For experimental purposes, the leaf water potential, preferably before dawn (pre-dawn, or base potential), is used when it is necessary to diagnose the crop's water state unequivocally and schedule irrigation or study evapotranspiration and render them in different water conditions (Fig. 2). The pre-dawn leaf potential represents an equilibrium between the water status of vegetation and of the soil layer colonized by the root system. Moreover, it is independent from the microclimatic conditions at the time of measurement and it depends uniquely on the soil water status.

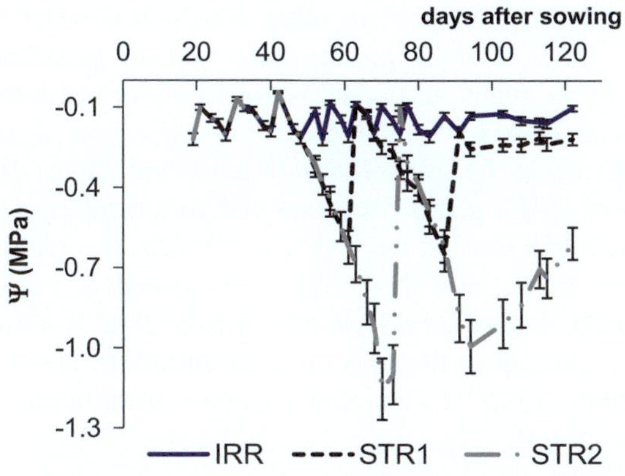

Fig. 2 Pre-dawn (or 'base') leaf potential (Ψ) monitored on corn in southern Italy thought the 'pressure chamber' (a simple tool, but requires manual and laborious measurements before dawn). Corn is grown under 3 water regimes (corresponding to the same degree of water stress): IRR, the corn is 'well-watered', and does not reduce gaseous exchanges through stomata regulation; STR1, the corn is sub-optimally irrigated and moderately reduces the stomatal conductance; STR2, insufficient irrigation and severe reduction in gaseous exchanges. Results show that same experimental protocol (consisting in scheduling irrigation on the basis of the measured Ψ values) repeated in two successive seasons, provided stable values of WUE ((IRR WUE was 1.8 in the first year and 1.7 kg m^{-3} in the second year, WUE in STR1 1.6 kg m^{-3} in both years and STR2 WUE 1.4 and 1.5 kg m^{-3}) and confirms that the values of WUE in corn decrease in proportion to the availability of water in the soil profile. (Source: Ben Nouna et al. (2000))

3.3.2 Mineral Fertilization

The effect of mineral fertilization on WUE depends on the soil's water content. In the case of dry wheat cultivated in a semi-arid environment (Oweis et al. 2000), to nitrogen doses greater than 50 kg ha^{-1} do not correspond to yield increases. Instead, in irrigated crops, grain productions increase in proportion to nitrogen doses supplied to wheat and, consequently, improves the efficiency of water use. This may not be the case if the phosphorous supply in the soil is limited, even though water and nitrogen in the soil are available for the crop during the whole cycle.

3.3.3 The Quality of the Water

In agronomy salinity is defined as the accumulation of salts in the root zone that damages the crops. The accumulation comes from irrigation practices to deal with drought and the main causes are: poor irrigation management and unconventional water use.

Salinity and water scarcity in the soil produce the same effect on plants. Saline water, even though it is in the soil, however it is not available for plants. To absorb water, the root must overcome the force with which the soil particles retain water. This force is defined as the potential matrix and increases – in absolute value – with the decrease of soil water content. In other words, less water is in the soil, the greater capillary ('cohesive force' linking water and soil particles), and the greater will be the work of the plants to uptake water from the soil into the roots. If soil water is saline, to the matrix potential, the osmotic potential of water is added, due to the presence of salts in the soil solution (Atzori et al. 2016). Increasing salinity, the availability of water for plants decreases and, as a result, the water status of the plants changes, as in the case of the soil water deficit. If a plant has less water, it reduces gaseous exchanges at a short time scale, growth at a mean, and yield at a long time scale. In the case of salinity, in addition the plant is subjected to the toxicity of specific ions present in the water in concentrations above certain tolerance thresholds. The relationship between toxicity and concentration of a particular ion is specific to each species.

3.3.4 The Species

The choice of the cultivated species plays a key role in the efficiency of brackish water use. Figure 3 shows the behaviour of ten species in response to the level of salinity of the soil. These are watered with water at increasing levels of salinity. Tolerance to the salinity of a species translates into the ability to maintain or improve WUE values when the crop is irrigated with brackish water (wheat, sunflower, potato, corn, beet). On the contrary, in sensitive species (legumes, tomatoes), WUE values tend to decrease if they are watered with low quality waters.

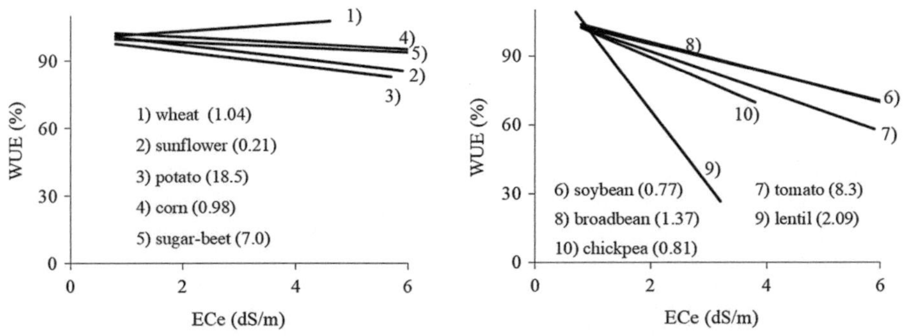

Fig. 3 Water use efficiency (WUE) values of ten species varying in salinity of soil. Species that tolerate salinity are numbered from 1 to 5, those sensitive to 5 to 10. WUE values are expressed as % of WUE measured in salt soil compared to the WUE value measured in non-saline soil. The value of WUE (kg m^{-3}) of control (non-saline) treatment for each species is given in brackets. The salinity of the soil is measured as ECe, electrical conductivity of the 'saturated paste' extract. (Source: Katerji et al. (2003))

3.3.5 The Genotypes

If the physiology of the cultivated species is sufficiently known, it is possible to obtain the same production levels even with reduced water volumes. If the same yield is obtained, but with less water, the efficiency of water use by the crop increases.

Knowledge on the physiology of abiotic stresses has allowed to identify genetic traits that regulate the relationship between tolerance to drought and production (Ceccarelli et al. 2004; Liu et al. 2017). The variety range is expanding with the application of this knowledge to genetic improvement. Today, the market offers a growing number of new cultivars with water stress resistance factors capable of maintaining high production standards even in poor water conditions. Genetic improvement should not aim at obtaining a variety with high WUE, if yield remains low: productivity and WUE are two complementary indicators. From the economic point of view, the ideal cultivar model is the one that enhances any condition of water availability and transforms it into production. Ultimately, the ideotype to be pursued by breeders, WUE and yield should be highly correlated (Trethowan 2014).

3.3.6 Sensitivity of the Phenological Phases

For the same species, every phenological stage responds differently to water. For many cultivated varieties, the sensitive stages to water stress have been defined. Only if it is known in which phenophase crop production is more vulnerable to water stress it is possible to reduce seasonal irrigated volumes and practice 'deficit irrigation'. 'Irrigation deficit' is a strategy for optimizing water by supplying water at the crop stages which are recognized to affect the final yield if conditions of soil water stress occur (Fereres and Soriano 2007; Hueso and Cuevas 2010; Katerji et al. 2013; Campi et al. 2014).

The crop cycle can be outlined as a succession of phenological phases. Each of them, if subjected to a water stress, has a peculiar effect on the final yield. The critical phenophase is that one which affects the yield formation processes more negatively than the other ones. Although limited to only one critical stage (i.e. emergency, flowering, or fruit-setting), a temporary soil water shortage hampers the yield. In order to carry out the 'deficit irrigation', critical phenophases must be identified *a priori*, species by species. As well as it must be known the relationship between a soil water stress occurring at each phenophase and the final yield. This physiological information, specific for each species, becomes crucial to schedule irrigation when the available water resources are not enough to fully cover the crop water needs. Knowing the ranking of the consequences of a temporary (in a specific phenophase) water stress on the final yield, the water should be supplied during the critical phenophase, and then supplied during those phenophases that in turn show a less sensitivity to the water stress.

3.3.7 Environmental Factors

Climate plays a key role in the consumption of water by cultivated species (Molden and Oweis 2007) and on the efficiency with which water is transformed into commercial agricultural products or grassland (Gang et al. 2016). Sadras and Angus (2006) collected the seasonal and yield evapotranspiration data of several semi-arid environments and showed that wheat WUE ranges from 1 to 0.53 kg m^{-3} and that this variability depends on the evapotranspiration demand of the atmosphere (ETref) during the flowering phase.

ETref is mainly governed by two climatic factors: net radiation and air vapor pressure deficit (VPD, difference between actual steam pressure and maximum steam pressure). The relationship between VPD and WUE has been the subject of many studies (Zwart and Bastiaanssen 2004; Hsiao et al. 2007). The VPD values measured during the cultivation cycle of species sown in winter (wheat) and in spring (corn, rice and cotton) at different latitudes (10–40° to the north and south of the equator) decreased by moving away from the equator. WUE values, on the other hand, increase with the distance from the equator. The linear relationship between WUE and latitude has also been confirmed in eastern Australia for wheat.

Practically, in field conditions, climatic parameters cannot be altered to reduce the causes of high levels of evapotranspiration. A few examples can be mentioned in which the microclimate of the crop is modified, but these are cases that do not refer to the full field. Solar radiation can be reduced by artificial shading (Losciale et al. 2011, Fig. 4) or self-shading (Apulian table grape or Sorrento citrus, Fig. 5) or by distributing radiation-reflecting substances (kaolin and synthetic products with

Fig. 4 Management of light energy in fruit trees: reducing irradiance improves water use efficiency. (Source: Losciale et al. (2011))

Fig. 5 An example of self-shading (Sorrento citrus)

reflective properties) or preventing evaporation from the soil with mulching materials. These are remedies that did not pass the scientific screen because the experimental trials provided partial results on the actual water saving and that, however, can hardly be applied to the open field.

In the field, an example of modification of the culture microclimate to control evapotranspiration is the distribution of field parcels. The contemporary presence of irrigated and dry fields in a landscape causes the oasis effect. Advection consists in the lateral transport of energy (in the form of sensitive heat) from dry to irrigated parcels. If they are close to a dry surface, irrigated crops are inevitably, in addition to solar energy, also subject to an energy surcharge that comes laterally, and which results in increased evapotranspiration (French et al. 2012).

The use of windbreaks is another technique that modifies the microclimate, reducing the aerodynamic component of evapotranspiration (Fig. 6).

A recent threat to the efficient use of water is represented by the photochemical oxidants (Rai et al. 2016). The typical Mediterranean climate conditions (high levels of temperature and radiation associated with stable air masses), combined with the emission of pollutants into the air, favour the formation of secondary pollutants such as ozone (O_3). Cultivated species show great variability in the ozone response (Fiscus et al. 2005). Very variable is also the ozone concentration between one season and the other (Bou Jaoudé et al. 2008). Field experiment demonstrated that soybean, despite being fully irrigated, if exposed to high levels of ozone, reduces evapotranspiration and grain yield compared to control (soybean not exposed to ozone pollution). Instead, if the crop is grown with sub-optimal irrigation volumes, the presence or absence of ozone has no significant effect on yield and evapotranspiration. This is because of soil water stress that, by regulating the stomatal opening,

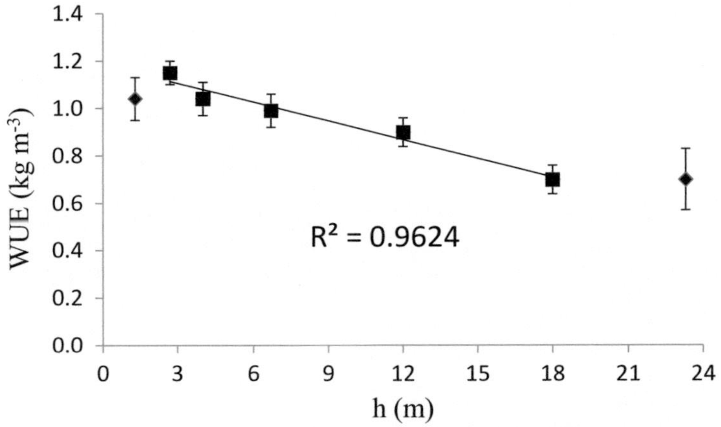

Fig. 6 Values of water use efficiency (WUE) measured in wheat at different distances from windbreak. The distance (h) is measured in m, as multiple of the height of the windbreak barrier. (Source: Campi et al. 2009)

Table 2 Soybean crop growing in a Mediterranean environment: seasonal evapotranspiration (ET), yield and water use efficiency (WUE). Crops were growing under to two irrigation treatments and two levels of AOT40 (0 or 10,000) which indicates the ozone concentration per hour

	Well watered crop		Stressed crop	
	AOT40 = 0 pbb.h	AOT40 = 10,000 pbb.h	AOT40 = 10,000 pbb.h	AOT40 = 0 pbb.h
ΣET ET (m^3 m^{-2})	0.38	0.28	0.27	0.28
Yield (kg m^{-2})	0.28	0.15	0.18	0.19
WUE (kg m^{-3})	0.74	0.53	0.68	0.67

Source: Bou Jaoudè et al. (2008)

reduces the ozone flow within the leaves. From this experience on soybean, it can be concluded that when the atmosphere is ozone-polluted, full watering a crop means wasting water. Table 2 shows that if the air contains high ozone levels, the WUE values of the irrigated crop are reduced by 30% compared to a growing crop in an ozone-free environment.

The soil type changes the ability of crops to use water. This is to say that the same cultivated crop in two different soils provides different productive performances and uses soil water differently. In Australia Turner (2004), simulating evapotranspiration and yields, shows that wheat grown in sandy soils better transforms water in grains than wheat does in a clay soil. At the same time, it points out how the nitrogen fertilizer of the soil plays an equally important role in determining the WUE values of wheat.

Multi-year simulation of corn yield and evapotranspiration, grown in different locations in southern Italy having comparable climatic conditions (evapotranspiration demand of the atmosphere during the maize cultivation season varying between 808 and 730 mm), but different water reserves (108–208 mm of available water in the soil profile explored by the roots), have shown that the success of the deficit irrigation is 'site-specific' (Katerji et al. 2010). In that it is determined by the soil water reserve which depends on soil properties (texture and available water). Under favourable irrigation conditions, the WUE values of corn do not show significant differences (between 1.3 and 1.4 kg m^{-3}) between the sites. Instead, applying the 'deficit irrigation' technique, WUE decreases significantly where the water reserve is low (from 1.42 to 1.15 kg m^{-3}) and increases (from 1.3 to 1.6 kg m^{-3}) where the water reserve of the soil is greater.

The measurements of seasonal evapotranspiration and yield of crops growing on different soil type and under experimental neutrality (obtained in a battery of large lysimeters containing soils of a different nature) have shown that soil texture affects evapotranspiration and yield, and consequently WUE. In general, in loam soil water is better transformed by crops into commercial products, while some species (soybean and tomato) have shown to be indifferent to soil texture (Table 3).

Table 3 Seasonal evapotranspiration, yield, and water use efficiency (WUE) measured in six crops growing without soil water constraints on two soil types

Species		Clay	Loam	
Potato	ET (mm)	363	415	*
	Yield (t of tuber ha^{-1})	5.8	8.6	**
	WUE (kg m^{-3})	16.1	21.0	*
Corn	ET (mm)	644	607	n.s.
	Yield (t of grain ha^{-1})	0.55	0.68	**
	WUE (kg m^{-3})	0.87	1.13	*
Sunflower	ET (mm)	1215	1450	*
	Yield (t of achene ha^{-1})	0.22	0.35	**
	WUE (kg m^{-3})	0.18	0.24	*
Sugar-beet	ET (mm)	731	836	*
	Yield (t of root ha^{-1})	4.47	6.56	**
	WUE (kg m^{-3})	6.11	7.85	**
Soybean	ET (mm)	430	410	n.s.
	Yield (t of grain ha^{-1})	0.31	0.33	n.s.
	WUE (kg m^{-3})	0.73	0.81	n.s.
Tomato	ET (mm)	667	708	n.s.
	Yield (t of fruit ha^{-1})	5.31	6.12	*
	WUE (kg m^{-3})	8.01	8.65	n.s.

n.s. not significant
$*p > 0.05$; $** p > 0.01$
Source: Katerji e Mastrorilli (2009)

4 Policy, Normative and Regulation Instruments for a Sustainable Water Management

Water is a key resource for many different needs, but mostly for civil, industrial, agricultural, environmental (water-related ecosystem services) uses. Also, water has cultural, religious and social values that are not tradable on a market.

Because of their importance, globally, agriculture and water as topics are always present into strategic programming and address documents. In fact, they play a substantial role in the 2030 Agenda for Sustainable Development, and this is clearly reflected in the Sustainable Development Goals (SDGs). Sustainable water management (SDG6) and sustainable agriculture (SDG2) are both primary goals, and neither one can be achieved independently of the other.

Water, as recognized in the 2017 G20 Agriculture Ministers' Declaration, is an essential production resource for agriculture, crucial for feeding the growing world population. Ministers confirmed their commitment to policies which boost agricultural productivity while ensuring that water and water-related ecosystems are protected, managed and used sustainably.

The multiple benefits that agriculture provides to society depend on the long-term sustainable management of natural resources, including water. However, a number of current pressures are affecting the quantity and quality of water supply, affecting its present and future sustainability. Promoting certain sustainable agriculture and forest management practices is of paramount importance for water ecosystems. Addressing the pressures while maximizing the beneficial effects of good agricultural land management will greatly enhance sustainable water management and sustainable agriculture.

4.1 One Step Back

To talk about the concept of efficiency and sustainability in water management from an economic point of view it is necessary to introduce some important economic concepts related to natural resources economics.

An economic good is defined as an item or a service suitable to satisfy human needs that is scarce if compared to its demand (Besanko and Braeutigam 2011). Water is universally known as the resource that plays a fundamental role in different fields of the society and, most of all, it allows the development of life in different forms, and also in the economic development, being included in the production of a large number of goods and services (UNESCO 2015).

According to the Principle n. 4 of Dublin Statement, 'Water has an economic value in all its competing uses and should be recognized as an economic good' (ICWE 1992) water has been considered as an economic good that can lead to an efficient and equitable use of resource.

Seeing water as an economic good started a debate about two main vision of water: as a private good, that implies that it should be priced letting the market ensure its allocation to its best uses (Van der Zaag and Savenije 2006); as a basic human need and also a social good which implies that it has to be kept outside the process of market pricing (Perry et al. 1997).

In general terms, economics distinguish between four main categories of goods: private goods (which are highly excludable and rival), public goods (which are non-excludable and non-rival), club goods (which are excludable and rival) and common pool resources. In the economics conception, water is identified for being non-excludable but rival, so as a common pool resources. In fact, there do not exist manners to prevent to some consumers free access to the good, while rival means that quantity of the good used by one consumer cannot be shared with another because, once consumed, it is no more available for others (FAO 2009).

When users of a scares resource are not excluded by the property right, then the resource is an open access resource. This kind of goods is like the common pool resources when there are no limitations. Open access resources impacted by agriculture include: ambient, air and water system. Common pool resources, unlike pure public goods, can face problems of congestion or overuse, because they are subtractable.

As will be described below, the economist's policy prescription for that kind of resources generally include regulation, marketable permits, or taxes, to offset the inefficiencies inherent in private management.

The Environmental economics literature considers the way in which producers and consumers use natural resources depends on the underlying set of property rights. Property rights refer to a bundle of entitlements that convey to the owner certain privileges and constrains. They can take the form of property rights, liability rules, or inalienable entitlements (Calabresi and Melamed 1972). For markets to lead to an efficient allocation of natural resources, the property rights to the resources generally must present four characteristics: ownership, exclusivity, transferability, and enforcement.

The economic incentive for efficient resource management is only possible if the right to use is guaranteed by the ownership. Ownership is a legal device that assigns the right to use a resource to a private owner. If there is exclusivity, then all benefits and costs from the use of the resource will accrue only to the owner, indeed the absence of exclusivity is the main distinction between a private property resource and an open access resource. Transferability implies that property rights are fully transferable between people through trade: restrictions on water transfers are often a source of inefficiency in water resource policy. Finally, to be effective, a system of rights must be enforceable. Well defined property rights coupled with competitive markets can lead to a set of inventive for efficient market exchange (Carlson et al. 1993).

When property rights are not well defined, externalities are generated. An economic externality requires both the environmental change and a human reaction to that change (Dwyer et al. 2006). Externalities related to water for agriculture are the off-farm effects of irrigation and drainage that impose costs or benefits on other farmers and/or the public. Negative externalities involve near-term and long-term damages caused by irrigation and drainage, include the waterlogging, salinization and downstream salinity that generate external cost. Positive externalities involve external benefits, such as the generation of usable surface runoff or the provision of water supply to a desirable wetland area, for which no payment or compensation is received.

Externality is a form of market failure, and the presence of an externality can lead to economic inefficiency. If private action fails to address externality, governments intervention potentially can correct the inefficiency thought a variety of policy tools, that are described below.

Finally, to define what is meant by optimal use of resources, one must consider resource use over time. The most widely used criterion for determining optimal resource use is dynamic efficiency. This criterion assumes that society's objective is to maximize the present value of net benefits from the use of the resource (Carlson et al. 1993).

4.2 Economic Efficiency and Sustainability

Agricultural system based on renewable resources (resources that can regenerate themselves within an acceptable time period) are required for long-run sustainability. So, an important issue in the economics of renewable resources is the identification of efficient and sustainable resource allocation. A related issue is the development of policy interventions associated with inefficient markets.

A feature that distinguishes renewable resources from non-renewable resources is the capability of the first type to reach sustainability. A renewable resource system is at sustainable state when the resource inventory does not change over time. In such situation the system results in constant yield, and extraction is equal to resource growth. There may be many possible sustainable states, but economists are interested in outcomes in which both physical variables and their economic counterparts stay constant over time. Which means that the whole system is in a steady state and resource-use levels and their prices do not change with time. Stable steady-state outcomes may be viewed as the dynamic equivalent of long-run equilibria. Natural resources management decisions are complex because they involve connections and trade-offs between the present and the future.

Economists often investigate the conditions under which optimal management of renewable resources result in steady-state outcomes and the stability of these outcomes. Economic theory demonstrates that optimal renewable resource management over time and the steady-state outcomes, depend mostly on the discount rate, extraction costs, and resource-growth function. Stronger preferences for present consumption, expressed in the form of higher discount rate, tend to result in smaller steady-state resource stocks. On the other hand, larger marginal costs of extraction, with respect to resource stock, tend to increase the steady-state resource stock (Carlson et al. 1993).

With respect to an efficient resource allocation, the absence of externalities is one of the sets of first best conditions required so that competitive markets will achieve a Pareto optimality.

Talking about water, a sustainable model of water provision includes the perfect correspondence between its value and its costs (Rogers et al. 1998). In particular, for a sustainable use of water its cost and its value have to be equal, maximizing the social welfare. The perfect balance between Full Cost and Full Value is achievable only in theoretical cases. Usually Water Value is higher than the Full Cost by reason of the impossibility to precisely calculate the number of externalities connected with the use of water (Rogers et al. 1998).

According to OECD, there are some distinctive economic features that make the supply and demand for water more complex than other economic goods and services; they include (Hanemann 2006; Thompson 2006): private (extraction) and public good (stewardship) characteristics of water imply different allocation mechanisms. When water is used on a farm it is a private good, but when left in situ, such as a lake or wetland, it is a public good for which private markets are generally absent. Moreover, water is largely used by the private sector (farms, households, industry) but its ownership and delivery are normally in the public domain.

Mobility of water, in that it flows, leaches, evaporates, and can be reused, which makes it distinctive as a commodity compared to land, for example. Moreover, agriculture can contribute positively to the hydrological cycle, for example, through groundwater recharge and water purification functions; it can, however, also contribute to surface water and groundwater pollution and through excessive extraction may lead to diversion of water from supporting ecosystems.

Heterogeneity of water in terms of space, quality and variability over time (seasonal and annual), which presents challenges in terms of matching supply and demand and structuring legal and institutional arrangements, as a given quantity of water is not the same as another available at a different location, point in time, quality and probability of occurrence.

Complex and multi-layered institutional and governance arrangements for water resources, reflected in the national institutions and governance of water resources (and in some cases cross national border structures) and sub-national regional and local governments (water user associations) management of water; sometimes the governance of surface water and groundwater are often separated.

Understanding the economics of water is difficult but very important because it can help inform decision makers of the full social costs of water use in agriculture and the full social value or benefits that agriculture's use of water can provide (Hanemann 2006). The usefulness of understanding these concepts for policy analysis is the transparency they bring in terms of how the value of water to society is more that just as an agricultural input, and to clarify what the costs are of agriculture's use of water resources (Malik 2008; Rogers et al. 1998; Rogers et al. 2002). The value and cost of water can be summarized as follows (Fig. 7).

- Value of water is the sum of the economic and intrinsic value. The economic value includes: adjustment for societal objectives, such as the additional increase

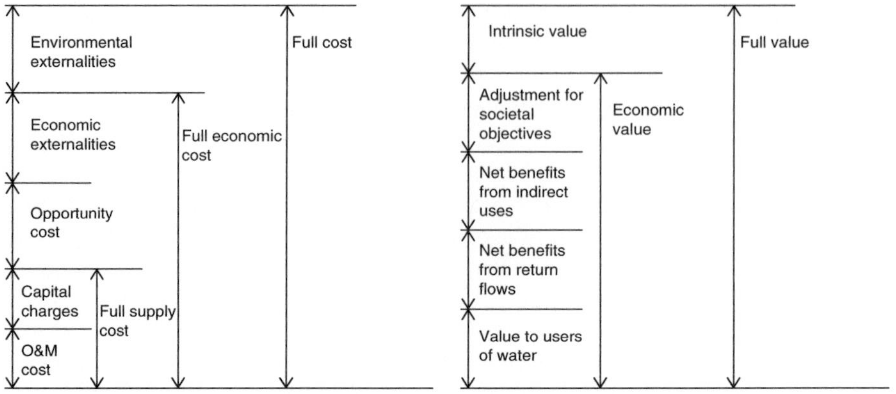

Fig. 7 A sustainable model of water provision. (Source: Rogers et al. (2002))

in commodity production gained from irrigation, higher employment and benefits for rural development;

- net benefits from indirect use such as drinking water for domestic purposes and providing habitat for flora and fauna, although these benefits can be offset by various negative environmental externalities, such as salinization of soils and pollution of water from farm chemicals used in irrigation;
- net benefits from return flows of water diverted for agriculture and other users, measuring the effects caused by water that returns flowing in nature and can be reused (Brouwer and Pearce 2005), which may also include groundwater recharge, although these benefits will depend on the lost to evapotranspiration;
- value to users of water for productive activities, such as irrigated farming, based on the marginal utility (i.e. the increase in utility received from the addition of another unit of the resource (Besanko and Braeutigam 2011).

The Intrinsic Value is linked to the attributes of water that are the most difficult to assign values, for example, the aesthetics of waterscapes and recreational attributes; it is not easy to estimate in monetary terms and comprehends all those aspects related to the existence of the resource, as cultural or aesthetic aspects (Rogers et al. 1998).

Cost of water consists of two elements, full economic cost, and environmental externalities, where the full economic costs are the sum of the supply costs, the opportunity costs and the economic externalities.

The full supply costs are included in full economic costs and are associated with supplying water to consumers without considering either the externalities of water consumption (positive or negative) or alternate uses of water (opportunity costs). These costs consist of two elements, which are very important in terms of measuring agricultural support for irrigation: operation and maintenance costs (O&M), associated with daily running of the water supply system, such as electricity for pumping, labor and repair costs; capital costs, are the cost of investment and cover both capital for renewal investment of existing infrastructure and new capital investment costs, such as building a new dam and canal network.

The opportunity (or resource) costs, address the cost of one consumer depriving another of the use of the water if that other use has a higher value for the water; although opportunity costs are zero when there is no alternate use, that is no shortage of water.

The economic cost of externalities is composed by positive externalities, such as the groundwater recharge benefits from irrigation; and negative externalities, typically upstream diversion of water or the release of pollutants downstream within an irrigation system.

While economic externalities cover costs to producers and consumers upstream and downstream, environmental externalities are associated with costs to public health and ecosystems.

Valuing the opportunity cost of water can be extremely difficult. The economic value of water, however, covers goods and services that are not usually marketed, such as the net benefits from return water flows (e.g. groundwater recharge) and indirect use (e.g. wetlands or pollution); social values (e.g. rural employment); and intrinsic values (e.g. recreational, scenic, and cultural attributes).

While economists have tools to provide proxy values for these non-marketed goods and services (e.g. contingent valuation) their application to guide policy decisions can be difficult.

It is possible to look at the economic efficiency in two different way: static efficiency and dynamic or intertemporal efficiency. In the first case a state that is efficient in the static sense is efficient strictly for a single time period, the present one; dynamic or intertemporal efficiency indicates a situation that is efficient when not only the present year is taken into account, but all future years as well, considering the future consequences flowing from today's decision. Intertemporal efficiency, which maximizes the welfare of the present generation, involves discounting in the future values of benefits and costs; this is a controversial issue, in particular with respect to the value to choose. Another way of thinking about the problem of balancing the interests of distant future generations with those of present generation is to talk about sustainability. Sustainability has become a principle for much of the subsequent politic discussion about natural resources and environmental policy.

4.3 *Efficiency of Irrigation*

Water is an extremely complex resource so address the issue of water resource management is complex. It is both a public and private good; it has multiple uses; the hydrology and externalities require to examine potential productivity gains at farm, system, and basin level; both quantity and quality are important in measuring availability and scarcity; and the institutions and policies that govern the use of water are typically fragmented.

Given these complexities, it is small wonder that there is little agreement among scientists, practitioners, and policy makers as to the most appropriate course of action to improve the management of water resources for the benefit of society.

In recent years several researchers have introduced new terms describing irrigation efficiency to enhance the information available when evaluating water policy alternatives.

Starting from the awareness that classical irrigation performance parameters failed to measure and differentiate between consumptive and beneficial uses, (Solomon and Burt 1999) proposed a new performance parameter, named Irrigation Sagacity (IS), in order to measure the irrigation performance addressing reasonable and beneficial water use. Reasonable uses are those that, while not directly benefiting agronomic production, are nonetheless reasonable under prevailing economic and physical conditions (i.e losses which contribute towards environmental goals). Beneficial uses are those that contribute directly to the agronomic production of the crop (i.e. crop evapotranspiration, water used for salt control). Sagacious uses are either beneficial, or non-beneficial but reasonable. Non-sagacious uses (non-beneficial and unreasonable) are those uses which are without economic, practical, or other justification (Solomon and Burt 1999).

Other definition is Effective Efficiency of irrigation water delivered to farms and the Basin Efficiency or Global Efficiency of water use within a river basin or irrigated area. Keller and Keller (1996) proposed the Effective Efficiency to account for the reuse of surface runoff and deep percolation by farmers along a watercourse. Basin or Global Efficiency describes the aggregate beneficial use of water in a river basin as a portion of the total volume of water available (Seckler 1996; Molden and de Fraiture 2000). Irrigated areas described by high levels of Effective Efficiency will be described also by high levels of Basin or Global efficiency.

Some authors suggest that when the estimate of Basin Efficiency approaches 100% there is little opportunity to save water by improving water management to achieve higher levels of classical, farm-level efficiencies in upstream positions of a river basin. They contend that such efforts largely would reduce the volume of surface runoff and deep percolation used by other farmers in the basin, while generating little or no gain in the amount of water available for irrigation.

Perry (1999) describes the water resources paradigm developed in recent years in the International Water Management Institute, highlighting how surface runoff and deep percolation from upstream irrigation projects often provide water supply to downstream projects, and he suggests that researcher must take a basin-wide perspective when considering policy alternatives for improving water management.

According to those studies efforts to improve classical farm-level irrigation efficiencies appears to be efforts that do not save water but create water to be moved to another location or allocated to another use outside the basin. However, in most river basin, saving water is not the ultimate policy goal, that is maximizing the social net benefits generated with limited water supply. This goal is consistent with achieving economic efficiency.

In some cases, thus, externalities and opportunity costs can prevent region or nation from achieving economic efficiency, even when irrigation is described by high measures of basin or effective efficiency. In this way, efforts to improve farm-level and regional water management will enhance the productivity of irrigation water and other inputs.

So it becomes important to distinguish between measures of irrigation efficiency and water productivity, which involve both physical variables and measures of economic efficiency as well as costs and revenue.

Aside irrigation efficiency (Ec), the concept of basin or global efficiency is an aggregate version of effective efficiencies, as sketched in the paragraph 3.1. Molden and de Fraiture (2000) also note the importance of accounting for drainage water reuse when evaluating irrigation efficiency, and to do this they introduced the concept of basin efficiency to account for the recycling of irrigation return flows. Therefore, some of the 'water saving' practices are not saving water but simply redistributing the water. The only real losses to the hydrological system are from bare soil and water evaporation or from flows to the sea or to the sinks.

Keller and Keller (1996) suggest that 'even if closed irrigation system were operating at nearly a 100% overall physical efficiency, substantial economic gains could be made by reallocating water from lower to higher valued uses'. For Molden and de Fraiture (2000) the concept of efficiency, even with basin efficiency, is that it refers only to physical quantities of water, and it does not capture differences in the value of water in alternative uses. Water productivity can be increased by obtaining more production per unit of water or by reallocating water from lower to higher valued crops. To talk about efficiency, it is important to consider economic dimensions.

4.4 Economic Irrigation Efficiency

Economic analysis of water use includes the value generated by production activities, the opportunity costs of inputs, and any pertinent costs or benefits that are external to producers and consumers (externalities). Economic efficiency describes the conditions that must be satisfied to ensure that resources are used to maximize net benefits, and it is achieved when limited resources are allocated and used giving greatest net value. Economic analysis can be useful in describing the private and public costs of an inefficient allocation of resources, and in determining strategies for moving towards an efficient allocation.

At the farm level the principal need is to maximize the profit, using water and other inputs in order to increase net returns e.g. choosing crops with a higher value or applying methods which consent to save water for its better uses. At a social level, it is necessary to considerate social net benefits. Maximization of social net benefits refers to the difference between farmers' returns and costs observed to produce their outputs, considering also the presence of externalities and opportunity costs in space and time (Wichelns 2010).

Economic efficiency in a production setting involves technical and allocative components (Allan 1999; Wichelns 2002). Production is technically efficient when the maximum possible output is generated with a given set of output, or when a selected output level is produced at minimum cost. Allocative efficiency describes the achievement of a specific goal with regard to the production process. Allocative efficiency concerns the minimization of production costs of a well-defined level of output using a precise set of inputs (Coelli et al. 2005).

Efficiency must be considered on different levels: even if a single user is not completely efficient, the entire system might gain benefits e.g. recycling those return flows coming from a particular use and employing them again for other ones (Molden and De Fraiture 2000). Total Basin Efficiency could be higher than the efficiency referred to a private use, or could be improved anyway, whether single users are totally efficient or not (FAO 2012).

Economic inefficiency is generated by a net loss that results from existing allocation and use decisions, and this loss has both private and public dimension, that can persist in irrigated area when production is not economically efficient. Policies that might encourage efficient production can be obtained by considering the consequences of poor water management at the farm-level and in regional water delivery systems.

Farm-level irrigation may be low because relative input and output prices require farmers to minimize water management expenditure or because farmers are prevented from achieving a desirable level of irrigation efficiency by constrains regarding the timing or availability of water and other key inputs. It is important to underline that the external (off-farm) economic consequences of low farm-level irrigation efficiencies include contamination of groundwater and surface water supplies with nutrient and other chemicals, waterlogging and salinization, and sediment loads entering streams and reservoirs. So, in this case higher regional expenditure may be required for operating and maintain regional water delivery system and for installing regional drainage system with greater capacity than might be required with higher farm-level irrigation efficiencies.

Necessary conditions for the optimal performance of regional water delivery system include well-defined water rights, infrastructure capable of providing the service embodied in the water rights and assigned responsibilities for all aspects of system operation (Perry 1995). Problems with cost recovery and inadequate maintenance also can reduce the efficiency of regional water delivery systems. The potential external consequences of poor delivery performances include the environmental problems generated by of low farm-level irrigation efficiencies. Furthermore, it can lead to inequitable distribution of water and income among farmers, while reducing aggregate production values.

4.5 *Policy Instruments for Efficiency and Sustainability*

There are many ways to define sustainability and sustainable development ranging from the very broad to the very narrow which creates potentials for misunderstanding (Dixon and Fallon 1989). We can define sustainability as the ability to continue extracting net positive social returns from a resource for an indefinite period of time.

The focus on ecosystems by environmentalists and on watersheds by hydrologists has carried the debate substantially above the commodity-based farm and farming systems level to land, water, and other highly valued natural and environmental resources. Lynam and Herdt (1999) argue that 'sustainability of common resource systems necessarily incorporates value judgements on multiple criteria over how the community wishes to utilize resources; moreover, sustainability of the system will depend more on social institutions controlling access and use than on production technologies.'

Many of the consequences of poor water management on farms and in regional delivery system will persist, over time, for a variety of reasons that involve inappropriate economic incentives, poorly defined property rights, incomplete information, lack of capital and individual planning horizons that are shorter than socially optimal perspectives.

Considering economic criteria is crucial when evaluating policy goals and instruments regarding water allocation issues and improvements in water management.

In general, policy instruments may operate in the agricultural product market or in factor markets. In any market, regulation may either be directed at the prices of goods or may dictate control over the quality, allocation, or allowable uses of output or factor goods. Each approach and each instrument have different implications for regulated sector, the legislators who develop the regulations, regulating agencies, and the consumers and other actors associated trough markets or external effects.

With respect to the specific case, policy recommendations for eliminating inefficiencies on farms and in regional delivery systems are derived from the reasons those inefficiencies persist. For example, policies that will motivate farm-level improvements in water management include:

- Improvements in the definition and enforcement of water rights in areas where those rights are uncertain or not secure;
- Water pricing or allocation strategies that reflect water scarcity;
- Water charges or restrictions that motivates reduction in the negative, off-farm impacts of irrigation and drainage;
- Removal of explicit and implicit taxation that reduces the portion of crop revenue retained by farmers;
- Low-interest loans and cost-share programs to support farm-level investments in water-saving irrigation methods;
- Programs that enhance farm-level access to complementary inputs, such as credit, fertilizer and pesticide.

Generally public policy instruments are classified in incentive-based policies and direct public action (Field 2008).

The incentive-based policies include market or property rights policies, that set the access to resources trough the institution of a system of property rights; taxes and subsidies for resources users are included. This second economic instruments are less diffused than the first ones. Between these are included the environmental taxation and water pricing. The only environmental taxation used in the water sector is represented by abstraction fees, that are due in exchange for the license. Subsidies are more frequently used. In many countries the public budget has subsidized, directly or indirectly, the most part of the water and sewerage infrastructure, as well as other environmental protection assets, not only for households, but for the productive sectors as well. To this category belong financial instruments that are used to cover costs of supplying water to irrigators, like water pricing. Water pricing might not be the ideal instrument to manage the most efficient allocation for the resource because responsiveness of farmers to changes in the price of water related to the elasticity of water demand is complex. But for many economists it is a way to obtain a financial income and cover at least part of the distribution costs (Savenije and Van der Zaag 2002).

The direct public action includes command-and-control policies and the direct public production. In the first case, public authorities establish direct controls on individual actions, enforcing these controls with standard legal enforcement practices. The basic policy instrument is represented by use license and authorizations. In the case of water, this instrument is applied for water abstractions, water discharges, works of any kind in the river territory; authorizations are also required for many activities that are potentially harmful for the water environment, such as the disposal of waste and the handling of pesticides. Direct public production occurs when public agencies themselves own natural resources and themselves pursue programs of production and distribution.

Generally, those instruments can be combined. And can be used by considering the nature of the resource to be ruled.

Finally, in some countries voluntary instruments are used, like 'management agreements', that normally involve subsidies. It is important to underlie that not always the existence of an externality requires government intervention. In many situations the involved parties may negotiate a solution that will address the externality problem and result in an efficient resource allocation. When this does not work, Government intervention, in the form of direct regulation, pollution charges, clean-up subsidies, etc., may be considered. If it is not guaranteed that intervention leads to improved efficiency. In some situation intervention is justified on distributional (equity) grounds. Even if an efficient solution could be reached through private or public means, that solution could be deemed to be suboptimal from a societal standpoint if it resulted in significant inequities in terms of income distribution or in the burden of regulation.

Economists argue that achieving efficiency should be the first objective of a policy since it results in the largest total benefit. If inequities result, then more benefits are available with which to achieve equity. Equity has to do with how the overall benefits and costs of natural resource use are distributed among subgroups of the overall population. And, as seen, one of the major issues is the balance among the generations, that is also the major focus of sustainability.

Economic efficiency and equity are important considerations in the allocation of water. Greater efficiency is required in the face of increasing water scarcity, and equity is a concern because of the importance of water to the livelihoods and well-being of rural communities.

Allocation of water can be socially suboptimal if there exist market failures and if government policy and associated institutional arrangements fail. Policy failure occurs where government regulatory instruments (e.g. taxes and exchange rates) or government policies create market price distortions that make it economically rational for individuals to use resources in a socially suboptimal manner (OECD 1999). Also, failure in sectoral policies can arise through inadequate consideration of impacts on other sectors, particularly about the environment (OECD 1994). Political failure can also occur through lack of government intervention and inadequate policy implementation.

5 Conclusions

Sustainable use of water depends on many natural factors, but mainly on man's choices and politics.

In this chapter, suggestions have been provided to manage water in a sustainable way.The correct determination of evapotranspiration occupies a fundamental role. This measurement is essential to dimension irrigation variables and to overcome the dangers arising from a water stress (loss of yield) or by overdose of watering volumes (water waste). More generally, the yield per unit of evapotranspirated water (WUE) is the indispensable indicator for assessing a crop or a farming system from an economic point of view (water productivity) and ecological (water footprint).

For practicing sustainable irrigation, monitoring of the soil's water status, or the use of plant water status indicators, is crucial to identifying the irrigation scheduling.

For any other strategy aimed at reducing water consumption, such as 'irrigation deficit', the crop physiology should be considered, by the knowledge of the phenological stages sensitivity to water scarcity. Plant physiology and plant water relationships guide agronomic options to improve the efficiency with which crops use water.

The sustainable water management in irrigated farming systems, and in semi-arid environments, is depending on the agronomic knowledge and on its transfer from the research centres down to the farm level. However, new research topics merit to be developed. Among them: the measure of the evapotranspiration of the multi-annual species (fruit trees are the most irrigated crops but those less studied in terms of water consumption), the exploitation of low-quality waters, the introduction of drought-resistant genes. The relationship between water use efficiency – root system architecture – physical and chemical characteristics of soil, the role of mycorrhizas, the consequences of air and soil pollution, adaptation to climate change.

Economic efficiency describes the conditions that must be satisfied to ensure that resources are used to maximize net benefits, and it is achieved when limited resources are allocated and used giving greatest net value. Economic inefficiency is generated by a net loss that results from existing allocation and use decisions, and this loss has both private and public dimension, that can persist in irrigated area when production is not economically efficient.

One of the most important institutional failures include inadequate availability of information for policy-makers (Burke et al. 1999). Therefore, in this contest the research also oriented to technical support to policy makers can help to achieve a more efficient and sustainable water management.

References

Allan T (1999) Productive efficiency and allocative efficiency: why better water management may not solve the problem. Agric Water Manag 40:31–35

Atzori G, Guidi Nissim W, Caparrotta S, Masi E, Azzarello E, Pandolfi C, Vignolini P, Gonnelli C, Mancuso S (2016) Potential and constraints of different seawater and freshwater blends as growing media for three vegetable crops. Agric Water Manag 176:255–262

Ballesteros R, Ortega JF, Hernández D, Moreno MA (2014) Applications of georeferenced high-resolution images obtained with unmanned aerial vehicles. Part I: description of image acquisition and processing. Precis Agric 15:579–592

Beltrán-Przekurat A, Pielke RA, Eastman JL, Coughenour MB (2012) Modelling the effects of land-use/land-cover changes on the near-surface atmosphere in southern South America. Int J Climatol 32:1206–1225

Ben Nouna B, Katerji N, Mastrorilli M (2000) Using the CERES-Maize model in a semi-arid Mediterranean environment. Evaluation of model performance. Eur J Agron 13:309–322

Besanko D, Braeutigam RR (2011) Microeconomics. Wiley, Hoboken, pp 26–42

Bethenod O, Le Corre M, Huber L, Sache I (2005) Modelling the impact of brown rust on wheat crop photosynthesis after flowering. Agric For Meteorol 131:41–53

Blum A (2009) Effective use of water (EUW) and not water-use efficiency (WUE) is the target of crop yield improvement under drought stress. Field Crop Res 112:119–123

Bou Jaoudé M, Katerji N, Mastrorilli M, Rana G (2008) Analysis of the ozone effect on soybean in the Mediterranean region. II – the consequences on growth, yield and water use efficiency. Eur J Agron 28:519–525

Brouwer R, Pearce DW (2005) Cost-benefit analysis and water resources management. Edward Elgar Publishing Lim, Cheltenham

Burke S, Mulligan M, Thomes JB (1999) Optimal irrigation efficiency for maximum plant productivity and minimum water loss. Agric Water Manag 40:377–391

Calabresi G, Melamed AD (1972) Property rules, liability rules, and inalienability: one view of the cathedral. Harv Law Rev 85:1089–1128

Campi P, Palumbo AD, Mastrorilli M (2009) Effects of tree windbreak on microclimate and wheat productivity in a Mediterranean environment. Eur J Agron 30:220–227

Campi P, Modugno F, Mastrorilli M, Tomei F, Villani G, Marletto V (2014) Evapotranspiration of tomato simulated with the CRITERIA model. Ital J Agron 2:93–98

Carlson GA, Zilberman D, Miranowski JA (1993) Agricultural and environmental resource economics. Oxford University Press, New York

Casa R, Castaldi F, Pascucci S, Pignatti S (2015) Chlorophyll estimation in field crops: an assessment of handheld leaf meters and spectral reflectance measurements. J Agric Sci 153:876–890

Castrignanò A, Buttafuoco G, Pisante M, Lopez N (2006) Estimating within-field variation using a nonparametric density algorithm. Environmetrics 17:465–481

Castrignanò A, Fiorentino C, Buttafuoco G, Troccoli A, Pisante M, Basso B (2008) A combined approach to delineating management zones for precision agriculture. Ital J Agron 3:147–148

Ceccarelli S, Grando S, Baum M, Udupa SM (2004) Breeding for drought resistance in a changing climate. In: Baker FWG (ed) Drought resistance in cereals. CAB International, Wallingford, pp 167–190

Chen DX, Coughenour MB (2004) Photosynthesis, transpiration, and primary productivity: scaling up from leaves to canopies and regions using process models and remotely sensed data. Glob Biogeochem Cycles 18:1–15

Chenoweth J, Hadjikakou M, Zoumides C (2014) Quantifying the human impact on water resources: a critical review of the water footprint concept. Hydrol Earth Syst Sci 18:2325–2342

Coelli TJ et al (2005) An introduction to efficiency and productivity analysis, 2nd edn. Springer, New York

Darzi-Naftchali A, Ritzema H, Karandish F, Mokhtassi-Bidgoli A, Ghasemi-Nasr M (2017) Alternate wetting and drying for different subsurface drainage systems to improve paddy yield and water productivity in Iran. Agric Water Manag 193:221–231

de Fraiture C, Wichelns D, (Coordinating Lead Authors) (2007) Looking ahead to 2050: scenarios of alternative investment approaches. In: Molden D (ed) Water for food, water for life: a comprehensive assessment of water management in agriculture. International Water Management Institute/Earthscan, Colombo/London, pp 91–145

De Groot RBA, Hermans LM (2009) Broadening the picture: negotiating payment schemes for water-related environmental services in the Netherlands. Ecol Econ 68:2760–2767

Deng XP, Shan L, Zhang H, Turner NC (2006) Improving agricultural water use efficiency in arid and semiarid areas of China. Agric Water Manag 80:23–40

Dixon JA, Fallon LA (1989) The concept of sustainability: origins, extensions, and usefulness for policy. Soc Nat Resour 2:73–84

Dwyer G, Heaney A, Beare S, Peterson D, Pechey L (2006) Third-party effects of water trading and potential policy responses. Aust J Agric Resour Econ 50:277–293

Fereres E, Soriano MA (2007) Deficit irrigation for reducing agricultural water use. J Exp Bot 58:147–159

Fereres E, Orgaz F, Gonzalez-Dugo V, Testi L, Villalobos FJ (2014) Balancing crop yield and water productivity tradeoffs in herbaceous and woody crops. Funct Plant Biol 41:1009–1018

Field CB (2008) Natural resource economics. Waveland Press

Fiscus EL, Booker FL, Burkey KO (2005) Crop responses to ozone: uptake, modes of action, carbon assimilation and partitioning. Plant Cell Environ 28:997–1011

Food and Agricultural Organization (FAO) (2009) Law for Water Management: a guide to concepts and effective approaches, FAO Legislative Study 101. http://www.fao.org/docrep/012/i1284e/i1284e.pdf

Food and Agricultural Organization (FAO) (2012) Coping with water scarcity. An action framework for agriculture and food security, FAO Water Rep 38. http://www.fao.org/docrep/016/i3015e/i3015e.pdf

French AN, Alfieri JG, Kustas WP, Prueger JH, Hipps LE, Chávez JL, Evett SR, Howell TA, Gowda PH, Hunsaker DJ, Thorp KR (2012) Estimation of surface energy fluxes using surface renewal and flux variance techniques over an advective irrigated agricultural site. Adv Water Resour 50:91–105

Gang C, Wang Z, Zhou W, Chen Y, Li J, Chen J, Qi J, Odeh I, Groisman PY (2016) Assessing the spatiotemporal dynamic of global grassland water use efficiency in response to climate change from 2000 to 2013. J Agron Crop Sci 202:343–354

Hanemann WH (2006) The economic conception of water. In: Rogers PP, Llamas MR, Martinez-Cortina L (eds) Water crisis: myth or reality. Taylor and Francis, London

Hsiao T, Steduto P, Fereres E (2007) A systematic and quantitative approach to improve water use efficiency in agriculture. Irrig Sci 25:209–231

Hueso JJ, Cuevas J (2010) Ten consecutive years of regulated deficit irrigation probe the sustainability and profitability of this water saving strategy in loquat. Agric Water Manag 97:645–650

International Conference on Water and the Environment (ICWE) (1992) The Dublin statement on water and sustainable development, Dublin UN (United Nations). http://www.un-documents.net/h2o-dub.htm

Katerji N, Mastrorilli M (2009) The effect of soil texture on the water use efficiency of irrigated crops: results of a multi-year experiment carried out in the Mediterranean region. Eur J Agron 30:95–100

Katerji N, Rana G (2006) Modelling evapotranspiration of six irrigated crops under Mediterranean climate conditions. Agric For Meteorol 138:142–155

Katerji N, Rana G (2014) FAO-56 methodology for determining water requirement of irrigated crops: critical examination of the concepts, alternative proposals and validation in Mediterranean region. Theor Appl Climatol 116:515–536

Katerji N, van Hoorn JW, Hamdy A, Mastrorilli M (2003) Salinity effect on crop development and yield, analysis of salt tolerance according to several classification methods. Agric Water Manag 62:37–66

Katerji N, Mastrorilli M, Rana G (2008) Water use efficiency of crops cultivated in the Mediterranean region: review and analysis. Eur J Agron 28:493–507

Katerji N, Mastrorilli M, Cherni HE (2010) Effects of corn deficit irrigation and soil properties on water use efficiency. A 25-year analysis of a Mediterranean environment using the STICS model. Eur J Agron 32:177–185

Katerji N, Campi P, Mastrorilli M (2013) Productivity, evapotranspiration, and water use efficiency of corn and tomato crops simulated by AquaCrop under contrasting water stress conditions in the Mediterranean region. Agric Water Manag 130:14–26

Kaune A, Werner M, Rodríguez E, Karimi P, de Fraiture C (2017) A novel tool to assess available hydrological information and the occurrence of sub-optimal water allocation decisions in large irrigation districts. Agric Water Manag 191:229–238

Keller A, Keller J (1996) Integrated water resources systems: theory and policy implications, Research report 3. International Irrigation Management Institute, Colombo

Liu Y, Bowman BC, Hu Y-G, Liang X, Zhao W, Wheeler J, Klassen N, Bockelman H, Bonman JM, Chen J (2017) Evaluation of agronomic traits and drought tolerance of winter wheat accessions from the USDA-ARS National Small Grains Collection. Agronomy 7:51

Losciale P, Chow WS, Corelli Grappadelli L (2010) Modulating the light environment with the peach 'asymmetric orchard': effects on gas exchange performances, photoprotection, and photoinhibition. J Exp Bot 61:1177–1192

Losciale P, Zibordi M, Manfrini L, Morandi B, Bastias RM, Corelli Grappadelli L (2011) Light management and photoinactivation under drought stress in peach. Acta Hortic 922:341–347

Lovelli S, Pizza S, Caponio T, Rivelli AR, Perniola M (2005) Lysimetric determination of muskmelon crop coefficients cultivated under plastic mulches. Agric Water Manag 72:147–159

Luquet D, Vidal A, Smith M, Dauzat J (2005) 'More crop per drop': how to make it acceptable for farmers? Agric Water Manag 76:108–119

Lynam JK, Herdt RW (1999) Sense and sustainability: sustainability as an objective in international agricultural research. Agric Econ 3:381–398

Malik RPS (2008) Towards a common methodology for measuring irrigation subsidies. The Global Subsidies Initiative, International Institute for Sustainable Development, Geneva. www.globalsubsidies.org

Mariolakos I (2007) Water resources management in the framework of sustainable development. Desalination 213:147–151

Mastrorilli M, Zucaro R (2016) Towards sustainable use of water in rainfed and irrigated cropping systems: review of some technical and policy issues. AIMS Agric Food 1:294–314. https://doi.org/10.3934/agrfood.2016.3.294

Mastrorilli M, Campi P, Colucci R, Di Bari V (2002) Double harvest: an agronomic strategy to increase the water use efficiency of biomass crops in Mediterranean Europe. In: Villalobos FJ, Testi L (eds) Book of proceedings VII congress of the European society for agronomy, Cordoba pp 113–114

Molden D, De Fraiture C (2000) Major paths to increasing the productivity of irrigation water. In: International Water Management Institute (IWMI). World Water Supply and Demand: 1995–2025. IWMI, Colombo, pp 41–50

Molden D, Oweis T, (coordinating leading authors) (2007) Pathways for increasing water productivity. In: Molden D (ed) Water for food, water for life: a comprehensive assessment of water management in agriculture. International Water Management Institute/Earthscan, London/Colombo, pp 279–310

Molden D, Oweis TY, Steduto P, Kijne JW, Hanjra MA, Bindraban PS, Bouman BAM, Cook S, Erenstein O, Farahani H, Hachum A, Hoogeveen J, Mahoo H, Nangia V, Peden D, Sikka A, Silva P, Turral H, Upadhyaya A, Zwart S (2013) Pathways for increasing agricultural water productivity. Water for food water for life: a comprehensive assessment of water management in agriculture, pp 279–314

OECD (1994) Managing the environment: the role of economic instruments. Organisation for Economic Co-operation and Development, Paris

OECD (1999) Agricultural water pricing in OECD countries. OECD, Paris. www.olis.oecd.org/olis/1998doc.nsf/LinkTo/env-epoc-geei(98)11-final

Oweis T (1997) Supplemental irrigation. A highly efficient water use practice. ICARDA editions

Oweis T, Zhang H, Pala M (2000) Water use efficiency of rainfed and irrigated bread wheat in a Mediterranean environment. Agron J 92:231–238

Pendergrass AG, Gerber EP (2016) The rain is askew: two idealized models relating vertical velocity and precipitation distributions in a warming world. J Clim 29:6445–6462

Perry CJ (1995) Determinants of function and dysfunction in irrigation performance, and implication for performance improvement. Int J Water Res Dev 11:25–35

Perry CJ (1999) The IWMI water resources paradigm: Definitions and implications. Agric Water Manag 40:45–50

Perry C (2014) Water footprints: path to enlightenment, or false trail? Agric Water Manag 134:119–125

Perry CJ, Rock M, Seckler D (1997) Water as an economic good: a solution, or a problem? Research report 14. International Irrigation Management Institute, Colombo

Pisante M, Stagnari F, Grant CA (2012) Agricultural innovations for sustainable crop production intensification. Ital J Agron 7:300–311

Pisante M, Stagnari F, Acutis M, Bindi M, Brilli L, Di Stefano V, Carozzi M (2015) Conservation agriculture and climate change. Conservation agriculture, pp 579–620

Rai R, Singh AA, Agrawal SB, Agrawal M (2016) Tropospheric O3: a cause of concern for terrestrial plants. Plant responses to air pollution, pp 165–195

Ramazzotti S, Stagnari F, Pisante M (2008) Integrated soil and water management for vineyards in southern Italy: a case study. Ital J Agron 3:117–118

Rana G, Katerji N (2000) Measurement and estimation of actual evapotranspiration in the field under Mediterranean climate: a review. Eur J Agron 13:125–153

Rinaldi M, Castrignanò A, De Benedetto D, Sollitto D, Ruggieri S, Garofalo P, Santoro F, Figorito B, Gualano S, Tamborrino R (2015) Discrimination of tomato plants under different irrigation regimes: analysis of hyperspectral sensor data. Environmetrics 26:77–88

Ritchie J, Basso B (2008) Water use efficiency is not constant when crop water supply is adequate or fixed: the role of agronomic management. Eur J Agron 28:273–281

Rockström J, Williams J, Daily G, Noble A, Matthews N, Gordon L, Wetterstrand H, DeClerck F, Shah M, Steduto P, de Fraiture C, Hatibu N, Unver O, Bird J, Sibanda L, Smith J (2017) Sustainable intensification of agriculture for human prosperity and global sustainability. Ambio 46:4–17

Rogers P, Bhatia R, Huber A (1998) Water as a social and economic good: how to put the principle into practice, Technical Advisory Committee (TAC) Background Papers No. 2. Global Water Partnership (GWP), Stockholm. www.gwpforum.org/gwp/library/TAC2.PDF

Rogers P, de Silva R, Bhatia R (2002) Water is an economic good: how to use prices to promote equity, efficiency, and sustainability. Water Policy 4:1–17

Rubino P, Stellacci AM, Rana RM, Catalano M, Caliandro A (2013) A non-linear model for optimal allocation of irrigation water and land under adequate and limited water supplies: a case study in southern Italy. Irrig Drain 62:145–155
Sadler EJ, Evans RG, Stone KC, Camp CR (2005) Opportunities for conservation with precision irrigation. J Soil Water Conserv 60:371–379
Sadras VO, Angus JF (2006) Benchmarking water use efficiency of rainfed wheat in dry environments. Aust J Agric Res 57:847–856
Savenije HHG, Van der Zaag P (2002) Water as an economic good and demand management paradigms with pitfalls. Water Int 27:98–104
Seckler D (1996) The new era of water resources management: from 'dry' to 'wet' water savings. Issues in agriculture; no. IAG 8. The World Bank, Washington, DC
Solomon KH, Burt CM (1999) Irrigation sagacity: a measure of prudent water use. Irrig Sci 18:135–140
Stagnari F, Jan S, Galieni A, Pisante M (2016) Sustainable agricultural practices for water quality protection. In: Ahmad P (ed) Water stress and crop plants: a sustainable approach. Wiley, Chichester, pp 75–85
Steduto P, Katerji N, Puertos-Molina H, Ünlü M, Mastrorilli M, Rana G (1997) Water-use efficiency of sweet sorghum under water stress conditions. Gas-exchange investigations at leaf and canopy scales. Field Crop Res 54:221–234
Stellacci AM, Castrignanò A, Diacono M, Troccoli A, Ciccarese A, Armenise E, Gallo A, De Vita P, Lonigro A, Mastro MA, Rubino P (2012) Combined approach based on principal component analysis and canonical discriminant analysis for investigating hyperspectral plant response. Ital J Agron 34:247–253
Thompson M (2006) National water initiative – the economics of water management in Australia – an overview. In: Water and agriculture: sustainability, markets and policies. OECD, Paris, pp 81–93. www.oecd.org/tad/env
Trethowan RM (2014) Delivering drought tolerance to those who need it: from genetic resource to cultivar. Crop Pasture Sci 65:645–654
Turner NC (2004) Agronomic options for improving rainfall-use efficiency of crops in dryland farming systems. J Exp Bot 55:2413–2425
Turral H, Svendsen M, Faures JM (2010) Investing in irrigation: reviewing the past and looking to the future. Agric Water Manag 97:551–560
United Nations Educational, Scientific and Cultural Organization (UNESCO) (2015) The United Nations World Water Development Report 2015: water for a sustainable world. WWAP (World Water Assessment Programme). UNESCO, Paris. http://unesdoc.unesco.org/images/0023/002318/231823E.pdf
Van der Zaag P, Savenije HHG (2006) Water as an economic good: the value of pricing and the failure of markets, Value of water research report series 19. UNESCO-IHE, Delft. http://waterfootprint.org/media/downloads/Report_19_Water_as_an_Econ_Good_1.pdf
Wichelns D (2002) An economic perspective on the potential gains from improvements in irrigation water management. Agric Water Manag 52:233–248
Wichelns D (2010) Agricultural water pricing: United States, OECD. www.oecd.org/water
Zwart SJ, Bastiaanssen WGM (2004) Review of measured crop water productivity values for irrigated wheat, rice, cotton, and maize. Agric Water Manag 69:115–133

Sustainable Nutrient Management

Abdul Rehman, Aman Ullah, Faisal Nadeem, and Muhammad Farooq

Abstract Agriculture production has substantially increased since green revolution due to introduction of modern cultivars and inputs. Organic fertilizers are key contributor to achieve high yield targets in sustainable way. From the last a few decades the uses of inorganic fertilizers have been increased to get higher yield. Low soil fertility is one of the major reasons for low crop production. However, under or over application of fertilizers and selection of wrong nutrient source causes nutrient imbalance in soil. Moreover, high application of inorganic fertilizers and unbalanced fertilization has reduced the nutrient use efficiencies (NUE) with high cost of production and environmental risks. Therefore, better NUE can reduce the fertilizer cost and environmental risks. This chapter discusses the challenges to sustainable nutrient management. Moreover, use of approaches for sustainable nutrient management including appropriate soil testing technique, fertilizer sources (organic, inorganic, biofertilziers and nanofertilizers) and application method in right combination using site specific nutrient management will reduce the fertilizer losses with high NUE and economic yield.

Keywords Organic · Fertilizer · Nutrient use efficiency · Biofertilizers and nanofertilizers

---

A. Rehman
Department of Crop Sciences and Biotechnology, Dankook University, Cheonan-si, South Korea

A. Ullah · F. Nadeem
Department of Agronomy, Faculty of Agriculture, University of Agriculture, Faisalabad, Pakistan

M. Farooq (✉)
Department of Crop Sciences, College of Agricultural and Marine Sciences, Sultan Qaboos University, Al-Khoud, Oman

Department of Agronomy, Faculty of Agriculture, University of Agriculture, Faisalabad, Pakistan

The UWA Institute of Agriculture, The University of Western Australia, Crawley, WA, Australia
e-mail: farooqcp@squ.edu.om

M. Farooq, M. Pisante (eds.), *Innovations in Sustainable Agriculture*,
https://doi.org/10.1007/978-3-030-23169-9_7

1 Introduction

Nutrient management involves the practices linked to plant, edaphic and environmental factors with irrigation, water and soil conservation practices to attain optimal crop yield, crop quality, nutrient use efficiency and economic benefits while decreasing the nutrient losses (Delgado and Lemunyon 2006). It includes matching of edaphic and environmental factors to rate, time, source and place of nutrient application. The rising population and consumption, and reduction in available land and other productive units are placing unprecedented pressure on the current agriculture and natural resources to meet the increasing food demand. Achieving food security under sustainable systems poses a significant challenge in the developing world and is highly critical for alleviating poverty. To circumvent this challenge, crop producers tended to overuse certain inputs such as chemical fertilizers and pesticides which in turn have already started deteriorating environment.

Arable soil usually lacks plant nutrient in sufficient quantity to achieve higher and sustainable yield goals. Application of fertilizers and nutrient availability are closely associated with higher crop yield (Kaur et al. 2008) as plant nutrition is very crucial to maintain the productivity and quality of soil (Jaga and Patel 2012). Chemical fertilizers help to maintain soil productivity by ensuring supply of vital plant nutrients and thus help in economic crop production. In most of the countries demand of chemical fertilizer is increasing due to introduction of new high yield and intensive input requiring crop cultivars. For instance, the use of major fertilizers (nitrogen (N), phosphorus (P), potassium (K)) has increased up to sixfolds since green revolution (FAO 2014). The use of fertilizers is tremendously increasing as the annual demand of N, P and K is rising by 1.4%, 2.2% and 2.6% annually (FAO 2015). Fertilizer use has also increased in developing world as overall the growth in annual use of fertilizer is higher in Africa (3.6%) and Sub-Saharan Africa (4.7%) than developed countries. Most of the fertilizer demand/consumption is higher in Asia. For instance, N, P and K fertilizer consumption is highest in South Asia (24.5%, 31.3 and 19.3%) and East Asia (29.1%, 19%, and 35.8%) respectively than rest of the world (FAO 2015).

The manufacturing of fertilizers causes serious threat to environment as from mining to manufacturing; different harmful chemicals are released into the air, water and soil. For instance, emission of ammonia, fluorine, oxides of sulphur and nitrogen, acid mists, fertilizer dust and harmful radiations are emitted from the fertilizer manufacturing units causing major environment pollution (Li et al. 2013; Ju et al. 2014). This high use of fertilizers has also posed serious threats to environment. Maintaining agricultural production, while minimizing pollution to water and air, is a global problem. Direct emissions from agriculture comprises roughly 11% of global greenhouse gas emissions and these emissions are projected to rise by 20% by 2030 (US-EPA 2011). Including indirect emissions increases the total emissions from agriculture to 19–29% of the global total (Vermeulen et al. 2012).

Anthropogenic activities have profoundly altered the global nitrogen and phosphorus cycles and will continue to do so (Bouwman et al. 2009). Net anthropogenic nitrogen inputs in China, US and Northern Europe are estimated at between 2 and 3.5 t ha^{-1} of which 15–30% is exported in rivers (Swaney et al. 2012). Indeed, studies across the globe have shown agriculture to be amongst the largest contributor of annual nitrate and phosphate loads to river waters (Liu et al. 2012).

Common field and farm management activities affecting diffuse pollution include the over-application of fertilizer (Withers et al. 2001), the inappropriate application of manure or slurry to land (Shepherd and Chambers 2007), or poor management of soil leading to erosion and surface runoff on both livestock and arable farms (Quinton et al. 2010). In this scenario, sustainable nutrient management approach will not only maintain the crop production but will also reduce the environmental pollution through over use of fertilizers. Sustainable nutrient management approach use the combination of well tested practices and principles of modern and traditional technologies in an integrated manner aimed at profitable crop production with better crop quality, nutrient use efficiencies and lower environmental pollution using crop management (crop rotation, intercropping), soil management (manures, green manures, organic fertilizers, nano fertilizers and crop residues), site specific nutrient application to fulfill the crop nutrient demand (Fig. 1). In this chapter, sustainable nutrient management approaches including soil management, crop management, fertilizer sources (organic, inorganic) and their application methods, site specific and integrated nutrient management practices and challenges to sustainable nutrient management are discussed.

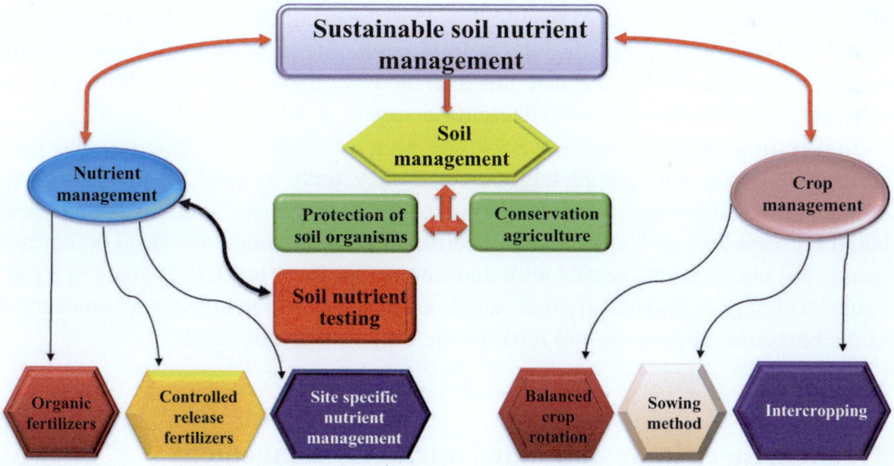

Fig. 1 Pillars for sustainable nutrient management

2 Soil-Testing for Sustainable Nutrient Management

Optimal crop growth and yield depends upon the availability of essential and some non-essential (Si, Se, Co etc.) crop nutrients. Multiplicity in the crop nutrient demand, fertilizer combination using specific formulation of nutrients can increase crop yield from one to tenfolds, depending upon the crop and nutrient (Dimkpa et al. 2017). For better crop production a certain concentration of these nutrients should be present in soil that can be taken up by plants. However, soil physiochemical properties and moisture availability influence the availability of these nutrients (Marschner 2012). Moreover, microbes present in rhizosphere also influence the nutrient availability. Therefore, a comprehensive soil testing system is very crucial to determine soil nutrient status considering the biotic and abiotic factors that can influence nutrient dynamics.

The nutrient dynamics keep changing in the soil from fixation to dissolution in soil solution and uptake and translocation to shoot. Classical soil testing methods usually predict nutrient uptake to their presence in the soil solution. However, this is not true in all the cases. For instance, in Zn deficient soils fractions of Zn interacting with root surface are better indicator of Zn availability than Zn present in soil solution (Duffner et al. 2013). After estimation of soil physiochemical properties then next step is fertilizer recommendation based on these tests. However, soils having more than one nutrient deficiency, the fertilizer recommendation are not easy (Oliver and Gregory 2015; Voortman and Bindraban 2015) as identification of right balance between quantity and composition and their availability to plants pose serious challenge. Moreover, nutrient ratios in soil are very critical as excess of one nutrient can dilute the other nutrient. For example, urea is alkaline in nature and can affect the micronutrient (Zn) availability (Milani et al. 2015). Lime treatment to acidic soil based on soil testing my help to overcome the problem of low pH and release magnesium (Mg) and calcium (Ca) also. Similarly, use of acidic fertilizers (HH_4SO_4) can enhance the iron (Fe) and Zn supply in alkaline soils.

Soil testing methods may not solve the issue of fertilizer availability and suitability to specific soils completely; however, they serve as basis for fertilizers recommendation and can help in formulation of suitable fertilizer selection and nutrient blend for a specific soil. Conclusively, fertilizer recommendations based on harmonizing soil chemical properties with nutrient products for nutrient balance in a particular soil may or may not right all the time. However, rapid nutrient testing serves as the basis to maintain the soil fertility and crop nutrient demand.

3 Challenges in Sustainable Nutrient Management

Plant nutrition is the key factor that influences soil quality and productivity (Jaga and Patel 2012). Fertilizers maintain the productivity and fertility of soil by furnishing essential nutrients and ultimately result in economic crop production. However, the rise in fertilizer demand and over use of fertilizer also pose serious threats to sustainable nutrient management and environmental health. Fertilizer use efficiency is low for most of the agriculture soils, therefor, for sustainable nutrient management the nutrient use efficiencies should be improved (Fixen 2009). For instance, about half of the applied N is only used by plants while remaining N is bound in organic form (15–25%) in soil, volatilization (2–20%) and leaching (2–10%) into ground water (Sonmez et al. 2007; Chien et al. 2009). The nitrogen use efficiency (NUE) is even lower in some parts of the world. For instance, in China, two decades ago NUE for major cereal crops was 28–41% (Zhu 1992), which has declined to 26.1%, 28.2% and 28.3% in maize, wheat and rice respectively during last decade (Wang 2007). The NUE in some of the farmer field's in north China plain is reported to be 15% and 18% for summer maize and winter wheat (Cui et al. 2008). Furthermore, P recovery is also very low as only one-fourth of the applied P was recovered during crop growing season. It also precipitates with oxides of Al and Fe in acidic soils (Vance et al. 2003) and with Ca and Mg in calcareous soils (Rahmatullah et al. 1994) with further decline P use efficiency.

Mismanagement or over use of chemical fertilizers has resulted in low nutrient uptake and use efficiencies. For instance, N losses through leaching (NO_3), volatilization (NH_3), nitrification/de-nitrification (N_2O/N_2) and emission of NO causing serious environmental issues (Zhu and Chen 2002; Ju et al. 2009). Nevertheless, P is most lost through surface runoff or erosion while losses due to subsurface leaching are very low. Organic P have more subsurface leaching when it is in inorganic form as it is more soluble (Aziz et al. 2015). Potassium use efficiency is also low due to K losses through drainage water in acidic and sandy soils receiving high rainfall (Havlin et al. 1999).

Nutrient budget calculation has showed that overuse of fertilizers have resulted in accumulation of nutrient in soil. Nitrogen and phosphorus budget calculation in China showed that N and P which were deficient in 1950s are now surplus. However, the budget of K and micronutrient is mostly negative around the globe which causes nutrient imbalance and also reduce the chances of yield improvement due to better N and P use efficiencies. Moreover, overuse of macronutrient particular N and P is due to high yield targets by the farmers, and unavailability of suitable nutrient sources. Mostly fertilizers are applied manually which reduces the fertilizer use efficiencies as most of the farmers have small land holdings and they don't afford soil testing and modern nutrient application technologies.

Application of organic fertilizer only is also not effective as; higher application of organic fertilizers can change the nutrient dynamics in soil and their availability to plants. For instance, FYM increase the level of P, K, Ca and organic matter in surface soil wile nitrate, Ca and Mg level rises in subsurface soil (Edmeades 2003), which can lead to higher N losses (Goulding et al. 2008). Moreover, it is difficult to predict the mineralization of nutrient from different types of manures in different cropping systems which can result in under or over fertilization (Zhao et al. 2010). Therefore, use of appropriate combination of fertilizers (organic, inorganic) and their application at right time, right place and suitable rates can help in reducing the nutrient losses with higher use efficiencies.

4 Fertilizer Source

Along with soil and crop management; selection of fertilizer source is very critical for sustainable nutrient management for long term ecosystem sustainability and food security. The presence of widespread nutrients deficiency in the soils causes great economic losses to farmers and considerably decreases the quantity and nutritional quality of grains both for human beings and livestock. The application of fertilizers can enhance the crop productivity; however, the available nutrients present in the bulk chemical forms are not fully accessible to plants and their utilization is very low owing to their inversion to insoluble form in the soil (Solanki et al. 2015). The use of chemical fertilizers in large quantity to increase crop productivity in long run is not suitable option as in one direction these increase crop production but on the other direction disturb the soil mineral balance, soil fertility, soil structure, mineral cycles, soil fauna and flora and food chains across ecosystems leading to heritable mutations (Solanki et al. 2015). There is need to adopt a system which has smart delivery system, targeted application and in long run should be sustainable.

4.1 Chemical Fertilizers

To increase and sustain food production the continuous fertilizer inputs are needed but there are problems with continued use of mineral fertilizers because of low nutrient uptake by crops in productive systems (Trenkel 1997). The high fertilizers application rates led to losses with negative impacts on atmospheric greenhouse gas concentration and water quality (Haygarth et al. 2013). Sustainable intensification with target to increase production on existing land area is a best option (Godfray et al. 2010). To keep the sustainability of agricultural and biogeochemical cycles there is need to develop nutrient efficient fertilizers which have high nutrient use efficiency. The nutrient efficient fertilizers include (i) controlled release fertilizers *vis* coated fertilizers, slow release fertilizers or uncoated fertilizers, (ii) nano fertilizers. These controlled release fertilizers have high efficiency owing to slow release of nutrients according to the crop demand and duration of the crop.

4.1.1 Coated Fertilizers

Excessive use of fertilizers causes problems especially with grown plants because roots are confined to small volumes, and the storage capacity of growth media for nutrients and water are limited. Frequent irrigation and fertilization are necessary to maintain the soil moisture and nutrient level, which may enhance leaching and runoff losses (Oertli 1980). Therefore, it is very important to select a proper fertilizer type, rate, and application technique to match the plant's nutrient and growth requirements as precisely as possible (Trenkel 2012). This can be achieved by using coated fertilizers compared to conventional fertilizers.

The population growth worldwide has increased the demand for food and to meet this demand a large area of fertile land is required to produce more food (Irfan et al. 2018). However, this fertile agricultural land is reduced owing to industrialization, urbanization and soil degradation (Chen et al. 2002; Brown et al. 2009; Gomiero 2016). To grow required food on diminished agricultural land a massive quantity of fertilizers are needed due to poor supply of nutrients (Irfan et al. 2018). The common solid fertilizers as uncoated or pristine granules have limitation as the release of nutrients from granules is fast and are vulnerable to losses in the form of leaching, volatilization and surface run-off (Bhat et al. 2011). Moreover, plants in their early/infancy stages cannot uptake all the supplied nutrients through fertilizers, so surplus nutrients are leached into the water table, causing problem for the aquatic life and cause economic losses (Vashishtha et al. 2010). To lessen these issues, one promising option is controlled-release fertilizers. There are two types of controlled-release fertilizers as (i) coated fertilizers (ii) uncoated fertilizers or slow-release fertilizers (Scherer 2005).

In coated fertilizers different types of impermeable coatings with small holes are used by which solubilized materials diffuse and semipermeable coatings through which water diffuses until the internal osmotic pressure raptures the coating (Scherer 2005). Coatings functions only as a physical barriers or a source of plant nutrient. The coating materials used in fertilizers are waxes, polymers, and sulfur. Osmocotes are covered with a plastic shell which allow the water to diffuse into the shell and tears the shell and nutrient diffuse into the soil. In sulfur-coated urea water vapor transfers through sulfur coating solubilizes the urea within the shell and builds sufficient osmotic pressure to disrupt the coating and urea is release (Scherer 2005).

The controlled release fertilizers are usually coated with organic polymers, modified biopolymers, natural macromolecule materials or nanocomposites. The coated film helps to achieve controlled, extended release rather than immediate release by providing the transport barrier to the fast dissolution of nutrients in the water when exposed without a coating (Salman 1989). The characteristics of coating materials are therefore important to get delayed or controlled release of nutrients (Table 1; Irfan et al. 2018).

The release process of coated fertilizers includes transport of water through coating, condensation of water molecules on the surface of nutrient core, dissolution of the active nutrient, development of osmotic pressure, swelling of controlled release fertilizers granule, and at the end the release of nutrient by transport through coating film via micro-pores (Irfan et al. 2018). The slow-release fertilizers (SRF) especially polymer-coated fertilizers improve the nutrient use efficiency and crop yield (Table 1; Shoji et al. 2001).

In a study, Tomaszewska and Jarosiewicz (2002) reported that the use of polysulfone as coating decrease the release rate of fertilizers and with the decrease of coating porosity the nutrient release rate further decrease. In case of coating with 38.5% porosity after 5 h 100% of NH_4^+ was released whereas in 11% porosity only 19% of NH_4^+ was released after 5 h.

Table 1 Effect of slow release fertilizer on the nutrient release and use efficiency

Nutrient	Compound name	Coating material	Formulation	Release time/ release amount	Increase in nutrient use efficiency (%)	Reference
Nitrogen	Sulfur-coated urea	Sulfur	–	–	–	Choi and Meisen (1997)
Nitrogen	Polyolefin-coated urea	Polyolefin	–	–	79	Shoji and Kanno (1994)
Nitrogen	Polyethylene- coated urea	Polyethylene	–	1–4 months	–	Wei et al. (2017)
NPK	Polysulfone coated NPK-fertilizers	Polysulfone	NPK 06-20-30 with 18% polysulfone coating	After 5 h 19% NH_4^+, 8.7% P_2O_5, and 3.8% K^+	–	Tomaszewska and Jarosiewicz (2002)
Nitrogen	Polymer coated urea fertilizers	DVB	N-DVB 20 g N and 0.38 mL DVB	82.3% 45th day	–	Abraham and Rajasekharan Pillai (1996)
Nitrogen	Polymer coated urea fertilizers	NNMBA	N-NNMBA 20 g N and 0.23 g NNMBA	85.6% 45th day	–	Abraham and Rajasekharan Pillai (1996)
Nitrogen	Polymer coated urea fertilizers	PETA	N-PETA 20 g N and 0.50 mL PETA	88.9% 45th day	–	Abraham and Rajasekharan Pillai (1996)
Nitrogen	Polymer coated urea fertilizers	TTEGDA	N-TTEGDA 20 g N and 0.41 mL TTEGDA	81.2% 45th day	–	Abraham and Rajasekharan Pillai (1996)
Nitrogen	Urea coated starch-g-PLLA fertilizers	Starch-g-PLLA	Urea 0.2 g and starch-g-PLLA 2.0 g	100% at 26 h	–	Chen et al. (2008)
Nitrogen	Urea coated starch-g-poly(vinyl acetate)	Starch-g-poly(vinyl acetate)	Urea 0.2 and starch-g-poly(vinyl acetate) 2.0 g	78% at 30 days	–	Niu and Li (2012)

Phosphate	Monoamonium phosphate coated polyethylene/paraffin waxes fertilizers	Polyethylene/ paraffin waxes	Monoamonium phosphate 18 g and polyethylene/ paraffin waxes 2 g	45 days paraffin wax,	–	Al-Zahrani (2000)
				52 days with polyethylene wax		
Phosphate	Diamonium phosphate coated polyethylene/paraffin waxes fertilizers	Polyethylene/ paraffin waxes	Diamonium phosphate 18 g and polyethylene/paraffin waxes 2 g	48 days paraffin wax,	–	Al-Zahrani (2000)
				58 days with polyethylene wax		
Mixture of nitrogen and phosphate	Compound nitrogen and phosphate coated polyethylene/paraffin waxes fertilizers	Polyethylene/ paraffin waxes	Compound nitrogen and phosphate 18 g and polyethylene/paraffin waxes 2 g	42 days paraffin wax,	–	Al-Zahrani (2000)
				48 days with polyethylene wax		
NPK	NPK coated polyethylene/ paraffin waxes fertilizers	Polyethylene/ paraffin waxes	NPK 18 g and polyethylene/ paraffin waxes 2 g	50 days paraffin wax,	–	Al-Zahrani (2000)
				64 days with polyethylene wax		
Phosphate	Triple superphosphate coated polyethylene/paraffin waxes fertilizers	Polyethylene/ paraffin waxes	Triple superphosphate 18 g and polyethylene/paraffin waxes 2 g	40 days paraffin wax,	–	Al-Zahrani (2000)
				48 days with polyethylene wax		
NPK	NPK-polymer coated fertilizers	Acrylate latex	NPK 80 kg and acrylate latex 16 kg	9 days	40	Cong et al. (2010)

DVB divinylbenzen, *NNMBA* N,~-methylenebisacrylamide, *PETA* pen-taerythritol triacrylate, *TTEGDA* tetraethyleneglycol diacrylate, *Starch-g-PLLA* starch-g-poly(L-lactide)

In conclusion, coated fertilizers are slow release fertilizers which provide the nutrients to the crop plants in a slow pattern; they slowly release with the passage of the time and fulfill the crop nutrients' demand with their growth pattern. Use of coated fertilizer can help in reducing the fertilizer application rates with higher NUE.

4.1.2 Slow/Controlled: Release Fertilizers

Controlled or slow release fertilizers are those fertilizers which contains plant nutrients in a form which either (i) delays the availability for plant uptake and use after its application (ii) or is available to the plant significantly longer than a "rapidly available nutrient fertilizers" (Table 1; AAPFCO 1995; Trenkel 1997).

Crops up take only 50–60% of the added N fertilizer to the soil in a growing season. This uptake of N fertilizer can be enhanced by controlling the rate of N fertilizer dissolution (Scherer 2005). On way to control the rate of N dissolution is controlled-release fertilizers and the aim of this slow release fertilizer is to provide the crops nutrients according to the demand (Scherer 2005). The slow-release fertilizers (SRF) release active nutrients in a controlled manner, extend the duration of release, and manipulate the rate of release so that they become compatible with the metabolic needs of plants (Irfan et al. 2018). The long term gradual release of nutrients from SRF is a solution to the current need of food (Trenkel 2010) and is necessary for the sustainability of the ecosystem. Un-coated urea fertilizers are readily soluble in water and quickly decomposed to release $NH^{+}_{4,}$ it forms several chemical reaction and products that are useful as slow-release N fertilizers (Scherer 2005).

Most of the studies (Yaseen et al. 2017; Trenkle 2010) have shown that by the application of P in the form of controlled-release fertilizers to citrus decrease potential losses and increase the fertilizer use efficiency compared with water soluble fertilizers (Zekri and Koo 1992). Conclusively, use of slow release fertilizers is effective approach for sustainable nutrient management as nutrients are available during the whole crop season. Moreover, it is also ecofriendly due to reduced nutrient losses through leaching and volatilization.

4.1.3 Nano Fertilizers

Nano-fertilizers are basically smart fertilizers which are designed to increase nutrient use efficiency and to reduce the adverse effects of conventional mineral fertilizers on the environment (Sharpley et al. 1992; Wurth 2007; Manjunatha et al. 2016). There are three types of nano-fertilizers as (i) nanoscale coating or host materials (nano-polymer), (ii) nanoscale additives and (iii) nanoscale fertilizers (synthesized nanoparticles) (Mastronardi et al. 2015). These nano fertilizers are most suitable alternatives to soluble fertilizers as they release nutrients at a slower rate during the crop growth so reduce nutrient losses (Table 2). In this regard, zeolites (natural clays) are best as they act as reservoir for nutrients that are slowly released (Manjunatha et al. 2016).

Table 2 Effect of nano-fertilizers application on nutrient uptake and crop yield

Nutrient	Compound name	Carrier	Grain nutrient contents	Technique	Crop	Release rate/time	Increase in yield (%)	Reference
Zinc	Zinc complexed chitosan nanoparticles (Zn-CNP)	Chitosan	36%	–	Wheat	–	–	Dapkekar et al. (2018)
Iron	Fe_2O_3 NPs	–	996 mg kg^{-1}	ICP-MS	Rice	–	–	Gui et al. (2015)
Iron	Nano-iron chelate	–	75 mg/g	–	Faba bean	–	–	Nadi et al. (2013)
Zinc	Zn-nano-zeolite	Zeolite	–	Atomic adsorption spectrophotometer	–	1176 hurs	–	Yuvaraj and Subramanian (2018)
Phosphorus	P-nano hydroxyapatite	Hydroxyapatite	–	–	Soybean	–	20	Liu and Lal (2014)
Nitrogen	Urea hydroxyapatite nanoparticles	Wood	–	Kjeldhal method and vnadomolybdate method	–	60 days	–	Kottegoda et al. (2011)
Zinc	MAP–nano-ZnO	Monoammonium phosphate	–	Inductively coupled plasma optical emission spectroscopy (ICP-OES)	–	48 hurs	–	Milani et al. (2012)
Zinc	Nano-scale zinc oxide	–	40 ppm	Transmission electron microscopy	Peanut	–	29	Prasad et al. (2012)
Zinc	Nano zinc	–	23 mg/kg	–	Rice	–	–	Apoorva et al. (2017)
Nitrogen	Nanozeourea	Zeolite	0.78 mg/kg	–	Maize	–	8	Manikandan and Subramanian (2016)
Iron and Zinc	Fe + Zn nano-fertilizers	–	–	–	Chickpea	–	34%	Drostkar et al. (2016)

The nano-fertilizers have high surface area, controlled release kinetics to targeted sites and sorption capacity called as smart delivery system (Fig. 2; Solanki et al. 2015). A nano-fertilizer is a product in nanometer regime that delivers nutrients to crops, for example encapsulation inside nanomaterials coated with a thin protective polymer film or in the form of particles or emulsions of nanoscale dimensions (DeRosa et al. 2010). The surface coatings of nanomaterials on fertilizer particle hold the material more strongly due to higher surface tension than conventional fertilizer surface and help in controlled release (Brady and Weil 1999). The nano-fertilizers have high solubility, effectiveness, stability, targeted activity, time-controlled release and less eco-toxicity, safe, easy mode of delivery and disposal (Tsuji 2000; Boehm et al. 2003; Green and Beestman 2007; Torney et al. 2007).

In a study, Corradini et al. (2010) evaluated the interaction and stability of chitosan nanoparticles suspensions containing N, P, and K fertilizers which can be useful for agricultural applications. In another study, Kottegoda et al. (2011) synthesized urea modified hydroxyapatite (HA) nanoparticles for gradual release of N to crop growth. These nano-fertilizers showed slow release of N up to 60 days of plant growth compared to commercial fertilizers which showed release only up to 30 days. The large surface area of HA facilitates the large amount of urea attachment on the HA surface. The strong interaction between HA and nanoparticles and urea contributes to slow and controlled release of urea. Few years back, Milani et al. (2012) compared the Zn solubility and dissolution kinetics of ZnO nanoparticles and bulk ZnO particles coated on macronutrient fertilizers (urea and monoammonium phosphate) and reported that coated monoammonium phosphate granules show faster dissolution rate.

Zeolite based nano-fertilizers are capable to release nutrient slowly to the crop plant which increase availability of nutrient to the crop throughout the growth period and prevent loss of nutrient from volatilization, leaching, denitrification and fixation in the soil especially NO_3-N and NH_4-N. The nutrient having particle size of below 100 nm nano-particles are used as efficient nutrient management which are more ecofriendly and reduce environmental pollution (Joseph and Morrisson 2006). The nano particles increased the NUE and minimized the costs of environmental protection (Naderi and Abedi 2012) and enhance plant growth by resisting the diseases and improving the stability of plants by deeper rooting and by anti-bending of crops (Fig. 2; Tarafdar et al. 2012).

In conclusion, nano fertilizers are ecofriendly can help in improving the agricultural productivity by improving the NUE with lower fertilizer requirement and better grain yield.

4.2 Organic Fertilizers

Organic fertilizers supply nutrients in slowly soluble organic with belief that plants will get balance nutrition through the actions of soil microbes, roots and weathering of minerals (Kirchmann et al. 2009) and these organic forms of nutrients are

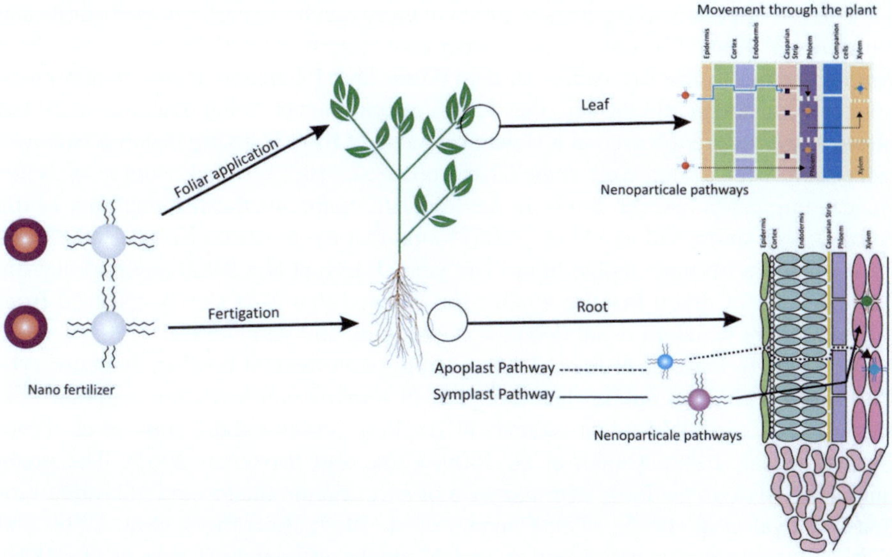

Fig. 2 Mechanism of nano-fertilizer uptake through foliar and fertigation in plant

available to the crop plants with longer time period. In inorganic fertilizers application the plants are directly fed owing to complete and high solubility of inorganic fertilizers in water (Kirchmann et al. 2009) compared with organic sources which release nutrients slowly which are available according to the crop need and has less or negligible losses to the environment.

The application of organic manures enhances build-up of soil organic matter, support soil structure, increase the cation exchange capacity, helps to chelate micronutrients, increase soil moisture retention while inorganic fertilizers supply crops with nutrients at times when their demand is large (Kirchmann et al. 2009). Organic materials are added to the soils to protect the productivity and sustainability of the land. The natural wastes are mostly used as organic fertilizers to increase the efficiency of nutrients and nutritional value of soils (Demir and Gulser 2015). Green manure/green manuring, farm yard manure and Compost are most widely used organic fertilizers.

4.2.1 Green Manuring

Quantity of agricultural production, crop yield, soil nutrient, and the environment all are influenced by fertilizer use. The increased mineral fertilizers prices and decreased soil fertility made legumes a popular option as organic fertilizer to improve the soil fertility in long run (Talgre et al. 2012). In a study, Talgre et al. (2012) found that after incorporation of green manure crops into the soil was effective in releasing nutrients into the soil even in the 3rd year. The application of green manures can replace the entire N requirement for non-leguminous succeeding crops (Guldan et al. 1997).

The use of perennial legumes as green manure (such as alfalfa) import additional nutrients (P, K and Ca) due to their deeper root system (Teit 1990) to the soil which are accessible to the succeeding crops (Witter and Johansson 2001). When green manures are added into the soil, they add large amounts of N and other nutrients, but these nutrients are released at a slower rate also N from N-fixing bacteria becomes accessible over a long time span. These processes supply steady source of N for succeeding crops (Freyer 2003), in the long run maintains the sustainability of the system. In a study, Viil and Vosa (2005) found that the positive effects (16–18%) of green manure become visible in the 2nd year. Talgre et al. (2012) reported that the yield results of green manure application showed that N is slowly released from green manure which in result decrease the lodging and yield losses.

The slow release of N from decomposing green manure residues is better synchronized with plant uptake than inorganic N sources as it increases N-uptake efficiency and crop yield while reduces N leaching losses (Abdul-Baki et al. 1996; Agustin et al. 1999; Aulakh et al. 2000; Cline and Silvernail 2002). The green manuring also drives long-term increase of soil organic matter and microbial biomass (Goyal et al. 1992, 1999; Chander et al. 1997; Biederbeck et al. 1998) and further improves nutrient retention and N-uptake efficiency (Cherr et al. 2006). Green manuring also offers habitat or resources for beneficial organisms (Bugg et al. 1991; Nicholls and Altieri 2001). The application of green manures reduced soil bulk density, increased soil organic matter and N, P, K, Ca and Mg (Adekiya et al. 2017). In conclusion, the incorporation of green manures improves the soil fertility, nutrients and crop growth and yield.

4.2.2 Farm Yard Manure

Farm yard manure is a decomposed mixture of urine and dung of the farm animals along with litter and left over material from roughages or fodder fed to the cattle. The application of farm yard manure improves the soil chemical, physical and biological properties (Bayu et al. 2006). Oswal (1994) reported that the application of farm yard manure (FYM) increased the electrical conductivity, cation exchange capacity, organic carbon and soil moisture contents. Likewise, Aggarwal et al. (1997) found that FYM increases water storage, crop yield and soil nutrient availability. Application of poultry manure (5 t ha^{-1}), FYM (10 t ha^{-1}) and piggery manure (2.5 t ha^{-1}) were equivalently effective and added 11.2 kg Zn ha^{-1} in maize-wheat cropping system (Nayyar et al. 1990). In another study, Alok and Yadav (1995) demonstrated that application of organic manure in rice wheat cropping system increased the Zn availability more than inorganic sources. Use of organic manures can meet the crop nutrient demand as they are rich in nutrients, improve physiochemical characteristics of soil and enhance nutrient uptake through formation of soluble nutrient complexes.

4.2.3 Compost

To satisfy the growing global food demand cereal production has increased (He et al. 2014) and this increase in production has in turn increased the amounts of harvested residues (straw, stubble) that can be a source of biomass feedstock or for animal feeding (Jiang et al. 2012; Habets et al. 2013). Unluckily, worst practice is removal of these residues by in situ burning with considerable environmental, human health and economic impacts (Singh et al. 2010; Gupta et al. 2016). These harvested residues can be a resource that can be utilized as organic raw material which improves the soil quality and productivity (Calabi-Floody et al. 2018).

One way to use these residues is their use as a composting agent (Roca-Perez et al. 2009; Medina et al. 2017). Compost is the final product which is obtained after composting and is rich and more stable than original material (feedstock) and can improve soil quality and productivity as well sustainability of the agricultural production (Farrell and Jones 2009; Barral et al. 2009). The application of compost slower the rate of mineralization (Bernal et al. 2009) and owing to this slow mineralization process the nutrients are available during the whole growing season and are more stable. With the application of compost soil structure is improved with the binding of soil organic matter and clay particles via cation bridges and through stimulation of microbial activity and root growth (Farrell and Jones 2009). In conclusion, the application of compost in the long run improves the soil structure, organic matter and fertility status of the soils.

4.3 Use of Soil Microbes

In intensive agriculture system, use of chemical fertilizer is necessary for getting good crop yield, however, the utilization efficiency of these nutrients remain low due to losses through leaching, volatilization and denitrification and fixation (Ayala and Rao 2002). These chemical fertilizers also increase the cost of production and are not ecofriendly (Adesemoye and Kloepper 2009). In this scenario, bio-fertilizers offer a better alternative to synthetic chemicals as they improve crop quality, yield and also increase resistance to abiotic stresses (Kumar et al. 2006). Integration of PGPR with traditional inorganic fertilizers in the field proved to be effective means to increase the availability of nutrients to plants with simultaneous reduction in diseases incidence of oil seed crop has been reported (Kumar et al. 2009). Plant growth promoting rhizobacteria (PGPR) and arbuscular mycorrhizae fungi (AMF) have been reported to increase the availability and uptake of nutrients in the soil.

4.3.1 Mycorhiza

Mycorhiza colonization enhances the absorption and uptake of nutrients limited to diffusion from soil solution to plant roots (Fageria et al. 2011). The AMF increase the phosphorus uptake by plants by enhancing the root surface area for absorption and by mineralizing the organic phosphorus (Wang et al. 2014). Few years back, Yang et al. (2012) reported that rice colonized with AMF received 70% of acquired pi from symbiotic fungi. The AM fungi also increases the nitrogen uptake as NH^{+4} transporters in AMF (GintAMT1, GintAMT2, GintAMT3) in *Rhizophagus irregularis* are involved in AM symbiosis (Lopez-Pedrosa et al. 2006; Perez-Tienda et al. 2011; Calabrese et al. 2016). These NH^{+4} transporters genes express under low NH^{+4} supply and enhance ammonium uptake from soil and surrounding media. Furthermore, AMF increase the availability of N to the plants by accelerating the decomposition of organic materials (Hodge et al. 2001; Leigh et al. 2009).

Mycorhizal colonization also increases micronutrient availability. For instance, soil inoculation with some *Penicillium sp. s*trains accelerates the uptake of Fe, Zn and Cu from soil and their accumulation in plants (Kucey 1988). Three decades ago, Meyer and Linderman (1986) established that co-inoculation of AMF and PGPR (*Pseudomonas putida*) enhanced the Al, Co, Cu, Fe and Ni uptake in shoots as *p. putida* release 2-ketogluconic acid which increase the micronutrient availability and uptake by plants through mycorrhizal colonization. However, plants grown on heavy metal polluted soils have low concentration of Cd, Cu, Mn and Zn with AMF inoculation than non-mycorrhizal association, indicating role of AMF in heavy metal stress tolerance (Weissenhorn et al. 1995; Guo et al. 1996). Furthermore, under enhanced Pd supply, AMF increase plant growth by accelerating the phosphorus uptake and protecting plants from Pd toxicity (Chen et al. 2005).

In conclusion, mycrorhizal associations improve the nutrient availability to plants under limited nutrient supply and can be used effectively as bio-fertilizers. Moreover, AMF regulate the metallic ions uptake thus protect plants from heavy metal toxicity.

4.3.2 Plant Growth Promoting Bacteria

Interaction of plants and soil bacteria play crucial role in maintaining the soil fertility. Plant growth promoting rhizobacteria improve the crop productivity when used in the form of biopesticides (Arora et al. 2008) and biofertilizer (Cakmakci et al. 2006). These PGPR improve the plant growth directly or indirectly through enhanced nutrient availability, root development and resistance against biotic and abiotic stresses (Glick 1995) by improving N_2 fixation, Fe sequestration, phosphate solublization, phytohormones synthesis, control on ethylene production and phytopathogens (Gamalero and Glick 2011). Plants can only utilize very small quantity of the applied phosphorus as >75% of applied P is precipitated with metallic cations and becomes fixed in the soil. In this scenario P solubilization and mineralization by phosphate solubilizing bacteria play key role in increasing phosphorus availability

(Jeffries et al. 2003). Phosphorus solubilizing bacterial synthesize organic acids like citric acid and gluconic acid which help in P solublization (Rodriguez et al. 2004).

Biological nitrogen fixation (BNF) almost accounts for 2/3 of the total N used in agriculture and it will be crucial for crop production sustainability in future (Matiru and Dakora 2004). Key BNF biochemical reaction occurs between legume and nitrogen fixing microbes that convert N_2 into NH_3 (Shiferaw et al. 2004). Nitrogen fixed by *Rhizobia* in legume crops also benefits the associated cereals or non-legume intercrops (Snapp et al. 1998) or subsequent crop. For instance, in many grass land systems with limited inputs, grasses depend on legume fixed N to meet the N requirement which is needed for better fodder quality (Paynel et al. 2001; Hayat and Ali 2010). Moreover, use of plant growth promoting bacterial can improve the micronutrient availability. Recently, Rehman et al. (2018a, b) reported an increase in Zn uptake and its translocation in wheat with Zn solubilizing bacteria due enhanced production of organic acids from the root exudates of wheat. Iron availability is also enhanced by PGPR as they release siderophores which help in Fe chelation. Fluorescent pseudomonads bacteria increase the iron sequestration by releasing iron-chelation siderophores (Dwivedi and Johri 2003).

Soil bacteria fix the atmospheric nitrogen and increase the availability of other macro and micronutrients through nutrient solubilization mineralization, siderophore production and root development. These PGPR can be used as bio fertilizers as this will be cost effective, ecofriendly and sustainable approach for nutrient management in crop production systems.

4.4 Fertilizer Application Methods

Fertilizer can be applied through several ways such as soil application, foliar application and through seed treatments (seed priming and seed coating). Each method of application has some limitations and advantages upon others.

4.4.1 Soil Application

Soil application of fertilizers is the most common approach to overcome the nutrient deficiencies. It can be done through broadcasting, band placement and fertigation. Mostly macronutrients are applied through soil application. For instance, soil incorporation/deep placement of urea minimizes the N losses with higher nitrogen use efficiency (Katyal et al. 1987). Nevertheless, soil incorporation of fertilizer depends on soil physiochemical properties but usually 5–10 cm depth is used for nutrient incorporation. Time of fertilizer application is very crucial for soil application as some fertilizers (phosphate fertilizers) are mostly applied at the time of sowing.

In a study, Rahim et al. (2012) found that band placement of P as basal application results in better phosphorus use efficiency than broadcasting of P. Contrary to this, Latif et al. (2001) reported that P application in splits as topdressing or fertigation was better than soil incorporation of P in wheat. In another study, Alam et al. (2005) established that application of N and P through fertigation enhanced the grain yield of wheat with better N and P uptake compared to topdressing. Furthermore, in wheat K is usually applied through broadcasting followed by incorporation in soil and drilling (Bijay-Singh et al. 2004).

Moreover, micronutrients are also supplied through soil fertilization. For instance, Zn fertilization through soil has increased the Zn uptake and bioavailability in wheat (Rehman et al. 2018a, b). In a recent study, Farooq et al. (2018) reported that Zn through soil application improved the grain Zn concentration in both conventional and conservation rice production systems. In another study, Zhao et al. (2018) demonstrated that band application of $ZnSO_4$ had little effect on grain Zn concentration but increase the loose organic matter bound Zn fraction in soil. Further, efficiency of Zn-EDTA and $ZnSO_4$ were higher when uniformly mixed rather than band application. However, both Zn sources have limited effect on grain Zn bioavailability due to higher fixation in calcareous soil.

Soil application of nutrient is very common approach to correct nutrient deficiencies and is most efficient method for macro nutrients (N, P and K) application. However, lack of soil plant nutrient status, higher rates of nutrient application, increased cost of production and unavailability of suitable nutrient sources are major bottleneck in this approach.

4.4.2 Foliar Application

Soil fertilization is mostly practiced for application of nutrients. However, higher plants also absorb nutrients through leaves when applied in suitable concentrations (Fageria et al. 2009). Foliar fertilization is mostly practiced for micronutrients as they are needed in small quantities. Otálora et al. (2018) conducted a study on foliar fertilization of urea and found that foliar applied urea enhances the N and other mineral uptake except Cu and Zn and enhanced the protein and amino acid accumulation. However, higher N rates reduced the sugar and phenolics accumulation in escarole (*Cichorium endivia* L. var. latifolium). Foliar fertilization has been found effective in improving the crop micronutrient demand. For instance Rehman et al. (2018a, b, c) reported that foliar Zn fertilization increases the grain Zn concentration in whole grain and endosperm with high Zn bioavailability. Likewise, a number of studies had reported increase in grain Zn and manganese (Mn) accumulation with foliar application (Zhao et al. 2014; Ullah et al. 2017a, b; Farooq et al. 2018; Rehman et al. 2018a, b).

Foliar application of $CaCl_2$ (1% solution) increased the Ca concentration of leaf while application of 2% $CaCl_2$ caused the leaf burn in pomegranate. Moreover, calcium fertilizer containing nanoparticles were not very effective in improving the leaf Ca concentration (Davarpanah et al. 2018). Likewise, application of B and Zn increased the both microelement concentrations in leaf of pomegranate (Davarpanah et al. 2016). Moreover, foliar fertilization of Fe at anthesis stage enhanced the grain Fe concentration and bioavailability. However, higher Fe accumulation was noticed for foliar application of Fe-EDTA (He et al. 2013). Foliar application of micronutrients along with endophytic bacteria improved the plant biomass, Fe and Zn concentration in wheat (Yaseen et al. 2018).

Foliar fertilization of macronutrients requires more number of sprays to fulfill crop nutrient demand. Moreover, there are chances of wash out by rain, leaf damage in case of higher solution concentration. Plant should also have sufficient leaf area for absorption of nutrient (Fageria et al. 2009). Despite these drawbacks in certain circumstances foliar fertilization is most effective method to overcome the nutrient deficiencies.

4.4.3 Seed Treatment

Nutrient application can be done through seed treatments. However, this practice mostly involves micronutrient application. Micronutrient delivery through seed treatment is economical and effective alternative to soil and foliar fertilization (Farooq et al. 2012, 2018). Seed treatments require very small amount of nutrient, hence are cost-effective and nutrients are available to the germinating seed (Singh et al. 2003).

4.4.3.1 Seed Priming

In nutrient priming seeds are soaked in aerated nutrient solution to initiate the metabolic activities prior to germination without radical protrusion. Primed seeds have better, and synchronized seedling emergence compared to dry seeds (Farooq et al. 2009). For instance, Zn seed priming in maize improved the maize performance (Harris et al. 2007). Likewise, seed priming with Zn increased the grain yield (19%) and seed Zn concentration respectively (Harris et al. 2008). Moreover, seed priming with Zn and plant growth promoting rhizobacteria (PGPR) enhanced the grain yield, grain and endosperm Zn accumulation (Rehman et al. 2018a, b). In maize, seed priming with Zn enhanced the seed Zn content (600%) compared to untreated seed and also improved the crop growth and biomass under normal and salt stressed condition (Imran et al. 2018). Seed priming with Mn increased the grain Mn content and also enhanced the grain yield of wheat and rice in both conventional and conservation rice production systems (Ullah et al. 2017a, b). Likewise in another study, Farooq et al. (2018) reported that Zn seed priming increased the grain yield and grain Zn concentration in rice compared with untreated control.

Boron application through seed priming substantially improved the rice yield and seed B concentration (Rehman et al. 2012). Recently, Ali et al. (2018) found that seed priming with B, Mn and Zn alone and in combination improves the concentration of respective nutrient in grain and straw of wheat. However, in nutrient seed priming solution concentration and duration of seed priming is very critical as priming in high nutrient solution may prove toxic and inhibit seedling germination and growth (Rehman et al. 2015). Moreover, for certain nutrients seed priming is better than soil application. For instance, Mo application through seed treatment was more effective than soil application (Johansen et al. 2006) as Mo application through seed priming increased the yield by 20–25% compared to soil Mo application (Johansen et al. 2007).

Seed priming with micronutrients is an eco-friendly and economical approach for nutrient delivery. This technique is help full under diverse climatic conditions as it helps in early stand establishment. However, selection of appropriate nutrient source and concentration are very critical for nutrient delivery through seed priming.

4.4.3.2 Seed Coating

In seed coating liquid or finely ground/suspended or dissolved solids are applied to the seed surface to cover the seed coat (Scott 1989). Seed coating mostly involve adhering of plant growth regulators, microorganisms, nutrients or other chemical on seed. Micronutrients are usually applied through seed coating as these are required in very small quantities. Seed coating of cowpea seeds substantially increased the grain yield. Moreover, seed coated with 250 mg $ZnSO_4$ kg^{-1} seed performed better than all other coating treatment and increased the yield by 32.1% than uncoated seeds (Masuthi et al. 2009). Similarly, seed coating with 1.25 g Zn kg^{-1} improved the stand establishment, grain yield, and grain Zn accumulation in wheat (Rehman et al. 2016). Application of Mn through seed coating improves the productivity and grain biofortification of rice (Ullah et al. 2017a). The application of Zn through seed coating improved the grain yield in direct seeded aerobic rice (Farooq et al. 2018).

Seed pelleting with boron i.e. 100 mg borax kg^{-1} seed improved the yield related traits and grain yield of cowpea than non-pelleted control (Masuthi et al. 2009). Molybdenum application through seed coating (80 g Mo ha^{-1}) enhanced the chlorophyll index, yield related traits and grain yield of common beans (Biscaro et al. 2009). In another study, soybean seed coating with 0.25 g ammonium molybdate ($(NH_4)6Mo_7O_24$) and 0.5 g ferrous sulphate kg^{-1} seed effectively improved the morphology, growth and dry matter production (Ramesh and Thirumurugan 2001).

Seed coating is very cost-effective approach of micronutrient application; however, success of seed coating depends on type of nutrients, coating materials, soil fertility status, soil type and seed to nutrient ratio. Moreover, seed coating is effective technique for nutrient supply during early stages of crop growth.

5 Site Specific Nutrient Management

Site specific nutrient management (SSNM) approach emphasize on application of nutrient to crop when needed. It does not thrive to limit or increase the fertilizer use but it helps to supply nutrient at optimal time and rate to obtain higher yield with better NUE. For instance, in South Asian countries fields are small with high spatial variability in crop management practices. Moreover, there are differences in soil inherent nutrient buildup, crop residue management, organic and inorganic use of fertilizers, fertilizer rate, time and application method and resource available to farmers which disturb the nutrient balance within a small piece of land. Moreover, nutrients are supplied on extensive recommendation based on large areas having similar climate and soil conditions. These recommendations are usually good but imbalanced use of fertilizer due to variable soil fertility and other soil characteristic of a field lower the NUE as higher nutrient application beyond a limit will not enhance the yield but will reduce the NUE. Therefore, SSNM offers nutrient management of crop according to its requirement in a specific field and environment (Table 3; Jat et al. 2014). Furthermore, SSNM help farmer to adjust fertilizer application in an accurate and efficient manner to fill the gap between nutrient demand of crop and supply of nutrient from soil, crop residues, organic and inorganic nutrient sources.

Site specific nutrient management was developed as INM strategy and is based on the quantitative relationship of crop demand and nutrient supply of each field which special and temporal variations (Dobermann et al. 2002, 2003) in different crop production systems. There are two types of SSNM approaches i.e. soil based (involves fertilizer recommendation of a specific field based on soil analysis and inherent capacity of soil to supply nutrients) and plant based which involves relationship between crop nutrient requirement and crop yield and usually determined from nutrient concentration at crop maturity (Witt et al. 1999).

In plant based SSNM crop nutrient demand is predicted by attainable yield target. Crop nutrient demand is fulfilled by inherent nutrient supply from soil, residual effect of previous crop and crop residues. A decade ago, Singh et al. (2008) evaluated SSNM in rice wheat crop rotation and reported average increase of 1.3 t ha^{-1} in rice yield than blanked recommendation. They further reported an increase of 0.39–1.92 t ha^{-1} across different locations in rice wheat cropping system. Recently, Banayo et al. (2018) conducted a study and found that site specific nutrient management using rice crop manager (RCM) software reduced the fertilizer application of N and P with an average increase of 6% in grain yield and average profit of 154 US$ ha^{-1}.

In conclusion: SSNM includes quantitative relationship between crop demand and nutrient requirement and it varies from field to field. It is very effective approach for sustainable nutrient management. However, success of this approach depends on rigorous plant and soil sampling and development of decision support system softwares.

6 Fertilizer Prediction Models

Fertilizer application to site specific field condition need estimation and understanding of crop nutrient status and soil spatial variability and its relation to plant response. However, use high resolution geo remote and proximal sensed data to quantify the approximate variation between management zone (Song et al. 2009).

Table 3 Effect of site specific nutrient management on crop yield and net economics returns

Cropping system	Crop	Blank fertilizer recommendation kg/ha	SSNM	Increase in yield (%)	Net return (USD) over control	Reference
Rice wheat	Rice	100 (N), 40 (P), 40 (K)	150 (N), 30 (P), 100 (K), 40 (S)	66	633	Singh et al. (2008)
	Wheat	120 (N), 60 (P), 40 (K)	150 (N), 30 (P), 100(K)	59	530	Singh et al. (2008)
Rice wheat	Rice	120 (N), 60 (P), 60 (K)	120 (N), 60 (P), 120 (K), 40 (S), 25 (Zn), 5 (B), 20 (Mn)	59	557	Singh et al. (2008)
	Wheat	120 (N), 60 (P), 60 (K)	150 (N), 60 (P), 120 (K)	65	475	Singh et al. (2008)
Rice wheat	Rice	150 (N), 75 (P), 75 (K), 2 (Zn)	150 (N), 30 (P), 80 (K), 20 (S), 25 (Zn), 5 (B), 20 (Mn)	45	678	Singh et al. (2008)
	Wheat	150 (N), 30 (P), 80 (K)	120 (N), 60 (P), 40 (K)	34	462	Singh et al. (2008)
	Rice	110 (N), 15 (P), 20 (K)	75 (N), 10 (P), 20 (K)	10	307	Banayo et al. (2018)
	Rice	75 (N), 8 (P), 18 (K)	75 (N), 8 (P), 18 (K)	11.2	275	Banayo et al. (2018)
	Cotton	312 (N), 312 (P), 180 (K)	225 (N), 105 (P), 150 (K), 45 (Mn), 30 (Zn)	19.8	561	Jin and Jiang (2002)
Maize-wheat-mungbean	Maize	150 (N), 60 (P), 60 (K)	144–170 (N), 46–50 (P), 63–105 (K)	5	68	Jat et al. (2018)
	Maize	150 (N), 60 (P), 60 (K)	144–170 (N), 46–50 (P), 63–105 (K)	7.4	130	Jat et al. (2018)
	Wheat	150 (N), 60 (P), 60 (K)	125–140 (N), 37–68 (P), 60–101 (K)	10.2	119	Jat et al. (2018)
	Wheat	150 (N), 60 (P), 60 (K)	125–140 (N), 37–68 (P), 60–101 (K)	12	184	Jat et al. (2018)

SSNM Site specific nutrient management, *N* nitrogen, *P* Phosphorus, *K* Potassium, *S* Sulphur, *B* Boron, *Mn* Manganese, *Zn* Zinc

These sensors generate and process large data set in real time to adopt precise management practices. For instance, site specific nutrient application using these sensors based on edaphic and soil condition increased the nitrogen use efficiency by 368% (Diacono et al. 2013).

Use of sensor and GPS technologies helped to monitor and identify plant and soil variability to specific inputs. Introduction of only GPS in farm machinery can improve the 5–10% efficiencies by decreasing overlaps and gaps during fertilizer application (Craighead and Yule 2001). Recently, Wang et al. (2014) studied the P losses from soil supplied with chemical fertilizers and cattle manures using SurPhos model. The model reliably predicted the losses of dissolved reactive P (DRP) from chemical fertilizer, liquid and solid cattle manure. Surphos also quantified the various sources of DRP loss and dynamics of labile P in soil which can help is adoption of appropriate P management practices to avoid P losses. Recently, Mahajan et al. (2014) used the hyperspectral remote sensing technique to predict wheat N, P, K and S requirement with very high accuracy. Efforts are going on to develop nutrient prediction models and technology for site specific nutrient management (Gregoret et al. 2011; Onoyama et al. 2015) however, there is still lot of work to be done on this aspect to achieve desired success.

6.1 Integrated Nutrient Management

Integrated nutrient management (INM) is soil fertility and plant nutrition management system according to soil properties with balanced fertilization using all possible nutrient sources (organic and inorganic) and biological agents in judicious and integrated manner (Janssen 1993; Roy et al. 2006). Moreover, INM consider nutrient cycling of macro and micronutrients to synchronize nutrient requirement of crop and its release in the environment (Table 4). All the INM approaches are aimed at reducing the nutrient losses through runoff, leaching, immobilization, volatilization and emission, and to increase the NUE (Zhang et al. 2012). The INM also helps in restoration of soil fertility and physiochemical properties with better soil organic carbon (C) and thus sustain the system productivity (Table 4; Das et al. 2014). In a study conducted by Das et al. (2014) on integrated nutrient management in rice-wheat cropping system, they found that incorporation of organic material improves the aggregation and structural stability of soil with better C accumulation in macro aggregates showing higher C sequestration of soil. They further reported that use of FYM in wheat and green gram residue (GR) in rice effectively improved the C accumulation in macro aggregates. Further, residue incorporation was more beneficial than 100% inorganic N application or GR to rice.

The INM substantially enhance rice yield by reducing nutrient losses and managing nutrient supply which help in cost reduction, better resource use efficiency and increased resistance to biotic and environmental stresses (Prasad et al. 2002; Zhang et al. 2012). Chemical fertilizer especially N fertilizers are being excessively used in China and other developing world (Peng et al. 2002; Zhang et al. 2012), which cause saturation of chemical nutrients in agro-ecosystems, thus leading to nutrient loss through runoff, leaching, volatilization, fixation, emissions with low NUE (Vitousek et al. 2009). For instance, in northern plains of China in maize wheat system about 227 kg N and 53 kg P ha^{-1} $year^{-1}$ surplus supply has been reported. Application of 120:26:37 kg NPK ha^{-1} in combination with green manures improved the grain yield of rice. Similarly, highest groundnut production was obtained with residual effect of green manure and 30:26:33 kg NPK ha^{-1} in combination with gypsum (Prasad et al. 2002).

Hossain et al. (2016) studied the INM in rice wheat cropping system by inoculation of legumes (mungbean, blackgram and dhaincha) and organic manures (poultry manure and cow dung). They reported that incorporation of legume residues enhanced the soil organic matter, N, extractable P and Zn, while all legume-based rotation with rice and wheat reduced the K and S concentration. Moreover, use of chemical fertilizer in combination with higher rates of organic manures increased the system productivity, showing that integrated approach is suitable option for balanced and sustainable nutrient management (Table 4). Likewise, maximum Zn concentration in grain and all seed fractions were recorded in wheat when chemical fertilizer was applied in through soil and foliar application in combination with Zn solubilizing microbial strain *Pseudomonas sp.* than sole application of chemical Zn fertilizer (Rehman et al. 2018a, b). Sharma et al. (2016) demonstrated that application of FYM along with recommended fertilizer dose substantially improved the physiochemical properties and biological activities of soil in finger-millet monocropping and groundnut finger millet crop rotation compared to sole inorganic fertilizer application.

In conclusion, INM approach is sustainable and ecofriendly approach as it reduces the chemical input by balanced fertilization and nutrient management using all possibly nutrient sources (crop rotation/intercropping, residue incorporation, organic manures and soil microbes) and minimize the greenhouse gas emissions.

7 Soil Management

The soil sustains all living organisms, being the ultimate source of their mineral nutrients. Good management of soils ensures that mineral elements do not become deficient or toxic to plants, and that appropriate mineral elements enter the food chain. Soil management is important, both directly and indirectly, to crop productivity, environmental sustainability, and human health. Because of the projected increase in world population and the consequent necessity for the intensification of food production, the management of soils will become increasingly important in the coming years. To achieve future food security, the management of soils in a sustainable manner will be the challenge, through proper nutrient management and appropriate soil conservation practices (White et al. 2012).

Table 4 Influence of long-term integrated nutrients management on the soil nutrients concentration at different soil depth

Nutrients combination	Study duration (years)	Soil depth (cm)	Increase in nutrients concentration (%)					Reference
			N	P	K	S	Zn	
N + FYM	41	0–15	27	–	–	–	–	Shahid et al. (2017)
N + FYM	41	15–30	5	–	–	–	–	
N + FYM	41	30–45	25	–	–	–	–	
NPK + FYM	41	0–15	18	–	–	–	–	
NPK + FYM	41	15–30	14	–	–	–	–	
NPK + FYM	41	30–45	24	–	–	–	–	
N fertilizers + cattle manure	26	20	51			–		Zhengchao et al. (2013)
P fertilizers + cattle manure	26	20	65	–	–	–	–	
N + P fertilizers + cattle manure	26	20	76	–	–	–	–	
25% RF + 75% RN (MOC)	2	–	23	46	11	–		Mondal et al. (2016)
100% RF + 25% RN (MOC) + 75% RF + 25% RN	2	–	9	17	5	–		
(MOC) + Biofertilizer	2	–	13	31	8	–		
100% RF + 25% RN (MOC) + Biofertilizer	2	–	15	39	11	–		
RF + Cow dung 5 t ha^{-1}	7	0–15	27	343	12	87	102	Saha et al. (2007)
RF + Cow dung 5 t ha^{-1}	7	16–30	75	25	–	17	34	
50 + 50% N (FYM)	23	0–15	46	566	63	–	201	Walia et al. (2010)
50 + 50% N (FYM)	23	15–30	79	428	61	–		
50 + 50% N (WCS)	23	0–15	40	226	49	–	188	
50 + 50% N (WCS)	23	15–30	73	214	47	–		
50 + 50% N (GM)	23	0–15	57	246	49	–	232	
50 + 50% N (GM)	23	15–30	81	228	44	–	–	
N + OM	33	–	116	–	–	–	–	Yang et al. (2015)
N + Straw	33	–	17	–	–	–	–	
N + green manure	33	–	9	–	–	–	–	
RF + VC at 2.5 t ha^{-1}	2	–	12	3	14	–		Kakraliya et al. (2017)
RF + FYM at 5 t ha^{-1}	2	–	12	21	15	–	–	
RF + FYM at 10 t ha^{-1}	2	–	19	46	19	–	–	
RF + VC at 2.5 t ha^{-1} + Azotobacter	2	–	14	17	14	–	–	
RF+ FYM at 5 t ha^{-1} + Azotobacter	2	–	15	22	15	–	–	
RF + VC at 2.5 t ha^{-1} + FYM at 5 t ha^{-1} + Azotobacter	2	–	19	40	19	–	–	

(continued)

Table 4 (continued)

Nutrients combination	Study duration (years)	Soil depth (cm)	Increase in nutrients concentration (%)					Reference
			N	P	K	S	Zn	
RF + 200 kg N ha^{-1} through FYM	7	0–15	–	42	–	–	–	Dhaliwal et al. (2014)
RF + 200 kg N ha^{-1} through FYM	7	15–30	–	62	–	–	–	
RF + 200 kg N ha^{-1} through FYM	7	30–45	–	51	–	–	–	
400 kg N ha^{-1} through VC	7	0–15	–	76	–	–	–	
400 kg N ha^{-1} through VC	7	15–30	–	110	–	–	–	
400 kg N ha^{-1} through VC	7	30–45	–	103	–	–	–	
400 kg N ha^{-1} through RSC	7	0–15	–	64	–	–	–	
400 kg N ha^{-1} through RSC	7	15–30	–	100	–	–	–	
400 kg N ha^{-1} through RSC	7	30–45	–	136	–	–	–	
20 kg N (crop residue) + 20 kg N (urea ha^{-1})	20	–	13	32	14	–	–	Maruthi Sankar et al. (2012)
10 kg N (FYM) + 10 kg N (urea ha^{-1})	20	–	27	51	20	–	–	
40 kg N (urea) + 20 kg P + 25 kg $ZnSO_4$ ha^{-1}	20	–	21	42	16	–	–	
25 kg N (Leucaena) + 25 kg N (urea) + 25 kg P ha^{-1}	20	–	15	26	24	–	–	
50% RF + 50% FYM	24	–	69	201	64	100	88	Gawde et al. (2017)
75% RF + 25% FYM	24	–	61	230	71	62	85	
50% RF+50% GM	24	–	77	188	57	63	70	
N + FYM + P + K	60	–	7	–	16	41	107	Verma (2017)
Lime + N	60	–	15	–	–	–	22	

N Nitrogen, *P* Phosphorus, *K* Potassium, *S* Sulphur, *Zn* Zinc, *RF* Recommended fertilizers application, *RN* Recommended N application, *FYM* Farmyard manure, *MOC* Mustard oil cake, *WCS* Wheat cut straw, *VC* Vermicompost, *RSC* Rice straw compost, *GM* Green manure

7.1 Mulching

Mulching is an agricultural technique which is used to cover soil surface around the plants to create congenial condition for the growth. Mulching reduces the deterioration of soil by preventing the runoff, soil loss and helps in the control of temperature fluctuations, improves physical, chemical and biological properties of soil, as it adds nutrients to the soil and ultimately enhances the growth and yield of crops

(Kumar et al. 1990). It reduces both the overland flow generation rates and velocity by increasing roughness (Jordán et al. 2010), and it cuts the sediment and nutrient concentrations in runoff (Gholami et al. 2013). It also enhances the activity of some species of earthworms as well as crop performance (Thierfelder et al. 2013), interactions with nutrients (Campiglia et al. 2014), the soil structure and the organic matter content within the soil (Karami et al. 2012).

The increases in the soil organic matter content can be particularly significant when vegetative residues are used as mulches, as shown by Jordán et al. (2010). Mulching has also been shown to reduce the topsoil temperature for more optimal germination and root development (Dahiya et al. 2007) which helps in enhanced nutrient uptake. Moreover, mulches also decrease evaporation (Vanlauwe et al. 2015) thus reduce the nutrient losses (particularly N) through volatilization. In conclusion, application of crop residue mulches helps in moisture conservation, soil and nutrient loss through runoff and volatilization. Thus, use of mulches can be helpful in sustainable crop management.

7.2 *Conservation Tillage and Residue Management*

Conservation tillage (CT), along with some complimentary practices such as soil cover and crop diversity (Corsi et al. 2012) has emerged as a viable option to ensure sustainable food production and maintain environmental integrity. Conservation tillage positively influence soil productivity and quality (Paul et al. 2013) by promoting the biological activities in top soil through maintaining soil organic matter (Dungait et al. 2012). Higher N, P and K concentration in soil was recorded for CT (Das et al. 2018) due to enhanced residue decomposition and nutrient mineralization. Increase in available soil P was observed in CT system (Das et al. 2018) as high soil organic carbon accelerate the conversion of immobile P into mobile form and also reduced losses due to erosion/runoff which maintained high applied P fertilizer on the soil (Falatah and Al-Derby 1993; Vincent et al. 2010).

Crop residue management also imparts significant impacts on crop productivity and soil fertility. Yield responses to crop residue retention are increased when the ratio of incorporation of inorganic N fertilizer at vegetative stage of crop plants are increased from 70% to 100% (Huang et al. 2013). Increases in soil productivity require balanced fertilization and residue retention (Whitbread et al. 2003). Residue management has significant effects on physical, chemical, and biological properties of soil. Biological nitrogen fixation by leguminous crops and the recycling of fixed nitrogen when leguminous crop residues are returned to the soil can prove to be a rich source of N to the soil organic N pool as well as for subsequent plant uptake (Mosier and Kroeze 1998).

Plant residue decomposition involves two processes: mineralization, humification of carbon compounds by microorganisms and the leaching downward in the soil in the form of soluble compounds (Couteaux et al. 1995). Moreover, incorporation of residues increases the soil microbial biomass carbon which accelerates the N mineralization from organic form (Das et al. 2014). Comparisons of N recoveries from crop residue N and inorganic N fertilizers have shown that, in general, N recoveries from leguminous and non-leguminous residues are about one-half and one-eighth, respectively, of that from various forms of N fertilizers. Also, more legume N than fertilizer N is retained in soil and enters the organic N pool, whereas losses of legume N and fertilizer N are generally similar. Thus, there is a need to minimize losses of N from both systems by devising proper management practices for all cropping systems so that N mineralization synchronizes with crop N demand (Kumar and Goh 1999).

In conclusion, conservation tillage practices and crop residue retention/incorporation improves the soil organic matter, microbial activity, moisture retention and nutrient availability. Therefore, CT and crop residue retention reduces the N fertilizer application through buildup in soil N pool.

7.3 *Use of High Intrinsic Nutrient Seeds*

Plant growth is not only affected by external factors, but maternal environmental condition and plant nutrient status influence the germination, seedling development and several other traits of crop plants (Aarssen and Burto 1990). For instance, seed vigor and biomass production during early vegetative growth are closely linked with intrinsic seed Zn (Rehman et al. 2018a). Seeds with low Zn concentration have reduced emergence and seedling growth in a Zn deficient soil (Yilmaz et al. 1998). Moreover, seed with high Zn concentration due to fertilization in maternal plants increased the dry matter production and grain yield (Rengel and Graham 1995; Yilmaz et al. 1998). Seed with lower Zn concentration may cause cellular damage, loss of food and nutrient reserves or disrupt vital biochemical process during germination and early seedling growth (Ozturk et al. 2006; Cakmak 2008).

Wulff and Bazzaz (1992) reported that *Abutilon theophrasti* seeds having high intrinsic nutrient concentration have resulted in higher leaf development, dry weight, seedling growth, cotyledon area and seed weight owing to enhanced maternal nutrient supply. However, in a study, addition of several nutrients applied to the maternal plants only increased one element in the progeny plants (Parrish and Bazzaz 1985). However, our knowledge on the effects of Zn biofortification on germination and crop performance of progeny is scarce. Nonetheless, under nutrient deficient condition high seed with high intrinsic nutrients can help in better crop stand and early plant growth.

In conclusion, initial seed nutrient concentration is crucial for germination and early seedling growth, especially in nutrient deficient condition. High initial nutrient reserve may help plant to cope with environmental stresses during early period of plant growth. Furthermore, nutrient dense seeds will increase agricultural productivity by enhancing the seed vigor, reduced fertilizer application and higher grain yield.

8 Crop Management

Crop management and soil cultivation practices can improve the nutrient availability in soil. Management practices which simultaneously improve soil properties and yield are mandatory to maintain high crop production and minimize deleterious impact on the environment. Retaining crop residues along with no-tillage improves soil properties and environment (Malhi et al. 2006). Selection of suitable planting technique, maintaining suitable plant population and crop rotation can influence the nutrient availability in soil.

8.1 Sowing Method and Planting Density

Nutrient losses from arable system can be minimized by adopting appropriate planting technique as it helps in adoption of appropriate fertilizer application method. Apart from balanced fertilizer and timely fertilizer application crop sowing method, crop sequence, crop root system and crop residue incorporation are very critical. For instance, top dressing and strip placement in maize-soybean improves NUE (Yong et al. 2018). Recently, Verma et al. (2018) found a decrease in weed dry mass with higher grain yield and increased availability of N, P, K, S and Zn in soil with raised bed sowing followed by furrow and ridge sowing of maize and these sowing methods were superior to flatbed sown maize.

The optimum plant population is very crucial for yield maximization in most of the field crops (Hiltbrunner et al. 2007). For instance, N uptake in wheat increased with optimal plant density while higher or lower seeding rated did not improve the N uptake (Blankenau and Olfs 2001). In a study, Dai et al. (2013) reported increase in N uptake, nitrogen use efficiency (NUE) and nitrogen uptake efficiency (UPE) due to increase uptake of above ground N when seedling rate increased from 135 to 405 m^{-2}, while, seedling rate higher than 405 m^{-2} did not improve the N uptake and use efficiency. Moreover, higher seedling rate is linked with reduce grain N concentration (Geleta et al. 2002), while no effect on grain N concentration of higher plant population has also been reported in wheat (Ozturk et al. 2006). Moreira et al. (2015) demonstrated that in soybean-wheat cropping system 50 cm spaced rows with no N application and 333,000 plants ha^{-1} are adequate for soybean as crop N supply is fulfilled with biological N fixation (BNF), while wheat N can also be fulfilled with BNF of soybean and N supply from organic matter.

Conclusively, plant population play key role in nutrient uptake and use efficiency. High and low planting density did not improve nutrient uptake. However, optimal planting density results in better grain yield with higher nutrient uptake and use efficiencies.

8.2 Crop Rotation and Intercropping

Crop rotation refers to the phenomenon of growing alternate crops in same field in order to avoid mono-cropping at a specific cropping season. Long term soil management practices affect soil pH, organic matter, bulk density, and nutrient availability. Different tillage and crop rotation practices require distinctly different soil fertility management strategies (Edwards et al. 1992). Changes in agricultural management can potentially increase the accumulation rate of soil organic carbon, thereby sequestering CO_2 from the external atmosphere (West and Post 2002). Enhanced monoculture production of cash grain crops and greater reliance on the import of chemical fertilizers and pesticides to maintain crop growth have resulted in greatly increased grain yields and labor efficiency. However, these conventional management practices have led to the decline in soil organic matter (SOM), increased soil erosion, nutrient depletion and surface and groundwater contamination (Reganold et al. 1987).

Legume based crop rotations reduced N leaching by 50% compared to conventional cropping systems (Drinkwater et al. 1998). For instance, reduced N fertilizer is needed for soybean as it meets 50–60% of N demand through N fixation (Salvagiotti et al. 2008). Few years back, Soltani et al. (2014) reported higher grain Zn concentration in wheat when grown after sunflower, safflower, clover and sudan grass. Likewise, wheat-cotton rotation increased the Zn accumulation of wheat (Khoshgoftar and Chaney 2007). Inclusion of legume crops in rice wheat cropping system enhanced the crop productivity with buildup in soil organic matter, N, P and Zn concentration (Hossain et al. 2016).

Intercropping can also improve the availability of nutrients to plants by altering the soil physiochemical properties. For instance, chickpea and wheat intercropping resulted in higher grain Zn accumulation in both crops than mono-cropping (Gunes et al. 2007). Similarly, barley-pea intercropping increased the N and C accumulation in both crops than monoculture (Chapagain and Riseman 2014). Intercropping of cereals with dicots is sustainable and effective Zn biofortification approach as it increases the Zn uptake in both crops (Zuo and Zhang 2009). In number of studies it has been reported that legume and cereal intercropping is efficient as it enhances N fixation, biodiversity, nutrient use efficiency with sustainable and higher grain yield (Awal et al. 2006; Hauggaard-Nielsen et al. 2008; Gao et al. 2010; Ghanbari et al. 2010; Wang et al. 2017; Yang et al. 2017).

In South West China, many studies have highlighted that relay intercropping of maize and soybean enhances the NUE, light use efficiency with higher crop productivity and is major planting pattern. Recently, Rehman et al. (2018a, b, c) concluded that intercropping of wheat with legumes augments Zn uptake more than mono-cropping of wheat due to formation of soluble Zn complexes. Recently, Gitari et al. (2018) demonstrated that potato intercropping with legumes improve the NUE. They reported that potato intercropping with *Pisum sativum* L. *Phaseolus vulgaris* L. and *Lablab purpureus* L. increased the NUE by 9%, 19% and 30% respectively while an increase of 21%, 14% and 6% in phosphorus use efficiency (PUE) was recorded respectively.

Intercropping of cereals and legume reduce N fertilizer input through enhanced N fixation. However, sowing method and fertilizer application are very crucial for legume N fixation (Li et al. 2001; Ghosh et al. 2006; Salvagiotti et al. 2008; Hauggaard-Nielsen et al. 2009; Wu et al. 2014). In conclusion, sequential cropping results in nutrient deficiencies. Intercropping of cereals with legumes enhances nutrient acquisition through enhanced N fixation, Zn availability, nutrient use efficiency and changing in the soil physiochemical properties through forming soluble nutrient complexes than mono-cropping.

9 Conclusion

Sustainable nutrient management includes optimization of all possible nutrients sources their special and temporal synchronization with plant nutrient demand with aiming at reducing nutrient losses and improving crop yield and soil nutrient balance. Use of appropriate chemical nutrient source with appropriate application method can help in meeting the crop demand. Recently, slow release fertilizer has found effective in improving nutrient use efficiency with significant increase in crop yield. However, excessive use of chemical fertilizers is serious threat to environmental health, use of organic nutrient sources and soil microbes (AMF and PGPR) can help in reducing the crop demand for chemical nutrient sources. Adoption of soil and crop management practices like minimum or no tillage, residue retention/incorporation, optimal plant density, crop rotation and planting methods reduce the nutrient losses and increase the soil N pool with high organic matter and increased soil microbial activities. Use of all possible nutrient sources (chemical, organic and biological) in an integrated manner using site specific nutrient management will help in improving the soil nutrient balance with better crop production and reduced environmental impact.

References

AAPFCO (1995) Official Publication No. 48. Published by Association of American Plant Food Control Officials. Association of American Plant Food Control Officials (AAPFCO). Inc.; West Lafayette, Indiana, USA

Aarssen LW, Burto SM (1990) Maternal effects at four levels in Senecio vulgaris (Asteraceae) grown on a soil nutrient gradient. Am J Bot 77:1231–1240

Abdul-Baki AA, Teasdale JR, Korcak R, Chitwood DJ, Huettel RN (1996) Fresh-market tomato production in a low-input alternative system using cover-crop mulch. Hortic Sci 31:65–69

Abraham J, Rajasekharan Pillai VN (1996) Membrane encapsulated controlled release urea fertilizers based on acrylamide copolymers. J Appl Polym Sci 60:2347–2351

Adekiya AO, Agbede TM, Aboyeji CM, Dunsin O, Ugbe JO (2017) Green manures and NPK fertilizer effects on soil properties, growth, yield, mineral and vitamin C composition of okra (*Abelmoschus esculentus* (L.) Moench). J Saudi Soc Agric Sci. https://doi.org/10.1016/j.jssas.2017.05.005

Adesemoye AO, Kloepper JW (2009) Plant–microbes interactions in enhanced fertilizer-use efficiency. Appl Microbiol Biotechnol 85:1–12

Aggarwal RK, Praveen K, Power JF (1997) Use of crop residue and manure to conserve water and enhance nutrient availability and pearl millet yields in an arid tropical region. Soil Tillage Res 41:43–51

Agustin EO, Ortal CI, Pascua SR, Santa Cruz PC Jr, Padre AT, Ventura WB, Obien SR, Ladha JK (1999) Role of indigo in improving the productivity of rainfed lowland rice-based cropping systems. Exp Agric 35:201–210

Alam SM, Shah SA, Iqbal MM (2005) Evaluation of method and time of fertilizer application for yield and optimum P-efficiency in wheat. Songklanakarin J Sci Technol 27:457–463

Ali N, Farooq M, Hassan MA, Arshad MS, Saleem MK, Faran M (2018) Micronutrient seed priming improves stand establishment, grain yield and biofortification of bread wheat. Crop Pasture Sci 69:479–487

Alok K, Yadav DS (1995) Use of organic manure and fertilizer in rice (*Oryza sativa*)-wheat (*Triticum aestivum*) cropping system for sustainability. Indian J Agric Sci 65:703–707

Al-Zahrani SM (2000) Utilization of polyethylene and paraffin waxes as controlled delivery systems for different fertilizers. Ind Eng Chem Res 39:367–371

Apoorva MR, Rao PC, Padmaja G (2017) Effect of zinc with special reference to nano zinc carrier on yield, nutrient content and uptake by rice (*Oryza sativa* L.). Int J Curr Microbiol App Sci 6:1057–1063

Arora S, Jain J, Rajwade JM, Paknikar KM (2008) Cellular responses induced by silver nanoparticles: in vitro studies. Toxicol Lett 179:93–100

Aulakh MS, Khera TS, Doran JW (2000) Yields and nitrogen dynamics in a rice–wheat system using green manure and inorganic fertilizer. Soil Sci Soc Am J 64:1867–1875

Awal MA, Koshi H, Ikeda T (2006) Radiation interception and use by maize/peanut intercrop canopy. Agric For Meteorol 139:74–83

Ayala S, Rao EP (2002) Perspectives of soil fertility management with a focus on fertilizer use for crop productivity. Curr Sci 82:797–807

Aziz T, Maqsood MA, Kanwal S, Hussain S, Ahmad HR, Sabir M (2015) Fertilizers and environment: issues and challenges. In: Crop production and global environmental issues. Springer, Cham, pp 575–598

Banayo NP, Haefele SM, Desamero NV, Kato Y (2018) On-farm assessment of site-specific nutrient management for rainfed lowland rice in the Philippines. Field Crop Res 220:88–96

Barral MT, Paradelo R, Moldes AB, Domínguez M, Díaz-Fierros F (2009) Utilization of MSW compost for organic matter conservation in agricultural soils of NW Spain. Resour Conserv Recycl 53:529–534

Bayu W, Rethman NF, Hammes PS, Alemu G (2006) Application of farmyard manure improved the chemical and physical properties of the soil in a semi-arid area in Ethiopia. Biol Agric Hortic 24:293–300

Bernal MP, Alburquerque JA, Moral R (2009) Composting of animal manures and chemical criteria for compost maturity assessment. A review. Bioresour Technol 100:5444–5453

Bhat MR, Murthy DVR, Saidutta MB (2011) Modeling of urea release from briquettes using semi infinite and shrinking core models. Chem Prod Process Model 6:1934–2659

Biederbeck VO, Campbell CA, Rasiah V, Zentner RP, Wen G (1998) Soil quality attributes as influenced by annual legumes used as green manure. Soil Biol Biochem 30:1177–1185

Bijay-Singh Y-S, Imas P, Jian-Chang X (2004) Potassium nutrition of the rice-wheat cropping system. Adv Agron 81:203–259

Biscaro GA, Junior G, Roseiro SA, Soratto RP, Freitas Júnior NA, Motomiya AV, Calado Filho GC (2009) Molybdenum applied to seeds and sidedressing nitrogen on irrigated common bean in cerrado soil. Ciênc Agrotecnol 33:1280–1287

Blankenau K, Olfs HW (2001) Effect of different crop densities of winter wheat on recovery of nitrogen in crop and soil within the growth period. J Agron Crop Sci 186:151–156

Boehm AL, Martinon I, Zerrouk R, Rump E, Fessi H (2003) Nanoprecipitation technique for the encapsulation of agrochemical active ingredients. J Microencapsul 20:433–441

Bouwman AF, Beusen AHW, Billen G (2009) Human alteration of the global nitrogen and phosphorus soil balances for the period 1970–2050. Glob Biogeochem Cycles 23:GB0A04

Brady NC, Weil RR (1999) Nitrogen and sulfur economy of soils. In: The nature and properties of soils. Prentice Hall, Upper Saddle River, pp 491–523

Brown ME, Hintermann B, Higgins N (2009) Markets, climate change, and food security in West Africa. Environ Sci Technol 43:21

Bugg RL, Wäckers FL, Brunson KE, Dutcher JD, Phatak SC (1991) Cool-season cover crops relay intercropped with cantaloupe: influence on a generalist predator, Geocoris punctipes (Hemiptera: Lygaeidae). J Econ Entomol 84:408–416

Cakmak I (2008) Enrichment of cereal grains with zinc: agronomic or genetic biofortification? Plant Soil 302:1–17

Cakmakçi R, Dönmez F, Aydın A, Şahin F (2006) Growth promotion of plants by plant growth-promoting rhizobacteria under greenhouse and two different field soil conditions. Soil Biol Biochem 38:1482–1487

Calabi-Floody M, Medina J, Rumpel C, Condron LM, Hernandez M, Dumont M, de la Luz Mora M (2018) Smart fertilizers as a strategy for sustainable agriculture. Adv Agron 147:119–157

Calabrese S, Pérez-Tienda J, Ellerbeck M, Arnould C, Chatagnier O, Boller T, Courty PE (2016) GintAMT3–a low-affinity ammonium transporter of the Arbuscular Mycorrhizal Rhizophagus Irregularis. Front Plant Sci 7:679

Campiglia E, Mancinelli R, Di Felice V, Radicetti E (2014) Long-term residual effects of the management of cover crop biomass on soil nitrogen and yield of endive (*Cichorium endivia* L.) and savoy cabbage (*Brassica oleracea* var. sabauda). Soil Tillage Res 139:1–7

Chander K, Goyal S, Mundra MC, Kapoor KK (1997) Organic matter, microbial biomass and enzyme activity of soils under different crop rotations in the tropics. Biol Fertil Soils 2224:306–310

Chapagain T, Riseman A (2014) Barley–pea intercropping: effects on land productivity, carbon and nitrogen transformations. Field Crop Res 166:18–25

Chen J, Chen J, Tan M, Gong Z (2002) Soil degradation: a global problem endangering sustainable development. J Geogr Sci 12:243–252

Chen C, Ridzon DA, Broomer AJ, Zhou Z, Lee DH, Nguyen JT, Barbisin M, Xu NL, Mahuvakar VR, Andersen MR, Lao KQ (2005) Real-time quantification of microRNAs by stem–loop RT–PCR. Nucleic Acids Res 33:179–179

Chen L, Xie Z, Zhuang X, Chen X, Jing X (2008) Controlled release of urea encapsulated by starch-g-poly (L-lactide). Carbohydr Polym 72:342–348

Cherr CM, Scholberg JM, McSorley R (2006) Green manure approaches to crop production. Agron J 98:302–319

Chien SH, Prochnow LI, Cantarella H (2009) Recent developments of fertilizer production and use to improve nutrient efficiency and minimize environmental impacts. Adv Agron 209:267–322

Choi MM, Meisen A (1997) Sulfur coating of urea in shallow spouted beds. Chem Eng Sci 52:1073–1086

Cline GR, Silvernail AF (2002) Effects of cover crops, nitrogen, and tillage on sweet corn. HortTechnology 12:118–125

Cong Z, Yazhen S, Changwen D, Jianmin Z, Huoyan W, Xiaoqin C (2010) Evaluation of water-borne coating for controlled-release fertilizer using Wurster fluidized bed. Ind Eng Chem Res 49:9644–9647

Corradini E, De Moura MR, Mattoso LHC (2010) A preliminary study of the incorporation of NPK fertilizer into chitosan nanoparticles. Express Polym Lett 4:509–515

Corsi S, Friedrich T, Kassam A, Pisante M, Sà JD (2012) Soil organic carbon accumulation and greenhouse gas emission reductions from conservation agriculture: a literature review. FAO United Nations, Rome

Couteaux MM, Bottner P, Berg B (1995) Litter decomposition, climate and liter quality. Trends Ecol Evol 10:63–66

Craighead MD, Yule IJ (2001) Opportunities for increased profitability from precision agriculture. Geospatial Information and Agriculture. http://www.regional.org.au/au/gia/13/439yule.htm. Accessed July 2018

Cui Z, Zhang F, Chen X, Miao Y, Li J, Shi L, Xu J, Ye Y, Liu C, Yang Z, Huang S (2008) On-farm evaluation of an in-season nitrogen management strategy based on soil Nmin test. Field Crop Res 105:48–55

Dahiya R, Ingwersen J, Streck T (2007) The effect of mulching and tillage on the water and temperature regimes of a loess soil: experimental findings and modeling. Soil Tillage Res 96:52–63

Dai X, Zhou X, Jia D, Xiao L, Kong H, He M (2013) Managing the seeding rate to improve nitrogen-use efficiency of winter wheat. Field Crop Res 154:100–109

Dapkekar A, Deshpande P, Oak MD, Paknikar KM, Rajwade JM (2018) Zinc use efficiency is enhanced in wheat through nanofertilization. Sci Rep 8:6832

Das B, Chakraborty D, Singh VK, Aggarwal P, Singh R, Dwivedi BS, Mishra RP (2014) Effect of integrated nutrient management practice on soil aggregate properties, its stability and aggregate-associated carbon content in an intensive rice–wheat system. Soil Tillage Res 136:9–18

Das S, Mohammad A, Sarode K, Pershwitz G, Hadidi O, Thomas R, Vinas A, Abu-Fadel M, Dieter RS, Ali MI, Shammas N (2018) CRT-213 comparison of revascularization outcomes for patients with lifestyle limiting claudication to Critical Limb Ischemia. J Am Coll Cardiol Intv 7:S37–S38

Davarpanah S, Tehranifar A, Davarynejad G, Abadía J, Khorasani R (2016) Effects of foliar applications of zinc and boron nano-fertilizers on pomegranate (*Punica granatum* cv. Ardestani) fruit yield and quality. Sci Hortic 210:57–64

Davarpanah S, Tehranifar A, Abadía J, Val J, Davarynejad G, Aran M, Khorassani R (2018) Foliar calcium fertilization reduces fruit cracking in pomegranate (*Punica granatum* cv. Ardestani). Sci Hortic 230:86–91

Delgado, Lemunyon (2006) Nutrient management. In: Lal R (ed) Encyclopedia of soil science. Boca Raton, CRC Press, pp 1157–1160

Demir Z, Gülser C (2015) Effects of rice husk compost application on soil quality parameters in greenhouse conditions. Eur J Soil Sci 4:185

DeRosa MC, Monreal C, Schnitzer M, Walsh R, Sultan Y (2010) Nanotechnology in fertilizers. Nat Nanotechnol 5:91

Dhaliwal JK, Saroa GS, Dhaliwal SS (2014) Effect of nutrient management techniques on phosphorus availability in basmati-wheat system. Int J Sci Environ Technol 3:1602–1608

Diacono M, Rubino P, Montemurro F (2013) Precision nitrogen management of wheat. A review. Agron Sustain Dev 33:219–241

Dimkpa C, Bindraban P, McLean JE, Gatere L, Singh U, Hellums D (2017) Methods for rapid testing of plant and soil nutrients. In: Sustainable agriculture reviews. Springer, Cham, pp 1–43

Dobermann A, Witt C, Dawe D, Abdulrachman S, Gines HC, Nagarajan R, Satawathananont S, Son TT, Tan PS, Wang GH, Chien NV (2002) Site-specific nutrient management for intensive rice cropping systems in Asia. Field Crop Res 74:37–66

Dobermann A, Ping JL, Adamchuk VI, Simbahan GC, Ferguson RB (2003) Classification of crop yield variability in irrigated production fields. Agron J 95:1105–1120

Drinkwater LE, Wagoner P, Sarrantonio M (1998) Legume-based cropping systems have reduced carbon and nitrogen losses. Nature 396:262

Drostkar E, Talebi R, Kanouni H (2016) Foliar application of Fe, Zn and NPK nano-fertilizers on seed yield and morphological traits in chickpea under rainfed condition. J Res Ecol 4:221–228

Duffner A, Hoffland E, Weng L, Van der Zee SEATM (2013) Predicting zinc bioavailability to wheat improves by integrating pH dependent nonlinear root surface adsorption. Plant Soil 373:919–930

Dungait JA, Cardenas LM, Blackwell S, Wu L, Withers PJ, Chadwick DR, Bol R, Murray PJ, Macdonald AJ, Whitmore AP, Goulding KW (2012) Advances in the understanding of nutrient dynamics and management in UK agriculture. Sci Total Environ 434:39–50

Dwivedi D, Johri BN (2003) Antifungals from fluorescent pseudomonads: biosynthesis and regulation. Curr Sci 25:1693–1703

Edmeades DC (2003) The long-term effects of manures and fertilizers on soil productivity and quality: a review. Nutr Cycl Agroecosyst 66:165–180

Edwards JH, Wood CW, Thurlow DL, Ruf ME (1992) Tillage and crop rotation effects on fertility status of a Hapludult soil. Soil Sci Soc Am J 56:1577–1582

Fageria NK, Filho MB, Moreira A, Guimarães CM (2009) Foliar fertilization of crop plants. J Plant Nutr 32:1044–1064

Fageria NK, Moreira A, Coelho AM (2011) Yield and yield components of upland rice as influenced by nitrogen sources. J Plant Nutr 34:361–370

Falatah AM, Al-Derby AM (1993) Chemical properties of a calcareous soil as affected by tillages practices in Saudi Arabia. Arid Soil Res Rehabil 7:347–353

FAO (2014) Fertilizer consumption database. http://faostat3.fao.org/faostat-gateway/go/to/download/R/*/E. Accessed 5 July 2018

FAO (2015) World fertilizer trends and outlook to 2018. Food and Agriculture organization of the United Nations (FAO), Rome, ISBN 978-92-5-108692-6, Annual Report 14

Farooq M, Wahid A, Kobayashi N, Fujita D, Basra SMA (2009) Plant drought stress: effects, mechanisms and management. In: Sustainable agriculture. Springer, Dordrecht, pp 153–188

Farooq M, Hussain M, Wahid A, Siddique KHM (2012) Drought stress in plants: an overview. In: Plant responses to drought stress. Springer, Berlin/Heidelberg, pp 1–33

Farooq M, Ullah A, Rehman A, Nawaz A, Nadeem A, Wakeel A, Nadeem F, Siddique KH (2018) Application of zinc improves the productivity and biofortification of fine grain aromatic rice grown in dry seeded and puddled transplanted production systems. Field Crop Res 216:53–62

Farrell M, Jones DL (2009) Critical evaluation of municipal solid waste composting and potential compost markets. Bioresour Technol 100:4301–4310

Fixen PE (2009) Nutrient use efficiency in the context of sustainable agriculture. In: IPNI symposium on nutrient use efficiency at the Latin American congress of soil science, San Jose, pp 1–10

Freyer B (2003) Crop rotation. Eugen Ulmer GmbH & Co, Stuttgart

Gamalero E, Glick BR (2011) Mechanisms used by plant growth-promoting bacteria. In: Bacteria in agrobiology: plant nutrient management. Springer, Berlin/Heidelberg, pp 17–46

Gao Y, Duan A, Qiu X, Liu Z, Sun J, Zhang J, Wang H (2010) Distribution of roots and root length density in a maize/soybean strip intercropping system. Agric Water Manag 98:199–212

Gawde N, Singh AK, Agrawal SK, Kumar R (2017) Long term effect of integrated nutrient management on soil nutrient status under rice-wheat cropping system in Inceptisols. Int J Chem Stud 5:1050–1057

Geleta B, Atak M, Baenziger PS, Nelson LA, Baltenesperger DD, Eskridge KM, Shipman MJ, Shelton DR (2002) Seeding rate and genotype effect on agronomic performance and end-use quality of winter wheat. Crop Sci 42:827–832

Ghanbari A, Dahmardeh M, Siahsar BA, Ramroudi M (2010) Effect of maize (*Zea mays* L.) cowpea (*Vigna unguiculata* L.) intercropping on light distribution, soil temperature and soil moisture in arid environment. J Food Agric Environ 8:102–108

Gholami L, Sadeghi SHR, Homaee M (2013) Straw mulching effect on splash erosion, runoff and sediment yield from eroded plots. Soil Sci Soc Am J 77:268–278

Ghosh PK, Manna MC, Bandyopadhyay KK, Tripathi AK, Wanjari RH, Hati KM, Misra AK, Acharya CL, Subba Rao A (2006) Interspecific interaction and nutrient use in soybean/sorghum intercropping system. Agron J 98:1097–1108

Gitari HI, Karanja NN, Gachene CK, Kamau S, Sharma K, Schulte-Geldermann E (2018) Nitrogen and phosphorous uptake by potato (*Solanum tuberosum* L.) and their use efficiency under potato-legume intercropping systems. Field Crop Res 222:78–84

Glick BR (1995) The enhancement of plant growth by free-living bacteria. Can J Microbiol 41:109–117

Godfray HCJ, Beddington JR, Crute IR, Haddad L, Lawrence D, Muir JF, Pretty J, Robinson S, Thomas SM, Toulmin C (2010) Food security: the challenge of feeding 9 billion people. Science 327:812–818

Gomiero T (2016) Soil degradation, land scarcity and food security: reviewing a complex challenge. Sustainability 8:281

Goulding K, Jarvis S, Whitmore A (2008) Optimizing nutrient management for farm systems. Philos Trans R Soc Sci 363:667–680

Goyal S, Mishra MM, Hooda IS, Singh R (1992) Organic matter–microbial biomass relationships in field experiments under tropical conditions: effects of inorganic fertilization and organic amendments. Soil Biol Biochem 24:1081–1084

Goyal S, Chander K, Mundra MC, Kapoor KK (1999) Influence of inorganic fertilizers and organic amendments on soil organic matter and soil microbial properties under tropical conditions. Biol Fertil Soils 29:196–200

Gregoret MC, Zorita MD, Dardanelli J, Bongiovanni RG (2011) Regional model for nitrogen fertilization of site-specific rainfed corn in haplustolls of the central pampas, Argentina. Precis Agric 12(6):831–849

Green JM, Beestman GB (2007) Recently patented and commercialized formulation and adjuvant technology. Crop Prot 26:320–327

Gui J, Zheng S, Shen J, Li L (2015) Grain setting defect1 (GSD1) function in rice depends on S-acylation and interacts with actin 1 (OsACT1) at its C-terminal. Front Plant Sci 6:804

Guldan SJ, Martin CA, Lindemann WC, Cueto-Wong J, Steiner RL (1997) Yield and green-manure benefits of inter seeded legumes in a high-desert environment. Agron J 89:757–762

Gunes A, Inal A, Alpaslan M, Eraslan F, Bagci EG, Cicek N (2007) Salicylic acid induced changes on some physiological parameters symptomatic for oxidative stress and mineral nutrition in maize (*Zea mays* L.) grown under salinity. J Plant Physiol 164:728–736

Guo Y, George E, Marschner H (1996) Contribution of an arbuscular mycorrhizal fungus to uptake of Cadmnium and Nickel in bean by maize plants. Plant Soil 184:195–205

Gupta S, Agarwal R, Mittal SK (2016) Respiratory health concerns in children at some strategic locations from high PM levels during crop residue burning episodes. Atmos Environ 137:127–134

Habets S, de Wild PJ, Huijgen WJJ, van Eck ERH (2013) The influence of thermochemical treatments on the lignocellulosic structure of wheat straw as studied by natural abundance 13 C NMR. Bioresour Technol 146:585–590

Harris L, Fritsche H, Mennel R, Norton L, Ravdin P, Taube S, Somerfield MR, Hayes DF, Bast RC Jr (2007) American society of clinical oncology 2007 update of recommendations for the use of tumor markers in breast cancer. J Clin Oncol 25:5287–5312

Harris JA, Livesey EJ, Gharaei S, Westbrook RF (2008) Negative patterning is easier than a biconditional discrimination. J Exp Psychol Anim Behav Process 34:494

Hauggaard-Nielsen H, Jørnsgaard B, Kinane J, Jensen ES (2008) Grain legume–cereal intercropping: The practical application of diversity, competition and facilitation in arable and organic cropping systems. Renewable Agric Food Syst 23:3–12

Hauggaard-Nielsen H, Gooding M, Ambus P, Corre-Hellou G, Crozat Y, Dahlmann C, Dibet A, Von Fragstein P, Pristeri A, Monti M, Jensen ES (2009) Pea–barley intercropping for efficient symbiotic N_2-fixation, soil N acquisition and use of other nutrients in European organic cropping systems. Field Crop Res 113:64–71

Havlin JL, Beaton JD, Tisdale SL, Nelson WL (1999) Soil fertility and fertilizers, 6th edn. Prentice Hall, Upper Saddle River, p 499

Hayat R, Ali S (2010) Nitrogen fixation of legumes and yield of wheat under legumes-wheat rotation in Pothwar. Pak J Bot 42:2317–2326

Haygarth PM, Bardgett RD, Condron LM (2013) Phosphorus and nitrogen cycles and their management. In: Russell's soil conditions and plant growth, 12th edn. Wiley-Blackwell, Hoboken, pp 132–158
He W, Shohag MJI, Wei Y, Feng Y, Yang X (2013) Iron concentration bioavailability, and nutritional quality of polished rice affected by different forms of foliar iron fertilizer. Food Chem 141:4122–4126
He Z, Xia X, Peng S, Lumpkin TA (2014) Meeting demands for increased cereal production in China. J Cereal Sci 59:235–244
Hiltbrunner J, Liedgens M, Bloch L, Stamp P, Streit B (2007) Legume cover crops as living mulches for winter wheat: components of biomass and the control of weeds. Eur J Agron 26:21–29
Hodge A, Campbell CD, Fitter AH (2001) An arbuscular mycorrhizal fungus accelerates decomposition and acquires nitrogen directly from organic material. Nature 413:297
Hossain MS, Hossain A, Sarkar MA, Jahiruddin M, da Silva JA, Hossain MI (2016) Productivity and soil fertility of the rice–wheat system in the High Ganges River Floodplain of Bangladesh is influenced by the inclusion of legumes and manure. Agric Ecosyst Environ 218:40–52
Huang S, Zeng Y, Wu J, Shi Q, Pan X (2013) Effect of crop residue retention on rice yield in China: a meta-analysis. Field Crop Res 154:188–194
Imran M, Boelt B, Mühling KH (2018) Zinc seed priming improves salt resistance in maize. J Agron Crop Sci 204:390–399
Irfan SA, Razali R, KuShaari K, Mansor N, Azeem B, Versypt AN (2018) A review of mathematical modeling and simulation of controlled-release fertilizers. J Control Release 271:45–54
Jaga PK, Patel Y (2012) An overview of fertilizers consumption in India: determinants and outlook for 2020–a review. Int J Sci Eng Technol 1:285–229
Janssen BH (1993) Integrated nutrient management: the use of organic and mineral fertilizers. In: van Reuler H, Prins WH (eds) The role of plant nutrients for sustainable crop production in Sub-Saharan Africa. Ponsen and Looijen, Wageningen, pp 89–105
Jat RK, Sapkota TB, Singh RG, Jat ML, Kumar M, Gupta RK (2014) Seven years of conservation agriculture in a rice–wheat rotation of Eastern Gangetic Plains of South Asia: yield trends and economic profitability. Field Crop Res 164:199–210
Jat RD, Jat HS, Nanwal RK, Yadav AK, Bana A, Choudhary KM, Kakraliya SK, Sutaliya JM, Sapkota TB, Jat ML (2018) Conservation agriculture and precision nutrient management practices in maize-wheat system: Effects on crop and water productivity and economic profitability. Field Crop Res 222:111–120
Jeffries P, Gianinazzi S, Perotto S, Turnau K, Barea JM (2003) The contribution of arbuscular mycorrhizal fungi in sustainable maintenance of plant health and soil fertility. Biol Fertil Soils 37:1–16
Jiang D, Zhuang D, Fu J, Huang Y, Wen K (2012) Bioenergy potential from crop residues in China: availability and distribution. Renew Sust Energ Rev 16:1377–1382
Jin J, Jiang C (2002) Spatial variability of soil nutrients and site-specific nutrient management in the PR China. Comput Electron Agric 36(2–3):165–172
Johansen C, Musa AM, Kumar Rao JVDK, Harris D, Shahidullah AKM, Lauren GJ (2006) Seed priming with molybdenum alleviates molybdenum deficiency and poor nitrogen fixation of chickpea in acid soils of Bangladesh and India. In: 18th world congress of soil science
Johansen C, Musa AM, Kumar Rao JVDK, Harris D, Yusuf AM, Shahidullah AKM, Lauren JG (2007) Correcting molybdenum deficiency of chickpea in the High Barind Tract of Bangladesh. J Plant Nutr Soil Sci 170:752–761
Jordán A, Zavala LM, Gil J (2010) Effects of mulching on soil physical properties and runoff under semi-arid conditions in southern Spain. Catena 81:77–85
Joseph T, Morrison M (2006) Nanotechnology in agriculture and food. Nanoforum Report, Institute of Nanotechnology, April 2006, pp 1–13
Ju XT, Xing GX, Chen XP, Zhang SL, Zhang LJ, Liu XJ, Cui ZL, Yin B, Christie P, Zhu ZL, Zhang FS (2009) Reducing environmental risk by improving N management in intensive Chinese agricultural systems. Proc Natl Acad USA 106:3041–3046

Ju XT, Kou CL, Christie P, Dou ZX, Zhang FS (2014) Changes in the soil environment from excessive application of fertilizers and manures to two contrasting intensive cropping systems on the North China Plain. Environ Pollut 145:497–506

Kakraliya SK, Jat RD, Kumar S, Choudhary KK, Prakash J, Singh LK (2017) Integrated nutrient management for improving, fertilizer use efficiency, soil biodiversity and productivity of wheat in irrigated rice wheat cropping system in indo-gangatic plains of India. Int J Curr Microbiol App Sci 6:152–163

Karami A, Homaee M, Afzalinia S, Ruhipour H, Basirat S (2012) Organic resource management: impacts on soil aggregate stability and other soil physico-chemical properties. Agric Ecosyst Environ 148:22–28

Katyal JC, Singh B, Vlek PLG, Buresh RJ (1987) Efficient nitrogen use as affected by urea application and irrigation sequence 1. Soil Sci Soc Am J 51:366–370

Kaur T, Brar BS, Dhillon NS (2008) Soil organic matter dynamics as affected by long-term use of organic and inorganic fertilizers under maize–wheat cropping system. Nutr Cycl Agroecosyst 81:59–69

Khoshgoftar AH, Chaney RL (2007) Preceding crop affects grain cadmium and zinc of wheat grown in saline soils of central Iran. J Environ Qual 36:1132–1136

Kirchmann H, Kätterer T, Bergström L (2009) Nutrient supply in organic agriculture–plant availability, sources and recycling. In: Organic crop production–ambitions and limitations. Springer, Dordrecht, pp 89–116

Kottegoda N, Munaweera I, Madusanka N, Karunaratne V (2011) A green slow-release fertilizer composition based on urea-modified hydroxyapatite nanoparticles encapsulated wood. Curr Sci 10:73–78

Kucey RMN (1988) Effect of Penicillium bilaji on the solubility and uptake of P and micronutrients from soil by wheat. Canad J Soil Sci 68:261–270

Kumar K, Goh KM (1999) Crop residues and management practices: effects on soil quality, soil nitrogen dynamics, crop yield, and nitrogen recovery. Adv Agron 68:197–319

Kumar DG, Sachin SS, Kumar R (1990) Importance of mulch in crop production. Indian J Soil Conserv 18:20–26

Kumar A, Roberts D, Wood KE, Light B, Parrillo JE, Sharma S, Suppes R, Feinstein D, Zanotti S, Taiberg L, Gurka D (2006) Duration of hypotension before initiation of effective antimicrobial therapy is the critical determinant of survival in human septic shock. Crit Care Med 34:1589–1596

Kumar P, Barrett DM, Delwiche MJ, Stroeve P (2009) Methods for pretreatment of lignocellulosic biomass for efficient hydrolysis and biofuel production. Ind Eng Chem Res 48:3713–3729

Latif A, Aslam SM, Iqbal Z, Shah SA (2001) Effect of fertigation applied nitrogen and phosphorus on yield and composition of maize. Pak J Soil Sci 19:23–26

Leigh J, Hodge A, Fitter AH (2009) Arbuscular mycorrhizal fungi can transfer substantial amounts of nitrogen to their host plant from organic material. New Phytol 181:199–207

Li L, Weinberg CR, Darden TA, Pedersen LG (2001) Gene selection for sample classification based on gene expression data: study of sensitivity to choice of parameters of the GA/KNN method. Bioinformatics 17:1131–1142

Li Y, Zhang W, Ma L, Huang G, Oenema O, Zhang F, Dou Z (2013) An analysis of China's fertilizer policies: impacts on the industry, food security, and the environment. J Environ Qual 42:972–981

Liu R, Lal R (2014) Synthetic apatite nanoparticles as a phosphorus fertilizer for soybean (*Glycine max*). Sci Rep 4:5686

Liu Z, Yang J, Yang Z, Zou J (2012) Effects of rainfall and fertilizer types on nitrogen and phosphorus concentrations in surface runoff from subtropical tea fields in Zhejiang, China. Nutr Cycl Agroecosyst 93:297–307

López-Pedrosa A, González-Guerrero M, Valderas A, Azcón-Aguilar C, Ferrol N (2006) GintAMT1 encodes a functional high-affinity ammonium transporter that is expressed in the extraradical mycelium of Glomus intraradices. Fungal Genet Biol 43:102–110

Mahajan GR, Sahoo RN, Pandey RN, Gupta VK, Kumar D (2014) Using hyperspectral remote sensing techniques to monitor nitrogen, phosphorus, sulphur and potassium in wheat (*Triticum aestivum* L.). Precis Agric 15:499–522

Malhi SS, Lemke R, Wang ZH, Chhabra BS (2006) Tillage, nitrogen and crop residue effects on crop yield, nutrient uptake, soil quality, and greenhouse gas emissions. Soil Tillage Res 90:171–183

Manikandan A, Subramanian KS (2016) Evaluation of zeolite based nitrogen nano-fertilizers on maize growth, yield and quality on inceptisols and alfisols. Int J Plant Soil Sci 9:1–9

Manjunatha SB, Biradar DP, Aladakatti YR (2016) Nanotechnology and its application to agriculture: a review. J Farm Sci 29:1–13

Marschner P (2012) Marschner's mineral nutrition of higher plants, 3rd edn. Academic, London

Maruthi Sankar GR, Sharma KL, Padmalatha Y, Bhargavi K, Babu MVS, Naga Sravaniv, Ramachandrappa BK, Dhanapal G, Sharma S, Thakur HS, Devi RA, Jawahar D, Ghabane VV, Abrol V, Singh B, Sharma P, Kumar NA, Girija A, Ravi P, Venkateswarlu B, Singh AK (2012) Efficient nutrient management practice for sustainable crop productivity and soil fertility maintenance based on permanent manorial experiments in different soil and agro-climatic conditions. In: Abrol V, Sharma P (eds) Resource management for sustainable agriculture. IntechOpen, ISBN: 978-953-51-0808-5. https://doi.org/10.5772/50572

Mastronardi E, Tsae P, Zhang X, Monreal CM, DeRosa MC (2015) Strategic role of nanotechnology in fertilizers: potential and limitations. In: Rai M, Ribeiro C, Mattoso L, Duran N (eds) Nanotechnologies in food and agriculture. Springer, Cham, pp 25–67

Masuthi DA, Vyakaranahal BS, Deshpande VK (2009) Influence of pelleting with micronutrients and botanical on growth, seed yield and quality of vegetable cowpea. Karnataka J Agric Sci 22:898–900

Matiru VN, Dakora FD (2004) Potential use of rhizobial bacteria as promoters of plant growth for increased yield in landraces of African cereal crops. Afr J Biotechnol 3:1–7

Medina M, Monreal C, Chabot D, Meier S, González ME, Morales E, Parillo R, Borie F, Cornejo P (2017) Microscopic and spectroscopic characterization of humic substances from a compost amended copper contaminated soil: main features and their potential effects on Cu immobilization. Environ Sci Pollut Res 24:14104–14116

Meyer JR, Linderman RG (1986) Selective influence on populations of rhizosphere or rhizoplane bacteria and actinomycetes by mycorrhizas formed by Glomus fasciculatum. Soil Biol Biochem 18:191–196

Milani N, McLaughlin MJ, Stacey SP, Kirby JK, Hettiarachchi GM, Beak DG, Cornelis G (2012) Dissolution kinetics of macronutrient fertilizers coated with manufactured zinc oxide nanoparticles. J Agric Food Chem 60:3991–3998

Milani N, Hettiarachchi GM, Kirby JK, Beak DG, Stacey SP, McLaughlin MJ (2015) Fate of zinc oxide nanoparticles coated onto macronutrient fertilizers in an alkaline calcareous soil. PLoS ONE 10:e0126275

Mondal S, Mallikarjun M, Ghosh M, Ghosh DC, Timsina J (2016) Influence of integrated nutrient management (INM) on nutrient use efficiency, soil fertility and productivity of hybrid rice. Arch Agron Soil Sci. https://doi.org/10.1080/03650340.2016.1148808

Moreira A, Moraes LA, Schroth G, Mandarino JM (2015) Effect of nitrogen, row spacing, and plant density on yield, yield components, and plant physiology in soybean–wheat intercropping. Agron J 107:2162–2170

Mosier A, Kroeze C (1998) A new approach to estimate emissions of nitrous oxide from agriculture and its implications for the global NzO budget. Glob Chang Newsl 34:8–13

Naderi MR, Abedi A (2012) Application of nanotechnology in agriculture and refinement of environmental pollutants. J Nanotechnol 11:18–26

Nadi E, Aynehband A, Mojaddam M (2013) Effect of nano-iron chelate fertilizer on grain yield, protein percent and chlorophyll content of Faba bean (*Vicia faba* L.). Int J Biosci 3:267–272

Nayyar VK, Takkar PN, Bansal RL, Singh SP, Kaur NP, Sadana US (1990) Micronutrients in soils and crops of Punjab. Research Bulletin, Department of Soils, Punjab Agricultural University, Ludhiana, India. Economic return, 1

Nicholls CI, Altieri MA (2001) Manipulating plant biodiversity to enhance biological control of insect pests: a case study of a northern California vineyard. In: Gliessman SR (ed) Agroecosystem sustainability: developing practical strategies. CRC Press, Boca Raton, pp 29–50

Niu Y, Li H (2012) Controlled release of urea encapsulated by starch-g-poly (vinyl acetate). Ind Eng Chem Res 51:12173–12177

Oertli JJ (1980) Controlled-release fertilizers. Fert Res 1:103–123

Oliver MA, Gregory PJ (2015) Soil, food security and human health: a review. Eur J Soil Sci 66:257–276

Onoyama H, Ryu C, Suguri M, Iida M (2015) Nitrogen prediction model of rice plant at panicle initiation stage using ground-based hyperspectral imaging: growing degree-days integrated model. Precis Agric 16:558–570

Oswal MC (1994) Soil water functional relationships of sandy soils and their modification. J Indian Soc Soil Sci 41:145–147

Otálora G, Piñero MC, López-Marín J, Varó P, del Amor FM (2018) Effects of foliar nitrogen fertilization on the phenolic, mineral, and amino acid composition of escarole (*Cichorium endivia* L. var. latifolium). Sci Hortic 239:87–92

Ozturk L, Yazici MA, Yucel C, Torun A, Cekic C, Bagci A, Ozkan H, Braun HJ, Sayers Z, Cakmak I (2006) Concentration and localization of zinc during seed development and germination in wheat. Physiol Plant 128:144–152

Parrish JA, Bazzaz FA (1985) Nutrient content of Abutilon theophrasti seeds and the competitive ability of the resulting plants. Oecologia 65:247–251

Paul BK, Vanlauwe B, Ayuke F, Gassner A, Hoogmoed M, Hurisso TT, Koala S, Lelei D, Ndabamenye T, Six J, Pulleman MM (2013) Medium-term impact of tillage and residue management on soil aggregate stability, soil carbon and crop productivity. Agric Ecosyst Environ 164:14–22

Paynel F, Murray PJ, Cliquet JB (2001) Root exudates: a pathway for short-term N transfer from clover and ryegrass. Plant Soil 229:235–243

Peng CYJ, Lee KL, Ingersoll GM (2002) An introduction to logistic regression analysis and reporting. J Educ Res 96:3–14

Pérez-Tienda J, Testillano PS, Balestrini R, Fiorilli V, Azcón-Aguilar C, Ferrol N (2011) GintAMT2, a new member of the ammonium transporter family in the arbuscular mycorrhizal fungus Glomus intraradices. Fungal Genet Biol 48:1044–1055

Prasad P, Satyanarayana V, Murthy V, Boote KJ (2002) Maximizing yields in rice-groundnut cropping sequence through integrated nutrient management. Field Crop Res 75:9–21

Prasad TNVKV, Sudhakar P, Sreenivasulu Y, Latha P, Munaswamy V, Reddy KR, Pradeep T (2012) Effect of nanoscale zinc oxide particles on the germination, growth and yield of peanut. J Plant Nutr 35:905–927

Quinton JN, Govers G, Van Oost K, Bardgett RD (2010) The impact of agricultural soil erosion on biogeochemical cycling. Nat Geosci 3:311–314

Rahim A, Hussain M, Abid M, Haq MZ, Ahmad S (2012) Phosphorus use efficiency of Trititicum aestivum as affected by band placement ofphosphorus and farmyard manure on calcareous soils. Pak J Bot 44:1391–1398

Rahmatullah, Gill MA, Shaikh BZ, Salim M (1994) Bioavailability and distribution of P among inorganic fractions in calcareous soils. Arid Soil Res Rehabil 8:227–234

Ramesh K, Thirumurugan V (2001) Effect of seed pelleting and foliar nutrition on growth of soybean. Madras Agric J 88:465–468

Reganold JP, Elliott LE, Unger YL (1987) Long-term effects of organic and conventional farming on soil erosion. Nature 330:370–372

Rehman J, Zhang HJ, Toth PT, Zhang Y, Marsboom G, Hong Z, Salgia R, Husain AN, Wietholt C, Archer SL (2012) Inhibition of mitochondrial fission prevents cell cycle progression in lung cancer. FASEB J 26:2175–2186

Rehman W, Milot RL, Eperon GE, Wehrenfennig C, Boland JL, Snaith HJ, Johnston MB, Herz LM (2015) Charge-carrier dynamics and mobilities in formamidinium lead mixed-halide perovskites. Adv Mater 27:7938–7944

Rehman S, Guo S, Hou Y (2016) Rational Design of Si/SiO2@ Hierarchical Porous Carbon Spheres as Efficient Polysulfide Reservoirs for High-Performance Li–S Battery. Adv Mater 28:3167–3172

Rehman A, Farooq M, Naveed M, Nawaz A, Shahzad B (2018a) Seed priming of Zn with endophytic bacteria improves the productivity and grain biofortification of bread wheat. Eur J Agron 94:98–107

Rehman A, Farooq M, Naveed M, Ozturk L, Nawaz A (2018b) Pseudomonas-aided zinc application improves the productivity and biofortification of bread wheat. Crop Pasture Sci 69:659–672

Rehman A, Farooq M, Ozturk L, Asif M, Siddique KH (2018c) Zinc nutrition in wheat-based cropping systems. Plant Soil 422:283–315

Rengel Z, Graham RD (1995) Wheat genotypes differ in zinc efficiency when grown in the chelate-buffered nutrient solution. I. Growth. Plant Soil 176:307–316

Roca-Perez L, Martínez C, Marcilla P, Boluda R (2009) Composting rice straw with sewage sludge and compost effects on the soil-plant system. Chemosphere 75:781–787

Rodriguez C, Mayo JC, Sainz RM, Antolín I, Herrera F, Martín V, Reiter RJ (2004) Regulation of antioxidant enzymes: a significant role for melatonin. J Pineal Res 36:1–9

Roy RN, Finck A, Blair GJ, Tandon HLS (2006) Nutrient management guidelines for some major field crops. In: Plant nutrition for food security. A guide for integrated nutrient management, FAO fertilizers and plant nutrition bulletin 16. FAO, United Nations, Rome, pp 235–226

Saha PK, Ishaque M, Saleque MA, Miah MAM, Panaullah GM, Bhuiyan NI (2007) Long-term integrated nutrient management for rice-based cropping pattern: effect on growth, yield, nutrient uptake, nutrient balance sheet, and soil fertility. Commun Soil Sci Plant Anal 38:579–610

Salman OA (1989) Polyethylene-coated urea. 1. Improved storage and handling properties. Ind Eng Chem Res 28:630–632

Salvagiotti F, Cassman KG, Specht JE, Walters DT, Weiss A, Dobermann A (2008) Nitrogen uptake, fixation and response to fertilizer N in soybeans: a review. Field Crop Res 108:1–13

Scherer HW (2005) Fertilizers and fertilization. In: Scow KM, Hillel D, Powlson DS, Hatfield JL (eds) Encyclopedia of soils in the environment, 4th volume

Scott JM (1989) Seed coatings and treatments and their effects on plant establishment. Adv Agron 42:43–83

Shahid M, Nayaka AK, Pureea C, Tripathia R, Lala B, Gautama P, Bhattacharyyaa P, Mohantya S, Kumara A, Pandaa BB, Kumara U, Shukla AK (2017) Carbon and nitrogen fractions and stocks under 41 years of chemical and organic fertilization in a sub-humid tropical rice soil. Soil Tillage Res 170:136–146

Sharma KL, Ramachandrappa BK, Chandrika DS, Sathish A, Dhanpal GN, Rao CS, Shankar MA, Grace JK, Sankar GM, Chary GR, Munnalal (2016) Effect of Organic Manure and Crop Residue Based Long-Term Nutrient Management Systems on Soil Quality Changes under Sole Finger Millet (*Eleusine coracana* (L.) Gaertn.) and Groundnut (*Arachis hypogaea* L.)–Finger Millet Rotation in Rainfed Alfisol. Commun Soil Sci Plant Anal 47:899–914

Sharpley AN, Meisinger JJ, Power JF, Suarez DL (1992) Root extraction of nutrients associated with long-term soil management. In: Hatfield JL, Stewart BA (eds) Limitations to plant root growth. Springer, New York, pp 151–217

Shepherd MA, Chambers B (2007) Managing nitrogen on the farm: the devil is in the detail. J Sci Food Agric 87:558–568

Shiferaw H, Teketay D, Nemomissa S, Assefa F (2004) Some biological characteristics that foster the invasion of Prosopis juliflora (Sw.) DC. at Middle Awash Rift Valley Area, north-eastern Ethiopia. J Arid Environ 58:135–154

Shoji S, Kanno H (1994) Use of polyolefin-coated fertilizers for increasing fertilizer efficiency and reducing nitrate leaching and nitrous oxide emissions. Fert Res 39:147–152
Shoji S, Delgado J, Mosier A, Miura Y (2001) Use of controlled release fertilizers and nitrification inhibitors to increase nitrogen use efficiency and to conserve air and water quality. Commun Soil Sci Plant Anal 32:1051–1070
Singh B, Natesan SKA, Singh BK, Usha K (2003) Improving zinc efficiency of cereals under zinc deficiency. Curr Sci 88:36–44
Singh VK, Tiwari KN, Gill MS, Sharma SK, Dwivedi BS, Shukla AK, Mishra PP (2008) Economic viability of site-specific nutrient management in rice wheat cropping system. Better Crops 92:28–30
Singh N, Agarwal R, Awasthi A, Gupta PK, Mittal SK (2010) Characterization of atmospheric aerosols for organic tarry matter and combustible matter during crop residue burning and non-crop residue burning months in north-western region of India. Atmos Environ 44:1292–1300
Snapp SS, Mafongoya PL, Waddington S (1998) Organic matter technologies for integrated nutrient management in smallholder cropping systems of southern Africa. Agric Ecosyst Environ 7:185–200
Solanki P, Bhargava A, Chhipa H, Jain N, Panwar J (2015) Nano-fertilizers and their smart delivery system. In: Rai M, Ribeiro C, Mattoso L, Duran N (eds) Nanotechnologies in food and agriculture. Springer, Cham/Heidelberg/New York/Dordrecht/London, pp 81–101
Soltani RDC, Rezaee A, Khataee AR, Safari M (2014) Photocatalytic process by immobilized carbon black/ZnO nanocomposite for dye removal from aqueous medium: Optimization by response surface methodology. J Ind Eng Chem 20:1861–1868
Song XY, Wang JH, Huang WJ, Liu LY, Yan GJ, Pu RL (2009) The delineation of agricultural management zones with high resolution remotely sensed data. Precis Agric 10:471–487
Sonmez S, Kaplan M, Somnez NK, Kaya H, Uz I (2007) Effect of both soil copper applications and foliar copper application frequencies on macronutrients contents of tomato plants. Asian J Chem 19:5372–5384
Swaney DP, Hong B, Ti C, Howarth RW, Humborg C (2012) Net anthropogenic nitrogen inputs to watersheds and riverine N export to coastal waters: a brief overview. Curr Opin Environ Sustain 4:203–211
Talgre L, Lauringson E, Roostalu H, Astover A, Makke A (2012) Green manure as a nutrient source for succeeding crops. Plant Soil Environ 58:275–281
Tarafdar JC, Raliya R, Rathore I (2012) Microbial synthesis of phosphorous nanoparticle from tri-calcium phosphate using Aspergillus tubingensis TFR-5. J Bionanosci 6:84–89
Teit R (1990) Soil organic matter biological and ecological effects. Wiley, New York, p 395
Thierfelder C, Mwila M, Rusinamhodzi L (2013) Conservation agriculture in eastern and southern provinces of Zambia: long-term effects on soil quality and maize productivity. Soil Tillage Res 126:246–258
Tomaszewska M, Jarosiewicz A (2002) Use of polysulfone in controlled-release NPK fertilizer formulations. J Agric Food Chem 50:4634–4639
Torney F, Trewyn BG, Lin VS-Y, Wang K (2007) Mesoporous silica nanoparticles deliver DNA and chemicals into plants. Nat Nanotechnol 2:295–300
Trenkel ME (1997) Improving Fertilizer Use Efficiency: Controlled-Release and Stabilized Fertilizers in Agriculture. International Fertilizer Industry Association, Paris, p 151
Trenkel ME (2010) Slow and controlled-release and stabilized fertilizers: an option for enhancing nutrient use efficiency in agriculture. International Fertilizer Industry Association (IFA), Paris
Trenkel ME (2012) Slow and controlled-release and stabilized fertilizers. An option for enchancing nutrient efficiency in agriculture, 2nd edn. IFA, Paris
Tsuji T (2000) Water in emission in the infrared space observatory spectrum of the early M supergiant Star μ Cephei. Astrophys J Lett 540:99
Ullah A, Farooq M, Nadeem A, Rehman A, Asad SA, Nawaz A (2017a) Manganese nutrition improves the productivity and grain biofortification of fine grain aromatic rice in conventional and conservation production systems. Paddy Water Environ 15(3):563–572

Ullah A, Farooq M, Rehman A, Arshad MS, Shoukat H, Nadeem A, Nawaz A, Wakeel A, Nadeem F (2017b) Manganese nutrition improves the productivity and grain biofortification of bread wheat in alkaline calcareous soil. Exp Agric. https://doi.org/10.1017/S0014479717000369
US-EPA, United States Environmental Protection Agency (2011) Global anthropogenic non-CO_2 greenhouse gas emissions: 1990–2030 EPA 430-D-11-003. (Draft.), vol. 10. Office of Atmospheric Programs, Climate Change Division, U.S. Environmental Protection Agency, Washington, DC, pp 110–121
Vance CP, Uhde-Stone C, Allan DL (2003) Phosphorus acquisition and use: critical adaptations by plants for recurring a non-renewable resources. New Phytol 157:423–447
Vanlauwe B, Descheemaeker K, Giller KE, Huising J, Merckx R, Nziguheba G, Wendt J, Zingore S (2015) Integrated soil fertility management in sub-Saharan Africa: unravelling local adaptation. Soil 1:491–508
Vashishtha M, Dongara P, Singh D (2010) Improvement in properties of urea by phosphogypsum coating. Int J Chem Tech Res 2:36–44
Verma S (2017) Long term effect of nutrient management practices on soil quality parameters under maize-wheat cropping system. Doctoral dissertation. Birsa Agricultural University, Kanke, Ranchi, Jharkhand
Verma GSK, Meena RS, Maurya AC, Kumar S (2018) Nutrients uptake and available nutrients status in soil as influenced by sowing methods and herbicides in Kharif Maize (*Zea mays* L.). Int J Agric Environ Biotechnol 11:17–24
Vermeulen SJ, Campbell BM, Ingram JSI (2012) Climate change and food systems. Annu Rev Environ Resour 37:195–222
Viil P, Võsa T (2005) Green manure crops. Trans Estonian Res Inst Agric 148:16
Vincent P, Larochelle H, Lajoie I, Bengio Y, Manzagol PA (2010) Stacked denoising autoencoders: learning useful representations in a deep network with a local denoising criterion. J Mach Learn Res 11:3371–3408
Vitousek PM, Naylor R, Crews T, David MB, Drinkwater LE, Holland E, Johnes PJ, Katzenberger J, Martinelli LA, Matson PA, Nziguheba G (2009) Nutrient imbalances in agricultural development. Science 324:1519–1520
Voortman R, Bindraban PS (2015) Beyond N and P: toward a land resource ecology perspective and impactful fertilizer interventions in Sub-Saharan Africa, VFRC report 2015/1. Virtual Fertilizer Research Center, Washington, DC, p 49
Walia MK, Walia SS, Dhaliwal SS (2010) Long-term effect of integrated nutrient management of properties of typic ustochrept after 23 cycles of an irrigated rice (*Oryza sativa* L.) – Wheat (*Triticum aestivum* L.) system. J Sustain Agric 34:724–743
Wang JQ (2007) Analysis and Evaluation of Yield Increase of Fertilization and Nutrient Utilization Efficiency for Major Cereal Crops in China, Ph.D. Dissertation. China Agricultural University, Beijing (in Chinese)
Wang YP, Bremer EC, Feng C, Gannon CP, Havens WH, Li J, Meier TP (2014) Inventors, Hand Held Products Inc, assignee. Indicia reading terminal processing plurality of frames of image data responsively to trigger signal activation. United States patent US 8,794,526
Wang XC, Deng XY, Pu T, Song C, Yong TW, Yang F, Sun X, Liu WG, Yan YH, Du J, Liu J, Su K, Yang WY (2017) Contribution of interspecific interactions and phosphorus application to increasing soil phosphorus availability in relay intercropping systems. Field Crop Res 204:12–22
Wei Y, Li J, Li Y, Zhao B, Zhang L, Yang X, Chang J (2017) Research on permeability coefficient of a polyethylene controlled-release film coating for urea and relevant nutrient release pathways. Polym Test 59:90–98
Weissenhorn I, Leyval C, Belgy G, Berthelin J (1995) Arbuscular mycorrhizal contribution to heavy metal uptake by maize (*Zea mays* L.) in pot culture with contaminated soil. Mycorrhiza 5:245–251
West TO, Post WM (2002) Soil organic carbon sequestration rates by tillage and crop rotation. Soil Sci Soc Am J 66:1930–1946

Whitbread A, Blair G, Konboon Y, Lefroy R, Naklang K (2003) Managing crop residues, fertilizers and leaf litters to improve soil C, nutrient balances, and the grain yield of rice and wheat cropping systems in Thailand and Australia. Agric Ecosyst Environ 100:251–263

White PJ, Crawford JW, Díaz Álvarez MC, García Moreno R (2012) Soil management for sustainable agriculture. Appl Environ Soil Sci. https://doi.org/10.1155/2012/850739

Withers PJA, Edwards AC, Foy RH (2001) Phosphorus cycling in UK agriculture and implications for phosphorus loss from soil. Soil Use Manag 17:139–149

Witt C, Dobermann A, Abdulrachman S, Gines HC, Guanghuo W, Nagarajan R, Satawatananont S, Son TT, Tan PS, Simbahan GC, Olk DC (1999) Internal nutrient efficiencies of irrigated lowland rice in tropical and subtropical Asia. Field Crop Res 63:113–138

Witter E, Johansson G (2001) Potassium uptake from the subsoil by green manure crops. Biol Agric Hortic 19:127–141

Wu Z, Wang S, Zhao J, Chen L, Meng H (2014) Synergistic effect on thermal behavior during co-pyrolysis of lignocellulosic biomass model components blend with bituminous coal. Bioresour Technol 169:220–228

Wulff RD, Bazzaz FA (1992) Effect of the parental nutrient regime on growth of the progeny in Abutilon theophrasti (Malvaceae). Am J Bot 79:1102–1107

Wurth B (2007) Emissions of engineered and unintentionally produced nanoparticles to the soil. MSc thesis, ETH Zurich

Yang Y, Hu C, Abu-Omar MM (2012) Conversion of glucose into furans in the presence of AlCl3 in an ethanol–water solvent system. Bioresour Technol 116:190–194

Yang J, Gao W, Ren S (2015) Long-term effects of combined application of chemical nitrogen with organic materials on crop yields, soil organic carbon and total nitrogen in fluvo-aquic soil. Soil Tillage Res 151:67–74

Yang WS, Park BW, Jung EH, Jeon NJ, Kim YC, Lee DU, Shin SS, Seo J, Kim EK, Noh JH, Seok SI (2017) Iodide management in formamidinium-lead-halide–based perovskite layers for efficient solar cells. Science 356:1376–1379

Yaseen M, Aziz MZ, Manzoor A, Naveed M, Hamid Y, Noor S, Khalid MA (2017) Promoting growth, yield, and phosphorus-use efficiency of crops in maize–wheat cropping system by using polymer-coated diammonium phosphate. Commun Soil Sci Plant Anal 48:646–655

Yaseen M, Abbas T, Aziz MZ, Wakeel A, Yasmeen H, Ahmed W, Ullah A, Naveed M (2018) Microbial assisted foliar feeding of micronutrients enhance growth, yield and biofortification of wheat. Int J Agric Biol 20:353–360

Yilmaz A, Ekiz H, Gültekin I, Torun B, Barut H, Karanlik S, Cakmak I (1998) Effect of seed zinc content on grain yield and zinc concentration of wheat grown in zinc-deficient calcareous soils. J Plant Nutr 21:2257–2264

Yong TW, Chen P, Dong Q, Du Q, Yang F, Wang XC, Liu WG, Yang WY (2018) Optimized nitrogen application methods to improve nitrogen use efficiency and nodule nitrogen fixation in a maize-soybean relay intercropping system. J Integr Agric 17:60345–60347

Yuvaraj M, Subramanian KS (2018) Development of slow release Zn fertilizer using nano-zeolite as carrier. J Plant Nutr 41:311–320

Zekri M, Koo RCJ (1992) Use of controlled release fertilizers for young citrus trees. Sci Hortic 49:233–241

Zhang F, Cui Z, Chen X, Ju X, Shen J, Chen Q, Liu X, Zhang W, Mi G, Fan M, Jiang R (2012) Integrated nutrient management for food security and environmental quality in China. Adv Agron 116:1–40

Zhao BQ, Li XY, Li XP, Shi XJ, Huang SM, Wang BR, Zhu P, Yang XY, Liu H, Chen Y, Poulton P (2010) Long-term fertilizer experiment network in China: crop yields and soil nutrient trends. Agron J 102:216–230

Zhao AQ, Tian XH, Cao YX, XC L, Liu T (2014) Comparison of soil and foliar zinc application for enhancing grain zinc content of wheat when grown on potentially zinc-deficient calcareous soils. J Sci Food Agric 94:2016–2022

Zhao T, Zhao W, Nowlin WC (2018) Inventors, Intuitive Surgical Operations Inc, assignee. Configuration marker design and detection for instrument tracking. United States patent US 9,867,669

Zhengchao Z, Zhuoting G, Zhouping S, Fuping Z (2013) Effects of long-term repeated mineral and organic fertilizer applications on soil organic carbon and total nitrogen in a semi-arid cropland. Eur J Agron 45:20–26

Zhu ZL (1992) Efficient management of nitrogen fertilizers for flooded rice in relation to nitrogen transformations in flooded soils. Pedosphere 2:97–114

Zhu ZL, Chen DL (2002) Nitrogen fertilizer use in China-contributions to food production, impacts on the environment and best management strategies. Nutr Cycl Agroecosyst 63:117–127

Zuo Y, Zhang F (2009) Iron and zinc biofortification strategies in dicot plants by intercropping with gramineous species: a review. Agron Sustain Dev 29:63–71

Alternative Fertilizers and Sustainable Agriculture

Muhammad Zahir Aziz, Muhammad Naveed, Tanveer Abbas, Sulman Siddique, and Muhammad Yaseen

Abstract Sustainable food production requires application of fertilizers including macro and micronutrients in arable land. The fertilizers application in agricultural practices has significantly increased the production of food, fiber and other plant products. However, a significant portion of nitrogen (40–60%), phosphorus (80–90%) and potash (30–50%) applied fertilizers in the agricultural field is not taken up by plants due to different soil dynamics. Such losses increase the cost of fertilizers that severely reduce crop yield. Yet future access to mineral fertilizers receives major attention of plant scientists to overcome the applied fertilizer losses for food security. To overcome these problems, different alternative fertilizers are developed to efficient utilization of nutrients including microbial inoculants, value-added compost and biochar, acidulated-microbial active products, formula-modified fertilizers, liquid macro and micro- nutrient fertilizers with different mode of application to partial or complete substitution of reputed chemical fertilizers. This chapter puts forward the case of different alternative fertilizers and their potential for sustainable crop production.

Keywords Nutrients · Foliar application · Crops · Yield

1 Introduction

Despite substantial increase in food production over the past half century, now a day, one of most important challenge facing the society is how to feed the expected 10 billion peoples by the mid of current century and to reach at peak 11 billion by the close of this century. Moreover, in light of decreasing the arable land, rapid global climate change, increasing water scarcity and rising prices of agricultural inputs, it will be the challenge to increase crop productivity in order to fulfil the food demand of expected population over the next 50 years. To meet the predictable food demand without substantial increase in food price, it has been forecasted to

M. Z. Aziz · M. Naveed (✉) · T. Abbas · S. Siddique · M. Yaseen
Institute of Soil and Environmental Sciences, University of Agriculture, Faisalabad, Pakistan
e-mail: muhammad.naveed@uaf.edu.pk

M. Farooq, M. Pisante (eds.), *Innovations in Sustainable Agriculture*,
https://doi.org/10.1007/978-3-030-23169-9_8

increase 70–100% more production than present productivity (UN 2011). Furthermore, high cereal production only possible through high inputs of chemical fertilizers, currently 40–60% cereal production depends upon chemical fertilizers and by the 2050 almost 110% grain production will have to depend on inorganic fertilizers (Tilman et al. 2011). However, intensive utilization of limited agricultural resources i.e. water and fertilizer, presently and in future is remarkably expensive and environmentally dangerous, cannot forever, due to increase demand of fresh water for non-agricultural uses and escalation of fertilizer production cast (Ghimire and Craven 2013).

Since the green revolution, a steady rise in agriculture production has been occurred. However, high inputs agricultural technologies and available crop varieties will not adequately feed the exponential increasing population in perspective of variable supply of agricultural inputs. This scenario particularly is observed truly in developing countries, where population will be increased rapidly but access to agricultural inputs will be limited (Den Herder et al. 2010). So, intensive utilization of chemical fertilizers in intensive cropping system is necessary for adequate nutrient supply and obtain optimum crop yield. However, when nutrients supply externally, only a small proportion of fertilizer applied to the soil is actually utilized by plants. Subsequently, up to 70% nitrogen, 90% phosphorus, 70% potash and 10% micronutrients (Fe and Zn) of applied conventional mineral fertilizers are lost to the environment due to different soil dynamics (Heffer and Prudehome 2012). This loss of agricultural nutrients leads to reduction in yield and quality, and valuable resources.

Therefore, as far as sustainable agriculture crop production and environmental protection are concerned, the traditional fertilizers and the associated manufacturing technology are inadequate, and further improvements of fertilizer properties and new innovations in fertilizers are required. Thus, there exists an imperative research need on alternative fertilizers for enhancing applied fertilizers use efficiency, increasing crop yield and quality, and minimizing environmental pollution. In fact, several research endeavors have been made in the development of alternative fertilizers and new fertilizers featuring in recent years. This chapter discusses the recent fertilizers formulations to enhance fertilizers use efficiency, to identify the related knowledge gaps, and to prioritize the research needs. Different alternative fertilizers including compost, biochar, acidulated organic fertilizer, biofertilizers and formula-modified fertilizers potential for crops production are reviewed in present chapters.

2 Conventional Fertilizers: Their Role in Food Production

Like human beings, animals and plants also require nutrients for their growth and development. There are 17 identified essential nutrients for plants out of which 6 are major nutrients, 8 are micronutrients and 3 are structural nutrients. Carbon, hydrogen and oxygen are structural nutrients and are usually obtained from air (through leaves) and water (through roots) whereas the mineral nutrients are taken from the soil (Whiting et al. 2011). These nutrients are required not only for healthy development of plant body but are also required for the development of healthy flowers, fruits and seeds. Therefore, the availability of nutrients in the soil determines the quantity and quality of crop produced. Thus, a regular replenishment of nutrients in the soil is necessary in order to maintain the fertility of the soil, maintain or increase the yield and or to improve the quality of the harvest (FAO 1965). The replenishment of nutrients is usually carried out by adding extraneous substances particularly synthetic chemical fertilizers. Food production estimates revealed that 30–50% increase in crop yield is due to the use of commercial fertilizers. To feed the linear increasing population, grain yield must increase double over current production in 2050 by farmer which will only possible by wise use of new fertilizers formulation along with commercial fertilizers (Stewart et al. 2005).

3 Why Alternative Fertilizers Are Needed?

The world consumption of nitrogen (N), phosphorus (P) and potash (K) fertilizers in 2010/2011 was reported to be 104.1, 40.5 and 27.6 million tons and forecast figures for these fertilizers in 2016/2017 are 114.7, 45.4 and 32.7 million tons, respectively (Heffer and Prudehome 2012). FAO (2011) has estimated that the world demand for total fertilizer nutrients was increased by 2.0% per annum from 2011 to 2015 which constitutes an increment of 19.3 million tons of fertilizer nutrients. Consequently, a major portion of macro- and micronutrients of the total applied conventional fertilizers are lost to the environment due to different soil and environmental conditions (Trenkel 2010). This level of loss in agricultural nutrients not only lead to the loss of valuable resources but also severely reduced the crop production by which farmer facing economic and marketing problems (Dave et al. 1999). Various alternative approaches have been studied to make effective fertilizers application such as organic waste management, priming techniques, next generation fertilizers formulation and use of biofertilizers that not only reduce the nutrients losses but also improve the soil and environment health.

3.1 Potential Benefits of Alternative Fertilizers to Agriculture and Environment

Alternative fertilizers relative to conventional or standard mineral fertilizers are those that counter conditions or environments and thus, increased nutrients use efficiency. There are different fertilizers technologies which have high nutrients use efficiency and can reduce nutrients losses as:

- Formula-modified fertilizers
- Crop residues/compost/biochar
- Biofertilizers
- Liquid fertilizers

Added nutrients are quickly adsorbed with Fe and Al oxides and Ca-P and volatilized/leached (Raven and Hossner 1994). The added alternative fertilizers application not only reduce itself losses but also remobilized the fixed nutrients in soil by modification of reactions and soil conditions (Abbas et al. 2017; Naveed et al. 2017; Noor et al. 2017; Yaseen et al. 2017, 2018; Khalid 2018). Their use benefits the crops and soils as under:

- Controlled-release of nutrients
- Reduce nutrients fixation and losses by denitrification and volatilization
- Increase nutrients availability
- Remobilized the fixed micro and micro-nutrients
- Improve soil health
- Increase water holding capacity of soil

4 Alternative Fertilizers for Sustainable Agriculture

4.1 Organic Residues Utilization and Their Potential as Fertilizer

Crop residues are plant parts left in the field after harvesting and threshing of crops. Agricultural wastes are classified into two categories; crop residues (i.e., plant residues such as leaves, stubbles, root and straw) and residues which are by-products of post-harvest and food processing, from agro-industrial activities. The annual production of dry crop residues is about 74 Tg in worldwide, of which 45 Tg are generated by rice-wheat crop rotation systems (Kim and Dale 2004). Generally, wheat residues after separating from grain used to feed the cattle, building materials, livestock bedding and composting in the South Asia. Leftover of crop considers as waste materials whereby they may become a potential environmental problem that requires disposal and management. Crop residues are the major source of nutrients (Azam 1990), which can positively influence the biological and physio-chemical properties of soil.

Straw incorporation of cereal crop contributes 40–70 kg N ha^{-1} $year^{-1}$ return in soil (Nicholson et al. 1997). Straw is broken down through microbial activity when incorporated into the soil, with N in mineral form (i.e., ammonium and nitrate) released in a process termed 'mineralization'. The mineralization of crop residue depends on several factors including the N content, residue C:N ratio, degree of contact with the soil matrix, residue placement, cropping practices and tillage, as well as microbial activity, soil temperature, aeration and moisture (Jensen et al. 1997; Silgram and Chambers 2002). In agricultural soils, the straw incorporation and other crop residues give a principal energy source for microbial nitrogen and carbon alterations and has an important effect on plant nutrients source-sink relationship. Retention of straw into soil not only gives a waste disposal but also gives an opportunity to improve soil conservation by reducing soil erosion and compaction because of improvements in soil structure (Cannell 1987). Few effects of straw retained on the soil are likely to develop only after numerous years (Ball et al. 1990). During winter, nitrogen fertilizer immobilized due to straw incorporation that's why the loss of nitrate reduced by leaching and the pollution of water systems decreased (Powlson et al. 1985).

The management of the above ground rice residues (here we called stubbles) needs a proper consideration. Livestock and local area industries do not use all the stubbles (Flinn and Marciano 1984). Nitrogen wastes during stubble burning (Raison 1979), and the resulting ash can reduce herbicide activity (Toth et al. 1981). Incorporation of rice stubbles increase the crop growth and cause N deficiency (Bacon et al. 1987). However, Californian studies with aerial-sown rice (Williams et al. 1972) specify that technique of stubble-management has minute effect on yield of subsequent rice crops, while research about rice transplanting determine that incorporation of stubbles can increase yield (Ponnamperuma 1984). Phosphorous and potassium are most valuable nutrients returns into soil when straw is incorporated (HGCA 2009). These nutrients are returned to the soil through incorporation of straw, thus reducing the requirement of inorganic fertilizers.

Potassium is not commonly considered an environmental pollutant because leaching of potassium losses from the soil and straw are more important in relation to the effects on plant quality and growth. Though, phosphorous lost from soils rather through surface run-off and leaching may severely disturb the water quality as it contributes to freshwater eutrophication. Loss of P following straw incorporation has been reported, however, the phosphorous is safe when straw is merged into soil (Addiscott and Dexter 1994). Later, Bailey et al. (2013) assumed that residues incorporation rather than removing and bailing may be an effective method for reducing P losses and protect the soil surface runoff.

Crop residues incorporation increase soil organic matter and change the nutrient status and physio-chemical properties of soil (Bhogal et al. 2009), and also influence the soil microbial biomass, which in turn affects nutrient cycling and soil stability (Powlson et al. 2011). Bhogal et al. (2009) reported two non-UK studies where organic carbon (OC) return from crop residues was reported to increase water holding capacity, infiltration, porosity and soil aggregation (Schjonning et al. 1994; Munkholm et al. 2002). Malhi and Lemke (2007) stated that repetitive addition of straw improved root growth, emergence of seedling and reduce fertilizer application rate and nutrients losses risk. Recently, Yasmeen et al. (2018) reported that wheat stubbles incorporation in soil improved soil fertility, nutrients status and succeeding crop growth with reducing rate of fertilizers.

4.2 Compost-as Fertilizer

Composting is a recycling process in which organic materials are biologically converted into amorphous and stable humus like substance that can be handled, stored and applied to land without environmental impacts. Recently, composting technology got attention due to many benefits associated with huge amount of solid organic waste management (Gallardo-Larva and Nogales 1987; Millner et al. 1998) as bellow:

- Reduces the bulk of organic material application
- Killing the pathogens
- Reduces the chance of weeds spreading as by conventional manuring
- Reduces the transportation cost
- Modified unpleasant smell of organic waste
- Narrow down C:N ratio
- Value added product (Nutrient enriched compost)

The use of organic amendments such as compost manure and other organic waste materials boost up bio-availability of nutrients i.e. nitrogen (31%), phosphorus (25%) and potassium (23%) as compared with commercial fertilizers and serve as conditioner for soil (Ahmad et al. 2008a, b; Naveed et al. 2008). Different soil dynamics and environmental conditions under climate change scenario lead to inadequate nutrient status and plants require bulk nutrients at certain growth stage that solely cannot be obtained from simple composting. The value addition of compost materials overcome the issue of low nutrients, and thus, enhanced the crop growth, yield and nutrient uptake with reduced fertilizer application rates (Zahir et al. 2007; Naveed et al. 2008).

Extensive use of mineral fertilizers declines organic matter and caused stagnant crop yield which is not tolerable under intensive agriculture. Integration of organic materials with inorganic fertilizers could enhance crop yield and improve soil health on sustainable basis. A novel approach is to convert composted organic matter into value-added product through the enrichment/blending of the compost with nutrients (inorganic fertilizer), plant growth regulators (PGRs) and plant growth promoting rhizobacteria (PGPR) (Ahmad et al. 2006, 2008a; Zahir et al. 2007). Results of various field trials revealed that blending of compost with nitrogen fertilizer improved uptake of nitrogen, phosphorus and potassium upto 20%, 10% and 31%, respectively, as compared to control. This wise manipulation of composted material, not only reduced application rates of compost but also helped in achieving the product of desired characteristics (Ahmad et al. 2007, 2008a, b; Zahir et al. 2007). In a study, Zahir et al. (2007) revealed that compost enriched with PGRs indole-3-acetic acid, gibberellic acid and kinetin enhanced growth, yield and nutrients NPK uptake compared with conventional chemical fertilizer. Likewise, PGPR enriched compost along with 75% recommended dose of nitrogen fertilizer enhanced growth, yield and N uptake under greenhouse and field conditions (Naveed et al. 2008; Ahmad et al. 2008b).

4.3 Crop Rotation

Crop rotation is defined as planned sequence of crops growing in a regularly recurring succession on same area of land (SSSA 2008). Growing crops in rotation has many positive effects on crop yields and soil health. These effects are control of diseases, insects, weeds and increase nutrients as well as water use efficiency. In an appropriate crop rotation legume are rotated with cereals. In irrigated or high-rainfall production regions, crop rotation with legumes have high NUE and can reduce the amount of available nutrients losses when compared with continuous crops (Halvorson et al. 2008; Yu et al. 2014). Likewise, enhanced nutrients use efficiency was recorded for corn grown in rotation than for continuous corn (Varvel 1994; Halvorson et al. 2008; Halvorson and Schlegel 2012). Unfortunately, rotations are not easily adopted by farmers who have become accustomed to monoculture production systems, since a new crop often requires purchase of additional equipment and learning to integrate new cultural practices. In irrigated agriculture, the use of high N rates as a substitute for more N-use- efficient rotation systems (such as corn-soybean) must be weighed against the increased potential for NO_3-N loss (Anderson et al. 1997; Yu et al. 2014). Nitrogen use efficiency for wheat following legumes is greater than that for wheat following fallow or continuous wheat (Badaruddin and Meyer 1994; Halvorson et al. 2008). Wheat-corn-fallow production systems are now promoted instead of the popular wheat-fallow where only 420 mm precipitation is received per year (Kolberg et al. 1996). The more intensive systems (growing more crops in a given period), require greater fertilizer inputs but are higher in total yield and thus can be economically advantageous in relation with legumes crop rotation (Kolberg et al. 1996; Halvorson et al. 2008). More intensive dry-land cropping systems lead to increased water use efficiency and better maintain soil quality (Halvorson et al. 2008). Alternative dry-land systems proposed include spring barley, corn, and winter wheat grown in rotation with adequate nutrients fertilization instead of continuous winter wheat-fallow (Halvorson and Schlegel 2012).

Including legumes in rotation with cereals is an appropriate rotation. However, many studies have shown that even growing cereals in rotation or including oil crops in rotation with cereals have positive effects on yields compared to growing cereals in monoculture (Yau and Ryan 2012). Yau and Ryan (2012) reported that growing wheat and barley following of safflower (*Carthamus tinctorius*) increased wheat yield by 30% and barley yield by 23%. It is studied that pea in rotation with spring wheat and barley can not only sustain their yields efficiently using soil water but reduce N fertilization rates by supplying supplemental N from pea residues due to its higher N concentration (Sainju et al. 2014). In fact, crop rotation gave promising results but unsuccessfully this practice is not common among the farming community.

4.4 Char Coal Farming

Biochar is a solid product resulted from pyrolysis of crop residues, animal manure, biological wastes and any other type of organic material. Pyrolysis is a thermochemical breakdown of substances in the absence of oxygen under high temperature and this process is considered as negative because carbon is converted in more stable form (Shenbagavalli and Mahimairaja 2012). Biochar is prepared in specially designed kiln. Previously, biochar has been produced in conventional kiln because apparatus and processes were very simple. Muffle furnace and special designed apparatus are now used for conduction of pyrolysis as well as for biochar production.

Biochar composition is greatly heterogenous, containing both labile and stable components (Sohi et al. 2009). Different components of biochar determine its chemical behaviour and functions as whole. Biochar produced from animal manure is generally nutrient rich, and therefore easily degraded by microbes present in soil environment (Brown 2009). Biochar produced from material like straw residues, grain husks and manures generally contain more ash contents (Demirabes 2004). Biochar composition varies by condition of pyrolysis because this process causes some nutrients present in original feedstock to volatilize (Deluca et al. 2009). During the production of wood-based biochar N, S, K and P start volatilizing at 200, 375, 700 and 800 °C, respectively (Neary et al. 1999). Nitrogen is therefore, heat sensitive among the essential nutrients. Carbon, N, P and K contents ranges between 17.2–90.5%, 0.18–5.6%, 0.27–48% and 0.1–5.8%, respectively (Chan et al. 2007). Biochar also contains varying concentration of other elements such as H, O, S and base cations (Preston and Schmidt 2006). Freshly produced biochar consists of an amorphous phase with aromatic structures and crystalline phase with graphene layers (Ofri et al. 2007). The outer surfaces of biochar contain various functional groups (i.e. O and H) and graphene sheets may contain various free radicals and oxygen groups (Bourke et al. 2007).

Pyrolysis of feedstock at lower temperature is beneficial than at high temperature (Ca and Harris 2010). Due to high cation exchange capacity of biochar (Chan et al. 2008) it can adsorb NH_4^+ present in soil solution (Lehmann et al. 2006). In agricultural soils NH_4^+ volatilization is favoured at high pH (Stevenson and Cole 1999). Increased nutrient retention by biochar may be most important factor for increased crop yield on infertile soils (Farrel et al. 2013). Eldridge et al. (2010) performed an experiment to check the NH_4^+ retention capacity of soil amended with biochar. Results showed that biochar amended soil increased NH_4^+-N concentration in soil compared to control treatment. Additions of biochar to soil have shown increased availability of P for plant uptake. The immediate positive effects of biochar additions on nutrients may be due to higher P availability (Lehmann et al. 2003). Many studies have shown improved P uptake because of biochar application. Various mechanisms including biochar as a source of available and exchangeable P, ameliorator of P complexing metals, modifier of soil pH and as promotor of P mineralization and microbial activity. Biochar is a store of P bound to surface sites through its anion exchange capacity (Deluca et al. 2009). In a field trial, biochar was used to investigate its potential for improving growth, yield and nutrient recovery of wheat at varying fertilizer rates (Hamdani et al. 2017). Two levels 0% and 1% (w/w) of biochar were used along with five fertilizer rates i.e., 0, 25%, 50%, 75% and 100% of recommended chemical fertilizer. Result revealed that 1% rate of biochar along with 75% of recommended chemical fertilizer enhanced plant growth, yield and NPK uptake of wheat as compared to control. Naeem et al. (2016) investigated the effect of biochar produced from maize straw at different temperatures (300, 400, and 500 °C), on growth and nutrient concentration of maize. Shoot and root dry matter of maize increased significantly with application of biochar produced at 300 and 400 °C and decreased significantly at 500 °C. Maximum shoot and root dry matter of maize was obtained at biochar produced at 300 °C. Phosphorus and K concentration in shoots and roots increased with biochar, and it was significantly more with fertilizer application. The findings of the study indicated that application of biochar produced at low pyrolysis temperature may be a practical approach to improve maize growth. Zhang et al. (2010) conducted a field experiment in flooded paddy rice in China and observed that biochar application @ 40 t ha^{-1} significantly improved NUE upto 130% compared with un-amended soil. Widowati and Asnah (2014) performed a field experiment to evaluate the effect of biochar on potassium leaching and uptake efficiency of maize while applying @ 30 t ha^{-1}. It was concluded that with application of biochar K leaching reduced significantly and uptake efficiency increased upto 18% compared to control. Carbonaceous soil amendment, comprising mixture of biosolids and biochar have been used to improve soil fertility and biofortification of vegetables (Gartler et al. 2013). Zinc concentration was maximum increased upto 172 and 1200 mg kg^{-1} in the bulbs and leaves of Beetroot, respectively.

Biochar improves soil water holding capacity, aggregates stability and soil fertility. It can also reduce pollutants problems and mitigate greenhouse gases emission. Biochar seems to be beneficial to soil, however, production, availability and cost can hinder its wider adoption among farming community.

4.5 Biodynamic Organic Product

For nutrient availability in alkaline calcareous soils, addition of fertilizers and acidifying amendments is a common practice to augment plant nutrient availability and promote plant growth. From previously practiced, elemental sulfur (S°) is of peculiar interest, it possesses a slow releasing characteristic as it is insoluble in water and only microbes oxidize it into acid which readily mobilize the fixed nutrients (Chien et al. 2011). The acidifying characteristic of S° is attributed to microbial activity having ability to convert it into H_2SO_4 over time (Vidyalakshmi et al. 2009). Each mole of S° after oxidation produces 2 moles of hydrogen and changed into SO_4^{-2} through multiple steps (Wainwright 1987; Modaihsh et al. 1989; Besharati 1998; Kaplan and Orman 1998) (Eq. 1). Sulfur oxidation is accomplished through both chemical and biological processes (Wainwright 1987) and cited as biochemical process.

$$S^{\circ} + 1.5O_2 + H_2O \rightarrow SO_4^{-2} + 2H \quad (\Delta Go = -587.1 kJ / reaction) \tag{1}$$

Successful oxidation of S° and in response, nutrient release from unavailable pools to soil solution was observed repeatedly for alkaline soils by different researchers (Cui et al. 2004; Wang et al. 2006). Furthermore, high concentration of H^+ produce during oxidation might raise the nutrient contents in soils by exchange of cations (Zn, Fe, Mn etc.) from soil exchange sites, enhance mineral weathering, change the oxidation state of nutrients, and additionally high concentrations of hydrogen also enhance plants nutrient uptake (Lambers et al. 2008; Viani et al. 2014). Nevertheless, it is hard to envisage the nutrient response (mobility, dissolution) to acidified amendments in soil but their interactions somehow disturb nutrient dynamics in soil and their uptake to plant. Because nutrient contents in calcareous soils are not a problem, issue is their availability due to fixation and precipitation (Mayer et al. 2008). Soil acidification through bioaugmentation of S° with sulfur oxidizing bacteria (SOB) could enhance the plant macronutrient especially phosphorus and micronutrients i.e., several folds increase in Zn (Cui et al. 2004; Klikocka 2011; Skwierawska et al. 2012; Safaa et al. 2013), more than fivefolds Fe (Karimizarchi et al. 2014) and Mn (Hopkins and Ellsworth 2005; Mostashari et al. 2008). Sulfur oxidation and acid production has a long history but unsuccessful oxidation is common due to absence or low population of SOB.

Integrated application of S° and SOB is better option to enhance nutrient availability as compared to exclusive S° application (Aria et al. 2010). Rezapour (2014) observed the simultaneous effect of S°, SOB and manure on nutrient availability in calcareous soils and revealed that bio-augmentation of S° and manure with SOB reduces soil pH from 0.1 to 0.9 units by enhancing the number of oxidizers and have the oxidation rate of S°. Sulfur oxidizing bacteria oxidize S° and convert it into H_2SO_4 which solubilize macro- and micronutrients. Meanwhile, due to heterogeneity, microbes require a food source and in soil, reason of their low activity is low organic matter contents. Inoculation of SOB, S° and cow manure is a good approach, but a huge quantity is required for soil application which is not economical. Very recently, Sattar et al. (2017) prepared an acidulated organic amendment though bioaugmentation of S° and manure with SOB under suitable conditions. By applying the product, higher growth, yield and nutrient uptake in maize were recorded under greenhouse and field conditions. The authors further stated that bio-augmentation with novel bacteria would enhance bio-availability of macro-micronutrients in soil through manipulation of pH which will not only lead to improve the crop productivity but also enhance the product quality and food security.

4.6 *Biofertilizers as Partial Replacer of Chemical Fertilizers*

This section is concerned with known mechanisms that different types of plant growth promoting microbes (PGPM) i.e. phosphorus solubilizing bacteria (PSB), mycorrhizae and endophytes use to improve plant growth, yield and fertilizer use efficiency. The PGPM play important for efficient fertilizer management in agriculture are as:

1. Rhizosphere microbes for nutrient management
2. Mycorrhizae for agricultural crops nutrition
3. Endophytes as biofertilizers

4.6.1 Rhizosphere Microbes for Sustainable Nutrient Management

Rhizosphere is the place of maximum microbial activity that enhances the plant health. Various organic compounds and root exudates are continuously metabolized by the rhizospheric microbes. For that reason, quantity and quality changes in the root exudates and organic compounds occur mostly through microbial activities. A bacterium in the rhizosphere meaningfully influences the nutrient supply of plants. As a result, PGPR in the root zone play important role on the revenue of nutrients in the soil (Robinson et al. 1989; Nadeem et al. 2014).

Phosphate solubilizing bacteria (PSB) plays vital role in acquisition of fixed P through break down of bonding in phosphatic compounds in alkaline calcareous and acidic soils by releasing different kinds of organic acids and enzymes (Table 1). Phosphorus solubilization totally depends on PSB strains, soil types, plants and environmental circumstances (Sharma et al. 2013). Microbes belonging to *Pseudomonas*, *Azotobacter, Burkholderia, Bacillus, Enterobacter, Pantoea* and *Rhizobium* genera solubilize the fixed phosphate in soil (Sharma et al. 2013; Song et al. 2008; Naveed et al. 2017). The phosphate solubilizing microbes release different kinds of organic acids i.e., citric acid, gluconic acid, lactate, keto-gluconic acids, succinate and phosphatases for lowering the pH that solubilized the fixed inorganic as well as organic phosphate (Grover et al. 2011). However, PSB released some assimilate that make complexes with phosphorus to prevent its fixation (Khan et al. 2010; Grover et al. 2011). Factors affecting PSB abilities for P solubilization including nutrients status of soil, biotic and abiotic stresses. Furthermore, bacteria and fungi produce anionic compound i.e. carboxylic anions that release fixed P up to 42 mg P mL^{-1} by anionic exchange process (Stephen and Jisha 2011). The P solubilization by PSB improves crop produce as a potential source of biofertilizers.

Table 1 List of organic acid producing important phosphorus solubilizing rhizobacteria for P solubilization

Bacteria	Acid(s) produced	References
Stenotrophomonas maltophilia	Propionic acid	Rojas-Solis et al. (2018)
Pseudomonas	Gluconic acid, propionic acid	Lopez et al. (2011)
Putida		
P. trivialis	Formic acid, lactic acid	Vyas and Gulati (2009)
P. fluorescens	Citric acid, tartaric acid, Gluconic malic acid	Fankem et al. (2006)
B. subtilisvar	Oxalic acid, malonic acid, formic acid	Puente et al. (2008)
Enterobacter intermedium	2-ketogluconic acid	Hwangbo et al. (2003)
Bacillu amyloliquefaciens, Chryseomonas Luteola	Lactic acid, itaconic acid, acetic acid, isovaleric acid, isobutyric acid,	Vazquez et al. (2000)
P. radicum	Gluconic acid	Whitelaw et al. (1999)
Enterobacter agglomerans	Citric acid, oxalic acid	Kim et al. (1998)
Arthrobacter sp.	Citric acid, lactic acid	Bajpai and Rao (1971)
Pseudomonas sp.	Lactic acid, malic acid	Taha et al. (1969)
Escherichia freundii	Lactic acid	Sperber (1958)
Phosphorus solubilizing bacteria	Not determined	Gerretsen (1948)
Phosphorus solubilizing bacteria	Not determined	Pikovskaya (1948)

4.6.2 Mycorrhizae for Agricultural Crops Nutrition

The mycorrhizal symbiosis is most common mutualistic association between fungi and plant roots. During this mutualistic association, the mycorrhizal companion enhances the plant access to nutrients and water due to increase surface area of root through extra radical hyphal network. Mutualistic fungal associates produce enzymes that involved in the breakdown of phosphorus and nitrogen complexes from the carbon-based material in the soil and play important role in the breakdown of minerals by production of organic acids (Nadeem et al. 2014).

The enhancement of P nourishment of plants has finally been recognized by mycorrhizae through wider physical exploration of the soil. Various secondary mechanisms have clearly been explained P uptake through mycorrhizal fungi (Joner and Jakobsen 1995) like

- P uptake kinetics by hyphae higher affinity (lower Km) than root system
- Explore microsites by collective roots and hyphae in soil
- More chemical changes for P solubility in rhizosphere by release of organic acid and enzymes through mycorrhizae association

The latter mechanism leads to admittance to organic and inorganic P sources which are inaccessible to uninoculated mycorrhizae plants. It is suggested that mycorrhizae are benefited to plant growth by increasing the accessibility of P from fixed sources. Mycorrhizal and non-mycorrhizal crop plants use identical labile P sources, but mycorrhizal plants acquired more P from similar sources through extended root system, more organic acid release and enzymes over non-mycorrhizal treatment (Nogueira et al. 2004; Giri et al. 2005; Nadeem et al. 2014). It has been estimated that 80% of phosphorus taken up by plants is supplied by fungus. In addition to their significant role in P acquisition fungi can also provide other macro and micro nutrients particularly in those soils where these are in less quantity. In nutshell, mycorrhizae increase the P availability through P solubilizing mechanism that has potential for agricultural industry.

4.6.3 Endophytes as Biofertilizers

The word “endophyte” is used for microbes that live in plant body for all or some time of life cycles and cause no disease symptoms to host (Naveed et al. 2017). Bacterial endophytes are classified as facultative or obligate according to their life strategies. Moreover, obligate endophytes microbe strictly needed host plant for survival and growth while facultative endophytic microbes have potential to survive outside and within the host plant body (Hardoim et al. 2008). The utmost frequently upraised question about endophytic microbes for nutrient managing, in what way they are superior than rhizosphere microbes for improving nutrient efficiency? To answer this question, endophytic microbes originate from vegetative plant part, seed and their intimate location or contact within roots that secure from adverse environment changes than rhizosphere microbes results gain in term of improved growth and final yield (Hardoim et al. 2015). Excitingly, various research findings had directed the uses of genetically modified and natural endophytic microbes for enhancing nutrient use efficiency and plant growth by different mechanisms (Fig. 1).

A fundamental mechanism use by endophyte for enhancement of phosphorus use efficiency is modulation of root architecture; they improve root hair production and branching that account 70% of total root length. However, root growth and development are heritably programmed but endophytes modulated root structure and size (Compant et al. 2012). Specially, plant growth promoting endophytes i.e., *Azotobacter, Gluconacetobacter, Bacillus, Azospirillum, Herbaspirillum, Pseudomonas* and *Burkholderia* species have previously been perceived to increase plant growth by root growth stimulation (Schulz 2006; Tadych et al. 2009; Naveed et al. 2017). Various endophytic microbes’ classes are documented for enhancing nutrient use efficiency by root growth stimulation resulting in efficient absorption of nutrients from soil (Malinowski et al. 1998; Rodriguez et al. 2008). In addition, these endophytes produce hormones i.e. auxins, cytokinins, gibberellins and ethylene that induce initiation of adventitious roots, which cause increase in yield and NUE (Naveed et al. 2017). Inoculation of fescue with endophyte *N. coenophialum* in phosphorus deficient soil increased roots diameter (11%) and root hair (17%) over uninoculated plants by the production of hormone (Malinowski et al. 1998). Furthermore, in field trials application of endophytic microbes to rice and wheat

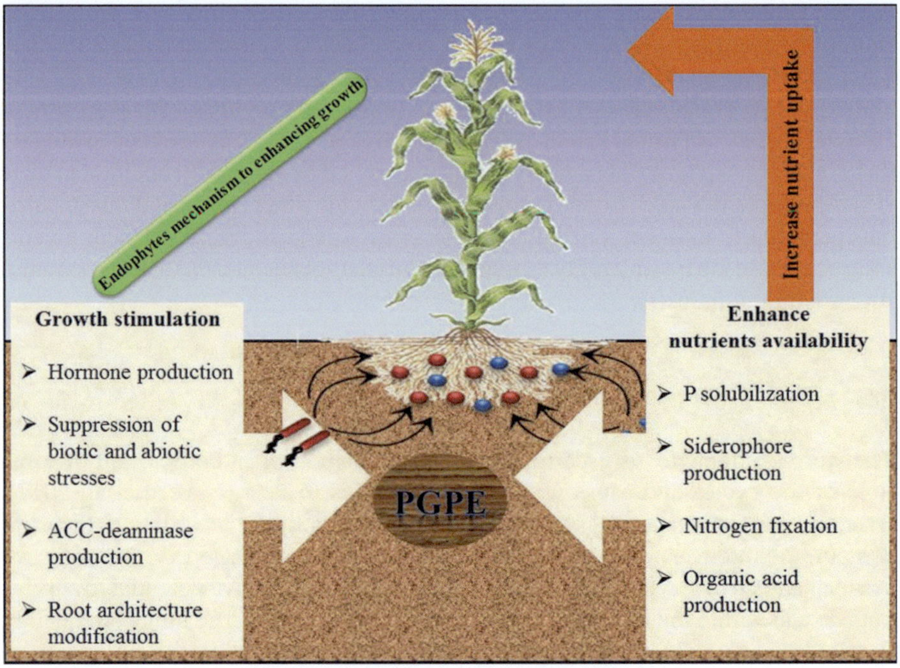

Fig. 1 Plant growth promoting endophytes (PGPE) mechanism to increase the nutrient uptake

increase P fertilizer use efficiency and grain yield by root system optimization and solubilization of fixed P (Yanni et al. 2001). Ethylene hormone is produced by plants on revelation to abiotic and biotic stresses, is known as stress ethylene that negatively affects plant physiology and inhibits lateral root growth that in turn decreases root surface area available for nutrient absorption (Glick et al. 2007). Endophytic bacterium *Burkholderia phytofirmans* PsJN secrete 1-aminocyclopropa ne-1-carboxylate (ACC) deaminase, which breaks down the ethylene precursor ACC into α-ketobutyrate and ammonia, the latter is then used as a reduced nitrogen source by these soil-inhabiting microbes (Glick 2005; Sessitsch et al. 2005; Naveed et al. 2013) and increased lateral root formation for effective nutrients absorption. Phosphorous is a major essential macronutrient for plant growth and absorbed by plant in two forms: dibasic (HPO_4^{2-}) and monobasic ($H_2PO_4^-$) ions. Availability of these ions to plant is reduced due to fixation of applied fertilizer in soil and this reduced availability limits the growth of plants (Yaseen et al. 2017). Potential of P solubilization of various endophytes is documented to transform the insoluble P to soluble form by secretion of organic acids and protons for enhancing crops growth (Hardoim et al. 2015; Naveed et al. 2017). Micronutrients deficiency reduce the yield and quality of crop produce. Very recently, Rehman et al. (2018a, b) illustrated the potential of endophytic bacteria *Pseudomonas* sp. MN12 with different methods of zinc (Zn) for enhancing growth, yield and grain Zn biofortification of bread wheat. Inoculation along with Zn source caused maximum increase in grain Zn bioavailable concentration upto 45% compared to control.

Due to increasing price and environmental anxieties regarding chemical fertilizers, use of microbial inoculants could be mandatory for sustainable agriculture and monetary benefits to farmer through reducing the application rate and losses of applied fertilizers and potential upturns in yields.

4.7 Seed Priming for Sustainable Nutrients Management

In agricultural market improved seed quality is important determinant to face the present demand of high standards food quality. Attaining quick and uniform seed emergence is essential for better crop production while slow seedling germination generally expose sprouts to soil-borne diseases and poor nutritious quality due to low nutrients uptake.

For enhancing seed quality 'priming' is an effective technology to achieve uniform and fast emergence. Primed seed show increased germination and resistance against biotic/abiotic stresses (Rajjou et al. 2012; Jisha et al. 2013). Seed priming is a water-based pre-sowing technique, which allows controlled rehydration of seeds to activate the metabolic process to a point of germination initiation but preventing conversion towards full germination by drying which stops radicle protrusion. Priming changes the protein content and nutrients concentration (Gonzalez-Zertuche et al. 2001; Lopez-Urrutia et al. 2014; Santini and Martorell 2013). The success of seed priming is strongly correlated to plant species/genotype and physiology, seed lot and vigour, as well as to the priming methods applied (Parera and Cantliffe 1994). Several priming techniques have been established which include hydro-priming, osmo-priming, halo-priming, hormonal priming, bio-priming and nutri-priming.

In nutri-priming, solution containing limiting nutrients are used for seed soaking instead of simple water. As nutrient solution increases the nutrient content of seed along with biochemical advantages of priming that improve seed quality for better germination and seedling establishment (Farooq et al. 2012). Zinc Seed priming enhanced productivity of wheat and chickpea (Arif et al. 2007), rice germination and early seedling growth (Abbas et al. 2014), while under saline condition potassium priming brought promising result on nutrient content of cotton seedling and increase growth (Shaheen et al. 2016). Harris et al. (2007) reported that increased in seed Zn content and grain yield of maize was observed by priming maize seed with 1% $ZnSO_4$. For seed preparation for growers some priming techniques are generally used by seed companies. Broad spectrum nutrient seed priming (BSN) is one of this method based on seed imbibing in mixture of minerals, such as copper, zinc, molybdenum, manganese and phosphorus, which provides nutrients for initial growth, seedling vigor, and root system development.

Most of the applied fertilizer are not accessible to the plants and lost through different soil dynamics. Low fertilizers use efficiency is not only cause high production cost but also lead to the serious environmental concerns such as ground water contamination, soil acidification and N_2O emission. Therefore, maintaining agricultural productivity in such a way that minimizes the harmful effects of fertilizers on environment is need of the hour. Microbes are the key component of soil health and nutrient transformation through mobilization and uptake. Soil microbial communities have been proved to be important constituent of biogeochemical cycling of materials affecting the composition, concentration and soil nutrient availability. Several microorganisms are potential rhizosphere colonizer used as seed dressers and support plant health. Various techniques such as soil application, seed coating and foliar treatment can be used to increase plant growth and as biocontrol agents. Among these techniques, seed treatment/coating is considered as the most effective method as it required less microbial inoculant dose with higher efficiency (Keswani 2015; Keswani et al. 2016a, b).

In biological priming mixture is integrated with bacterial inoculant and their bioactive molecules (Callan et al. 1990). It is well known that the association of microorganisms with plant result into very convenient outcomes, meanwhile they begin endophytic relations with the plant, leading to phytohormones production and provide resistance against biotic/abiotic stress (Waller et al. 2005). The strains mostly used for bio-priming belong to *Enterobacter* spp., *Bacillus* spp., *Pseudomonas* spp. and *Trichoderma* spp. (Niranjan et al. 2004). Furthermore, some bacteria directly or indirectly support plant growth and act as biocontrol agent by colonization in rhizosphere after germination (Table 2) (Callan et al. 1997). It was found that for disease management bio-priming is more effective than other approaches such as film coating and pelleting (Müller and Berg 2008). Nowadays, the use of plant

Table 2 Bio priming increase nutrient use efficiency

Biological agent	Crop	Nutrient use efficiency			References
		Primary (N, P, K)	Secondary (Ca, Mg)	Micro (Fe, Cu Mn, Zn)	
T. harzianum	Maize (*Zea mays*)	3.5% N in shoot; 8.8–9.76% N in root	–	–	Akladious and Abbas (2012)
Pseudomonas fluorescens (strains R62 + R81) + Natural mycorrhiza consortium	Wheat-rice and wheat-black gram rotations	0.695 PUE (kg P grain kg^{-1} P fertilizer)	–	–	Mäder et al. (2011)
Trichoderma asperellum strain T 34	Cucumber (*Cucumis sativus*)	–	–	Fe (85.7%); Zn (29.5%); Mn (58.6%); Cu (25%)	de Santiago et al. (2013)
Fluorescent Pseudomonas strains R62 + R81	Sugarcane (*Saccharum officinarum*)	0.719 PUE (kg P grain kg^{-1} P fertilizer)	–	–	Yadav et al. (2013)
T. harzianum T22	Tomato (*Lycopersicon esculentus*)	N (2.5%); P (38%); K (9.7%)	Ca (22%); Mg (20%)	Zn (27%); Fe (46%)	Molla et al. (2012)

growth-promoting bacteria (PGPB) as bio-priming agent shows great potential in agricultural practice (Timmusk et al. 2014). Under saline condition, radish seed bio-primed with rhizobacteria enhanced germination parameters (Kaymak et al. 2009). Akladious and Abbas (2012) found that maize seed priming with *T. harzianum* increase N content 8.8–9.76% in root as compared to convention fertilizer treatment without bioagent. Bio-priming in pearl millet with *Pseudomonas fluorescens* increased resistance against downy mildew and improved plant growth (Raj et al. 2004). The commercialization of biological agents is a problem that will need to be addressed. If technology is refined that allows storage for longer period than there is an opportunity for seed producers to develop a value-added product (Bio-primed seeds).

4.8 Formula Modified Fertilizers: As Alternative Fertilizers

4.8.1 Controlled Release: Coated Fertilizers

To enhance nutrient use efficiency with reduction in pollution, modification in fertilizer technology has been suggested which reduces environmental hazards. Many studies proved that controlled/slow-release fertilizers, which are actually "enhanced efficient fertilizers", when added in part or during the fertilization scheme have great ability to decrease losses of nutrients in soils (Trenkel 2012; Yaseen et al. 2017) and increase yield by releasing nutrients in controlled fashion as compared to the water-soluble fertilizers (Trenkel 2012). Quantity and release pattern of these fertilizers with time have been measured within certain limits to record enhancement in nutrient use efficiency of crops. Such fertilizers are being developed by coating of suitable natural or artificial barrier that switches controlled discharge (release) of nutrients from granule fertilizer (Du et al. 2004; Du et al. 2007; Trenkel 2010). Analogous term is also being used for controlled-release fertilizer and slow-release fertilizers. But the clear difference between two analogues words have been reported (Shaviv 2001); slow-release fertilizers have unpredictable release pattern of nutrient from coated granules. On the other hand, controlled-release fertilizers follow release pattern under expected/unexpected certain biotic and abiotic conditions. According to earlier literature, controlled/slow-release fertilizers concept has origin in the early 1960 (Fujii and Yazawa 1989). In early era, polyethylene and sulphur materials were used to develop these fertilizers and thus gave a way for many natural coating agents, multi-functional super-absorbent, even nano-composites with inconsistent results in term of controlled-release of nutrients due to surface cracking (Trenkel 2012). However, controlled-release fertilizers use for crops was limited due to high cost and these were suggested to use where more frequent application of conventional fertilizers was not practicable.

Difficulty faced in split application and cost problems is being tackled by polymers; hence providing resolution of the nutrients upturn problem, especially P. Polymers have unique characteristics that they are being attached to the nutrients and water molecules provided them distinctive controlled release property (Yaseen et al. 2017) and polymer improved soil structure by binding soil separates (Basak et al. 2012). Soluble fertilizers have best replaced by coated/slow-release fertilizer since they reduce cost of labour, enhance nutrient uptake via a single application of coated fertilizer, and save economy as well as reduce the environmental pollution (Sanders et al. 2007). Presently, coated fertilizers are being used at very low rate (only 1% of the total conventional fertilizer used) but have been attaining much attention for high value crop production gradually (Trenkel 2012). However, in future their use would be enhanced due to low use efficiency of conventional fertilizers and pollution.

Currently, polymer coated fertilizer gaining more attention of researchers to enhance nutrients use efficiency especially multilayer polymer coated diammonium phosphate (Yaseen et al. 2017). Controlled release of nutrients from polymer-coated fertilizers (PCF) is due to penetration of water (especially vapor) through the coating material. The mechanism is driven by concentration gradient from higher concentration to lower concentration across the coating material and by mass flow driven through the pressure gradient, or by the combination of the these two (Shaviv 2001). PCFs are the most sophisticated and propelled means of controlling fertilizer stability and nutrient release.

Many benefits are associated with the use of polymer coated P fertilizer such as growth and yield enhancement, higher fertilizer use efficiency, reduced nutrient losses via fixation and modification of soil chemical processes and consistent supply of nutrients (Table 3) (Trenkel 2010). However, the use of controlled-release fertilizers is limited due to dilemma associated to the adoption of controlled-release fertilizer due to its cost (Obreza and Rouse 1992). Many options are being used to reduce the cost of controlled-release fertilizer. Among these, polymer coated fertilizers can play important role to reduce not only cost of production but also

Table 3 Enhancing nutrients use efficiency by coated fertilizers

Coating material	Fertilizer	Fertilizers use efficiency (%)	References
Alginate	DAP	57–107	Aziz et al. (2018)
Polyacrylamide	DAP	30–40	Yaseen et al. (2017)
Polyacrylamide	DAP	45	Noor et al. (2017)
Polyacrylamide	Urea	60	Khalid (2018)
Alginate	DAP	53	Aziz et al. (2016a)
Carboxymethyl cellulose	DAP	49	Aziz et al. (2016b)
Starch	$NH_4H_2PO_4$	60	Jin et al. (2013)
Chitosan	N	60	Jin et al. (2011)
Polyacrylic acid	N and K	60	Wu et al. (2008)
Lignin	N	200	Entry and Sojka (2008)
Cellulose acetate,	NPK	69	Wu and Liu (2008)
Polyacrylic acid,			
Polyacrylamide,			

TSP Triple super phosphate, *DAP* Di-ammonium phosphate, *NPK* Nitrogen, phosphorus and potassium

some extra features like water holding, soil pulverization and improvement in soil aeration. Phosphorus uptake and recovery efficiency of applied polymer-coated fertilizer has been increased because polymer has high cation exchange capacity which is used to hold competing ions responsible for nutrients fixation/precipitation. This results in least losses and increased availability to crop plant for longer periods (Sanders et al. 2007). Further research on coating strength and number of coating layers suggests that polymer coated fertilizer can become successful and the best replacer of common fertilizers due to high use efficiency, reduction in labor and production cost (Basak et al. 2012). Sanders et al. (2007) applied phosphate fertilizer coated with 0.5–1.0% organic polymer with broadcast and band placement methods to different crops for enhanced P uptake. The coated fertilizers showed 6–25% increase in yield as compared to conventional MAP. Recently, Yaseen et al. (2017) synthesized polymer coated P fertilizer to assess its effect on wheat-maize crops under field experiment. The data revealed that polymer coated P fertilizer improved growth and yield upto 10% and 36%, respectively over conventional P fertilizer (DAP). However, biological yield and P agronomic efficiency was increased by 36% and 72%, respectively in treatment where recommended polymer coated P fertilizer was applied as compared to uncoated P fertilizer. Similarly, application of polymer coated P fertilizer at recommended rate increased plant height, grains and biological yield and P agronomic efficiency up to 4%, 29%, 39% and 58% as compared to commercial un-coated phosphatic fertilizer (Noor et al. 2017).

4.8.2 Polymer-Entrapped Microbe's Coated Fertilizer: A Novel Approach for Nutrient Management

Carriers are the main delivery vehicle of viable microorganisms from factory to the field (Bashan 1998). These carriers represent the key portion of the inoculants (volume/weight) and this must have suitable conditions for viable cells in right number (Fig. 2). These carriers, either physically, creating protective microhabitats or nutrition by providing the specific substrate to the inoculant (Bashan et al. 2014).

Various types of polymers might be taken part for encapsulation including natural or synthetic and homo-, hetero-, or co-polymers. Natural polymer includes protein material and polysaccharides while synthetic include polyurethane and polyacrylamide. Moreover, 1350 polymers are reported for encapsulation in different combinations based on chemical composition ability (John et al. 2011) but alginate and polyacrylamide (PAM) are commonly used for microbial encapsulation with more preference to alginate over PAM due to acrylamide toxicity (Bashan et al. 2014). Alginate is nontoxic, natural and biodegradable polymer makes three dimensional gels when mixed with (Bashan 1998; Bashan et al. 2014) by which microbes and additives easily dispersed into alginate to extend shelf life of inoculation.

Recently, Aziz et al. (2016a) evaluated the effect of polymer immobilized microbe's coated P fertilizer application on growth, yield and P use efficiency on wheat. Two techniques were joined with the help of polymer to load microbes on DAP fertilizer successfully. By this method, release pattern of both nutrients and

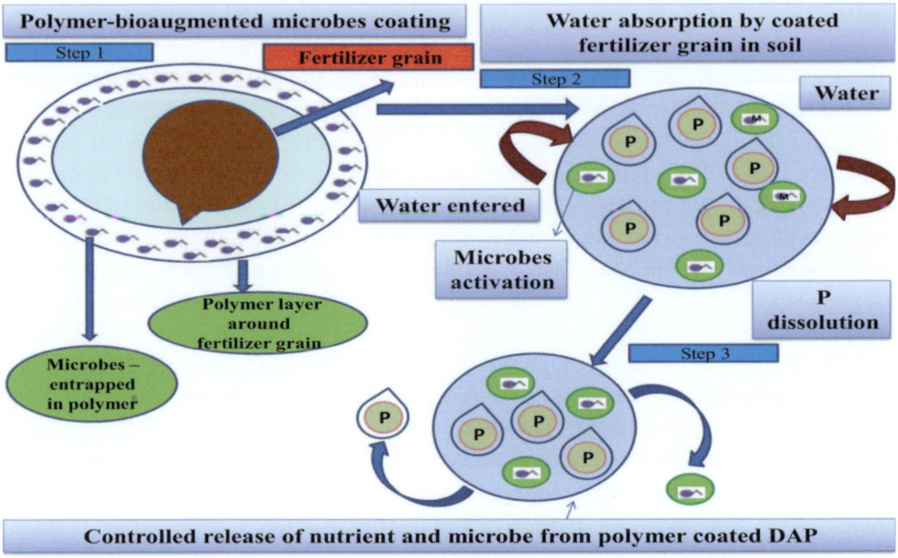

Fig. 2 Nutrients and microbes release mechanism from polymer bioaugmented microbes coated fertilizer

microbes was controlled successfully according to requirement of crop plants with microbial survival for long span from biotic and abiotic stresses. Results exposed that these recent methods successfully provide microbes to target place and enhanced crop growth, yield and P use efficiency of wheat. In other studies, Aziz et al. (2016a, 2018) reported alginate as a microbial carrier for survival of microbe on polymer-coated DAP fertilizer. They screened the different concentration of alginate i.e. 0.5%, 1% and 1.5% on the base of microbial survival in liquid alginate formulation. Best selected alginate concentration of 1.5% was amended with different organic carbon sources in different combination of glucose and glycerol. The prototype solution of screened alginate concentration along with organic carbon and microbe was coated on DAP to evaluate the microbial survival on DAP surface. However, viable number of microbes was recovered from DAP surface during different time periods. Results regarding the fertilizer release pattern showed that controlled release of P fertilizer i.e. 79% and microbes 80×10^8 CFU g^{-1} soil was recorded at 60 days sampling. Moreover, wheat growth and yield attributes were improved along with P use efficiency. Similarly, wheat phosphorus use efficiency was increased by the application of carboxymethyl cellulose entrapped microbe coated DAP. Results revealed that after application of coated DAP deliver microbes in rhizosphere i.e. 60×10^8 CFU g^{-1} soil and released P 85% upto 2 month along with improved P use efficiency.

4.8.3 Nano-fertilizers for Crop Production

Materials having nano size particles at least one dimension is called nano-particles (NPs). Nano-fertilizers are supplied nutrients to the crop plants and improve growth and development as well as fertilizers use efficiency. Moreover, nano-fertilizers are classified like macronutrient and micronutrient nano-fertilizers. Nano-fertilizers are estimated to significantly increase crop yields, fertilizer use efficiency, reduce nutrient losses and environmental hazards compared to conventional fertilizers (Lui and Lal 2015). However, NPs to unswervingly pass in plant cells because particle sizes are smaller than cell wall pores. But auxiliary proceeding of NPs enters through cell membrane in to cytoplasm and use of NP is also complicated (Nair et al. 2010). However, lack of research regarding nanoparticle nutrient elements absorbance by plant root system from soil solution is cited. In short, dissolve NPs in solution simply release the nutrient in soluble form and plants absorb the soluble nutrient but the dissolution rate of NPs in water and soil varies (Lui and Lal 2015).

Nano-fertilizers of macronutrient are comprised of one/more elements being able to supply essential macro-nutrients to plants. Global fertilizers i.e. nitrogen (N), phosphorus (P_2O_5) and potassium (K_2O) consumption was 177 million ton in 2011 and is estimated to upsurge 263 Mt in mid of running century (Alexandratos and Bruinsma 2012). It is projected fertilizers contributed 40% share in food production currently and 110% food production depends on fertilizer end of current century due to low efficiency of applied fertilizer particularly P, remaining amount is lost into environments creating problems for human being and aquatic life (Smil 2002). Therefore, there is need to develop next generation technologies to enhance nutrient use efficiency to overcome above mention problems.

Apatite (P), calcite (Ca), magnesium, zinc and iron nano-fertilizer were synthesized to assess its effects on different agronomic crops under different conditions (Liu and Lal 2014). The data revealed that nano-fertilizer improved growth and yield up to 33% and 20%, respectively over conventional fertilizer. The biological yield and below ground biomass were increased by 18% and 41%, respectively. The results directed that roots absorb nano-fertilizer effectively as nutrient source and improved growth and yield. In addition, nano-fertilizer as innovative class of fertilizer has potential to enhance agronomical yield, nutrients use efficiency and reduce environmental risks. NPs showed weaker interactions with soil components and Ca/Mg as significant number of NPs remain in soil solution for roots to absorb and NPs is less bioavailable to microbes that pose low risks of environment. Additionally, no phytotoxicity of the NPs was observed through a seed germination test. Similarly, wheat nutrient uptake was improved upto 61% using nano-fertilizer as compared to control (Montalvo et al. 2015). Moreover, results on germination trial of vegetables and cereals showed that nano-fertilizer is not significantly affected the germination (Sharonova et al. 2015). However, research thrust regarding merits and demerits, interaction with soil and microbes along with different crops of nano-fertilizer are needed under different conditions.

4.9 *Liquid Fertilizers-Crop Production*

4.9.1 Foliar Feeding of Nutrients in Sustainable Agriculture

Fertilizer is vital component in agricultural practices to increase crop yield. Soil application of nutrients is most common practice but attention toward use of foliar fertilizer is increasing day by day. In modern crop management, foliar spray is widely adopted when soil is unable to supply nutrients to plant due to unavailability of specific element and when plant nutrients demand exceeds the capacity. An advantage of foliar fertilizer application is that it provides nutrients directly to the metabolizing parts of the plant. Cereals crops are sprayed several times for different purpose during the season, like growth regulators and pesticide treatment (Yaseen et al. 2018). In such cases pesticides mixtures with compatible foliar fertilizer can increase the pesticide activity and recover fertilizer cost. To attain optimal utilization through leaves nutrients must dissolved in solution before to spray on the leave surface. Spray in evening and overcast weather gives better result than bright sunlight and dry days (Wittwer and Teubner 1959). Nutrients follow three paths from leaf surface to the leaf cells. The cutin swells when nutrient spray wet the cuticle, as a result distances increase between the wax plates in the cuticle. As the distance increases nutrients can pass through the cuticle and into the cell wall. The nutrients can then have absorbed in to cell membrane or they can diffuse deeper into the leaves along the epidermal cell and absorbed by parenchymal cell. In second pathway cytoplasmic strands transport the absorbed nutrient on epidermal cell to other cells, the plasma-desmata. If nutrient spray includes surfactants then nutrients can diffuse through the stomata into the air spaces of the leaf, third pathway (Marschner 2012). Nutrients applied to the foliage are generally absorbed more rapidly than when applied to the soil. Foliar application provides a means of quickly correcting plant nutrient deficiencies, when identified on the plant. It often provides a convenient method of applying fertilizer materials, especially those required in very small amounts and the highly soluble materials. Moreover, foliar feeding of micronutrients increased the nutrients concentration in edible parts as well as non-edible parts of the cereal crop up to 33% over non-treated crop plant. Recently, Yaseen et al. (2018) demonstrated the microbe-assisted foliar application of micronutrients for enhancing growth, yield and nutrient concentration of wheat. Result revealed that combined application of endophyte and micronutrients as foliar enhanced the nutrients concentration in grains up to 15% as compared to individual application of foliar nutrients. The disadvantages of foliar feeding are that the nutrient delivered is quite small and their action time is short, and there are questions of compatibility in relation to concentration and prevailing weather conditions.

4.9.2 Fertigation-Substitute Mineral Fertilizers

In arid and semi-arid areas specifically in developing countries higher crop yield depends mainly on the sustained use of energy resources and limited water. Moreover, strengthening of agricultural production to encounter market demand needs the simultaneous application of fertilizer and irrigation water. To increase yield and reduce environmental pollution, fertigation is an excellent opportunity by minimizing fertilizer use, increasing fertilizer use efficiency and return on investment (Hagin et al. 2002). The practice of providing nutrients to crops by dissolving fertilizer in water and applied with irrigation water is called fertigation. By this method nutrients and water amounts and concentration are easily controlled in the root zone. Fertigation lets the crop to absorb applied nutrients up to 90%, while 10–40% absorption was observed with dry or granular fertilizer application. Fertigation saves fertilizer up to 40–60%, due to reduction in leaching and improved nutrient uptake (Kumar and Singh 2002; Sathya et al. 2008). The advantage of fertigation over conventional system is that it assures uniform application of nutrients where the active roots are concentrated. Fertigation saves time, labour and ensure yield, which makes fertigation economically profitable (Singh 2002). Now a day's drip irrigation is preferred due to high water-application efficiency, less surface evaporation and deep percolation over other irrigation means. The dripper provides controlled supplies of water that not only affect the plant shoot and root growth but also increase the fertilizer use efficiency. Wastage of chemical fertilizer and water reduces by fertigation through drip irrigation, nutrient use optimizes by applying them at proper place, time and critical stages, which increase nutrient use efficiency. Moreover, it is well known as the most convenient approach of sustaining ideal nutrient level and water according to specific requirements of each crop and type of soil. Kaushal et al. (2011) reported that fertigation reduces 20–33% fertilizer requirement of mineral fertilizers and increased 7–25% yield as compared to conventional fertilizers. Rekha and Mahavishnan (2008) reported that drip fertigation saves water 40–70% and fertilizer 30–50%.

5 Concluding Remarks and Outlooks

Low use efficiency of applied fertilizers and risk of eutrophication can pollute environment – a striking challenge for scientific community for long time with little success. Moreover, limited and uneven distribution of conventional fertilizers resources has been recognized as major bottleneck for sustainable agriculture and economy. Crop economy is largely based on fertilizer applied to use ratio. At present, this ratio is very wide. This is reason that fertilizer use efficiency is emerged out as one of possible solution to narrow down gap between increasing population and food demand. So, increasing nutrients use efficiency by different scientific approaches has got significant importance in this regard. The synthesis of available information from the literature review presented herein shows potential of next generation fertilizers formulation as alternative fertilizer over commercial mineral fertilizers; however, cost effective strategy/approach is very much needed here for sustainable crop production. Potential of polymer coated P fertilizer for keeping nutrient soluble form, increase in diffusion shell and release of nutrient according to crop demand and intervention of entrapped microbes in polymer to further ensure the supply of nutrients through different mechanism i.e., hormone production, organic acid secretion and siderophore production seem viable cost-effective approach for enhancing nutrient use efficiency particularly P, Fe and Zn. Nano-fertilizer could be a good fertilizer replacer and has ability to enhance nutrient use efficiency many folds. Taking advantages of organic residue management as fertilizer by conducting series of laboratory and pilot scales experiments that gave encouraging results in combination of reduce rate of mineral fertilizers without compromising the crop yield.

Pertinent literature highlights the gaps and missing links between the controlled conditions and field results. This may be probable reason of least acceptance of new/modern fertilizers. However, further research data at field scale related different aspect especially improving yield is needed to encouraging adoption of new fertilizer intervention in agriculture. Some future outlooks are given below:

- Remodulation of agronomic approaches and screening of crops with the use of alternative fertilizers
- A greater understanding of nutrients dynamics in the soil/rhizosphere/plant continuum is of utmost importance after the application of alternative fertilizers as it can provide a significant basis for optimizing nutrients management to improve NUE in crop production.
- To know the mechanisms to understand insight microbial activity behavior by application of alternative fertilizers, as well as different dynamics related to alternative fertilizers and microbe's interactions.
- The research thrust should be directed toward the improvement of nutrients efficiency. Developed new fertilizer and formulation should be more efficient regarding nutrient uptake, soil and plant botany.

References

Abbas A, Yaseen M, Khalid M, Naveed M, Aziz MZ, Hamid Y, Saleem M (2017) Effect of biochar-amended urea on nitrogen economy of soil for improving the growth and yield of wheat (*Triticum Aestivum* L.) under field condition. J Plant Nutr 40(16):2303–2311

Abbas SQ, Hassan MU, Hussain B, Rasool T, Ali Q (2014) Optimization of zinc seed priming treatments for improving the germination and early seedling growth of *Oryza sativa*. Adv Life Sci 2:31–37

Addiscott TM, Dexter AR (1994) Tillage and crop residue management effects on losses of chemicals from soils. Soil Till Res 30:125–168

Ahmad R, Khalid A, Arshad M, Zahir ZA, Naveed M (2006) Effect of raw (un-composted) and composted organic waste material on growth and yield of maize (*Zea mays* L.). Soil Environ 25:135–142

Ahmad R, Shehzad SM, Khalid A, Arshad M, Mahmood MH (2007) Growth and yield response of wheat (*Triticum aestivum* L.) and maize (*Zea mays* L.) to nitrogen and L-Tryptophan enriched compost. Pak J Bot 39:541–549

Ahmad R, Arshad M, Khalidand A, Zahir ZA (2008a) Effectiveness of organic-/bio-fertilizer supplemented with chemical fertilizers for improving soil water retention, aggregate stability, growth and nutrient uptake of maize (*Zea mays* L.). J Sustain Agric 31:57–74

Ahmad R, Arshad M, Khalid A, Zahir ZA (2008b) Effect of compost enriched with N and L-tryptophan on soil and maize. Agron Sustain Dev 28:299–305

Akladious SA, Abbas SM (2012) Application of Trichoderma harzianum T22 as a biofertilizer supporting maize growth. Afr J Biotechnol 11:8672–8683

Alexandratos N, Bruinsma J (2012) World agriculture towards 2030/2050: the 2012 revision, ESA Working Paper No. 12-03. FAO, Rome

Anderson IC, Buxton DR, Karlen DL, Cambardella C (1997) Cropping system effects on nitrogen removal, soil nitrogen, aggregate stability, and subsequent corn grain yield. Agron J 89(6):881

Aria MM, Lakzian A, Haghnia GH, Berenji AR, Besharati H, Fotovat A (2010) Effect of *Thiobacillus*, sulfur, and vermicompost on the water-soluble phosphorous of hard rock phosphate. Bioresour Technol 101:551–554

Arif M, Waqas M, Nawab K, Shahid M (2007) Effect of seed priming in Zn solutions on chickpea and wheat. Afr Crop Sci Conf Proc 8:237–240

Azam F (1990) Comparative effect of organic and inorganic nitrogen sources applied to a flooded soil on rice yield and availability of Nitrogen. Plant Soil 125:255–263

Aziz MZ, Yaseen M, Naveed M (2016a) Efficacy of carboxymethyl cellulose coated diammonium phosphate bioaugmented with *Burkholderia phytofirmans* to promote phosphorus use efficiency of wheat via control release of microbe. 1st National conference on Recent Trends in Microbiology, 19–21 December, Abbottabad University of Science and Technology, Pakistan

Aziz MZ, Yaseen M, Naveed M, Shahid M (2016b) Improving growth and phosphorus use efficiency of wheat via controlled release of bacteria-immobilized in alginate coated on diammonium phosphate. 3rd conference of the world association of soil and water conservation, August 22–26, Belgrade, Republic of Serbia

Aziz MZ, Yaseen M, Naveed M, Shahid M (2018) Alginate-entrapped *Enterobacter* spp. MN17 coated diammonium phosphate improves growth, yield and phosphorus use efficiency of wheat. Int J Agric Biol. https://doi.org/10.17957/IJAB/15.0770

Bacon PE, Welsh T, Heenan DP, McCaffery D, Batten GD, Williams A (1987) Effect of N application rate and timing on growth and yield of Pelde and M7 at Yanco. Adv Agron 51:33–76

Badaruddin M, Meyer DW (1994) Grain legume effects on soil nitrogen, grain yield, and nitrogen nutrition of wheat. Crop Sci 34:1304–1309

Bailey A, Deasy C, Quinton J, Silgram M, Jackson B, Stevens C (2013) Determining the cost of in-field mitigation options to reduce sediment and phosphorus loss. Land Use Policy 30:234–242

Bajpai PD, Rao WVBS (1971) Phosphate solubilising bacteria. Soil Sci Plant Nutr 17(2):46–53

Ball BC, Bicker DC, Roberison EAG (1990) Straw incorporation and tillage for winter barley: soil structural effects. Soil Till Res 15:309–327

Basak BB, Pal S, Datta SC (2012) Use of modified clays for retention and supply of water and nutrients. Curr Sci 102:12–72

Bashan Y (1998) Inoculants of plant growth promoting bacteria for use in agriculture. Biotechnol Adv 16:729–770

Bashan Y, de-Bashan EL, Prabhu SR, Hernandez J (2014) Advances in plant growth-promoting bacterial inoculant technology: formulations and practical perspectives (1998–2013). Plant Soil 378:1–33

Besharati H (1998) Effects of the sulfur applications with Thiobacillus species in the ability to absorb certain nutrients in the soil. PhD diss., M. Sc. Thesis, Faculty of Agriculture, Tehran University. (In Farsi)

Bhogal A, Nicholson FA, Chambers BJ (2009) Organic carbon additions: effects on soil biophysical and physicochemical properties. Eur J Soil Sci 60:276–286

Bourke J, Harris MM, Fushimi C, Dowaki K, Nunoura T, Antal MJJ (2007) Do all the carbonised charcoal have same structure? A model of chemical structure of carbonized charcoal. Ind Eng Chem Res 50:169–175

Brown R (2009) Biochar production technology. In: Lehmann J, Joseph S (eds) Biochar for environment management. Science and Technology. Earthscan, London, pp 127–146

Ca X, Harris W (2010) Properties of dairy manure derived biochar pertinent to its potential use in remediation. Bioresour Technol 101:5222–5228

Callan NW, Mathre DE, Miller JB (1990) Bio-priming seed treatment for biological control of *Pythium ultimum* pre-emergence damping off in sh2 sweet corn. Plant Dis 74:368–372

Callan NW, Mathre DE, Miller JB, Vavrina CS (1997) Biological seed treatments: factors involved in efficacy. Hortic Sci 32:179–183

Cannell RQ (1987) Straw incorporation in relation to soil conditions and crop growth. Outlook Agric 13:130–135

Chan KY, Zwieten LV, Meshzaros I, Downie A, Joseph S (2007) Agronomic value of greenwaste biochar as soil amendment. Aust J Soil Res 45:629–634

Chan KY, Zwieten LV, Meshzaros I, Downie A, Joseph S (2008) Using poultry litter biochar as soil amendments. Aust J Soil Res 46:437–444

Chien SH, Gearhart MM, Villagarcķa S (2011) Comparison of ammonium sulfate with other nitrogen and sulfur fertilizers in increasing crop production and minimizing environmental impact. Soil Sci 176(7):327–335

Compant PF, Sessitsch A, Mathieu F (2012) The 125th anniversary of the first postulation of the soil origin of endophytic bacteria-a tribute to M.L.V. Galippe. Plant Soil 356:299–301

Cui Y, Dong Y, Li H, Wang Q (2004) Effect of elemental Sulphur on solubility of soil heavy metals and their uptake by maize. Environ Int 30(3):323–328

Dave AM, Mehta MH, Aminabhavi TM, Kulkarni AR, Soppimath KS (1999) A review on controlled release of nitrogen fertilizers through polymeric membrane devices. Polym-Plast Technol Eng 38(4):675–711

de Santiago A, García-López AM, Quintero JM, Avilés M, Delgado A (2013) Effect of Trichoderma asperellum strain T34 and glucose addition on iron nutrition in cucumber grown on calcareous soils. Soil Biol Biochem 57:598–605

Deluca TH, Mackenzie MD, Dundale MJ (2009) Biochar effects on soil nutrient transformations. In: Lehmann J, Joseph S (eds) Biochar for environment management. Science and Technology. Earthscan, London, pp 251–270

Demirabes A (2004) Effects of temperature and particle size on biochar yield from pyrolysis of agricultural residues. J Anal Appl Pyrolysis 72:243–248

Den Herder G, Van Isterdael G, Beeckman T, De Smet I (2010) The roots of a new green revolution. Trends Plant Sci 15(11):600–607

Du C, Zhou J, Wang H, Li S (2004) A preliminary study on natural matrix materials for controlled release nitrogen fertilizers. Pedosphere 14:45–52

Du C, Tang D, Zhou J, Wang H, Shaviv A (2007) Prediction of nitrate release from polymer coated fertilizers using an artificial neural network model. Bio Syst Eng 99:478–486

Eldridge S, Chen C, Xu Z, Meszaros I, Chan, KY (2010) Green waste biochar potentially reduces nitrogen fertiliser losses. In Proceedings of the 19th World Congress of Soil Science: Soil solutions for a changing world, Brisbane, Australia, 1–6 August 2010. Division Symposium 3.2 Nutrient best management practices. International Union of Soil Sciences (IUSS), c/o Institut für Bodenforschung, Universität für Bodenkultur, pp 100–103

Entry JA, Sojka RE (2008) Matrix based fertilizers reduce nitrogen and phosphorus leaching in three soils. J Envir Manag 87:364–372

Fankem H, Nwaga D, Deubel A, Dieng L, Merbach W, Etoa FX (2006) Occurrence and functioning of phosphate solubilizing microorganisms from oil palm tree (*Elaeis guineensis*) rhizosphere in Cameroon. Afri J Biotech 5:24

FAO (1965) Fertilizers and their use. FAO, Rome

FAO (2011) Plant nutrition for food security: a guide for integrated nutrient management, FAO fertilizer and plant nutrition bulletin 16. Food and Agriculture Organization of the United Nations, Rome

Farooq M, Wahid A, Siddique KHM (2012) Micronutrients application through seed treatments a review. J Soil Sci Plant Nutr 12:125–142

Farrel M, Kuhn TK, Macdonald LM, Maddern TM, Murphy DV, Hall PA, Singh BP, Baumann K, Krull ES Baldock JA (2013) Microbial utilization of biochar derived carbon. Sci Total Environ 465:288–297

Flinn JC, Marciano, VP (1984) Rice straw and stubble management

Fujii T, Yazawa F (1989) Development of coated fertilizer. Jpn Soc Soil Sci Plant Nutr 34:101–109

Gallardo-Larva F, Nogales R (1987) Effect of application of town refuse compost on the soil plant system – a review. Biol Wastes 19:35–62

Gartler J, Robinson B, Burton K, Clucas L (2013) Carbonaceous soil amendments to biofortify crop plants with zinc. Sci Total Environ 465:308–313

Gerretsen FC (1948) The influence of microorganisms on the phosphate intake by the plant. Plant Soil 1(1):51–81

Ghimire SR, Craven KD (2013) Endophytes in low-input agriculture and plant biomass production. In: Saha MC, Bhandari HS, Bouton JH (eds) Bioenergy feedstocks: breeding and genetics. Wiley, Oxford, pp 249–265

Giri B, Kapoor R, Mukerji KG (2005) Effect of the arbuscular mycorrhizae *Glomus fasciculatum* and *G. macrocarpum* on the growth and nutrient content of *Cassia siamea* in a semi-arid Indian wasteland soil. New For 29:63–73

Glick BR (2005) Modulation of plant ethylene levels by the bacterial enzyme ACC deaminase. FEMS Microbiol Lett 251(1):1–7

Glick B, Cheng Z, Czarny J, Duan J (2007) Promotion of plant growth by ACC deaminase-producing soil bacteria. Eur J Plant Pathol 119:329–339

González-Zertuche L, Vázquez-Yanes C, Gamboa A, Sánchez-Coronado ME, Aguilera P, Orozco-Segovia A (2001) Natural priming of *Wigandia urens* seeds during burial: effects on germination, growth and protein expression. Seed Sci Res 11(01):27–34

Grover M, Ali SZ, Sandhya V, Rasul A, Venkateswarlu B (2011) Role of microorganisms in adaptation of agriculture crops to abiotic stresses. World J Microbiol Biotechnol 27:1231–1240

Hagin J, Sneh M, Lowengart-Aycicegi A (2002) In: Johnston AE (ed) Fertigation fertilization through irrigation, IPI Research Topics No. 23. International Potash Institute, Basel

Halvorson AD, Schlegel AJ (2012) Crop rotation effect on soil carbon and nitrogen stocks under limited irrigation. Agron J 104(5):1265

Halvorson AD, Del Grosso SJ, Reule CA (2008) Nitrogen, tillage, and crop rotation effects on nitrous oxide emissions from irrigated cropping systems. J Environ Qual 37(4):1337

Hamdani SAF, Aon M, Khalid M, Ali L, Aslam Z, Naveed M (2017) Application of *Dalbergia sissoo* biochar enhanced wheat growth, yield and nutrient recovery under reduced fertilizer doses in calcareous soil. Pak J Agric Sci 54:107–115

Hardoim P, Van LO, Van JE (2008) Properties of bacterial endophytes and their proposed role in plant growth. Trends Microbiol 16:463–471

Hardoim PR, van-Overbeek LS, Berg G, Pirttila AM, Compant S, Campisano A, Doring M, Sessitsch A (2015) The hidden world within plants: ecological and evolutionary considerations for defining functioning of microbial endophytes. Microbiol Mol Biol Rev 79(3):293–320

Harris D, Rashid A, Miraj G, Arif M, Shah H (2007) On-farm seed priming with zinc sulphate solution, a cost effective way to increase the maize yields of resource poor farmers. Field Crop Res 110:119–127

Heffer P, Prudehome M (2012) Fertilizer outlook 2012–2016. International Fertilizer Industry Association (IFA), Paris

HGCA (2009) Information Sheet 05: assessing the nutrient content of cereal straw. HGCA, London. High value chemicals. J Environ Sci Health B 49:51–67

Hopkins B, Ellsworth J (2005) Phosphorus availability with alkaline/calcareous soil. In Western Nutrient Management Conference 6:88–93

Hwangbo H, Park RD, Kim YW, Rim YS, Park KH, Kim TH, Jang SS, Jim, KY (2003) 2-Ketogluconic acid production and phosphate solubilization by Enterobacter intermedium. Current microbiology 47(2): 87–92

Jensen LS, Mueller T, Magid J, Nielsen NE (1997) Temporal variation of C and N mineralization, microbial biomass and extractable organic pools in soil after oilseed rape straw incorporation in the field. Soil Biol Biochem 29:1043–1055

Jin S, Yue G, Feng L, Han Y, Yu X, Zhang Z (2011) Preparation and properties of a coated slow-release and water-retention biuret phosphoramide fertilizer with superabsorbent. J Agric Food Chem 59(1):322–327

Jin S, Wang Y, He J, Yang Y, Yu X, Yue G (2013) Preparation and properties of a degradable interpenetrating polymer networks based on starch with water retention, amelioration of soil, and slow release of nitrogen and phosphorus fertilizer. J Appl Polym Sci 128(1):407–415

Jisha KC, Vijayakumari K, Puthur JT (2013) Seed priming for abiotic stress tolerance: an overview. Acta Physiol Plant 35:1381–1396

John RP, Tyagi RD, Brar SK, Surampalli RY, Prevost D (2011) Bio-encapsulation of microbial cells for targeted agricultural delivery. Crit Rev Biotechnol 31:211–226

Joner EJ, Jakobsen I (1995) Uptake of P^{32} from labeled organic matter by mycorrhizal and non-mycorrhizal subterranean clover (*Trifolium subterraneum* L). Plant Soil 172:221–227

Kaplan M, Orman Ş (1998) Effect of elemental Sulphur and Sulphur containing waste in a calcareous soil in Turkey. J Plant Nutr 21(8):1655–1665

Karimizarchi M, Aminuddin H, Khanif MY, Radziah O (2014) Elemental Sulphur application effects on nutrient availability and sweet maize (*Zea mays* L.) response in a high pH soil of Malaysia. Malaysian J Soil Sci 18:75–86

Kaushal A, Lodhi AS, Singh KG (2011) Economics of growing sweet pepper under low tunnels. Prog Agric 24:67–72

Kaymak HC, Guvenc I, Yarali F, Donmez MF (2009) The effects of bio-priming with PGPR on germination of radish (*Raphanus sativus* L.) seeds under saline conditions. Turk J Agric For 33:173–179

Keswani C (2015). Proteomics studies of thermotolerant strain of Trichoderma spp. PhD thesis, Banaras Hindu University, Varanasi, India, p 126

Keswani C, Bisen K, Singh V, Sarma BK, Singh HB (2016a) Formulation technology of biocontrol agents: present status and future prospects. In: Arora NK, Mehnaz S, Balestrini R (eds) Bioformulations: for sustainable agriculture. Springer, New Delhi, pp 35–52

Keswani C, Bisen K, Singh SP, Sarma BK, Singh HB (2016b) A proteomic approach to understand the tripartite interactions between plant-Trichodermapathogen: investigating the potential for efficient biological control. In: Hakeem KR, Akhtar MS (eds) Plant, soil and microbes, vol. 2. Mechanisms and molecular interactions. Springer, Cham, pp 79–93

Khalid MA (2018) Synchronized nitrogen release from polymer coated nitrochalk enhances nitrogen use efficiency and yield of wheat. Pak J Agric Sci 55(02):367–373

Khan MS, Zaidi A, Ahemad M, Oves M, Wani PA (2010) Plant growth promotion by phosphate solubilizing fungi – current perspective. Arch Agron Soil Sci 56:73–98

Kim S, Dale B (2004) Global potential bioethanol production from wasted crops and crop residues. Biomass Bioenergy 26:361–375

Kim KY, Jordan D, McDonald GA (1998) Enterobacter agglomerans, phosphate solubilizing bacteria, and microbial activity in soil: effect of carbon sources. Soil Biol Biochem 30(8–9):995–1003

Klikocka H (2011) The effect of Sulphur kind and dose on content and uptake of micronutrients by potato tubers (*Solanum tuberosum* L.). Acta Sci Pol Hort Cult 10:137–151

Kolberg RL, Kitchen NR, Westfall DG, Peterson GA (1996) Cropping intensity and nitrogen management impact of dryland no-till rotations in the semi-arid Western Great Plains. J Prod Agric 9(4):517

Kumar A, Singh AK (2002) Improving nutrient and water use efficiency through fertigation. J Water Manage 10:42–48
Lambers H, Raven J, Shaver G, Smith S (2008) Plant nutrient-acquisition strategies change with soil age. Trends Ecol Evol 23(2):95–103
Lehmann J, Silva JPD, Seiner C, Nehls T, Zech W, Glaser B (2003) Nutrient availability and leaching in an archaeological Anthrosol and a Ferrasol of the central Amazon basin: Fertilizer, manure and charcoal amendments. Plant Sci 249:343–357
Lehmann J, Gaunt J, Rondon M (2006) Biochar sequestration in terrestrial ecosystem- a review mitigation and adaptation. Strateg Glob Chang 11:403–427
Liu R, Lal R (2014) Synthetic apatite nanoparticles as a phosphorus fertilizer for soybean (Glycine max). Sci Rep 4:5686–5691
Lopez BR, Bashan Y, Bacilio M (2011) Endophytic bacteria of *Mammillaria fraileana*, an endemic rock-colonizing cactus of the Southern Sonoran Desert. Arch Microbiol 193:527–541
López-Urrutia E, Martínez-García M, Monsalvo-Reyes A, Salazar-Rojas V, Montoya R, Campos JE (2014) Differential RNA- and protein-expression profiles of cactus seeds capable of hydration memory. Seed Sci Res 24(2):91–99
Lui R, Lal R (2015) Potentials of engineered nanoparticles as fertilizers for increasing agronomic productions. Sci Total Environ 514:131–139
Mäder P, Kaiser F, Adholeya A, Singh R, Uppal HS, Sharma AK, Srivastava R, Sahai V, Aragno M, Wiemken A, Johri BN, Fried PM (2011) Inoculation of root microorganisms for sustainable wheat–rice and wheat–black gram rotations in India. Soil Biol Biochem 43(3):609–619
Malhi SS, Lemke R (2007) Tillage, crop residue and N fertilizer effects on crop yield, nutrient uptake, soil quality and nitrous oxide gas emissions in a second 4-yr rotation cycle. Soil Till Res 96:269–283
Malinowski DP, Alloush GA, Belesky DP (1998) Evidence for chemical changes on the root surface of tall fescue in response to infection with the fungal endophyte *Neotyphodium coenophialum*. Plant Soil 205:1–12
Marschner P (2012) Marschner's mineral nutrition of higher plants, 3rd edn. Academic
Mayer JE, Pfeiffer WH, Beyer P (2008) Biofortified crops to alleviate micronutrient malnutrition. Curr Opin Plant Biol 11(2):166–170
Millner PD, Sikora LJ, Kaufman DD, Simpson ME (1998) Agricultural uses of biosolids and other recyclable municipal residues. In: Wright RJ, Kemper WD, Millner PD, Power JF, Korcack RF (eds) Agricultural uses of municipal, animal, and industrial byproducts, Conservation research reports 44. USDA Agricultural Research Survey, Washington, DC, pp 9–44
Modaihsh AS, Al-Mustafa WA, Metwally AI (1989) Effect of elemental Sulphur on chemical changes and nutrient availability in calcareous soils. Plant Soil 116(1):95–101
Molla AH, Haque MM, Haque MA, Ilias GNM (2012) Trichoderma-enriched biofertilizer enhances production and nutritional quality of tomato (*Lycopersicon esculentum* mill.) and minimizes NPK fertilizer use. Agric Res 1(3):265–272
Montalvo D, McLaughlin MJ, Degryse F (2015) Efficacy of hydroxyapatite nanoparticles as phosphorus fertilizer in andisols and oxisols. Soil Sci Soc Am 79:551–558
Mostashari M, Muazardalan M, Karimian N, Hosseini HM, Rezai H (2008) Phosphorus fractions of selected calcareous soils of Qazvin province and their relationships with soil characteristics. Am Eurasian J Agric Environ Sci 3:547–553
Müller H, Berg G (2008) Impact of formulation procedures on the effect of the biocontrol agent Serratia plymuthica HRO-C48 on Verticillium wilt in oilseed rape. BioControl 53(6):905–916
Munkholm LJ, Schjonning P, Debrosz K, Jensen HE, Christensen BT (2002) Aggregate strength and mechanical behaviour of a sandy loam soil under long term fertilization treatments. Eur J Soil Sci 53:129–137
Nadeem SM, Ahmad M, Zahir ZA, Javaid A, Ashraf M (2014) The role of mycorrhizae and plant growth promoting rhizobacteria (PGPR) in improving crop productivity under stressful environments. Biotechnol Adv 32(2):429–448

Naeem MA, Khalid M, Ahmad Z, Naveed M (2016) Low pyrolytic temperature biochar improve growth and nutrient availability of maize on Typic Calciargid. Commun Soil Sci Plant Anal 47(1):41–51

Nair R, Varghese SH, Nair BG, Maekawa T, Yoshida Y, Kumar DS (2010) Nanoparticulate material delivery to plants. Plant Sci 179:154–163

Naveed M, Khalid M, Jones DL, Ahmad R, Zahir ZA (2008) Relative efficacy of Pseudomonas spp., containing ACC-deaminase for improving growth and yield of maize (*Zea mays* L.) in the presence of organic fertilizer. Pak J Bot 40:43–51

Naveed M, Mitter B, Yousaf S, Pastar M, Afzal M, Sessitsch A (2013) The endophyte *Enterobacter* sp. FD17: a maize growth enhancer selected based on rigorous testing of plant beneficial traits and colonization characteristics. Biol Fertil Soils 50:249–262

Naveed M, Aziz MZ, Yaseen M (2017) Perspectives of endophytic microbes for legume improvement. In: Zaidi A et al (eds) Microbes for legume improvement. Springer, Switzerland, pp 277–299

Neary D, Klopatek C, Debano L, Ffolliott P (1999) Fire effects on belowground sustainability. A review and synthesis. For Ecol Manag 122:51–71

Nicholson FA, Chambers BJ, Mills AR, Strachan PJ (1997) Effects of repeated straw incorporation on crop fertilizer nitrogen requirements, soil mineral nitrogen and nitrate leaching losses. Soil Use Manag 36:136–142

Niranjan RS, Shetty NP, Shetty HS (2004) Seed bio-priming with Pseudomonas fluorescens isolates enhances growth of pearl millet plants and induces resistance against downy mildew. Int J Pest Manag 50:41–48

Nogueira MA, Magelhaes GC, Cardoso EJBN (2004) Manganese toxicity in mycorrhizal and phosphorus-fertilized soybean plants. J Plant Nutr 27:141–156

Noor S, Yaseen M, Naveed M, Ahmad R (2017) Use of controlled release phosphatic fertilizer to improve growth, yield and phosphorus use efficiency of wheat crop. Pak J Agric Sci 54:xxx, xxx; ISSN (print) 0552-9034, ISSN (online) 2076-0906. http://www.pakjas.com.pk

Obreza TA, Rouse RE (1992) Controlled-release fertilizers use on young Hamlin orange trees. Proc Soil Crop Sci Soc Fla 51:64–68

Ofri IC, Niro RP, Weiner S (2007) Structural characterization of modern and fossilized charcoal produced in natural fires as determined by using electron energy loss spectroscopy. Chem Eur J 13:2306–2310

Parera CA, Cantliffe DJ (1994) Pre-sowing seed priming. Hortic Rev 16:109–141

Pikovskaya RI (1948) Mobilization of phosphorus in soil in connection with vital activity of some microbial species. Mikrobiologiya 17:362–370

Ponnamperuma FN (1984) Straw as a source of nutrients for wetland rice. Organic Matter Rice 117:136

Powlson DS, Jenkinson DS, Pruden G, Johnston AE (1985) The effect of straw incorporation on the uptake of nitrogen by winter wheat. J Sci Food Agric 36:26–30

Powlson DS, Glendining MJ, Coleman K, Whitmore AP (2011) Implications for soil properties of removing cereal straw: results from long-term studies. Agron J 103:279–287

Preston CM, Schmidit MWI (2006) Black (Pyrogenic) carbon. A synthesis of current knowledge and uncertainties with special consideration of boreal region. Biogeosciences 3:397–420

Puente ME, Bashan Y, Li CY, Lebsky VK (2008) Microbial populations and activities in the rhizoplane of rock-Weathering Desert plants. I. Root colonization and weathering of igneous rocks. Plant Biol 6(5):629–642

Raison RJ (1979) Modification of the soil environment by vegetation fires, with particular reference to nitrogen transformations: a review. Plant Soil 51:73–108

Raj N, Shetty N, Shetty H (2004) Seed biopriming with Pseudomonas fluorescens strains enhances growth of pearl millet plants and induces resistance against downy mildew. Int J Pest Manag 50:41–48

Rajjou L, Duval M, Gallardo K, Catusse J, Bally J, Job C, Job D (2012) Seed germination and vigor. Annu Rev Plant Biol 63:507–533

Raven KP, Hossner LR (1994) Soil phosphorus desorption kinetics and its relationship with plant growth. Soil Sci Soc Am J 58(2):416

Rehman A, Farooq M, Naveed M, Ozturk L, Nawaz A (2018a) *Pseudomonas*-aided zinc application improves the productivity and biofortification of bread wheat. Crop Past Sci 69:659–672

Rehman A, Farooq M, Naveed M, Nawaz A, Shahzad B (2018b) Seed priming of Zn with endophytic bacteria improves the productivity and grain biofortification of bread wheat. Eur J Agron 94:98–107

Rekha BK, Mahavishnan K (2008) Drip fertigation in vegetable crops with emphasis on lady's finger (*Abelmoschus esculentus* (L.) Moench) – a review. Agric Rev 29:298–305

Rezapour S (2014) Effect of sulfur and composted manure on SO4-S, P and micronutrient availability in a calcareous saline–sodic soil. Chem Ecol 30(2):147–155

Robinson D, Griffiths B, Ritz K, Wheatley R (1989) Root-induced nitrogen mineralisation: a theoretical analysis. Plant and Soil 117(2): 185–193

Rodriguez RJ, Henson J, Van-Volkenburgh E (2008) Stress tolerance in plants via habitat-adapted symbiosis. ISME J 2:404–416

Rojas-Solís D, Zetter-Salmón E, Contreras-Pérez M, Rocha-Granados M d C, Macías-Rodríguez L, Santoyo G (2018) *Pseudomonas stutzeri* E25 and Stenotrophomonas maltophilia CR71 endophytes produce antifungal volatile organic compounds and exhibit additive plant growth-promoting effects. Biocatal Agric Biotechnol 13:46–52

Safaa MM, Khaled SM, Hanan SS (2013) Effect of elemental Sulphur on solubility of soil nutrients and soil heavy metals and their uptake by maize plants. J Am Sci 9:19–24

Sainju UM, Stevens WB, Caesar-TonThat T, Liebig MA, Wang J (2014) Net global warming potential and greenhouse gas intensity influenced by irrigation, tillage, crop rotation, and nitrogen fertilization. J Environ Qual 43(3):777

Sanders L, Kimmerly M, Murphy L (2007). A new method for influencing phosphate availability to plants. Fertilizer industry round table, Winston-Salem

Santini BA, Martorell C (2013) Does retained-seed priming drive the evolution of serotiny in drylands? An assessment using the cactus *Mammilaria hernandezii*. Am J Bot 100:365–373

Sathya S, Pitchai JG, Indirani R, Kannathasan M (2008) Effect of fertigation on availability of nutrients (N, P & K) in soil – a review. Agric Rev 29:214–219

Sattar A, Asghar HN, Zahir ZA, Asghar M (2017) Bioactivation of indigenous and exogenously applied micronutrients through acidified organic amendment for improving yield and biofortification of maize in calcareous soil. Int J Agric Biol 19:1039–1046

Schjonning P, Christensen BT, Carstensen B (1994) Physical and chemical properties of a sandy loam receiving animal manure, mineral fertilizer or no fertilizer for 90 years. Eur J Soil Sci 45:257–268

Schulz B (2006) Mutualistic interactions with fungal root endophytes. In: Schulz BJE, Boyle CJC, Sieber TN (eds) Microbial root endophytes. Springer, New York, pp 261–279

Sessitsch A, Coenye T, Sturz AV (2005) *Burkholderia phytofirmans* sp. a novel plant-associated bacterium with plant-beneficial properties. Int J Syst Evol Microbiol 55:1187–1192

Shaheen HL, Iqbal M, Azeem M, Shahbaz M, Shehzadi M (2016) K-priming positively modulates growth and nutrient status of salt-stressed cotton (*Gossypium hirsutum*) seedlings. Arch Agron Soil Sci 62:759–768

Sharma SB, Sayyed RZ, Trivedi MH, Gobi TA (2013) Phosphate solubilizing microbes: sustainable approach for managing phosphorus deficiency in agricultural soils. Springer Plus 2:587–600

Sharonova NL, Yapparov AK, Khisamutdinov NS, Ezhkova AM, Yapparov IA, Ezhkov VO, Degtyareva IA, Babynin EV (2015) Nanostructured water-phosphorite suspension is a new promising fertilizer. Nanotech Russia 10:651–661

Shaviv A (2001) Preparation methods and release mechanism of the controlled release fertilizer: agronomic efficiency and the environmental significances. Int Fertil Soc Proc 431:1–35

Shenbagavalli S, Mahimairaja S (2012) Characterization and effect of biochar on nitrogen and carbon dynamics in soil. Int J Adv Biol Res 2:249–255

Silgram M, Chambers BJ (2002) Effects of long term straw management and fertilizer nitrogen additions on soil nitrogen supply and crop yields at two sites in eastern England. J Agric Sci (Camb) 139:115–127
Singh HP (2002) Precision farming in horticulture. In: Proceedings of the national seminar cum workshop on Hi-tech horticulture and precision farming, 26–28 July 2012, Lucknow, pp 1–20
Skwierawska M, Zawartka L, Skwierawski A, Nogalska A (2012) The effect of different sulfur doses and forms on changes of soil heavy metals. Plant Soil Environ 58(3):135–140
Smil V (2002) Nitrogen and food production: proteins for human diets. Ambio 31:126–131
Sohi S, Cpel EL, Krull E, Bol R (2009) Biochar, climate change and soil: a review of guide future research. Land Water Sci Res:1–56
Song OR, Lee SJ, Lee YS, Lee SC, Kim KK, Choi YL (2008) Solubilization of insoluble inorganic phosphate by *Burkholderia cepacia* Da23 isolated from cultivated soil. Braz J Microbiol 39:151–156
Sperber JI (1958) Solution of apatite by soil microorganisms producing organic acids. Aust J Agric Res 9(6):782
SSSA (2008) Glossary of soil science terms. Soil Science Society of America, Inc., Madison
Stephen J, Jisha MS (2011) Gluconic acid production as the principal mechanism of mineral phosphate solubilization by *Burkholderia* sp. J Trop Agric 49:99–103
Stevenson FJ, Cole MA (1999) Cycles of soil, carbon, nitrogen, phosphorus, sulphur micronutrients. Wiley, New York
Stewart WM, Dibb DW, Johnston AE, Smyth TJ (2005) The contribution of commercial fertilizer nutrients to food production. Agron J 97(1):1–6
Tadych M, White JF, Moselio S (2009) Endophytic microbes. In: Schaechter M (ed) Encyclopedia of microbiology, 3rd edn. Academic, Oxford, pp 431–442
Taha SM, Mahmoud SAZ, Halim El-Damaty A, Abd El-Hafez AM (1969) Activity of phosphate-dissolving bacteria in Egyptian soils. Plant Soil 31(1):149–160
Tilman D, Balzer C, Hill J, Befort BL (2011) Global food demand and the sustainable intensification of agriculture. Proc Natl Acad Sci USA 108:20260–20264
Timmusk S, Abd El-Daim IA, Copolovici L, Tanilas T, Kännaste A, Behers L, Nevo E, Seisenbaeva G, Stenström E, Niinemets Ü (2014) Drought-tolerance of wheat improved by rhizosphere bacteria from harsh environments: enhanced biomass production and reduced emissions of stress volatiles. PLoS One 9:960–986. https://doi.org/10.1371/journal.pone.0096086
Toth J, Milham PJ, Raison JM (1981) Ash from rice stubble inactivates Thiobencarb and Molinate. Weed Res 21:113–117
Trenkel ME (2010) Slow and controlled-release and stabilized fertilizers: an option for enhancing nutrient use efficiency in agriculture. International Fertilizer Industry Association (IFA), Paris
Trenkel ME (2012) Slow and Controlled-release and stabilized fertilizers. An option for enhancing nutrient efficiency in agriculture, 2nd edn. IFA, Paris
United Nations (2011) World population prospects: the 2010 revision, Standard variants. Updated: 28 June 2011. http://esa.un.org/wpp/Excel-Data/population.htm. Accessed 15 May 2013
Varvel GE (1994) Rotation and nitrogen fertilization effects on changes in soil carbon and nitrogen. Agron J 86(2):319
Vazquez P, Holguin G, Puente ME, Lopez-Cortes A, Bashan Y (2000) Phosphate-solubilizing microorganisms associated with the rhizosphere of mangroves in a semiarid coastal lagoon. Biol Fertil Soils 30(5–6):460–468
Viani RAG, Rodrigues RR, Dawson TE, Lambers H, Oliveira RS (2014) Soil pH accounts for differences in species distribution and leaf nutrient concentrations of Brazilian woodland savannah and seasonally dry forest species. Perspect Plant Ecol Evol Syst 16(2):64–74
Vidyalakshmi R, Paranthaman R, Bhakyaraj R (2009) Sulphur oxidizing bacteria and pulse nutrition- a review. World J Agric Sci 5:270–278
Vyas P, Gulati A (2009) Organic acid production in vitro and plant growth promotion in maize under controlled environment by phosphate-solubilizing fluorescent Pseudomonas. BMC Microbiol 9(1):174

Wainwright M (1987) Microbial sulfur oxidation in soil. Soil Prog (Oxford) 65:459–475
Waller F, Achatz B, Baltruschat H, Fodor J, Becker K, Fischer M, Heier T, Huckelhoven R, Neumann C, Von-Wettstein D (2005) The endophytic fungus Piriformis indica reprograms barley to salt-stress tolerance, disease resistance, and higher yield. Proc Natl Acad Sci USA 102:13386–13391
Wang AS, Angle JS, Chaney RL, Delorme TA, Reeves RD (2006) Soil pH effects on uptake of cd and Zn by Thlaspi caerulescens. Plant Soil 281(1–2):325–337
Whitelaw MA, Harden TJ, Helyar KR (1999) Phosphate solubilisation in solution culture by the soil fungus Penicillium radicum. Soil Biol Biochem 31(5):655–665
Whiting SJ, Langlois KA, Vatanparast H, Greene-Finestone LS (2011) The vitamin D status of Canadians relative to the 2011 dietary reference intakes: an examination in children and adults with and without supplement use. Am J Clin Nutr 94(1):128–135
Widowati, Asnah (2014) Biochar can enhance potassium fertilization efficiency and economic feasibility of maize cultivation. J Agric Sci 6:24–32
Williams WA, Morse MD, Ruckman JE (1972) Burning vs incorporation of rice straw residues. Agron J 64:467–468
Wittwer SH, Teubner FG (1959) Foliar absorption of mineral nutrients. Annu Rev Plant Physiol Plant Mol Biol 10:13–32
Wu L, Liu M (2008) Preparation and characterization of cellulose acetate-coated compound fertilizer with controlled-release and water-retention. Polymer Adv Tech 19:785–792
Wu L, Liu M, Liang R (2008) Preparation and properties of a double-coated slow-release NPK compound fertilizer with superabsorbent and water-retention. Bioresour Technol 99(3):547–554
Yadav SK, Dave A, Sarkar A, Singh HB, Sarma BK (2013) Co-inoculated biopriming with Trichoderma, Pseudomonas and Rhizobium improves crop growth in Cicer arietinum and Phaseolus vulgaris. Inter J Agric Environ Biotechnol 6(2):255–259
Yanni YG, Rizk RY, El-Fattah FK (2001) The beneficial plant growth-promoting association of *Rhizobium leguminosarum* bv. trifolii with rice roots. Aust J Plant Physiol 28:845–870
Yaseen M, Aziz MZ, Manzoor A, Naveed M, Hamid Y, Noor S, Khalid MA (2017) Promoting growth, yield, and phosphorus-use efficiency of crops in maize–wheat cropping system by using polymer-coated diammonium phosphate. Commun Soil Sci Plant Anal 48:646–655
Yaseen M, Abbas T, Aziz MZ, Wakeel A, Humaira Y, Wazir A, Ullah A, Naveed M (2018) Microbial assisted foliar feeding of micronutrients enhance growth, yield and biofortification of wheat. Int J Agric Biol 20:353–360
Yasmeen H, Yaseen M, Aziz MZ, Naveed M, Arfan-ul-Haq M, Jilani G, Qadeer A, Abbas T (2018) Wheat residue management improves soil fertility and productivity of maize. Int J Agric Biol 20:2181–2188
Yau S-K, Ryan J (2012) Does growing safflower before barley reduce barley yields under Mediterranean conditions? Agron J 104(6):1493
Yu Y, Xue L, Yang L (2014) Winter legumes in rice crop rotations reduces nitrogen loss, and improves rice yield and soil nitrogen supply. Agron Sustain Dev 34(3):633–640
Zahir ZA, Iqbal M, Arshad M, Naveed M, Khalid M (2007) Effectiveness of IAA, GA3 and kinetin blended with recycled organic waste for improving growth and yield of wheat (*Triticum aestivum* L.). Pak J Bot 39(3):761–768
Zhang A, Cui L, Pan G, Li L, Hussain Q, Zhang X, Zheng J, Crowley D (2010) Effect of biochar amendment on yield and methane and nitrous oxide emissions from a rice paddy from tai Lake plain, China. Agric Ecosyst Environ 139(4):469–475

Part IV
Sustainable Pest Management

Sustainable Weed Management

Ali Ahsan Bajwa, Muhammad Jamal Khan, Prasanta Chitta Bhowmik, Michael Walsh, and Bhagirath Singh Chauhan

Abstract Weed management is an essential element of successful crop production. In recent times, an exponential rise in human population and drastic changes in climate and production techniques have intensified the crop production systems and increased the weed infestations. The evolution of herbicide resistance in a large number of weed species across the world has further aggravated the situation. These circumstances require sustainable weed management tools that can be used effectively to achieve decent crop yields without affecting the environment and ecosystem services negatively. Some of the conventional weed control methods, including the use of preventive measures, tillage and mechanical control, crop competition, soil coverage, crop rotations and crop diversification, are still effective and eco-friendly. The ecological phenomenon of allelopathy could also be explored in different ways for sustainable weed management. Recent advances in the fields of renewable energy, remote sensing, modelling, automation and robotics have opened new windows for more physical weed control methods such as thermal weed control, precision weed control and harvest weed seed control. These methods are quite

A. A. Bajwa (✉)
School of Agriculture and Food Sciences, The University of Queensland, Gatton, QLD, Australia

The Centre for Crop Science, Queensland Alliance for Agriculture and Food Innovation (QAAFI), The University of Queensland, Gatton, QLD, Australia
e-mail: ali.bajwa@dpi.nsw.gov.au

M. J. Khan
Faculty of Veterinary and Agricultural Sciences, The University of Melbourne, Dookie, VIC, Australia

P. C. Bhowmik
Stockbridge School of Agriculture, University of Massachusetts, Amherst, MA, USA

M. Walsh
The University of Sydney, Narrabri, NSW, Australia

B. S. Chauhan
The Centre for Crop Science, Queensland Alliance for Agriculture and Food Innovation (QAAFI), The University of Queensland, Gatton, QLD, Australia

M. Farooq, M. Pisante (eds.), *Innovations in Sustainable Agriculture*,
https://doi.org/10.1007/978-3-030-23169-9_9

expensive and do not suit all geographical conditions, but these are going through rapid evolution and modifications that will make them more affordable, precise and pragmatic in the near future. Although the ecologically-based, cultural and physical weed control methods have great potential for sustainable weed management, herbicides cannot be completely left out. In fact, herbicides could be an effective tool in an integrated weed management kit. However, the true integrated weed management strategies must have a balance between chemical and non-chemical options that can be used judiciously in order to achieve a reasonable weed control. Sustainable weed management is an achievable goal provided that efforts are made to diversify the management.

Keywords Integrated weed management · Sustainable agriculture · Conservation agriculture · Herbicide resistance · Weed control · Crop productivity

1 Introduction

Weeds are among the most important challenges to sustainable crop production and global food security (Oerke 2006). These prolific competitors not only cause significant yield losses but also deteriorate the normal crop husbandry practices (Zimdahl 2013). The negative impact and economic losses become incredibly remarkable when the environmental damage, ecosystem deterioration and health problems caused by crop weeds and invasive plant species are considered (Liebman et al. 2016). Therefore, weed management is essential for sustainable crop production and environmental safety. Unfortunately, not all the weed management tactics are sustainable or safe for the environment. For instance, the most reliable and efficient method to control weeds is the use of synthetic herbicides. It has worked perfectly over the years, but sole dependence on chemicals and their continuous use has led to the evolution of herbicide resistance in a large number of weed species around the world (Bagavathiannan and Davis 2018). The herbicide tolerant crops were introduced to simplify weed control (essentially narrowing it down to just herbicides), but nature retaliated in the form of pronounced herbicide resistance in weeds (Owen et al. 2015). Similarly, the gene flow from herbicide tolerant crop species or resistant weed populations to the closely related wild species have exacerbated this problem (Ohadi et al. 2017). So, it is evident that we cannot beat the problem of herbicide resistance with the use of more herbicides and therefore, there is an urgent need for sustainable weed management options.

Climate change poses a threat to agricultural productivity in many different ways. Crops suffer abiotic stresses more often (Fahad et al. 2017) and rapid fluctuations in climatic elements also promote disease and pest outbreaks and the spread of invasive weed species (Ziska et al. 2011; Nguyen et al. 2017). Increasing global temperature and carbon dioxide (CO_2) levels not only favour the invasion, growth and reproduction of many problematic weed species but also decrease the efficacy of herbicides (Bajwa et al. 2016; Ziska 2016). The intensification of crop production systems and the adoption of conservation tillage systems have changed the weed dynamics, infestation patterns and weed-crop interactions. These challenges restrict the realization of sustainable crop production and require pragmatic weed management strategies.

This chapter gives a comprehensive account of several weed management options that can be sustainable if used in a site-specific manner. Different conventional and modern weed control methods and their efficacy in different scenarios have been discussed in detail. This chapter provides the readers an in-depth analysis of potential sustainable weed management tools that can be used to achieve high crop productivity while ensuring the environmental safety and long-term sustainability.

2 Conventional Approaches

Since the inception of agriculture, weeds have been infesting the crop production systems and numerous management approaches have been practiced to control them. Some of the old and established weed control methods are still the most sustainable ones. These include the use of preventive measures, tillage and mechanical control, and a range of cultural methods.

2.1 Preventive Measures

Weeds are often aggressive invaders, which can dominate any landscape once after their successful introduction and establishment. Therefore, the foremost weed management approach in sustainable agriculture is the prevention (Christoffoleti et al. 2007; Bajwa 2014). The preventive measures include but are not limited to the deliberate efforts to avoid the contamination of crop seed lots with weed seeds, the use of clean farm machinery, containment of the weed populations in landscapes adjacent to the crop fields and restricting the weed seed dispersal (Jordan 1996; Bond and Grundy 2001; Christoffoleti et al. 2007; Bajwa et al. 2018). These precautionary measures avoid the spread of existing weed flora and restrict the introduction of new weed species in any crop production system (Anderson 2007). Proactive control methods keep the weed pressure below economic thresholds and therefore, offer more sustainable weed management (Jordan 1996). Often these approaches are not applied solely and sometime may get less attention in an integrated weed management (IWM) strategy.

Sometimes weed seeds are similar to those of crops which make the identification difficult and deteriorate the quality of crop produce due to their adulteration (Jabran et al. 2017). Similarly, some of the most problematic weeds mimic the associated crops in terms of morphology, phenology and seed production. Therefore, careful crop husbandry practices from sowing to harvesting of any crop may reduce the weed infestations and weed seedbank build-up (Anderson 2007). Controlling the weed populations during fallow seasons and in non-cropped areas such as water channels, field bunds, pathways and along the fences is also a very important preventive measure for sustainable weed management.

2.2 *Tillage Interventions and Mechanical Weed Control*

Tillage plays a major role in weed control in conventional cropping systems (Bond and Grundy 2001; Zimdahl 2013). It is used for mechanical weed control as interculture. Similarly, the tillage operations involved in seedbed preparation also serve the purpose of weed management (Reicosky and Allmaras 2003). Tillage influences weed dynamics and, thus, weed management through cutting, burial, uprooting and other disturbances. Weed germination, stand establishment and subsequent growth is affected by tillage process (Clements et al. 1996). In conventional tillage systems, the use of the stale seedbed technique has been very effective in controlling various weeds at the time of sowing (Buhler 2002). In this method, the field is tilled and levelled for planting the crop, but actual planting is delayed allowing the emergence of most weed species in seedbank. The crop is then planted after killing the emerged cohort of weeds by herbicides or another round of tillage (Singh et al. 2015). It reduces the weed seedbank strength as well as the weed pressure in crops. Safdar et al. (2011) reported significant reductions in the densities of *Avena fatua* L., *Chenopodium album* L. and *Phalaris minor* Retz. in wheat (*Triticum aestivum* L.) with the use of the stale seedbed technique.

Inter-culture using different tools and tillage implements also provide effective weed control in different crops. The use of different types of harrows have provided variable control of different weeds (Jabran et al. 2017). However, harrowing does not provide effective control of tall stature weeds with a tap root system. Inter-row cultivation has proved more effective in controlling most of the annual weed species in different row crops. However, it may not eradicate some perennial weed species such as *Cirsium arvense* (L.) Scop. (Graglia et al. 2006). Hoeing, slashing, mowing and earthing-up are some other forms of mechanical weed control, which are effective for weed management (Bajwa et al. 2015).

Different tillage systems affect weed dynamics and weed control measures depending upon the soil type, cropping system and weed flora (Chauhan et al. 2006; Bajwa 2014). Modified tillage in conservation agriculture itself offers an opportunity to manage some weeds. Most of the times, weed seeds persist in the upper soil layer in conservation tillage systems. Pareja et al. (1985) studied that 85% of all weed seeds were present in top 5 cm soil whereas, 28% seeds were prevailing in soil depth approached by mouldboard plough. Similarly, Staricka et al. (1990) found that 50% weed seeds were present in the top 4 cm soil layer and 11% in the mouldboard plough layer. No-till systems concentrate weed seeds near the soil surface due to less soil disturbance (Torresen et al. 2003). Weed infestation is a serious problem during initial years under conservation tillage which cause hindrance in getting good crop yield (Blackshaw et al. 2001). Under conservation tillage, newly produced seeds of weeds may germinate immediately due to fewer disturbances; however, buried seeds become dormant due to less manipulations. Relatively higher amounts of residues on the soil surface in conservation tillage systems not only affect weed distribution but also hamper herbicide efficacy. In such systems, weeds producing vegetative reproductive parts may be more difficult to control.

On the other hand, heavy tillage may invert soil layers and bury weed seeds deeper which reduce the weed density in following season, but in the long run, a high weed seedbank is developed. Tillage system may also affect weed seedbank persistence in different seasons. For example, in Australia, the seedbank of *Lolium rigidum* Gaud. was below 1% of the soil seedbank before the adoption of conservation tillage (McGowan 1970) but 20–30% after that (Peltzer and Matson 2002). Such shifts were attributed to changes in cropping patterns, cropping intensity and tillage systems. Tillage may affect weed seedling recruitment directly depending upon type, amount, frequency and equipment. Weeds respond differently under different tillage regimes. For instance, intensive tillage may affect small seeded weeds like *Vulpia bromoides* (L.) S.F. Gray. through deep burial (Chauhan et al. 2006). The timing of tillage operations also affects weed emergence and therefore, alters the seedbank dynamics (Chauhan and Johnson 2010a). For example, Myers et al. (2005) reported that a spring cultivation restricted the emergence of *Ambrosia artemisiifolia* L., *Amaranthus albus* L. and *Setaria faberi* Herrm. These changes in weed germination and emergence patterns make the herbicide selection and application time more difficult (Bullied et al. 2003). Tillage systems influence perennial species more as compared to annual (Chauhan et al. 2006).

Since tillage and related crop management practices are changing continuously, the way of using mechanical approaches to control weeds is also evolving. With the large-scale adoption of conservation tillage systems in the United States of America (USA), Australia, Europe and parts of Asia, conventional tillage is no more a pragmatic option for weed control (Bajwa 2014). In this regard, targeted or strip tillage not only conserves the precious resources but also control stubborn perennial weeds such as *Solanum carolinense* L. and *Taraxacum officinale* G.H. Weber ex Wiggers (Brainard et al. 2013). However, the infestations of annual weeds like *Stellaria media* (L.) Vill. and *Lamium amplexicaule* L. may increase in this system (Brainard et al. 2012). The introduction of a strategic deep tillage in a conservation tillage system has also shown some merits in breaking the continuity of weed infestations by sending a portion of viable weed seeds to the deeper soil layers. In a long-term study, the rotation of zero-tillage with a full-width tillage suppressed the density of several summer annual weed species (Peachey et al. 2006). Hence, tillage interventions play a major role in weed management.

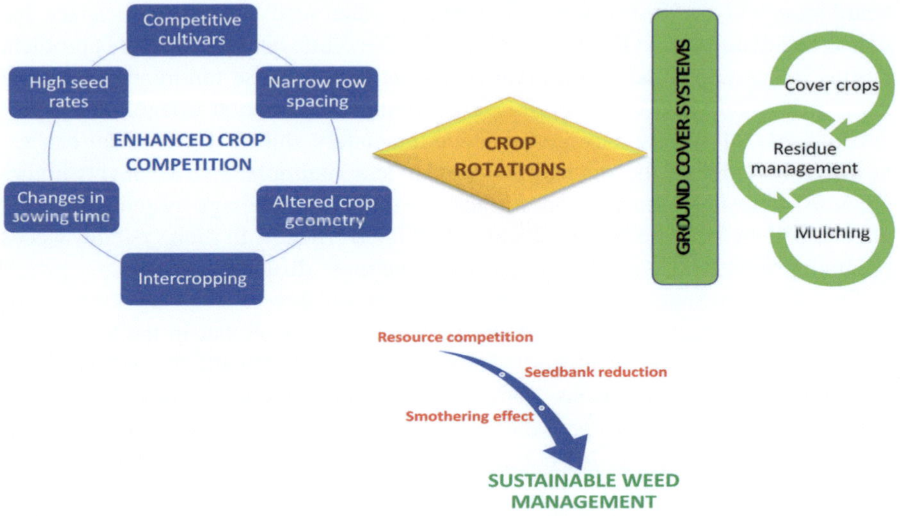

Fig. 1 An illustration of the use of cultural methods for sustainable weed management (enhanced crop competition, ground cover systems and diversified crop rotations suppress weeds through resource competition, physical smothering, and seedbank reduction)

2.3 Cultural Methods

A range of cultural practices has been used to control weeds since ancient times. These practices often have multi-faceted impacts on crop production, but weed control is a key benefit in most cases. Most of the cultural practices are inter-related and sometimes overlap (Fig. 1).

2.3.1 Crop Competition

There are numerous cultural methods to enhance the competitiveness of crops, which suppress the weed emergence, establishment, growth, reproduction and in some cases also the weed seed dispersal (Sardana et al. 2017). Crop competition offers effective and sustainable weed control. Different approaches to enhance the competitive ability of crops include the use of competitive crop cultivars, higher seed rates or planting densities, modified sowing methods, early or delayed sowing, narrow row spacing, altered row orientation or cropping geometry, intercropping, and fertilizer management (Table 1). All these practices work on the basic principle of giving advantage to crops over weed species by enabling them to acquire more resources efficiently (Bajwa et al. 2017).

Table 1 The use of enhanced crop competition for weed management in different crops

Cultural approach	Crop	Weeds controlled	References
Competitive cultivars	Barley (*Hordeum vulgare* L.)	*Bromus diandrus* Roth., *Lolium rigidum* Gaud.	Lemerle et al. (1995) and Paynter and Hills (2009)
	Canola (*Brassica napus* L.)	*Avena fatua* L., *Lolium multiflorum* Lam., *L. rigidum,* volunteer wheat	Zand and Beckie (2002), Beckie et al. (2008), Asaduzzaman et al. (2014) and Lemerle et al. (2014, 2016)
	Cotton (*Gossypium hirsutum* L.)	*Anoda cristata* (L.) Schlecht, *Xanthium strumarium* L.	Chandler and Meredith (1983) and Rezakhanlou et al. (2013)
	Maize (*Zea mays* L.)	*Abutilon theophrasti* Medik, *Panicum miliaceum* L., *Setaria glauca* (L.) Beauv	Staniforth (1961) and Lindquist and Mortensen (1998)
	Pea (*Pisum sativum* L.)	*L. rigidum* and other grass weed species	McDonald (2003) and Lemerle et al. (2006)
	Rice (*Oryza sativa* L.)	*Cyperus* spp., *Cynodon dactylon* (L.) Pers., *Echinochloa* spp.	Mahajan et al. (2014)
	Sorghum [*Sorghum bicolor* (L.) Moench]	*Echinochloa esculenta* (A. Braun) H. Scholtz and mixed weed flora	Wu et al. (2010) and Mishra et al. (2015)
	Soybean [*Glycine max* (L.) Merr.]	*A. theophrasti, Ipomoea lacunosa* L., *Senna obtusifolia* (L.) H.S. Irwin & Barneby, *Setaria italica* (L.) Beauv and mixed weed flora	Rose et al. (1984), Bennett and Shaw (2000), Nordby et al. (2007) and Rezvani et al. (2013)
	Wheat (*Triticum aestivum* L.)	*B. diandrus, Bromus tectorum* L., *L. multiflorum, L. rigidum, Setaria viridis* (L.) Beauv.	Gill et al. (1987), Blackshaw (1994a) and Lemerle et al. (1996)

(continued)

Table 1 (continued)

Cultural approach	Crop	Weeds controlled	References
High seed rate/ planting density	Canola	*A. fatua, L. rigidum, Polygonum convolvulus* L., *Sinapis arvensis* L. and mixed weed flora	O'Donovan et al. (2004) and Lemerle et al. (2017)
	Cotton	*Commelina benghalensis* L., *S. obtusifolia*	Stephenson and Brecke (2010)
	Lentil (*Lens culinaris* L.)	*Brassica* spp.	McDonald et al. (2007)
	Maize	*A. theophrasti, Amaranthus retroflexus* L., *P. miliaceum*	McLachlan et al. (1993), Lindquist et al. (1998) and Williams and Boydston (2013)
	Rice	*Cyperus rotundus* L., *Echinochloa crus-galli* (L.) Beauv and mixed weed flora	Chauhan et al. (2011) and Kaur and Surjit (2014)
	Sorghum	*E. esculenta* and mixed weed flora	Stahlman and Wicks (2000) and Wu et al. (2010)
	Soybean	*S. obtusifolia*	Norsworthy and Oliver (2002)
	Wheat	*A. fatua, Avena ludoviciana* Durieu, *Erodium cicutarium* L., *Galium aparine* L., *L. rigidum, L. multiflorum, Phalaris paradoxa* L.	Blackshaw et al. (2000), Walker et al. (2002), Mennan and Zandstra (2005), Lemerle et al. (2006) and Paynter and Hills (2009)

(continued)

Table 1 (continued)

Cultural approach	Crop	Weeds controlled	References
Narrow row spacing	Barley	*L. rigidum*	Borger et al. (2016a)
	Canola	*L. rigidum* and Mixed weed flora	Borger et al. (2010, 2016a)
	Chickpea (*Cicer arietinum* L.)	*L. rigidum* and Mixed weed flora	Borger et al. (2010, 2016b)
	Cotton	*A. retroflexus, C. benghalensis, Euphorbia hyssopifolia* L., *Jacquemontia tamnifolia* (L.) Griseb. Iaqta, *S. obtusifolia, Sida spinosa* L.	Molin et al. (2004), Reddy and Boykin (2010) and Stephenson and Brecke (2010)
	Maize	*Chenopodium album* L., *Setaria faberi* Herrm. and Mixed weed flora	Shrestha et al. (2001) and Dalley et al. (2006)
	Mungbean [*Vigna radiata* (L.) Wilczek]	*Chloris gayana* Kunth	Chauhan et al. (2017)
	Pea	*L. rigidum*	Lemerle et al. (2006)
	Rice	*Cyperus* spp., *Echinochloa colona* (L.) Link, *E. crus-galli*	Chauhan and Johnson (2010b) and Kaur and Surjit (2014)
	Sorghum	*Digitaria sanguinalis* (L.) Scop., *E. crus-galli, Panicum texanum* Buckl. and Mixed weed flora	Smith et al. (1990) and Holland and McNamara (1982)
	Soybean	*A. theophrasti, S. obtusifolia, Sorghum halepense* (L.) Pers., *X. strumarium* and Mixed weed flora	Felton (1976), Shaw et al. (1991) and Hock et al. (2006)
	Sunflower (*Helianthus annuus* L.)	Mixed weed flora	Osten et al. (2006)
	Wheat	*Bromus secalinus* L., *G. aparine, Lepidium sativum* L., *L. rigidum, P. minor*	Solie et al. (1991), Mahajan and Brar (2002), Peltzer et al. (2009) and Fahad et al. (2015)

(continued)

Table 1 (continued)

Cultural approach	Crop	Weeds controlled	References
Intercropping	Maize + soybean	Mixed weed flora	Tripathi and Singh (1983)
	Persian clover (*Trifolium resupinatum* L.) and white clover (*Trifolium repens* L.) + wheat	Mixed weed flora	Hartl (1989)
	Oats (*Avena sativa* L.) + alfalfa (*Medicago sativa* L.)	Mixed weed flora	Lanini et al. (1991)
	Chickpea + wheat	Mixed weed flora	Banik et al. (2006)
	Cotton + sorghum	*Convolvulus arvensis* L., *Trianthema portulacastrum* L.	Iqbal et al. (2007)
	Pea + wheat	*A. fatua, Rumex dentatus* L.	Khan et al. (2013)
East-west row orientation	Barley, canola, lupin (*Lupinus angustifolius* L.), pea, wheat	*Arctotheca calendula* (L.) Levyns., *Emex australis* Steinheil, *L. rigidum*	Borger et al. (2010, 2016b)
Altered spatial pattern	Wheat	*Matricaria perforata* Mérat, *Papaver rhoeas* L.	Olsen et al. (2005)
Early sowing	Wheat	*P. minor*	Singh et al. (1999)
Late sowing	Wheat	*P. minor, C. album*	Farooq and Cheema (2013)
Bed planting	Wheat	Mixed weed flora	Farooq and Cheema (2013)
Twin row system	Cotton	*C. benghalensis, J. tamnifolia, S. obtusifolia*	Stephenson and Brecke (2010)
Fertilizer management	Canola	*A. fatua*	Blackshaw et al. (2003) and Harker et al. (2013)
	Wheat	*A. fatua, Eleusine indica* (L.) Gaertn., *Geranium carolinianum* L., *L. multiflorum, S. arvensis, Veronica persica* Poir., *Vicia sativa* L.,	Melander et al. (2003), Blackshaw et al. (2003), Blackshaw and Brandt (2008), Scursoni et al. (2012) and Tang et al. (2013)

The development of crop cultivars with superior yield potential and tolerance to biotic and abiotic stresses has long been a pivotal research goal for breeders and agronomists (Lemerle et al. 2014). There are a number of studies reporting the weed suppression by competitive cultivars of major cereals and pulses crops (Table 1). Competitive cultivars possess multiple morphological and physiological attributes, which enable them to absorb water and nutrients efficiently and grow vigorously to suppress the co-existing weed species (Lemerle et al. 1996). The use of high seed rates provides dense crop populations which affect weed growth and development (O'Donovan et al. 2004; Lemerle et al. 2017). Narrow row spacing allows early canopy closure which restricts the light interception to weeds growing between the rows (Blackshaw et al. 2000). In this scenario, inter-specific competition is often more severe than intra-specific competition, which often avoids crop growth suppression but suppresses weed growth and development (Bajwa et al. 2017; Chauhan et al. 2017). Changing the planting time, method or the crop row orientation have also shown remarkable weed control in different crop production systems (Liebman and Dyck 1993; Singh et al. 1999; Olsen et al. 2005; Borger et al. 2010, 2016a, b; Farooq and Cheema 2013).

Intercropping is another important cultural practice which is used to improve the land use efficiency, soil health, system productivity and weed control (Liebman and Dyck 1993; Farooq et al. 2011a; Liebman et al. 2016). Sowing of two or more compatible crops together not only enhances the crop competition to beat the existing weeds but also helps in reducing the weed infestations in subsequent crops (Bajwa 2014; Singh et al. 2015; Jabran et al. 2017). Proper fertilizer management is often considered as a good management practice to enhance crop yields and net profits, and it also plays a vital role in weed management (Blackshaw et al. 2008; Bajwa et al. 2014). The research has shown that weeds are efficient in nutrient absorption from the soil in cropping situations depriving the crop plants of essential nutrients (Liebman and Dyck 1993). However, adjusting the fertilizer type, dose, application time and application method in a way that could enhance the nutrient uptake and fertilizer efficiency of crops may help in suppressing the weed species (Yin et al. 2005; Blackshaw and Brandt 2008; Bajwa et al. 2014; Tang et al. 2013; Jabran et al. 2017). Therefore, an improved fertilizer management could effectively be used to enhance the crop competitive ability for suppressing the weed species.

2.3.2 Ground Cover Systems

Protecting the soil by having a permanent vegetative, crop residue or artificial cover is a key principle of conservation agriculture (Farooq et al. 2011b). The same principle applies for sustainable crop production in modern-day agriculture. Like other cultural practices, soil cover also serves as a great tool for weed suppression in addition to its benefits for soil health, crop productivity and so on (Teasdale et al. 2007). The ground cover systems include the use of cover crops and retention of the previous crop residues for mulching (Bajwa 2014).

The practice of mulching has proved successful in managing weeds in conventional as well as organic crop productions systems (Bond and Grundy 2001; Bajwa 2014). Especially, the use of natural and artificial mulches has provided significant yield gains and weed control in vegetable crops (Boz et al. 2009; Bakht and Khan 2014). Mulching works on the principle of continuous ground cover which inhibits the light penetration to the lower layers of soil, and therefore reduce the germination of several weed species. Moreover, the natural mulches have smothering effects on the germination and growth of weeds. Natural mulching could be achieved by leaving the vegetative matter of cover crops on the soil surface or by retaining the previous crop residues (Erenstein 2003). The polyethylene and biodegradable mulches are gaining popularity in organic and sustainable production systems due to their low environmental impacts and efficiency (Kasirajan and Ngouajio 2012). The artificial materials such as plastic not only restrict weed emergence due to physical hindrance and moisture unavailability but also cause weed seed destruction by increasing the soil temperature (Katan 2015). Mulching not only provides effective weed control but also improves the water use efficiency, soil health and system productivity (Jabran et al. 2016). Therefore, it is a win-win approach for sustainable crop production systems.

There are numerous studies reporting the weed suppressive potential of multiple cover crops. Some of the important cover crops with high weed suppressive potential have been enlisted in Table 2. These crops provided an effective weed control during their growing season and lowered the weed pressure in subsequent crops (Weston 1996; Lowry and Smith 2018). Long ago, Teasdale (1996) described cover crops as a viable weed management option for sustainable agriculture due to their related benefits to soil health and crop performance. Such crops occupy the land surface during the period when main crops are not sown, restricting the weed emergence from the soil seedbank and suppressing the existing weed flora due to the resource competition and smothering effect (Teasdale et al. 2007; Hodgdon et al. 2016; Lowry and Smith 2018). However, the weed suppressive potential depends upon the growth and biomass accumulation of the selected cover crop (Liebman and Davis 2000; Teasdale et al. 2007). Usually the crop species with rapid germination ability, vigorous growth, high biomass accumulation capacity, efficient nutrient uptake and short growth period are selected for this purpose (Hodgdon et al. 2016; Lowry and Smith 2018).

The incorporation of previous crop residues or the vegetative matter of cover crops into the soil has proved effective in reducing the weed emergence and growth in subsequent crops (Derksen et al. 2002). However, this approach is not suitable in conservation systems and therefore, the trend has been shifted towards the surface retention of crop residues while crops are planted with minimal soil disturbance/reduced tillage (Bajwa 2014). Franke et al. (2007) reported a 50% reduction in *P. minor* emergence in zero-tilled wheat crop due to the retention of rice (*Oryza sativa* L.) crop residues. Similarly, the surface retention of crop residues also resulted in significant reduction in the emergence of *Anagallis arvensis* L. and *C. album* in wheat (Chhokar et al. 2007).

Ground cover systems provide an effective weed control in sustainable crop production systems with least impact on the environment. The consistent soil cover could usually be achieved by natural inputs from within the system and therefore, offers an economical and environment-friendly solution for weed problems.

Table 2 Different cover crops with demonstrated in-season and/or carry-over weed suppressive potential

Cover crop			Weeds suppressed	Referencess
Common name	Scientific name	Family		
Alfalfa	*Medicago sativa* L.	Fabaceae	*C. album, E. crus-galli, Poa annua* L., *Polygonum* spp., *Stellaria media* (L.) Vill, *V. sativa*	Kruidhof et al. (2008)
			A. retroflexus, Digitaria spp., *P. oleracea, S. media*	Hodgdon et al. (2016)
Barley	*Hordeum vulgare* L.	Poaceae	*A. retroflexus, C. album, L. multiflorum*	Singh et al. (2003)
Buckwheat	*Fagopyrum sagittatum* Moench.	Polygonaceae	*A. theophrasti, A. retroflexus, C. album, S. viridis*	Wortman et al. (2013)
Chickling vetch	*Lathyrus sativus* L.	Fabaceae	*A. theophrasti, A. retroflexus, C. album, S. viridis*	Wortman et al. (2013)
Crimson clover	*Trifolium incarnatum* L.	Fabaceae	*A. theophrasti, A. retroflexus, C. album, S. viridis*	Wortman et al. (2013)
			A. retroflexus, Digitaria spp., *P. oleracea, S. media*	Hodgdon et al. (2016)
Finger millet	*Eleucine corocana* (L.) Gaertn.	Poaceae	*E. crus-galli, Eclipta prostrata* L., *Isachne globose* L.	Samarajeewa et al. (2006)
Fodder radish	*Raphanus sativus* L.	Cruciferae	*C. album, E. crus-galli, S. media, V. sativa*	Kruidhof et al. (2008)
			A. retroflexus, Digitaria spp., *P. oleracea*	Hodgdon et al. (2016)
Hairy vetch	*Vicia villosa* Roth.	Fabaceae	*A. theophrasti, A. retroflexus, C. album, E. crus-galli, Panicum capillare* L., *Setaria viridis* var. major (Gaudin) Pospichel, *S. media*	Mohler and Asdale (1993)
			Lolium temulentum L., *S. media*	Campiglia et al. (2009)
			A. retroflexus, Digitaria spp., *P. oleracea, S. media*	Hodgdon et al. (2016)
Oats	*Avena sativa* L.	Poaceae	*A. retroflexus, C. album, L. multiflorum*	Singh et al. (2003)

(continued)

Table 2 (continued)

Cover crop			Weeds suppressed	Referencess
Common name	Scientific name	Family		
Pea	*Pisum sativum* L.	Fabaceae	*A. theophrasti, A. retroflexus, C. album, S. viridis*	Wortman et al. (2013)
Rapeseed	*Brassica napus* L.	Cruciferae	*Lolium temulentum* L., *S. media*	Campiglia et al. (2009)
			C. album, E. crus-galli, S. media, V. sativa	Kruidhof et al. (2008)
			A. theophrasti, A. retroflexus, C. album, S. viridis	Wortman et al. (2013)
Rye	*Secale cereal* L.	Poaceae	*Amaranthus* spp., *Portulaca oleracea* L.	Nagabhushana et al. (2001)
			A. retroflexus, C. album, L. multiflorum	Singh et al. (2003)
			C. album, E. crus-galli, S. media, V. sativa	Kruidhof et al. (2008)
			A. retroflexus, Digitaria spp., *P. oleracea, S. media*	Hodgdon et al. (2016)
Snail medick	*Medicago scutellate* Mill.	Fabaceae	*Lolium temulentum* L., *S. media*	Campiglia et al. (2009)
Subclover	*Trifolium subterraneum* L.	Fabaceae	*Lolium temulentum* L., *S. media*	Campiglia et al. (2009)
White clover	*Trifolium repens* L.	Fabaceae	*A. retroflexus, Digitaria* spp., *P. oleracea, S. media*	Hodgdon et al. (2016)
White lupin	*Lupinus albus* L.	Fabaceae	*C. album, E. crus-galli, S. media, V. sativa*	Kruidhof et al. (2008)
White mustard	*Sinapis alba* L.	Cruciferae	*A. theophrasti, A. retroflexus, C. album, S. viridis*	Wortman et al. (2013)

2.3.3 Crop Rotations and Diversification

Crop rotations play an important role in weed management by having a direct impact on long-term weed dynamics of any cropping system (Liebman and Dyck 1993; Bond and Grundy 2001; Blackshaw et al. 2008; Jabran et al. 2017). Weed control may not be the most important factor while planning a crop rotation but it often appears as the most rewarding aspect of a well-planned rotation (Liebman and Davis 2000). Kotile and Martin (2000) reported crop rotation as one of the most important sustainable weed management practices in terms of its impact and adoption in diverse farming communities. Changing the cropping sequence and diversifying the cropping system break the cycle of reproduction and emergence of noxious weeds by altering the microclimate at a particular site (Anderson 2007).

Diverse crop rotations restrict certain weed species to build strong association to a specific crop (Blackshaw et al. 2007). Rotating less competitive crops with more competitive crops, non-allelopathic crops with allelopathic crops, exhaustive crops with the crop requiring low inputs, long-duration crops with short-duration crops, cereals with legumes, and tap-root crops with crops having fibrous root-systems provides effective weed control (Liebman and Dyck 1993; Liebman and Davis 2000; Anderson 2007; Blackshaw et al. 2007, 2008). On the other hand, skipping one crop in rotation and keeping the land fallow has proved effective in controlling the perennial weeds (Bond and Grundy 2001).

Several studies have reported the potential of planned crop rotations for weed management. For instance, introducing rice in place of maize (*Zea mays* L.) in a rotation with wheat, almost eliminated *A. fatua* in the following seasons (Gill and Brar 1975). Similarly, rotating canola (*Brassica napus* L.) with wheat reduced *Bromus tectorum* L. to just 50 plants m^{-2} as compared to wheat monoculture where the density of *B. tectorum* was above 700 plants m^{-2} (Blackshaw 1994b). Similarly, introducing maize or sunflower (*Helianthus annuus* L.) in wheat-fallow rotation caused remarkable reductions in *Aegilops cylindrica* Host and *B. tectorum* infestations (Daugovish et al. 1999). In a rice-wheat cropping system, the introduction of barseem (*Trifolium alexandrinum* L.) or oat (*Avena sativa* L.) in place of wheat once in 3 years significantly reduced the *P. minor* infestations (Chhokar and Malik 2002). In another study, replacing a typical rice-wheat rotation with rice-barseem-sunflower-wheat and cotton (*Gossypium hirsutum* L.)-pigeon pea [*Cajanus cajan* (L.) Millsp.]-wheat rotations reduced isoproturon resistance in *A. fatua* by alleviating the selection pressure (Malik and Singh 1995). Koocheki et al. (2009) reported that different crop rotations involving wheat had significantly different weed flora, infestation levels and weed seedbank dynamics. The weed flora of a wheat-wheat rotation comprised of about 90% grass weeds compared to wheat-sugar beet (*Beta vulgaris* var. *saccharifera*) rotation consisting of only 43% of grass weeds (Koocheki et al. 2009). Moreover, the wheat-sugar beet and wheat-maize rotations provided 28 and 12% reductions in weed seedbank as compared to continuous wheat. Chauhan (2012) reported that changing the crop sequence provided effective weed control in dry direct-seeded rice. Recently, Shahzad et al. (2016) reported that sorghum-wheat rotation reduced the weed infestation in conventional as well as conservation tillage systems.

Crop diversification in long-term rotations has been identified as a robust weed management tool in sustainable agriculture. In fact, the cropping system diversity has been suggested as a functional tool for agro ecological weed control (Gaba et al. 2014). A diversified crop rotation involving wheat, maize and soybean [*Glycine max* (L.) Merr.] resulted in a substantial decline in broadleaf weed infestations and their seedbank (Teasdale et al. 2004). In Canada, a shift from cereal-based crop rotations to more diversified rotations including legumes and forage crops such as canola, flax (*Linum usitatissimum* L.), pea (*Pisum sativum* L.), lentil (*Lens culinaris* Medik.), alfalfa (*Medicago sativa* L.) and red clover (*Trifolium pretense* L.) significantly reduced weed densities (Blackshaw et al. 2008). Anderson (2015) reported that diversified crop rotations integrated with no-till significantly improved weed management in organic crop production systems. A complex rotation involving several cool and warm season annual crops and perennial legumes reduced weed emergence in most of the annual crops, delayed the weed emergence overall, reduced the infestations of *Taraxacum officinale* G.H. Weber ex Wiggers, and reduced the yield penalties due to weed interference (Anderson 2015).

Diversified crop rotations have immense potential for weed management in sustainable agriculture. It is one of the most natural and eco-friendly way to control noxious weed species. The disruptive rotations not only enhance the soil and crop productivity but also create harsh conditions for most of the weed species associated to certain monocultures. Crop rotations also restrict the weed seedbank development which avoids the long-term weed infestations.

3 Innovative Approaches

There are some innovative and novel approaches to manage weed species in sustainable production systems. These approaches are based on enhancing the ecological service of the system or technological advances. These techniques are coherent with the existing conventional methods and provide more flexibility and reliability to the IWM suite.

3.1 Allelopathy

Allelopathy is an ecological phenomenon in which some plant species release certain secondary metabolites/allelochemicals in their above- and or below-ground surroundings to inhibit or promote the germination, growth and development of their neighbouring plant species (Singh et al. 2003; Farooq et al. 2011a). Generally, allelochemicals have inhibitory effects at their high concentrations while they may stimulate plant growth at low concentrations (Farooq et al. 2013). The inhibitory aspect of allelopathy has great potential for weed suppression in sustainable crop production systems (Cheema et al. 2004; Narwal and Haouala 2013; Jabran et al. 2015). These allelochemicals could be released in the form of root exudates or volatiles directly from the living plants or may be expressed in the form of leachates from plant residues or decaying parts (Farooq et al. 2013). Therefore, the allelopathic effects of different crops can be exploited for weed management by (i) using the crop cultivars with superior allelopathic potential, (ii) using allelopathic crops as cover or inter-crops or simply by including them in rotations, (iii) mulching or residue incorporation, and (iv) foliage application of allelopathic crop extracts (Bhowmik and Inderjit 2003; Farooq et al. 2013; Nawaz et al. 2014).

The use of allelopathy for weed management has been extensively studied and reported over the last three decades. Several studies have reported that weeds can be effectively managed through allelopathic effects of various field crops (Table 3).

Table 3 Weed suppressive potential of different allelopathic crops

Mode of application	Weeds suppressed	Crops benefited	References
Allelopathic cultivars			
Rice	*E. crus-galli, E. colona*	Rice	Masum et al. 2016
Allelopathic extracts			
Sorghum	*C. album, C. arvensis, C. dactylon, Cyperus iria* L., *C. rotundus, E. colona, P. minor, R. dentatus, T. portulacastrum*	Cotton, mungbean, rice, wheat	Cheema and Khaliq (2000), Cheema et al. (2001, 2002) and Wazir et al. (2011)
Sunflower	*A. fatua, Melilotus officinalis* (L.) Lam., *P. minor, Rumex obtusifolius* L.	Wheat	Cheema et al. (2003) and Naseem et al. (2010)
Sorghum + Sunflower, Sorghum + Canola, Sorghum + Tobacco (*Nicotiana tabacum* L.)	*A. fatua, P. minor*	Wheat	Jamil et al. (2009)
Soil incorporation of residues			
Sorghum	*C. album, Fumaria officinalis* L., *P. minor, R. dentatus*	Wheat	Cheema and Khaliq (2000)
Sunflower + Canola + Rice	*T. portulacastrum*	Maize	Khaliq et al. (2010)
Mulching			
Canola	*A. fatua*	Wheat	Farooq et al. (2011a)
Sorghum	*C. arvensis, C. dactylon, C. rotundus, T. portulacastrum,*	Cotton	Cheema et al. (2000)
Cover crops			
Rye	*Amaranthus* spp., *P. oleracea*	–[a]	Nagabhushana et al. (2001)
Velvet bean [*Mucuna pruriens* (L.) DC.]	*E. crus-galli*	–	Peters et al. (2003)
Barley	*Digitaria ciliaris* (Retz.) Koel.	–	Farooq et al. (2011a)
Intercrops			
Chickpea	*Anagallis arvensis* L., *A. fatua, Coronopus didymus* (L.) Sm.	Wheat	Banik et al. (2006)
Sorghum	*C. arvensis, T. portulacastrum*	Cotton	Iqbal et al. (2007)
Soybean	*C. rotundus*	Cotton	Iqbal et al. (2007)
Crop rotations			
Oat and pearl millet [*Pennisetum glaucum* (L.) R.Br.] in place of wheat	*C. rotundus, E. crus-galli*	Rice	Kobayashi et al. (2004)
Wheat	*Orobanche minor* Sm.	Red clover (*Trifolium pretense* L.)	Lins et al. (2006)

[a]Data not available or applicable

The allelopathic potential of several field crops including, different *Brassica* spp., wheat, maize, soybean, sunflower, rice, sorghum [*Sorghum bicolor* (L.) Moench], rye (*Secale cereal* L.) and buckwheat (*Fagopyrum esculentum* Moench.) has been explored in different ways to control noxious weeds in agro-ecosystems (Bhowmik and Doll 1982; Putnam et al. 1983; Weston 1996; Cheema and Khaliq 2000; Cheema et al. 2004; Jamil et al. 2009; Rehman et al. 2018). Imported residues (applied as surface mulch or incorporated shallowly) of maize, soybean and sunflower were effective in controlling early emerging weed species such as *Amaranthus retroflexus* L., *C. album*, and *Galinsoga parviflora* Cav. (Barker and Bhowmik 2001). However, using allelopathic cover crops or rotational crops for weed management is challenging in field applications. There are limitations in using cover crops for various cropping systems. Delayed planting, delayed crop emergence, phytotoxic effects to major crop, and increased pest pressure are some of the limitations. Under the best management practices, it is possible to integrate allelopathic crop residues and chemical control strategies (such as pre- or post-emergence herbicides). Cover crops in general are killed to leave the crop residues on the soil surface to the following crop. In the majority of cases, herbicides are used to kill cover crops. Some crops such as oat and *Brassica* spp. when used as cover crops can be killed naturally by winter snow or severe frost (Putnam et al. 1983; Parish 1990). Residue from a fall-planted oat cover crop resulted in reduced herbicide use in the following spring in reduced-tillage maize (Bhowmik, 1992). However, oat cover crop residue did not make any impact in reducing herbicide treatments under the no-tillage system.

Significant research has been conducted to screen crop cultivars with high allelopathic potential (Putnam et al. 1983; Wu et al. 1999; Olofsdotter et al. 1995; Olofsdotter 2001). In general, monocot crop species have been searched for allelopathy. Several members of the family Poaceae have been identified as allelopathic. Over the last two decades, rice accessions or cultivars have been extensively examined for their allelopathic activity in suppressing weed species (Dilday et al. 1994; Olofsdotter et al. 1995; Fischer et al. 1997; Kim et al. 1999; Olofsdotter 2001; Kato-Noguchi et al. 2008; Khanh et al. 2009; Masum et al. 2016, 2018). Salam and Kato-Noguchi (2009) compared the allelopathic activity among 102 modern and traditional cultivars and reported that BR17 (modern variety) was the most allelopathic. The authors also reported that Kartiksail (an indigenous cultivar) might have a great inhibitory activity against *Echinochloa crus-galli* (L.) Beauv. Masum et al. (2016) screened 50 rice cultivars from Bangladesh for allelopathic activity against *E. crus-galli* and *Echinochloa colona* (L.) Link. It was reported that different cultivars caused 7–37% suppression of both weed species. Recently, Masum et al. (2018) identified four potential allelochemicals from four indigenous rice cultivars.

There are limitations for using allelochemicals for successful weed management. Some of these constraints in implementing natural products for effective weed management include i) natural products have generally short half-lives, (ii) compounds are present in very low concentration, (iii) narrow spectrum selectivity and (iv) high cost of production (Bhowmik and Inderjit 2003; Inderjit and Bhowmik

2004; Tharayil et al. 2008). Continued research on these areas is important and we must invest our resources in exploring allelopathy as a complimentary component in successful weed management practices.

3.2 *Thermal Weed Control*

The use of different technologies delivering the thermal energy to kill weeds has shown promise in sustainable crop production systems. This section will demonstrate the practicability and suitability of the distinct heat sources for weed management.

3.2.1 Flaming

Flame weeding has been successfully used to control weeds. Of the several kinds of applicators, tractor-mounted flame weeder has been extensively used in this regard. Ascard (1994) tested the flame weeder under field conditions for 3 years to develop a dose-response relationship between applied liquid petroleum gas (LPG) doses and percent weed control. A total load of 159 kg LPG ha^{-1} was needed for 95% reduction in high density (607 plants m^{-2}) of *Sinapis alba* L. at the 4–6 leaf stage (Ascard 1994). This technique has an advantage over chemical application and can effectively control herbicide-resistant weeds (Wszelaki et al. 2007; Ulloa et al. 2011).

3.2.2 Steaming and Solarisation

Soil steaming has been reconsidered for weed management in recent years. Sheet steaming is the most common steam applicator which involves covering of the soil with a heat resistance membrane. Saturated steam is pumped to penetrate soil surface for weed control (Gay et al. 2010). The application of steam for 5 min achieved equilibrium in a temperature gradient, with a maximum threshold of 100 °C (Gay et al. 2010). With this exposure, the average weed density was less than 50 plants m^{-2} compared to un-steamed soil where it was 400 plants m^{-2}. The authors proposed that soil steaming can effectively control weeds and disinfect the soil; however, it requires sufficient energy investment for a steam generation.

Soil solarisation has also been reported to suppress soil-borne pathogens and weeds (McGovern et al. 2013). The use of double-layer polyethylene sheets has increased the efficacy of this weed control method by increasing a temperature gradient of 3–10 °C compared to a single layer sheet (Barakat and AL-Masri 2012). Solarisation can kill a wide range of weed species by breaking seed dormancy and heating-induced devitalization of young seedlings. Soil solarisation is the best suit to IWM (McGovern et al. 2013).

3.2.3 Electrocuting

The use of electric current for soil disinfestation has long been proposed and attempted (Diprose et al. 1984). Electrical methods of killing weeds fall into two main types; spark discharge and direct contact, and either of these methods needs high voltages (20 kV) for effective weed control (Diprose and Benson 1984). Diprose et al. (1980) reported that short exposure (20 s) of the high alternating voltage of 5 kV can potentially kill annual weeds. Various plant species require a different range of current; however, 0.5–1.0 A was enough threshold level for killing 1.0–1.4 m high vegetation. A tractor-driven system with a capacity of 8 kV voltage and a covering speed of 1.6 km ha^{-1} was found to kill 75% of weed flora. Electrocuting had been successfully used to kill weeds both in laboratory and field conditions and offers advantages over chemical and mechanical weed control methods (Diprose et al. 1980).

3.2.4 Radiations

Radiative heat transfer refers to electromagnetic energy transfer from adjacent hot material to the target (Brodie 2018). Infrared radiation systems use gas burners to heat up ceramic and metal surfaces, which then transfer the heat to the ground for disinfection scenarios. Parish (1990) identified that a "medium-wave tubular fused quartz infrared emitter" was an effective tool for weed management. Compared to flame weeder, there is no profound effect of wind on the infrared burners, and the target area coverage is much higher. In either burners, for 100% weed control, a total load of the 120 kg ha^{-1} of propane was required and relatively much higher temperature (1350 °C) was needed in case of flame weeder compared to the infrared radiator (770 °C) (Ascard, 1998). For efficient weed control, an energy density of 200–400 kJ m^{-2} was needed in case of medium-wave infrared energy (Parish 1990).

Ultraviolet radiation falls in the range of 100–400 nm and can be distinguished into three groups based on wavelength (Brodie 2018). When plants are subjected to ultraviolet radiation, most of the energy obsorbed into the thin outer layer (0.1–0.2 mm) of plant tissues and, thus, results in heating of plant material in similar pattern to other thermal weed control methods like flame weeding (Andreasen et al. 1999). Moreover, regrowth after ultraviolet radiation was observed by Andreasen et al. (1999), suggesting multiple treatments for effective weed control.

3.2.5 Microwaves

Microwaves (MW) are non-ionizing electromagnetic waves occupying a frequency range of 300 MHz $< f <$ 300 GHz and the wavelength range of 1 mm $< \lambda <$ 1 m, respectively (Banik et al. 2003). In several studies, the MW application has been proposed for pre- and post-emergence weed control. Wayland et al. (1973) buried permeable seed packets, imbibed for 10 h before burial, of wheat and radish up to 2.5 cm in Lufkin fine sandy loam soil (6.8% moisture content). Microwaves (1.5 kW; 2.45 GHz) applied to soil via a stainless steel radiator. It was found that for a 50% seed mortality of wheat and radish, 100 J cm^{-2} and 180 J cm^{-2} MW energy, respectively, was required. Sartorato et al. (2006) tested a MW-based (2.45 GHz; 900 W) weed control system in the field (loamy soil) on the suppression of *Abutilon theophrasti* Medik. and *Setaria viridis* (L.) Beauv. The estimated MW dose of 1015–3433 kJ m^{-2} gave up to 90% reduction of *A. theophrasti* and *S. viridis* dry weight.

In Australia, Brodie et al. (2009) reported the complete reduction in the germination capacity of *L. rigidum* seedbank at 80–100 °C soil temperature, this was achieved with MW heating time of 12 min. This (8–12 min) exposure in the top 0–5 cm of wet sandy soil completely inhibited the germination of *L. rigidum*. Brodie (2018) reported that the post-emergence application of 400–500 J cm^{-2} killed *L. rigidum*, *E. crus-galli*, *Conyza canadensis* (L.) Cronq. and *Malva parviflora* L. In addition, the field test of prototype MW weed control device killer, reduced (60–80%) the weed establishment in rice crop and sustained crop productivity. Khan et al. (2017) applied microwave energy of 560 J cm^{-2} via open structured horn antenna (aperture dimensions: 110 × 55 mm) for weed suppression in temperate rice crop before crop planting. This energy application achieved a maximum temperature of 80 °C and ultimately reduced the weed pressure by up to 70–80%. Additionally, Khan et al. (2018) tested the pre-sowing effect of microwave soil treatment for weed suppression in the dryland wheat production system of Australia. It was found that the temperature range of 70–80 °C significantly reduced weed establishment under field conditions.

3.3 *Precision Weed Management*

Precision agriculture is becoming increasingly popular with the advancements in the fields of remote sensing, modelling, robotics and artificial intelligence (Westwood et al. 2018). These modern technologies also offer viable weed management options for sustainable agriculture (Bajwa et al. 2015). Modelling techniques help to estimate the weed infestation levels and to determine the weed dynamics in a particular field. Modelling the population dynamics and the efficacy of control measures has become a strong tool for precision weed management in sustainable agriculture (Freckleton and Stephens 2009). There are several models available to measure the weed populations and the efficacy of herbicides in actual field scenarios, which help in decision making in weed management. Remote sensing is a powerful tool to mark the weed patches in the fields and then apply control options to specific sites (Medlin and Shaw 2000). Weed mapping based on high-resolution imagery and remote sensing aided by modelling approaches is a precise way to tackle weeds on a broad scale (Thorp and Tian 2004). These methods are not only time, labour and resource efficient but also have minimal environmental footprints. Unmanned aerial vehicles have been used successfully in taking remote images of different weeds in different row crops (Torres-Sánchez et al. 2013).

The precise digital surveillance, identification and mapping of weeds have paved the way for a highly efficient and precise weed control through robots (Young et al. 2014). Robotic weed control is a real time integration of sensing and mechanical technologies (Slaughter et al. 2008). There are already some success stories such as *See & Spray* robotic weed management technology. Such high-tech systems sense and identify weeds with extreme precision and then spot spray them in real time in fallow lands and row crops (Westwood et al. 2018). These systems also have the ability to identify a specific problematic weed species such as *Amaranthus palmeri* S. Wats. and only spray that. Similarly, machine vision has also been used successfully to aid the robots to kill weeds mechanically or by thermal methods (Astrand and Baerveldt 2002; Blasco et al. 2002).

Precision tools are undoubtedly the future of sustainable weed management being efficient and eco-friendly (Westwood et al. 2018). However, these technologies are very expensive and still need fine tuning to be used in diverse crop production systems.

3.4 Harvest Weed Seed Control

Herbicide weed seed control (HWSC) is an innovative system developed in Australia where the herbicide resistance threatened the future of highly productive conservation cropping systems that relied almost entirely on chemical weed control. This was substantial motivation for growers to develop alternate systems that can be used within conservation cropping systems. Therefore, systems were developed that target the weed seed passing through the harvester during harvesting (Walsh et al. 2013). In Australia, researchers have optimized HWSC systems over the last 20 years and now this technology is being rapidly adopted. A major portion of weed seeds (e.g. > 95% for *L. rigidum*), collected during harvest, exit the combine in the chaff fraction of the harvest residues (Walsh and Powles 2007; Broster et al. 2016). The residue-dispersal systems on modern combines effectively redistribute the weed seeds across the field during harvest. This process exacerbates the weed control problems by further spreading the resistant seeds of weed populations. Due to this reason, chaff material became the focus of innovative Australian growers when they began the development of HWSC systems.

3.4.1 Chaff Carts

Chaff carts were the first HWSC tools to be introduced in Australia in the 1980s. This system consists of a trailing cart attached to the rear of the harvester to collect chaff material during harvest (Fig. 2a). Chaff cart collection systems can collect and remove high proportions of *A. fatua, L. rigidum* and *Raphanus raphanistrum* L. seeds during harvest (Shirtliffe and Entz 2005; Walsh and Powles 2007). Because of the large volume of chaff produced during harvest, this residue fraction is typically placed in piles, lined up across the field (Fig. 2b) in preparation for subsequent burning to destroy the weed seeds. This chaff material can also be used to graze or feed the livestock (Table 4).

3.4.2 Narrow Windrow Burning

First used in the mid-1990s, narrow windrow burning was developed as an inexpensive and simple approach to HWSC. This system involves the attachment of a chute to the rear of the harvester that, during harvest, concentrates the chaff into a narrow-windrow (50–60 cm wide). These windrows are later burnt when weather is suitable to contain the fire within the windrows. Windrow burning has higher temperature and burning duration due to concentrated chaff and straw compared to a typical whole field burn (Walsh and Newman 2007). The simplicity and efficacy of this approach to HWSC has led to its widespread adoption in Australia (Table 4; Walsh et al. 2017).

Fig. 2 (**a**) Chaff cart system operating in wheat crop harvest in Australia and (**b**) resulting chaff piles across the field that will be later managed to destroy weed seeds

Table 4 The disadvantages and advantages associated with the use of currently available HWSC systems

HWSC system	Disadvantages	Advantages (in addition to weed seed control)
Chaff cart	Cart and transfer system need to be attached to and towed by the harvester.	Collection of a valuable source of livestock feed
	Management (burning/grazing/removal) of collected chaff to ensure weed seed control can be time consuming, with chaff burning a fire risk in cereal stubbles.	
	Chaff management results in the loss of organic matter and nutrients from across the field.	
Narrow windrow burning	Burning windrows is time consuming and a	Easy to attach a chute and use during harvest Inexpensive and simple to manufacture chutes
	Significant fire risk in cereal stubbles.	
	Concentration of straw and chaff residues in windrows results in the removal of nutrients and organic matter from across the field	
Bale direct system	Baler is attached to and powered by harvester	Baled straw and chaff material is a marketable livestock feed
	Comparatively expensive	
	Removal of nutrients and organic matter in the baled material	Stubble removal allows easier crop planting
Chaff lining and tramlining	Allows some weed seed survival in low chaff or disturbed chaff line areas	Can create chaff lines with easy to attach chutes Simple and inexpensive to manufacture chutes Chaff on tramlines reduces vehicle dust during spraying
	Concentration of chaff material into narrow rows results in removal of nutrients and organic matter from across the field.	
	Build-up of chaff material placed on dedicated lines	
Impact mills (e.g. HSD, iHSD, seed terminator)	Attached to and powered by harvester	Allows retention of all crop residues
	Processing chaff creates additional dust during harvest	No after harvest management of residues
	Comparatively expensive	

3.4.3 Bale Direct System

This system was developed in the early 2000s to realise a commercial opportunity of using the baled harvest residues for livestock feed. It has a large square baler trailing and powered by the combine that makes bales from the chaff and straw residues. It works to capture weed seeds and bale the residues for livestock feed. Bale direct system captured >95% seeds of *L. rigidum* at the harvest and therefore, prevented the seedbank build-up of this noxious weed (Walsh and Powles 2007). This approach is suitable for use in the fields where high straw residue levels impede subsequent crop planting. A large amount of baled material is produced during crop harvest, e.g. ~ 2 t ha^{-1} for every 1 t ha^{-1} of harvested grain, therefore securing a market for this material is critical (Table 4).

3.4.4 Chaff Impact Mills

Since the initial development of HWSC, there has been the demand for a system that controls weed seeds during grain harvest without the need for follow-up operations. In 2005, the development began on the Harrington Seed Destructor (HSD; Fig. 3a), a trailer mounted cage mill, with chaff and straw transfer systems, and a diesel motor as a power source. Although the HSD had proven weed seed destruction efficacy, the demand was for a harvester integrated system. Subsequently, the integrated HSD (iHSD; Fig. 3b) and seed terminator have been developed and commercially released. Extensive testing of the iHSD has proven its seed destruction efficacy on a range of weed species (Table 5).

3.4.5 Chaff Lining and Chaff Tramlining

The confinement of the weed seed bearing chaff material into narrow (20–30 cm) rows on dedicated wheel tracks (Chaff tramlining) or between stubble rows (chaff lining) during harvest was developed by Western Australian growers in the

a b

Fig. 3 (**a**) HSD and (**b**) iHSD systems developed for chaff processing and weed seed destruction during harvest

Table 5 Individual seed weight and percent seed destruction of 11 weed species by Integrated Harrington Seed Destructor in Australia

Weed species	Seed weight (mg/seed)	Seed kill (%)[a]
L. rigidum	2.8	96 (0.9)
Avena spp.	26.8	99 (0.1)
R. raphanistrum	5.2	99 (0.1)
Hordeum glaucum Steud.	10.0	99 (0.1)
Bromus spp.	15.7	98 (1.0)
Echinochloa spp.	2.2	99 (0.8)
Sisymbrium orientale Torn.	0.20	99 (0.4)
Conyza bonariensis (L.) Cronq.	0.05	99 (0.2)
Chloris truncata R. Br.	0.28	97 (0.4)
Sonchus oleraceus L.	0.33	99 (0.5)
Chloris virgata Sw.	0.33	98 (0.3)

[a]Figures in brackets are the standard errors for the mean of eight replicates (Walsh et al. 2018)

mid-2000s. In this approach, collected weed seeds are placed in an inhospitable environment where the combination of physical and chemical influences of the chaff material prevent the germination and emergence of weed seeds. Chaff rows are established by a chute attached on the rear of the combine that concentrate chaff material into narrow rows. The placement of chaff material on tramlines has the added advantage of reducing the amount of dust, created by vehicle movement during summer spraying, which can interfere with herbicide efficacy (Table 4).

The HWSC systems have evolved considerably since they were first introduced about 30 years ago. As the adoption of HWSC systems continues to grow there will be the increasing demand for more refined systems that are easier to use and have minimal impact on harvest efficiency as well as maintaining weed control efficacy. Estimates are that HWSC will grow from 42% in 2014 to 82% adoption by 2019, indicating a strong demand for a range of HWSC systems.

4 The Role of Herbicides in Sustainable Weed Management

Chemical weed control is the most efficient, economical and adopted method of weed management. However, the increasing problems of herbicide resistance, environmental pollution, health issues and off-target application are making this option less suitable (Liebman et al. 2016). Although different non-chemical weed control methods enlisted above are best fit for sustainable crop production, herbicides cannot be abandoned completely due to lower efficacy of other methods. Therefore, it would be unrealistic to say that herbicides have no place in sustainable agriculture at present and in the near future. In fact, herbicides are essential to manage weeds on a large scale in order to secure crop yields. These chemicals have served humanity a great deal during the last 40 years, but the reliance on herbicides has led to the development of herbicide resistance and environmental concerns (Westwood et al. 2018). For instance, the consistent use of herbicides with the same mode of action over a long period of time increased the selection pressure and the evolution of resistance in weeds (Mortensen et al. 2012). Similarly, mono-cropping encouraged the use of the same herbicides repeatedly leading to similar consequences. The use of herbicides in isolation, without integrating other suitable weed control methods has not been sustainable.

The future of herbicides lies in their judicious and integrated use in order to preserve the existing chemistry (Westwood et al. 2018). For example, glyphosate is a "one in a century" discovery, but we are losing this great chemical to widespread resistance across the globe (Mortensen et al. 2012). Still, efforts to minimize and optimize the use of glyphosate can preserve this herbicide for more targeted applications in future. Farmers must be educated and encouraged to use herbicides according to the local recommendations, keeping in view the long-term sustainability of the system and environmental protection (Senseman and Grey 2014). Herbicides with different modes of action should be used in rotation within the same crops. The reliance on only herbicides should be ruled out by including diverse weed control methods in IWM strategies (Mortensen et al. 2012; Liebman et al. 2016). On the other hand, research efforts need to expedited to discover new chemicals with novel modes of action. These strategies may provide a durable solution for herbicide-related problems and keep herbicides as a viable and efficient weed control method in sustainable agriculture (Westwood et al. 2018). Research effort should also be dedicated to reduce the off-target herbicide applications and the environmental footprint of chemicals.

5 Conclusions and Future Directions

Sustainable weed management is strongly linked with the use of ecologically-based weed control methods. However, none of the methods discussed above is a silver bullet when it comes to reliable weed control and therefore, key to success is IWM (Jordan and Davis 2015; Neve et al. 2018). A mix of traditional and innovative approaches discussed in this chapter may provide a suitable IWM package for most of the crop production systems. It is important to choose the right methods and then finding a balance between typical herbicide weed control and cultural or ecological methods. The motivation to develop and adopt diverse IWM strategies necessitates the trans-disciplinary research focus, which is essential to achieve broader sustainability goals (Neve et al. 2018; Westwood et al. 2018).

In future, weed management should be oriented around sustainability. Most of the cultural weed control practices such as crop rotations, cover crops and ground cover techniques should be revisited in relation to the changing crop production systems. The reduction of the weed seedbank should be a prime goal as is the control of above-ground weed infestations (Norsworthy et al. 2018). Research should also be focused on developing innovative weed control tools with lowest possible environmental footprint. For instance, crop cultivars with more competition ability and higher allelopathic potential should be developed. Research efforts should be expedited in the areas of modelling, robotics and thermal weed control. Researchers should endeavour to develop true IWM packages (1) by introducing non-conventional crops into the cropping systems which may provide effective weed control, (2) by introducing site-specific weed control technologies, (3) by testing the right combinations of management tactics, and (4) by evaluating the efficacy of cultural, mechanical or ecological weed control methods. Work on technology adoption is also crucial in this regard. Multidisciplinary research should focus on the adoption of advanced weed management tools such as HWSC, robotic weed control and thermal weed control across the world. Research in relatively neglected areas such as the biology and management of invasive weed species should also be accelerated. Moreover, the research investigating the role and impact of globalization and climate change on weed biology, dispersal and management should be prioritized.

Acknowledgements Ali Ahsan Bajwa acknowledges the financial support from The University of Queensland, Australia for his PhD studies. Authors are also thankful to the editors for the invitation to contribute this chapter.

References

Anderson RL (2007) Managing weeds with a dualistic approach of prevention and control: a review. Agron Sustain Dev 27:13–18

Anderson RL (2015) Integrating a complex rotation with no-till improves weed management in organic farming: a review. Agron Sustain Dev 35:967–974

Andreasen C, Hansen L, Streibig JC (1999) The effect of ultraviolet radiation on the fresh weight of some weeds and crops. Weed Technol 13:554–560

Asaduzzaman M, Pratley JE, An M, Luckett DJ, Lemerle D (2014) Canola interference for weed control. Spring Sci Rev 2:63–74

Ascard J (1994) Dose-response models for flame weeding in relation to plant size and density. Weed Res 34:377–385

Ascard J (1998) Comparison of flaming and infrared radiation techniques for thermal weed control. Weed Res 38:69–76

Astrand B, Baerveldt AJ (2002) An agricultural mobile robot with vision-based perception for mechanical weed control. Auton Robots 13:21–35

Bagavathiannan MV, Davis AS (2018) An ecological perspective on managing weeds during the great selection for herbicide resistance. Pest Manag Sci. https://doi.org/10.1002/ps.4920. In press

Bajwa AA (2014) Sustainable weed management in conservation agriculture. Crop Prot 65:105–113

Bajwa AA, Anjum SA, Nafees W, Tanveer M, Saeed HS (2014) Impact of fertilizer use on weed management in conservation agriculture-a review. Pak J Agric Res 27:70–78

Bajwa AA, Mahajan G, Chauhan BS (2015) Nonconventional weed management strategies for modern agriculture. Weed Sci 63:723–747

Bajwa AA, Chauhan BS, Farooq M, Shabbir A, Adkins SW (2016) What do we really know about alien plant invasion? a review of the invasion mechanism of one of the world's worst weeds. Planta 244:39–57

Bajwa AA, Walsh M, Chauhan BS (2017) Weed management using crop competition in Australia. Crop Prot 95:8–13

Bajwa AA, Nguyen T, Navie S, O'Donnell C, Adkins S (2018) Weed seed spread and its prevention: the role of roadside wash down. J Environ Manage 208:8–14

Bakht T, Khan IA (2014) Weed control in tomato (*Lycopersicon esculentum* Mill.) through mulching and herbicides. Pak J Bot 46:289–292

Banik S, Bandyopadhyay S, Ganguly S (2003) Bioeffects of microwave-a brief review. Bioresour Technol 87:155–159

Banik P, Midya A, Sarkar B, Ghose S (2006) Wheat and chickpea intercropping systems in an additive series experiment: advantages and weed smothering. Eur J Agron 24:325–332

Barakat RM, Al-Masri MI (2012) Enhanced soil solarization against *Fusarium oxysporum* f. sp. lycopersici in the uplands. Int J Agron. https://doi.org/10.1155/2012/368654

Barker AV, Bhowmik PC (2001) Weed control with crop residues in vegetable cropping systems. J Crop Prod 4:163–183

Beckie HJ, Johnson EN, Blackshaw RE, Gan Y (2008) Weed suppression by canola and mustard cultivars. Weed Technol 22:182–185

Bennett AC, Shaw DR (2000) Effect of Glycine max cultivar and weed control on weed seed characteristics. Weed Sci 48:431–435

Bhowmik PC (1992) Performance of reduced rates of herbicides with low and high residues in minimum-tillage corn. Mass Weed Sci Res Resul 11:37–42

Bhowmik PC, Inderjit (2003) Challenges and opportunities in implementing allelopathy for natural weed management. Crop Prot 22:661–671

Bhowmik PC, Doll JD (1982) Corn and soybean response to allelopathic effects of weed and crop residues. Agron J 74:601–606

Blackshaw RE (1994a) Differential competitive ability of winter wheat cultivars against downy brome. Agron J 86:649–654

Blackshaw RE (1994b) Rotation affects downy brome (*Bromus tectorum*) in winter wheat (*Triticum aestivum*). Weed Technol 8:728–732

Blackshaw RE, Brandt RN (2008) Nitrogen fertilizer rate effects on weed competitiveness is species dependent. Weed Sci 56:743–747

Blackshaw RE, Semach GP, O'Donovan JT (2000) Utilization of wheat seed rate to manage redstem filaree (*Erodium cicutarium*) in a zero-till cropping system. Weed Technol 14:389–396

Blackshaw RE, Larney FJ, Lindwall CW, Watson PR, Derksen DA (2001) Tillage intensity and crop rotation affect weed community dynamics in a winter wheat cropping System. Can J Plant Sci 81:805–813

Blackshaw RE, Brandt RN, Janzen HH, Entz T, Grant CA, Derksen DA (2003) Differential response of weed species to added nitrogen. Weed Sci 51:532–539

Blackshaw RE, Anderson RL, Lemerle D (2007) Cultural weed management. In: Upadhyaya MK, Blackshaw RE (eds) Non-chemical weed management: principles, concepts, and technology. CAB International, Cambridge, MA, pp 35–47

Blackshaw RE, Harker KN, O'donovan JT, Beckie HJ, Smith EG (2008) Ongoing development of integrated weed management systems on the Canadian prairies. Weed Sci 56:146–150

Blasco J, Aleixos N, Roger JM, Rabatel G, Molto E (2002) Robotic weed control using machine vision. Biosyst Eng 83:149–157

Bond W, Grundy AC (2001) Non-chemical weed management in organic farming systems. Weed Res 41:383–405

Borger CP, Hashem A, Pathan S (2010) Manipulating crop row orientation to suppress weeds and increase crop yield. Weed Sci 58:174–178

Borger CP, Riethmuller G, D'Antuono M (2016a) Eleven years of integrated weed management: long-term impacts of row spacing and harvest weed seed destruction on *Lolium rigidum* control. Weed Res 56:359–366

Borger CP, Hashem A, Powles SB (2016b) Manipulating crop row orientation and crop density to suppress *Lolium rigidum*. Weed Res 56:22–30

Boz O, Ogut D, Kir K, Doğan MN (2009) Olive processing waste as a method of weed control for okra, faba bean, and onion. Weed Technol 23:569–573

Brainard DC, Henshaw B, Snapp S (2012) Hairy vetch varieties and bicultures influence cover crop services in strip-tilled sweet corn. Agron J 104:629–638

Brainard DC, Peachey RE, Haramoto ER, Luna JM, Rangarajan A (2013) Weed ecology and non-chemical management under strip-tillage: implications for Northern US vegetable cropping systems. Weed Technol 27:218–230

Brodie G (2018) The use of physics in weed control. In: Jabran K, Chauhan BS (eds) Non-chemical weed control. Academic Press, San Diego, pp 33–59

Brodie G, Harris G, Pasma L, Travers A, Leyson D, Lancaster C, Woodworth J (2009) Microwave soil heating for controlling ryegrass seed germination. Trans ASABE 52:295–302

Broster JC, Walsh MJ, Chambers AJ (2016) Harvest weed seed control: the influence of harvester set up and speed on efficacy in south-eastern Australia wheat crops. In: Randall R, Lloyd S, Borger C (eds) Proceedings of the 20th Australasian Weeds conference. 11–15 Sep 2016, Perth, pp 38–41

Buhler DD (2002) Challenges and opportunities for integrated weed management. Weed Sci 50:273–280

Bullied WJ, Marginet AM, Van Acker RC (2003) Conventional- and conservation-tillage systems influence emergence periodicity of annual weed species in canola. Weed Sci 51:886–897

Campiglia E, Paolini R, Colla G, Mancinelli R (2009) The effects of cover cropping on yield and weed control of potato in a transitional system. Field Crop Res 112:16–23

Chandler JM, Meredith WR (1983) Yields of 3 cotton (*Gossypium hirsutum*) cultivars as influenced by spurred anoda (*Anoda cristata*) competition. Weed Sci 31:303–307

Chauhan BS (2012) Weed ecology and weed management strategies for dry-seeded rice in Asia. Weed Technol 26:1–13

Chauhan BS, Johnson DE (2010a) The role of seed ecology in improving weed management strategies in the tropics. Adv Agron 105:221–262

Chauhan BS, Johnson DE (2010b) Implications of narrow crop row spacing and delayed *Echinochloa colona* and *Echinochloa crus-galli* emergence for weed growth and crop yield loss in aerobic rice. Field Crop Res 117:177–182

Chauhan BS, Gill GS, Preston C (2006) Influence of tillage systems on vertical distribution, seedling recruitment and persistence of rigid ryegrass (*Lolium rigidum*). Weed Sci 54:669–676

Chauhan BS, Singh VP, Kumar A, Johnson DE (2011) Relations of rice seeding rates to crop and weed growth in aerobic rice. Field Crop Res 121:105–115

Chauhan BS, Florentine SK, Ferguson JC, Chechetto RG (2017) Implications of narrow crop row spacing in managing weeds in mungbean (*Vigna radiata*). Crop Prot 95:116–119

Cheema ZA, Khaliq A (2000) Use of sorghum allelopathic properties to control weeds in irrigated wheat in semiarid region of Punjab. Agric Ecosyst Environ 79:105–112

Cheema ZA, Asim M, Khaliq A (2000) Sorghum allelopathy for weed control in cotton (*Gossypium arboreum* L.). Int J Agric Biol 2:37–41

Cheema ZA, Khaliq A, Akhtar S (2001) Use of sorghum water extract as a natural weed inhibitor in spring mungbean. Int J Agric Biol 3:515–518

Cheema ZA, Khaliq A, Tariq M (2002) Evaluation of concentrated sorghum water extract alone and in combination with reduced rates of three pre-emergence herbicides for weed control in cotton (*Gossypium hirsutum* L.). Int J Agric Biol 4:549–552

Cheema ZA, Khaliq A, Mubeen K (2003) Response of wheat and winter weeds to foliar application of different plant water extracts of sorghum (*Sorghum bicolor*). Pak J Weed Sci Res 9:89–97

Cheema ZA, Khaliq A, Saeed S (2004) Weed control in maize (*Zea mays* L.) through sorghum allelopathy. J Sustain Agric 23:73–86

Chhokar RS, Malik RK (2002) Isoproturon-resistant littleseed canarygrass (*Phalaris minor*) and its response to alternate herbicides. Weed Technol 16:116–123

Chhokar RS, Sharma RK, Jat GR, Pundir AK, Gathala MK (2007) Effect of tillage and herbicides on weeds and productivity of wheat under rice-wheat growing system. Crop Prot 26:1689–1696

Christoffoleti PJ, Carvalho SJP, Nicolai M, Doohan D, VanGessel M (2007) Prevention strategies in weed management. In: Upadhyaya MK, Blackshaw RE (eds) Non-chemical weed management: principles, concepts, and technology. CAB International, Cambridge, MA, pp 1–16
Clements DR, Benoit DL, Murphy SD, Swanton CJ (1996) Tillage effects on weed seed return and seed bank composition. Weed Sci 44:314–322
Dalley CD, Bernards ML, Kells JJ (2006) Effect of weed removal timing and row spacing on soil moisture in corn (*Zea mays*). Weed Technol 20:399–409
Daugovish O, Lyon DJ, Baltensperger DD (1999) Cropping systems to control winter annual grasses in winter wheat (*Triticum aestivum*). Weed Technol 13:120–126
Derksen DA, Anderson RL, Blackshaw RE, Maxwell B (2002) Weed dynamics and management strategies for cropping systems in the northern Great Plains. Agron J 94:174–185
Dilday RH, Lin J, Yan W (1994) Identification of allelopathy in the USDA-ARS rice germplasms collection. Aust J Exp Agric 34:907–910
Diprose M, Benson F (1984) Electrical methods of killing plants. J Agric Eng Res 30:197–209
Diprose M, Benson F, Hackam R (1980) Electrothermal control of weed beet and bolting sugar beet. Weed Res 20:311–322
Diprose M, Benson F, Willis A (1984) The effect of externally applied electrostatic fields, microwave radiation and electric currents on plants and other organisms, with special reference to weed control. Bot Rev 50:171–223
Erenstein O (2003) Smallholder conservation farming in the tropics and sub-tropics: a guide to the development and dissemination of mulching with crop residues and cover crops. Agric Ecosyst Environ 100:17–37
Fahad S, Hussain S, Chauhan BS, Saud S, Wu C, Hassan S, Tanveer M, Jan A, Huang J (2015) Weed growth and crop yield loss in wheat as influenced by row spacing and weed emergence times. Crop Prot 71:101–108
Fahad S, Bajwa AA, Nazir U, Anjum SA, Farooq A, Zohaib A, Sehrish S, Nasim W, Adkins S, Saud S, Ihsan MZ, Alharby H, Wu C, Wang D, Huang J (2017) Crop production under drought and heat stress: plant responses and management options. Front Plant Sci. https://doi.org/10.3389/fpls.2017.01147
Farooq O, Cheema ZA (2013) Impact of planting time and seedbed conditions on little seed canary grass and lambsquarters dynamics in wheat. Int J Agric Biol 15:1003–1007
Farooq M, Jabran K, Cheema ZA, Wahid A, Siddique KH (2011a) The role of allelopathy in agricultural pest management. Pest Manag Sci 67:493–506
Farooq M, Flower KC, Jabran K, Wahid A, Siddique KH (2011b) Crop yield and weed management in rainfed conservation agriculture. Soil Tillage Res 117:172–183
Farooq M, Bajwa AA, Cheema SA, Cheema ZA (2013) Application of allelopathy in crop production. Int J Agric Biol 15:1367–1378
Felton WL (1976) The influence of row spacing and plant population on the effect of weed competition in soybeans. Anim Prod Sci 16:926–931
Fischer A, Ramirez HV, Lozano J (1997) Suppression of junglerice (*Echnochloa colona* L. Link.) by irrigated rice cultivars in Central America. Agron J 89:516–521
Franke A, Singh S, McRoberts N, Nehra A, Godara S, Malik R, Marshall G (2007) *Phalaris minor* seedbank studies: longevity, seedling emergence and seed production as affected by tillage regime. Weed Res 47:73–83
Freckleton RP, Stephens PA (2009) Predictive models of weed population dynamics. Weed Res 49:225–232
Gaba S, Fried G, Kazakou E, Chauvel B, Navas ML (2014) Agroecological weed control using a functional approach: a review of cropping systems diversity. Agron Sustain Dev 34:103–119
Gay P, Piccarolo P, Aimonino DR, Tortia C (2010) A high efficiency steam soil disinfestation system, part I: Physical background and steam supply optimisation. Biosyst Eng 107:74–85
Gill HS, Brar LS (1975) Importance of weedicides in the agriculture of Punjab and Haryana. Pesticides 9:20–24

Gill GS, Poole ML, Holmes JE (1987) Competition between wheat and brome grass in Western Australia. Aust J Exp Agric 27:291–294

Graglia E, Melander B, Jensen RK (2006) Mechanical and cultural strategies to control *Cirsium arvense* in organic arable cropping systems. Weed Res 46:304–312

Harker KN, O'Donovan JT, Turkington TJ, Blackshaw RE, Johnson EN, Brandt S, Kutcher HR, Clayton GW (2013) Weed interference impacts on yield recovery after four years of variable crop inputs in no-till barley and canola. Weed Technol 27:281–290

Hartl W (1989) Influence of under sown clovers on weeds and on yield of winter wheat in organic farming. Agric Ecosyst Environ 27:389–396

Hock SM, Knezevic SZ, Martin AR, Lindquist JL (2006) Soybean row spacing and weed emergence time influence weed competitiveness and competitive indices. Weed Sci 54:38–46

Hodgdon EA, Warren ND, Smith RG, Sideman RG (2016) In-season and carry-over effects of cover crops on productivity and weed suppression. Agron J 108:1624–1635

Holland JF, McNamara DW (1982) Weed control and row-spacing in dry-land sorghum in northern New South Wales. Anim Prod Sci 22:310–316

Inderjit, Bhowmik PC (2004) Sorption of benzoic acid onto soil colloids and its implications for allelopthy studies. Biol Fertil Soils 40:345–348

Iqbal J, Cheema ZA, An M (2007) Intercropping of field crops in cotton for the management of purple nutsedge (*Cyperus rotundus* L.). Plant and Soil 300:163–171

Jabran K, Mahajan G, Sardana V, Chauhan BS (2015) Allelopathy for weed control in agricultural systems. Crop Prot 72:57–65

Jabran K, Hussain M, Fahad S, Farooq M, Bajwa AA, Alharrby H, Nasim W (2016) Economic assessment of different mulches in conventional and water-saving rice production systems. Environ Sci Pollut Res 23:9156–9163

Jabran K, Mahmood K, Melander B, Bajwa AA, Kudsk P (2017) Weed Dynamics and Management in Wheat. Adv Agron 145:97–166

Jamil M, Cheema ZA, Mushtaq MN, Farooq M, Cheema MA (2009) Alternative control of wild oat and canary grass in wheat fields by allelopathic plant water extracts. Agron Sustain Dev 29:475–482

Jordan N (1996) Weed prevention: priority research for alternative weed management. J Prod Agric 9:485–490

Jordan NR, Davis AS (2015) Middle-way strategies for sustainable intensification of agriculture. BioScience 65:513–519

Kasirajan S, Ngouajio M (2012) Polyethylene and biodegradable mulches for agricultural applications: a review. Agron Sustain Dev 32:501–529

Katan J (2015) Soil solarization: the idea, the research and its development. Phytoparasitica 43:1–4

Kato-Noguchi H, Ino T, Ota K (2008) Secretion of momilactone A from rice roots to the rhizosphere. J Plant Physiol 165:691–696

Kaur S, Surjit S (2014) Influence of crop density on weeds, growth and yield of direct-seeded rice. Indian J Weed Sci 46:318–321

Khaliq A, Matloob A, Irshad MS, Tanveer A, Zamir MSI (2010) Organic weed management in maize through integration of allelopathic crop residues. Pak J Weed Sci Res 16:409–420

Khan MA, Marwat KB, Umm-E-Kulsoom, Hussain Z, Hashim S, Rab A, Nawab K (2013) Weed control effects on the wheat-pea intercropping. Pak J Bot 45:1743–1748

Khan MJ, Brodie G, Gupta D (2017) The effect of microwave soil treatment on rice production under field conditions. Trans ASABE 60:517–525

Khan MJ, Brodie GI, Gupta D, Foletta S (2018) Microwave soil treatment improves weed management in Australian dryland wheat. Trans ASABE 61:671–680

Khanh TD, Cong LC, Chung IM, Xuan TD, Shinkichi T (2009) Variation of weed-suppressing potential of Vietnamese rice cultivars against barnyardgrass (*Echinochloa crus-galli*) in laboratory, greenhouse and field screenings. J Plant Interact 3:209–218

Kim KU, Shin DH, Kim HY, Lee ZL, Olofsdotter M (1999) Evaluation of allelopathic potential in rice germplasm. Korean J Weed Sci 19:1–9

Kobayashi H, Miura S, Oyanagi A (2004) Effects of winter barley as a cover crop on the weed vegetation in a no-tillage soybean. Weed Biol Manage 4:195–205

Koocheki A, Nassiri M, Alimoradi L, Ghorbani R (2009) Effect of cropping systems and crop rotations on weeds. Agron Sustain Dev 29:401–408

Kotile DG, Martin RA (2000) Sustainable agricultural practices for weed management: implications to agricultural extension education. J Sustain Agric 16:31–51

Kruidhof HM, Bastiaans L, Kropff MJ (2008) Ecological weed management by cover cropping: effects on weed growth in autumn and weed establishment in spring. Weed Res 48:492–502

Lanini WT, Orloff SB, Vargas RN, Orr JP, Marble VL, Grattan SR (1991) Oat companion crop seeding rate effect on alfalfa establishment, yield and weed control. Agron J 83:330–333

Lemerle D, Verbeek B, Coombes N (1995) Losses in grain yield of winter crops from *Lolium rigidum* competition depend on crop species, cultivar and season. Weed Res 35:503–509

Lemerle D, Verbeek B, Cousens RD, Coombes NE (1996) The potential for selecting wheat varieties strongly competitive against weeds. Weed Res 36:505–513

Lemerle D, Verbeek B, Diffey S (2006) Influences of field pea (*Pisum sativum*) density on grain yield and competitiveness with annual ryegrass (*Lolium rigidum*) in south-eastern Australia. Anim Prod Sci 46:1465–1472

Lemerle D, Luckett DJ, Lockley P, Koetz E, Wu H (2014) Competitive ability of Australian canola (*Brassica napus*) genotypes for weed management. Crop Past Sci 65:1300–1310

Lemerle D, Luckett DJ, Koetz EA, Potter T, Wu H (2016) Seeding rate and cultivar effects on canola (*Brassica napus*) competition with volunteer wheat (*Triticum aestivum*). Crop Past Sci 67:857–863

Lemerle D, Luckett DJ, Wu H, Widderick MJ (2017) Agronomic interventions for weed management in canola (Brassica napus L.)-a review. Crop Prot 95:69–73

Liebman M, Davis AS (2000) Integration of soil, crop and weed management in low-external-input farming systems. Weed Res 40:27–47

Liebman M, Dyck E (1993) Crop rotation and intercropping strategies for weed management. Ecol Appl 3:92–122

Liebman M, Baraibar B, Buckley Y, Childs D, Christensen S, Cousens R, Eizenberg H, Heijting S, Loddo D, Merotto A, Renton M, Riemens M (2016) Ecologically sustainable weed management: how do we get from proof-of-concept to adoption? Ecol Appl 26:1352–1369

Lindquist JL, Mortensen DA (1998) Tolerance and velvetleaf (*Abutilon theophrasti*) suppressive ability of two old and two modern corn (*Zea mays*) hybrids. Weed Sci 46:569–574

Lindquist JL, Mortensen DA, Johnson BE (1998) Mechanisms of corn tolerance and velvetleaf suppressive ability. Agron J 90:787–792

Lins RD, Colquhoun JB, Mallory-Smith C (2006) Investigation of wheat as a trap crop for control of Orobanche minor. Weed Res 46:313–318

Lowry CJ, Smith RG (2018) Weed control through crop plant manipulations. In: Jabran K, Chauhan BS (eds) Non-chemical weed control. Academic, San Diego, pp 73–96

Mahajan G, Brar LS (2002) Integrated management of *Phalaris minor* in wheat: rationale and approaches-a review. Agric Rev 23:241–251

Mahajan G, Poonia V, Chauhan BS (2014) Integrated weed management using planting pattern, cultivar, and herbicide in dry-seeded rice in northwest India. Weed Sci 62:350–359

Malik RK, Singh S (1995) Littleseed canarygrass (*Phalaris minor*) resistance to isoproturon in India. Weed Technol 9:419–425

Masum S, Hossain A, Akamine H, Jun-Ichi S, Bhowmik PC (2016) Allelopathic potential of indigenous Bangladeshi rice varieties. Weed Biol Manage 16:119–131

Masum S, Hossain A, Akamine H, Jun-Ichi S, Ishii T, Gima S, Kensaku T, Bhowmik PC (2018) Isolation and characterization of allelopathic compounds from the indigenous rice variety 'Boterswar' and their biological activity against *Eichornia crus-galli* L. Allelo J 43:31–42

McDonald GK (2003) Competitiveness against grass weeds in field pea genotypes. Weed Res 43:48–58

McDonald GK, Hollaway KL, McMurray L (2007) Increasing plant density improves weed competition in lentil (*Lens culinaris*). Anim Prod Sci 47:48–56

Mcgovern RJ, Chaleeprom W, Chaleeprom W, Mcgovern P, To-Anun C (2013) Evaluation of soil solarization and amendments as production practices for lettuce and vegetable soybean in northern Thailand. J Agric Technol 9:1863–1872

McGowan A (1970) Comparative germination patterns of annual grasses in north-eastern Victoria. Aust J Exp Agric Anim Husb 10:401–404

McLachlan SM, Tollenaar M, Swanton CJ, Weise SF (1993) Effect of corn induced shading on dry matter accumulation, distribution and architecture of redroot pigweed (*Amaranthus retroflexus*). Weed Sci 41:568–573

Medlin CR, Shaw DR (2000) Economic comparison of broadcast and site-specific herbicide applications in nontransgenic and glyphosate-tolerant *Glycine max*. Weed Sci 48:653–661

Melander B, Cirujeda A, Jørgensen MH (2003) Effects of inter-row hoeing and fertilizer placement on weed growth and yield of winter wheat. Weed Res 43:428–438

Mennan H, Zandstra BH (2005) Effect of wheat (*Triticum aestivum*) cultivars and seeding rate on yield loss from *Galium aparine* (cleavers). Crop Prot 24:1061–1067

Mishra J, Rao S, Patil J (2015) Response of grain sorghum (*Sorghum bicolor*) cultivars to weed competition in semi-arid tropical India. Indian J Agric Sci 85:688–694

Mohler CL, Asdale JR (1993) Response of weed emergence to rate of *Vicia villosa* Roth and *Secale cereale* L. residue. Weed Res 33:487–499

Molin WT, Hugie JA, Hirase K (2004) Prickly sida (*Sida spinosa* L.) and spurge (*Euphorbia hyssopifolia* L.) response to wide row and ultra-narrow row cotton (*Gossypium hirsutum* L.) management systems. Weed Biol Manag 4:222–229

Mortensen DA, Egan JF, Maxwell BD, Ryan MR, Smith RG (2012) Navigating a critical juncture for sustainable weed management. BioScience 62:75–84

Myers MW, Curran WS, Van Gessel MJ, Majek BA, Mortensen DA, Calvin DD, Karsten HD, Roth GW (2005) Effect of soil disturbance on annual weed emergence in the Northeastern United States. Weed Technol 19:274–282

Nagabhushana GG, Worsham AD, Yenish JP (2001) Allelopathic cover crops to reduce herbicide use in sustainable agricultural systems. Allelo J 8:133–146

Narwal SS, Haouala R (2013) Role of allelopathy in weed management for sustainable agriculture. In: Cheema Z, Farooq M, Wahid A (eds) Allelopathy: current trend and future applications. Springer, Berlin/Heidelberg, pp 217–249

Naseem M, Aslam M, Ansar M, Azhar M (2010) Allelopathic effects of sunflower water extract on weed control and wheat productivity. Pak J Weed Sci Res 15:107–116

Nawaz A, Farooq M, Cheema S, Cheema Z (2014) Role of allelopathy in weed management. In: Chauhan BS, Mahajan G (eds) Recent advances in weed management. Springer, New York, pp 39–61

Neve P, Barney JN, Buckley Y et al (2018) Reviewing research priorities in weed ecology, evolution and management: a horizon scan. Weed Res. https://doi.org/10.1111/wre.12304. In press

Nguyen T, Bajwa AA, Navie S, O'Donnell C, Adkins S (2017) Parthenium weed (*Parthenium hysterophorus* L.) and climate change: the effect of CO2 concentration, temperature, and water deficit on growth and reproduction of two biotypes. Environ Sci Pollut Res 24:10727–10739

Nordby DE, Alderks DL, Nafziger ED (2007) Competitiveness with weeds of soybean cultivars with different maturity and canopy width characteristics. Weed Technol 21:1082–1088

Norsworthy JK, Oliver LR (2002) Pitted morningglory interference in drillseeded glyphosate-resistant soybean. Weed Sci 50:26–33

Norsworthy JK, Korres NE, Bagavathiannan MV (2018) Weed seedbank management: revisiting how herbicides are evaluated. Weed Sci. https://doi.org/10.1017/wsc.2018.11. In press

O'Donovan JT, Newman JC, Harker KN, Clayton GW (2004) Crop seeding rate influences the performance of variable herbicide rates in a canola-barley-canola rotation. Weed Technol 18:733–741

Oerke EC (2006) Crop losses to pests. J Agric Sci 144:31–43

Ohadi S, Hodnett G, Rooney W, Bagavathiannan M (2017) Gene flow and its consequences in Sorghum spp. Crit Rev Plant Sci 36:367–385

Olofsdotter M (2001) Rice – a step toward use of allelopathy. Agron J 93:3–8

Olofsdotter M, Navarej DC, Moody K (1995) Allelopathic potential of rice (*Oryza sativa* L.) germplasm. Ann Appl Biol 127:543–560

Olsen J, Kristensen L, Weiner J (2005) Effects of density and spatial pattern of winter wheat on suppression of different weed species. Weed Sci 53:690–694

Osten V, Wu H, Walker S, Wright G, Shields A (2006) Weeds and summer crop row spacing studies in Queensland. In: Preston C, Watts JH, Crossman ND (eds) Proceedings of the 15th Australian weeds conference. 24–28 Sep 2006, Adelaide, pp 347–350

Owen MD, Beckie HJ, Leeson JY, Norsworthy JK, Steckel LE (2015) Integrated pest management and weed management in the United States and Canada. Pest Manag Sci 71:357–376

Pareja MR, Staniforth DW, Pareja GP (1985) Distribution of weed seeds among soil structural units. Weed Sci 33:182–189

Parish S (1990) A review of non-chemical control technologies. Biol Agric Hortic 7:117–137

Paynter BH, Hills AL (2009) Barley and rigid ryegrass (*Lolium rigidum*) competition is influenced by crop cultivar and density. Weed Technol 23:40–48

Peachey RE, William RD, Mallory-Smith C (2006) Effect of spring tillage sequence on summer annual weeds in vegetable row crop. Weed Technol 20:204–214

Peltzer SC, Matson PT (2002) How fast do the seedbanks of five annual cropping weeds deplete in the absence of weed seed input? In: Jacob HS, Dodd J, Moore J (eds) Proceedings of the 13th Australian weeds conference. Plant Protection Society of WA, Perth, pp 553–555

Peltzer SC, Hashem A, Osten VA, Gupta ML, Diggle AJ, Riethmuller GP, Douglas A, Moore JA, Koetz EA (2009) Weed management in wide-row cropping systems: a review of current practices and risks for Australian farming systems. Crop Pasture Sci 60:395–406

Peters RD, Sturz AV, Carter MR, Sanderson JB (2003) Developing disease-suppressive soils through crop rotation and tillage management practices. Soil Tillage Res 72:181–192

Putnam AR, DeFrank J, Barnes JP (1983) Exploitation of allelopathy for weed control in annual and perennial cropping systems. J Chem Ecol 9:1001–1010

Reddy KN, Boykin JC (2010) Weed control and yield comparisons of twin- and single-row glyphosate-resistant cotton production systems. Weed Technol 24:95–101

Rehman S, Shahzad B, Bajwa AA, Hussain S, Rehman A, Cheema SA, Abbas T, Ali A, Shah L, Adkins S, Li P (2018) Utilizing the allelopathic potential of Brassica species for sustainable crop production: a review. J Plant Growth Regul. https://doi.org/10.1007/s00344-018-9798-7. In press

Reicosky DC, Allmaras RR (2003) Advances in tillage research in North American cropping systems. J Crop Prod 8:75–125

Rezakhanlou A, Mirshekari B, Zand E, Farahvash F, Baghestani MA (2013) Evaluation of competitiveness of cotton varieties to cocklebur (*Xanthium strumarium* L.). J Food Agric Environ 11:308–311

Rezvani M, Zaefarian F, Jovieni M (2013) Weed suppression ability of six soybean [*Glycine max* (L.) Merr.] varieties under natural weed development conditions. Acta Agron Hung 61:43–53

Rose SJ, Burnside OC, Specht JE, Swisher BA (1984) Competition and allelopathy between soybean and weeds. Agron J 76:523–528

Safdar ME, Asif M, Ali A, Aziz A, Yasin M, Aziz M, Afzal M, Ali A (2011) Comparative efficacy of different weed management strategies in wheat. Chil J Agric Res 71:195–204

Salam MD, Kato-Noguchi H (2009) Screening of allelopathic potential Bangladesh rice cultivars by donor-receiver bioassay. Asian J Plant Sci 8:20–27

Samarajeewa K, Horiuchi T, Oba S (2006) Finger millet (*Eleucine corocana* L. Gaertn.) as a cover crop on weed control, growth and yield of soybean under different tillage systems. Soil Tillage Res 90:93–99

Sardana V, Mahajan G, Jabran K, Chauhan BS (2017) Role of competition in managing weeds: an introduction to the special issue. Crop Prot 95:1–7

Sartorato I, Zanin G, Baldoin C, De Zanche C (2006) Observations on the potential of microwaves for weed control. Weed Res 46:1–9

Scursoni J, Palmano M, De Notta A, Delfino D (2012) Italian ryegrass (*Lolium multiflorum* Lam.) density and N fertilization on wheat (*Triticum aestivum* L.) yield in Argentina. Crop Prot 32:36–40

Senseman SA, Grey TL (2014) The future of herbicides and genetic technology: ramifications for environmental stewardship. Weed Sci 62:382–384

Shahzad M, Farooq M, Jabran K, Hussain M (2016) Impact of different crop rotations and tillage systems on weed infestation and productivity of bread wheat. Crop Prot 89:161–169

Shaw DR, Bruff SA, Smith CA (1991) Effect of soybean (Glycine max) row spacing on chemical control of sicklepod (*Cassia obtusifolia*). Weed Technol 5:286–290

Shirtliffe SJ, Entz MH (2005) Chaff collection reduces seed dispersal of wild oat (*Avena fatua*) by a combine harvester. Weed Sci 53:465–470

Shrestha A, Rajcan I, Chandler K, Swanton CJ (2001) An integrated weed management strategy for glufosinate-resistant corn (*Zea mays*). Weed Technol 15:517–522

Singh S, Kirkwood RC, Marshall G (1999) Biology and control of *Phalaris minor* Retz. (littleseed canarygrass) in wheat. Crop Prot 18:1–16

Singh HP, Batish DR, Kohli RK (2003) Allelopathic interactions and allelochemicals: new possibilities for sustainable weed management. Cr Rev Plant Sci 22:239–311

Singh VP, Barman KK, Singh R, Sharma AR (2015) Weed management in conservation agriculture systems. In: Farooq M, Siddique KHM (eds) Conservation agriculture. Springer, Berlin/Heidelberg, pp 39–77

Slaughter DC, Giles DK, Downey D (2008) Autonomous robotic weed control systems: a review. Comp Elec Agric 61:63–78

Smith BS, Murray DS, Green JD, Wanyahaya WM, Weeks DL (1990) Interference of three annual grasses with grain sorghum (*Sorghum bicolor*). Weed Technol 4:245–249

Solie JB, Solomon SG, Self KP, Peeper TF, Koscelny JA (1991) Reduced row spacing for improved wheat yields in weed-free and weed-infested fields. Am Soc Agric Eng 34:1654–1660

Stahlman PW, Wicks GA (2000) Weeds and their control in grain sorghum. In: Smith CW, Frederikson RA (eds) Sorghum: origin, history, technology, and production. Wiley, New York, pp 535–582

Staniforth DW (1961) Responses of corn hybrids to yellow foxtail competition. Weeds 9:132–136

Staricka JA, Burford PM, Allmaras RR, Nelson WW (1990) Tracing the vertical distribution of simulated shattered seeds as related to tillage. Agron J 82:1131–1134

Stephenson DO, Brecke BJ (2010) Weed management in single- vs. twin-row cotton (*Gossypium hirsutum*). Weed Technol 24:275–280

Tang L, Wan K, Cheng C, Li R, Wang D, Pan J, Tao Y, Xie J, Chen F (2013) Effect of fertilization patterns on the assemblage of weed communities in an upland winter wheat field. J Plant Ecol 7:39–50

Teasdale JR (1996) Contribution of cover crops to weed management in sustainable agricultural systems. J Prod Agric 9:475–479

Teasdale JR, Mangum RW, Radhakrishnan J, Cavigelli MA (2004) Weed seedbank dynamics in three organic farming crop rotations. Agron J 96:1429–1435

Teasdale JR, Brandsaeter LO, Calegari A, Skora Neto F (2007) Cover crops and weed management. In: Upadhyaya MK, Blackshaw RE (eds) Non-chemical weed management: principles, concepts, and technology. CAB International, Cambridge, MA, pp 49–64

Tharayil N, Bhowmik PC, Xing B (2008) Bioavailability of allelochemicals as affected by companion compounds in soil matrices. J Agric Food Chem 56:3706–3713

Thorp KR, Tian LF (2004) A review on remote sensing of weeds in agriculture. Prec Agric 5:477–508

Torresen KS, Skuterud R, Tandsaether HJ, Hagemo MB (2003) Long-term experiments with reduced tillage in spring cereals. I. Effects on weed flora, weed seedbank and grain yield. Crop Prot 22:185–200

Torres-Sánchez J, Lopez-Granados F, De Castro AI, Pena-Barragan JM (2013) Configuration and specifications of an unmanned aerial vehicle (UAV) for early site specific weed management. Plos One 8:e58210
Tripathi B, Singh CM (1983) Weed and fertility management using maize/soyabean intercropping in the northwestern Himalayas. Trop Pest Manage 29:267–270
Ulloa SM, Datta A, Bruening C, Neilson B, Miller J, Gogos G, Knezevic SZ (2011) Maize response to broadcast flaming at different growth stages: effects on growth, yield and yield components. Eur J Agron 34:10–19
Walker SR, Medd RW, Robinson GR, Cullis BR (2002) Improved management of *Avena ludoviciana* and *Phalaris paradoxa* with more densely sown wheat and less herbicide. Weed Res 42:257–270
Walsh MJ, Newman P (2007) Burning narrow windrows for weed seed destruction. Field Crop Res 104:24–40
Walsh MJ, Powles SB (2007) Management strategies for herbicide-resistant weed populations in Australian dryland crop production systems. Weed Technol 21:332–338
Walsh MJ, Newman P, Powles SB (2013) Targeting weed seeds in-crop: a new weed control paradigm for global agriculture. Weed Technol 27:431–436
Walsh M, Ouzman J, Newman P, Powles S, Llewellyn R (2017) High levels of adoption indicate that harvest weed seed control is now an established weed control practice in Australian cropping. Weed Technol 31:1–7
Walsh MJ, Broster JC, Powles SB (2018) iHSD Mill Efficacy on the Seeds of Australian Cropping System Weeds. Weed Technol 32:103–108
Wayland J, Davis F, Merkle M (1973) Toxicity of an UHF device to plant seeds in soil. Weed Sci 21:161–162
Wazir I, Sadiq M, Baloch MS, Awan IU, Khan EA, Shah IH, Nadim MA, Khakwani AA, Bakhsh I (2011) Application of bio-herbicide alternatives for chemical weed control in rice. Pak J Weed Sci Res 17:245–252
Weston LA (1996) Utilization of allelopathy for weed management in agroecosystems. Agron J 88:860–866
Westwood JH, Charudattan R, Duke SO, Fennimore SA, Marrone P, Slaughter DC, Swanton C, Zollinger R (2018) Weed management in 2050: perspectives on the future of weed science. Weed Sci 66:275–285
Williams II, Boydston R (2013) Crop seeding level: implications for weed management in sweet corn. Weed Sci 61:437–442
Wortman SE, Francis CA, Bernards MA, Blankenship EE, Lindquist JL (2013) Mechanical termination of diverse cover crop mixtures for improved weed suppression in organic cropping systems. Weed Sci 61:162–170
Wszelaki AL, Doohan DJ, Alexandrou A (2007) Weed control and crop quality in cabbage (*Brassica oleracea* capitata group) and tomato (*Lycopersicon lycopersicum*) using a propane flamer. Crop Prot 26:134–144
Wu H, Pratley J, Lemerle D, Haig T (1999) Crop cultivation with allelopathic activity. Weed Res 39:171–180
Wu H, Walker SR, Osten VA, Robinson G (2010) Competition of sorghum cultivars and densities with Japanese millet (*Echinochloa esculenta*). Weed Biol Manage 10:185–193
Yin L, Cai Z, Zhong W (2005) Changes in weed composition of winter wheat crops due to long-term fertilization. Agric Ecosyst Environ 107:181–186
Young SL, Meyer GE, Woldt WE (2014) Future directions for automated weed management in precision agriculture. In: Young SL, Pierce FJ (eds) Automation: the future of weed control in cropping systems. Springer, Dordrecht, pp 249–259
Zand E, Beckie HJ (2002) Competitive ability of hybrid and open-pollinated canola (*Brassica napus*) with wild oat (*Avena fatua*). Can J Plant Sci 82:473–480

Zimdahl RL (2013) Fundamentals of weed science. Elsevier/Academic Academic Press, Cambridge, MA

Ziska LH (2016) The role of climate change and increasing atmospheric carbon dioxide on weed management: herbicide efficacy. Agric Ecosyst Environ 231:304–309

Ziska LH, Blumenthal DM, Runion GB, Hunt ER Jr, Diaz-Soltero H (2011) Invasive species and climate change: an agronomic perspective. Clim Change 105:13–42

Sustainable Management of Insect-Pests

Ahmad Nawaz, Muhammad Sufyan, Muhammad Dildar Gogi, and Muhammad Wajid Javed

Abstract The conception of 'insect-pests' has ascended from human crop cultivation practice and desire of food security from ubiquitous insects. They are also regarded by some as the main competitors of humans for dominance on the earth. The management of insect-pests is hammered by both biotic and abiotic factors. Sustainable pest management is a two-strand approach which requires complete information about control strategy, pest biology and ecology which helps to determine the most appropriate procedure/method (how), timing (when) and place (where) for effective use of any control technology of any pest. In this context, IPM (Integrated Pest Management), ICM (Integrated Crop Management) and IRM (Integrated resistance Management) can help to reduce crop yield losses while managing insect pests without causing harm to non-target organisms. However, the global implementation of these practices has been slow down due to different factors. Conclusively, integration of non-chemical control methods including new technologies with synthetic insecticides will be a promising option for sustainable insect pest management. This chapter will highlight the issues hampering sustainable insect-pests management and suggest ways to overcome these factors. Furthermore, the potential role of different stakeholders is also discussed which can be integrated for fruitful solutions of common problems of insect pest management. Finally, the integration of different therapeutic tools (IPM, ICM and IRM etc.) is underscored to increase crop production without harming the environment.

Keywords IMP · ICP · IRM · Sustainable management · Pesticides

A. Nawaz (✉) · M. D. Gogi
Integrated Pest Management Laboratory, Department of Entomology, University of Agriculture, Faisalabad, Pakistan
e-mail: ahmad.nawaz1793@uaf.edu.pk

M. Sufyan
Biological Control Laboratory, Department of Entomology, University of Agriculture, Faisalabad, Pakistan

M. W. Javed
Integrated Genomics Cellular Developmental and Biotechnology Laboratory, Department of Entomology, University of Agriculture, Faisalabad, Pakistan

M. Farooq, M. Pisante (eds.), *Innovations in Sustainable Agriculture*,
https://doi.org/10.1007/978-3-030-23169-9_10

1 Introduction

Insects belong to phylum arthropoda of kingdom Animalia. They have been evolving from about 350 million years, compared to human beings (2 million years). They are remarkable biological organisms and regarded by some as the main competitors of humans for dominance on the earth. Humans depend on insects in many ways like crop pollination, honey and silk production, organic matter decomposition and carbon recycling, and various other vital ecological roles. Honey bees estimated annual contribution in crop production is US$ 15 billion in the United States alone. Similarly, the total estimated pollination services globally exceed US$200 billion for one hundred crops used directly for human food. In addition, the predators and parasitic wasps controlling other pests often go unrecognized but worth billions of US$ annually (Gullan and Cranston 2014). Therefore, insects are inevitable for human survival on earth. In spite of benefits, insects have been of greatest concerns to human race by causing negative impact on their valuable resources. Among the one million described species of insect pests, more than 10,000 species are involved in food losses and considered as major pests (Dhaliwal et al. 2007). The estimated global annual loss of major crops (cotton, wheat, rice, maize, soyabeen, potatoes) due to insect pests ranges from 7.9% to 15.2% (Dhaliwal et al. 2010). The crop damage percentage increases in developing countries due to the lack of knowledge and availability of new technologies for the management of insect pests.

The insect pest management is an essential part of agriculture to compensate the ever increasing food demands of humans. Therefore, the prime-most objective of all countries is the increase in food production. The annual increase in world population is around 97 million per year. In this way, the world population is likely to grow from 7.6 billion to nearly 10 billion by 2050 and further predicted to be 11.2 billion by the end of this century (UN 2017; Saravi and Shokrzadeh 2011). Similarly, it is predicted by Food and Agricultural Organization the United Nations that to keep pace with the demand of increasing population, the world food production needs to increase by seventy percent. Therefore, the existing agricultural systems are under tremendous pressure of ever increasing global population and its food requirements from the same current resources (Saravi and Shokrzadeh 2011). The pests of agricultural crops are actually the bottlenecks for the increase in world food production and the insect pests are of prime importance. Before the introduction of synthetic chemicals, different sustainable practices (cultural, mechanical, and physical control strategies etc.) were used to manage these insect pests in the process of increasing crop production.

Previously, the agriculture intensification investments were often increasing in external inputs of fertilizers and synthetic chemicals. The toxic residues, pest resistance, secondary pest outbreaks, and pest resurgence are the four major glitches encountered with conventional chemical pesticides. It has been reported that more than 90% of the arthropod species (Sukhoruchenko and Dolzhenko 2008) with resistant populations belong to Lepidoptera (15%), Diptera (35%), Coleoptera (14%), Hemiptera (14%). Similarly, the microbial community is also essential for better yield of crops as reported by Allison and Martiny that microbial organisms are sensitive to nitrogen (N), phosphorus (P), and potassium (K) fertilization (Allison and Martiny 2008). In addition, similar to pesticides the concept of mono-culture or clean culture crop cultivation also has negative impact on biodiversity, insect pest natural enemies and other non-target organisms.

In agricultural systems, the basic principle for undesired variables management is similar to other systems, including the human body and social systems. The direct application of a corrective measure for an undesired entity never produces sustainable desired effects. Instead, as a matter of fundamental principle, external counter force application into a system can be effective only for short term relief from the problem. The sustainable and long term solutions should be achieved through restructuring the system. Therefore, the underpinning for pest management approach should be a full composite of innate plant defenses, plant mixtures, soil, pest natural enemies, and other components of the agricultural system. These natural "built in" managers are interconnected and renewable in sustainable manner. The "treat-the-symptoms" approaches like synthetic chemicals and other tactics should be the last option of defense.

The conception of 'insect pest' has ascended from human crop cultivation practice and desire of food security from ubiquitous insects. A pest management plan should constantly start with the question "Why is the pest a pest?" and in response, it should seek weaknesses of the cultivation system and poor agronomic practice (s) that allowed organism (s) to stretch pest status and finally address underlying issues to manage that pest (s). However, the global implementation of these practices has been slow down due to lack of knowledge, limited technical capacity, dearth of priority in agriculture sector and low demand by small land holder farmers. The sustainable pest management system enhances those pest management methods that support crop production sustainability and do not pose risk to farmers' incomes, health and environment. Therefore, the current chapter will discuss the factors affecting insect-pest management and their management. The role of stakeholders including academia, industry, research institutes, farmers, government agencies, etc. regarding legislation of crop production and protection, natural enemies conservation, genetically modified crops management, resistance management etc., is also highlighted. Finally, the chapter will focus the recent investments in crop production which are based on economically sound, socially acceptable, and environment friendly inputs. In this context, IPM, ICM and IRM are considered comprehensive agricultural practices that help to reduce crop yield losses while protecting environment as well as human health.

2 Issues of Sustainable Insect-Pest Management

Insects are ubiquitous in nature and known as most successful animals on planet earth due to several aspects like diversification, high reproductive potential, malleable exoskeleton and metamorphosis etc. The sustainable management of insect pests is quite difficult task due to both biotic and abiotic factors mainly climate change, biodiversity management, misuse of resources, and most importantly the resistance development against pesticides. Following are the major constraints for sustainable management of insect pests.

2.1 Climate Change

The ongoing changes in climatic conditions and regular weather effects can alter development and dispersal of different insect species. It is evident that fluctuations in surrounding temperature regimes are involved in modifications in development rates, insect survival, voltinism and consequently effect the size, density, genetic variability of populations and host plant interactions (Table 1) (Bale et al. 2002). The change in temperature thresholds is also pre-requisite for insect flight and can also vary among insect species, with season as well as with region. Black bean aphid (*Aphis fabae* Scopoli) require different temperatures for wing beating (6.5°C), horizontal flight (13°C), sustained upward flight (15°C) and for take-off (17°C) (Cockbain 1961). There might be both positive and negative reproduction and developmental responses of insects to temperature conditions. Some economically important multivoitine insects like bark beetle (*Ips typographus* L.) can get benefit by an earlier completion of life cycles and establishment of additional generations within a season due to increase in temperature (Jönsson et al. 2009). Increase in temperature frequencies can lead to decreased growth rates and fecundity for the multitude of species. The increased mortality rates are also observed due to increase in temperature. Similar effects (*Operophthera brumata* L., *Epirrita autumnata* Borkhausen and *Lymantria dispar* L.) are possible with decrease in temperature extremes (Moore and Allard 2008). The gradually changing climate scenario is likely to influence distribution and severity of crop pests and diseases (Oerke 2006), impact the sustainability of the crop production and protection system (Lamichhane et al. 2015), complicate the use of reduced-risk-pesticides (Hossard et al. 2014) and ultimately limit global food production (Foley et al. 2011). Climate change may enhance the adaptability of pests in previously detrimental areas (Chakraborty 2013) and accelerate spatio-temporal pest pressure due to resurgence and replacement phenomena (Chakraborty and Newton 2011). Adaption of indigenous and exotic pest species to changing climate, better-adapted pest genotypes, lack of stable and predictable cropping system, resurged impact of pest status and their losses, comprehensive revision in plant health strategies, climate-resilient production and protection system, future legislation to increasingly stringent climate and human-health concerns and augmented pressure on high-yielding cropping system and food-security are the major operational and practical challenges generated by climate change (Lamichhane et al. 2015).

No doubt, temperature is important for survival, growth, development, dispersal and voltinism but other factors like draught and precipitation also share a vital role in insect abundance. Overall, climate transformation might effects population

Table 1 Examples depicting impact of climate change on various life parameters of insects

Insect	Technical name	Climate change associates	Effects and insect associated change	References
Stinkbug in England and Japan	*Acrosternumhilae* (Say)	Temperature increase of only 2 °C	Distribution range shifted to 300 km northward	Trumble and Butler (2009)
Mountain pine beetle in the USA and Canada	*Dendroctonus ponderosae* (Hopkins)	Temperature increase of only 2 °C	Distribution range shifted to 30–400 km northward	Logan and Powell (2001)
European corn borer	*Ostrinia nubilalis* (Hübner)	Temperature variation	Distribution range shifted to maize growing areas which previously free of infestation	Lamichhane et al. (2015)
Codling moth	*Cydia pomonella* (L.)	Temperature variation	Phenological changes occur and formerly univoltine species have become bi- or multivoltine	Stoeckli et al. (2012)
Aphid species	*Myzus ascalonicus* (Doncaster)	Mild change in winter temperature	Survival and colonization patterns of aphids shifted from holocyclic to anholocyclic form	Radcliffe and Ragsdale (2002)
Aphids	*Brachycaudus helichrysi* (Kaltenbach), *Myzus persicae* (Sulzer) and *Sitobion avenae* (Fabricius)	1°C rise in winter Temperature	Radical change in migration phenology by 19 days	Zhou et al. (1995)
Potato psyllid	*Bactericera cockerelli* (Sulc)	Warmer temperature of winter	Previously not establishing species since centuries has migrated, introduced, established and colonized successfully in California	Liu and Trumble (2007)
Poly voltine species of bark beetles	*Ips typographus* (L.)	Temperature increase	Augmented development and reproduction rates, earlier completion of life cycle and establishment of additional generations	Jönsson et al. (2009)

(continued)

Table 1 (continued)

Insect	Technical name	Climate change associates	Effects and insect associated change	References
Brown plant hopper and rice leaf folder	*Nilaparvatha lugens* (Stal) and *Cnaphalocrocismedinalis* (Guen)	Temperature increase	Declined survival rate, alteration in voltinism and changed geographical distribution	Karuppaiah and Sujayanad (2012)
Phloem-feeders, ants, chewing herbivores and parasitoids,	Leaf miner	Elevated level of CO_2	Reduction in abundance of phloem-feeders and ants, while increase in abundance of chewing herbivores and parasitoids,	Hillstrom and Lindroth (2008)
Ichneumonids, Brachonids and Chalcidoids parasitoids	Ichneumonoidea (Latrelle)	Elevated level of O_3	41%, 33% and 26% reduction in abundance of Ichneumonids, Brachonids and Chalcidoids, respectively	
Several butterflies, beetles, dragonflies and grasshoppers	*Carterocephalus palaemon* (Pallas)	Isothermal shift (increase in temperature)	Expansion in their geographical range to higher latitudes and altitudes	Parmesan et al. (1999)
Migratory butterflies in Europe		Isothermal shift (increase in temperature)	≈ 60% of non-migratory butterflies in Europe have extended their geographical distributions by 35–240 km northwards	
Monophagous butterfly	*Boloria titania* (Esper)	Climate changes in term of temperature increase	Incongruity in synchronization and disturbance in trophic interactions between *B. titania* and its larval host plant *Polygonum bistorta*	Schweiger et al. (2008)

(continued)

Table 1 (continued)

Insect	Technical name	Climate change associates	Effects and insect associated change	References
Winter moth	*Operophtera* brumata (L.)	Climate changes in term of temperature increase	Asynchronization in insect-plant and disturbance in their trophic interactions eg. egg hatching (> 90%) before oak (*Quercus robur*) bud burst	Visser and Holleman (2001)
Marsh fritillary butterfly and its parasitoid	*Euphydrias aurinia* (Rottermburg) and *Cotesia bignellii* (Marshal)	Climate change (increase in temperatue)	Decrease in developmental times of the host Alteration in dynamics of host-parasitoid system and synchronization host-parasitoid interactions	Klapwijk et al. (2010)
Herbivore, pollinators, seed-dispersing insects	*Plutella xylostella* (L.) and the generalist predator *Podisus maculiventris* (Say)	Changes in temperature, rainfall patterns and atmospheric concentration of gases	Positive and negative insect-plant interactions; 2–3 fold increase in emission in plant volatile organic compounds (VOCs) (such as methyl jasmonate or methyl salicylate), more fragrant environment, reduction in future herbivory rates and disruption in pollination and seed-dispersal causing reduction in reproduction and fitness of plants	Constable et al. (1999); Penuelas and Staudt (2010)
Parasitoid wasp	*Cotesia marginiventris* (Cresson)	Increasing temperature	Effects on fecundity and 90% reduction of off-spring production	Dukes and Mooney (1999)
Argentine ants	*Linepithema humile* (Mayr)	Increasing temperature	Distribution dissemination to northward and fecundity disruption of more inborn ant species	

(continued)

Table 1 (continued)

Insect	Technical name	Climate change associates	Effects and insect associated change	References
Spruce budworm	*Choristoneura fumiferana* (Clemens)	Increasing temperature	50% increase in fecundity	Régnière (1983)
Moth in Norway birch forests	*Argyresthia retinella* (Zeller)	High temperatures and droughts	Severe outbreak of epizootics	Tenow et al. (1999)
Winter moth	*Operophtera brumata* (L.)	High temperatures	Increased epizootic and range in in Norway birch forests	Hagen et al. (2007)
Pine processionary	*Thaumetopoea pityocampa* (Denis & Schiffermüller)	Warmer winters due to rising temperatures	Epizootics outbreak on Scot pine	Buffo et al. (2007)
Oak dieback disease and ambrosia beetle	*Platypus quercivorus* (Murayama)	Global warming	Increase in range of ambrosia beetle and epidemic outbreak of oak dieback disease in Japan due to encounter of beetle with fungus.	Kamata et al. (2002)
European pine sawfly and shoot beetle	*Neodiprion sertifer* (Geoffroy) and *Tomicus destruens* (Wollaston)	Global warming	Epizootics outbreak and severe damage on pines	Faccoli (2007)

dynamics of insect pests differently in different agro-ecological zones and agro-ecosystems. Therefore, it is obligatory to understand and address these issues through more research on different aspects (metabolic alterations, prediction models, evolutionary changes) of insect pests.

2.2 *Insecticide Resistance*

The indiscriminate use of pesticides posed a major challenge to the targeted pests by forcing them to either disperse to novel environment and/or adapt newfangled conditions. Such adaptations could be attributed as gene mutation, alteration in population growth rates, and escalation of generations etc., which ultimately cause pest resurgence and pest resistant incidents. Pest resistance can be defined as "Heritable change in the sensitivity of a pest population that is reflected in the repeated failure of a product to achieve the expected level of control when used according to the label recommendation for that pest species" (IRAC 2013). The development of pesticide resistance is a most serious bottleneck in the sustainable pest management because resistant individuals continue to reproduce and eventually become the dominant part of the population with the passage of time. In 2001, the estimated number of resistant insects and mites was around 700 which were calculated 600 by the end of 1990 and this trend is likely to be continued. Pesticide resistance has been reported against a number of insecticide groups. According to Rai and Ingle (2012), a large number of weed species (270), plant pathogens (150) and insect species (more than 500) acquired resistance against herbicides, fungicides and insecticides respectively (Rai and Ingle 2012). There are many insect species who developed resistance against oganochlorine insecticides like cyclodiene (291 species) and DDT (263 species). Similarly resistance has been found in other insecticide groups e.g., 260 species have developed resistance against organophosphates followed by carbamates (85 species), pyrethroids (48 species) and fumigants (12 species) (Dhaliwal et al. 2006). Almost 330 cases of imidacloprid resistance have been reported by APRD (Arthropod Pesticide Resistance Database) followed thiamethoxam (130 cases) and acetamiprid (50 cases) resistance (Bass et al. 2015). The insect pests who showed resistance was mainly white fly (*Bemisia tabaci* Gennadius) followed by the green peach aphid (*Myzus persicae* Sulzer), the cotton aphid (*Aphis gossypii* Glover) and the rice brown plant hopper (*Nilaparvata lugens* Stål). The genetically modified crops with insecticidal toxins from *Bacillus thuringiensis* (Bt) showed resistant against a number of insect pests. But some reports (Tabashnik et al. 2003) showed resistance development in major insect pests like diamondback moth (*Plutella xylostella* L), pink bollworm (*Pectinophora gossypiella* Saunders) and American bollworm (*Helicoverpa armigera* Hübner). Some examples of resistance development in insect pests of economically important crops are given in Table 2. Thus, the issue of insecticide resistance always causes pressure to develop novel compounds to avoid resistant development in insect pests. Therefore, every year around one million insecticidal compounds have been screened out (Resh and Cardé 2009) for better management of insect pests.

2.3 *Pest Resurgence*

Besides resistance development, the pest resurgence is another phenomenon which hampers the efforts of sustainable pest management. Pest resurgence is known to occur due to many reasons (Dhaliwal et al. 2006) but the use of broad spectrum and persistent pesticides is considered as leading cause because of their toxic effect on non-target organism especially the insect natural enemies. In literature, many pesticide-induced pest outbreaks have been reported (Gill and Garg 2014) but brown plant hopper (*Nilaparvata lugens* Stal) outbreak in rice gained major importance. Generally, BPH population was kept under control by different entomophagous insects (mirid bugs, ladybird beetles, spiders) but pesticide exposure destroyed the BHP natural enemies. Pesticides influenced the fecundity of females (Wang et al. 2010) which further enhanced BHP resurgence. The other examples of pest resurgence are bed bug (*Cimex lectularius* Latreille) and cotton bollworm (*Helicoverpa armigera* Hübner) which occur due to indiscriminate use of pesticides and resistance development (Davies et al. 2012; Mironidis et al. 2013).

Table 2 Incidence of insecticide resistance development in insect pests of economically important crops

C. Name	T. Name	Family & order	Insecticide	Crop	Reference
Cotton leaf hopper	*Amrasca devastans* (Dist.)	Hemiptera: Cicadellidae	Neonicotinoids	Cotton	Rabia et al. (2017)
Dusky Cotton Bug	*Oxycarenus byalinipennis* (Costa)	Hemiptera: Lygaeidae	Pyrethroids, organophosphate, spinosad, benzoate, nitenpyram and imidacloprid.	Cotton	Ullah et al. (2016)
Cotton aphid	*Aphis gossypii* (Glover)	Hemiptera: Pseudococcidae	Neonicotinoid (thiamethoxam, acetamiprid and clothianidin)	Cotton	Herron and Wilson (2011)
Cotton Whitefly	*Bemisia tabaci* (Gennadius)	Hemiptera: Aleyrodidae	lambda-cyhalothrin, cypermethrin, deltamethrin fenpropathrin	Cotton	He et al. (2007)
Cowpea Aphid	*Aphis craccivora* (C.L. Koch)	Hemiptera: Aphididae	Malathion, chlorpyrifos, thiamethoxam and carbosulfan	Cowpea and pulses	Kandil et al. (2017)
Cabbage aphid	*Brevicoryne brassicae* (Linnaeus)	Hemiptera: Aphididae	Organophosphate, Neonicotinoids	Cabbage	Ahmad and Akhtar (2013)
Cotton Whitefly	*Bemisia tabaci* (Gennadius)	Hemiptera: aleyrodidae	Acetamiprid	Cotton	Basit et al. (2011)
Mustard Aphid	*Lipaphis erysimi* (Kalt.)	Hemiptera: Aphididae	neemarin	Mustard	Kumar et al. (2007)
Green peach aphid	*Myzus persicae* (Sulzer)	Hemiptera: Pseudococcidae	Pyrethroids, carbametes, organophosphate	Canola and cabbage	Needham and Sawicki (1971)
Spotted bollworm	*Earias vittella* (Fabricius)	Lepidoptera: noctuidae	Organophosphate, pyrethroid and biorational Insecticides	Cotton and okra	Jan et al. (2015)

(continued)

Table 2 (continued)

C. Name	T. Name	Family & order	Insecticide	Crop	Reference
Cotton bollworm	*Helicoverpa armigera* (Hübner)	Lepidoptera: noctuidae	Endosulfan and chlorpyriphos	Cotton	Kranthi et al. (2002)
Pink bollworm	*Pectinophora gossypiella* (Saunders)	Lepidoptera: noctuidae	Pyrethroids and endosulfan		
Spotted bollworm	*Earias vittella* (Fabricius)	Lepidoptera: noctuidae	Cypermethrin		
Tobacco caterpillar	*Spodoptera litura* (Fabricius)	Lepidoptera: noctuidae	Chlorpyriphos		
Whitefly	*Bemisia tabaci* (Gennadius)	Hemiptera: aleyrodidae	Cypermethrin		
Cotton mealybug	*Phenacoccus Solenopsis*	Homoptera: pseudococcidae	Bifenthrin	Cotton	Mansoor et al. (2016)
Green peach aphid	*Myzus persicae* (Sulzer)	Hemiptera: Aphididae	Imidacloprid		Nauen and elbert (2003)
Corn earworm	*Helicoverpa zea* (Boddie)	Lepidoptera: noctuidae	Pyrethroids	Maize	SEMINIS (2018)
European corn borer	*Ostrinia nubilalis* (Hubner)	Lepidoptera: crsmbidae	Pyrethroids		
Fall army worm	*Spodoptera frugiperda*	Lepidoptera: noctuidae	Organophosphates or carbamate		
Whitefly	*Bemisia tabaci* (Gennadius)	Hemiptera: aleyrodidae	Neonicotinoid insecticide	Cotton, vegetables, and ornamental crops	Nauen and Denholm (2005)
Colorado potato beetle	*Leptinotarsa decemlineata* (Say)	Coleoptera: chrysomelidae	Neonicotinoid insecticide	Potato	
Army worm	*Spodoptera litura* (Fabricius)	Lepidoptera: noctuidae	Emamectin and Indoxacarb	Multiple crops	Ahmad et al. (2008)
Whitefly (biotype Q)	*Bemisia tabaci* (Gennadius)	Hemiptera: aleyrodidae	Acetamiprid, Imidacloprid, and Thiamethoxam	cotton, vegetables, and ornamental crops	Luo et al. (2010)
Tomato leafminer	*Tuta absoluta* (Meyrick)	Lepidoptera: Gelechiidae	Abamectin	tomator	Lietti et al. (2005)
Anopheles spp.	*Anopheles culicifacies* (Giles)	Diptera: Culicidae	malathion		Raghavendra et al. 1991
Stem borer	*Chilo suppressalis (Walker)*	Lepidoptera: Crambidae	Fipronil	Rice	Li et al. (2007)

2.4 Pesticides Impact on Non-target Organisms

Biological control is an effective strategy for controlling arthropod pests. Natural enemies such as predators, parasitoids and entomopathogens have been utilized for crop pest management for centuries. Today the biological control schemes have been operated successfully across the globe for the management of pests in agriculture, forestry and horticulture. But it is evident that the pesticides severely affect the non-target organisms especially predators, parasitoids, pollinators, earthworms, humans, birds, fishes, amphibians (Gill and Garg 2014). Unfortunately, biological control agents (parasitoids and predators) are most susceptible to negative effects of insecticides. These pesticides are severely affected their viability and efficiency to reduce the pest infestation. There are a number of studies which showed the negative impacts of pesticides on insect natural enemies. Ghananand et al. reported that spraying with cypermethrin and imidacloprid caused higher mortality of insect natural enemies. On the other hand, spraying with bio-pesticides and neem based insecticides were less toxic to coccinellids, braconid wasps and predatory spiders (Ghananand et al. 2011). Similarly, more number of arthropods such as coccinellids was present in non-sprayed fields compared to fields sprayed with insecticides and herbicides to control the insect pests (Amalin et al. 2009). In addition, different pesticides may have different levels of toxic effects on natural enemies. The foliar application of spirotetramat, buprofezin and fipronil were significantly less toxic to insect natural enemies in comparison with imidacloprid, clothianidin, admire, thiamethoxam and acetamiprid (Kumar et al. 2012).

2.5 Biodiversity

Before the introduction of synthetic insecticides in agriculture, the forces of potential creation (biotic potential) of insects and forces of potential destruction (environmental resistance) were playing their roles to maintain a balance (biological equilibrium) among the insect population. The use of chemical insecticides caused a huge imbalance among phytophagous and entomophagous insect diversity. The practice of monoculture for crop production also caused effects on plant diversity similar to insecticides. Recent investigation revealed the negative impacts on the ecosystem biodiversity as well as the biological control of pests (Geiger et al. 2010). Plant biodiversity plays an important role in sustainable pest management. An analysis of 22 case studies dedicated plant diversity services for orchard pest control revealed positive impact on pest control (16 cases) or null (9 cases), but also negative (5 cases) (Simon et al. 2010). The negative impact could be due to many reasons which need further research to identify the processes involved at different levels of natural control. Therefore, the biodiversity management could yield effective control of insect pests on sustainable basis.

The above mentioned issues are the bottlenecks for the sustainable pest management of insect pests in different agro-ecosystems. The climate change impacts can be minimized by adapting innovative technological developments, modifications in farm production practices and financial management. In addition, the government programs and insurance policies are also important. The insecticide resistance and resurgence issues are addressable by integration of different pest control options like reduce risk pesticides, biopesticides, botanical insecticides, pheromones. Furthermore, the use of nanotechnology based nanopesticides also has potential to tackle the issues of pest resistance and resurgence. Similarly, the biodiversity and genetic diversity at farm level can be maintained by practicing zero tillage and conservation agriculture techniques for sustainable pest management of different agricultural insect pests.

3 Pre-requisites of Sustainable Management of Insect Pests

Pest management is a two-strand approach which mainly relies on the knowledge of the strategy, pest biology and pest ecology in agroecosystem. The selection of appropriate pest control technology as well as its effective and efficient application mainly depends upon a comprehensive knowledge about it. The biological and ecological knowledge of pest helps to determine the most appropriate procedure/method (how), timing (when) and place (where) for effective use of any technology and economically effective management of any pest (Buurma 2008; Pedigo and Rice 2009).

The knowledge of various aspects of biology and ecology of pests lays the foundation of an efficient and economical pest control strategy and is important for achieving key objectives of pest management. For examples, such kind of knowledge reduces the threat of crop failure by endemic or epidemic pest outbreak. Such knowledge also strengthens the effectiveness of pest control strategies, reduces operational cost of technique used, enhances productivity and profitability by reducing the amount of inputs and ultimately eliminates or reduces the threats of environmental degradation and hazards of human health (Norris et al. 2002; Pedigo and Rice 2009).

Integrated application of multiple and highly compatible tactics; reduction in number or effects of pest below defined economic decision levels (EIL and ETL); and conservation of environmental quality are the key characteristics/elements of sustainable pest management (Knipling 1979; Pedigo and Rice 2009). However, Geier (1966) suggests some supplementary characteristics/elements of sustainable pest management system that a pest management technology/system should be: (1) highly target specific i.e., very selective for pest and safe for non-target organisms; (2) Comprehensive and conducive for crop productivity (not be phytotoxic and enhance plant-growth and yield); (3) highly compatible with the key principles of ecology and (4) tolerant to potential pests but within economically tolerable limit. A comprehensive and practical knowledge of above-mentioned elements guarantees the development of an ecofriendly, economical and efficient, crop production and protection program (Dhaliwal et al. 2006; Buurma 2008; Pedigo and Rice 2009; Alam 2010; Schowalter 2011).

Effective and sustainable insect pest management also depends on economic decision levels which are mandatory for determining the course of action, ensuring sensible pesticide application, reducing unacceptable economic damages, safeguarding the profits of producer and conserving the environmental quality in any pest situation (Knipling 1979; Inayatullah 1995; Alam 2010; Jha 2010).

3.1 *Information of Insect Pest Biology*

Various aspects of pest's biology that can be helpful in devising efficient pest management strategies include:

- What kind of habitat does the pest prefer? (Darkness, indoor, outdoor, humid, warm, temperate, aquatic, terrestrial etc.)
- What kind of food does the pest prefer?
- What is the total life span of pest?
- What is longevity of incubation period of the pest?
- Where is different life stages found?
- What is the breeding place and season of the pest?
- What kind of behavior does the pest exhibit in its life? (Knipling 1979; Sorby et al. 2005; Dhaliwal et al. 2006; Pedigo and Rice 2009; Jha 2010)

An efficient, effective and successful management of insect pests is always founded on a comprehensive knowledge of the biology, morphology, internal anatomy, behavior, growth (metamorphosis), life history and ecology of any insect pest. The morphological knowledge of an insect helps to develop an appropriate technology and selection of appropriate insecticide. Chemotropism based techniques involving attractant or repellents have been developed for various insect pests. The development of such techniques depends upon knowledge about chemoreceptors like, gustatory, olfactory, sensory receptors etc. Development and selection of color of light for light-traps depend on the knowledge of structural components and physiology of compound eyes of insects (Pedigo and Rice 2009) which provide information about the type of color which is highly attractive for any insect. For example, yellow sticky traps are used for the control of aphids as aphids are attracted to yellow color. The knowledge about the types of mouthparts of insect pests helps to decide that what type of insecticides should be selected for successful control of insect pests. For example, if the infesting insect pests have sucking type of mouthparts, then, insecticides with systemic and contact action will be the most appropriate selection of insecticides. Unlikely, if the infesting insect pests have chewing type of mouthparts then stomach poisons will give effective control. Unawareness of the knowledge of the mouthparts of insect pests leads to wrong selection of insecticides and ineffective management of insect pest in spite of investment of money in form of insecticides application (Dhaliwal et al. 2006; Saha and Dhaliwal 2012).

Knowledge of internal anatomy and physiology is also very advantageous in devising pest management tactic. For example, so many insecticide molecules based on growth and development based hormones, peptides from sting gland of parasitoids and pheromones- having IGR or karomone activity have been discovered and their analogs have been synthesized for commercialization and management of insect pests. The knowledge of spiracle respiration can be helpful in controlling the insect pests with fumigation.

The knowledge of insect metamorphosis and its physiology provides so many useful cues about the weak links of insect growth stages and their activity periods and sites which if targeted can ensure effective management of any pest. Such knowledge can also be useful in synchronizing the timing of application of pest management tactics with weak-link or susceptible growth stage of insect; thus, ultimately would be helpful in reducing blind use and application intensity of pesticide on crop. These facts would lay the foundation of decision on when, where and how to use available and recommended insecticides or other pest management tactics. Information on the metamorphic stages like, eggs, larvae/nymphs/naiads, pupae and adults of insects comprehend the facts that which stage is notorious, devastating and damaging one and which are not.

Incorporating pest controls at many different stages and limiting pests' abilities in many small ways are the foundation of ecological pest management (Schowalter 2011). Production systems that use ecological principles to imitate nature, along with multiple tactics and the right information, can: (i) synergize individual impacts of strategies when used together, (ii) reduce the risk of crop failure by distributing the burden of crop protection across many tactics, (iii) minimize environmental disruptions and threats to human health, (iv) slow the rate at which pests adapt or evolve resistance to a given management tactic because and reduce operating costs and ultimately improve profitability by minimizing inputs (Dhaliwal et al. 2006; Pedigo and Rice 2009)

The study of the behavior of insects also laid the foundation of successful control of insect's pests. Insect's behavioral studies figure out following important facts of their life that can be helpful in controlling them.

3.1.1 Egg Laying Behavior

The some insects are endophytic (fruit flies) and some are exophytic (most of the bollworms, borers etc.). Most of the insects deposit exposed eggs while some deposit the covered egg masses. Depending on egg laying behavior, pest management tactic is decided to control insect pest at egg stage.

3.1.2 Behavior of Newly Emerged Young Ones

The young ones of most of the borers just after hatching enter into the leaf whorls or stem of the plants, avoid the direct exposure of insecticides and become very difficult to kill with contact insecticides. Similarly, leaf miners just after hatching enter into the cortex tissues forming mines and cannot be controlled with contact insecticides. Young ones as well as later instar larvae of cutworm remain hidden in cracks and crevices and insecticides direct application on plants during day time will not yield effective control. Their effective management can be ensured if chemigation of insecticides or bait application is employed.

3.1.3 Feeding Behavior in Young Ones

Feeding habit of insect pest also help in deciding the types of tactics and method of their application for effective management of any insect pest. The insect pest which prefer to feed underside the leaf can be controlled effectively by application of systemic and translaminar insecticides. Similarly, borers (Insecta: Lepidoptera and Coleoptera) exhibit concealed feeding inside the stem which cannot be killed with contact insecticides; rather systemic insecticides will be the most appropriate borer management tactic.

3.1.4 Breeding Place

Nipping the evil in the bud for insect pests is possible only if their exact breeding sites are known. It is possible only through comprehensive studies of their biology. The breeding places of mosquitoes are stagnant water and their treatment with larvicides, ovicides or oils help in controlling the breakout of adult population. Cockroaches breed in gutter or filthy places which should be targeted with insecticides treatment for their management at bud/root level (breeding places) for terminating their further population buildup and outbreak. For fruit flies, the breeding substrates are dropped fruits which should be collected and destroyed for their population management.

3.2 *Insect Ecology*

The study of insect ecology provides the conceptual and theoretical framework which offers the practical ground for the application of pest management discipline. Recent advances in understanding the complex effects of insects and their interactions with other organisms on ecosystem services have influenced evaluation of the need for insect management (Schowalter 2011). The solution of insect problems majorly depends on ecological management which is considered as one of the oldest, least expensive and ecologically the most compatible tactics. Ecological studies of insects help in identifying and exploiting the weak links of seasonal life cycle of insects. Such studies also help to explore the food and physical factors which impact insect's life negatively. By manipulating of such factors unfavorable for insect survival, insect pest's outbreak, population buildup and damage impacts can be avoided in an ecofriendly way (Pedigo and Rice 2009). Study of insect ecology also laid the foundation of plant-insect pest-entomophagous insect interactions which help to frame out the pest management strategy for any insect pest. In vegetable system, combining minimum tillage with cover crops and cover crop mulch creates enough biological diversity to pests. Such integrated practices resulted in conservation of field and increased beneficial insect populations 14 times higher than in the conventional fields. Leaving some undisturbed areas on a farm can help maintain the balance between beneficial and pest organisms. Many predators and parasites that attack crop pests thrive in the less-disturbed areas provided by hedgerows, weedy borders, woodlots and riparian buffers on the farm; in grassed alleyways in orchards and grassed waterways in field crops; and even in the small areas left between crop rows by zone tillage. Small sites allow natural enemies to persist and migrate into crop fields to keep pest populations in check. Maintenance of diversity in agroecosystem based on ecological studies of insect life and then diversity maintained in the crops grown can reduce pest problems. Maintenance of dissimilar types of crops growing at various stages and under diverse management practices will results in an encounter of pests with a broader range of stresses, they will face difficulties in locating their hosts in both space and time and their resistance to control measures also will be hampered (Schillhorn et al. 1997; Pattison 2005).

Insect ecological studies also help to select various alternate host plants which can serve as trap and cover crops. Such crops, when intercropped or border-cultivated, not only recruit entomophagous insects in their battle against insect pests on major crops but also create a nice habitat for feeding and overwintering of beneficial insects. The dandelion flowers serve as source of food for nectar- and pollen-seeking insects before mowing them down. Insect ecological studies also laid the foundation of insect chemical ecology that yielded the discovery of so many semiochemials and their potential implementation in pest management program of so many insect pests (Pattison 2005). For example, discovery of pheromones (methyl euginol for fruit flies, gossyplure for pink bollworm etc.), allomones, kairomones and synomones are based on insect chemical ecology studies (Dhaliwal and Arora 2003; Dhaliwal et al. 2006; Saha and Dhaliwal 2012).

3.3 *Information of Control Methods*

The various aspects of the knowledge of technology help to select and use an appropriate pest management tool (insecticides, equipment). The knowledge of biological aspects of pest life highlights the appropriate place (where), timing (when) and procedure/method (how) for efficient application of any technology as well as for economically effective management of any pest (Buurma 2008; Pedigo and Rice 2009). Various aspects of any technology which lead towards its proper and effective application (Pedigo and Rice 2009) are given below:

- Nature and type of technology
- Mode of its application (aerial, foliar, chemigation, baits, traps etc.)
- Bio/shelf life of the technology
- Equipment required for its application
- Factors affecting the performance of technology
- Compatibility with other management tactics
- Target specific or broad spectrum
- Mode of action

3.4 Economic Decision Levels

The decision staircase of pest management program shows that successful and sustainable pest management depend on certain pillars that basically stand on the foundation of six slabs (biology, ecology, threshold, models, sampling and taxonomy) and one of those is economic decision levels (thresholds). Economic decision levels (EDLs) are indispensible for devising and implementing insect pest management program in an effective and economical way (Dhaliwal and Arora 2003; Dhaliwal et al. 2006; Pedigo and Rice 2009). The comprehensive and true practical knowledge of such decision levels ensure the sensible and timely use of insecticides because these levels highlight the exact density of insect population that may cause economic damage if insecticides are not used. An ignorance of these economic decision levels leads to ridiculous economic gaffes spending more cost on pest management and crop protection. A comprehensive and proper knowledge, understanding and use of these economic decision levels can enhance the profit ratio of the growers and ensure the conservation of the environment and biodiversity (Knipling 1979; Inayatullah 1995; Pedigo and Rice 2009). Briefly, proper and sensible utilization of EDLs has following plus-points (Knipling 1979; Pedigo and Rice 2009):

1. Sensible use of insecticides and avoidance from the indiscriminate use of insecticides
2. Reduction in insecticides use
3. Increase producer's profit ratio
4. Conserve natural biodiversity
5. Conserve the environment quality
6. Solution of some problems like ecological backlash (resistance, resurgence and replacement), pesticide residues and negative impacts on non-target organisms.

These EDLs include EIL (Economic Injury Level), ETL (Economic Threshold Level), GT (Gain Threshold) and DB (Damage Boundary). Among these, ETL is the practical operational level which is recommended to and being practiced by the growers for making pest management decisions in many situations. ETL is mostly used for making decision about the strategic implementation of curative/therapeutic management tactics including insecticides.

3.5 Climate Change Management

Climate is changing due to global-warming associated with anthropogenic activities (Pachauri and Reisinger 2007). A comprehensive and long-term monitoring data and imperial-approach for feeding in modeling system is required to determine the impacts of climate change on distribution, outbreak, and dynamics of pests (Shaw and Osborne 2011). Gradual or abrupt increase in seasonal temperature, rise in the level of CO_2 and higher precipitation intensity are the major climate change manipulating factors (Pachauri and Reisinger 2007) which affect biology, development, physiology, epizootiology, phenology, distribution, invasion, population dynamics, (Willmer et al. 2000; Lamichhane et al. 2015), life history patterns, evolutionary adaptation (Bradshaw and Holzapfel 2011), distribution range (to suitable altitudes) (Parmesan 2006), and traveling speed (increases) (Aluja et al. 2011) of indigenous and invasive insects pest species (Fig. 1). Global changes also modify the host-pest-natural enemies' synchronization and interactions (both bi- or tri-trophic interactions), Synchronization in mutualistic interactions (pollination and seed dispersion) among species intensify losses potentials of pests (VanAsch and Visser 2007; Lamichhanc ct al. 2015), enhance adaptability of pest to changing climate (Trumble and Butler 2009), alter pest management protocols and strategies (low pesticide-residue IPM) and accelerate pest resistance buildup against control measure in practice (Lamichhane et al. 2015). A mild elevation in winter temperature due to climate change enhances the survival and colonization of previously low-temperature tolerant insect species which change their mode of reproduction from sexual to new asexual generation (Lamichhane et al. 2015). It has been reported that an increase of 2°C in temperature might result in 1–5 additional biological cycles per season in insects depending on insect species (Yamamura and Kiritani 1998).

Atmosphere enriched with CO_2 and O_3 influences plant quality, host plant selection and herbivory behavior of insects (Peltonem et al. 2006). Elevated ozone level in atmosphere indirectly influences insects by regulating insects associated bottom-up and top-down factors (Hillstrom and Lindroth 2008). Elevation in the level of CO_2 and O_3 will entirely change the insect community structure (abundance and diversity), population dynamics of hosts and biological control system due to substantial change in abiotic, bottom-up (resource concentrations) and top-down factors (predation, parasitism, pathogens etc.) (Dermody et al. 2008; Hillstrom and Lindroth 2008). Changes in rainfall, hurricanes and flooding can affect food-web dynamics, herbivory patterns, and insect-plant interactions due to alteration in the biochemical based plant-defenses (Koptur et al. 2002; Angulo-Sandoval et al. 2004). Climate change can also affect the production of plant volatile organic compounds (VOCs) which will influence the positive (e.g., pollination and seed dispersal) to

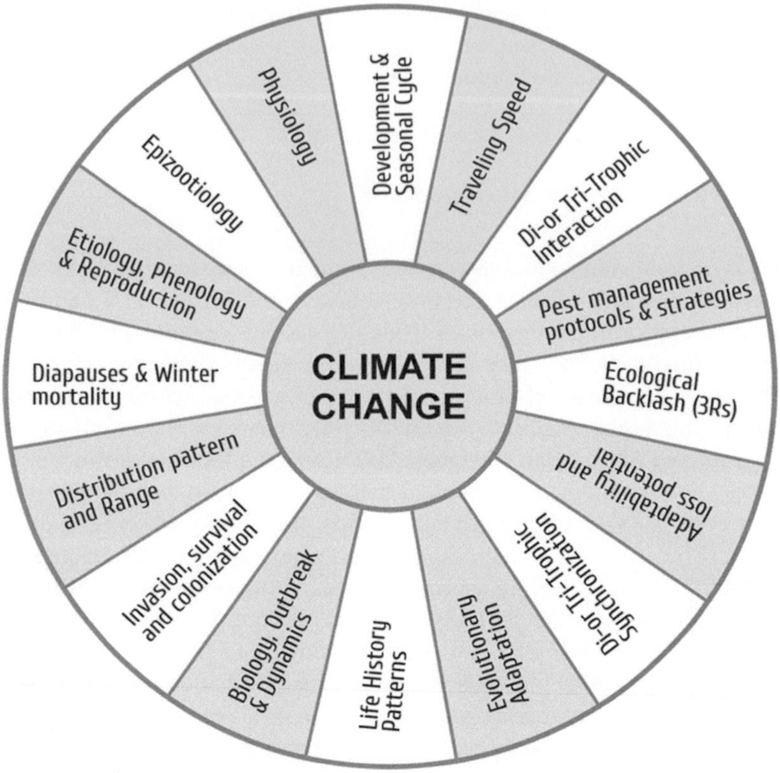

Fig. 1 Diagrammatic illustration of climate change impact on insect pest associated parameters including tri-trophic interaction, pest management practices, biotic potential, evolution and adaptations

negative (e.g., defenses against herbivory) insect-plant interactions (Penuelas and Staudt 2010).

3.5.1 Implications of Climate Change and Insect Pest Management

Agriculture and climate resilient sustainable pest management is crucial for sustainability of any pest management program. This imperativeness is attributed to the fact that climate change has been declared as one of the imperative factors which not only directly regulate agricultural productivity but also indirectly influence it by affecting regional and marginal distribution of indigenous and invasive pest species. Climate change also has significant effects on the biodiversity of pests, pollinators, crops and decomposers etc. It is, therefore, indispensable to execute modeling for biodiversity and climate change not only to devise climate resilient IPM but also to ensure sustainability of crop production and protection system (Fig. 2). Climate change has made the previously detrimental environmental conditions highly

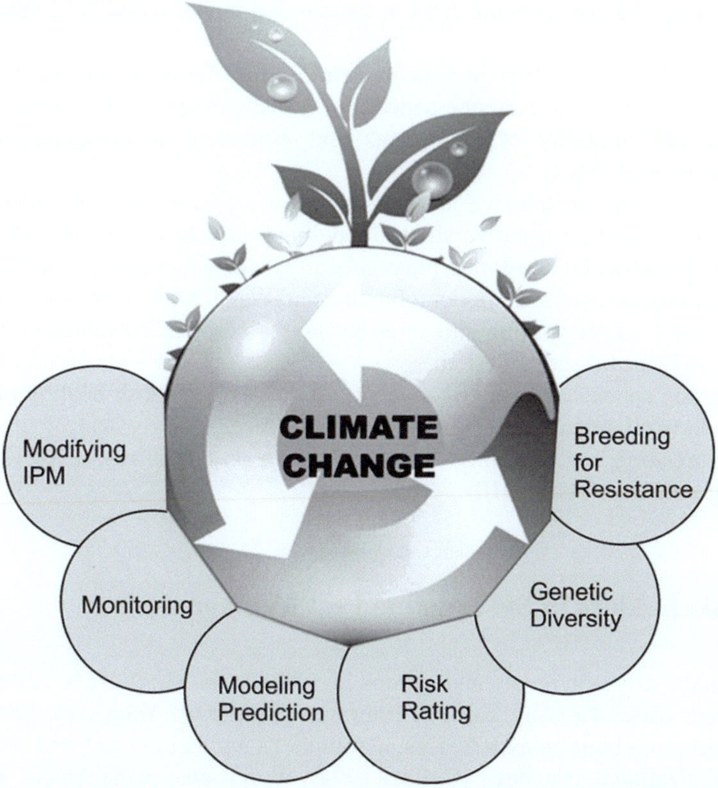

Fig. 2 Strategies for promising program of crop-health and sustainable ecosystem management against insect pests under changing climate conditions

favourable and conducive for less adapted indigenous as well as for exotic/invasive pest species. The least adapted pest species have become successful survivors under conducive climate changes than highly adapted species (Chakraborty 2013; Lamichhane et al. 2015). Extreme and excessive climate changes like severe and prolong dry, wet and warm condition may hamper crop productivity and effectiveness of the protection methods and technologies (Chakraborty and Newton 2011). Climate change may also cause severe increase in crop losses by pests and threaten food security (Lamichhane et al. 2015). Various models like CLIMEX have been utilized to predict the insects' responses to climate changes (Desprez-Loustau et al. 2007). However, insufficient knowledge on the interactions between climate change and disturbances (magnitude, severity or frequency insects' ecosystem disturbances) and inadequate information on climate-change induced modifications in insect-biology, host-resistance and phenology/physiology of host–insect interactions are major barriers in the modification of IPM strategies/programs. Following strategies can be exploited to address the issue of crop-health and sustainable-ecosystem management against insects under climate change.

- Modifying the conventional IPM to degree-day-model based IPM (Stacey and Fellowes 2002)
- Ensuring long-term efficient management of crops by coordination of monitoring data on spatial occurrence patterns of insects, ranges of host crops, growth, health and mortality of host-plants and efficiency of management tactics (Sturrock et al. 2011)
- Forecasting the prospective array of changes across a landscape and insects-attack outcomes using a diversity of modeling tools like climate models coupled with environmental envelopes, phenomenally diverse model and insects/climate-envelopes/host-reactions model (Hamann and Wang 2006; Sturrock et al. 2011)
- Application of crop-health strategies before the outbreak (epidemics) of insects using risk/hazard-rating system
- Increasing species and genetic diversity of crops to establish healthy plantation resilient to climate change (O'Neill 2008)
- Crop breeding for resistance and tolerance to pests and resilience to climate changes (Woods et al. 2010).

4 Stakeholders in Sustainable Pest Management

Stakeholders are individuals and groups of individuals who have a vested/staked interest in a particular issue, cause or enterprise (Dent 2000; Yang et al. 2009). Their expectations are built on experiences, assumptions and beliefs and will reflect specific organizational structures (Collins 1994). Within pest management, there are numerous stakeholders (Bryan et al. 2010), who can include, for instance, shareholders, managers, employees, suppliers, customers and communities who are all linked to different degrees to a commercial company that produces a chemical insecticide. In followings, there is an abridge description of the stakeholders categories with respective importance and roles.

4.1 Public Sector Agents and Agencies

The stakeholders in government need to establish policies that will work for the benefit of all other stakeholders (farmer, private sector, middle man etc.,) (Dent 2000; Freeman 2010). As stakeholders in pest management are so diverse in their interests and needs, henceforth governments perform balancing acts (Dent 2000). Public sector is responsible for funding public research interests in their institutes and universities; they are also responsible for ensuring that commercial companies generate new products which need to be manufactured, employing people, generating wealth and paying taxes (Altbach 2009; Parker 2009). Government policies including tax incentives made the production and use of pesticides attractive to business and farmers (Hall and Matos 2010). Pesticide development, production and use became institutionalized and in the same fashion, farmers became increasingly dependent (Zalom 1993; Alford 2000; Lamine 2011). However, with increasingly obvious environmental concerns raised about pesticides, there is also increased awareness of the public for these issues; thus, they became politically more important (Pimentel 2007; Damalas and Eleftherohorinos 2011). Henceforward, a balance of policy, which allowed environmental issues to be addressed but that, did not influence too heavily on the agribusiness, is required.

4.2 Research and Academia

The development of an IPM program, detailed knowledge of agro ecosystem and its components and how they interact in pest management is the job of scientists (Dent 2000; Feder and Savastano 2006). It has been argued that IPM is the creation of scientists, and these are scientists who have largely controlled its evolution, notwithstanding subject to pressures (Morse and Buhler 1997). The development of transgenic crop plants is one such example currently receiving a great deal of interest and, of course, funding (Marris 2008). The changes in seasonal abundance of a pest are easily described but much less easily explained. The understanding that is central to the philosophy of IPM necessitates an in-depth enquiry by scientists into the complexities and subtleties of insect biology and ecology (Dent 2000; Walter 2005). Despite the obvious role for interdisciplinary research in integrating control measures at a research level, the statement made by Pimentel in 1982 and 1985, still remains largely true today that: 'most remain *ad hoc* efforts by individual pest control specialists, each developing so called integrated pest management programs independently of one another' (Pimentel 1982; Dent 2000).

4.3 Industry

Commercial enterprises generate income through the provision of services, products or a combination of the two (Sievers and Vandenberg 2007). Within agribusiness, there is a greater emphasis on manufacturing and sale of products rather than the service side of the industry. Growers expect to budget for tangible items such as machinery, pesticides and fertilizer but the concept of purchasing, for example, is less acceptable (Zalom 1993). Whereas chemical pesticides were the predominate type of control product in the 1960s, since that time there has been a proliferation of different types of pest management products (de Faria and Wraight 2007). Proclamation made by Dent (2000) should be written with distinguished marks depicting that commercial companies are not in the business of alleviating the world's pest problems, but rather, providing solutions that will generate a viable income and maintain the long-term prospects of the individual companies. The pest control business is worth billions of dollars worldwide each year, its presence influences the whole philosophy of pest management, continually driving for 'its' products (Dent 2000; Pimentel 2007). The commercial company stakeholders are major players in pest management affecting agricultural policy, R&D and also farmers' expectations and needs. The wealth and taxes, the employment and the assurance they generate, provide a powerful incentive for their continued role in the future.

4.4 Growers and Farmers

Farmers have often been viewed as passive recipients of pest management technologies (Pannell et al. 2006), however, this view is changing and farmers tend now to be seen as an integral part of the pest management stakeholder network, with a role in defining pest control needs, evaluating their effectiveness and influencing their wider adoption (Dent 2000). Farmers, more than any other group are sensitive to customer needs and the more competitive and intensive farming becomes the more consumers dependent leading to dictate the pest control practices adopted by farmers (Pimentel 2007). Farmers' objectives may vary. They may, for example, be interested in the maximization of profit or alternatively the minimization of risk (Zadoks 1987). Nevertheless, on both respect, their role remains essential in sustainable pest declines.

4.5 Final Product Users or Consumers

Consumers in developed countries have increasingly high expectations concerning food quality (Grunert 2005). In addition; there is increasing concern about pesticide residues on food (Boobis et al. 2008). It will be the need to maintain consumer confidence in the food industry that will continue to drive other stakeholders to invest in 'safe' technologies (Brunsø et al. 2002). This approach is being mirrored in developing countries that demands high quality standards (Napolitano et al. 2010). The concerns first expressed in *Silent Spring* have been maintained in the public arena by vociferous groups committed to environmentalism (Dent 2000). These groups, which initially campaigned successfully to maintain a high profile on the problems with pesticide use, are now equally vigilant and vocal concerning the potential hazards posed by genetically manipulated crop plants. Public concern may yet significantly influence the widespread use of these and other novel control pest measures (Alford 2000; Pimentel 2007).

5 Sustainable Legislature, Governance and Other Agricultural Regulations

Legislatures are very significant entities in controlling and regulating anything working in state area (Hopper 2016). In pest management, governing bodies and agricultural ordinances are appearing to be employed at national and international fronts to monitor pests, pest-mitigating products, food commodity; and henceforth; monitoring of invasive or out-placed pests (WHO 2015). These laws or codes have varying forms and formats (Peters and Law 2017). One of the fundamental forms, at country scale, is the establishment of quarantine ordinances and departments (Topinka 2009). Simultaneously, WTO inspections and agendas are at international arenas (Black 2017). Restrictions implied on the use of pyrethroids, on cotton in Columbia and in European areas, against *Helicoverpa virescens*, due to insecticidal resistance, is a prominent example at that time (Dent 2000). Ministry of National Food Security and Research is the propositional element under Government of Pakistan, working to frame Agricultural Pesticide Ordinances and Acts time-to-time (Ahmad and Farooq 2010).

5.1 *Pesticide Ordinances and Orders*

Regulations of pesticides and their products i.e., of biological or synthetic origins, are mainly done by government and its related institutes (DPP 2014a). They are relating not only to commercial pesticides but also to phytohormones (Alberto et al. 2016; Javed 2016) in plant protection. EPA and FDA with respective titles of 'Environmental Protection Agency' and 'Food and Drug Administration', are the basic governing bodies in US addressing these issues (Miller 2015). In Pakistan, Plant Protection Institutes-PPIs and PWQCPs 'Pest Warning and Quality Control of Pesticides' are playing those pivotal roles (DPP 2014a). At international sites, the Organization of Economic Co-operation and Development (OECD; an official governing body with different nations), United Nations Food and Agriculture Organisation (FAO) and WHO-the World Health Organisation, are the main working groups (Haya et al. 2015; WHO 2017).

FAO Article 6.11 in an 'International code of conduct on the distribution and use of Pesticides' has signify and intensify the considerations of legislations in words as: 'governments should take action to introduce the necessary legislation for regulation, including registration of pesticides and make provisions for its effective enforcement, including the establishment of appropriate educational, advisory, extension and health care services' (Dent 2000). Agricultural Pesticide Ordinance 1971 of Pakistan is in compliance with FAO, encloses directives for the importation, manufacture, preparations, trade, delivery and use of pesticides in Pakistan (DPP 2014a).

5.2 *GM Crops Regulations*

Since the introduction of biotechnology and other genetic tools in living world, scientists and others research institutes have to face huge controversies on the aspects of their GM crops regular incorporations in human globe (Qiu 2014). The removal of direct gene inductions, and making crops capable of self-defense by mediated inductive interference resistance, can also yield the equivalent crop protection and productions results (Javed 2016) without GM debates. Here, the aspects of supervisory measures mainly focusing on environmental safety, implications of gene shifting from GM to wild and also pollen contaminations of non-GM from GM, is prevailing (Qiu 2014).

However, an acceptance to GM crop can be granted, in US and EU, if GM crop satisfies and appears at par with the criteria for nearest conventional crop/product, i.e., in both botanical attributes and chemical constituency, depicting safety ranks for human and ecology (OECD 1993; FAO/WHO 1996). Animal and Plant Health Inspection Service (APHIS) are taking directives in US; and in Europe, the European Union implemented Directive 90/220/EEC in 1990, contracts with the proclamation and commerce of GMOs (Hygnstrom et al. 2014).

5.3 *Quarantine Conducts and Orders*

With the enhancements in global links and increasing trades across continents, have intensified the threat of pest transfer and introduction of potential pests in other countries (Dowell and Gill 1989; DPP 2014b) as by nature pests can have a limited expansion power by flight etc. Nevertheless fast trades and transports have facilitated them to move beyond the hemisphere (Hurley et al. 2016). Such activities preventing the introduction of any across border pests and are denoted under 'quarantine' with support of ministry of agriculture on legislative grounds (Mittinty et al. 2015). The things with considerations are plants, crop germ plasms, plant constituents, agronomic consignments, soil, vessels, stuffing, budding media, or any article that theoretically provide anchorage to exotic pests (DPP 2014b).

Government of Pakistan has formulated such conducts under the Department of Plant Protection with title 'Plant Quarantine Act 1976', the standing body is Ministry of Food, Agriculture and Co-Operatives (DPP 2014a, b). International Plant Protection Convention or IPPC has already loaded varying instructions and rules to prohibit any such cases (Hallman 2017). Any violation or mismanagement of such rules/orders lead to havoc as indicated from the plant products importation from USA in the form of American native *Helicoverpa armigera* (Hübner) or commonly statured American bollworm, with the present status of destructive key-pests in Pakistan and other regional countries (Kriticos et al. 2015). Most important quarantine pests in Pakistan, 'Plant Quarantine Act 1976' representing, are Black wart (*Synchytrium endobioticum* Schilbersky), Golden Nematode (*Globodera rosto-*

chiensis Wollenweber), Colorado potato beetle (*Leptinotarsa decemlineata* Say) and South American leaf blight (*Dothidella ulei* Hennings) (DPP 2014b).

5.4 Crop Production and Protection Legislations

Most of the crops are being infested by the pests when they are sown without paying any considerations to the prevailing pests' occurrence time, and thus, crops become susceptible to such pests (Sarwar 2012; Javed 2016). Any alternations in crop sowing time, cultural practices (Javed 2016) and changing cropping schemes can produce healthy crops with profitable results (Medvedev et al. 2015). But making and maintain the crop production rules are the sole responsibility of government to avoid any such pest threats (Lazpoulos Friedman and Van Camp 2016). Approval of warranted crop varieties, improved pesticides, plant defense mediators are need to be addressed for pest control. Sowing of rice nursery with planting time manipulation in Pakistan, to avoid yellow stem borer (*Scirpophaga incertulas* Walker) disposing off the double seeds of cotton to restrict pink bollworm and timely burning of crop residues and stubbles to prohibit litter dwelling insect pests, are few pest managing legislative examples (Attique et al. 2001). Similarly, on further perspectives, changing the planting geometries and varietal nature can also be helpful for such aspects (Sarwar 2012).

5.5 Biological Diversity Conservations

The case of sustainable pest control should not only assimilate the measures to mitigate or reduce the pest population/pesticides, rather, for instance, with reference to biocontrol agents (de Melo et al. 2018), must involves and encompass the measures to conserve biological diversifications of fascinating world (Ong et al. 2016). Biocontrol agents are the natural non-paid labor and farmer assistant with no proper attention of crop protection community (de Melo et al. 2018). There should be such legislative endeavors to conserve those (Sutherland et al. 2017). The supreme imperative international episode was 'United Nations Conference on Environment and Development (UNCED)', Rio de Janeiro in 1992 leading to scientific documentation title 'the Convention on Biological Diversity, that was later ratified in 1995 by 142 countries around the world with 'Agenda 21', embraces a subdivision/chapter 14 on 'Promoting Sustainable Agriculture and Rural Development'. This compacts utterly with the glitches of pesticide over employment, thus flashing, Integrated Pest Management along with, further, to launch operative and collaborative linkages among farmers, academics and extension personnel lead facilities to uphold IPM task (Dent 2000; Sherman et al. 2017).

6 Sustainable Management Approaches for Insect Pests

Since the first introduction of integrated control term, integrated pest management strategies has increasingly received attention as a practical solution to pests without ecological backlashes (Stern et al. 1959). Sustainable pest management techniques in crop protection emphasizing systematic approaches focusing on preventive and curative methods drawn from a wide array of connotations (Fig. 3). It mostly encompasses physical, agronomic, mechanical and biological principles resorting to selective reduced risk pesticide when addressing situations cannot be effectively managed with other control tactics (Gadanakis et al. 2015). Sustainable management approaches aims to mitigate the input of pesticides and lessen detrimental effects of chemicals on non-target organisms and the environment. Today sustainable management approaches has become a fundamental strategy of sustainable agricultural arthropod pest management in developed and developing countries (Peshin et al. 2009). Moreover, durability of sustainable control approaches relies on the diverse array of solution, rather than repeated use of a single management approach (Barzman et al. 2014).

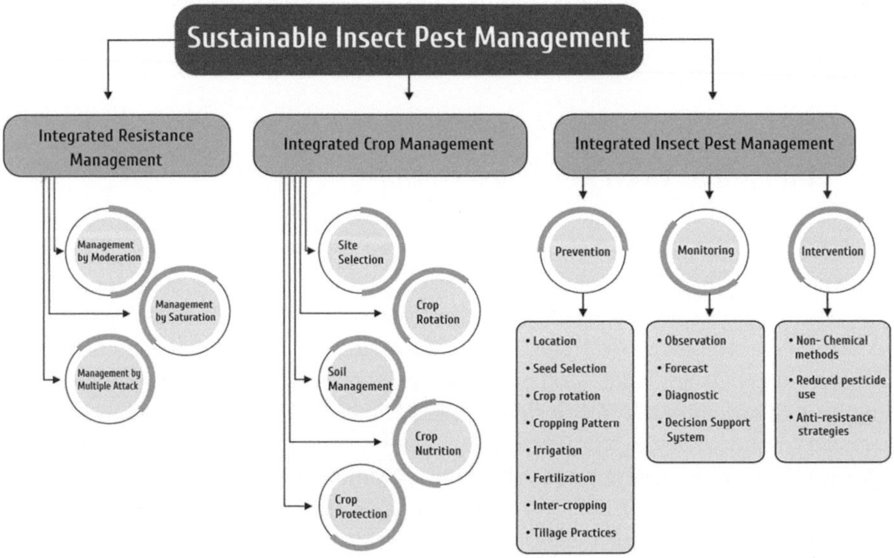

Fig. 3 Sustainable insect pest management techniques (IIPM, IIRM, ICM) in crop protection emphasizing systematic approaches focusing on preventive and curative methods drawn from a wide array of connotations

6.1 *Integrated Insecticide Resistance Management (IIRM)*

Insecticides, being the necessary entities are neither accepted nor denied in sustainable agricultural productive mechanics (Rother 2018). Additively, indiscrimination or undesirability in employments are making theme to be strongly deterred among eco-activists (Joshua 2017; Rowell 2017). But the most prevalent problem of present pest management is accompanied to be linked with the pervasiveness of insecticides so there is some sorted 'tilt at the wind mill' ignoring judicious strategies. No doubt, excessiveness in utilization of pesticides, henceforth creating the insecticidal resistance predicaments (Pedigo and Rice 2014; Rowell 2017), is the dire focal considerations to discourage insecticides but the blunders on behalf of insecticides applicator/farmers need to addressed too (Rother 2018; Sudo et al. 2018). The forum of discussion not only can assist to solve the prevailing resistance issues but pave a glaring way toward Integrated Insecticide Resistance Management (Fig. 4). IRAC (2018) has asserted a well accomplished integrated approach in this regard. This being apparently a single topic, is overwhelmed and encompassed broader ranges of resistance responsible representatives (Fig. 4). These may be either of operational (prolonged exposure/use of single active ingredient insecticide, high lethality/causality pressures, immediate knockdowns, no/zero percent refugia, agri-advisories insufficiencies) or biological ones (monophagy of pest, multi-volatility and high mobility) (Pedigo and Rice 2014; Elahi et al. 2018; Arain et al. 2018; Sudo et al. 2018).

Nowadays, the core theme of focus for management and mitigation of pest/insecticides resistance is incorporating the integrative measures at multiple dimensional strategies (Fig. 4) i.e., inculcation of moderation, saturation and multiple attacks as depicted by Pedigo and Rice (2014).

6.1.1 Mitigation by Moderation

That is basically preventing any paradigm gene shift from vulnerable to resistance genes acquisitions by the insects hence making population more insecticide hazard free (Pedigo and Rice 2014). On the other, conservation and increments of bio-control agents along with natural eco-arena is also achieved under 'moderation' (Arain et al. 2018; Joshua 2017; Rowell 2017). On a broader approach, it uses less dosage, leaving some vulnerable pest population, less persistent agro-chemicals and localized application as the dire flashing points, i.e., prohibition of all extremes (Pedigo and Rice 2014; Rother 2018).

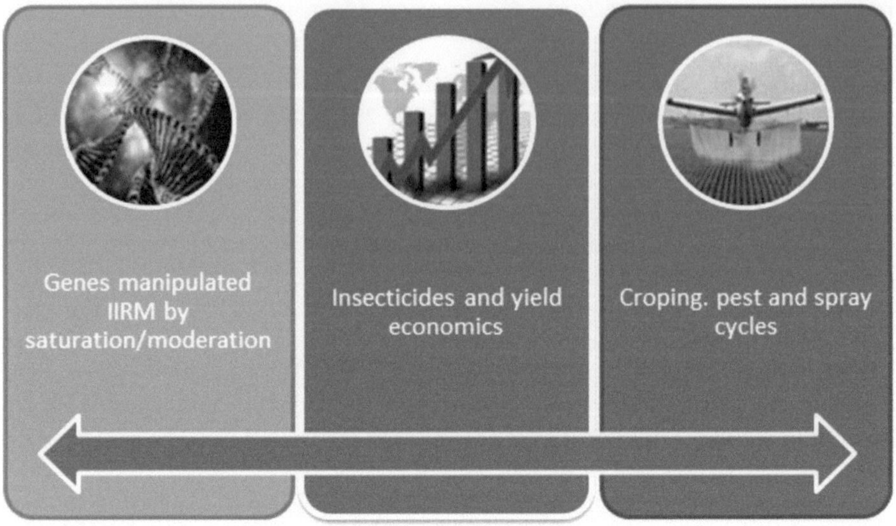

Fig. 4 Schematics of integrated insect resistance management occupying sub/genes level to main/field levels depicting integrative approaches in all aspects

6.1.2 Mitigation by Saturation

Saturation is basically involving intermingling the pest built-in defenses in such a manner that it cannot remain further capable of coping the insecticide exertive pressures (Sudo et al. 2018). This either be achieved by imposing the resistance genes to be in susceptible categories especially by making *R* gene as recessive or may be by repressions of detoxifying in insects, e.g., UDP-glycosyltransferases/metabolized systematizations in Drosophila (Arain et al. 2018) and by the synergism of piperonyl butoxide in spray mixes (Pedigo and Rice 2014).

Mitigation by multiple-attacks: Multiple-invasions in any cases render the target organism to be in confusion, henceforth decline its ability or total collapse to counter-act. The same scheme can be applied on field levels, but on lower intensive scales for resistance integrated management (Pedigo and Rice 2014; Sudo et al. 2018). These may be applying the insecticides in permitted mixtures; compatible insecticide sprays in assortment patterns, and of course, insecticides replacements/alternations with novel modalities (Arain et al. 2018; Sudo et al. 2018).

The other commonly prevailing strategies in normally nominated 'IIRM' escorted by the fundamental theme point of coordinated techniques with furthermore reconciliation of various administration strategies (Fig. 4), by IRAC (2018) and relevant researches, can be elaborated as follows.

6.1.2.1 Economic Considerations on Pest Levels

Option of insecticides utilization as final assertion is only if pests square-up adequately to cause financial calamities/monetary loss (Pedigo and Rice 2014). Otherwise, on the far side retrospect and prospect, the estimation of insecticides abuses not only far reaching but may be aggravated. Aggravation may be apparent as deformed fruits, pre-mature crop evacuations or sever defoliation with burnt phyto-textures along with pest resurgences (Arain et al. 2018; Sudo et al. 2018).

6.1.2.2 Integrated Insecticide Administration Strategies

The fusion of the most critical possible scope of different insecticidal administration techniques (Pedigo and Rice 2014; Rother 2018) with incorporation of the working efficiencies of chemicals/multiple toxins, botanicals and bio-pesticides with simultaneous tendency of parasitoids/entomo-pathogens has always yielded commendable outcomes (IRAC 2018; Sudo et al. 2018). But importantly, the inclination toward eco-friendly tactics must not sacrifice the pest decrements, henceforth must be assimilated with other cultural crop sanitation programs. It is a key crucially to maintain the action of sphere with pesticides timing, pest dynamicity and over-dose despondencies (Pedigo and Rice 2014; IRAC 2018).

Integration of altered insecticide class sprays with cropping/pest cycles: One of the key components of powerful resistance-mitigated program and so forth pest protection framework, is the utilization of pivoted grouping of altered classes synthetic cum novel pesticides (Arain et al. 2018; Rother 2018). Additionally perpetual spray cycles, all through the cropping yield and/or pest cycles, should be rendered active by 1 year or so, rather than focusing on single insecticide/seasonal pest optimized management (Sudo et al. 2018). It is of utmost considerations that pest can naturally integrate them with cropping patterns/cycles but the integration of spray programs must be executed on personal willpower for encouraging pest/resistance predicaments hold under proper agri-advisories and ecological safties (Rowell 2017; Elahi et al. 2018).

6.2 Integrated Crop Management

The development of agriculture production system has advanced into a variety of socio-economic and socio-ecological conditions especially in the developing countries. The wise use of natural resources is the key for the long term sustainable production levels in different cropping systems (Meerman et al. 2008). In the recent years it has become obvious that the pest management requires selective use of control methods for the sustainable agriculture. In the days to come farming system is shifting towards a blend of traditional and modern insect pest management approaches. Integrated pest management is a part of integrated crop management

(ICM) that is not only a form of crop production but a dynamic system that adapts the wise use of latest technologies for agriculture productivity. Integrated crop management is the key in response to farmer's concern and related economic liabilities (Leake 2000). Integrated crop management is a holistic approach that reflects the prevailing conditions on the farm with a consideration of sound economic and environmental factors for the sustainable agriculture. There are several components that involve in the integrated crop management e.g. site selection, crop rotation, soil management, crop nutrition and the crop protection (Kumar and Shivay 2008).

6.2.1 Site Selection

Optimal growing condition is the key component of integrated crop management that provides appropriate plant growth features from seedling onwards. Adjacent crops and environmental conditions should be considered while selecting the location of planting new crop. Moreover considerations should be given to overwintering pests that can move from neighbouring ignored habitats (Pedigo and Rice 2009).

6.2.2 Crop Rotation

Growing different crops in a rotation pattern helps to reduce pest buildup both above and below ground regimens especially when the pest has narrow host range, egg laying before new crop plantation and the damaging stage not very mobile. Rotations have been successful for arthropods that target roots and cannot move out the area to obtain their food requisites. Crop rotations with longer period of time and with the adding of more new crops in the area can better manage the soil pests of several crops (Pedigo and Rice 2009). Furthermore diverse rotation can also reduce the impact of weeds and involves breaking the life cycle of antagonists to keep them below threshold levels which would ultimately require pesticide application (Leake 2000).

6.2.3 Soil Management

Soil is the essential natural resource provides stability, structure and fertility to the crops; that needs to be managed properly and is vital component of any ICM plan. Sometimes erosion caused by various factors (wind, water) triggered unhealthy effects on some soil types and it is quite important to minimize those factors. Measures might be taken by planting permanent grasses however care should be taken while establishing rotation pattern. Moreover alternate ploughing and non-inversion techniques should have been adapted in the rotation while establishing integrated crop management.

6.2.4 Crop Nutrition

Diverse soil types exhibit different amounts of nutrients required for healthy plants. Plant nutrition management is also another key component of ICM that provides necessary nutrients at right time for proper growth of plants. Moreover, planned nutrients inputs are the key to enhance the crop production and maintain the economically and environmentally sound soil fertility for the longer period of time. However care should be taken while fertilizers application that might create unhealthy effects on beneficial fauna of the soil.

6.2.5 Crop Protection

The invasion of pests and other diseases is inevitable in any farming systems. Much can be done under the umbrella of ICM programs for the effective control of damaging pests without disturbing other practices in the holistic management of the farm. One of the essential aspects is the adaptation of prevention strategies as a first line of defense to keep the pests below economic threshold levels. Although severity of pest and disease may varies depending on the agro-ecology, genotype susceptibility, crop growth stage and locality etc. Adaptation of modern crop protection strategies using a combination of cultural, physical biological and chemical methods within the requirements of ICM can play a pivotal role in sustainable agriculture.

6.3 Integrated Insect Pest Management

Man has to compete with the insect pests from the pre-historic days that cause severe economic losses to agricultural crops worldwide. Insect pests are considered a major constraint to achieve global food security and poverty alleviation especially in the developing countries due to lack of management technology. To overcome this problem the use of pesticides for pest management presents additional negative impacts on ecosystem; and it is now clear that alternative holistic control methods needs to be applied for sustainable agriculture. Integrated insect pest management is an important substitute method of rationalizing synthetic pesticides use to avoid or delay pest resurgence and to protect the natural enemies in the agriculture ecosystem (Alastair 2003). Integrated pest management has been used in varied inferences and the term was first used as "integrated control" by Bartlett in 1956 and was further elaborated in 1959 by Stern and his co-workers.

Integrated pest management system is the socioeconomic in the perspective of farming system, the associated environment and the population dynamics of pest species, utilizes all suitable possible techniques to keep pest below economic thresholds (Pretty et al. 2011; Pretty and Bharucha 2014).

Integrated pest management highlights pest problems followed by simultaneous integration of different tactics, the regular monitoring of insect pests and natural enemies and a thresholds assessment for decisions. After the proper identification of pest damage and the responsible pests there are different tactics i.e. cultural, physical, genetic, biological, chemical and regulatory methods to suppress the pests (Ehi-Eromosele et al. 2013). Integration or compatibility of these management practices is the key in integrated pest management programs for sustainable development. Dependency on a single pest management method may have undesirable effects on ecosystem. Moreover reliance on a single management practice might favor pests that can cause resurgence in future cropping system. So for that IPM takes the advantage of using all appropriate management methods in integration with judicial use of pesticides. Integrated pest management practices include; planning, regular monitoring and timely decision that play a key role for sustainable crop production. Some of the important IPM control methods are;

- **Cultural methods** – these are the practices that make the less favorable conditions for pests' establishment, their reproduction, dispersal and survival. Moreover by adjusting crop location, time of sowing or crop rotation and cultivation techniques also destroy their food, shelter, breeding habitats and exposing them to predators play a significant role to keep pest population below threshold levels.
- **Mechanical and physical methods** – using these methods pests can be controlled directly or make the conditions unsuitable for them. The insect pests can be kept away using barriers and traps or physically remove them from the target area. Moreover hot or cold treatments make the environment unsuitable for insect pest developments provide control at key times.
- **Genetic methods** – insect-resistant varieties developed by classical breeding or via genetic engineering suppress pest population or elevate plant tolerance level through insect movement interfering their feeding behavior or reproduction on or in the plant. The resistance can be generated by change in color, thickness of the cell walls or plant tissues, surface wax, trichomes (hairs) or spines etc.
- **Biological methods** – it's a self-perpetuating use of natural enemies including predators, parasites or microbial pathogens to suppress pests for an extended period of time than other methods of pest control. Biological control agents can help to suppress pest populations by competing with the same pest resources or by parasitism or predation of the target pest species.
- **Chemical methods** – chemical control is a cure-all for pest problems and is the best practical and cost effective technique to bring population below economic threshold levels. Regardless of the synthetic pesticides used, there are several considerations like mode of action, delivery, selectivity and resistance should be addressed before application.
- **Regulatory methods** – include all forms of legislation and regulations that might prevent, establish or the entry or spread of pests and restrict the movement from one area to another. Additionally, regulatory control gives growers and producers a short reprieve before invasion of the pest and provides a cushion time for better management.

7 Integration of BC with Insecticide Application

As an alternate for the sustainable pest management in the 21st century, the opportunities and demand for the effective biological control are greater than ever before (Bale et al. 2008). Although, the level of pest suppression using bio-control agents have never been exceeded from 50% (Hall and Ehler 1979; Hall et al. 1980) and the success rates may vary depending on environmental conditions. For the sustainable insect pest management, biological control is more effective when coupled with other integrated pest management (IPM) tactics. Among these strategies, making pesticides more compatible with bio-control agents is one of the key combinations keeping pest population below economic threshold level with least disturbance of ecosystem.

Although pesticides have variety of unpredictable negative impacts on closely related beneficial organisms; however pesticides still remains an integral component of sustainable pest management strategies (Guillebeau 2004). However pesticides can be used in a variety of modified manner (e.g. selective use of active ingredients and formulations, only when economic thresholds dictates, temporal and spatial separation of natural enemies and pesticides, use of lowest effective rates of pesticides) to protect the natural enemies in the ecosystem (Hull and Beers 1985; Poehling 1989; Ruberson et al. 1998). Moreover it is worth to know that natural enemies can recuperate quickly even when broad spectrum pesticides have been used, particularly if they are easily degradable, and recolonization of population in the refuge areas at margins. Differences regarding susceptibility among taxonomically close species have been documented in some studies and even within the same species strains. Adults of *Eretmocerus mundus* Mercet parasitoid were less susceptible to cypermethrin, amitraz and thiodicarb residues compared to *Encarsia formosa* or *E. pergandiella* Howard (Jones et al. 1995). Additionally there are several studies that showed predator/parasitoid tolerance against some insecticides even when there is no indication of resistance in the natural enemies (Guillebeau 2004) resulted unpredictable impact of broad spectrum pesticides on beneficial insects. For instance, *Chrysoperla rufilabris* (Bermeister) adults and larvae showed toxic susceptibility towards organophosphates and carbamates; while pyrethroids were non-toxic to this natural enemy (Mizzell and Schiffhauer 1990). On the contrary, organophosphates were non-toxic to some predatory beetle compared to pyrethroids. There are clear indications of varying impacts on natural enemies within a single group of pesticides. In another study, cypermethrin was recorded less toxic than permethrin to parasitoid; but in case of predators reverse effect was observed (Wright and Verkerk 1995). Furthermore, the sublethal effects of pesticides on natural enemies make more complications. For example, some Braconidae minute wasp females lay fewer viable eggs when exposed to sublethal doses of carbaryl (Grosch 1975). Similarly decreased fecundity has been recoded for some coccinellids after sublethal effects of organophosphates (Parker et al. 1976). Similarly formulations improve the selectivity of pesticides to protect the natural enemies. For example dust formulations that are more toxic to beneficial than powders or emulsifiable concentrates that can cause mortality of some parasitioids even in the absence of pesticides. Irrespective of the chemical used there are few considerations that should be addressed before applications.

- **Mode of action:** generally mode of action needs to be designed so that the pest can be targeted at weaker stage of their life cycle.
- **Delivery:** pesticides application should supremely be multi-disciplinary in order to minimize their impacts on ecosystem and must have target specific.
- **Selectivity:** while making a treatment decision careful selection of pesticide is quite necessary to spare natural enemies. However selectivity sometime differs from specificity that is the ability of a compound results higher mortality for the particular target pests (Fisher et al. 1999).
- **Resistance:** information regarding resistance has been observed in either the pest or natural enemy or the exposure of target pest to a particular pesticide before and the potential for the resistance buildup of in the target pest population against a particular chemical should have been carefully examined.

Making pesticides more compatible with biological control system; placement and careful timing of pesticide application can minimize the pesticide contact with natural enemies. Additionally selective treatment with non-persistent pesticides can limit overall negative effect on beneficial population of insects in the cropping ecosystem and these types of pesticides should also be considered in integrated pest management programs.

8 Use of Insect Behavior Modifiers

The natural phenomena regarding an organism's behavior in response to external or internal stimuli released by other organisms of the same or other species (or different phylogeny) play a key role in insect-plant interactions. Eco-physiological, biochemical and behavioral processes involve in insect plant interactions in which secondary metabolites play a significant role. For their development insects have adapted to these phytochemicals; using them as cues for host recognition or other biological activities. Change in structural diversity of plants with the passage of time has resulted in synthesize of substances like, phenolic compounds, non-protein aminoacids, terpenes, alkaloids and flavonoids that ultimately led to behavioral and biochemical adaptations in insects towards plants. Primarily most of the insects rely on olfactory receptors to contact with the external environment (Krieger and Breer 1999). Insect's attraction towards plants or other host organisms involves the recognition of specific semiochemicals (Fig. 5) or specific ratios of these compounds (Bruce et al. 2005). For the sustainable management of insect pests, there is an opportunity to develop interventions using semiochemicals that influence the behavior of noxious insect pests in agriculture, forestry, horticulture, stored food products and insect vectors of several diseases. Semiochemicals are signaling chemicals naturally produced by insects that transmit chemical messages. Semiochemicals provide environmentally safe, non-toxic and species specific alternative solutions for pest management in different cropping systems. Moreover, semiochemical's role as pest repellents and natural enemies attractant can be helpful keeping pest populations below economic threshold levels without harming agro-ecosystem. Semiochemicals can be classified into allelochmeicals which have interspecific interactions and pheromones with intraspecific interactions (Fig. 5). Furthermore allelochemicals divided into allomones signals in which emitting species benefits, kairomones when receptor species get benefit and synomones when both species have the advantage (Nordlund et al. 1981). Antimone are harmful for both emitter and the receiver, e.g. non-host chemicals arrest parasitic wasps (honeybees); while in case of apneumone chemical signals from nonliving sources like salt and there is no benefit damage to the emitter.

One of the most widespread and successful application of semiochemicals is the detection and monitoring of pest populations (Witzgall et al. 2010) to justify the pesticide use before exceeding the economic thresholds. Pest sampling/scouting is always a laborious and expensive especially on large scale areas. In this regards sex pheromones are key and have the potential to suppress or eradicate low density populations and are effective for tracking invasive species in the establishment phase (El-Sayed et al. 2006; Liebhold and Tobin 2008). Sex pheromones based threshold action was taken first time by monitoring pea moth *Cydia nigricana* in England from 1980–1985 (Wall et al. 1987). Similarly, a pheromone-based monitoring system was developed at Rothamsted, UK for the orange wheat blossom midge (OWBM), *Sitodiplosis mosellana*; a serious pest of wheat in Northern Hemisphere causing severe losses to crop yield (Bruce et al. 2007). In another studies pheromone based field monitoring for Agriotes spp. have been successfully conducted in Europe and North America (Vernon and Toth 2007; Toth et al. 2008; Sufyan et al. 2013). Apart from pest monitoring, pheromone-based mass trapping,

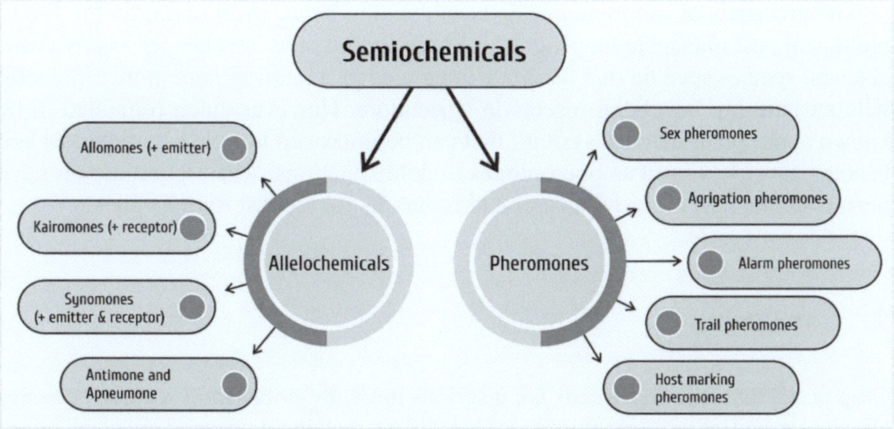

Fig. 5 Schematic diagram of different types of semiochemicals (allelochmeicals and pheromones) which can be exploited for the management of different insect pests in different crops

mating disruption and lure and kill techniques has been applied in the integrated pest management programs successfully for several decades. From the last four decades more than 200 studies have been conducted for mass trapping of pests from Lepidoptea, Homoptera, Diptera and Coleoptera (El-Sayed 2012; Alpizar et al. 2012). Some of the most successful mass trapping attempts of pest management against *Leucinodes orbonalis* Guenee (Cork et al. 2001, 2003, 2005), *Dendroctonus* spp. and *Ips duplicatus* Sahlberg (Silverstein et al. 1968; Schlyter et al. 2001) and *Ephestia kuehniella* Zeller (Trematerra and Gentile 2010) has been recoded with a significant reduction of pesticides consumption and resulted significant increase in crop yield. The blend of mass trapping with insecticides composites into "lure and kill" or attracticides (Jones 1998); the approach have been used in integrated pest management programs for the last several decades documented in numerous studies (El-Sayed et al. 2009). Similarly attract and kill technique has been successfully employed for the management of cotton boll weevil (*Anthonomus grandis*) in USA and South America on several thousand hectares (Ridgeway et al. 1990; Smith 1998), fruit flies *Bacterocera* spp. in USA (Cunningham et al. 1990; Hee and Tan 2004; Vargas et al. 2010; El-Sayed et al. 2009). Correspondingly pheromone mediated mating disruption is also an alternate sustainable pest management strategy by disrupting chemical communication among organism that reduce chances of organisms reproduction and ultimately reduce future pest population. The area under mating disruption has been significantly increased from the last 2–3 decades with 770,000 ha globally in 2010 (Ioriatti et al. 2011; Witzgall et al. 2010). Mating disruption is more effective on large areas because the large areas permeated with synthetic pheromones reduce the impact of gravid females to immigrate into treated areas. A wide range area under the infestation of gypsy moth (Lymantria dispar) in North America forests, codling moth (*Cydia pomonella*) in apple trees globally and the grape vine moth (*Lobesia botrana*) in the EU and Chile grapes fields has been managed by permeation of synthetic sex pheromones in the infested areas (Witzgall et al. 2010).

The prospects of semiochemicals considered to be an encouraging in the future biological pest management programs. Since pheromones are cheaper, easily available and species specific that facilitate integrated pest management more efficiently without harming beneficial insects in agriculture. However much remained to be known about plant defense system; that can be improved in future by breeders and become as widely used as pheromones in field situations; where pheromones have been essential part of pest management programs for the last four decades.

9 Conclusion

Crop pests, diseases and weeds are a serious threat to global food security, poverty alleviation and other agricultural products. The sustainable management of insect pests is quite challenging due to interference of some biotic and abiotic factors especially the climate change and global warming. The extent of losses due to insect pests is expected to increase in future due to changes in crop diversity, pest types and changing environmental conditions. Moreover, the changing environment also interferes with the normal functioning of various pest management strategies like host-plant resistance, biological control methods and chemical control.

Sustainable insect pest management strategies including integrated pest management and integrated crop management is much more than just a simple resource-conserving technology.

A successful adaptation of IPM plan against certain insect pests also accounts for the protection of beneficial insects, secondary pest outbreaks, pest resurgence and ecological backlashes. Moreover, the use of non-chemical control methods based on the philosophy of integration of indigenous natural enemies with other biological control techniques to partially replace the synthetic chemicals is worth considering. Although non-chemical control methods of insect pests can be utilized for longer period of time; however these measures may be insufficient to manage the outbreaks of migratory pests. Therefore, integration of non-chemical control methods with synthetic pesticides will be a promising option for sustainable insect pest management.

Moreover the use of therapeutic tools (biological, chemical, physical, mechanical and resistance management) are considered primary means of regulating pests rather than as occasional supplements to keep them below economic thresholds. Additionally the focus should be on the development of farming practices that are more compatible with ecological systems and cropping patterns that naturally limit an organism to attain pest status. Furthermore, there is need to understand and address pest management issues by keeping on board the other crop producing stakeholders for the sustainable insect pest management.

References

Ahmad M, Akhtar S (2013) Development of insecticide resistance in field populations of *Brevicoryne brassicae* (Hemiptera: Aphididae) in Pakistan. J Econ Entomol 106:954–958

Ahmad M, Farooq U (2010) The state of food security in Pakistan: Future challenges and coping strategies. Pak Dev Rev:903–923

Ahmad M, Sayyed AH, Saleem MA, Ahmad M (2008) Evidence for field evolved resistance to newer insecticides in *Spodoptera litura* (Lepidoptera: Noctuidae) from Pakistan. Crop Prot 27:1367–1372

Alam MA (2010) Encyclopedia of Applied Entomology. Anmol Publications Pvt. Ltd, New Delhi

Alastair O (2003) Integrated Pest Management for Resource-Poor African Farmers: Is the Emperor Naked? World Dev 31:831–845

Alberto D, Serra AA, Couée I, Sulmon C, Pateyron S, Gouesbet G (2016) Signalling and regulation pathways involved in cryptic effects induced by residual levels of pesticide degradation products on higher plants. In: SETAC Europe Meeting

Alford DV (2000) Pest and disease management handbook. Wiley, New York

Allison SD, Martiny JB (2008) Resistance, resilience, and redundancy in microbial communities. Proc Natl Acad Sci 105:11512–11519

Alpizar D, Fallas M, Oehlschlager AC, Gonzalez LM (2012) Management of cosmopolites *sordidus* and *Metamasius hemipterus* in banana by pheromone-based mass trapping. J Chem Ecol 38:245–252

Altbach PG (2009) Peripheries and centers: Research universities in developing countries. As Pac Educat Rev 10(1):15–27

Aluja M, Guillén L, Rull J, Höhn H, Frey J, Graf B, Samietz J (2011) Is the alpine divide becomingmore permeable to biological invasions? Insights on the invasion and establishment of the Walnut Husk Fly, Rhagoletis complete (Diptera: Tephritidae) in Switzerland. Bull Entomol Res 101:451–465

Amalin DM, Peña JE, Duncan R, Leavengood J, Koptur S (2009) Effects of pesticides on the arthropod community in the agricultural areas near the Everglades National Park. Pro Florida State Hort Soc 122:429–437

Angulo-Sandoval P, Fernandez-Marin H, Zimmerman JK, Alde TM (2004) Changes in patterns of understory leaf phenology and herbivory following hurricane damage. Biotropica 36:60–67

Arain MS, Shakeel M, Elzaki MEA, Farooq M, Hafeez M, Shahid MR, Li GQ (2018) Association of detoxification enzymes with butene-fipronil in larvae and adults of Drosophila melanogaster. Environ Sci Poll Res 25:1–8

Attique M, Ahmad M, Ahmad Z, Rafiq M (2001) Sources of carry-over and possibilities of cultural control of *Pectinophora gossypiella* (Saunders) in the Punjab, Pakistan. Crop Prot 20(5):421–426

Bale JS, Masters GJ, Hodkinson ID, Awmack C, Bezemer TM, Brown VK, Butterfield J, Buse A, Coulson JC, Farrar J (2002) Herbivory in global climate change research: direct effects of rising temperature on insect herbivores. Global Change Biol 8:1–16

Bale JS, van Lenteren JC, Bigler F (2008) Biological control and sustainable food production. Philos Trans R Soc B 363:761–776. https://doi.org/10.1098/rstb.2007.2182

Bartlett BR (1956) Natural predators. Can selective insecticides help to preserve biotic control? Agric Chem 11(2):42–44

Barzman MS, Bertschinger L, Dachbrodt-Saaydeh S, Graf B, Jensen JE, Jorgensen LN, Kudsk P, Messéan A, Moonen AC, Ratnadass A, Sarah JL, Sattin M (2014) IPM policy, research and implementation: European initiatives. In: Peshin R, Pimentel D (eds) Integrated pest management, experiences with implementation, global overview, vol 4. Springer, London, pp 415–428

Basit M, Sayyed AH, Saleem MA, Saeed S (2011) Cross-resistance, inheritance and stability of resistance to acetamiprid in cotton whitefly, *Bemisia tabaci* Genn (Hemiptera: Aleyrodidae). Crop Prot 30(6):705–712

Bass C, Denholm I, Williamson MS, Nauen R (2015) The global status of insect resistance to neonicotinoid insecticides. Pestic Biochem Physiol 121:78–87

Black R (2017) CAREC common agenda for SPS modernization. In: C5+1 Expert-level meeting on regional phytosanitary cooperation, 24 August 2017, Dushanbe, Tajikistan

Boobis AR, Ossendorp BC, Banasiak U, Hamey PY, Sebestyen I, Moretto A (2008) Cumulative risk assessment of pesticide residues in food. Toxicol Lett 180(2):137–150

Bradshaw WE, Holzapfel CM (2011) Genetic shift in photoperiodic response correlated with global warming. Proc Natl Acad Sci U S A 98:14509–14511

Bruce TJA, Wadhams LJ, Woodcock CM (2005) Insect host location: a volatile situation. Trends Plant Sci 10:269–274

Bruce TJ, Hooper AM, Ireland L, Jones OT, Martin JL, Smart LE, Oakley J, Wadhams LJ (2007) Development of a pheromone trap monitoring system for orange wheat blossom midge, *Sitodiplosis mosellana* in the UK. Pest Manag Sci 63:49–56

Brunsø K, Fjord TA, Grunert KG (2002) Consumers' food choice and quality perception. The Aarhus School of Business Publications, Aarhus

Bryan BA, Grandgirard A, Ward JR (2010) Quantifying and exploring strategic regional priorities for managing natural capital and ecosystem services given multiple stakeholder perspectives. Ecosystems 13:539–555

Buffo E, Battisti A, Stastny M, Larsson S (2007) Temperature as a predictor of survival of the pine processionary moth in the Italian Alps. Agric Entomol 9:65–72

Buurma JS (2008) Stakeholder involvement in crop protection policy planning in the Netherlands. ENDURE – RA3.5/SA4.5 Working paper. LEI Wageningen UR, The Hague, The Netherland.

Chakraborty S (2013) Migrate or evolve: options for plant pathogens underclimate change. Glob Chang Biol 19:1985–2000

Chakraborty S, Newton AC (2011) Climate change, plant diseases and food security: an overview. Plant Pathol 60:2–14

Cockbain A (1961) Low temperature thresholds for flight in *Aphis fabae* Scop. Entomol Exp Appl 4:211–219

Collins SA (1994) Strategic management in the development of integrated pest and disease management programmes. In: Proceedings 1994 British crop protection conference – pests and diseases. British crop protection council, Farnham, UK, pp 115–124

Constable JVH, Litvak ME, Greenberg JP, Monson RK (1999) Monoterpene emission from coniferous trees in response to elevated CO_2 concentrations and climate warming. Glob Chang Biol 5:255–267

Cork A, Alam SN, Das A, Das CS, Ghosh GC, Phythian S, Farman DI, Hall DR, Maslen NR, Vedham K, Rouf FMA, Srinivasan K (2001) Female sex pheromone of brinjal fruit and shoot borer, *Leucinodes orbonalis* (Lepidoptera: Pyralidae): blend optimization. J Chem Ecol 27:1867–1877

Cork A, Alam SN, Rouf FMA, Talekar NS (2003) Female sex pheromone of brinjal fruit and shoot borer, *Leucinodes orbonalis* (Lepidoptera: Pyralidae) trap optimization and application in IPM trials. Bull Entomol Res 93:107–113

Cork A, Alam SN, Rouf FMA, Talekar NS (2005) Development of mass trapping technique for control of brinjal shoot and fruit borer, *Leucinodes orbonalis* (Lepidoptera: Pyralidae). Bull Entomol Res 95:589–596

Cunningham RT, Kobayashi RM, Miyashita DH (1990) The male lures of tephritid fruit flies. In: Ridgway RL, Silverstein RM, Inscoe MN (eds) Behavior modifying Chemicals for Insect Management: Applications of Pheromones and Other Attractants. Marcel Dekker, New York, pp 113–129

Damalas CA, Eleftherohorinos IG (2011) Pesticide exposure, safety issues, and risk assessment indicators. Int J Environ Res Public Health 8:1402–1419

Davies T, Field L, Williamson M (2012) The re-emergence of the bed bug as a nuisance pest: implications of resistance to the pyrethroid insecticides. Med Vet Entomol 26:241–254

de Faria MR, Wraight SP (2007) Mycoinsecticides and mycoacaricides: a comprehensive list with worldwide coverage and international classification of formulation types. Biol Control 43:237–256

de Melo AG, Levesque S, Moineau S (2018) Phages as friends and enemies in food processing. Curr Opin Biotechnol 49:185–190

Dent D (2000) Insect pest management. CABI Bioscience, Wallingford

Dermody O, O'Neill BF, Zangerl AR, Berenbaum MR, DeLucia EH (2008) Effects of elevated CO_2 and O_3 on leaf damage and insect abundance in a soybean agroecosystem. Arhtrop Plant Int 2:125–135

Desprez-Loustau ML, Robin C, Bue'e M, Courtecuisse R, Garbaye J, Suffert F, Sache I, Rizzo D (2007) The fungal dimension of biological invasions. Trends Ecol Evol 22:472–480

Dhaliwal GS, Arora R (2003) Principles of Insect Pest Management, 2nd edn. Kalyani Publishers, Ludhiana, pp 90–94

Dhaliwal GS, Singh R, Chhillar BS (2006) Essentials of Agricultural Entomology. Kalyani Publishers, Ludhiana/New Delhi

Dhaliwal GS, Dhawan AK, Singh R (2007) Biodiversity and ecological agriculture: Issues and perspectives. Indian J Ecol 34:100–109

Dhaliwal G, Jindal V, Dhawan A (2010) Insect pest problems and crop losses: changing trends. Indian J Ecol 37:1–7

Dowell RV, Gill R (1989) Exotic invertebrates and their effects on California. Pan-Pac Entomol 65:132–145
DPP (2014a) Pakistan agricultural pesticide ordinance/act. Department of Plant Protection (Edition). Ministry of National Food Security & Research. Government of Pakistan
DPP (2014b) Pakistan plant quarantine act 1976. Department of Plant Protection (Edition). Ministry of National Food Security & Research. Government of Pakistan
Dukes JS, Mooney HA (1999) Does global change increase the success of biological invaders? Trends Ecol Evol 14:135–139
Ehi-Eromosele CO, Nwinyi OC, Ajani OO (2013) Integrated Pest Management. In: Sonia S, Marcelo L (eds) Weed and Pest Control – Conventional and New Challenges, pp 105–115
Elahi E, Abid M, Zhang L, ul Haq S, Sahito JGM (2018) Agricultural advisory and financial services; farm level access, outreach and impact in a mixed cropping district of Punjab, Pakistan. Land Use Policy 71:249–260
El-Sayed AM (2012) The pherobase: database of insect pheromones and semiochemicals. http://www.pherobase.com. Accessed 15 Oct 2012
El-Sayed AM, Suckling DM, Wearing CH, Byers JA (2006) Potential of mass trapping for long-term pest management and eradication of invasive species. J Econ Entomol 99:1550–1564
El-Sayed AM, Suckling DM, Byers JA, Jang EB, Wearing CH (2009) Potential of "lure and kill" in long-term pest management and eradication of invasive species. J Econ Entomol 102:815–835
Faccoli M (2007) Breeding performance and longevity of *Tomicus destruens* on mediterranean and continental pine species. Entomol Exp Appl 123:263–269
FAO/WHO (1996) Biotechnology and food safety. Report of a joint fao/who consultation. FAO food and nutrition paper 61. Food and Agriculture Organisation of the United Nations, Rome, Italy.
Feder G, Savastano S (2006) The role of opinion leaders in the diffusion of new knowledge: The case of integrated pest management. World Dev 34:1287–1300
Fisher TW, Bellows TS, Caltagirone LE, Dahlsten DL, Huffaker CB, Gordh G (1999) Handbook of biological control: principles and applications of biological control. Academic, San Diego
Foley JA, Ramankutty N, Brauman KA, Cassidy ES, Gerber JS, Johnston M, Mueller ND, O'Connell C, Ray DK, West PC, Balzer C, Bennett EM, Carpenter SR, Hill J, Monfred C, Polasky S, Rockström J, Sheehan J, Siebert S, Tilman D, Zaks DPM (2011) Solutions for a cultivated planet. Nature 478:337–342
Freeman RE (2010) Strategic management: a stakeholder approach. Cambridge University Press, Cambridge, MA
Gadanakis Y, Bennett R, Park J, Areal FJ (2015) Evaluating the sustainable intensification of arable farms. J Environ Manag 150:288–298
Geier PW (1966) Management of insect pests. Annu Rev Entomol 11:471–490
Geiger F, Bengtsson J, Berendse F, Weisser WW, Emmerson M, Morales MB, Ceryngier P, Liira J, Tscharntke T, Winqvist C (2010) Persistent negative effects of pesticides on biodiversity and biological control potential on European farmland. Basic Appl Ecol 11:97–105
Ghananand T, Prasad CS, Lok N (2011) Effect of insecticides, bio-pesticides and botanicals on the population of natural enemies in brinjal ecosystem. Vegetos- Int J Plant Res 24:40–44
Gill HK, Garg H (2014) Pesticide: Environmental impacts and management strategies. In: Solenski S, Larramenday ML (eds) Pesticides—toxic effects. Intech, Rijeka, pp 187–230
Grosch DS (1975) Reproductive performance of *Bracon hebetor* after sublethal doses of carbaryl. J Econ Entomol 68:659–662
Grunert KG (2005) Food quality and safety: consumer perception and demand. Eur Rev Agric Econ 32:369–391
Guillebeau P (2004) The pesticide paradox in IPM: risk-benefit analysis. In: Koul O (ed) Integrated Pest Management: Potential, Constraints and Challenges. CABI Publishing, Cambridge, MA, pp 169–184
Gullan PJ, Cranston PS (2014) The insects: an outline of entomology. Wiley, Chichester

Hagen SB, Jepsen JU, Ims RA, Yoccoz NG (2007) Shifting altitudinal distribution of outbreak zones of winter moth, *Operophtera brumata* in sub-arctic birch forest a response to global warming? Ecography 30:299–307
Hall RW, Ehler LE (1979) Rate of establishment of natural enemies in classical biological control. Bull Entomol Soci America 25:280–282
Hall J, Matos S (2010) Incorporating impoverished communities in sustainable supply chains. Int J Phys Distrib Logist Manag 40:124–147
Hall RW, Ehler LE, Bisabri-Ershadi B (1980) Rate of success in classical biological control of arthropods. Bull Entomol Soci America 26:111–114
Hallman GJ (2017) Process control in phytosanitary irradiation of fresh fruits and vegetables as a model for other phytosanitary treatment processes. Food Control 72:372–377
Hamann A, Wang T (2006) Potential effects of climate change on ecosystem and tree species distribution in British Columbia. Ecology 87:2773–2786
Haya N, Baessler K, Christmann-Schmid C, De Tayrac R, Dietz V, Guldberg R, Mascarenhas T, Nussler E, Ballard E, Ankardal M (2015) Prolapse and continence surgery in countries of the organization for economic cooperation and development in 2012. Am J Obstet Gynecol 212:755.e1–755.e27
He YX, Weng YQ, Huang J, Liang SZ, Lin JG, Wu DD (2007) Insecticide resistance of *Bemisia tabaci* field populations. J Appl Ecol 18:1578–1582
Hee AKW, Tan KH (2004) Male sex pheromonal components derived from methyl eugenol in the haemolymph of fruit fly *Bactrocera papaya*. J Chem Ecol 30:2127–2138
Herron GA, Wilson JL (2011) Neonicotinoid resistance in *Aphis gossypii* Glover (Aphididae: Hemiptera) from Australian cotton. Aust Entomol 50:93–98
Hillstrom ML, Lindroth RL (2008) Elevated atmospheric carbon dioxide and ozone alter forest insect abundance and community composition. Insect Conserv Div 1:233–241
Hopper JAS (2016) The regulation that comes from combination: the effects of combining multiple policy areas within a single environmental protection agency in the states. Ph.D. Dissertation. Department of Plitical Science, University of Missouri – Columbia
Hossard L, Philibert A, Bertrand M, Colnenne-David C, Debaeke P, Munier-Jolain N, Jeuffroy MH, Richard G, Makowski D (2014) Effects of halving pesticide use on wheat production. Sci Rep 4:4405
Hull LA, Beers EH (1985) Ecological selectivity: modifying chemical control practices to preserve natural enemies. In: Hoy MA, Herzog DC (eds) Biological Control in Agricultural IPM Systems. Academic, Orlando, pp 103–112
Hurley BP, Garnas J, Wingfield MJ, Branco M, Richardson DM, Slippers B (2016) Increasing numbers and intercontinental spread of invasive insects on eucalypts. Biol Invasions 18:921–933
Hygnstrom SE, Timm RM, Curtis PD, Nolte DL, Tobin ME, VerCauteren KC (2014) Prevention and control of wildlife damage. Univerisity of California, Davis, pp 440–442
Inayatullah C (1995) Training manual: integrated insect pest management. Entomological Research Laboratories and NARC Training Institute, Islamabad
Ioriatti C, Anfora G, Tasin M, De Cristofaro A, Witzgall P, Lucchi A (2011) Chemical ecology and management of *Lobesia botrana* (Lepidoptera: Tortricidae). J Econ Entomol 104:1125–1137
IRAC (2013) Resistance management for sustainable agriculture and improved public health. http://www.irac-online.org/. Accessed 12 Mar 2018
IRAC (2018) Insecticide resistance action committee, IRAC. http://www.irac-online.org. Accessed 23 Apr 2018
Jan MT, Abbas N, Shad SA, Saleem MA (2015) Resistance to organophosphate, pyrethroid and biorational insecticides in populations of spotted bollworm, *Earias vittella* (Fabricius) (Lepidoptera: Noctuidae), in Pakistan. Crop Prot 78:247–252
Javed MW (2016) Study on induced resistance based defense genes expression to control consequential insect pests of tomato (*Lycopersicon esculentum*). MSc. (Hons.) Thesis, Department of Entomology, University of Agriculture Faisalabad

Jha LK (2010) Applied agricultural entomology. New Central Book Agency (Pvt.) Ltd, Kolkata/Pune/Delhi
Jones OT (1998) The commercial exploitation of pheromones and other semiochemicals. Pestic Sci 54:293–296
Jones WA, Wolfenbarger DA, Kirk AA (1995) Response of adult parasitoids of *Bemisia tabaci* (Homoptera: Aleyrodidae) to leaf residues of selected cotton insecticides. Entomophaga 40:153–162
Jonsson A, Appelberg G, Harding S, Barring L (2009) Spatio-temporal impact of climate change on the activity and voltinism of the spruce bark beetle, *Ips typographus*. Glob Chang Biol 15:486–499
Joshua J (2017) The Environmental Effects of Overconsumption. In: Jashua J (ed) The Economics of Addictive Behaviours, vol IV. Palgrave Macmillan, Cham, pp 67–73
Kamata N, Esaki K, Kato K, Igeta Y, Wada K (2002) Potential impact of global warming on deciduous oak dieback caused by ambrosia fungus *Raffaelea* sp. carried by ambrosia beetle *Platypus quercivorus* (Coleoptera: Platypodidae) in Japan. Bull Entomol Res 92:119–126
Kandil MA, Abdallah IS, Abou-Yousef HM, Abdallah NA, Fouad EA (2017) Mechanism of resistance to pirimicarb in the cowpea aphid *Aphis craccivora*. Crop Prot 94:173–177
Karuppaiah V, Sujayanad GK (2012) Impact of climate change on population dynamics of insect pests. World J Agric Sci 8:240–246
Klapwijk MJ, Grobler C, Ward K, Wheeler D, Lewis O (2010) Influence of experimental warming and shading on host-parasitoid synchrony. Glob Chang Biol 16:102–112
Knipling (1979) The basic principles of insect population suppression and management, Agriculture handbook no. 512. US Government Printing Office, Department of Agriculture, Washington, DC
Koptur S, Rodriguez C, Oberbauer SF, Weekley C, Herndon A (2002) Herbivore-free time? Damage to new leaves of woody plants after Hurricane Andrew. Biotropica 34:547–554
Kranthi KR, Jadhav DR, Kranthi S, Wanjari RR, Ali SS, Russell DA (2002) Insecticide resistance in five major insect pests of cotton in India. Crop Prot 21(6):449–460
Krieger J, Breer H (1999) Olfactory reception in invertebrates. Science 286:720–723
Kriticos DJ, Ota N, Hutchison WD, Beddow J, Walsh T, Tay WT, Borchert DM, Paula-Moreas SV, Czepak C, Zalucki MP (2015) The potential distribution of invading *Helicoverpa armigera* in North America: is it just a matter of time? PLoS One 10:e0119618. https://doi.org/10.1371/journal.pone.0119618
Kumar D, Shivay YS (2008) Integrated Crop Management. Indian Agricultural Research Institute, New Delhi
Kumar A, Jandial VK, Parihar SBS (2007) Efficacy of different insecticides against mustard aphid, *Lipaphis erysimi* (Kalt.) on mustard under field conditions. Int J Agric Sci 3:90–91
Kumar R, Kranthi S, Nitharwal M, Jat SL, Monga D (2012) Influence of pesticides and application methods on pest and predatory arthropods associated with cotton. Phytoparasitica 40:417–424
Lamichhane JR, Barzman M, Booij K, Boonekamp P, Desneux N, Huber L, Kudsk P, Langrell SRH, Ratnadass A, Ricci P, Sarah JL, Messéan A (2015) Robust cropping systems to tackle pests under climate change: A review. Agron Sustain Dev 35:443–459
Lamine C (2011) Transition pathways towards a robust ecologization of agriculture and the need for system redesign. Cases from organic farming and IPM. J Rural Stud 27:209–219
Lazpoulos FL, Van Camp W (2016) Pitfalls of the Food Safety Modernization Act: Enhanced Regulation, Minimal Consumer Benefit, and Zero Tolerance Levels for Naturally-Occurring Trace Pathogens. U Miami Inter-Am L Rev 48:13
Leak A (2000) The development of integrated crop management in agricultural crops: comparisons with conventional methods. Pest Manag Sci 56:950–953
Li X, Huang Q, Yuan J, Tang Z (2007) Fipronil resistance mechanisms in the rice stem borer, *Chilo suppressalis* Walker. Pestic Biochem Physiol 89(3):169–174
Liebhold AM, Tobin PC (2008) Population ecology of insect invasions and their management. Annu Rev Entomol 53:387–408

Lietti MM, Botto E, Alzogaray RA (2005) Insecticide resistance in argentine populations of Tuta absoluta (Meyrick)(Lepidoptera: Gelechiidae). Neotrop Entomol 34:113–119

Liu D, Trumble JT (2007) Comparative fitness of invasive and native populations of the potato psyllid (Bactericera cockerelli). Entomol Exp Appl 123:35–42

Logan JA, Powell JA (2001) Ghost forests, global warming and the mountain pine beetle. Am Entomol 47:160–173

Luo C, Jones C, Devine G, Zhang F, Denholm I, Gorman K (2010) Insecticide resistance in *Bemisia tabaci* biotype Q (Hemiptera: Aleyrodidae) from China. Crop Prot 29:429–434

Mansoor MM, Afzal MBS, Basoalto E, Raza ABM, Banazeer A (2016) Selection of bifenthrin resistance in cotton mealybug *Phenacoccus solenopsis* Tinsley (Homoptera: Pseudococcidae): Cross-resistance, realized heritability and possible resistance mechanism. Crop Prot 87:55–59

Marris E (2008) More crop per drop: farmers' yields in the developing world are often limited by unreliable rains. Improving their harvests will require plant breeders, agronomists and geneticists to pull together--but can these experts work out their differences? Nature 452:273–278

Medvedev S, Topaj A, Badenko V, Terleev V (2015) Medium-term analysis of agroecosystem sustainability under different land use practices by means of dynamic crop simulation. In: International symposium on environmental software systems. Springer, New York, pp 252–261

Meerman F, Van De Ven GWJ, Van Keulen H, Breman H (2008) Integrated crop management: An approach to sustainable agricultural development. Int J Pest Manag 42:13–24

Miller JS (2015) Multi-agency revision and development of a standardized template for assessing occupational and residential risks of pesticide exposure through fumigation and spray drift pathways. Master of Public Health Student Report. Kansas State University, USA

Mironidis GK, Kapantaidaki D, Bentila M, Morou E, Savopoulou-Soultani M, Vontas J (2013) Resurgence of the cotton bollworm *Helicoverpa armigera* in northern Greece associated with insecticide resistance. Insect Sci 20:505–512

Mittinty M, Whittle P, Burgman M, Mengersen K (2015) The Role of Surveillance in Evaluating and Comparing International Quarantine Systems. In: Jarrad F, Low-Choy S, Mengersen K (eds) Biosecurity Surveillance: Quantitative Approaches, p 137

Mizell RF, Schiffhauer DE (1990) Effects of pesticides on pecan aphid predators *Chrysoperla rufilabris* (Neuroptera: Chrysopidae), *Hippodamia convergens*, *Cycloneda sanguinea* (L.), *Olla v-nigrum* (Coleoptera: Coccinellidae), and *Aphelinus perpallidus* (Hymenoptera: Encyrtidae). J Econ Entomol 83:1806–1812

Moore B, Allard G (2008) Climate change impacts on forest health. Forest health and biosecurity working paper FBS/34E. Rome, FAO

Morse S, Buhler W (1997) IPM in developing countries: the danger of an ideal. Integr Pest Manag Rev 2:175–185

Napolitano F, Girolami A, Braghieri A (2010) Consumer liking and willingness to pay for high welfare animal-based products. Trends Food Sci Technol 21:537–543

Nauen R, Denholm I (2005) Resistance of insect pests to neonicotinoid insecticides: current status and future prospects. Arch Insect Biochem Physiol 58:200–215

Nauen R, Elbert A (2003) European monitoring of resistance to common classes of insecticides in *Myzus persicae* and *Aphis gossypii* (Homoptera: Aphididae) with special reference to imidacloprid. Bull Entomol Res 93:47–54

Needham PH, Sawicki RM (1971) Diagnosis of resistance to organophosphorus insecticides in *Myzus persicae* (Sulz.). Nature 230:125–126

Nordlund DA, Jones RL, Lewis WJ (1981) Semiochemicals: their role in pest control. John Wiley and Sons, New York

Norris RF, Caswell-Chen EP, Kogan M (2002) Concepts in integrated pest management. Prentice-Hall of India Private Limited, New Delhi

O'Neill GA (2008) Assisted migration to address climate change in British Columbia: recommendations for interim seed transfer standards. Ministry of Forests and Range, Forest Science Program, Victoria

OECD (1993) Safety evaluation of foods produced by modern biotechnology: concepts and and principles. Organization of Economic Co-operation and Development, OECD, Paris, pp 339–347

Oerke EC (2006) Crop losses to pests. J Agric Sci 144:31–43

Ong B, Lye LH, Chun J (2016) Biological diversity conservation laws in South East Asia and Singapore: a regional approach in pursuit of the United Nations' sustainable development goals? Asia Pac J Environ Law 19:14–23

Pachauri RK, Reisinger A (2007) Climate change 2007: synthesis report. Contribution of working groups I, II and III to the fourth assessment report of the intergovernmental panel on climate change. Geneva (IPCC), Switzerland

Pannell DJ, Marshall GR, Barr N, Curtis A, Vanclay F, Wilkinson R (2006) Understanding and promoting adoption of conservation practices by rural landholders. Aust J Exp Agric 46:1407–1424

Parker SC (2009) The economics of entrepreneurship. Cambridge University Press, Cambridge

Parker BL, Ming NS, Peng TS, Singh G (1976) The effect of malathion on fecundity, longevity and geotropism of *Menochilus sexmaculatus*. Environ Entomol 5:495–501

Parmesan C (2006) Ecological and evolutionary responses to recent climate change. Annu Rev Ecol Evol Syst 37:637–669

Parmesan C, Ryrholm N, Stefanescu C, Hill JK, Thomas CD (1999) Poleward shifts in geographical ranges of butterfly species associated with regional warming. Nature 399:579–583

Pattison M (2005) The Quiet Revolution: Push–Pull Technology and the African Farmer. Gatsby Charitable Foundation, London

Pedigo LP, Rice ME (2009) Entomology and pest management, 6th edn. PHI Learning Private Limited, New Delhi

Pedigo LP, Rice ME (2014) Entomology and pest management. Waveland Press, Long Grove

Peltonen PA, Julkunen TR, Vapaavuori E, Holopainen JK (2006) Effects of elevated carbon dioxide and ozone on aphid oviposition preference and birch bud exudate phenolics. Glob Chang Biol 12:1670–1679

Penuelas J, Staudt M (2010) BVOCs and global change. Trends Plant Sci 15:133–144

Peshin R, Bandral RS, Zhang W, Wilson L, Dhawan AK (2009) Integrated pest management: A global overview of history, programs and adoption. In: Peshin R, Dhawan AK (eds) Integrated Pest Management: Innovation-Development Process. Springer Science + Business Media, Berlin, pp 1–49

Peters D, Law A (2017) Statement of Senator Deb Peters, South Dakota Legislature President-elect, National Conference of State Legislatures. Quart Theol Quart 41:133–135

Pimentel D (1982) Perspectives of integrated pest management. Crop Prot 1:5–26

Pimentel D (2007) Encyclopedia of pest management. CRC press, Boca Raton

Poehling HM (1989) Selective application strategies for insecticides in agricultural crops. In: Jepsen PC (ed) Pesticides and non-target invertebrates. Intercept, Wimborne, pp 151–176

Pretty J, Bharucha ZP (2014) Foresight and African agriculture: innovations and policy opportunities. Report to the UK Government Foresight Project. https://www.gov.uk/government/uploads/system/uploads/attachment_data/file/300277/14-533-future-african-agriculture.pdf

Pretty J, Toulmin C, Williams S (2011) Sustainable intensification in African agriculture. Int J Agric Sustain 9:5–24

Qiu J (2014) Controversy of GM crops in China. Nat Sci Rev 1:466–470

Rabia S, Muhammad R, Naeem A, Tahir JM, Muhammad N (2017) Toxicity and resistance of the cotton leaf hopper, *Amrasca devastans* (Distant) to neonicotinoid insecticides in Punjab, Pakistan. Crop Prot 93:143–147

Radcliffe EB, Ragsdale DW (2002) Aphid-transmitted potato viruses: the importance of understanding vector biology. Am J Potato Res 79:353–386

Raghavendra K, Vasantha K, Subbarao SK, Pillai MKK, Sharma VP (1991) Resistance in Anopheles culicifacies sibling species B and C to malathion in Andhra Pradesh and Gujarat States, India. J Am Mosq Control Assoc 7(2):255–259

Rai M, Ingle A (2012) Role of nanotechnology in agriculture with special reference to management of insect pests. Appl Microbiol Biotechnol 94:287–293
Régnière J (1983) An oviposition model for the spruce budworm, *Choristoneura fumiferana* (Lepidoptera: Tortricidae). Can Entomol 115:1371–1382
Resh VH, Carde RT (2009) Encyclopedia of insects. Academic, New York
Ridgway RL, Inscoe MN, Dickerson WA (1990) Role of the boll weevil pheromone in pest management. In: Ridgway RL, Silverstein RM, Inscoe MN (eds) Behavior-modifying chemicals for insect management, pp 437–471
Rother HA (2018) Pesticide Labels: Protecting Liability or Health?–Unpacking "misuse" of pesticides. Curr Opin Environ Sci Health 4:10–15
Rowell A (2017) Green backlash: Global subversion of the environment movement. Routledge, London
Ruberson JR, Nemoto H, Herose Y (1998) Pesticides and conservation of natural enemies in pest management. In: Barbosa P (ed) Conservation biological control. Academic, San Diego
Saha LR, Dhaliwal GS (2012) Handbook of Plant Protection, 2nd edn. Kalyani Publishers, New Delhi
Saravi SSS, Shokrzadeh M (2011) Role of pesticides in human life in the modern age: a review. In: Stoytcheva M (ed) Pesticides in the modern world: risks and benefits. In-Tech, Rijeka, pp 4–11
Sarwar M (2012) Management of rice stem borers (Lepidoptera: Pyralidae) through host plant resistance in early, medium and late plantings of rice (*Oryza sativa* L.). J Cer Oils 3:10–14
Schillhorn VVT, Forno D, Joffe S, Umali-Deininger D, Cooke S (1997) Integrated pest management: strategies and policies for effective implementation, Environmentally sustainable development studies and monographs, no. 13. World Bank, Washington, DC
Schlyter F, Zhang QH, Liu GT, Ji LZ (2001) A successful case of pheromone mass trapping of the bark beetle *Ips duplicatus* in a forest Island, Analysed by 20-year time-series data. Integr Pest Manag Rev 6:185–196
Schowalter TD (2011) Insect Ecology: An Ecosystem Approach. Academic, London
Schweiger O, Settele J, Kudrna O, Klotz S, Kuhn I (2008) Climate change can cause spatial mismatch of tropically interacting species. Ecology 89:3472–3479
SEMINIS (2018) Resistance to insecticides in sweet corn insects. http://www.seminis-us.com/resources/agronomic-spotlights/resistance-to-insecticides-in-sweet-corn-insects/. Accessed 03 June 2018
Shaw MW, Osborne TM (2011) Geographic distribution of plant pathogens in response to climate change. Plant Pathol 60:31–43
Sherman K, Sevilla NPM, Torres PÁ, Peterson B (2017) Sustainable development of latin american and the caribbean large marine ecosystems. Elsevier
Sievers M, Vandenberg P (2007) Synergies through linkages: Who benefits from linking microfinance and business development services? World Dev 35:1341–1358
Silverstein RM, Brownlee RG, Bellas TE, Wood DL, Browne LE (1968) Brevicomin: Principal sex attractant in the frass of the female western pine beetle. Science 159:889–891
Simon S, Bouvier JC, Debras JF, Sauphanor B (2010) Biodiversity and pest management in orchard systems. A review. Agron Sustsain Dev 30:139–152
Smith JW (1998) Boll weevil eradication: area-wide pest management. Ann Entomol Soc Am 91:239–247
Sorby K, Fleischer G, Pehu E (2005) Integrated pest management in development: review of trends and implementation strategies. Agriculture and rural development working paper 5, World Bank, Washington, DC. http://documents.worldbank.org/curated/en/ 2003/04/2455449/integrated-pestmanagement-development-eviewtrendsimplementation- strategies. Accessed 25 Dec 2016
Stacey D, Fellowes M (2002) Influence of temperature on pea aphid *Acyrthosiphon pisum* (Hemiptera: Aphididae) resistance to natural enemy attack. Bull Entomol Res 92:351–358
Stern VM, Smith RF, van den Bosch R, Hagen KS (1959) The integrated control concept. Hilgardia 29:81–101

Stoeckli S, Hirschi M, Spirig C, Christoph S, Pierluigi C, Mathias WR, Jörg S (2012) Impact of climate change on voltinism and prospective diapause induction of a global pest insect *Cydia pomonella* (L.). PLoS One 7:e35723

Sturrock R, Frankel S, Brown A, Hennon P, Kliejunas J, Lewis K, Worrall J, Woods A (2011) Climate change and forest diseases. Plant Pathol 60:133–149

Sudo M, Takahashi D, Andow DA, Suzuki Y, Yamanaka T (2018) Optimal management strategy of insecticide resistance under various insect life histories: Heterogeneous timing of selection and interpatch dispersal. Evol Appl 11:271–283

Sufyan M, Neuhoff D, Furlan L (2013) Effect of male mass trapping of *Agriotes* species on wireworm abundance and potato tuber damage. Bull Insect Contr 66:135–142

Sukhoruchenko GI, Dolzhenko VI (2008) Problems of resistance development in arthropod pests of agricultural crops in Russia. EPPO Bull 38:119–126

Sutherland WJ, Barnard P, Broad S, Clout M, Connor B, Côté IM, Dicks LV, Doran H, Entwistle AC, Fleishman E (2017) A 2017 horizon scan of emerging issues for global conservation and biological diversity. Trends Ecol Evol 32:31–40

Tabashnik BE, Carriere Y, Dennehy TJ, Morin S, Sisterson MS, Roush RT, Shelton AM, Zhao JZ (2003) Insect resistance to transgenic Bt crops: lessons from the laboratory and field. J Econ Entomol 96:1031–1038

Tenow O, Nilssen A, Holmgren B, Elverum F (1999) An insect (*Argyresthia retinella*, Lep., Yponomeutidae) outbreak in northern birch forests, released by climatic changes? J Appl Ecol 36:111–122

Topinka JB (2009) Yaw, pitch, and roll: quarantine and isolation at United States airports. J Legal Med 30:51–81

Tóth M, Furlan L, Xavier A, Vuts J, Toshova T, Subchev M, Szarukán I, Yatsynin V (2008) New sex attractant composition for the click beetle *Agriotes proximus*: Similarity to the pheromone of *Agriotes lineatus*. J Chem Ecol 34:107–111

Trematerra P, Gentile P (2010) Five years of mass trapping of *Ephestia kuehniella* Zeller: a component of IPM in a flour mill. J Appl Entomol 134:149–156

Trumble JT, Butler CD (2009) Climate change will exacerbate California's insect pest problems. Calif Agric 63:73–78

Ullah S, Sarfaraz SA, Abbas N (2016) Resistance of dusky cotton bug, *Oxycarenus yalinipennis* Costa (Lygaidae: Hemiptera), to conventional and novel chemistry insecticides. J Econ Entomol 109:345–351

UN (2017) World population prospects 2017. https://esa.un.org/unpd/wpp/DataQuery/. Retrieved 31 Oct 2017

VanAsch M, Visser M (2007) Phenology of forest caterpillars and their host trees: the importance of synchrony. Annu Rev Entomol 52:37–55

Vargas RI, Shelly TE, Leblanc L, Pinero JC (2010) Recent advances in methyl eugenol and cue-lure technologies for fruit fly detection, monitoring, and control in Hawaii. Vitam Horm 83:575–596

Vernon RS, Toth M (2007) Evaluation of pheromones and a new trap for monitoring *Agriotes lineatus* and *Agriotes obscurus* in the Fraser Valley of British Columbia. J Chem Ecol 33:345–351

Visser ME, Holleman LJ (2001) Warmer springs disrupt the synchrony of oak and winter moth phenology. Proc R Soc Lond B Biol Sci 268:289–294

Wall C, Garthwaite DG, Blood Smyth JA, Sherwood A (1987) The efficacy of sex-attractant monitoring for the pea moth, *Cydia nigricana* in England 1980–1985. Ann Appl Biol 110:223–229

Walter GH (2005) Insect pest management and ecological research. Cambridge University Press, Cambridge

Wang LP, Shen J, Ge LQ, Wu JC, Yang GQ, Jahn GC (2010) Insecticide-induced increase in the protein content of male accessory glands and its effect on the fecundity of females in the brown planthopper *Nilaparvata lugens* Stål (Hemiptera: Delphacidae). Crop Prot 29:1280–1285

WHO (2015) International code of conduct on pesticide management: guidelines on pesticide legislation. World Health Organization, United Nation Organization, Geneva

WHO (2017) Safety evaluation of certain food additives and contaminants. World Health Organization, Geneva
Willmer P, Stone G, Johnston I (2000) Environmental physiology of animals. Blackwell, Oxford
Witzgall P, Kirsch P, Cork A (2010) Sex pheromones and their impact on pest management. J Chem Ecol 36:80–100
Woods AJ, Heppner D, Kope HH, Burleigh J, Maclauchlan L (2010) Forest health and climate change: a British Columbia perspective. Chronicle 86:412–422
Wright DJ, Verkerk RHJ (1995) Integration of chemical and biological control systems for arthropods: evaluation in a multitrophic context. Pestic Sci 44:207–218
Yamamura K, Kiritani K (1998) A simple method to estimate the potential increase in the number of generations under global warming in temperate zones. Appl Entomol Zool 33:289–298
Yang J, Shen Q, Ho M (2009) An overview of previous studies in stakeholder management and its implications for the construction industry. J Facil Manag 7:159–175
Zadoks J (1987) Rationale and concepts of crop loss assessment for improving pest management and crop protection. In: Teng PS (ed) Crop loss assessment and pest management. APS Press, St Paul, pp 1–5
Zalom F (1993) Reorganizing to facilitate the development and use of integrated pest management. Agri Eco Environ (Netherlands) 46:1–4
Zhou XL, Harrington R, Woiwod IP, Joe NP, Jeffrey SB, Suzanne JC (1995) Effects of temperature on aphid phenology. Glob Chang Biol 1:303–313

Sustainable Management of Plant Diseases

Davide Spadaro and Maria Lodovica Gullino

Abstract The disease management strategy represents an important contribution to the sustainability of the farming systems. Plant disease management attempts to maintain disease levels below economic thresholds because complete elimination of disease is unnecessary and may result in unacceptable costs, labour and environmental impacts. Integrated disease management intends to manage plant diseases by assembling complementary approaches, depending on the pathosystem involved, the geographical location and the pedoclimatic conditions. The current chapter provides several examples of sustainable disease management, with particular reference to the control of soilborne diseases of vegetable and ornamentals crops. Healthy soils are fundamental to sustainable disease management. Most practices designed to improve soil health, such as organic matter supplementation also help to suppress the disease development. The use of healthy or treated propagation material is an effective tool to prevent native or alien pathogens. Chemical control with fumigants and fungicides should be considered when other approaches do not achieve the required pathogen control. Rapid and reliable diagnostic methods allow a rational and efficient choice of the management options. Decision support systems should be developed through forecasting models. The choice of the appropriate plant disease management strategy should not only integrate the impact on the soil and crop health, but also on the agricultural and non-agricultural environments, the natural resources, and human health. Economic, social, legislative and political issues should be considered together with regional, national and international regulations.

Keywords Biocontrol agents · Chemical control · Diagnostics · Induced resistance · Integrated disease management · Plant pathogen · Seed health · Soilborne disease · Soil health

D. Spadaro (✉) · M. L. Gullino
Center of Competence AGROINNOVA, University of Torino, Grugliasco, Italy

DISAFA, University of Torino, Grugliasco, Italy
e-mail: davide.spadaro@unito.it; marialodovica.gullino@unito.it

M. Farooq, M. Pisante (eds.), *Innovations in Sustainable Agriculture*,
https://doi.org/10.1007/978-3-030-23169-9_11

1 Introduction

An important goal of sustainable agriculture is the development of integrated farming systems with reduced use of natural resources (water, soil, energy), as well as of chemical fertilizers and pesticides. Sustainable farming systems should maintain and possibly enhance the quantity and quality of crop production, improve the farmer's income, and balance the economic, environmental and social consequences of human interventions. An important contribution to the sustainability of the farming systems is the choice of the disease management strategy. In fact, despite the use of pesticides, 20–30% of production is estimated to be lost due to plant diseases every year (Oerke 2006). Such figures would be even higher without any intervention for reducing losses causes by plant diseases (Esker et al. 2012). Crop losses due to plant diseases affect the potential production in industrialized countries, but in developing countries they are even costly in terms of food security, foreign exchange requirements for food imports, and income losses to farmers (Oerke et al. 1994).

Plant diseases result from complex interactions among host, pathogen and the environment. The disease triangle represents the main elements required for plant diseases: a susceptible host plant, a virulent pathogen able to cause disease, and a favourable environment. Moreover, time can influence a disease, so the disease triangle could become a tri-dimensional disease pyramid, by including this element. Other elements important for some disease could be vectors and human activities, which modify the interaction through agricultural practices, genetic resistance and fungicide application (Burdon and Thrall 2009).

2 Plant Disease Management

Plant disease management attempts to maintain disease levels below economic thresholds because complete elimination of disease is unnecessary and may result in unacceptable costs, labour and environmental impacts. Plant disease management faces significant challenges due to increasing demands for safe and diversified food (Flood 2010); reducing the production potential due to land competition in fertile areas; depletion of natural resources; reduction of biodiversity in the agroecosystems; and increased risk of disease epidemics due to agricultural intensification, monoculture, and climate change (Dun-chun et al. 2016). The pathogen spread is facilitated by human transportation, but there is an increasing evidence that global warming can drive pathogen movement towards the pole, by altering their latitudinal range (Bebber et al. 2013).

In the late 1960s and 1970s, the commercialization of many broad-spectrum pesticides of novel structure and mostly with systemic activity marked an era characterized by intensified agricultural production. After some years of intensive chemical control, new pathogens became dominant once their competitors were eliminated and fungicide resistance developed (Delp and Dekker 1985). To address these problems, growers intensified the use of fungicides, which increased production costs and increased the risk of fungicide residues on crops (Oliver and Hewitt 2014).

Integrated disease management intends to manage plant diseases by assembling complementary approaches, depending on the pathosystem involved, the geographical location and the pedoclimatic conditions. As stated by the European Directive on the Sustainable Use of Pesticides, integrated pest management carefully considers all available plant protection methods and subsequent integration of appropriate measures that discourage the development of pathogen populations and keep the use of fungicides and other forms of intervention to economically and ecologically justified levels, by minimising the risks for human health and the environment. Plant pathogens are difficult to control partly due to their spatial-temporal dynamics and rapid evolution (Strange and Scott 2005), associated with high genetic diversity and short generation times that favour their ability to overcome effective disease control approaches. Integrated disease management emphasises the growth of a healthy crop with the least possible disruption of the agro-ecosystems and encourages natural disease control mechanisms.

3 Sustainable Plant Disease Management

Sustainable management of plant diseases aims to create environments adverse for the pathogens and suitable for healthy plants, by ensuring high yield through the efficient use of natural resources (Zhan et al. 2015). An agroecological approach should be used for the management of diseases, leading to solutions serving the public good by simultaneously fostering agrifood system productivity and resilience, reducing energy consumption and supporting bioenergy production, as well as conserving water resources (Kremen and Miles 2012). Agroecology is the science of applying ecological concepts and principles to the design and management of sustainable food systems (Gliessman 2014). In addition, economic and societal impacts should be evaluated for each plant disease management scheme. An agroecological system approach to plant disease management consists of four pillars: (i) prevention of pathogen introduction and spread in the cropping system; (ii) reduction of pathogen populations to levels which can be controlled through natural mechanisms; (iii) introduction of practices into the cropping system designed to promote beneficial microbiota; and (iv) reduction of fungicide use through the adoption of integrated disease management (Chellemi et al. 2016). To achieve the goal of sustainable plant disease management, multidisciplinary collaboration between disciplines, such as plant pathology, plant breeding, agronomy, horticulture, agricultural entomology, soil science, environmental science, economics and social sciences is needed. Agroecology, besides being multidisciplinary, is also transdisciplinary, as it incorporates elements of practice and collective action, which enable the scaling of agricultural practices from individual farms to larger landscape-level (DeLonge and Basche 2017).

The current chapter provides several examples of sustainable disease management, with particular reference to the control of soilborne diseases of vegetable and ornamentals crops. Soilborne pathogens can cause heavy losses in vegetable production, by affecting both yield and quality. Soilborne pathogens can occur from the initial nursey stage, to the harvest. In vegetable production, crop rotations are minimal and soilborne pathogen propagules may accumulate in the soil, which is the primary inoculum. Soilborne pathogens are particularly favoured in vegetables, which are an intensive and dynamic system, characterized by a wide range of crop species and varieties, a continuous introduction of innovative technologies and the use of intensive cultivation techniques. For the above-mentioned reasons, the management of soilborne diseases in vegetable production represents a very interesting case study, both in terms of phytopathological issues and innovative strategies adopted for their control (Colla et al. 2012).

3.1 Maintaining Healthy Soils

Healthy soils are fundamental to sustainable disease management, as they affect the density of pathogens, particularly of the soilborne ones (Janvier et al. 2007), the structure of beneficial microbiota, and the availability of organic and inorganic nutrition for plants (Larkin 2015; van Bruggen et al. 2016). Agricultural management strategies can have a major impact on soil quality with consequent effects on disease incidence. Soil organic matter, one of the primary indicators of soil health, is fundamental to the long-term sustainability of agroecosystems. Managing soil health is a matter of maintaining a suitable habitat for the soil (micro)-organisms. The aim of the practices adopted is to achieve the resilience (the capacity to self-organize into desirable steady states) and homeostasis (the maintenance of desirable steady states) of the soil microbiota. In most cases, regular additions of organic matter are necessary to replenish soil resources and improve soil health.

3.1.1 Suppressive Soils

Suppressive soils are those where the disease development is naturally controlled, even in the presence of a virulent pathogen, a susceptible plant host, and with environmental conditions conducive for the development of the disease. Soil suppressiveness is a complex system of biotic and abiotic factors, such as soil structure, nutrient and water availability, microbiota (including pathogens and symbionts), and plant genotype. Natural soils have a general disease suppression compared to the same pasteurised soils, and it is directly related to the microbial activity (Schlatter et al. 2017). In cropping systems, a specific suppression is present when a group of microorganisms, selected for their antagonistic activity, is directly responsible for disease suppression. Soil bacteria and fungi, as *Pseudomonas* spp. and *Alcaligenes* spp. in the USA (Kloepper et al. 1980; Yuen et al. 1985) and *Fusarium* spp. in

France and Italy (Janvier et al. 2007; Garibaldi and Gullino 1987), have been shown to be involved in Fusarium wilt suppression. Antagonistic *Fusarium* spp., isolated from the rhizosphere of carnation grown in suppressive soils, showed high rhizosphere competence. When applied to soil and substrates they controlled Fusarium wilts on different crops, such as tomato, basil, carnation, cyclamen, and bulb crops (Gullino and Garibaldi 2007). Soils suppressive to *Rhizoctonia solani* are correlated with the presence of large amounts of *Trichoderma* spp. (Chet 1987).

3.1.2 Soil Management for Disease Suppression

Organic matter can be added through agronomic practices, such as crop residues, rotations, and cover crops. Crop rotations are one of the most interesting agronomic practices, as they are able to combine the optimal use of nutrients with the reduction of soilborne pathogens. The evolution of agriculture has led to the abandonment of rotations in favour of monoculture, with consequent negative plant disease profile. Monoculture, in fact, leads to the progressive soil accumulation of propagules of plant pathogens to unacceptable levels, which force the adoption of disinfestation practices. Some pathogens (*Fusarium* spp., *Verticillium* spp., or *Rhizoctonia spp.*), which show high competitiveness at saprophytic level or differentiate survival structures, tend to accumulate in the soil. The mechanism underlying the beneficial effects of rotation is starving the pathogen when the susceptible host is not cultivated. This occurs in the case of organisms with narrow host spectrum, modest saprophyte capacity, and lack of survival structures. The level of specialization of the parasite is important: crop rotation has higher effect on species-specific pathogens (i.e. *formae speciales* of *Fusarium oxysporum*) than on the polyphagous ones (*Sclerotinia sclerotiorum*, *Verticillium dahliae*). Crop rotation can also include the alternation of cultivars of the same species with different levels of pathogen susceptibility.

Crop rotations are associated with increasing soil microbial activity and diversity, due to the cultivation of different plant species in the soil (Garbeva et al. 2004; Welbaum et al. 2004). Crop rotations that maximize diversity of plant and root systems (mixing legumes, cereals, solanaceous, cucurbits, brassica, etc.) may significantly modify soil microbiota and their disease suppression potential.

Cover crops are grown primarily to cover the soil, to protect it from erosion and nutrient losses when production crops are not present. Benefits of cover crops may include disease control (Larkin 2015).

Green manuring is the incorporation of fresh plant material to enrich the soil organic matter. Green manuring results in higher organic matter inputs than traditional crop rotations or cover crops, producing improvements in soil fertility, structure, and microbiota, with an effect on disease suppression (Collins et al. 2006; Stark et al. 2007). Most practices designed to improve soil health, such as organic matter supplementation also help to suppress the disease development (Welbaum et al. 2004; Bonilla et al. 2012a, b; Page et al. 2013).

3.1.3 Suppressive Substrates

Suppressiveness has been found for several substrates used in horticulture. Sphagnum peat mixes can naturally suppress soilborne pathogens, but few weeks after potting, they become conducive to diseases (Hoitink and Boehm 1999). Peat mixes well tolerate the introduction of biocontrol agents or the addition of composts (Hoitink and Locke 2012). When hardwood bark is used, improved plant vigour and disease suppressiveness, from richer microbiota, are observed in potted plants (Hoitink and Boehm 1999).

Increasing the use of compost as a potting substrate would contribute to waste recycling and reduction of chemical fertilizers. Compost is interesting as a peat substitute, for the lower production cost and for the increasing concern about the environmental impact of peat extraction (Silva et al. 2007). Some composts, particularly those amended with composted bark, suppress most soilborne plant pathogens (Hoitink and Boehm 1999; Noble and Coventry 2005; Termorshuizen et al. 2006). Composts were demonstrated to be more suppressive than crop residues and peat (Bonanomi et al. 2007). Low amounts of compost in growing media avoid the lower growth and the phytotoxicity caused by high pH and electrical conductivity (Sullivan and Miller 2001). Composts originating from green wastes or municipal biowastes, blended with a peat substrate effectively reduced Fusarium wilt on basil, *Pythium ultimum* on cucumber, *Phytophthora nicotianae* on tomato and *Phytophthora capsici* on pepper (Pugliese et al. 2014). On the contrary, saline composts were reported to enhance *Pythium* and *Phytophthora* diseases, while high nitrogen composts could enhance *Fusarium* wilts (Hoitink et al. 2001). The efficacy of compost for disease control depends on the raw materials from which the compost was prepared, the composting process used, and the compost maturity and quality (Termorshuizen et al. 2006). Of particular interest is the use of disease suppressive composts, thanks to the introduction of selected antagonists: their use is particularly interesting in the case of nurseries (Garibaldi 1988; Hadar 2011; Hoitink and Fahy 1986). In other cases, composts have been identified as a potential source of antagonistic microorganisms (Pugliese et al. 2008). In some cases, it is interesting to combine the use of compost with that of resistant rootstocks (Pugliese et al. 2014).

Although interesting for field crops and vegetables, the use of organic amendments for disease control is still not widespread, due to many factors such as the lack of standardization, the inconsistency in their efficacy, and the complexity of their use.

3.1.4 Soilless Media

Soilless cultivation is realized in inert or cation exchange capacity substrates (rock wool, perlite, peat), used as a mechanical support for the plant, replacing the soil. Soilless cultivation requires a continuous feeding of the plants with a complete nutrient solution. This technique offers numerous advantages, such as better control of soilborne pathogens and more effective planning of crop cycles. Soilless cultivation could permit the production cycle completely free of pathogens. It also permits eradication of the soilborne pathogens in the recirculating nutrient solutions (Van Os et al. 2012). Soilless cultivation allows excluding soilborne pathogens: the possibility of contact between the host and pathogen is avoided by growing the plant in a pathogen-free environment (Postma 2004; Garibaldi and Gullino 2010). Soilless systems, while they strongly limit some pathogens, they could favour pathogens that find favourable conditions for their diffusion in the nutrient solution. *Pythium* and *Phytophthora* are the most frequent pathogen genera in the root system of soilless vegetables and ornamentals. Many pathogens (*Pythium aphanidermatum, P. myriotylum, Phytophthora cryptogea, P. nicotianae*) found in hydroponics are the same present in normal soil conditions, while others affect plant hosts which are resistant when grown in soil. *Phytophthora cryptogea* in soilless systems becomes strongly virulent on lettuce. *Pythium dissotocum* becomes extremely virulent in soilless cultivation of spinach and lettuce. Other pathogens are specific for soilless crops, such as *Plasmopara radicis-lactucae*, reported on lettuce roots.

Among the potential sources of pathogen infection in soilless crops, there are the substrates; perlite, vermiculite, rock wool, polyurethane, and polystyrene are generally considered sterile, but organic materials, such as peat, coconut fibre or non-composted bark, represent the main source of infection of *Pythium* spp., *Fusarium* spp., *Olpidium* spp. and *Thielaviopsis* spp. (Van Os 2010). On the other hand, the cultivation substrate could show a natural suppressiveness, depending both on chemical and microbiological factors. By comparing different substrates, there are substantial differences in the microflora established, which generate a different degree of suppressiveness.

In closed systems, higher electrical conductivity of the nutrient solution, amendment with potassium silicate, and their combination were effective against powdery mildews, downy mildews, leaf spots, and Fusarium wilts (Gullino et al. 2015a, b). Silicon provided partial control of powdery mildews on greenhouse crops and soilborne diseases on turfgrass (Bélanger et al. 1995; Brecht et al. 2004; Uriarte et al. 2004): in addition to the deposition of amorphous silica in the cell wall, there is an increased lignin production, which could limit the pathogen penetration in the plant cell (Gullino et al. 2015a, b).

Soilless systems also permit microbial optimization, thanks to the application of microorganisms able to colonize the plant rhizosphere. Slow sand filtration combined with the application of different antagonistic strains of *Fusarium* spp. and *Trichoderma* spp. was effective against *Phytophthora cryptogea* in gerbera (Garibaldi et al. 2004a).

Pathogen diffusion in soilless cropping systems can be greatly reduced by adopting proper disinfection methods for the recirculating solution, such as slow sand filtration (Van Os 2010). Moreover, preventative methods to increase the plant resistance to diseases and the use of diagnostic tools constitute an integrated approach for soilless systems (Van Os et al. 2012).

3.1.5 Organic Amendments

Organic amendments include manure, crop residues, compost, and organic fertilisers. The application of organic amendments is commonly adopted in traditional agricultural systems to provide nutrients to the crop and to improve the soil fertility and structure (Bailey and Lazarovits 2003; Bonanomi et al. 2007; Bonilla et al. 2012a, b). Suppressiveness has been found for organic amendments used in agriculture. Several chemical and physical changes in the soil are due to the incorporation of amendments and result in control of soilborne pathogens, with reduced application of chemicals (Pugliese et al. 2015).

A proper nutritional status makes plants more easily able to react to any kind of stress. High nitrogen fertilization, by favouring the vegetative growth of the host and the tissue turgidity, is conducive to the pathogen attack. Generally, adequate potassium fertilization makes the host resistant to several parasites. Soil amendments can be useful to modify the soil pH. For example, pH values above 7 reduce the incidence of *Plasmodiophora brassicae* on cabbage (Webster and Dixon 1991), though at these pH values the occurrence of *Erwinia carotovora* increases (Bain et al. 1996). Alkaline soils are conducive to the spread of the scab of potatoes, as *Streptomyces scabies* usually develops between pH 5.2 and 8.0 (Hooker 1981). It is, however, difficult to generalize and to choose a unique intervention practice. For example, on carnation, soil pH reduction reduces the attacks of *Phytophthora nicotianae* (Spencer and Benson 1981) and increases the wilts caused by *Fusarium oxysporum* (Jones et al. 1993).

When added to soil, amendments, such as cow or poultry manure and brassica residues, are subjected to microbial degradation that releases toxic and volatile compounds directly affecting soilborne pathogens or indirectly increasing microbial soil suppressiveness. Organic amendments can promote the re-establishment of a more balanced and suppressive microflora. Furthermore, the development of plant disease is reduced thanks to the extended root systems growing in a rich soil (Chellemi 2010).

Composts and Brassica pellets are considered among the most promising organic amendments. A growing interest is directed to the use of isothiocyanate precursors, contained in selected brassicaceae (*Brassica juncea* and *B. carinata*), used as alternating species and then applied as green manure or as flour or pellets (Larkin and Griffin 2007). The use of Brassica species as green manure is a type of biofumigation that involves the release of volatile compounds able to control a wide array of soilborne pathogens (Larkin and Griffin 2007). Biofumigation, however, provides results that are not always univocal: promising efficacy was obtained against *Colletotrichum coccodes* on tomato, *Fusarium oxysporum* f. sp. on cucumber, *Verticillium dahliae* on eggplant grafted onto *Solanum torvum,* and Fusarium wilt of lettuce, rocket and basil (Garibaldi et al. 2010, 2014a, b). Partial or negative results have been observed in other crops, such as *Brassica* spp., where the inoculum of soilborne pathogens could be favoured (Lu et al. 2010). The combination of green manure with soil solarisation is also very effective and reduces the period of soil mulching with plastic films.

Organic amendments for disease control are not yet widespread, due to lack of standardisation of production parameter, inconsistent efficacy and difficult application. Control of soilborne diseases with organic amendments must be considered a component of a system approach, where the impact of crop production practices on resident soil microflora is addressed.

3.1.6 Soil Solarisation

Solarization is the soil covering with plastic film during the summer. The method has been widely exploited in warm and temperate countries (Katan and DeVay 1991). Farmers are generally sceptical about its adoption, as it requires soil free of cultivation for at least 4 weeks. An integration strategy, often adopted to increase soil solarization efficacy, is its combination with biocontrol agents, to reduce the solarisation period and to permit its use in marginal areas (Minuto et al. 2006). The combination of soil solarization and *Streptomyces griseoviridis* is effective against fusarium and verticillium wilts and corky root, and it increases the range of pathogens controlled with respect to the single treatments. Significant increases in yield and fruit weight were observed, confirming the potential additive effect caused by biocontrol agent and solarization in terms of yield increase.

3.2 Planting Material

3.2.1 Healthy Propagation Material

Considering the losses caused by most emerging pathogens, the first preventative strategy that should be considered by seed producers and farmers is the use of healthy seeds and propagation material. The use of healthy or treated propagation material is an effective tool to prevent native or alien pathogens from being introduced in the agricultural environment. It is estimated that almost 800 fungi, over 150 viruses, 100 bacteria and 20 phytopathogenic nematodes are transmitted through propagation material. To avoid this risk, programs have been activated for the most important crops aimed at certifying the health of the seed or propagation material. This requires specific phytosanitary assays, which consist in estimating the possible presence of the pathogen using different biological and molecular methods.

The control of propagation material is important for clonal species (carnation, geranium, strawberry) for which the use of uncontrolled material could facilitate disease outbreaks. The importance of the use of healthy or treated material is particularly evident in the case of pathogens (viruses, bacteria) with few or ineffective control strategies (Gullino and Munkvold 2014). On strawberry, the use of certified propagation material, obtained by thermotherapy, meristem cultivation and subsequent indexing is a consolidated practice.

Another important aspect is seed health. Stock seeds should be produced in locations with low disease risk, characterized by low humidity and dry summer climate, to reduce fungal or bacterial epidemics (Munkvold 2009). The choice of proper geographical areas, possibly isolating seed and seedling production from the environment, and the application of good agricultural practices are critical for producing high-quality, pathogen-free seed.

As it is unrealistic to pursue an absolute seed health of the seed, a certain tolerance is admitted. Very common is the diffusion of fungal and bacterial seedborne pathogens on vegetables (Koch and Roberts 2014). The production of virus-free seed must follow appropriate production and certification schemes, which involve the controlled cultivation of the mother plants and diagnostic tests both on the mother plants and the seed produced (Gullino and Bonants 2014).

To reduce the risks of fungal and bacterial seedborne diseases, it is recommended that stock seeds undergo precautionary chemical or physical treatments. Chemical seed treatments have successfully been applied to vegetable seeds and are in commercial use for a wide range of crops against different seedborne pathogens (Munkvold 2009). Several surface disinfectants (bleach, hydrogen peroxide, ethanol) can be applied to remove pathogen inoculum from seed coats (Mancini and Romanazzi 2014). Chemical treatments are effective, but they can also negatively affect germination and cause phytotoxicity (Axelrood et al. 1995; du Toit 2004), besides having negative effects on human health and the environment (Lamichhane et al. 2016). Alternative strategies for the control of seedborne pathogens include physical seed treatments, treatments with natural compounds, antagonistic microorganisms, and resistance inducers. Physical strategies include mechanical (sorting and brushing), heat, ultrasonic, radiations (with microwaves resulting in elevated temperatures), UV-C light, and redox treatments (cold plasma and electrons (Spadaro et al. 2017). Thermal treatments with hot water, aerated steam or dry heat can be very effective, but they need to be optimised for the pathosystems, due to the different temperature and time required (Koch and Roberts 2014). Although alternative seed treatments have been intensively investigated, there are few examples of commercial application (Koch and Roberts 2014; Gullino et al. 2014).

Seed treatments can also be an effective means to increase seedling emergence, particularly when done on seeds of low vigour and when the seed coat has been damaged (Mancini and Romanazzi 2014). In general, the use of healthy or disinfected seed is a very useful practice for plant disease management.

3.2.2 Resistant Varieties and Grafting

Host resistance, which is the use of resistant and/or tolerant plant varieties, is one of the most effective strategies against pathogens. Varieties, which are resistant or at least tolerant to one or more pathogens, are available for many crops and the industry is investing on research in this field. Resistant cultivars of lettuce can control Fusarium wilt. Lettuce varieties that are resistant, or at least tolerant, to race 1 of Fusarium wilt are available (Garibaldi et al. 2004b, 2014a, b), but their use is complicated by the presence of different races of the pathogen. Seed breeding companies are currently working hard in order to develop planting material resistant to the recently detected race 4 (Gilardi et al. 2017a).

Host resistance, and the integration of such varieties with other management strategies is fundamental within the framework of IPM, but few researches focused on the integration of plant resistance with other IPM strategies (Stout and Davis 2009). Moreover, the breeding approach used to date to develop resistant and/or tolerant crop varieties should be revised, as most crop cultivars bred to date are based on a market-driven approach focused on high yield and remunerative crop varieties. This trend has facilitated the adoption of short rotations or monoculture practices and ignored the potential that minor side crops may have for IPM. The limited range of available minor crop varieties is one obstacle to crop diversification, thereby confining certain beneficial practices such as multiple cropping or intercropping. Sustainable disease management should develop crop breeding based on the competitiveness of crops and their adaptation to diversified cropping systems (Lamichhane et al. 2017).

Grafting is used to reduce susceptibility against pests, root rots and wilts, and to increase yield (Rouphael et al. 2010). Despite disadvantages associated with grafting, including the additional cost and physiological disorders due to incompatibility between rootstocks and scions, the use of resistant rootstock strongly increased, mainly for vegetable crops. Despite disadvantages associated with grafting, including the additional cost and physiological disorders due to incompatibility between rootstocks and scions, the use of resistant rootstock, despite its high cost, strongly increased. Grafting on resistant rootstock is becoming popular on pepper and some of the commercially available rootstock provide a good control of Phytophthora blight (Gilardi et al. 2013). In the case of *P. capsici* on bell pepper, due to the lack of commercial cultivars with resistance, growers are interested in grafting. Grafted plants are popular in the case of tomato, to control soilborne pests and pathogens and to increase yield (Chellemi 2002; Lee and Oda 2003; Gilardi et al. 2013). Grafting susceptible crops onto resistant rootstocks is interesting also for cucumber (*Cucurbita vicifolia* as rootstock resistant to Fusarium wilt) and melon (*Benincasa cerifera* resistant to Fusarium wilt) (King et al. 2008).

3.3 Chemical and Biological Control Methods

3.3.1 Chemical Control: Fumigants and Fungicides

Chemical control with fumigants and fungicides is an inseparable component of plant disease management, and it should be considered when other approaches cannot achieve the required level of pathogen population density reduction.

Soil disinfestation with fumigants is becoming very difficult due to the loss of registered fumigants due to recent regulation strongly limiting their availability (Colla et al. 2014). Among the fumigants available, dimethyl disulphide, metham sodium, and dazomet provide significant control of Fusarium wilt of lettuce (Gilardi et al. 2017b). Covering the soil with low-density polyethylene film (LPDE) permits the reduction of fumigant dosage, with interesting results, both under greenhouse conditions and in the open field. Combination of fumigants with alternative methods, notably solarization, are promising. The combination of solarisation for 2 weeks and fumigation with reduced dosage of fumigants was effective, and allowed a shortening of solarization, permitting a reduction in the non-cultivation period (Gullino et al. 2003).

Fungicides are not used to control soilborne pathogens in open field, because of their relative high cost, but they could be used for seed dressing, in nursery to protect the plantlets from damping off and other soilborne diseases, and in potted plants. Mechanisms of action and risk of pathogen resistance development should be considered, when selecting the active ingredient (Siegwart et al. 2015). Diversity of fungicides, concerning their chemistry and mode of action, is essential to ensure effective crop protection, to control new threats and to manage fungicide resistance (Leadbeater and Gisi 2010). Overuse of many organic fungicides can result in resistant fungal populations, so it is important to use fungicides as part of an overall resistance management plan. In the case of Pythium damping off, control is mainly accomplished by treatments with fungicides, such as strobilurins and phenylamides. However, *Pythium* spp. can develop resistance to common fungicides, such as azoxystrobin or mefenoxam. This further suggests the necessity of using other fungicides and alternative means for damping off control, and an accurate identification of *Pythium* spp. before choosing the appropriate control strategy (Matic et al. 2018).

The use of fungicides in integrated disease management is not aimed at eradicating the disease but to reduce it at ecological and economical thresholds.

3.3.2 Induced Resistance

Plants have constitutive and induces responses to defend themselves against pathogens. Two main types of induced resistance are known: systemic acquired resistance (SAR) and induced systemic resistance (ISR) (Vallad and Goodman 2004). SAR elicits the death of one or a few cells, known as the hypersensitive response (HR) and the production of pathogenicity-related (PR) proteins, such as glucanases, chitinases and thaumatin-like proteins (Shoresh et al. 2010). New growth occurs following HR and salicylic acid plays a role in triggering the signal. SAR is often related to the induction via aerial plant parts and it usually takes a certain amount of time to be fully expressed in plants. ISR is often triggered by rhizosphere bacteria in the soil, it involves jasmonic acid and ethylene, but not salicylic acid and PR-proteins.

Induced resistance, mostly SAR, can be triggered by a variety of natural and chemical compounds (Walters et al. 2005). The increasing interest in their use depends on their broad spectrum of activity, and on the possibility of reducing the number of fungicide sprays (Walters et al. 2013). Very interesting results have been observed against Fusarium wilt of lettuce and crown and root rot of zucchini, caused by *Phytophthora capsici*, using resistant inducers, based on either phosphites or acibenzolar-S-methyl, applied as pre-plant treatment in the nursery. Phosphite-based products also show a very positive effect on plant biomass (Gilardi et al. 2015, 2016). The benefits of preventive and repeated treatments with silicates to reduce the attacks of *P. aphanidermatum* (Heine et al. 2007) and *Fusarium oxysporum* f. sp. *radicis-lycopersici* on tomato (Huang et al. 2011) were demonstrated. The commercial biocontrol agents (BCAs) were able to reduce Fusarium wilt of lettuce, particularly when their application starts at nursery (Gilardi et al. 2016) while they were not effective against crown and root rot on zucchini (Gilardi et al. 2015). BCAs can also be effectively applied, alone or combined with heat treatments, for seed dressing, in the case of seed-transmitted pathogens, such as *F. lactucae* (Lopez-Reyes et al. 2016). The efficacy of resistance inducers is seldom complete, as it is generally influenced by several factors (target pathogen, plant genotype, phenotype, environmental conditions, application timing, and formulation) (Walters et al. 2013).

3.3.3 Biocontrol Agents

Many laboratories around the world have developed their own microorganisms and this allowed the collection of important contributions about the biology of pathogens and antagonists. Biocontrol agents may act in various ways but have specific modes of action, including antibiosis, competition, mycoparasitism and induced resistance.

Among the antagonists studied, saprophytic *Fusarium oxysporum*, often isolated from Fusarium suppressive soils, have been widely exploited for their activity against several Fusarium wilts (Garibaldi et al. 1994; Spadaro and Gullino 2005; Gullino et al. 2015a, b). The good antagonistic attitude of strains belonging to *Trichoderma* spp. has been proved against Fusarium wilts in vegetables and ornamental crops (Harman 2006; Gilardi et al. 2016). Plant growth-promoting rhizobacteria, such as *Pseudomonas* spp. and *Bacillus* spp., can induce host systemic resistance against several diseases (Clematis et al. 2009; Lopez et al. 2014).

However, despite the initial great optimism and extensive research efforts, progress in achieving commercial, large-scale usage of biological control has been slow. When trials move towards the farm scale, many antagonists show inconsistent efficacy and lack reliability (Mathre et al. 1999).

Biofungicides still face significant constraints, but there are many possibilities for combining various biocontrol agents, with each other, or with agronomical, physical or chemical control methods (Spadaro and Gullino 2005). In particular, by combining different methods of control, the aim is to obtain a synergistic rather than additive effect. For that reason, a complete comprehension of the mechanism of control is needed. Combining a biocontrol agent with a fungicide improves the biofungicide efficacy and enables the reduction of the fungicide dosage. Moreover, the combination of control methods provides a wider spectrum of control, which is needed to replace fumigants.

3.4 Additional Tools for Sustainable Disease Management

3.4.1 Diagnostics

Rapid and reliable diagnostic methods allow a rational and efficient choice of the management options. The easy spread of fungal spores, virus and bacteria combined with the intense trading globalization are key factors to allow the movement of pathogens around the world, which can become invasive in new areas and even cause the destruction of the crop. Traditional detection methods based on visual assessment of plant symptoms, isolation, culturing in selective media, and direct microscopic observation of pathogens are frequently laborious, time-consuming and require extensive knowledge of classical taxonomy. For many diseases, the observation under microscope or stereoscopic microscope is used to determine the causal agent, taking into consideration pathogenicity tests and morphological features such as size and shape of the propagules and colony characteristics, such as colour. However, many microorganisms (including viruses) can produce the same symptoms in the plant, making difficult the correct identification of the causal agent. As many plant pathogens remain latent in the planting material, and may be present in very low numbers, high sensitivity, specificity, and reliability methods are required. The impossibility or difficulty of culturing some species *in vitro* and the inability for accurate quantification of the pathogen are other limitations. Early detection of pathogens in seeds and plant materials is of key importance to avoid further spreading and introduction of new pathogens into growing areas where they are not present yet. These limitations have led to the development of molecular approaches with improved accuracy and reliability. Molecular techniques are faster, more specific, sensitive, and accurate than traditional techniques and they can identify non-culturable microorganisms and facilitate early disease management decisions.

The combination of traditional and molecular techniques permits to characterize, detect, identify and quantify different pathogens. In the case of fungal pathogens, the Internal Transcribed Spacer region (rDNA ITS) has been selected by the Consortium for the Barcode of Life (CBOL) as the primary fungal barcode for species identification (Begerow et al. 2010). Other genomic regions are interesting for the fungal identification at species level, or even at subspecies level (Srinivasan et al. 2010). The 16S rRNA has been selected as universal barcode for bacteria identification (Weisburg et al. 1991).

An early pathogen detection represents the best preventative measure in several pathosystems, as in the case *formae speciales* and races of *Fusarium oxysporum* from seeds, plants and soil samples (Pasquali et al. 2007; Mbofung and Pryor 2010; Thomas et al. 2017; Gilardi et al. 2017a).

Loop-mediated isothermal amplification (LAMP) is a DNA amplification method that can be used to amplify nucleic acid in a target specific way without the need for thermal cycling (Notomi et al. 2000). LAMP is particularly promising for plant pathogen detection, as it is easier and quicker to perform than PCR, it can be performed on hand-held platforms, and it is well suited for in field use. The LAMP method has been demonstrated for the detection of bacteria (Hodgetts et al. 2015), fungi (Franco Ortega et al. 2018), phytoplasmas (Hodgetts et al. 2011) and viruses (Tomlinson et al. 2013).

The limit of detection of pathogens, by comparing the molecular techniques, can reach nanograms of DNA for PCR, picograms of DNA for biosensors, and femtograms of DNA for qPCR and digital PCR. NGS technologies are having an enormous impact on biological sciences, allowing the determination of genome variation within a species or a population. Comparative analysis of the genome sequences allows the identification of highly conserved gene families, conserved regulatory elements, repeated elements, uncultured pathogens, new species, symbionts, etc., on which new markers could be designed. On the other side, the use of field techniques, such as LAMP and portable platforms, is a promising tool to early and quickly detect pests and a useful decision support system for appropriate pest and disease management. The choice of the diagnostic technique depends on the balance between the reliability and the cost per sample. Microbiological techniques are generally cheap, but time-consuming, while molecular technologies have a higher cost, which is counterbalanced by the higher performance. PCR, qPCR and LAMP have a progressively lower cost per sample in the order of 2–10 € sample, while NGS are more expensive and they are not yet used for routine analysis (Spadaro et al. 2018). The development of new instruments and platforms and the continuous increase of bioinformatics-data have allowed the use of bioinformatics-based techniques such as metagenomics, comparative genomics and genome sequencing as routine analysis tools. The dramatic decrease of the cost of the new sequencing technologies permits to foresee a higher adoption rate in diagnostic laboratories in the near future.

3.4.2 Forecasting Models

Research tried to develop disease predictions models, also called forecasts or warnings, to help the farmers determine whether and when preventive management measures are needed. Plant disease models are simplifications of the relationships between pathogens, crops, and the environment that cause epidemics to develop over time and space. Plant disease models produce predictions about epidemics or single epidemic components that can be used as risk indicators. Such models also produce predictions about plant disease epidemics that allows growers to respond in timely and efficient ways by adjusting crop management practices. A prediction of low disease risk may result in reduced fungicide treatments with positive economic and environmental effects (Rossi et al. 2010). Disease prediction is most useful for economically important, sporadic diseases for which effective management measures are available. It is also important that growers or technicians be able to operate the prediction system themselves, or that there is a good communication tool between those who monitor and those who manage the disease.

3.4.3 Decision Support Systems

Decision support systems (DSS) should be developed through forecasting models, results of the early detection tools, as well as pathway, establishment and spread models. Data from various sources are interpolated using spatial statistics methods, making the DSS able to provide prediction data with high accuracy at field and site-specific scale. The DSS should have a user-friendly interface, having Geographic Information System (GIS)/mapping functionalities to project the pathogen occurrence. They also could provide alerts when a new pathogen has been identified and could provide recommendations for treatment applications (ideal timing and dosage, optimal sprayer calibration, real-time indicator for tractor speed).

Recently developed DSSs are characterised by holistic treatment of crop management problems (including pests, diseases, fertilisation, canopy management and irrigation); conversion of complex decision processes into simple and easy-to-understand 'decision supports'; easy and rapid access through the Internet; two-way communication between users and providers that make it possible to consider context-specific information (Rossi et al. 2012).

4 Conclusions

Attempts to control soilborne pathogen populations include the use of pesticides, genetic resistance, crop rotations and a variety of cultural practices, aimed at reducing plant infections. Since these measures not always provide adequate disease control, fumigants and fungicides are sometimes needed, as part of an integrated disease management. Adopting preventative and combined methods of disease management has become the choice for the control of soilborne pathogens on economically important crops. The management of soilborne pathogen represents a real challenge.

The implementation of the concepts of soil health and soil health management into agricultural production is essential for sustainable crop production and environmental quality (Larkin 2015). The choice of the appropriate plant disease management strategy should not only integrate the impact on the soil and crop health, but also on the agricultural and non-agricultural environments, the natural resources, and human health. Economic, social, legislative and political issues should be considered together with regional, national and international regulations.

New disease outbreaks emerge and will emerge, requiring continuous changes to the disease management system and reprioritization of goals and objectives. Globalization of trade, new consumption habits, shifts in diets, and climate change are among the factors influencing the occurrence, frequency and severity of new plant diseases, with an important impact on decision-making tools for the related disease management measures that should be adopted. Effort for a continuous monitoring and disease surveillance is necessary. Strategies to produce healthy seeds and seed treatment methods need to be investigated and made available to seed companies and growers. Plant disease management should be adapted to the geographical areas, to the crops and to the pathogens. Future plant disease management should continue to strengthen food security for a stable society, but also safeguard the health of associated ecosystems and reduce dependency on natural resources.

Acknowledgements This research has received funding from the European Union's Horizon 2020 research and innovation programme under grants agreements No. 634179. Effective Management of Pests and Harmful Alien Species – Integrated Solutions (EMPHASIS) and EUCLID EU-CHINA Lever for IPM Demonstration, No. 633999 (EUCLID).

References

Axelrood PE, Neumann M, Trotter D, Radley R, Shrimpton G, Dennis JJ (1995) Seedborne Fusarium on Douglas-fir: pathogenicity and seed stratification method to decrease Fusarium contamination. New For 9:35–51

Bailey KL, Lazarovits G (2003) Suppressing soil-borne diseases with residue management and organic amendments. Soil Tillage Res 72:169–180

Bain RA, Millard P, Perombelon MCM (1996) The resistance of potato plants to *Erwinia carotovora* subsp. *atroseptica* in relation to their calcium and magnesium content. Potato Res 39:185–193

Bebber DP, Ramotowski MAT, Gurr SJ (2013) Crop pests and pathogens move polewards in a warming world. Nat Clim Chang 3:985–988

Begerow D, Nilsson H, Unterseher M, Maier W (2010) Current state and perspectives of fungal DNA barcoding and rapid identification procedures. Appl Microbiol Biotechnol 87(1):99–108

Bélanger RR, Bowen PA, Ehret DL, Menzies JG (1995) Soluble silicon: its role in crop and disease management of greenhouse crops. Plant Dis 79:329–336

Bonanomi G, Antignani V, Pane C, Scala F (2007) Suppression of soilborne fungal diseases with organic amendments. J Plant Pathol 89:311–324

Bonilla N, Cazorla FM, Martinez-Alonso M, Hermoso JM, González-Fernández JJ, Gaju N, Landa BB, de Vicente A (2012a) Organic amendments and land management affect bacterial community composition, diversity and biomass in avocado crop soils. Plant Soil 357:215–226

Bonilla N, Gutierrez-Barranquero JA, de Vicente A, Cazorla FM (2012b) Enhancing soil quality and plant health through suppressive organic amendments. Diversity 4:475–491

Brecht MO, Datnoff LE, Kucharek TA, Nagata RT (2004) Influence of silicon and chlorothalonil on the suppression of gray leaf spot and increase plant growth in St. Augustine grass. Plant Dis 88:338–344

Burdon JJ, Thrall PH (2009) Coevolution of plants and their pathogens in natural habitats. Science 324:755–756

Chellemi DO (2002) Nonchemical management of soilborne pests in fresh market vegetable production systems. Phytopathology 92:1367–1372

Chellemi DO (2010) Back to the future: total system management (organic, sustainable). In: Gisi U, Chet I, Gullino ML (eds) Recent developments in management of plant diseases. Plant pathology in the 21st century. Springer, Dordrecht, pp 285–292

Chellemi DO, Gamliel A, Katan J, Subbarao KV (2016) Development and deployment of systems-based approaches for the management of soilborne plant pathogens. Phytopathology 106:216–225

Chet I (1987) *Trichoderma* – application, mode of action, and potential as a biocontrol agent of soilborne plant pathogenic fungi. In: Chet I (ed) Innovative approaches to plant disease control. Wiley, New York, pp 137–160
Clematis F, Minuto A, Gullino ML, Garibaldi A (2009) Suppressiveness to *Fusarium oxysporum* f. sp. *radicis-lycopersici* in re-used perlite and perlite-peat substrates in soilless tomatoes. Biol Control 48:108–114
Colla P, Gilardi G, Gullino ML (2012) A review and critical analysis of the European situation of soilborne disease management in the vegetable sector. Phytoparasitica 40:515–523
Colla P, Gilardi G, Garibaldi A, Pugliese M, Gullino ML (2014) Knowledge and technology transfer on sustainable use of pesticides: the life+ sustuse project experience on soil disinfestation sector. Acta Hortic 1044:359–361
Collins HP, Alva AK, Boydston RA, Cochran RL, Hamm PB, McGuire A, Riga E (2006) Soil microbial, fungal, and nematode responses to soil fumigation and cover crops under potato production. Biol Fertil Soils 42:247–257
DeLonge M, Basche A (2017) Leveraging agroecology for solutions in food, energy, and water. Elem Sci Anth 5:6. https://doi.org/10.1525/elementa.211
Delp CJ, Dekker J (1985) Fungicide resistance: definitions and use of terms. EPPO Bull 15:333–335
du Toit LJ (2004) Management of diseases in seed crops. In: Dekker GRM (ed) Encyclopedia of plant and crop science, Encyclopedia of plant and crop science. M. Dekker, New York, pp 675–677
Dun-chun H, Jia-sui Z, Lian-hui X (2016) Problems, challenges and future of plant disease management: from an ecological point of view. J Integr Agric 15(4):705–715
Esker PD, Savary S, McRoberts N (2012) Crop loss analysis and global food supply: focusing now on required harvests. CAB reviews: perspectives in agriculture, veterinary science. Nutr Nat Resour 052:1–14
Flood J (2010) The importance of plant health to food security. Food Sec 2:215–231
Franco-Ortega S, Tomlinson J, Gilardi G, Spadaro D, Gullino ML, Garibaldi A, Boonham N (2018) Rapid detection of *Fusarium oxysporum* f. sp. *lactucae* on soil, lettuce seeds and plants using loop-mediated isothermal amplification. Plant Pathol 67:1462–1473
Garbeva P, vanVeen JA, van Elsas JD (2004) Microbial diversity in soil: selection of microbial populations by plant and soil type and implications for disease suppressiveness. Annu Rev Phytopathol 42:243–270
Garibaldi A (1988) Research on substrates suppressive to *Fusarium oxysporum* and *Rhizoctonia solani*. Acta Hortic 221:271–277
Garibaldi A, Gullino ML (1987) Fusarium wilt of carnation: present situation, problems and perspectives. Acta Hortic 216:45–54
Garibaldi A, Gullino ML (2010) Le colture fuori suolo come approccio alla difesa delle colture minori. Protezione delle colture 3(2):9–12
Garibaldi A, Minuto G, Gullino ML (1994) Biological control of Fusarium wilt of basil (*Ocimum basilicum* L.). Brighton Crop Prot Conf – Pest Dis 2:811–815
Garibaldi A, Minuto A, Bertetti D, Gullino ML (2004a) Fusarium wilt of Gerbera in soil and soilless crops in Italy. Plant Dis 88:311
Garibaldi A, Gilardi G, Gullino ML (2004b) Seed transmission of *Fusarium oxysporum* f.sp. *lactucae*. Phytoparasitica 32:61–65
Garibaldi A, Gilardi G, Clematis F, Gullino ML, Lazzeri L, Malaguti L (2010) Effect of green *Brassica* manure and *Brassica* defatted seed meals in combination with grafting and soil solarization against Verticillium wilt of eggplant and Fusarium wilt of lettuce and basil. Acta Hortic 883:295–302
Gilardi G, Gullino ML, Garibaldi A (2013) Critical aspects of grafting as a possible strategy to manage soil-borne pathogens. Crop Prot 149:19–21

Gilardi G, Demarchi S, Gullino ML, Garibaldi A (2014a) Effect of simulated soil solarization and organic amendments on *Fusarium* wilt of rocket and basil under controlled conditions. J Phytopathol 162:557–566

Gilardi G, Demarchi S, Gullino ML, Garibaldi A (2014b) Varietal resistance to control fusarium wilts of leafy vegetables under greenhouse. Commun Agric Appl Biol Sci 79(2):21–27

Gilardi G, Demarchi S, Gullino ML, Garibaldi A (2015) Nursery treatments with non-conventional products against crown and root rot, caused by *Phytophthora capsici*, on zucchini. Phytoparasitica 43:501–508

Gilardi G, Demarchi S, Gullino ML, Garibaldi A (2016) Evaluation of the short term effect of nursery treatments with phosphite-based products, acibenzolar-S-methyl, pelleted *Brassica carinata* and biocontrol agents, against lettuce and cultivated rocket Fusarium wilt under artificial inoculation and greenhouse conditions. Crop Prot 85:23–32

Gilardi G, Franco Ortega S, van Rijswick PCJ, Ortu G, Gullino ML, Garibaldi A (2017a) A new race of *Fusarium oxysporum* f. sp. *lactucae* of lettuce. Plant Pathol 66(4):677–688

Gilardi G, Gullino ML, Garibaldi A (2017b) Soil disinfestation with dimethyl disulfide for management of Fusarium wilt of lettuce. J Plant Dis Protect 124:361–370

Gliessman SR (2014) Agroecology: the ecology of food systems, 3rd edn. CRC Press, Boca Raton, 405 pp

Gullino ML, Bonants P (2014) Detection and diagnostics of plant pathogens. Springer, Dordrecht, 200 pp

Gullino ML, Garibaldi A (2007) Critical aspects in management of fungal diseases of ornamental plants and directions in research. Phytopathol Mediterr 46:135–149

Gullino ML, Munkvold G (2014) Global perspectives on the health of seeds and plant propagation material. Springer, Dordrecht, 136 pp

Gullino ML, Camponogara A, Gasparrini G, Rizzo V, Clini C, Garibaldi A (2003) Replacing methyl bromide for soil disinfestation: the Italian experience and implications for other countries. Plant Dis 87:1012–1021

Gullino ML, Gilardi G, Garibaldi A (2014) Seed-borne pathogens of leafy vegetable crops. In: Gullino ML, Munkvold G (eds) Global perspectives on the health of seeds and plant propagation material. Springer, Dordrecht, pp 47–58

Gullino ML, Pugliese M, Garibaldi A (2015a) Use of silicon amendments against foliar and vascular diseases of vegetables grown soilless. In: Sangeetha G, Kurucheve V, Jayaraj J (eds) Sustainable crop disease management using natural products. CABI, Oxfordshire, pp 293–306

Gullino ML, Daughtreu ML, Garibaldi A, Elmer WH (2015b) Fusarium wilts of ornamental crops and their management. Crop Prot 73:50–59

Hadar Y (2011) Suppressive compost: when plant pathology met microbial ecology. Phytoparasitica 39:311–314

Harman GE (2006) Overview of mechanisms and uses of *Trichoderma* spp. Phytopathology 96:190–194

Heine G, Tikum G, Horst WJ (2007) The effect of silicon on the infection by and spread of Pythium aphanidermatum in single roots of tomato and bitter gourd. J Exp Bot 58:569–577

Hodgetts J, Tomlinson J, Boonham N, Gonzlzález-Martín I, Nikolić P, Swarbrick P, Yankey EN, Dickinson M (2011) Development of rapid in-field loop-mediated isothermal amplification (LAMP) assays for phytoplasmas. Bull Insectol 64:S41–S42

Hodgetts J, Karamura G, Johnson G, Hall J, Perkins K, Beed F, Nakato V, Grant M, Studholme DJ, Boonham N, Smith J (2015) Development of a lateral flow device for in-field detection and evaluation of PCR-based diagnostic methods for *Xanthomonas campestris* pv. *musacearum*, the causal agent of banana Xanthomonas wilt. Plant Pathol 64:559–567

Hoitink HAJ, Boehm MJ (1999) Biocontrol within the context of soil microbial communities: a soil-dependent phenomenon. Annu Rev Phytopathol 37:427–446

Hoitink HAJ, Fahy PC (1986) Basis for the control of soilborne plant pathogens with composts. Annu Rev Phytopathol 24:93–114

Hoitink HAJ, Locke JC (2012) An integrated approach to biological control of Fusarium species in containerized crops. In: Gullino ML, Katan J, Garibaldi A (eds) Fusarium wilts of greenhouse vegetable and ornamental crops. APS Press, St Paul, pp 109–115

Hoitink HAJ, Krause MS, Han DY (2001) Spectrum and mechanisms of plant disease control with composts. In: Stofella PJ, Kahn BA (eds) Compost utilization in horticultural cropping systems. Lewis Publishers, Boca Raton, pp 263–274

Hooker WJ (1981) Common scab. In: Compendium of potato diseases. American Phytopathological Society Press, St. Paul, pp 33–34

Huang CH, Roberts PD, Datnoff LE (2011) Silicon suppresses Fusarium crown and root rot of tomato. J Phytopathol 159:546–554

Janvier C, Villeneuve F, Alabouvette A, Edel-Hermann V, Mateille T, Steinberg C (2007) Soil health through soil disease suppression: which strategy from descriptors to indicators? Soil Biol Biochem 39:1–23

Jones JP, Engelhard AW, Woltz SS (1993) Management of Fusarium wilt of vegetables and ornamentals by macro-and microelement nutrition. In: Engelhard WA (ed) Soilborne plant pathogens: management of diseases with macro- and microelements. American Phytopathological Society Press, St Paul, pp 18–32

Katan J, DeVay JE (1991) Soil solarization. CRC Press, Boca Raton, 267 pp

King SR, Davis AR, Liu WG, Levi A (2008) Grafting for disease resistance. HortScience 43:1673–1676

Kloepper JW, Leong J, Teintze M, Schroth MN (1980) Pseudomonas siderophores: a mechanism explaining disease-suppressive soils. Curr Microbiol 4:317–320

Koch EGND, Roberts SJ (2014) Non-chemical seed treatment in the control of seed-borne pathogens. In: Gullino ML, Munkvold G (eds) Global perspectives on the health of seeds and plant propagation material. Springer, Dordrecht, pp 105–123

Kremen C, Miles A (2012) Ecosystem services in biologically diversified versus conventional farming systems: benefits, externalities, and trade-offs. Ecol Soc 17(4):40. https://doi.org/10.5751/ES-05035-170440

Lamichhane JR, Dachbrodt-Saaydeh S, Kudsk P, Messéan A (2016) Toward a reduced reliance on conventional pesticides in European agriculture. Plant Dis 100:10–24

Lamichhane JR, Dürr C, Schwanck AA, Robin MH, Sarthou JP, Cellier V, Messéan A, Aubertot JN (2017) Integrated management of damping-off diseases. A review. Agron Sustain Dev 37:10. https://doi.org/10.1007/s13593-017-0417-y

Larkin RP (2015) Soil health paradigms and implications for disease management. Annu Rev Phytopathol 53:199–221

Larkin RP, Griffin TS (2007) Control of soilborne potato diseases using Brassica green manures. Crop Prot 26:1067–1077

Leadbeater A, Gisi U (2010) The challenges of chemical control of plant diseases. In: Gisi U, Chet I, Gullino ML (eds) Recent developments in management of plant diseases, plant pathology in the 21th century. Springer, Dordrecht, pp 3–17

Lee JM, Oda M (2003) Grafting of herbaceous vegetable and ornamental crops. Hortic Rev 28:61–124

Lopez-Reyes GG, Gilardi G, Garibaldi A, Gullino ML (2014) Efficacy of bacterial and fungal biocontrol agents as seed treatments against *Fusarium oxysporum* f. sp. *lactucae* on lettuce. J Plant Pathol 96:535–539

Lopez-Reyes GJ, Gilardi G, Garibaldi A, Gullino ML (2016) *In vivo* evaluation of essential oils and biocontrol agents combined with heat treatments on basil cv Genovese Gigante seeds against *Fusarium oxysporum* f. sp. *basilici*. Phytoparasitica 44:35–45

Lu P, Gilardi G, Gullino ML, Garibaldi A (2010) Biofumigation with *Brassica* plants and its effect on the inoculum potential of *Fusarium* yellows of Brassica crops. Eur J Plant Pathol 126:387–402

Mancini V, Romanazzi G (2014) Seed treatments to control seedborne fungal pathogens of vegetable crops. Pest Manag Sci 70(6):860–868

Mathre DE, Cook RJ, Callan NW (1999) From discovery to use – traversing the world of commercializing biocontrol agents for plant disease control. Plant Dis 83:972–980

Matić S, Gilardi G, Gisi U, Gullino ML, Garibaldi A (2018) Differentiation of *Pythium* spp. from vegetable crops with molecular markers and sensitivity to azoxystrobin and mefenoxam. Pest Manag Sci, in press

Mbofung GY, Pryor BM (2010) A PCR-based assay for detection of *Fusarium oxysporum* f. sp. *lactucae* in lettuce seed. Plant Dis 94:860–866

Minuto A, Spadaro D, Garibaldi A, Gullino ML (2006) Control of soilborne pathogens of tomato with a commercial formulate of *Streptomyces griseoviridis* and solarization. Crop Prot 25:468–475

Munkvold GP (2009) Seed pathology progress in academia and industry. Annu Rev Phytopathol 47:285–311

Noble R, Coventry E (2005) Suppression of soil-borne plant diseases with composts: a review. Biocontrol Sci Tech 15:3–20

Notomi T, Okayama H, Masubuchi H, Yonekawa T, Watanabe K, Amino N, Hase T (2000) Loop-mediated isothermal amplification of DNA. Nucleic Acids Res 28:E63

Oerke EC (2006) Crop losses to pests. J Agric Sci 144:31–43

Oerke EC, Dehne HW, Schonbeck F, Weber A (1994) Crop production and crop protection: estimated losses in major food and cash crops. Elsevier, Amsterdam, p 808

Oliver R, Hewitt H (2014) Fungicides in crop protection. CABI Publishing, Wallingford, 200 pp

Page K, Dang Y, Dalal R (2013) Impacts of conservation tillage on soil quality, including soil-borne diseases, with a focus on semi-arid grain cropping systems. Australas Plant Pathol 42:363–377

Pasquali M, Dematheis F, Gullino ML, Garibaldi A (2007) Identification of race 1 of *Fusarium oxysporum* f. sp. *lactucae* on lettuce by inter-retrotransposon sequence-characterized amplified region technique. Phytopathology 97:987–996

Postma J (2004) Suppressiveness of root pathogens in closed cultivation systems. Acta Hortic 644:503–510

Pugliese M, Liu BP, Gullino ML, Garibaldi A (2008) Selection of antagonists from compost to control soil-borne pathogens. J Plant Dis Prot 115:220–228

Pugliese M, Benetti A, Garibaldi A, Gullino ML (2014) Use of compost from different origins to control soilborne pathogens in potted vegetables. Acta Hortic 1044:145–148

Pugliese M, Gilardi G, Garibaldi A, Gullino ML (2015) Organic amendments and soil suppressiveness: results with vegetable and ornamental crops. In: Meghvansi MK, Varma A (eds) Organic amendments and soil suppressiveness in plant disease management. Springer, Dordrecht, pp 495–509

Rossi V, Giosuè S, Caffi T (2010) Modelling plant diseases for decision making in crop protection. In: Oerke EC, Gerhards R, Menz G, Sikora RA (eds) Precision crop protection – the challenge and use of heterogeneity. Springer, Dordrecht, pp 241–258

Rossi V, Caffi T, Salinari F (2012) Helping farmers face the increasing complexity of decision-making for crop protection. Phytopathol Mediterr 51:457–479

Rouphael Y, Schwarz D, Krumbein A, Colla G (2010) Impact of grafting on product quality of fruit vegetables. Sci Hortic 127:172–179

Schlatter D, Kinkel L, Thomashow L, Weller D, Paulitz T (2017) Disease suppressive soils: new insights from the soil microbiome. Phytopathology 107:1284–1297

Shoresh M, Harman GE, Mastouri F (2010) Induced systemic resistance and plant responses to fungal biocontrol agents. Annu Rev Phytopathol 48:21–43

Siegwart M, Graillot B, Blachere Lopez C, Besse S, Bardin M, Nicot PC, Lopez-Ferber M (2015) Resistance to bio-insecticides or how to enhance their sustainability: a review. Front Plant Sci 6:381. https://doi.org/10.3389/fpls.2015.00381

Silva MTB, Menduina AM, Seijo YC, Viquiera FDF (2007) Assessment of municipal solid waste compost quality using standardized methods before preparation of plant growth media. Waste Manag Res 25:99–108

Spadaro D, Gullino ML (2005) Improving the efficacy of biocontrol agents against soilborne pathogens. Crop Prot 23:601–613

Spadaro D, Herforth-Rahmé J, Van Der Wolf J (2017) Organic seed treatments of vegetables to prevent seedborne diseases. Acta Hortic 1164:23–32

Spadaro D, Agustí N, Franco Ortega S, Hurtado Ruiz MA (2018) Diagnostics and identification of diseases, insects and mites. In: Gullino ML, Albajes R, Nicot P, van Lenteren JC (eds) Integrated pest and disease management in greenhouse crops. Springer, Dordrecht, in press

Spencer S, Benson DM (1981) Root rot of *Aucuba japonica* caused by Phytophthora cinnamomi and P. citricola and suppressed with bark media. Plant Dis 65:918–921

Srinivasan K, Gilardi G, Spadaro D, Garibaldi A, Gullino ML (2010) Molecular characterization through IGS sequencing of *formae speciales* of *Fusarium oxysporum* pathogenic on lamb's lettuce. Phytopathol Mediterr 49:309–320

Stark C, Condron LM, Stewart A, Di HJ, O'Callaghan M (2007) Influence of organic and mineral amendments on microbial soil properties and processes. Appl Soil Ecol 35:79–93

Stout M, Davis J (2009) Keys to the increased use of host plant resistance in integrated pest management. In: Peshin R, Dhawan AK (eds) Integrated pest management innovation process. Springer, Dordrecht, pp 163–181

Strange RN, Scott PR (2005) Plant disease: a threat to global food security. Annu Rev Phytopathol 43:83–116

Sullivan DM, Miller RO (2001) Compost quality attributes, measurements, and variability. In: Stofella PJ, Kahn BA (eds) Compost utilization in horticultural cropping systems. Lewis Publishers, Boca Raton, pp 95–120

Termorshuizen AJ, van Rijn E, van der Gaag DJ, Alabouvette C, Chen Y, Lagerlöf J, Malandrakis AA, Paplomatas EJ, Rämert B, Ryckeboer J, Steinberg C, Zmora-Nahum S (2006) Suppressiveness of 18 composts against 7 pathosystems: variability in pathogen response. Soil Biol Biochem 38:2461–2477

Thomas JE, Wood TA, Gullino ML, Ortu G (2017) Diagnostic tools for plant biosecurity. In: Gullino ML, Stack J, Fletcher J, Mumford J (eds) Practical tools for plant and food biosecurity. Springer, Dordrecht, pp 209–226

Tomlinson JA, Ostoja-Starzewska S, Adams IP, Miano DW, Abidrabo P, Kinyua Z, Alicai T, Dickinson MJ, Peters D, Boonham N, Smith J (2013) Loop-mediated isothermal amplification for rapid detection of the causal agents of cassava brown streak disease. J Virol Methods 191:148–154

Uriarte RF, Shew HD, Bowman DC (2004) Effect of soluble silica on brown patch and dollar spot of creeping bentgrass. J Plant Nutr 27:325–339

Vallad GE, Goodman RM (2004) Systemic acquired resistance and induced resistance in conventional agriculture. Crop Sci 44:1920–1934

Van Bruggen AHC, Gamliel A, Finckh MR (2016) Plant disease management in organic farming systems. Pest Manag Sci 72:30–44

Van Os EA (2010) Disease management in soilless culture system. Acta Hortic 883:385–393

Van Os EA, Hofland-Zijlstra JD, Postma J (2012) Management of irrigation water as a tool for disease management. In: Gullino ML, Katan J, Garibaldi A (eds) Fusarium wilts of greenhouse vegetable and ornamental crops. APS Press, St Paul, pp 75–82

Walters D, Walsh D, Newton A, Lyon G (2005) Induced resistance for plant disease control: maximizing the efficacy of resistance elicitors. Phytopathology 95:1368–1373

Walters DR, Ratsep J, Havis ND (2013) Controlling crop diseases using induced resistance: challenges for the future. J Exp Bot 64:1263–1280

Webster MA, Dixon GR (1991) Calcium, pH and inoculum concentration influencing colonization by *Plasmodiophora brassicae*. Mycol Res 95:64–73

Weisburg WG, Barns SM, Pelletier DA, Lane DJ (1991) 16S ribosomal DNA amplification for phylogenetic study. J Bacteriol 173(1):697–703

Welbaum GE, Sturz AV, Dong Z, Nowak J (2004) Managing soil microorganisms to improve productivity of agro-ecosystems. Crit Rev Plant Sci 23:175–193

Yuen GY, Schroth MN, McCain AH (1985) Reduction of Fusarium wilt of carnation with suppressive soils and antagonistic bacteria. Plant Dis 69:1071–1075

Zhan J, Thrall PH, Papaïx J, Xie LH, Burdon JJ (2015) Playing on a pathogen's weakness: using evolution to guide sustainable plant disease control strategies. Annu Rev Phytopathol 53:19–43

Part V
Genetic Resources and Crop Improvement for Sustainable Agriculture

Conservation of Biodiversity and Genetic Resources for Sustainable Agriculture

Mehmet Karaca and Ayse Gul Ince

Abstract The aim of conservation of biodiversity and genetic resources is to secure existing genetic biodiversity while allowing evolution and to build a wide base of genetic resources that meet demands of present and future uses not only for human kind but also for all livings forms on earth. Genetic resources for sustainable agriculture are irreplaceable natural sources for food, spice, medicine, fuel, fodder and building materials. Genetic diversity is an essential natural resource, like soil, water and the sun, without it life may not exist. Unfortunately, the most dramatic decline in the genetic diversity occurred with dramatic yield improvement of modern crops due to the development of hybrid technologies, synthetic fertilizers, irrigation, pest managements and farm machinery. Among 500,000 land species on earth, 100,000–160,000 are estimated to be under threats or about to enter the red list. It is estimated that today 15% of the earth land surface is protected for conservation, however, coverage varies widely among ecosystems and countries. Today approximately 7.4 million germplasm accessions representing more than 16,500 plant species are conserved in approximately 1750 gene banks worldwide, and more than two million accessions are estimated to be added soon. However, most gene banks around the world lack facilities, sufficient funds and staff to successful regeneration of gene bank collections and maintenances. Conservation of biodiversity and genetic resources is needed more than ever, given the cumulative effects of exploitation and destruction that is compounded by climate change. This chapter focuses on a brief history of public awareness on biodiversity and genetic resources for sustainable agriculture with specific highlights on next generational high-throughput techniques. The application of high-throughput phenotyping genomics and phenomics

M. Karaca (✉)
Department of Field Crops, Akdeniz University, Antalya, Turkey
e-mail: mkaraca@akdeniz.edu.tr

A. G. Ince
Vocational School of Technical Sciences, Akdeniz University, Antalya, Turkey
e-mail: aince@akdeniz.edu.tr

M. Farooq, M. Pisante (eds.), *Innovations in Sustainable Agriculture*,
https://doi.org/10.1007/978-3-030-23169-9_12

opens new ways for a substantial enhancement of plant conservation activities for sustainable agriculture. Monitoring tools utilizing machine and deep learning approaches coupled with traditional plant breeding could not only inform us about the risk of genetic erosion through genetic drift and nonrandom viability selection within gene banks, but could help us to fight current pest and disease outbreaks, would also have the dual effects of contributing to enhanced food production and to the conservation of plant genetic resources.

Keywords Crop biodiversity · Gene discovery · Next generation genotyping · Phenotyping · Sources of diversity · Threats

Abbreviations

2b-RAD	Type IIB Restriction Enzyme Digestion Restriction Site Associated DNA
2D	Two Dimensional
3'-UTRs	3'-Untranslated Regions
3D	Three Dimensional
5'-UTRs	5'-Untranslated Regions
CBD	The Convention on Biological Diversity
CCD	Charge-coupled Device
cDNA	Complementary DNA
CFI	Chlorophyll Fluorescence Imaging
CGIAR	Consultative Group on International Agricultural Research
CGRFA	The Commission on Plant Genetic Resources for Food and Agriculture
CMOS	Complementary Metal Oxide Semiconductor
CNVs	Copy Number Variations
CPC	Centre for Plant Conservation
CRoPS	Complexity Reduction of Polymorphic Sequences
ddRAD	Double-digest RAD Sequencing
ECOSOC	The United Nations Economic and Social Council
FAO	Food and Agriculture Organization
GBS	Genotyping by Sequencing
GPS	The Global Positioning System
GRs	Genetic Resources
GT-seq	Genotyping-in-Thousands by Sequencing
IARCs	International Agricultural Research Centres
IBP	International Biological Program
IBPGR	International Board for Plant Genetic Resources
ICARDA	The International Center for Agricultural Research in the Dry Areas
INDELs	Insertions or Deletions

IPGRI	International Plant Genetic Resources Institute
IR	Infrared
iRRL	Improved RRLs
ITPGR	International Treaty for Plant Genetic Resources
IUCN	The World Conservation Union
LiDAR	Light Detection and Ranging
LMOs	Living Modified Organisms
loRNA	Long RNA
MIG-seq	Multiplexed Inter-SSR Genotyping by Sequencing
miRNA	Micro RNA
MSG	Multiplexed Shotgun Genotyping
NGS	Next Generation Sequencing
NIR	Near-infrared
NMR	Nuclear Magnetic Resonance
NRC/NAS	National Research Council/National Academy of Sciences
PAV	Presence/Absence Variations
PCR	Polymerase Chain Reaction
pERPLs	Paired-end Reduced Representation Libraries
RAD	Restriction Site-associated DNA
RAD-seq	Restriction Site-associated DNA Sequencing
RAM-seq	Random Amplicon Sequencing
Rapture	RAD Capture
RESTseq	Restriction Fragment Sequencing
RF	Radio Frequency
RNA-seq	High-throughput RNA Sequencing Analysis
RRL	Reduced-representation Library
RRS	Reduced Representation Shotgun Sequencing
SAR	Synthetic Aperture Radar
SBG	Sequence-based Genotyping
SEQ	Sequencing
sGBS	Spiked Genotyping-by-sequencing
SkimGBS	Skim Genotyping-by-sequencing
SLAF-Seq	Specific Length Amplified Fragment Sequencing
SMRT	Single Molecule Real-Time
SMTA	Standard Material Transfer Agreement
SNP	Single Nucleotide Polymorphism
sRNA	Small RNA
SSRs	Simple Sequence Repeats
tGBS	Tunable Genotyping-By-Sequencing
ToF	Time of Flight
UAV	Unmanned Aerial Vehicles
UN	United Nations
UNCED	The United Nations Conference on Environment and Development
UNCSD	United Nations Conference on Sustainable Development
UNFCCC	United Nations Framework Convention on Climate Change

UV	Ultraviolet
VIS	Visible
WES	Whole Genome Exome Sequencing
WGR	Whole Genome Resequencing
WGS	Whole Genome Sequencing
X-ray CT	X-ray Computed Tomography

1 Introduction

There exist approximately 500,000 land plant species on earth. Although 10–20% of earth's land plant species are not scientifically studied, about 100,000–160,000 land species are estimated to be threatened. About 15% of the earth land surface is protected for conservation, however, coverage varies widely within ecosystem (Sharrock et al. 2014; Coetzee et al. 2014; Geldmann et al. 2015; Sajjadi et al. 2015; Corlett 2016; Yousef et al. 2018). Conservation of plant biodiversity and genetic resources probably begun some 13,000–15,000 years ago when hunter-gatherers started to collect plants. Gatherers turned into farmers and discovered to save crop seeds they found easiest to process or tasted the best (Zhang et al. 2017). During the history of crop domestication, more than 7000 plant species have been cultivated. However, today in modern agriculture no more than 500 plant species are being grown, among the most widely cultivated crop species include cereals such as wheat, maize, rice (Duvick 2005), fiber crops such as cotton, flax, hemp (Uygur Gocer and Karaca 2016a), oil crops such as sunflower, canola, soybean (Edgerton 2009), pulses such as chickpea, lentil, bean (Ince and Karaca 2011a), forage crops such as bermudagrass, alfalfa, common vetch (Cakmakci et al. 2006; Karaca and Ince 2008), vegetables such as tomato, pepper, lettuce, mushroom, carrot, asparagus, celery, turmeric, artichoke, (Ince et al. 2009a, b, 2010a, b, c; Polat et al. 2010; Ince and Karaca 2011b), starch and sugar plants such as potato, sugar beet, cassava (Peroni and Hanazaki 2002), medicinal and aromatic plants such as sage, oregano, thyme, ginger, jojoba, chicory (Karaca et al. 2008, 2015; Ince et al. 2010d, e, 2011a; Ince 2012), ornamental crops such as carnation, rose, lily (Ince et al. 2009c; Ince and Karaca 2015a).

During the cultivation periods, genomes of the most cultivated species mentioned above experienced intense selection for desirable characteristics, many of which are not found in wild and crop wild relatives. On the other hand, wild species and crop wild relatives are rich in genes against biotic and abiotic stress factors and secondary metabolites such as phenolic compound many of which have economic importance (Elmasulu et al. 2009; Ince and Karaca 2009; Karaca et al. 2011). During the domestication periods cultivated crop species received long-standing events include the domestication bottleneck (occurs when a subset of the wild populations is brought into cultivation), directional selection (diversity can subsequently be lost through selective breeding for desirable traits during crop improvement), dispersal bottlenecks and gradual increase of genetic diversity as a consequence of gene transfer within and between the domesticated species or crop wild relatives. Also cultivated species gained or lost allelic combinations via mutations and recombination events which could affect conservation of biodiversity and genetic resources for sustainable agriculture (Shepherd et al. 2016; Zhang et al. 2017; Kopnina et al. 2018).

In the agricultural history, the most dramatic decline in the genetic diversity occurred with dramatic yield improvement due to the development and widespread use of new farming technologies such as hybrid technologies, synthetic fertilizers, irrigation, pest managements and farm machinery (especially during the period of Green Revolution). Modernization of agriculture started in the middle of the nineteenth century in Europe and North America leading to the irreversible loss of innumerable heterogeneous landraces and other genetic materials. In Asia and other developing countries, the Green Revolution started in the beginning of the twentieth century and gained momentum in the 1960s. It is important to note that modernization of agriculture was evident in Europe and North America long before World War II indicating that Green Revolution started during nineteenth century in today's most developed countries (Baur 1914; Harlan 1975).

Biodiversity and genetic resources of plant species and their wild relatives are not equally and evenly distributed on earth. Significant amounts of *in situ* genetic resources are within the developing countries while developed countries have *ex situ* genetic resources. It is known that significant amounts of landraces in North America and northwestern Europe were lost due to genetic erosion. It is also interesting to note that loss in genetic variations due to genetic erosion has been less intensive in remote areas where these traditional varieties are still grown in small cultivation and patches of land (Evenson and Gollin 2003; Ince et al. 2009d; Karaca and Ince 2011a, b).

Landraces, heirlooms and traditional varieties are old cultivars selected by farmers over hundreds years to best fit their needs. These genetic resources generally display greater diversity and many desired metabolomic traits than modern cultivars as they have been selected to adapt to local, sometimes hostile environments. Metabolomic traits especially taste and health-promoting related traits are contained in landraces, heirloom and traditional varieties and thus serve as a good source of the best alleles for organoleptic quality improvement (Ince and Karaca 2011a; Gascuel et al. 2017; Vlk and Repkova 2017). Therefore, conservation of biodiversity and genetic resources of these valuable resources are very important for sustainable agriculture.

Landraces, heirloom and traditional varieties are also beneficial crops for soil fertility characteristics that save soil's organic matter and protect it from soil erosion. Also many of heirloom and traditional varieties contribute to healthy human nutrition and potential sources of resistance to abiotic and biotic stresses. However, they are less suited for new agricultural technologies and do not provide high yields as high as modern cultivars. Due to their less productivity in terms of yield, they are not widely cultivated in most parts of the world at present, thus, genetic diversity within landraces, heirloom and traditional varieties is seriously reduced. The rapid expansion of plant breeding applications during the second half of the twentieth century brought the introduction of a big number of improved varieties, which progressively replaced old landraces, especially in developing countries resulted in genetic erosion. Also the release of a large number of commercial varieties into traditional farming systems caused a reduction in the number of varieties cultivated in a given area. Unfortunately, almost due to the same reasons biodiversity in other valuable genetic materials such breeding lines, genetic stocks, obsolete cultivars, landraces, accessions, heirlooms, traditional or heritage varieties of crops along with wild species and crop wild relatives (please see the glossary at the end of this chapter for short descriptions) is narrowed. In addition, due to the climatic changes and monoculture agricultural practices genetic diversity on earth is being eroded (Elmasulu et al. 2011; Ince et al. 2010f). Therefore, conservation of biodiversity and genetic resources for sustainable agriculture is required to secure existing genetic biodiversity and to build a wide base of genetic resources that meet demands of present and future utilizations (Evenson and Gollin 2003; Ince et al. 2009e; Karaca et al. 2013).

According to the central dogma of conservation genetics, genetic variability is beneficial and thus it is worth preserving to the greatest extent (Pertoldi et al. 2007). Therefore, conservation of biodiversity is important for maintaining the adaptive potential of species and populations for sustainable agriculture. In turn, conservation of biodiversity ultimately depends on the conservation of genetic diversity and increasing genetic variation enhances the probability of population survival (Aravanopoulos 2016). Almost everybody on earth agrees that biodiversity is at risk from multiple threats including increasing human population and the genetic diversity contained within plant genetic resources needs to be conserved.

There have been two different concepts on what types of biodiversity and genetic resources should be conserved. Also important concepts on how biodiversity and genetic resources are conserved and how genetic erosion is reduced, and what kinds of technologies can be implemented to enhance their conservation and use still need attentions. During the second half of twentieth century germplasm collection expeditions adopted an approach called "mission-oriented approach" (Dulloo et al. 2013; Buse et al. 2015; Kopnina et al. 2018). This approach focuses on targets specific to plant breeding projects. Therefore, collected genetic materials are mainly used in plant breeding stations. However, this approach, while responding to immediate individual or organizational needs, have limited effects on reducing the genetic erosion. Opposite to the mission-oriented approach is the "generalist approach" that directs towards collecting and conservation of much possible genetic materials in plant centers of origin (Bayerl et al. 2017; Fu 2017).

Today approximately 7.4 million germplasm resources representing more than 16,500 plant species are being conserved in 1750 gene banks worldwide, and more than two million accessions are estimated to be added (Shepherd et al. 2016; Fu 2017). Conservation of these genetic resources uses *in situ* strategy and *ex situ* strategy or both strategies. Although *ex situ* and *in situ* conservations are two main strategies for conserving genetic resources for sustainable agriculture, they are equally important and should be utilized at the same time as complementary approaches (Dulloo et al. 2013; van Kleunen et al. 2015).

Ex situ conservation is the conservation of genetic resources outside their natural habitats and it is generally used to conserve populations that potentially under threat. *Ex situ* conservations in gene banks are in the form of seeds, live plants, tissues, cells and/or DNA materials. On the other hand, *in situ* conservation is the conservation of populations of species at their natural habitats or close their gene centers including maintenance and recovery of viable populations of species. *In situ* conservation can be either on farm, requiring the maintenance of the agro-ecosystem along with the cultivation and selection processes on local varieties and landraces, or in the wild, which involves the maintenance of the ecological functions that allow species to evolve under natural conditions (Ince and Karaca 2011a; Korun et al. 2013; Buse et al. 2015; Hernandez-Suarez 2018; Kopnina et al. 2018; Manhaes et al. 2018).

We know that identification of genetic resources is also as important as conservation of biodiversity for sustainable agriculture. In this chapter, we also discuss the novel, and emerging approaches such as next generation phenotyping and next generation genotyping systems for detecting and conservation of biodiversity and genetic resources for sustainable agriculture after a brief historical view on global biodiversity and genetic resources, values and current status of genetic resources and diversity.

2 A Brief Historical View on Global Biodiversity and Genetic Resources

During 1845 and 1849, Irish Potato Famine caused about one million people death and a million more migrated from Ireland. One of the main causes of famine was potato blight disease that ravaged potato throughout the Europe during 1840s. Probably this was the first well know indication of the result of genetic erosions in cultivated crops. Second indication of genetic erosion was noted in 1970 with considerable yield loss in United States of America corn production caused by fungus *Helminthosporium maydis* race T, known as the southern corn leaf blight. These two events are good examples showing the consequences of the lack of genetic diversity and the use of monoculture modern varieties instead of landraces (Fu 2017).

Accumulated scientific, political and public awareness on conservation of biodiversity and genetic resources for sustainable agriculture were internationally sounded in 1960s and the landmark conferences sponsored by international organizations such as the Food and Agriculture Organization of the United Nations (FAO), the International Biological Program (IBP) and World Bank were held. The IBP and FAO in 1967 helped to lay the foundation for modern genetic resources conservation efforts (Goulart et al. 2018; Li et al. 2018). Table 1 summarizes some events, conferences and establishments concerned conservation of biodiversity and genetic resources of the world.

Table 1 Events relevant to the establishment and evolution of international instruments related to the conservation and sustainable utilization of plant genetic resources during the period 1961–2018

Event	Some underpinning principle(s)/agreements
1961–1973	Rising concern about formulating criteria for the conservation, diversity and genetic resources for sustainable agriculture
FAO technical meeting on plant exploration and introduction, (Rome, 1961)	The use of *ex situ* and *in situ* conservation strategies as a complementary strategy to conserve gene erosion of landraces and wild relatives
FAO/IBP technical conference plant exploration, utilization and conservation of plant genetic resources, (Rome, 1967)	Priority geographic areas for exploration and conservation of plant genetic resources.
	Establishment of the technical advisory committee for conservation of plant genetic resources
	Establishment of the world network of genetic resources centers
Third session of the FAO panel of experts on plant exploration and introduction, (Rome, 1969)	Establishment of a coordinating center support to gene banks already existing in international agricultural research centers (IARCs) of the consultative group on international agricultural research (CGIAR)
Founding meeting of the consultative group on international agricultural research (CGIAR), (Washington, DC, 1971)	
UN conference on human environment, (Stockholm, 1972)	
Establishment of the international board for plant genetic resources (IBPGR) group, (Beltsville, 1972)	
FAO/IBP technical conference on genetic resources, (Rome, 1973)	
1981–1991	Suggested clarity regarding the legal situation of the *ex situ* collections
FAO/IBP technical conference on genetic resources, (Rome, 1981)	Suggesting the need for an international agreement to ensure the conservation, maintenance and free exchange of plant genetic resources
21st session of the FAO conference, (Rome, 1981)	Adoption of the international undertaking on plant genetic resources establishment of the commission on plant genetic resources for food and agriculture (CGRFA) and of the global system on plant genetic resources
22nd session of the FAO conference, (Rome, 1983)	
	Plant breeders' rights are not inconsistent with the international undertaking, recognition of farmers' rights

(continued)

Table 1 (continued)

Event	Some underpinning principle(s)/agreements
National forum on biodiversity (Washington 1986)	Requested FAO a code of conduct for biotechnology to be used in conservation of genetic resources
25th session of the FAO conference, (Rome, 1989)	International board for plant genetic resources (IBPGR) was transformed into the international plant genetic resources institute (IPGRI)
3rd regular session of CGRFA, (Rome, 1989)	
26th session of the FAO conference, (Rome, 1991)	
1992–1994	Plant genetic resources of nations are recognized the sovereign rights of nations
The United Nations conference on environment and development (UNCED) (Rio de Janeiro, 1992)	Agreement on the development of the 1st state of the world's plant genetic resources and global plan of action on plant genetic resources
1st extraordinary session of the CGRFA, (Rome, 1994)	Agreement on risk assessment and management of all aspects of biotechnology
The convention on biological diversity (CBD, 1992)	Agreement on international policy framework for the conservation of plant genetic diversity
Establishment of the Scarascia Mugnozza community genetic resources center, (Chennai, 1994)	Agreement to hold the designated germplasm in trust for the benefit of the international community
1995–1999	Stating that biodiversity loss is not only an important environmental problem, but also a socio-economic, political and ethical problem
28th session of the FAO conference, (Rome, 1995)	
Science academies summit at the M. S. Swami Nathan research foundation, (Madras, 1996)	*Ex situ* conservation of plant genetic resources are essential but must be integrated by in situ, on farm, a community level conservation strategy
4th international technical conference on plant genetic resources, (Leipzig, 1996)	Genetic resources should be conserved and made available to scientists and farmers, but access should be regulated by international agreements
World Food Summit, (Rome, 1996)	
1st extraordinary meeting of the conference of the parties to the CBD, (Cartagena, 1999)	

(continued)

Table 1 (continued)

Event	Some underpinning principle(s)/agreements
2000–2004	Protecting biological diversity from the potential risks posed by living modified organisms (LMOs).
Resumed session of the conference of the parties to the CBD, (Montreal, 2000)	Recognition of farmers' rights access to plant genetic resources and equitable sharing of the benefits derived from their use
31st session of the FAO conference, (Rome, 2001)	
6th ordinary meeting of the conference of the parties to the convention on biological diversity, (Hague, 2002)	The rate of loss is still accelerating and the threats must be addressed
UN world summit on sustainable development, (Johannesburg, 2002)	Ensuring an absolutely requirement of funding for the conservation of plant genetic resources
Convention on biological diversity (CBD, Paris 2002)	The world conservation union (IUCN) red list categories and criteria were preliminary constructed to assess the threatened status of species or lower taxa on a global scale. Red data lists can play a crucial role by focusing attention on species most in need of conservation action
Establishment of the global crop diversity trust (now renamed crop trust) (2004)	
2005–2009	Stated that over the past 50 years, humans have changed ecosystems and have substantial net gains for their well-being, but at growing environmental costs
Publication of the millennium ecosystem assessment (2005)	
1st meeting of the governing body of international treaty for plant genetic resources (ITPGR), (Madrid, 2006)	Degradation of ecosystem services could grow significantly worse during the first half of this century
Establishment of the Svalbard global seed vault, (Svalbard, 2008)	Standard material transfer agreement (SMTA) is the legal instrument through which the multilateral system of access and benefit sharing operates
12th Regular session of the CGRFA, (Rome, 2009)	Recognition of the crop trust as an essential element of the treaty's funding strategy, in regard to *ex situ* conservation and availability of plant genetic resources
36th session of the FAO conference, (Rome, 2009)	*Ex situ* gene bank collections are put under the international treaty for plant genetic resources (ITPGR)
2010–2012	Establishing more predictable conditions for access to genetic resources, helping to ensure benefit-sharing when genetic resources leave the country providing the genetic resources
International technical FAO Conference on agricultural biotechnologies in developing countries (Guadalajara, 2010)	Target for 2020, establishment of an online flora of all known plants
10th meeting of the conference of the parties to the convention on biological diversity, (Nagoya, 2010)	Status and trends of biotechnologies applied to the conservation and utilization of genetic resources for food and agriculture and matters relevant for their future development
13rd regular session of the CGRFA, (Rome, 2011)	Biotechnologies largely used for conservation and use of plant genetic resources in developed countries but many developing countries do not have biotechnological capacities
143rd session of the FAO council, (Rome, 2011)	Need for a roadmap on climate change and genetic resources for food and agriculture

(continued)

Table 1 (continued)

Event	Some underpinning principle(s)/agreements
2012–2018	Need to promote, enhance and support more sustainable agriculture that improves food security, eradicates hunger
United nations conference on sustainable development (UNCSD), (Rio de Janeiro, 2012)	
14th regular session of the CGRFA (Rome, 2013)	Conserving land, water, and genetic resources, biodiversity and ecosystems and enhancing resilience to climate change and natural disasters
International symposium on forest biotechnology for smallholders, (Foz do Iguacu, 2015)	Importance of genetic resources for food and agriculture for coping with climate change
39th session of the FAO conference, (Rome, 2015)	Biotechnologies can be used in production systems, based on agro-ecological principles, to enhance productivity while ensuring sustainability, conservation of genetic resources and use of indigenous knowledge
FAO international symposium on the role of agricultural biotechnologies in sustainable food systems and nutrition, (Rome, 2016)	Although in June 2017, United States announced the withdrawal for the Paris agreement, in May 2018, 195 UNFCCC members have signed the Paris agreement, and 177 have become party to it
United Nations Framework Convention on Climate Change (May 2018)	The Paris agreement aims long-term goal of keeping the increase in global average temperature to well below 2 °C above pre-industrial levels, and to aim to limit the increase to 1.5 °C, since this would significantly reduce risks and the impacts of climate change

During the period of 1961 to 1991, a large number of scientific and technical conferences, and workshops were held mainly in Europe and United States of America. These activities were among the initiatives on rising the concern about conservation of biodiversity, sustainable use of its components and a fair and equitable sharing of its benefits. For instance, the term genetic erosion was coined at the Technical Conference Plant Exploration, Utilization and Conservation of Plant Genetic Resources of FAO/International Biological Program, held in Rome in 1967 to describe this loss of individual genes and combinations of genes. The concept of biodiversity was first conceived by Walter G. Rosen from the National Research Council/National Academy of Sciences (NRC/NAS) in 1985, while planning to conduct a forum on biological diversity (Sonnino 2017; Kopnina et al. 2018).

During the last decades of twentieth century, plant genetic resources-related activities were primarily focused on the collection and *ex situ* conservation of germplasm. E*x situ* conservation strategy, which was suggested in 1973, was reinforced in 1981 by the FAO/IBP Technical Conference on Genetic Resources. However, several scholars expressed concern that the storage of seeds in gene banks (*ex situ* conservation) not allowed natural evolution to proceed (Brown and Hodgkin 2015). Centre for Plant Conservation (CPC, 1991) set *ex situ* guidelines range from relatively small targets (e.g. collection of seed from 10 individuals in each of five populations) to comprehensive collections of germplasm. Furthermore, in 1996, the Leipzig Declaration appropriated both the *ex situ* and *in situ* approaches, considering them as not mutually exclusive, but complementary components of conservation programs. The Convention on Biological Diversity (CBD) promotes *ex situ* conservation, via the establishment of protected areas and natural parks. In addition, on-farm conservation is often adopted to grow, utilize and conserve landraces, native varieties and other local materials, within their original landscapes and traditional farming systems (Kopnina et al. 2018).

The United Nations Conference on Environment and Development (UNCED) (also known as the Rio de Janeiro Earth Summit) was a major United Nations conference held in Rio de Janeiro from 3 to 14 June 1992. In 2012, the United Nations Conference on Sustainable Development was also held in Rio de Janeiro from 13 to 22 June, and is commonly called Rio + 20 or Rio Earth Summit 2012. Among issues addressed in these two conferences included systematic scrutiny of patterns of production of toxic components, such as lead in gasoline, or poisonous waste including radioactive chemicals, alternative sources of energy to replace the use of fossil fuels which delegate linked to global climate change, new reliance on public transportation systems in order to reduce vehicle emissions, congestion in cities and the health problems caused by polluted air and smoke and the growing usage and limited supply of water (Ogwu et al. 2014; Kopnina et al. 2018). The Convention on Biological Diversity was opened for signature at the Earth Summit, and made a start towards redefinition of measures that did not inherently encourage destruction of natural eco-regions and so called uneconomic growth. USA failed to sign the proposed Convention on Biological Diversity. In order to ensure compliance to the agreements at Rio delegates to the Earth Summit established the Commission on Sustainable Development. In 2013, the Commission on Sustainable Development was replaced by the High-level Political Forum on Sustainable Development that meets every year as part of the United Nations Economic and Social Council (ECOSOC) meetings, and every fourth year as part of the General Assembly meetings. Critics point that many of the agreements made in Rio have not been realized regarding such fundamental issues as fighting poverty and cleaning up the environment (Sonnino 2017; Kopnina et al. 2018).

3 Genetic Resources, Conservation and Values

3.1 Genetic Resources

Genetic resources are genetic materials of actual or potential values that are used in the future improvement of crops utilized in food, spice, medicine, fuel, fodder and building material production. Genetic resources consist of genotypes or population of landraces, advanced cultivars, domestically bred cultivars, old local cultivars, genetic stocks, wild relatives and weedy species which are maintained in the form of seeds, plants, tissues etc. Genetic resources should be properly monitored in order to reduce the risk of within-gene bank erosion or *in situ* conservations. In the absence of such monitoring, some unique germplasm accessions are lost and this reduces the biodiversity coverage present in a gene bank collection or *in situ* conservations (Krishnan et al. 2013; Brown and Hodgkin 2015; Kopnina et al. 2018).

Genetic resources of crop wild relatives on earth are not evenly distributed geographically, therefore, there exist great differences among the gene banks in terms of number and biodiversity. Due to global warming conditions efforts to collect germplasm of crop wild relatives have gradually increased (Castaneda-Alvarez et al. 2016). However, gene bank capacities and low funding limit the success of these efforts. Crop wild relatives usually have low germination rate, and require taxonomic evaluation, specialized pollinators and different life cycle. Many gene banks have insufficient capacity to maintain both old and newly acquired germplasm, affecting the efficacy of long-term germplasm conservation (Ogwu et al. 2014; Kopnina et al. 2018).

Conservation of genetic resources for sustainable agriculture is the art and science for the benefit of genetic improvement of crops in present and future generations. Researchers and staff involved in germplasm conservation through *ex situ* and *in situ* methods are expected to have knowledge and experiences in a variety of fields including biology, molecular biology, molecular genetics, plant systematics, population genetics, plant pathology, plant physiology, plant ecology, biochemistry, computer science, legal science, economics, and political science. This indicates that conservation teams should have special training, however, there is no specific institutes providing comprehensive professional training in the germplasm conservation. In the developing countries many researchers and staff working in gene banks since the 1970s have retired or will retire soon (Fu et al. 2015). That means some useful knowledge and experience in germplasm conservation for sustainable agriculture are being lost without replacing young and dynamic researchers due to restricted financial supports.

Gene banks in different parts of the world may suffer catastrophic events and can collapse. For instance, the N. I. Vavilov Institute of Plant Genetic Resources was damaged during World War II. Genetic resources in gene banks of Guinea-Bissau, Liberia and Sierra Leone have been damaged due to civil wars. Afghanistan's gene bank in Kabul and Iraq's Abu Ghraib national gene bank in Baghdad were looted. Syria's gene bank at the International Center for Agricultural Research in the Dry Areas (ICARDA) in Aleppo was probably damaged during civil war although the center has been relocated to Terbol, Lebanon (Bhattacharya 2016). Some gene banks have been destroyed due to natural disasters too. For instance, national seed bank of Nicaragua was lost in the 1971 earthquake. The national seed bank of Honduras was demolished by hurricane Mitch in 1998. The Thai gene bank was flooded and some of the 20,000 unique rice accessions were lost forever in 2011. The typhoon Milenyo damaged the Philippines' national gene bank in 2006 and was again destroyed by fire in January 2012. Destruction due to increased frequency of flooding, typhoons and civil war disasters rationalized for constructing the Svalbard Global Seed Vault in 2006 for long-term safety backup of valuable germplasm (Fu et al. 2015; Fu 2017).

3.2 Conservation of Genetic Resources

Genetic resources can be conserved either *in situ* (in their natural setting) or *ex situ* (outside their natural setting). *Ex situ* conservation is the dominant method of conserving natural ecosystem. On the other hand, *ex situ* conservation is commonly used by plant breeders. However, agricultural resources can also be held *in situ*. Many farmers developed landraces contain significant diversity and encouraging use of these varieties is one method to conserve agro-biodiversity *in situ*. Wild relatives of cultivated varieties may also be conserved *in situ* on wild land. More recent approaches view the *ex situ* and *in situ* forms of conservation as complimentary, rather than as substitutes (Ogwu et al. 2014; Kopnina et al. 2018).

The *ex situ* conservation method needs to obtain genetic materials from their ecological environment and grow them in different environment for long-term conservation. *Ex situ* examples include gene banks, national parks, botanic gardens, arboretums, museums, zoos and protected areas. Compared to *in situ* conservation certain methods of *ex situ* conservation can be used to store large amounts of genetic material at relatively low cost. Gene banks can hold a large amount of germplasm resources, for instance, the world's gene banks presently hold more than four million accessions, or specific samples of certain crop varieties (Ogwu et al. 2014). On the other hand, the *in situ* conservation method does not remove the genetic materials from their environment, instead genetic materials remain in their natural habitats. Most of the world's genetic diversity is found *in situ*. For agriculturally important species, the greatest diversity in landraces and in wild relatives may be found in or near their primer and seconder centers of origin, or the places in which they were first domesticated (Uygur Gocer and Karaca 2016b; Kopnina et al. 2018; Li et al. 2018).

Conservation of *ex situ* genetic resources is an efficient and economical way by seed conservation (seed bank). Seed bank represents the most cost-effective *ex situ* conservation strategy. Although most crop seeds can be stored for long periods under low relative humidity and low temperature conditions, it is not feasible for some species that cannot stand desiccation below a relatively high critical water content value (10–12% or 20%). Therefore, seed bank conservation is for the storage of predominantly orthodox seeds to maintain the allelic integrity and identity of a sample (Chandrakant et al. 2017). For those non-orthodox seeds gene bank requires essential infrastructure for short- and long-term seed storage, but also the efficient management of germplasm from safety backup to regeneration and characterization, germplasm distribution, and data management (Hernandez-Suarez 2018; Li et al. 2018).

Conservation of *ex situ* genetic resources may consist of following procedures; (i) collection, identification and characterization, (ii) regeneration, (iii) conservation, (iv) data management, (v), distribution, (vi) evaluation for subsequent use, (vii) acquisition, (viii) characterization, (ix) re-evaluation for supportive research and germplasm enhancement. All newly arrived plant seed samples are controlled for health and purity. Primarily a seed viability and some other related tests are made and seed samples having the required standards are dried and stored. Seed drying is an important step to assure a long viability of seeds and it is carried out gently using temperatures below 25 °C. Seeds are dried to 4–8% moisture, filled into glass containers with vapor proof covers and placed into moving shelves in cooled chambers. The storage temperature is below −18 °C for the base collections. Seed viability as well as seed supply are regularly monitored during long-term storage. Regeneration of multiplied accessions is initiated when one of these parameters drops below the standard level. Recent years witnessed dramatic development in freezer technologies and thus ultra-low temperature freezers became much cheaper for preserving excised embryos, embryonic axes, or dormant buds of many non-tropical and some tropical species, avoiding the need to replace liquid nitrogen as it evaporates. It is likely that cryopreservation will become cheaper and easier and, become more widespread in next decades (Ogwu et al. 2014; Fu et al. 2015; Fu 2017; Li et al. 2018).

The information about the level of allelic diversity of a species is very important to capture a high proportion of the total genetic diversity in *ex situ* conservation such as gene bank. Genetic screening studies would provide information about the genetic diversity of the population. It is possible that a large number of individuals can be represented by a few genotypes or a few individuals can represent a large number of genotypes. Because genetic drift in gene banks is caused by the use of inadequate sample sizes, greater numbers of accessions are stored to guarantee that a particular proportion of possible genotypes has been preserved, but this can lead to prohibitively large sample sizes. Therefore, coverage of a sample is very important for conservation of biodiversity for sustainable agriculture. In simpler terms, the coverage of a sample can be defined as the fraction of individuals in the population that is represented in the sample. Clearly, the goal of conservation of biodiversity and genetic resources for sustainable agriculture is to achieve high coverage at all loci and accurately estimate the proportion of existing alleles in a genome that is included in an accession. The use of next generation phenotyping and genotyping method could be very useful on screening of accessions (monitoring) for high coverage and high level conservation of biodiversity in *ex situ* approaches (Truong et al. 2012; Karaca and Ince 2017).

3.3 Values of Genetic Resources

Today the agriculture of virtually every country is heavily dependent on a supply of genetic resources from other parts of the world. The United States of America and Australia, for example, place considerable reliance on many species originating in other regions of the world for their major food and industrial crops. Sub-Saharan Africa is estimated to be 87% dependent upon other parts of the world for the plant genetic resources, and the figure is estimated to be about 90% in Europe and 62% in East and Southeast Asia. Many countries hold a significant amount of plant genetic diversity in their gene banks, farmers' fields and natural habitats for food and agriculture. In the medium- to long-term, however, these countries are likely to require additional genetic resources from the crop species' centers of diversity, the majority of which are restricted to eight crop diversity hotspots identified early in the last century (Vavilov 1926). Although developing countries contain many of the Vavilov centers of crop diversity and therefore have much of the world's genetic resources. However, many of developing countries struggle to conserve genetic resources, and they have limited technologies of advanced molecular and genomic tools and the corresponding expertise to use the genetic wealth for their own benefit. The onus is on the developed countries to work with those developing countries to conserve agricultural plant genetic resources, including diversity of crop wild relatives for sustainable agriculture (Phelps and Webb 2015; Castaneda-Alvarez et al. 2016; Vavilov 1926; Karaca and Ince 2017).

It is not possible to determine an organism's value if it is not used in direct or indirect by humans. However, if the organism is valued in agriculture (including all types of activities such as landscaping, forestry, arboriculture, horticulture, floriculture, viticulture, aquaculture etc.), its value arises from the direct use of genetic resources for sustainable agriculture. Direct use values include the use of genetic resources to produce food and fiber, or to help create new varieties of crops. Otherwise the value of genetic resources is not typically revealed by markets, because genetic resources are not directly traded in the markets. Conserved genetic resources could have more economic value in the future even if the resources are not currently being used or known. Therefore, an organism that is not presently economically valued, may have considerable value in the future, though this value is difficult to measure at the present. For instance, prior to 1980s, the economic value of bacterium *Thermus aquaticus* that lives in hot springs and hydrothermal vents was not known. *Taq* DNA polymerase of this bacterium and polymerase chain reaction brought $2 billion in royalties (Polat et al. 2010; Korun et al. 2013; Timmermann and Robaey 2016; Pavan et al. 2017).

Plant genetic resources could be used by breeders to develop new and improved varieties for agricultural production. This process of genetic enhancement has produced substantial economic benefits for the producers and plant breeders. Plant breeders are requested to provide genetic diversity to farmers' fields in order farmers produce the agricultural products. Therefore, well conserved genetic diversity and genetic resources such as crop wild relatives and landraces have replicable values in the development of new varieties resistant to biotic and abiotic stresses, as they are drought, insect, pest, disease tolerant and resistant to biotic and abiotic stresses (Sharrock et al. 2014; Timmermann and Robaey 2016; Li et al. 2018).

Due to the genetic recombination, gene flow, mutations and the intensities of pest pressure in the field farmers need to replace their seeds used for production. Although depending on many factors it was estimated that new varieties are resistant for an average of about 5 years, while it generally takes about 10 years to breed new varieties (Karaca 2001; Ince and Karaca 2011c). Breeders often rely on landraces, old cultivars or crop wild relatives as a last resort to gain alleles of interest from these materials (Ince et al. 2010g). These genetic resources are constantly required as repertories into the continuing process of enhancement through selective breeding

(Ince et al. 2010h). Because pests and diseases evolve over time, new alleles and epialleles are continually needed to transfer from outside the utilized stock, landraces, and wild relatives to maintain or improve yields (Ince and Karaca 2011a, 2016). Genetic resources are not only utilized to transfer resistance to pests and diseases, and tolerance to non-biological stresses, such as drought but also include rapid and simultaneous germination, flowering, and maturation of crops (Ince et al. 2011b; Ogwu et al. 2014; Ince and Karaca 2015a, b; van Kleunen et al. 2015; Uygur Gocer and Karaca 2016b).

4 Biodiversity

Biodiversity refers to variation in number and frequency within the natural system. In another word it refers to the variety of all forms of life in the world. Reduction or decline in genetic diversity, also called genetic erosion, in many commercially important plant species has been observed (Ince and Karaca 2012; Aavik et al. 2017; Dorey and Walker 2018). One reason for this decline in diversity has been the loss of genetic resources such as landraces and wild relatives of cultivated crops. It is a widely held belief that modern agriculture, particularly the transition from landraces to modern varieties as exemplified in the twentieth century's green revolution, has profoundly narrowed the genetic base of modern crop varieties. In the broadest sense, however, genetic alteration and narrowing began with the first domestication of wild plants (Karaca and Ince 2008, 2016; Ince et al. 2010i; Ogwu et al. 2014; Aavik et al. 2017; Manhaes et al. 2018).

Principle threats causing adverse effects on the status or sustainable use of any component of biological diversity include habitat alteration or destruction, overharvesting or over-exploitation of biological resources, weather, water or soil or biological pollution, introduced or invading species, climatic changes, and expanding human population (Karaca et al. 2015). Increased demand for resources results to land use changes hence loss to genetic diversity, species reduction and increased ecosystem changes such as random population changes, habitat fragmentation and many others resulted in biodiversity losses. Population size and habitat fragmentation differ in response to inbreeding depression and environmental adaptations. Levels of genetic diversity are affected from population size along with habitat fragmentation. Therefore, if not well conserved, *ex situ* populations can become genetically different from their original source populations, may lose adaptation to their source environment, and may become inbred (Karaca et al. 2015; Aavik et al. 2017; Li et al. 2018).

Major sources of threats for genetic diversity include (i) plant loss, fragmentation, and degradation, (ii) over-exploitation (including over-collection and overgrazing), (iii) invasive species, (iv) increased air, soil water or biological pollution and nitrogen deposition, (v) severe climate change and (vi) wrong land uses and urbanization (Shearman et al. 2012; Ogwu et al. 2014; Sharrock et al. 2014; Goettsch et al. 2015; Phelps and Webb 2015; Specht et al. 2015; ter Steege et al. 2015;

Thomas and Palmer 2015; Corlett 2016; Timmermann and Robaey 2016; Dorey and Walker 2018). Plant loss, fragmentation, and degradation are among most important threats to plant diversity particularly in the tropics. Tropical forests have been replaced with monoculture of oil palm, rubber, soybean (ter Steege et al. 2015). Also logging, fire, and other impacts, including fuelwood harvesting in densely populated areas are among the main threats in developing countries and regions (Specht et al. 2015). Furthermore, mining of stone, construction of wide roads in the regions of endemic plant species are among other threats (Ince et al. 2014; Ogwu et al. 2014; Dorey and Walker 2018).

Over-exploitation (including over-collection, over-harvesting and over-grazing) of the whole plant, seeds or reproduction systems reduces the chance of survival. Over-harvesting results when individuals of a particular species are harvested at a higher rate than that they can be sustained by the natural reproductive. This could lead to extinction of the biological resources, eventually leading to loss of species. It is known that over-exploitation is usually species-specific and it is well correlated with its value. Horticultural trade for private collections is important threat to some plant species such as cacti, orchids as well as cycads and ornamental species (Ogwu et al. 2014; Sharrock et al. 2014; Goettsch et al. 2015; Phelps and Webb 2015; Dorey and Walker 2018; Li et al. 2018). In some other plant species overexploitation of animals may also threaten plant species in the long term, by restricting seed dispersal or pollination. Also in the tropical forests, it is known that over-logging is the main threat factor (Shearman et al. 2012; Dorey and Walker 2018).

Species that are not habitats but were introduced in an ecosystem may cause changes in the host (existing) ecosystem. Introduced species are those species arising in areas/habitats in which they were previously not native. Such some introduced species could refer to as biological pollutants. These kinds are also called as invasive alien species that have potentially threat to native species. A study of van Kleunen et al. (2015) showed that more than 13,000 species of the world's vascular plant flora naturalized somewhere outside their native range as a result of human activity. It is known that invasive plant species can reduce native plant diversity by changing hybridization, out competition, disruption of original ecosystem, plant pathogenic influences, disease transmission, fire regimes, nutrient cycling, pollen transfer and some other physiological requirements, disruption of food webs and to some situations extinction (Thomas and Palmer 2015; Dorey and Walker 2018; Manhaes et al. 2018).

Any chemical, thermal, air, soil, water or biological pollution is a threat to biodiversity. Species in habitats are increasingly being harmed by industrial activities and pollution from excessive use of chemicals such as dichlorodiphenyltrichloroethane (DDT), oil spills, acid precipitation etc. Due to human activities the concentration of the major greenhouse gases (CO_2, CH_4, N_2O) changed, thus, plants on earth today are exposed to an atmosphere that differed significantly in composition from any that their ancestors experienced. Burning of fossil fuels is the major source of air pollution and primary pollutants are SO_2 and NO. Ozone produced from hydrocarbons and nitrogen oxides in the presence of sunlight is the most important secondary pollutant. According to Corlett (2016) air pollution is declining in Europe and other developed regions, but increasing in much of Asia. Zhu et al. (2015) informed that wet and dry deposition of nitrogen compounds dramatically changed nutrient cycles in southern China due to its acidifying effects on the soils.

Impacts of climate change caused by humans are complex and mostly unpredictable, and even more pervasive. The rate of evolution of plant species is driven by their genetic makeup and mainly climate. When the climate changes in a local area in where plant populations adapt, a plant can either adjust physiologically within the lifetime (acclimation), or evolves by evolutionary changes over multiple generations (evolution), or move to some other places where with a more suitable climate (migration), or vanish (extinction). Although the problem is not directly attributable to climate change, it has been reported that crop yields of wheat, barley, and canola have been reduced by over 40% in Australia due to drought (Hijioka et al. 2014). It has been estimated that rising temperatures and reduced precipitation would affect semiarid regions and reduce yields of maize, wheat, and rice over the next 20 years. It is known that extinctions of some species at local have occurred at the climatic margins of species ranges (Buse et al. 2015) and some species have extended their ranges escaping regions where the high temperature and water are limiting factors (Hijioka et al. 2014; ter Steege et al. 2015; Castaneda-Alvarez et al. 2016).

Land uses, urbanization, hydroelectric dam construction, road construction for transportation and competitions in global market economies do strongly contribute indirect negative effects on biodiversity. Also the use of alien species and chronic weed infestation have increased the number of threatened species. A significant of damage for the biodiversity is also caused by collection for local and global markets, often by professional collectors. Many countries have laws against inappropriate collections, but often commercial collection makes use of legal loopholes which urgently need to be closed (Corlett 2016). Furthermore, it is known that many protected areas fail to prevent over-exploitation of valuable plants. In these regions *ex situ* conservation requires effective monitoring to ensure that viable plant populations of threatened species persist within protected areas. Unfortunately, in many developing countries many protected areas are subject to encroachment by farmers or their fires. For instance, it has been reported that the expansion of rubber plantations and the promotion of biofuel crops such as physic nut (*Jatropha curcas*) in southern Yunnan's Xishuangbanna region in China was reduced two-thirds of a unique rainforest (Heywood 2015; Corlett 2016; Chandrakant et al. 2017).

5 Next Generation High-Throughput Phenotyping

Next generation high-throughput phenotyping consists of collection of huge quantities of image data and safe storage, fast and well-organized working flow, economical and time-saving analysis procedures, and dissection of objective data (without influence of human perception). The lack of accession-level information on conserved germplasm is one of the major limitations to wider germplasm utilization and conservation for sustainable agriculture. Varying among the gene banks around the world a majority of the conserved genetic resources show only with basic germplasm description records such as passport data. However, due to a lack of detail in the passport data, factors influencing genetic diversity like sampling strategy, regeneration procedures and selection during regeneration could not be well reconstructed, fortunately next generation phenotyping approaches could be used to add valuable and detailed information to passport data that could be used in sustainable agriculture (Cobb et al. 2013; Song et al. 2015; Afonnikov et al. 2016; Walter et al. 2018).

Many activities may be classified as genomic research or simply genomics, including mapping the genome of an organism; sequencing a single individual or several individuals from a given species; identifying a large number of genes; studying genetic variability within species; studying genetic similarities and differences within and among species; discovering a large number of genes' function, and the relationship between gene structure, protein synthesis, and metabolic pathways; studying gene regulation, including gene activation and silencing; studying gene interaction and phenomena dependent on many genes (Ogwu et al. 2014; Karaca and Ince 2017).

Genomics is expected to provide a comprehensive view of genetic capacity, however, the information it contains is cryptic and does not directly explain the differences between cells and all plant phenotypes. On the other hand, some phenotypic traits provide more direct information about plant production and health than genomic data. The recent improvement in phenotyping methods enable us to broaden the concept of phenotyping and include both molecular mechanisms (proteomics and metabolomics) and all intermediate layers that result in macroscopic, physiological and phenological traits (Ubbens and Stavness 2017).

Phenotyping can be performed at different depth scales such as high or low resolution, and high or low throughput volumes. High-throughput techniques in general involve the analysis of the whole plant with medium-low resolution, therefore, it is suitable for conservation of biodiversity and genetic resources for sustainable agriculture. High-throughput techniques could be used in *ex situ* and *in situ* conservation fields for phenotyping and crop monitoring which could allow the screening of hundreds of accessions per day in a nondestructive manner with automated systems. The integration of genomics and phenomics has promised to revolutionize the field of plant breeding indicating that these high-throughput methods can be used in conservation of biodiversity and genetic resources for sustainable agriculture (Breccia and Nestares 2014; Lobos et al. 2017; Sun et al. 2017.

Phenomics is driven by large-scale and economical generation of phenotype differences coupled with increasingly sophisticated and comprehensive sensors and cameras called high-throughput phenotyping (phenomics). The aim of plant phenomics is to characterize all the possible phenotypes under different environmental conditions of a given genotype or species. For that purpose, phenomics includes phenotyping at multiple levels of organization (ranging from cellular components to whole plant and canopy level) and comprises structural, physiological, and performance-related traits (Busemeyer et al. 2013; Breccia and Nestares 2014; Lobos et al. 2017; Sun et al. 2017; Crain et al. 2018; Thompson et al. 2018; Walter et al. 2018).

Satellite imaging technologies have become an extremely useful tool for collecting data for various agricultural applications including conservation of biodiversity and genetic resources. However, the major limitations of using the currently available satellite sensors include the high cost, the lack of spatial resolution for the identification of desirable traits, the risk of cloudy scenes and the long revisit periods. To data many field-based high-throughput phenotyping methods and platforms have been developed for high-throughput phenotyping. Some of the platforms use push carts, tractor mounted systems and aerial vehicles (Crain et al. 2018). Advanced plant phenotyping platforms include phenomobiles, phenotowers and blimps equipped with a global positioning system, navigation device and sensors, however, although performing well, they are quite costly, data acquisition and handling needs for specialized personnel (Araus and Cairns 2014; Lausch et al. 2017; Habib et al. 2018; Tripodi et al. 2018). Therefore, these platforms are not easily affordable in many developing countries where they are needed most. Although still often used in practical phenotyping, manual measurements of crop traits have significant limitations and drawbacks such as they are time-consuming, labor intensive and subject to human error due to fatigue and distractions during data collection (Arend et al. 2016; Yang et al. 2017; Jimenez-Berni et al. 2018).

Studies of conservation of biodiversity and genetic resources for sustainable agriculture could utilize phenotyping and monitoring technologies. These technologies include spectral laboratory and phenomics facilities, close-range, airborne and satellite approaches (Lausch et al. 2017). For phenotyping purposes, images obtained from satellite, manned and unmanned aerial vehicles typically have a low spatial resolution (in the context of *ex situ* and *in situ* conservation), poor sensitivity under cloudy conditions, and slow data transmission and expensive. Most long-distance remote sensing technologies could sufficiently capture the fine data suitable for the studies of conservation of biodiversity and genetic resources for sustainable agriculture. On the other hand, spectral laboratory and plant phenomics facilities provide biochemical-biophysical, structural variables in organs (roots, leaf, stem) and whole plants. These close-range remote sensing methods or platforms include field spectrometers, wireless sensor networks, towers and next generation unmanned aerial vehicles provide taxonomic, phylogenetic, genetic, epigenetic or morphological-functional features. Therefore, spectral laboratory and phenomics facilities could be used to detect biochemical-biophysical and morphological traits (Lausch et al. 2017; Jimenez-Berni et al. 2018; Thompson et al. 2018).

Among the phenotyping and monitoring technologies proximal sensing carts and unmanned aerial vehicles (UAV), both can be called as phenomobiles, are suitable in the application of conservation of biodiversity and genetic resources for sustainable agriculture. UAVs are equipped with multiple sensors [some of visible light imaging sensors, spectral sensors, infrared thermal sensors, fluorescence sensors, digital camera (RGB), multispectral camera, infrared thermal imager, hyperspectral camera, Light Detection and Ranging (LiDAR), three-dimensional camera and synthetic aperture radar (SAR)], using communication technology and GPS positioning technology to rapidly and non-destructively acquire high-resolution images (Table 2). The typical UAVs used for phenomics include multi-rotors, helicopters,

Table 2 Some sensing and imaging techniques used next generation phenotyping

Technique	Short description of the technique
Magnetic resonance imaging (MRI)	It is based on nuclear magnetic resonance (NMR) analysis, allows measuring resonance signals produced from H, C, and N isotopes. With this technology, 3D acquisition can be accomplished to acquire information on plant phenotype with high resolution
Tomography imaging	It uses radio frequency (RF) magnetic fields to construct tomographic images. It produces imagines by sections or sectioning, through the use of any kind of penetrating wave. For instance, x-ray computed tomography (x-ray CT) employs x-rays to produce tomographic images of specific areas of the scanned object. The process of attenuation of rays together with a rotation and axial movement over objects produces 3D images
Light detection and ranging (LiDAR)	It uses short pulses of laser light distributed from a scanning device across a wide area and their reflections from different objects are recorded by the sensor. It produces set of 3D points, which represent the scanned surfaces from where the pulses were reflected. LiDAR provides an alternative approach for 3D plant model reconstruction.
Synthetic aperture RADAR (SAR)	SAR uses a receiver (an antenna) to transmit microwave pulses in a specific waveband (or frequency) at an oblique angle to the target area. Radio waves that are reflected off the object back (from the target area) to the source can be acquired in a variety of modes
The time of flight camera (ToF camera)	It is one of the recent imaging devices to be incorporated into automatic plant phenotyping. ToF has as a general principle the measurement of the distance between the objective of the camera and each pixel. This is achieved by measuring the time it takes for a signal emitted in near infrared (NIR) to come back, reflected by the object. This allows a precision 3D reconstruction.
Multispectral imaging sensor Hyperspectral imaging sensor	Multispectral imaging sensors are defined as hardware that are capable of sensing and recording radiation from invisible as well as visible parts of the electromagnetic spectrum, which have been widely used for crop phenotyping due to the advantages of low cost, fast frame imaging and high work efficiency; however, they are limited by the low number of bands, low spectral resolution, and discontinuous spectrum. Hyperspectral imaging sensors are cameras that can obtain a large number of very narrow bands and continuous spectra. Compared with multispectral imagers, hyperspectral imagers have the advantages of more band information and higher spectral resolution and can accurately reflect the spectral characteristics of the crop in the field and the spectral differences between crops
Thermography imaging/Thermal imaging	Thermographic cameras are able to acquire images at wavelengths ranging from 300 to 14,000 nm allowing the conversion of the irradiated energy into temperature values once the environmental temperature is assessed. Thermal imaging uses infrared detectors and an optical imaging lens to receive infrared radiation and produces time series or single-time-point analysis based data
Fluorescence imaging	It belongs to spectroscopy but differs greatly from reflectance, absorbance, and transmittance measurements in the way by which plant tissues interact with electromagnetic radiation. It uses a low-light camera/sensor and appropriate filters to collect fluorescence emission light from samples

fixed-wing, blimps and flying wing. Among this one or more are selected based on the purpose and budget (Busemeyer et al. 2013; Sun et al. 2017; Virlet et al. 2017; Thompson et al. 2018).

Past three decades witnessed the development of large number of phenotyping technologies. Those phenotyping techniques using high-throughput and high resolution are called next generation phenotyping methods that contain more than one sensing (multi-sensor) approaches or platforms. It has been shown that sensor characteristics (spatial, radiometric, spectral, temporal or angular resolution) and the sensing approaches (hyperspectral, multispectral, digital (RGB), LiDAR, SAR and passive microwave sensors) show different level of discriminate between certain plant species, populations, communities, habitats. Advanced phenotyping methods and platforms based on multi-sensor remote sensing would be able to discriminate and monitor threatened plant species or invasive species, bio-pollutants, the pattern and spatial distribution and diversity of plant species and communities as well as natural disasters and disturbance regimes, i.e., volcano eruptions, wildfires, beetle infestations, and the global carbon cycles (Perez-Sanz et al. 2017; Sun et al. 2017; Virlet et al. 2017; Thompson et al. 2018).

Various studies have shown that the implementation of multi-sensor approaches improves the discrimination of plant properties over time and thus the accuracy of estimation of population indicators. Multi-sensor systems on single platform enable the simultaneous acquisition of information related to different spectral traits and ensure the same illumination conditions, weather conditions and flight parameters for all mounted sensors. The package of a platform may include RGB camera, infrared thermometers, active spectral reflectance, and light or ultrasonic sensors. Next generation phenotyping platforms can be classified considering many different characteristics. For instance, they can roughly be divided into the categories of point sensors (spectro-radiometers and fluorimeters) and imaging sensors that allow the acquisition of spatial information of the detected data. We classified some phenotyping techniques based on the sensors and platform and depicted in Table 2.

Recent developments in remote and proximal sensing for high-throughput field phenotyping have led to proposed alternatives to the destructive sampling, including the use of digital photography and sensors, across multiple scales, using both aerial and ground platforms. High-throughput phenotyping spectral traits (Table 3) suitable for conservation of biodiversity and genetic resources for sustainable agriculture include plant structure and morphogenetic traits; abiotic and biotic stresses, adaptation to abiotic and biotic limiting conditions, metabolomics traits, quality traits and physiological traits (Perez-Sanz et al. 2017; Yang et al. 2017; Espina et al. 2018; Jimenez-Berni et al. 2018).

Next generation phenomobiles equipped with infrared thermal imagers can quickly and non-destructively acquire the crop canopy temperature, which can effectively identify the temperature differences in the crop canopy under different environmental conditions. The canopy temperature can be used to predict plant yield when a significant positive correlation between lower canopy temperature and higher yield under conditions of high temperature and drought exists. Leaf water potential could be estimated since the stomatal closure results in the leaf tempera-

Table 3 Spectral traits suitable for conservation of biodiversity and genetic resources

Category	Traits suitable for analysis
Structure and morphogenetic traits	These traits include plant height, chlorophyll content, biomass, yield, length of the growth period, flowering, crop canopy cover, canopy spectral texture
Plant physiological traits	These traits include chlorophyll, pigment content, carotenoids, pigment indices photosynthesis, protein content, malnutrition, crop vigor and water status
Plant yield and quality traits	These traits include total oil, protein, starch, moisture content, fatty acid and amino acid compositions. Yield prediction is defined as building the relationship between the canopy spectra and crop yield based on the biological characteristics of crops for yield prediction using spectral data at different crop growth stages
Plant geometric traits	These traits include crop height, vegetation cover fraction, fraction of intercepted radiation, leaf area, leaf area index, lodging, 3D structure, leaf angle distribution, tiller densities, and emergence
Plant biotic and abiotic stress	These traits include water stress and deficit, low temperature, high temperature, high salinity, environmental pollution, susceptibility to pests and diseases, stomatal conductance, canopy temperature difference, leaf water potential, senescence index
Metabolomic traits	These traits include flavor, phenolic, vitamins, sugars, organic acids, and volatile compounds. Metabolomics plays a remarkable role in assessing genotypic and phenotypic diversity in plants, in defining biochemical changes associated with developmental changes during plant growth and, increasingly, in compositional comparisons
Quality traits	These traits include fatty acid and amino acid compositions, fiber quality, nitrogen concentration and protein content, seed traits such as total oil, protein, starch, moisture content
Ground canopy cover	It is an important parameter related to the crop photosynthesis and transpiration. It is dynamic during the crop growth stages and is reduced as a result of leaf rolling or wilting under drought stress conditions, which can be used for studying the response of crop varieties under abiotic/biotic stress.
Qualification and selection	These traits include leaf/pod/fruit counting, vigor ratings, injury ratings, disease detection, age estimation, and mutant classification

ture increase under osmotic stress caused by excess salinity and high temperature. Also drought and salinity can induce the same effects on stomatal conductance and photosynthesis (Hoyos-Villegas et al. 2014; Tripodi et al. 2018).

Crop yield of conserved genetic resources could be estimated using next generation phenotyping approaches such as using phenomobiles. Since the crop canopy temperature is related to photosynthesis, the canopy air temperature difference, which is the ratio of the canopy temperature and air temperature, can be used to predict crop yield when there is a significant negative correlation between the air temperature difference and yield of plant as seen in sorghum. In wheat it has been seen that there existed a significant positive correlation between air temperature difference and wheat yield under water stress conditions. The water deficit index obtained from thermal imaging data can be used to determine the water status of crop leaves and to estimate the stomatal conductance (Rascher et al. 2011; Ecarnot et al. 2013; Simko et al. 2016; Padilla et al. 2017; Crain et al. 2018; Tripodi et al. 2018).

Cell structures could be estimated using next generation phenotyping methods. The reflectance of plant leaves in visible light (about 390–700 nm) is affected by the contents of chlorophyll, carotene and lutein in the palisade tissue. The reflectance of plant leaves in the near-infrared (NIR) band is closely related to the cell structure and can be used to estimate several spectral traits including plant physiological trait, geometric traits and ground canopy cover (Perez-Sanz et al. 2017; Espina et al. 2018; Jimenez-Berni et al. 2018). Biodiversity and genetic resources could be validated, monitored or conserved using plant cell structures based on the next generation phenotyping techniques and platforms such as phenomobiles. Phenotypic information plays an important role in revealing the resistance of crops to stress, therefore, rapid phenotyping is also essential for conservation of biodiversity and genetic resources for sustainable agriculture. Infrared canopy temperatures provide an efficient method for rapid, non-destructive monitoring of whole plant or population response to water stress, which has been widely used to screen drought tolerance in domesticated plant species. Biotic and abiotic stress factors, including water deficit, low temperature, heavy metals, high temperature, high salinity, environmental pollution, pests and diseases, can have significantly adverse effects on plant growth and development. Abiotic stress during early canopy development can decrease plant biomass and height, reduce leaf area, and abbreviate green area duration. Effects of most of these biotic and abiotic stress factors affect plant's membrane permeability, the chlorophyll content, hormone and enzymatic activities under stress conditions, thus, they can be detected by spectroscopy at early growth stage if an effective correlation or regression method is available (Liebisch et al. 2015; Crain et al. 2017, 2018).

The absorption and reflection characteristics differ between spectral bands in the crop leaves, with strong absorption in the visible band and strong reflection in the near-infrared band, providing the physical basis of crop growth monitoring by remote sensing suitable for conservation of biodiversity and genetic resources. Digital cameras in the range of visible spectrum (400–700 nm, VIS) allow capturing 2D images in which raw data are recorded in the red (about 600 nm), green (about 550 nm), and blue (about 450 nm) array using charge coupled device (CCD) or complementary metal oxide semiconductor (CMOS) silicon-made sensors. These kinds of 2D images show many limitations, especially when used for plants that have a high degree of structure complexity, therefore, 3D images are preferred for the estimation of plant biomass, leaf area and leaf area index, and plant morphology. The use of stereo cameras and computer programs produce 3D images taken by multiple angulations allow drawing sophisticated models for the reconstruction of plant structures. Also digital cameras offer further characteristics that deal with plant color analysis, however, to the specific purpose of plant structure and biomass analysis, the most widely adopted technologies are based on light detection and ranging (LiDAR) by using laser-scanner sensors. LiDAR provides direct measurements of canopy architecture and organ distribution for the estimation of plant volume, leaf area index, and biomass. LiDAR allows plant growth analyses from the vegetative to reproductive stages (Jin et al. 2017; Tripodi et al. 2018).

Measurements for different data can be obtained in the range of ultraviolet (UV), visible (VIS), near-infrared (NIR), and infrared (IR) radiation using the electromagnetic spectrum. Instruments working in the hyperspectral range (from tens to hundreds of wavelengths) offer more flexibility analysis than multispectral analysis (from two to tens of wavelengths) or single-wavelength measurements since the broader the covered wavelength range and the number of measured wavelengths, the higher the detection capabilities are obtained. Crop growth rates based on changes in crop height could be used to assess the efficiency and effectiveness of management strategies. Ultrasonic sensors are most commonly used sensors to measure crop height in agriculture applications. However, the main disadvantage of an ultrasonic sensor is that the field of view becomes larger as the distance between the sensor and the object increases due to the sensor's relatively wide angle divergence of ultrasonic waves. This reduces the accuracy of ultrasonic measurements. Furthermore, the ultrasonic sensor is sensitive to temperature variations, which limits its outdoor use (Sun et al. 2017).

LiDAR equipped on an airborne vehicle could detect fallen dead trees and the remains of large branches on the ground in forests indicating that LiDAR and similar remote sensing techniques could be used in conservation of biodiversity and genetic resources. For instance, they provide opportunities to monitor endangered plant and animal species for conservation purposes. However, the application of LiDAR is costly because it is limited to airborne missions covering local to regional areas (Lausch et al. 2017). Aerial LiDAR has been successfully used to obtain forest structure attributes such as tree height, leaf area, and branch detection. However, aerial LiDAR was not as effective in annual crops phenotyping activities since it has limited capability to provide high resolution information for crops which are much smaller than trees. This indicates that aerial LiDAR is not suitable for conservation of biodiversity and genetic resources for sustainable agriculture. On the other hand, terrestrial LiDAR has the potential to provide denser point over a relatively small area, from which high resolution information could be extracted. Therefore, it has been increasingly used in field phenotyping. Comparison studies of ultrasonic sensors and LiDAR indicated that LiDAR was generally more precise than data obtained with ultrasonic sensors (Sun et al. 2017). One of major limitations of image based methods is that data quality can be significantly affected by the variable environment, since shadows and sunlight can result in under or over exposure and limit automatic data processing (Araus and Cairns 2014; Walter et al. 2018).

Fluorescence imaging has been used in a large number of experimental setups, as ultraviolet (UV) light in the range of 340–360 nm is reflected by different plant components as discrete wavelengths. Chlorophyll fluorescence emits in red and far-red (690–740 nm). Chlorophyll fluorescence imaging (CFI) is a step forward in fluorescence analysis, accomplished by the support of CCD cameras. In CFI, different lamps are used to induce fluorescence excitation while the plant response is monitored by the digital camera measuring fluorescence at different wavelengths in the typical spectral ranges of blue (440 nm), green (520–550 nm), red (690 nm), far-red (740 nm), and NIR (800 nm). Fluorescence imaging can be utilized in phenotyping of crops to asses biotic and abiotic stresses, tissue chemical composition and characterization, and different plant physiological conditions (Zarco-Tejada et al. 2012; Hoffmann et al. 2015; Virlet et al. 2017; Yang et al. 2017; Tripodi et al. 2018).

Thermography is a widely used technology in plant phenotyping. Plants are induced to open stomata in response to environmental cues and circadian clock depending on the type of photosynthetic metabolism they have. With this imaging method the evapotranspiration can be assessed with thermography, and quantification can be made at different scales, such as a leaf, a tree, a field, or a complete region. Thermography imaging provides monitoring and detecting water stress, irrigation management and plant diseases where all the specimens are located under strict control conditions: However, temperature, wind velocity, irradiance, leaf angle, and canopy leaf structures are potential issues for quality image acquisition. Both thermographic and fluorescent images capture a single component, and images are in principle easy to analyze but require sophisticated data analysis methods to obtain quality data, but it is an emerging solution (Prashar and Jones 2014; Perez-Sanz et al. 2017; Tripodi et al. 2018).

Synthetic Aperture RADAR (SAR) is an imaging radar used for conducting coherent processing of the obtained echo in different fields or area locations to obtain high-resolution data. SAR is a type of active microwave sensor and high-resolution radar images can obtain in a fashion similar to optical sensor. RADAR data can be acquired in a variety of modes, including standard polarizations (horizontal (H)- vertical (V), HH, VV, VH), polar metric and interferometric way (two signals at slightly different incident angles). This technique can obtain images in very low visibility weather conditions and can work around the clock, which can be used for crop identification, crop acreage monitoring, key crop trait estimation and yield prediction, providing strong technical support for large-scale crop growth monitoring by remote sensing. It is suitable in in tropical areas where persistent cloud cover, or in northern boreal areas where low sun angle effects can reduce the quality of optical model estimates. SAR is very effective in the determination of above ground biomass, fire impacts and forest inundation. It is clear that forest removal, disturbance and degradation analysis and monitoring using RADAR is very important for conservation of forest biodiversity and genetic resources (Perez-Sanz et al. 2017; Thompson et al. 2018; Tripodi et al. 2018).

Thermal infrared imaging sensors equipped with infrared detectors and optical imaging lens receive infrared radiation energy and can produce time series or single-time-point analysis (Gonzalez-Dugo et al. 2015). As the stomatal conductance, photosynthetic characteristics and transpiration rate are closely related to canopy temperature. Canopy temperature in the infrared thermal imaging technology can be used to determine the response of crops under stress conditions, to estimate leaf water potential and stomatal conductance, the cell structure and can be used to estimate plant physiological trait, geometric traits and ground canopy cover (Thompson et al. 2018; Tripodi et al. 2018).

The digital camera, multispectral camera, hyperspectral camera, thermal infrared imager and LiDAR have been widely used to field-based phenotyping. The use of phenomobiles in the studies of conservation of biodiversity and genetic resources will enhance our ability to conserve and widen genetic resources on earth since they provide the advantages of high operation efficiency, low cost, suitability for complex field environments, and high resolution. The limiting factors for phenomobiles based phenotyping for conservation of biodiversity and genetic resources include the strict airspace regulations and higher costs in many countries, the lack of methods and researchers for fast data processing and models for estimating complex traits under different environmental conditions. Also low payload and short endurance in air are among disadvantages. Improving the phenomobiles with machine learning approaches, reducing the cost of sensors, speeding up data processing and developing strategies for analyzing crop phenotype by remote sensing are future trends to be used in conservation of biodiversity and genetic resources. Fortunately, it is expected that with the advancement of new technologies with larger payload and longer endurance, low-cost sensors, improved image processing methods and effective airspace regulations, phenomobiles will find wider applications in high-throughput phenotyping and would be very suitable in conservation of biodiversity and genetic resources for sustainable agriculture (Perez-Sanz et al. 2017; Thompson et al. 2018; Tripodi et al. 2018).

6 Next Generation High-Throughput Genotyping

A DNA marker may be defined as a DNA sequence or fragment that is detected and its inheritance can be monitored. A DNA marker can be as small as a single base or it can be as long as several hundred or more bases. A marker must show at least two different forms (polymorphism) so that genotype carrying a form can be distinguished from other genotype with the other forms. Following the first DNA marker technology developed in the 1980s, a larger number of polymerase chain reaction (PCR) based DNA markers were developed and acted as versatile tools in fingerprinting of varieties, mapping of genes and quantitative trait loci, marker assisted breeding, positional cloning of genes, identification of chromosomes or/and chromosome segments, inferring and establishing phylogenetic relationships among species, building and detection of gene pyramiding; and maintenance and utilization of genetic resources (Bostein et al. 1980; Jeffreys et al. 1985; Bilgen et al. 2004; Ince et al. 2008; Karaca et al. 2008; Wang et al. 2009a, b; Zhang et al. 2009; Ince and Karaca 2011a; Ince et al. 2011c; Jonah et al. 2011; Ince and Karaca 2012; Olarte et al. 2013; Saebnazar and Rahmani 2013; Erbano et al. 2015; Ince and Karaca 2015b; Will et al. 2015; Aydin and Karaca 2016; Karaca and Ince 2017; Song et al. 2017).

Traditional (Karaca et al. 2005a, b; Ince et al. 2007, 2010j; Karaca and Ince 2017) and next generation sequence (NGS)-based DNA markers (Ali et al. 2016; Jiang et al. 2016; Du et al. 2017; Karaca and Ince 2017) are single (such as single nucleotide polymorphism, SNP) or larger nucleotide sequences (fragments) that are located within or between regulator sequences (promoters, enhancers and silencers) and gene bodies (5'-UTRs, exons, introns and 3'UTRs). DNA marker polymorphisms could result from substitutions, insertions or deletions (INDELs), variation in repeats (such as simple sequence repeats, SSRs) and copy number variations (CNVs). Those DNA markers associated with phenotypic/physiologic trait variations are called functional DNA markers, gene based markers or perfect markers. Functional DNA markers are divided into two main groups. Those functional markers that closely associated with the phenotypic trait variations are called direct functional markers whereas those functional markers that less or not directly associated with the phenotypic traits due to recombination and genetic interaction are called indirect functional markers (Karaca and Ince 2017). Functional DNA markers have advantages over general DNA markers including: (i) not lost due to the recombination between marker and gene of interest; (ii) more meaningful in plant breeding; (iii) more useful in determination of population dynamics, germplasm collections, and monitoring evolutionary changes (Ince et al. 2007, 2010j, 2011d; Salgotra et al. 2014; Michael and van Buren 2015; Kage et al. 2016; Karaca and Ince 2017).

High-throughput sequencing technologies opened new ways for development of novel types of DNA markers, increased our ability to genotype larger numbers of genomes and individuals, and dramatically improved our understanding of how evolutionary processes shape genetic variation across populations, species, and genomes of plant species. High-throughput approaches provide great help and monitor the transfer of genes from distantly related species into breeding programs. Wild species and crop wild relatives have already contributed significantly to improving food production using traditional DNA markers. For instance, Asian rice is one of the clearest examples on application of biotechnological techniques for the genetic improvement of crops. More than 7000 lines were screened to find one from wild *Oryza nivara* that possessed a resistance to the grassy stunt virus; this resistance can now be found in most rice crop germplasm (Li et al. 2018). It has been some time that plant breeding has been supplemented with newer processes involving chromosomal manipulation, embryo rescue, alien introgression lines, mapping populations, marker-assisted selection, and the use of doubled haploids to create inbred lines, allele mining, map-based cloning, the analysis of quantitative trait loci, gene isolation, and genetic modification. Many of these approaches can be used in conservation of biodiversity and genetic resources for sustainable agriculture (Ogwu et al. 2014; Li et al. 2018).

Sequencing of whole genomes involves considerable time, labor, and financial and other resources. In order to reduce time, labor and cost of whole genome sequencing, genotyping by sequencing methods have been developed (Huang et al. 2009; Rife et al. 2015; Rowan et al. 2015). Although the term genotyping by sequencing (GBS) method was first introduced to plant science by Elshire et al. (2011) it had been already available since the earliest form of GBS methods such as complexity reduction of polymorphic sequences (CRoPS), restriction site-associated DNA sequencing (RAD-seq) and reduced-representation library (RRL). Whole genome sequencing and resequencing (WGS and WGR) along with GBS methods produce polymorphisms of SNPs, insertions/deletions (InDels), microsatellites (SSRs) and copy number variation (Kozarewa et al. 2009; Andolfatto et al. 2011; Mascher et al. 2013; He et al. 2014; Yang et al. 2015; Voss-Fels and Rod 2016; Zhu et al. 2016; Furuta et al. 2017; Scheben et al. 2017; Stetter and Schmid 2017).

High-throughput sequencing methods could be mainly divided in two approaches; reduced representation sequencing (RRS) and whole-genome resequencing (WGR) approaches (Table 4). Although both RRS and WGR methods profit from prior genomic information, reference sequence is a prerequisite only for WGR methods. This relative independence from prior genomic information means that RRS shows particular promise for characterizing the genomes of non-model species. The sequencing read depth can be affected by some biological factors of a target species, including: genome size, genome complexity, ploidy, and expected heterozygosity. Read depth differs between RRS and WGR. Low read depth in WGR methods is typically less than 1x and this low read depth can cause problems when genotyping heterozygotes. On the other hands, read depth in most GBS methods is grater but varies from 1× to 15× depending on the type of GBS methods used (Table 4). Read depth in GBS methods can be increased by reduced numbers of genotypes per library, use of rare cutting restriction enzymes, double digestion, and multiple sequencing runs for a library (Deschamps et al. 2012; Stolle and Moritz 2013; Beissinger et al. 2013; Rife et al. 2015; Du et al. 2017; Karaca and Ince 2017).

GBS methods are derivatives or improvement of approaches that have mainly evolved from reduced representation library (RRL) or reduced representation sequencing (RRS). The use of RRL for single nucleotide polymorphism discovery was first based on Sanger sequencing (Altshuler et al. 2000). In this method, pools of DNA from multiple individuals are reduced in complexity by the type II DNA restriction enzyme digestion and fragments produced by complete digestion of enzymes are size selected. The use of restriction enzyme digestion has the advantages of reducing the fraction of the genome present in the RRL by one to two orders of magnitude and ensuring that independently constructed libraries contain nearly identical fragment populations. Other strategies for genome reduction such as multiplexed amplification of target sequences, molecular inversion probes or the use of probes to capture DNA fragments by direct hybridization prior to sequencing are available but in comparison to the use of restriction enzyme they can be labor intensive. RRS approach is suitable for simultaneous de novo discovery of high-quality SNPs and population characterization of allele frequencies of any species with at least a partially sequenced genome. RRS is a general category of techniques that sequence a subset of the genome following different strategies and can be obtained using restriction enzymes, mechanical shearing or amplification, or natural resources such mRNA populations. High-throughput sequencing RRS can be classified in three major approaches: Restriction site Associate DNA sequencing (RAD-seq) and related method collectively called genotyping by sequencing (GBS), sequencing of cDNA obtained from mRNA and other non-coding RNA (RNA-seq) and whole exome sequencing (WES) (van Orsouw et al. 2007; Baird et al. 2008; van Tassell et al. 2008; Ali et al. 2016; Karaca and Ince 2017).

Table 4 Some high-throughput sequencing (next generation) methods currently available, divided into reduced-representation sequencing (RRS) and whole genome resequencing (WGR)

Reduced-representation sequencing based methods	References
Reduced representation shotgun sequencing (RRS)	Altshuler et al. (2000)
Complexity reduction of polymorphic sequences (CRoPS)	van Orsouw et al. (2007)
Restriction site-associated DNA sequencing (RAD-seq)	Baird et al. (2008)
Reduced-representation library (RRL)	van Tassell et al. (2008)
Paired-end reduced representation libraries (pERPLs)	Kerstens et al. (2011)
Multiplexed shotgun genotyping (MSG)	Andolfatto et al. (2011)
Simple genotyping-by-sequencing (GBS)	Elshire et al. (2011)
Two-enzyme genotyping-by-sequencing (GBS)	Poland and Rife (2012)
Double-digest RAD sequencing (ddRAD)	Peterson et al. (2012)
Sequence-based genotyping (SBG)	Truong et al. (2012)
Paired-end reduced representation libraries	Deschamps et al. (2012)
Type IIB endonucleases restriction-site associated DNA (2b-RAD)	Wang et al. (2012)
ezRAD	Toonen et al. (2013)
Restriction fragment sequencing (RESTseq)	Stolle and Moritz (2013)
Specific length amplified fragment sequencing (SLAF-Seq)	Sun et al. (2013)
Scalable genotyping by sequencing (GBS)	Sonah et al. (2013)
Genotyping by genome reducing and sequencing	Chen et al. (2013)
GBS with one enzyme digest	Beissinger et al. (2013)
Ion torrent genotyping by sequencing	Mascher et al. (2013)
Flexible and scalable GBS	Heffelfinger et al. (2014)
GBS with two enzyme digests	Gardner et al. (2014)
Improved RRLs (iRRL)	Greminger et al. (2014)
Genotyping-in-thousands by sequencing (GT-seq)	Campbell et al. (2015)
Spiked genotyping-by-sequencing (sGBS),	Rife et al. (2015)
Multiplexed inter-SSR genotyping by sequencing (MIG-seq)	Suyama and Matsuki (2015)
RAD capture (Rapture)	Ali et al. (2016)
Tunable genotyping-by-sequencing (tGBS)	Ott et al. (2017)
Random amplicon sequencing (RAM-seq)	Bayerl et al. (2017)
Whole genome resequencing (WGR) methods	**References**
Sliding window WGR	Huang et al. (2009)
Parental inference WGR	Xie et al. (2010)
Parental inference WGR with individualized model	Rowan et al. (2015)
Skim genotyping-by-sequencing (SkimGBS)	Bayer et al. (2015)
Whole-genome shotgun (WGS) SMRT sequencing	Du et al. (2017)

The reduced representation sequencing approaches select a fraction of the whole genome for sequencing and reduce the cost and labor for high-throughput genotyping. For instance, hypo-methylated regions of a genome can be obtained (selected) for sequencing. The genomic DNA of the target individual is digested with a 5-methylcytosine-sensitive restriction enzyme and the digest is subjected to electrophoresis; fragments of 100–600 bp are separated and used for sequencing using a suitable platform of NGS technologies. Alternatively, WES or RNA-seq (also called transcriptome sequences) could be used for genotyping studies. There are several different strategies or approaches for DNA and RNA studies such as sequence capture approach of NimbleGen SeqCap, Agilent SureSelect method and RainDance Targeted Sequencing System (Cui et al. 2011; Levy and Myers 2016; Karaca and Ince 2017).

Knowledge of the biologic system and genomic resources can assist in selecting among RRS (RAD-seq and other GBS methods), RNA-seq, WES or WGR. It is important to select correct high-throughput method to be used in conservation of biodiversity and genetic resources. Clearly it depends on the aim of study, biological system, genomic resources available, the genetic architecture of phenotypic traits, background of the researchers and ultimately on funding. For example, if selection is operating on a specific tissue, stage or development time, RNA-seq would be very appropriate for assessing genetic variation in the genomic regions expressed at time of sampling. On the other hand, if the genes of interest are already known, then GBS such as target capture could be the best strategy. However, if no candidate genes are known, a higher density screening methods such as WES or WGR could be preferable. When selection acts on protein-coding parts of the genome, the use of WES would be a cost-effective approach than WGR. On the other hand, if selection could be acting in regulatory elements or could be mediated by large structural variations and the research focus is the analysis of neutral processes, then WGR could be the best choice because it provides the highest DNA marker density. When, WGR would not be necessary as RRS methods would excel for an affordable price (Bayerl et al. 2017; Fuentes-Pardo and Ruzzante 2017; Karaca and Ince 2017).

In a typical high-throughput RRS method, different samples from the related organisms are pooled and pooled samples are then digested with a type II DNA restriction enzyme. Enzyme treated DNA samples are size selected and selected DNA fragments are ligated with adapters required for sequencing on a NGS platform. Ligated fragments are again size selected and purified. Purified DNA fragments are amplified and the PCR products are sequenced using an Illumina platform (van Tassell et al. 2008; Kerstens et al. 2011). One of the main limitations of RRS method is that it requires reference sequence of the species under study. A reference genome sequence is used to order SNPs within the sequence assembly. However, this challenge may be overcome by genotyping linkage mapping populations or by using comparative genomic information to infer likely or relative genome position (Elshire et al. 2011; Deschamps et al. 2012; Karaca and Ince 2017).

RAD-seq refers to a group of RRS methods such as RAD, ddRAD, ezRAD, RAD-cap that evaluate the genetic variation present within and at the restriction cut sites. The selection of frequent or rare cutter restriction enzyme determines marker density making these methods flexible and customizable. RAD-seq typically examine thousands of low-density genome wide SNPs located in neutral and putatively functional loci that can be genotyped by sequencing in multiple individuals and populations for a relatively low cost. A typical RAD-seq is performed as follows: genomic DNA samples are individually digested with a restriction enzyme and adaptors with nucleotide barcodes for unique identification of each sample are ligated to DNA fragments. Fragments with 300–700 bp are size selected and different DNA samples are pooled. Pooled DNA fragments with adapters are randomly sheared by sonication, and ends are ligated with a second type adapters. Purified fragments are PCR amplified and sequenced using a high-throughput sequencing (NGS) such as reversible dideoxy based Illumina sequencing which uses either sequencing one (one read, single end) or both (two reads, paired end) ends of each fragment and currently gives reads of up to 300 bp in length (Karaca and Ince 2017).

High-throughput RNA sequencing analysis (RNA-seq) focuses on genetic variation of genome transcribed in a particular time/tissue. RNA-seq is able to reveal genes that are being actively expressed in specific tissue and species of interest, and facilitate the discovery of potential molecular marker of SNPs, microsatellites or InDels markers, some of which could be functional DNA markers. This type of analysis is useful in non-model organisms where the full genome data is still not available for comparison. Sequences that are targets for RNA-seq analysis do not contain repetitive genomic regions and rich in regulatory sequences 5'-UTR, 3'-UTR, miRNA and gene bodies. Furthermore, these regulatory sequences and genes are present in only those genes that are transcribed in a particular tissue/organ during the given developmental stage and under the environmental conditions. RNA-seq is mostly used as a cost-effective approach for gene expression quantification research (Li et al. 2010; Yan et al. 2010; Fuentes-Pardo and Ruzzante 2017; Yamanaka et al. 2018).

Although RNA-seq provides abundant information on gene expression, gene regulation and amino acid content of proteins, it is limited to only those genes that are transcribed in the concerned tissue/organ during the given developmental stage and under the environmental conditions prevailing at the time of sample collection. Therefore, a fair number of organs/tissues, developmental stages should be sampled to ensure the representation of most, if not all, of the genes present in the genome of the concerned species. For a typical RNA-seq analysis, mRNA, RNA with polyA tails is isolated from total RNA and reverse transcribed to cDNA with reverse transcriptase and polyT or polyU primers (Wang et al. 2009a, b; Hua et al. 2011; Du et al. 2015; Waiho et al. 2017). To isolate micro (miRNA), small (sRNA), and long (loRNA), these non-coding RNA molecules are selectively ligated to 3′ and 5′ adapters and reverse transcribed to cDNA (Li et al. 2010; Yan et al. 2010; Batovska et al. 2017; Waiho et al. 2017; Wei et al. 2017).

Whole genome exome sequencing (WES) provides a cost and time effective alternate to whole genome sequencing. The goal of WES is to determine DNA sequence information for regions of a genome that code for proteins. Target regions are referred to as exons. WES selects exonic regions of interest and separating them from non-exon regions of the genome. It is fast and cost effective approach to identification of variants (SNPs, copy number variations (CNVs), small InDels), linkage, association and conservation pedigree studies. WES is often chosen as a substitute for whole genome sequencing because of its lower cost, lower data storage and analysis requirements. RNA-seq and WES differ in the first steps of creating a sequencing library. WES uses genomic DNA regions while RNA-seq utilizes

RNA molecules. WES is a cost-effective alternative to RAD-seq, RNA-seq and whole genome resequencing (Elshire et al. 2011; Altmann et al. 2012; Krumm et al. 2012; He et al. 2014; Suyama and Matsuki 2015; Yamanaka et al. 2018).

Whole genome sequencing and resequencing (WGR) could produce complete or nearly complete genomic DNA sequences of an organism using and assembling numerous shotgun reads that cover the genome multiple times. WGR studies could use four different approaches such as the sequencing of individuals to a high depth of coverage with resolved haplotypes and unresolved haplotypes, the sequencing of population genomes to a high depth by pooling the same amounts of individual DNA, the sequencing of multiple individuals from a population to a low depth. WGR allows the discovery of a huge number of DNA markers such as SNPs, InDels, copy number variations, and presence/absence variations (PAV) in crops and provides deep insight into genome evolution. Moreover, the combination of WGR with bulked segregant analysis allows rapid identification of genes and causal mutations in crops (Huang et al. 2009; Xie et al. 2010; Bayer et al. 2015; Du et al. 2017; Fuentes-Pardo and Ruzzante 2017). Unfortunately, WGR is not currently cost-effective for particularly those species with large genomes, or for those studies requiring large numbers of individuals (Jamann et al. 2017; Karaca and Ince 2017; Vlk and Repkova 2017; Parchman et al. 2018).

A typical WGR method is performed as follows: genomic DNA is fragmented to about 500 bp by sonication and the fragments are end repaired before adding dATPs to generate a protruding 3′ A for ligating with the adaptor carrying a three-base index. Three based indexes are linked to adapters and the indexed DNA samples are run on 2% agarose gels to purify fragments of 150–180 bp. Each sample is amplified by PCR for about 18 cycles and DNA samples of individuals with different indexes are mixed in an equal molar concentration and are loaded into one lane of the Illumina GA for 36-cycle sequencing, with the Illumina PhiX sample used as control. Image analysis and base calling are performed using Illumina GA processing pipeline (Huang et al. 2009; Xie et al. 2010; Bayer et al. 2015; Du et al. 2017; Karaca and Ince 2017).

WGR methods based on NGS technologies and platforms are theoretically capable of identification all genetic variants among individuals of populations. WGR is more robust than WES for the detection of exome variants as it provides a more homogeneous sequence read coverage and a better sequencing quality overall. Another advantage of WGR approaches is that they examine multiple types of genetic variations simultaneously including structural variations (deletions, insertions, substitution, rearrangements, and copy number variation) and mutations in regulatory elements. In contrast, RRS techniques are mostly restricted to one base changes (i.e., SNPs), and RNA-seq and WES are for detection of variation within coding sequences (Fuentes-Pardo and Ruzzante 2017). Although WGR provides complete resolution of any genome it is cost-prohibitive for researchers in developing countries and indeed WGR may be unnecessary for many studies involving a large number of individuals. The parental genomes with high-quality sequences and a reference sequence are often required for WGR. It differs from RRS, in the lack of complexity reduction steps before sequencing. WGR is well suited to genotyping biparental cross populations with complex, small and moderate sized genomes. It provides the lowest cost per marker data point. Compared to WGR methods, RRS approaches differ in their suitability for various tasks, but demonstrate similar costs per marker data point. However, RRS approaches are generally better suited for de novo applications and more cost-effective when genotyping populations with large genomes or high heterozygosity. On the other hands, WGR offers the greatest cost-efficiency per marker data point, and is particularly useful when recombination is high and many markers are needed for a well-resolved genetic map in a species with a small or moderate sized genome. WGR has the added benefit of increasing the chances of finding causative SNPs, InDels or genes, which allow development of "perfect" or "functional" markers. In the light of the decreasing costs of sequencing, the use of WGR to increase the resolution of mapping studies is likely to become more common in the future (Huang et al. 2009; Rife et al. 2015; Rowan et al. 2015).

WGR could be used in the detection of biodiversity, selection of genetic resources and the characterization of the genetic basis of phenotypic traits and diseases affecting survivor. RRS approaches can also be used for this purpose at the fraction of the genome screened, although their success may depend on the proportion of the genome covered. With the help of high resolution of high-throughput genotyping approaches (high-throughput sequencing) measures of nucleotide diversity and divergence can be estimated. For instance, deviation from neutrality can be readily tested, and identification of thousands of genes altered can be achieved. In typical genetic conservation studies about 10–50 variables are used but conservation genomics based on high-throughput sequencing involves tens of thousands of genes. Conservation genomics, in particular the availability of genome-wide sequences permits the simultaneous study of the effects of demographic history, migration and selection (Bayerl et al. 2017).

High-throughput sequencing based genotyping provides higher resolution for phylogenomics, hybridization and taxonomical studies, all of which relate with conservation of biodiversity and genetic resources. The successful implementation of conservation plans relies on the correct identification of the taxonomic status of organisms that are targeted for conservation of biodiversity and genetic resources. Whole or nearly whole-genome data provide a complete record of a species evolutionary process. However, more works are required to be done to resolve algorithm limitations associated with the analysis of such large amount of genomic data. In some cases, genome rearrangements, lateral gene transfer, incomplete lineage sorting make analyses more difficult (Aravanopoulos 2016; Karaca and Ince 2017; Yousef et al. 2018).

High-throughput sequencing based genotyping could provide data on species demographic history, migration patterns, range expansion and changes in historical effective population size. Such data also allow obtaining information regarding barriers to gene flow, anthropogenic disturbance, climate change, historical demographic processes, population structure and admixture. It is very important to maintain high genetic diversity in vulnerable species with lower population size for genetic conservation. Because most natural populations are structured in local subpopulations, genetic differences may occur among subpopulations over time as a result of gene flow, genetic drift and local adaptation. Because high-throughput sequencing based GBS and WGS approaches provide the highest marker density, these methods allow the simultaneous evaluation of genome wide patterns in neutral and functional loci that act as a record of demographic and historical events, and adaptation. GBS and WGS provide data on the identification of genomic regions, which involved in adaptation to local environmental conditions. These data are crucial for conservation biology because of the importance of functional genetic diversity. Furthermore, these data provide connection between genotype, phenotype and fitness (Fuentes-Pardo and Ruzzante 2017; Karaca and Ince 2017).

High-throughput sequencing based genotyping provides valuable data that could be used assessment of genetic diversity, which is essential for the organization, conservation and use of genetic resources to develop strategies for optimal germplasm collection, evaluation and seed regeneration. Next generation genotyping methods have advantages for characterizing gene bank accessions such as a major advantage is their applicability to any species. These do not cost much per individual data, but provide sufficient power for genome-wide analyses of population structure and genetic relationships. The main disadvantage of high-throughput sequencing is the presence of a high proportion of missing data that may reduce the power for correct estimation of population parameters. Also, high cost of high-throughput sequencing and the elevated demand for computing resources limit their implementations in conservation of biodiversity and genetic resources (Aravanopoulos 2016; Yamanaka et al. 2018; Yousef et al. 2018).

Genomics provides an unprecedented level of resolution for population genetic studies since next-generation sequencing data will be more powerful and accurate, especially in cases where significant adaptive differentiation is expected among evolutionary significant units considered as candidates for gene conservation. Today because of high-throughput resequencing platforms, it is feasible to substantially increase the numbers of populations, individuals per population and loci per individual studied at a fraction of earlier experimental costs (Pertoldi et al. 2007; Karaca and Ince 2017). With the use of NGS based genotyping approaches genomics offers high precision estimates of genetic and demographic parameters and could result in high-resolution characterization of adaptive genetic variation in nature. Therefore, studies dealing with conservation of biodiversity and genetic resources would provide considerable benefits for humankind (Bayerl et al. 2017; Jamann et al. 2017; Scheben et al. 2017).

The use of genomics in genetic monitoring of biodiversity is very important since genetic monitoring provides valuable information regarding an early detection mechanism that leads to management decisions aimed to lessen potential harmful effects before permanent damage occurs. In another word, genetic monitoring is an effective prognostic tool to secure genetic diversity in natural populations. It could provide plenty information on natural selection, genetic drift, mating system, migration, gene flow and health of population. For instance, the effects of natural selection may lead to differentiation associated with local or regional adaptation, while genetic drift can lead to genetic erosion (Ali et al. 2016; Jiang et al. 2016; Du et al. 2017). Second generation based GBS technologies use DNA enrichment methods prior to amplification, resulting in relatively short sequencing templates. Third generation sequencing platforms are capable of producing significantly larger read lengths and sequencing through traditionally difficult sequence templates with high GC content (Du et al. 2017). Third generation sequencing platforms seem best suitable method for conservation of biodiversity and genetic resources when several associated disadvantages are mitigated (Beissinger et al. 2013; Sun et al. 2013; Heffelfinger et al. 2014; Karaca and Ince 2017; Scheben et al. 2017; Elbasyoni et al. 2018).

7 Conclusions and Future Prospects

It is estimated that the global population is approaching to nine billion by 2050, and demand for food and fiber crops is expected to increase by about 60% (Sun et al. 2018). Although phenotypic, metabolomics, proteomics and genetic diversity are more heavily reduced in cultivated germplasm, fortunately international movement on conservation and sustainable use of biodiversity and genetic resources for agriculture have greatly sounded during the last 50–60 years. Today approximately 7.4 million germplasm accessions, representing more than 16,500 plant species are secured in 1750 gene banks worldwide. However, unfortunately conservation programs are chronically underfunded and the impact of climate change on crop genetic diversity is not completely understood. In many parts of the world, appropriate capacities and adequate infrastructures to explore biodiversity are still lacking and genetic erosion is far from being stopped (Sari et al. 2005; Davey et al. 2011; Fu et al. 2015).

The genetic drift in gene banks is caused by the use of inadequate sample sizes. Also regeneration delays cause genetic integrity loss for some cross-pollinating species in gene banks. Furthermore, gene bank conservation gets less strengthen political support in todays' capitalist world. In many developing countries there exist inadequate germplasm evaluation and characterization. Efficient conservation of genetic resources requires efficient and effective global networking of gene banks around the world. Effectively upgrading gene bank information systems is also important and required. In many countries there exist low diversity coverage and inadequate gene bank capacities. Unfortunately, private sectors not interested in conservation of biodiversity and there is inadequate gene bank support from stakeholders. Many stakeholders are mainly interested in germplasm for economic potential and do not provide supports for management of gene banks and establishment facilities for long-term conservation (Fu 2017).

Over-grazing, over-exploitation, urbanization, hydroelectric dam construction, roads and global market economies have caused the impoverishment of many native forests and grasslands. For instance, heavy collection of aromatic and medicinal plant species narrowed genetic diversity in the Mediterranean basin of Turkey. It is known that over-exploitation in some other parts of the world threatened genetic biodiversity of many plant species. Increasing water and air pollution along with deforestation and biologic pollution contribute to the genetic erosion of both cultivated and wild species. In turn, the unsustainable use of natural resources such as forests and ponds has resulted into disturbed water balance and severe erosion. Also in many countries current legislations discourage the use of landraces and also have a strong negative impact on their conservation. For instance, Italy reports that out of 41 farms growing landraces of forage legumes in the 1970s only one now carries through this activity (Fu et al. 2015; Manhaes et al. 2018).

In order to conserve the biodiversity and genetic resources for sustainable agriculture next generation based genomic and phenomic monitoring should be considered and used simultaneously. NGS based phenotyping and genotyping could be effectively used in monitoring of genetic diversity during seed regeneration and plantation. These technologies would allow to manage diversity within accessions to mitigate some disadvantages of small population sizes of *ex situ* conservation (Davey et al. 2011; Poczai et al. 2013; Jia et al. 2016; Tsai et al. 2015). But most of these techniques are not yet widely available in developing countries where they are most needed such as in tropics regions including many threatened species (Fu et al. 2015; Kang et al. 2016; Manhaes et al. 2018). Most phenotypic traits involved in local adaptation survival are polygenic, and the importance of epistasis, transposable element activity or epigenetics plays significant roles. Since polygenic traits could be effectively analyzed using GBS technologies, genomic monitoring based on GBS is very suitable for conservation of biodiversity and genetic resources for sustainable agriculture. GBS could be used to estimate population parameters including allelic richness, expected heterozygosity and the total and the effective number of alleles, outcrossing and inbreeding rate, out coming gene flow and effective population size (Kang et al. 2016; Watanabe et al. 2017).

The presence of dramatic climate changes and the direct adverse anthropogenic influence and activity are two major issues that are driving the need for immediate, extensive and comprehensive conservation of genetic resources of world. It is expected that global temperature will rise about 1.8–4.0 °C during the twenty-first century and this will cause a shift of species spatial distributions more than 6 km towards the poles and 1 m in elevation, per decade. This may result in population spatial shifts, fragmentation, reduction of population size or even extinction in mountainous ranges (Aravanopoulos 2016). As a final sentence, we believe that a well-designed, genomic and phenomic tools-monitored, and well-managed systems coupled with *ex situ* and *in situ* conservation strategies (seed banks, cryogenic storage, living collections in botanical gardens, arboreta, and similar facilities where necessary) is enough to protect many endangered plant species and conserve biodiversity through the several decades of rapid global change. However, people on earth should learn to live with nature as a part of nature not against to nature for long-term conservation for themselves and for the nature surrounding them, and should listen the nature while it is still able to speak.

Acknowledgements We thank Dr. Ayşe Gümüş Karaca, for her valuable comments and suggestions on this study. We also would like to thank all researchers and scientists whose works were cited in this study. We believe that knowledge should be shared and only shared knowledge gets stronger and bigger for humanity.

References

Aavik T, Talvea T, Thetloff M, Uuemaab E, Ojaa T (2017) Genetic consequences of landscape change for rare endemic plants – A case study of *Rhinanthus osiliensis*. Biol Conserv 210:125–135

Afonnikov DA, Genaev MA, Doroshkov AV, Komyshev EG, Pshenichnikova TA (2016) Methods of high-throughput plant phenotyping for large-scale breeding and genetic experiments. Russ J Genet 52:688–701

Ali OA, O'Rourke SM, Amish SJ, Meek MH, Luikart G, Jeffres C, Miller MR (2016) RAD capture (Rapture): flexible and efficient sequence-based genotyping genetics. Genetics 202:389–400

Altmann A, Weber P, Bader D, Preuss M, Binder E, Muller-Myhsok B (2012) A beginner's guide to SNP calling from high throughput DNA-sequencing data. Hum Genet 131:1541–1554

Altshuler D, Pollara VJ, Cowles CR, van Etten WJ, Baldwin J, Linton L, Lander ES (2000) An SNP map of the human genome generated by reduced representation shotgun sequencing. Nature 407:513–516

Andolfatto P, Davison D, Erezyilmaz D, Hu TT, Mast J, Sunayama-Morita T, Stern DL (2011) Multiplexed shotgun genotyping for rapid and efficient genetic mapping. Genome Res 21:610–617

Araus JL, Cairns JE (2014) Field high-throughput phenotyping: the new crop breeding frontier. Trends Plant Sci 19:52–61

Aravanopoulos FA (2016) Conservation and monitoring of tree genetic resources in temperate forests. Curr Forestry Rep 2:119–129

Arend D, Junker A, Scholz U, Schuler D, Wylie J, Lange M (2016) PGP repository: a plant phenomics and genomics data publication infrastructure. Database 1:1–10

Aydin A, Karaca M (2016) An alternative way to assign the chromosome location of genes/markers in cotton. FEBS J 283:335

Baird NA, Etter PD, Atwood TS, Currey MC, Shiver AL, Lewis ZA, Selker EU, Cresko WA, Johnson EA (2008) Rapid SNP discovery and genetic mapping using sequenced RAD markers. PLoS One 3:3376

Batovska J, Cogan NOI, Lynch SE, Blacket MJ (2017) Using next-generation sequencing for DNA barcoding: capturing allelic variation in ITS2. Gene Genome Genet 7:19–29

Baur E (1914) Die Bedeutung der primitiven kulturrassen und der wilden verwandten unserer kulturpflanzen fur die pflanzenzuchtung. (In German). Jahrbuch der DLG 29:104

Bayer PE, Ruperao P, Mason AS, Stiller J, Chan CK, Hayashi S et al (2015) High-resolution skim genotyping by sequencing reveals the distribution of crossovers and gene conversions in *Cicer arietinum* and *Brassica napus*. Theor Appl Genet 128:1039–1047

Bayerl H, Kraus RKS, Nowak C, Foerster DW, Fickel J, Kuehn R (2017) Fast and cost-effective single nucleotide polymorphism (SNP) detection in the absence of a reference genome using semi deep next-generation Random Amplicon Sequencing (RAMseq). Mol Ecol Resour 18:107–117

Beissinger TM, Hirsch CN, Sekhon RS, Foerster JM, Johnson JM et al (2013) Marker density and read depth for genotyping populations using genotyping-by-sequencing. Genetics 193:1073–1081

Bhattacharya S (2016) Syrian seed bank gets new home away from war. Nature 538:16–17

Bilgen M, Karaca M, Onus AN, Ince AG (2004) A software program combining sequence motif searches with keywords for finding repeats containing DNA sequences. Bioinformatics 20:3379–3386

Bostein D, White RL, Skolnick M, Davis RW (1980) Construction of a genetic link-age map in man using restriction fragment length polymorphisms. Am J Hum Genet 32:314–333

Breccia G, Nestares G (2014) Next-generation phenotyping in plants: old problems, new promises. J Basic App Genet 25:5–8

Brown AHD, Hodgkin T (2015) Indicators of genetic diversity, genetic erosion, and genetic vulnerability for plant genetic resources. In: Ahuja MR, Jain SM (eds) Genetic diversity and erosion in plants. Springer, Cham, pp 25–53

Buse J, Boch S, Hilgersd J, Griebeler EM (2015) Conservation of threatened habitat types under future climate change lessons from plant-distribution models and current extinction trends in southern Germany. J Nat Conserv 27:18–25

Busemeyer L, Ruckelshausen A, Moller K, Melchinger AE, Alheit KV, Maurer HP, Hahn V, Weissmann EA, Reif JC, Wurschum T (2013) Precision phenotyping of biomass accumulation in triticale reveals temporal genetic patterns of regulation. Sci Rep 3:2442

Cakmakci S, Aydinoglu B, Karaca M, Bilgen M (2006) Heritability of yield components in common vetch (*Vicia sativa* L.). Acta Agric Scand Sect B Soil Plant Sci 56:54–59

Campbell NR, Stephanie AH, Shawn RN (2015) Genotyping-in-thousands by sequencing (GT-seq): a cost effective SNP genotyping method based on custom amplicon sequencing. Mol Ecol Resour 15:855–867

Castaneda-Alvarez NP, Khoury CK, Achicanoy HA, Beernau V, Dempewolf H, Eastwood RJ (2016) Global conservation priorities for crop wild relatives. Nat Plants 2:16022

Chandrakant S, Poornima R, Rajkumar B (2017) Recent advances in conservation of plant genetic resources. Agri Res Tech 7:555718

Chen Q, Ma Y, Yang Y, Chen Z, Liao R, Xie X, Wang Z, He P et al (2013) Genotyping by genome reducing and sequencing for outbred animals. PLoS One 8:e67500

Cobb JN, DeClerck G, Greenberg A, Clark R, McCouch S (2013) Next-generation phenotyping: requirements and strategies for enhancing our understanding of genotype–phenotype relationships and its relevance to crop improvement. Theor Appl Genet 126:867–887

Coetzee BWT, Gaston KJ, Chown SL (2014) Local scale comparisons of biodiversity as a test for global protected area ecological performance: a meta-analysis. PLoS One 9:e105824

Corlett RT (2016) Plant diversity in a changing world: status, trends, and conservation needs. Plant Diversity 38:10–16

Crain JL, Reynolds MP, Poland JA (2017) Utilizing high-throughput phenotypic data for improved phenotypic selection of stress adaptive traits in wheat. Crop Sci 57:648–659

Crain J, Mondal S, Rutkoski J, Singh RP, Poland J (2018) Combining high-throughput phenotyping and genomic information to increase prediction and selection accuracy in wheat breeding. Plant Genome 11:170043

Cui G, Huang L, Tang X, Zhao J (2011) Candidate genes involved in tanshinone biosynthesis in hairy roots of *Salvia miltiorrhiza* revealed by cDNA microarray. Mol Biol Rep 38:2471–2478

Davey JW, Hohenlohe PA, Etter PD, Boone JQ, Catchen JM, Blaxter ML (2011) Genome-wide genetic marker discovery and genotyping using next-generation sequencing. Nat Rev Genet 12:499–510

Deschamps S, Llaca V, May GD (2012) Genotyping-by-sequencing in plants. Biol 1:460–483

Dorey K, Walker TR (2018) Limitations of threatened species lists in Canada: a federal and provincial perspective. Biol Conserv 217:259–268

Du Q, Li C, Li D, Lu S (2015) Genome-wide analysis, molecular cloning and expression profiling reveal tissuespecifically expressed, feedback-regulated, stress-responsive and alternatively spliced novel genes involved in gibberellin metabolism in *Salvia miltiorrhiza*. BMC Genomics 16:1087

Du H, Yu Y, Ma Y, Gao Q, Cao Y, Zhuo Chen Z et al (2017) Sequencing and de novo assembly of a near complete indica rice genome. Nat Commun 8:15324

Dulloo ME, Thormann I, Fiorino E, de Felice S, Rao VR, Snook L (2013) Trends in research using plant genetic resources from germplasm collections: from 1996 to 2006. Crop Sci 53:1217–1227

Duvick DN (2005) The contribution of breeding to yield advances in maize (*Zea mays* L.). Adv Agron 86:83–145

Ecarnot M, Baczyk P, Tessarotto L, Chervin C (2013) Rapid phenotyping of the tomato fruit model, Micro-Tom, with a portable VIS-NIR spectrometer. Plant Physiol Biochem 70:159–163

Edgerton MD (2009) Increasing crop productivity to meet global needs for feed, food, and fuel. Plant Physiol 149:7–13

Elbasyoni IS, Lorenz AJ, Guttieri M, Frels K, Baenziger PS, Poland J, Akhunov E (2018) A comparison between genotyping-by-sequencing and array-based scoring of SNPs for genomic prediction accuracy in winter wheat. Plant Sci 270:123–130

Elmasulu SY, Çınar A, Ince AG, Karaca M, Onus AN, Turgut K (2009) Identification of genotypes with high essential oil contents in *Origanum*, *Thymus* and *Sideritis*. Planta Med 75:932

Elmasulu S, Kürkçüoğlu M, Ince AG, Karaca M, Çınar A, Onus AN, Başer KHC, Turgut K (2011) Classification of 63 *Origanum* taxa based on microsatellite markers and essential oil composition. Planta Med 77:1296

Elshire RJ, Glaubitz JC, Sun Q, Poland JA, Kawamoto K, Buckler ES, Mitchell SE (2011) A robust, simple genotyping-by-sequencing (GBS) approach for high diversity species. PLoS One 6:e19379

Erbano M, Schuhli GSE, Santos EPD (2015) Genetic variability and population structure of *Salvia lachnostachys*: implications for breeding and conservation programs. Int J Mol Sci 16:7839–7850

Espina MJ, Ahmed CMS, Bernardini A, Adeleke E, Yadegari Z, Arelli P, Pantalone V, Taheri A (2018) Development and phenotypic screening of an ethyl methane sulfonate mutant population in soybean. Front Plant Sci 9:394

Evenson RE, Gollin D (2003) Assessing the impact of the green revolution, 1960 to 2000. Science 300:758

Fu YB (2017) The vulnerability of plant genetic resources conserved *ex situ*. Crop Sci 57:2314–2328

Fu YB, Ahmed Z, Diederichsen A (2015) Towards a better monitoring of seed ageing under *ex situ* seed conservation. Conserv Physiol 3:cov026

Fuentes-Pardo AP, Ruzzante DE (2017) Whole-genome sequencing approaches for conservation biology: advantages, limitations and practical recommendations. Mol Ecol 26:5369–5406

Furuta T, Ashikari M, Jena KK, Doi K, Reuscher S (2017) Adapting genotyping-by-sequencing for rice F_2 populations. Gene Genom Genet 7:881–893

Gardner KM, Brown P, Cooke TF, Cann S, Costa F, Bustamante C, Velasco R, Troggio M, Myles S (2014) Fast and cost-effective genetic mapping in apple using next-generation sequencing. Genes Genome Genet 4:1681–1687

Gascuel Q, Diretto G, Monforte AJ, Fortes AM, Granell A (2017) Use of natural diversity and biotechnology to increase the quality and nutritional content of tomato and grape. Front Plant Sci 8:652

Geldmann J, Coad L, Barnes M, Craigie ID, Hockings M, Knights K, Leverington F, Cuadros IC, Zamora C, Woodley S, Burgess ND (2015) Changes in protected area management effectiveness over time: a global analysis. Biol Conserv 191:692–699

Goettsch B, Hilton-Taylor C, Cruz-Pinon G, Dufy JP, Frances A et al (2015) High proportion of cactus species threatened with extinction. Nat Plants 1:15142

Gonzalez-Dugo V, Goldhamer D, Zarco-Tejada PJ, Fereres E (2015) Improving the precision of irrigation in a pistachio farm using an unmanned airborne thermal system. Irrigation Sci 33:43–52

Goulart F, Galanc AL, Nelsond E, Soares-Filh B (2018) Conservation lessons from Cuba: connecting science and policy. Biol Conserv 217:280–288

Greminger MP, Stölting KN, Nater A, Goossens B, Arora N, Bruggmann R et al (2014) Generation of SNP datasets for orangutan population genomics using improved reduced-representation sequencing and direct comparisons of SNP calling algorithms. BMC Genomics 15:16

Habib A, Zhou T, Masjedi A, Zhang Z, Flatt JE, Crawford M (2018) Boresight calibration of GNSS/INS-assisted push-broom hyperspectral scanners on UAV platforms. IEEE J 11:1734–1749

Harlan JR (1975) Our vanishing genetic resources. Science 188:617–621

He J, Zhao X, Laroche A, Lu ZX, Liu HK, Li Z (2014) Genotyping-by-sequencing (GBS), an ultimate marker-assisted selection (MAS) tool to accelerate plant breeding. Front Plant Genet Genomics 5:1–8

Heffelfinger C, Fragoso CA, Moreno MA, Overton JD, Mottinger JP, Zhao H, Tohme J, Dellaporta SL (2014) Flexible and scalable genotyping-by-sequencing strategies for population studies. BMC Genomics 15:979

Hernandez-Suarez C (2018) Measuring the representativeness of a germplasm collection. Biodivers Conserv 27:1471–1486

Heywood VH (2015) *In situ* conservation of plant species – an unattainable goal? Israel J Plant Sci 4:211–231

Hijioka Y, Lin E, Pereira JJ, Corlett RT, Cui X, Insarov G, Lasco R, Lindgren E, Surjan A, Barros VR (2014) Climate Change 2014: impacts adaptation, and vulnerability. Part B. Cambridge University Press, Cambridge/New York, pp 1327–1370

Hoffmann AM, Noga G, Hunsche M (2015) Fluorescence indices for monitoring the ripening of tomatoes in pre- and postharvest phases. Sci Hortic 191:74–81

Hoyos-Villegas V, Houx JH, Singh SK, Fritschi FB (2014) Ground-based digital imaging as a tool to assess soybean growth and yield. Crop Sci 54:1756–1768

Hua WP, Zhang Y, Song J, Zhao LJ, Wang ZZ (2011) De novo transcriptome sequencing in *Salvia miltiorrhiza* to identify genes involved in the biosynthesis of active ingredients. Genomics 98:272–279

Huang X, Feng Q, Qian Q, Zhao Q, Wang L, Wang A, Guan J, Fan D, Weng Q, Huang T, Dong G, Sang T, Han B (2009) High-throughput genotyping by whole-genome resequencing. Genome Res 19:1068–1076

Ince AG (2012) A contig-based microsatellite marker approach and its application in *Cichorium* ESTs. Rom Biotech Lett 17:7177–7186

Ince AG, Karaca M (2009) The MAGi RNA extraction method: a highly efficient and simple procedure for fresh and dry plant tissues. J Sci Food Agric 89:168–176

Ince AG, Karaca M (2011a) Genetic variation in common bean landraces efficiently revealed by Td-DAMD-PCR markers. Plant Omics J 4:220–227

Ince AG, Karaca M (2011b) Development of new genomic and genic SSR primer pairs for carrot. Planta Med 77:1359–1360

Ince AG, Karaca M (2011c) Early determination of sex in jojoba plant by CAPS assay. J Agric Sci 149:327–336

Ince AG, Karaca M (2012) Species-specific touch-down DAMD PCR markers for *Salvia* species. J Med Plant Res 6:1590–1595

Ince AG, Karaca M (2015a) Td-DAMD-PCR assays for fingerprinting of commercial carnations. Turk J Bot 39:147–159

Ince AG, Karaca M (2015b) E-microsatellite markers for some naturally occurring *Salvia* species in the Mediterranean region. Turk J Biol 39:69–77

Ince AG, Karaca M (2016) Association between cytosine methylation and tissue specific expression of microsatellites. FEBS J 283:334–335

Ince AG, Onus AN, Elmasulu SY, Bilgen M, Karaca M (2007) *In silico* data mining for development of Capsicum microsatellites. Acta Horticulturae 729:123–127

Ince AG, Karaca M, Bilgen M, Onus AN (2008) Digital differential display tools for mining microsatellite containing organism, organ and tissue. Plant Cell Tiss Org Cult 94:281–290

Ince AG, Karaca M, Onus AN (2009a) Development of microsatellite primer pairs for *Cynara cardunculus* var. *scolymus* (L.) Fiori. Planta Med 75:930

Ince AG, Karaca M, Onus AN (2009b) Differential expression of microsatellites in leaves and rhizomes of turmeric (*Curcuma longa* Linn.). Planta Med 75:929

Ince AG, Karaca M, Turgut K (2009c) PCR based minisatellites are useful in *Origanum*, *Thymus*, *Sideritis* and *Salvia* genetic studies. Planta Med 75:932

Ince AG, Yaşak Elmasulu S, Yılmaz Çınar A, Karaca M, Onus AN, Turgut K (2009d) Comparison of DNA marker techniques for Lamiaceae. Acta Hortic 826:431–438

Ince AG, Karaca M, Onus AN (2009e) Development and utilization of diagnostic DAMD-PCR markers for *Capsicum* accessions. Genet Res Crop Evol 56:211–221

Ince AG, Karaca M, Onus AN (2010a) Mono- and dinucleotide repeat DNA content differs in dioecious Asparagus (*Asparagus officinalis* L.). Planta Med 76:1203–1204

Ince AG, Karaca M, Turgut K (2010b) Development of new set of EST-SSR primer pairs for celery (*Apium graveolens* L.). Planta Med 76:1202

Ince AG, Karaca M, Onus AN (2010c) Genetic relationships within and between *Capsicum* species. Biochem Genet 48:83–95

Ince AG, Karaca M, Onus AN (2010d) The first report of microsatellite primer pairs for genetic studies in jojoba (*Simmondsia chinensis* (Link) Schneider). Planta Med 76:1201

Ince AG, Karaca M, Onus AN (2010e) CAPS-microsatellites are suitable for Ginger (*Zingiber officinale* Roscoe) genetic studies. Planta Med 76:1203

Ince AG, Karaca M, Turgut K (2010f) Screening of nine organelle DNA loci for polymorphism in twenty-two medicinal and aromatic plant species. Planta Med 76:1202

Ince AG, Karaca M, Onus AN (2010g) Differential expression patterns of genes containing microsatellites in *Capsicum annuum* L. Mol Breed 25:645–658

Ince AG, Karaca M, Onus AN (2010h) Polymorphic microsatellite markers transferable across *Capsicum* species. Plant Mol Biol Rep 28:285–291

Ince AG, Karaca M, Onus AN (2010i) CAPS-microsatellites: use of CAPS method to convert nonpolymorphic microsatellites into useful markers. Mol Breed 25:491–499

Ince AG, Karaca M, Onus AN (2010j) A reliable gender diagnostic PCR assay for jojoba (*Simmondsia chinensis* (Link) Schneider). Genet Res Crop Evol 57:773–779

Ince AG, Karaca M, Aydin A (2011a) Comparison of multiple DNA alignment algorithms for Labiatae molecular phylogeny inferences. Planta Med 77:1280

Ince AG, Karaca M, Onus AN (2011b) Exact microsatellite density differences among *Capsicum* tissues and development stages. J Agric Sci 17:291–299

Ince AG, Karaca M, Ay ST (2011c) Transferability of EST-Microsatellite markers to some Labiatae genera. Planta Med 77:1360–1360

Ince AG, Yildiz F, Karaca M (2011d) The MAGi DNA extraction method for fresh tissues and dry seeds. J Med Plant Res 5:5458–5464

Ince AG, Karaca M, Elmasulu SY (2014) New microsatellite and CAPS-microsatellite markers for clarifying taxonomic and phylogenetic relationships within *Origanum* L. Mol Breed 34:643–654

Jamann TM, Sood S, Wisser RJ, Holland JB (2017) High-throughput resequencing of maize landraces at genomic regions associated with flowering time. PLoS One 12:e0168910

Jeffreys AJ, Wilson V, Thein SL (1985) Hypervariable "minisatellite" regions in human DNA. Nature 314:67–73

Jia Q, Tan C, Wang J, Zhang XQ, Zhu J, Luo H, Yang J, Westcott S, Broughton S, Moody D, Li C (2016) Marker development using SLAF-seq and whole-genome shotgun strategy to fine map the semi-dwarf gene aria in barley. BMC Genomics 17:911

Jiang Z, Wang H, Michal JJ, Zhou X, Liu B, Woods LCS, Fuchs RT (2016) Genome wide sampling sequencing for SNP genotyping: methods, challenges and future development. Int J Biol Sci 12:100–108

Jimenez-Berni JA, Deery DM, Rozas-Larraondo P, Condon AG, Rebetzke GJ, James RA, Bovill WD, Furbank RT, Sirault XRR (2018) High throughput determination of plant height, ground cover, and above-ground biomass in wheat with LiDAR. Front Plant Sci 9:237

Jin X, Liu S, Baret F, Hemerle M, Comar A (2017) Estimates of plant density of wheat crops at emergence from very low altitude UAV imagery. Remote Sens Environ 198:105–114

Jonah PM, Bello LL, Lucky O, Midau A, Moruppa SM (2011) Review: the importance of molecular markers in plant breeding programs. Global J Sci Frontier Res 11:0975–5896

Kage U, Kumar A, Dhokane D, Karre S, Kushalappa AC (2016) Functional molecular markers for crop improvement. Crit Rev Biotechnol 36:917–930

Kang YJ, Lee T, Lee J, Shim S, Jeong H, Satyawan D, Kim MY, Lee SH (2016) Translational genomics for plant breeding with the genome sequence explosion. Plant Biotechnol J 14:1057–1069

Karaca M (2001) Characterization of *Cynodon* spp. and *Gossypium* spp. genomes using molecular and cytological techniques, PhD dissertation. Mississippi State, MS: Mississippi State University

Karaca M, Ince AG (2008) Minisatellites as DNA markers to classify bermudagrasses (*Cynodon* spp.): confirmation of minisatellite in amplified products. J Genet 87:83–86

Karaca M, Ince AG (2011a) Another CAPS DNA marker for sex identification in jojoba seedlings. Planta Med 77:1280–1280

Karaca M, Ince AG (2011b) New non-redundant microsatellite and CAPS-microsatellite markers for cotton (*Gossypium* L.). Turk J Field Crops 16:172–178

Karaca M, Ince AG (2016) Preliminary studies on differential methylation in A and D sub-genomes of Upland cotton (*Gossypium hirsutum* L.). FEBS J 283:334

Karaca M, Ince AG (2017) Molecular markers in *Salvia* L.: past, present and future. In: Georgiev V, Pavlov A (eds) Salvia biotechnology. Springer, Cham, pp 291–398

Karaca M, Ince AG, Elmasulu SY, Onus AN, Turgut K (2005a) Coisolation of genomic and organelle DNAs from 15 genera and 31 species of plants. Anal Biochem 343:353–355

Karaca M, Bilgen M, Onus AN, Ince AG, Elmasulu SY (2005b) Exact Tandem Repeats Analyzer (E-TRA): a new program for DNA sequence mining. J Genet 84:49–54

Karaca M, Ince AG, Ay ST, Turgut K, Onus AN (2008) PCR-RFLP and DAMD-PCR genotyping for *Salvia* species. J Sci Food Agric 88:2508–2516

Karaca M, Elmasulu S, Kürkçüoğlu M, Ince AG, Çınar A, Onus AN, Başer KHC, Turgut K (2011) Major volatile compounds of 50 *Thymus* taxa naturally grown in Antalya region of Turkey. Planta Med 77:1295–1296

Karaca M, Ince AG, Aydin A, Ay ST (2013) Cross-genera transferable e-microsatellite markers for 12 genera of the Lamiaceae family. J Sci Food Agric 93:1869–1879

Karaca M, Ince AG, Aydin A, Elmasulu SY, Turgut K (2015) Microsatellites for genetic and taxonomic research in thyme (*Thymus* L.). Turk J Biol 39:1406–1420

Kerstens HH, Crooijmans RP, Dibbits BW, Vereijken A, Okimoto R, Groenen MA (2011) Structural variation in the chicken genome identified by paired-end next-generation DNA sequencing of reduced representation libraries. BMC Genomics 12:94

Kopnina H, Washington H, Gray J, Taylor B (2018) The ‘future of conservation’ debate: defending ecocentrism and the nature needs half movement. Biol Conserv 217:140–148

Korun J, Ince AG, Karaca M (2013) Antibiotic resistance and plasmid profile of *Vibrio alginolyticus* strains isolated from cultured European sea bass (*Dicentrarchus labrax* L.). Bull Vet Inst Pulawy 57:173–177

Kozarewa I, Kozarewa I, Ning Z, Sanders MJ, Berriman M, Turner DJ (2009) Amplification-free Illumina sequencing-library preparation facilitates improved mapping and assembly of (G+C)-biased genomes. Nat Methods 6:291–295

Krishnan S, Ranker TA, Davis AP, Rakotomalala JJ (2013) An assessment of the genetic integrity of ex situ germplasm collections of three endangered species of coffea from Madagascar: implications for the management of field germplasm collections. Genet Resour Crop Evol 60:1021–1036

Krumm N, Sudmant PH, Ko A, O’Roak BJ, Malig M, Coe BP, Quinlan AR, Nickerson DA, Eichler EE (2012) Copy number variation detection and genotyping from exome sequence data. Genome Res 22:1525–1532

Lausch A, Erasmi S, King DJ, Magdon P, Heurich M (2017) Understanding forest health with remote sensing-part II-a review of approaches and data models. Remote Sens (Basel) 9:129

Levy SE, Myers RM (2016) Advancements in next-generation sequencing. Ann Rev Genom Human Genet 17:95–115

Li Y, Sun C, Luo HM, Li XW, Niu YY, Chen SL (2010) Transcriptome characterization for *Salvia miltiorrhiza* using 454 GS FLX. Acta Pharm Sin 45:524–529

Li D, Monahan WB, Baiser B (2018) Species richness and phylogenetic diversity of native and non-native species respond differently to area and environmental factors. Divers Distrib 24:853–864

Liebisch F, Kirchgessner N, Schneider D, Walter A, Hund A (2015) Remote, aerial phenotyping of maize traits with a mobile multi-sensor approach. Plant Methods 11:9

Lobos GA, Camargo AV, del Pozo A, Araus JL, Ortiz R, Doonan JH (2017) Editorial: plant phenotyping and phenomics for plant reeding. Front Plant Sci 8:2181

Manhaes AP, Loyolac R, Mazzochinia GG, Ganadea G, Oliveira-Filhob AT, Carvalho AR (2018) Low-cost strategies for protecting ecosystem services and biodiversity. Biol Conserv 217:187–194

Mascher M, Wu S, Amand PS, Stein N, Poland J (2013) Application of genotyping-by-sequencing on semiconductor sequencing platforms: a comparison of genetic and reference-based marker ordering in barley. PLoS One 8:e76925

Michael TP, van Buren R (2015) Progress, challenges and the future of crop genomes. Curr Opin Plant Biol 24:71–81

Ogwu MC, Osawaru ME, Ahana CM (2014) Challenges in conserving and utilizing plant genetic resources (PGR). Int J Genet Mol Biol 6:16–22

Olarte A, Mantri N, Nugent G, Wohlmuth H, Li CG, Xue C, Pang E (2013) A gDNA microarray for genotyping *Salvia* species. Mol Biotechnol 54:770–783

Ott A, Liu S, Schnable JC, Yeh CT, Wang C, Schnable PS (2017) Tunable genotyping-by-sequencing (tGBS®) enables reliable genotyping of heterozygous loci. BioRxiv. https://doi.org/10.1101/100461

Padilla FM, Pena-Fleitas MT, Gallardo M, Thompson RB (2017) Determination of sufficiency values of canopy reflectance vegetation indices for maximum growth and yield of cucumber. Eur J Agron 84:1–15

Parchman TL, Jahner JP, Uckele KA, Galland LM, Eckert AJ (2018) RADseq approaches and applications for forest tree genetics. Tree Genet Genomes 14:39

Pavan S, Marcotrigiano AR, Ciani E, Mazzeo R, Zonno V, Ruggieri V, Lotti C, Ricciardi L (2017) Genotyping-by-sequencing of a melon (*Cucumis melo* L.) germplasm collection from a secondary center of diversity highlights patterns of genetic variation and genomic features of different gene pools. BMC Genomics 18:59

Perez-Sanz F, Navarro PJ, Egea-Cortines M (2017) Plant phenomics: an overview of image acquisition technologies and image data analysis algorithms. GigaSci 6:1–18

Peroni N, Hanazaki N (2002) Current and lost diversity of cultivated varieties, especially cassava, under swidden cultivation systems. Agric Ecosyst Environ 92:171–183

Pertoldi C, Bijlsma R, Loeschcke V (2007) Conservation genetics in a globally changing environment: present problems, paradoxes and future challenges. Biodivers Conserv 16:4147–4163

Peterson BK, Weber JN, Kay EH, Fisher HS, Hoekstra HE (2012) Double digest RADseq: an inexpensive method for de novo SNP discovery and genotyping in model and non-model species. PLoS One 7:e37135

Phelps J, Webb EL (2015) Invisible wildlife trades: Southeast Asia's undocumented illegal trade in wild ornamental plants. Biol Conserv 186:296–305

Poczai P, Varga I, Laos M, Cseh A, Bell N, Valkonen JPT, Hyvonen J (2013) Advances in plant gene-targeted and functional markers: a review. Plant Methods 9:6

Poland JA, Rife TW (2012) Genotyping-by-sequencing for plant breeding and genetics. Plant Genome 5:92–102

Polat E, Ince AG, Karaca M, Onus AN (2010) Mining and utilization of mushroom ESTs for microsatellites. Conserv Genet 11:1123–1126

Prashar A, Jones HG (2014) Infra-red thermography as a high-throughput tool for field phenotyping. Agronomy 4:397–417

Rascher U, Blossfeld S, Fiorani F, Jahnke S, Jansen M, Kuhn AJ, Matsubara S, Martin LLA, Merchant A, Metzner R et al (2011) Non-invasive approaches for phenotyping of enhanced performance traits in bean. Funct Plant Biol 38:968

Rife TW, Wu S, Bowden RL, Poland JA (2015) Spiked GBS: a unified, open platform for single marker genotyping and whole-genome profiling. BMC Genomics 16:248

Rowan BA, Patel V, Weigel D, Schneeberger K (2015) Rapid and inexpensive whole-genome genotyping-by-sequencing for crossover localization and fine-scale genetic mapping. Genes Genome Genet 5:385–398

Saebnazar A, Rahmani F (2013) Genetic variation among *Salvia* species based on se-quence-related amplified polymorphism (SRAP) marker. J Plant Physiol Breed 3:71–78

Sajjadi SE, Ghanadian M, Haghighi M, Mouhebat L (2015) Cytotoxic effect of *Cousinia verbascifolia* Bunge against OVCAR-3 and HT-29 cancer cells. J Herb Med Pharmacol 4:15–19

Salgotra RK, Gupta BB, Stewart CN Jr (2014) From genomics to functional markers in the era of next-generation sequencing. Biotechnol Lett 36:417–426

Sari M, Sönmez NK, Karaca M (2005) Relationship between chlorophyll content and canopy reflectance in Washington navel orange trees (*Citrus sinensis* (L.) Osbeck). Pakistan J Bot 4:1093–1102

Scheben A, Batley J, Edwards D (2017) Genotyping-by-sequencing approaches to characterize crop genomes: choosing the right tool for the right application. Plant Biotechnol J 15:149–161

Sharrock S, Oldfield S, Wilson O (2014) Plant Conservation Report 2014: a review of progress in implementation of the global strategy for plant conservation 2011–2020. Secretariat of the convention on biological diversity and botanic gardens conservation international, Richmond

Shearman P, Bryan J, Laurance WF (2012) Are we approaching 'peak timber' in the tropics? Biol Conserv 151:17–21

Shepherd LD, de Lange PJ, Cox S, McLenachan PA, Roskruge NR, Lockhart PJ (2016) Evidence of a strong domestication bottleneck in the recently cultivated New Zealand endemic root crop, *Arthropodium cirratum* (Asparagaceae). PLoS One 11:e0152455

Simko I, Hayes RJ, Furbankm RT (2016) Non-destructive phenotyping of lettuce plants in early stages of development with optical sensors. Front Plant Sci 7:1985

Sonah H, Bastien M, Iquira E, Tardivel A, Legare G, Boyle B et al (2013) An improved genotyping by sequencing (GBS) approach offering increased versatility and efficiency of SNP discovery and genotyping. PLoS One 8:e54603

Song QJ, Hyten DL, Jia G, Quigley CV, Fickus EW, Nelson RL, Cregan PB (2015) Fingerprinting soybean germplasm and its utility in genomic research. G3 (Bethesda) 5:1999–2006

Song Z, Guo L, Liu T, Lin C, Wang J, Li X (2017) Comparative RNA-sequence transcriptome analysis of phenolic acid metabolism in *Salvia miltiorrhiza*, a traditional Chinese medicine model plant. Int J Genomics 2017:9364594

Sonnino A (2017) International instruments for conservation and sustainable use of plant genetic resources for food and agriculture: an historical appraisal. Diversity 9:50

Specht MJ, Pinto SRR, Albuquerque UP, Tabarelli M, Melo FPL (2015) Burning biodiversity: fuelwood harvesting causes forest degradation in human-dominated tropical landscapes. Glob Ecol Conserv 3:200–209

Stetter MG, Schmid KJ (2017) Analysis of phylogenetic relationships and genome size evolution of the *Amaranthus* genus using GBS indicates the ancestors of an ancient crop. Mol Phylogenet Evol 109:80–92

Stolle E, Moritz RFA (2013) RESTseq efficient benchtop population genomics with RESTriction fragment SEQuencing. PLoS One 8:e63960

Sun X, Liu D, Zhang X, Li W, Liu H, Hong W, Jiang C, Guan N, Ma C et al (2013) SLAF-seq: an efficient method of large-scale de novo SNP discovery and genotyping using high-throughput sequencing. PLoS One 8:e58700

Sun S, Li C, Paterson AH (2017) In-field high-throughput phenotyping of cotton plant height using LiDAR. Remote Sens (Basel) 9:377

Sun S, Li C, Paterson AH, Jiang Y, Xu R, Robertson JS, Snider JL, Chee PW (2018) In-field high throughput phenotyping and cotton plant growth analysis using LiDAR. Front Plant Sci 9:16

Suyama Y, Matsuki Y (2015) MIG-seq: an effective PCR-based method for genome-wide single nucleotide polymorphism genotyping using the next-generation sequencing platform. Sci Rep 5:16963

ter Steege H, Pitman NCA, Killeen TJ, Laurance WF, Peres CA et al (2015) Estimating the global conservation status of more than 15,000 Amazonian tree species. Sci Adv 1:e1500936

Thomas CD, Palmer G (2015) Non-native plants add to the British flora without negative consequences for native diversity. Proc Natl Acad Sci USA 112:4387–4392

Thompson AL, Thorp KR, Conley M, Andrade-Sanchez P, Heun JT, Dyer JM, White JW (2018) Deploying a proximal sensing cart to identify drought-adaptive traits in upland cotton for high-throughput phenotyping. Front Plant Sci 9:507

Timmermann C, Robaey Z (2016) Agrobiodiversity under different property Regimes. J Agric Environ Ethics 29:285–303

Toonen RJ, Puritz JB, Forsman ZH, Whitney JL, Fernandez-Silva I, Andrews KR, Bird CE (2013) ezRAD: a simplified method for genomic genotyping in non-model organisms. Peer J 1:e203

Tripodi P, Massa D, Venezia A, Cardi T (2018) Sensing technologies for precision phenotyping in vegetable crops: current status and future challenges. Agronomy 8:57

Truong HT, Ramos AM, Yalcin F, de Ruiter M, van der Poel HJA (2012) Sequence-based genotyping for marker discovery and co-dominant scoring in germplasm and populations. PLoS One 7:e37565

Tsai CC, Shih HC, Wang HV, Lin YS, Chang CH, Chiang YC, Chou CH (2015) RNA-Seq SSRs of moth orchid and screening for molecular markers across genus *Phalaenopsis* (Orchidaceae). PLoS One 10:e0141761

Ubbens JR, Stavness I (2017) Deep plant phenomics: a deep learning platform for complex plant phenotyping tasks. Front Plant Sci 8:1190

Uygur Gocer E, Karaca M (2016a) Cleaved minisatellite locus (CML) markers for fingerprinting of cotton cultivars grown in Turkey. FEBS J 283:334

Uygur Gocer E, Karaca M (2016b) Genetic characterization of some commercial cotton varieties using Td-DAMD-PCR markers. J Sci Eng Res 3:487–494

van Kleunen M, Dawson W, Essl F et al (2015) Global exchange and accumulation of non-native plants. Nature 525:100e103

van Orsouw NJ, Hogers RCJ, Janssen A, Yalcin F, Snoeijers S, Verstege E et al (2007) Complexity reduction of polymorphic sequences (CRoPS™): a novel approach for large-scale polymorphism discovery in complex genomes. PLoS One 2:e1172

van Tassell CP, Smith TP, Matukumalli LK, Taylor JF, Schnabel RD, Lawley CT et al (2008) SNP discovery and allele frequency estimation by deep sequencing of reduced representation libraries. Nat Methods 5:247–252

Vavilov NI (1926) Studies on the origin of cultivated plants. (Russian). Bull Appl Bot Plant Breed 14:1–245

Virlet N, Sabermanesh K, Sadeghi-Tehran P, Hawkesford MJ (2017) Field Scanalyzer: an automated robotic field phenotyping platform for detailed crop monitoring. Funct Plant Biol 44:143–153

Vlk D, Repkova J (2017) Application of Next-Generation Sequencing in plant breeding. Czech J Genet Plant Breed 53:89–96

Voss-Fels K, Rod J (2016) Snowdon understanding and utilizing crop genome diversity via high-resolution genotyping. Plant Biotechnol J 14:1086–1094

Waiho K, Fazhan H, Shahreza MS, Moh JHZ, Noorbaiduri S, Wong LL et al (2017) Transcriptome analysis and differential gene expression on the testis of orange mud crab, *Scylla olivacea*, during sexual maturation. PLoS One 12:e0171095

Walter J, Edwards J, McDonald G, Kuchela H (2018) Photogrammetry for the estimation of wheat biomass and harvest index. Field Crop Res 216:165–174

Wang Q, Zhang B, Lu Q (2009a) Conserved region amplification polymorphism (CoRAP), a novel marker technique for plant genotyping in *Salvia miltiorrhiza*. Plant Mol Biol Rep 27:139–143

Wang Z, Gerstein M, Snyder M (2009b) RNA-Seq: a revolutionary tool for transcriptomics. Nat Rev Genet 10:57–63

Wang S, Meyer E, McKay JK, Matz MV (2012) 2b-RAD: a simple and flexible method for genome-wide genotyping. Nat Methods 9:808–810

Watanabe K, Guo W, Arai K, Takanashi H, Kajiya-Kanegae H, Kobayashi M, Yano K, Tokunaga T, Fujiwara T, Tsutsumi N, Iwata H (2017) High-throughput phenotyping of sorghum plant height using an unmanned aerial vehicle and its application to genomic prediction modeling. Front Plant Sci 8:421

Wei T, Deng K, Zhang Q, Gao Y, Liu Y, Yang M, Zhang L, Zheng X, Wang C, Liu Z, Chen C, Zhang Y (2017) Modulating AtDREB1C expression improves drought tolerance in *Salvia miltiorrhiza*. Front Plant Sci 8:52

Will M, Schmalz N, Classen Bockhoff R (2015) Towards a new classification of *Salvia* s.l.: (re) establishing the genus Pleudia Raf. Turk J Bot 39:693–707

Xie W, Feng Q, Yu H, Huang X, Zhao Q, Xing Y, Yu S, Han B, Zhang Q (2010) Parent-independent genotyping for constructing an ultrahigh-density linkage map based on population sequencing. Proc Natl Acad Sci USA 107:10578–10583

Yamanaka ES, Tortajada-Genaro LA, Pastor N, Maquieira A (2018) Polymorphism genotyping based on loop-mediated isothermal amplification and smartphone detection. Biosens Bioelectron 109:177–183

Yan YP, Wang ZZ, Tian W, ZhongMin D, David S (2010) Generation and analysis of expressed sequence tags from the medicinal plant *Salvia miltiorrhiza*. Sci China Life Sci 53:273–285

Yang H, Li C, Lam HM, Clements J, Yan G, Zhao S (2015) Sequencing consolidates molecular markers with plant breeding practice. Theor Appl Genet 128:779–795

Yang G, Liu J, Zhao C, Li Z, Huang Y, Yu H, Xu B, Yang X, Zhu D, Zhang X, Zhang R, Feng H, Zhao X, Li Z, Li H, Yang H (2017) Unmanned aerial vehicle remote sensing for field-based crop phenotyping: current status and perspectives. Front Plant Sci 8:1111

Yousef EAA, MuEller T, BoErner A, Schmid KJ (2018) Comparative analysis of genetic diversity and differentiation of cauliflower (*Brassica oleracea* var. *botrytis*) accessions from two ex situ gene banks. PLoS One 13:e0192062

Zarco-Tejada PJ, Gonzalez-Dugo V, Berni JA (2012) Fluorescence, temperature and narrow-band indices acquired from a UAV platform for water stress detection using a micro-hyperspectral imager and a thermal camera. Remote Sens Environ 117:322–337

Zhang Y, Deng K, Xie L, Li X, Wang B, Chen L (2009) Genetic diversity analysis and conservation of the Chinese herb *Salvia miltiorrhiza* collected from different geographic origins in China. Afr J Biotechnol 8:4849–4855

Zhang J, Wang X, Yao J, Li Q, Liu F, Yotsukura N, Krupnova TN, Duan D (2017) Effect of domestication on the genetic diversity and structure of *Saccharina japonica* populations in China. Sci Rep 7:42158

Zhu JX, He NP, Wang QF, Yuan GF, Wen D, Yu GR, Jia YL (2015) The composition, spatial patterns, and influencing factors of atmospheric wet nitrogen deposition in Chinese terrestrial ecosystems. Sci Total Environ 511:777–785

Zhu F, Cui QQ, Hou ZC (2016) SNP discovery and genotyping using genotyping-by-sequencing in Pekin ducks. Sci Rep 6:36223

New Breeding Techniques for Sustainable Agriculture

Alessandra Gentile and Stefano La Malfa

Abstract In the scenario of a new agriculture, breeding techniques are asked to play an important role in order to make productive processes more sustainable under environmental, economic and social point of view. New crops need to be more efficient as for use of water and other resources, resilient and adaptable to different environments, able to produce healthy food products, but also other renewable material from their biomass to be used with an eco-logical approach in the different ambits of human activity. Plants will have to resist to new biotic and abiotic stresses where a drastic reduction of chemicals for their efficient production is expected. Also, plants will be more and more asked to fit new human needs such as the production of medicines and vaccines, or the detoxification of water and soils. The challenges facing breeders are highly demanding. However, breeding has always guided agriculture, improving performance and enabling the achievement of important goals, particularly during the "green revolution" age. The integration of different techniques and the development of both *in vitro* techniques and molecular strategies have accompanied the development of innovative breeding strategies during the last 50 years. New techniques have been utilized either directly for breeding or indirectly to obtain a more thorough understanding of the traits to be improved. Plant genetic resources have played a key role in this process. The use of New Breeding Techniques (NBTs) based on exhaustive knowledge of the genome of species and varieties will enable the development of new results that overcome the limitations of classical breeding techniques and their length and limiting the risks of the first generation of molecular breeding tools. In such a scenario, plant genetic resources are once again motivating breeders to achieve new results.

Keywords Genetic improvement · Plant genetic resources · Stress resistance · Yield · Resilience

A. Gentile
Hunan Agricultural University, Horticulture and Landscape College, Changsha, China

Department of Agriculture, Food and Environment, University of Catania, Catania, Italy

S. La Malfa (✉)
Department of Agriculture, Food and Environment, University of Catania, Catania, Italy
e-mail: stefano.lamalfa@unict.it

M. Farooq, M. Pisante (eds.), *Innovations in Sustainable Agriculture*,
https://doi.org/10.1007/978-3-030-23169-9_13

1 Introduction

Plant breeding (synonym: plant genetic improvement) represents one of the approaches to develop agriculture in different areas (Tester and Langridge 2010). This particular sector of plant science encompasses the study and the manipulation of plant heredity to develop novel plants with enhanced characteristics of interest that can be readily employed in agriculture with direct benefits to human society. Unlike other agronomic techniques, plant breeding approaches to improve the yield and quality of agricultural products rely on the inner characteristics of the plant and not on the external factors that can be modified with the adoption of agronomic techniques (fertilization, irrigation, the use of chemicals for pest and disease control, and the use of protected cultivation). In such a context, modifications made into plants through breeding are permanent and due to propagation, they can be spread in time and space (Acquaah 2007).

The needs for manipulating plant features and their behaviour are different and, at some level, they can change during time (Blum 2018). Generally, plants provide humans with food, feed, energy, shelter, fibers, and several other products, including pharmaceuticals. In addition, plants are often used for their ornamental value in varying contexts. However, plant breeding has been particularly important to satisfy the need for food by improving the yield and quality of several major crops, including cereals, corn, potato, vegetables and fruits (Edgerton 2009). The integration of traditional and modern techniques has improved the effectiveness of breeding methods and the achievement of new results in more efficient ways (Fig. 1). This challenge, particularly important during the "green revolution" period, is still needed to meet the world population increase expected over the next decades and to manage land scarcity and the competition for alternative uses. As a consequence, agriculture is expected to produce more food on less land such that agronomy and breeding are

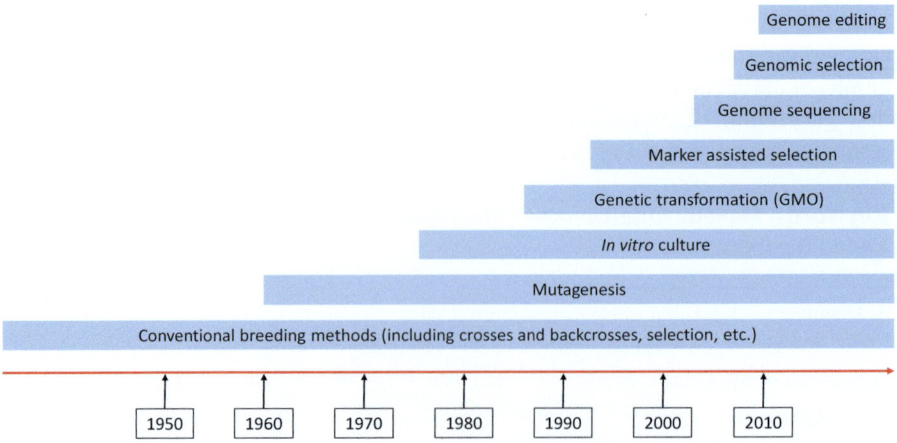

Fig. 1 Plant breeding: evolution of methods and role of biotechnologies. (Source: Cattivelli and Valè 2017)

required to contribute to this challenge, especially in developing countries, where most of the population growth is estimated (Acquaah 2012; Tester and Langridge 2010).

2 Role and Primary Objectives of Plant Breeding

The scenario in which plant breeders work changes during time and space. Global climate change requires the development of new varieties that can be successfully cultivated in different environments and that could face new and somewhat unpredictable climatic constrains (Pereira et al. 2012). Therefore, new cultivars are needed for environments that are becoming warmer, colder, drier or saltier (Blum 1985). In addition, pests and diseases are important stress conditions for crops, and their control by chemicals is often difficult or environmentally hazardous. Thus, there is increasing urgency to develop plant types that can also resist these stresses (Hussain 2015). Breeding is also required to develop new varieties that can be adapted to new cultivation areas, such as low-chilling varieties of pome and stone fruits that can be adapted to warmer climates or photoperiod-insensitive varieties of different seed and grain crops (Khush 2001). Another important input to the activity of breeders comes from the economic sustainability of the production process. Due to labour and other input costs, there is an increasing requirement to facilitate crop production and optimize the whole process. An example is the search for varieties more adapted to mechanical harvest or, particularly for fruit trees, dwarf plants that enable more intensive plantation systems and higher water use efficiency (Condon et al. 2004). The demand for new varieties of ornamentals is highly acute to help meet the consumer requirements for new colours, shapes, and other morphological features with increased appeal (Cadic and Widehem 2001; Smulders and Arens 2018). The agro-food industry seeks the improvement of specific traits, since quality requirements are often different from those of products that are consumed fresh. In addition, the particular uses of the products indicate the important diversification of breeding goals to be considered (final consumers or processers). This is the case, among the others, for grapes (table grape *versus* wine grape) and tomato or potato for specific uses (Murphy 2007).

Of course, some of the objectives of the breeding programmes are broad and common to different species. However, many other goals are specialized and specific for single crops or designed for particular needs. Increasing yield has been and continues to be one of the most important (Annichiarico 2002). There are several biochemical and physiological processes involved in yield mechanisms, including photosynthesis, respiration and transpiration. In addition, a distinction must be made between biological yield and economic yield. Biological yield generally corresponds to the biomass produced, while economic yield depends on how the biomass is partitioned in a crop (Hay 1995; Zhu et al. 2010). Breeders consider also other aspects when selecting new cultivars, such as the yield stability along time to avoid losses due to alternate bearing (for fruit species), lodging (for cereals) or shattering (for grains or dry fruits). Other traits linked to reproductive biology have been successfully modified in fruit tree species. In fact, the procurement self-fertile varieties that produce without being pollinated from different varieties is a goal of paramount importance, especially for cherry and almond but also in pear. Different sources of this character are available for each species and have been used for breeding programmes, especially those conducted via hybridization (Company et al. 2015).

The reduction of plant height is another trait that is often considered when selecting for yield. The use of dwarf or semi-dwarf cultivars (or rootstocks) is common in several annual and perennial crops and is often associated with lodging resistance. Dwarf genotypes can be used for more closely spaced fields, thus allowing more efficient orchard management. Photoperiod and temperature strongly affect flowering in plants and influence the shift from vegetative to reproductive phases and for some species, their adaptability to different growing areas. This is the case for pome and stone fruits in which the accumulation of a certain number of hours below a temperature is necessary to induce flowering (Byrne et al. 2000; Darnell 2000). New varieties with low chill requirements have enabled the diffusion of these species into hotter climates (Lyrene 2005). Alternatively, photoperiod insensitivity is a very desirable trait to grow crops at different latitudes, such as for cotton or tomato (Wallace et al. 1993). The synchronization of flowering time is also important for breeders when they plan to use different varieties in crossing-based breeding programs.

Maturity is another complex phenomenon in plants. For some fresh fruit species in particular, such as peach, mandarin, apricot, and grape, the search for early or late ripening varieties has driven the breeding programmes. In the first case, early maturity, also coupled with temporary or permanent protected cultivation, allows cultivars to escape environmental stresses and to attain better performance in the market. In contrast, late maturing cultivars fully exploit the growing season, allowing optimal yield and the display of qualitative features in terms of colour, sugar and vitamin content (Laurens et al. 2018).

Quality in agricultural products is a very complex issue whose comprehension includes several aspects. Of course, an indirect but relevant appraisal of quality is provided by the market value of a product, which is, in turn, given by its correspondence with the needs of final users, including consumers and transformers. As an example, seedlessness in fresh fruits, such as mandarins or table grape, has become a must in the last decades and has driven the breeding of these species (Perl et al. 2000; Cruz-Hernández and Paredes-López 2012). In the case of citrus, seedless fruit formation is due to the particular form of sterility coupled with parthenocarpy (the ability of a plant to set fruit without fecundation or with a precocious seed abort).

In the case of grain crops, milling quality, cooking and processing aptitude, and nutritional quality are the most important traits that determine their value as perceived by bakers and other product transformers (Battenfield et al. 2016; Rutkoski et al. 2017). Of course, both agricultural production regions and the production method (i.e. organic *versus* conventional) play an important role. Based on the different parts or organs of a plant used for food, the nutritional quality can have very different effects, and as a consequence, also the aims of the breeding programmes. The increase of the protein content (or of some essential amino acids) remains one of the most important goals for important crops, such as cereals and tubers (Jansen and Flamme 2006; Alvarez and Guzmán 2018). For rice, the primary approach to increase nutritional value consists of the augmentation of vitamin A, particularly using genetic transformation techniques (Paine et al. 2005). The improvement of fatty acid composition is also being considered as an important objective in those species cultivated for oil production (Fehr 2007).

For those species that supply fresh products, such as fruit and vegetables, the extension of shelf life represents a high priority objective for breeders (White 2002; Anwar et al. 2018). The most important results have been attained so far studying ethylene biosynthesis and regulating the cell wall modifying enzymes in tomatoes, apples and peaches (Brummel and Harpster 2001).

Apart from food production, plants are important for human for many other specific products, including biomass for energy, gums, cellulose, amylopectin, pharmaceuticals, enzymes, and vaccines, or functions, such as soil detoxification and phytoremediation (Gomes 2012). Of course for these specific uses, breeding, and in particular molecular breeding, is utilized to modify specific pathways to enable the production or functioning of specific mechanisms (Grabowski et al. 2014; Ho et al. 2014). Thus, most of the results achieved to date involve species that can be easily manipulated *in vitro* for regeneration (corn and tobacco *in primis*).

As already discussed in this book, insect-pests and disease represent important limiting factors to the sustainable cultivation of many species and a threat to safe post-harvest product storage. The presence of plant pathogens and pests often requires the adoption of chemical, biological or agronomic strategies for their control with the use of chemicals increasingly being debated for the risks associated with their residues (Bebber et al. 2014; Pertot et al. 2017). Biological control includes the use of resistant or tolerant varieties. Therefore, obtaining these varieties through breeding is among the high priority goals for several crops. In addition, for some virus and bacterial diseases, particularly for those affecting fruit tree species (*Citrus Tristeza Virus* in the case of citrus, *Plum Pox Virus* in the case of stone fruits, *Xylella fastidiosa* in the case of olive trees *inter alia*), the adoption of tolerant rootstocks and/or varieties remains the sole way to control disease and avoid important losses and complete plant death (Basile and DeJong 2018; Cuenca et al. 2018; Marini and Fazio 2018). Of course, quarantine laws are also important methods for disease control in these and other cases. The defence mechanisms that plants possess differ substantially based on avoidance, resistance and tolerance strategies and based on the availability of natural resistance, several breeding strategies could be adopted. The wild relatives of cultivated crops are a critically important source of resistance genes for most of the crops. The inclusion of different resistant genes in one variety is an important strategy to manage different races of one pathogen, a strategy known as gene pyramiding (Fuchs 2017).

Plant performance depends largely on the interaction between the genotype and the conditions of the environment where cultivation occurs (Hall 2018). Stressing conditions determine a reduction in plant growth and productivity. Within some limits, the environment in which a plant is cultivated can be modified, such as in protected cultivation. However, the environment can determine several stressing conditions for crops that can hamper their genetic yield potential. This implies that prior to their release in the market, novel varieties are tested on different environments, and their response in terms of production and/or trait stability is evaluated. The most relevant results in this topic to date have been achieved for salt tolerance (Arzani and Ashraf 2016) and thermal stress (both cold resistance and heat stress conditions) (Schwarz et al. 2010). In perennial crops, the late frost susceptibility in various species has been partially overcome by the use of late sprouting and flowering varieties. Resistance to waterlogging conditions, oxidative stress and heavy metal toxicity (also for plants to be used for phytoremediation) represent other important traits that are pursued in specific breeding programmes (Hasanuzzaman et al. 2012).

3 Selection and Other Breeding Strategies

The selection of living organisms, including plants, has been happening since very ancient times and determined slow but important changes in gene frequencies and, in turn, contributed to the diversity of plant life now present (Gepts 2004). Along their history, plants have been changing, and genotypes with higher survival potential or fitness became predominant. With the advent of agriculture, humans started to select for traits considered desirable. Selection is effective only when the observed variation in the population is due to a genetic component (Chen et al. 2017). In natural conditions, each species has a specific level of genetic diversity distributed at different hierarchical levels until that of individuals, each with a specific allelic composition. In a population, the allelic distribution in different groups is responsible for the adaptability of species to different conditions and environments (Lopes et al. 2015). In a cultivated species, the genetic diversity level is higher because of their exploitation as crops that determined relevant impacts on their genetic structure for different processes, including domestication, diffusion from centres of origin, selection, adaptation to different environments, hybridization, and breeding. However, the biodiversity level of most species in many cases has been substantially eroding (van de Wouw et al. 2009). The aspects related to the preservation of such an important heritage from erosion risk, the identification of the most suitable strategies for its conservation, and the sustainable and fair use of genetic resources are discussed in Chap. 12 of this volume.

The success of a breeding programme largely relies on selection methods. This assumption is particularly true for traditional breeding programmes, while molecular breeding strategies are more precise and can allow the acquirement or modification of desired traits in a faster and more precise manner (Al-Khayri et al. 2015). The reproductive mechanisms of each species dramatically affect the choice of the genetic improvement methods that can be applied, and hybridization plays a key role in crop development and evolution (Soltis and Soltis 2009). Plants species exhibit a wide range of reproductive systems spanning from cleistogamy (strict self-pollination) to obligate outcross. Hybridization for plants has played an important

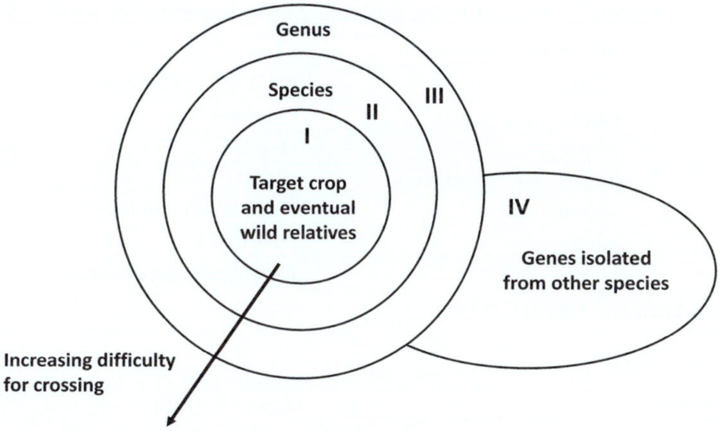

Fig. 2 The "Gene Pools" model. (Source: Gepts 2000)

role for crop development and modification, and breeders often refer to gene pool models to assess crop germplasm cross compatibility (Fig. 2).

The first gene pool refers to cultivated varieties and wild progenitors of a given crop. Within this group, crosses are easily performed and result in fertile progenies. The second and the third pools include related species that can still be crossed with the target one but with an increasing loss of fertility and/or with an increasing difficulty, rendering the use of embryo rescue or other *in vitro* techniques for embryo recovery. The fourth gene pool includes species that cannot be crossed but can represent sources of useful genes to be transferred with molecular breeding and new breeding techniques (Lee 1998).

Classical breeding has historically been performed by crossing. The choice of parents with superior traits and selection within the progeny obtained from the performed crosses of those individuals with a suitable combination of the desired characters of both parents until the release of a new variety are the most important steps of a breeding programme (Johnson 2000). However, the choice of the selection scheme and the type of variety to be released strongly relies on the reproductive method of each species. Schematically, we can consider different approaches that can be used for self-pollinated species, cross-pollinated species and those species, which are mostly propagated vegetatively. However, the distinction between these categories is not strict, and some of the breeding schemes that have been initially set up for one of these categories can be applied for another one (Table 1). The genetic structure of self-pollinated species requires the exploitation of the genetic diversity that can be found in natural populations or local varieties (Duc et al. 2015). Mass selection, pure-line selection and pedigree selection are the most common procedures that can be applied with the latter based on a selection scheme that follows the application of a breeding strategy aimed at generating new variability, such as hybridization or, to a lesser extent, the use of polyploidy or induced mutagenesis (Fig. 3).

Table 1 Types of varieties released by breeders according to crop and pollination methods

Crop	Pollination method	Main released cultivar typology
Rice	Self	Purelines and hybrids
Wheat	Self	Purelines
Maize	Cross	Hybrids
Potato	Cross (not for cultivar)	Clonal
Sorghum	Mainly self	Hybrids and purelines
Barley	Self	Purelines
Sunflower	Cross	Hybrids and populations

Source: Chrispeels and Sadava (2003)

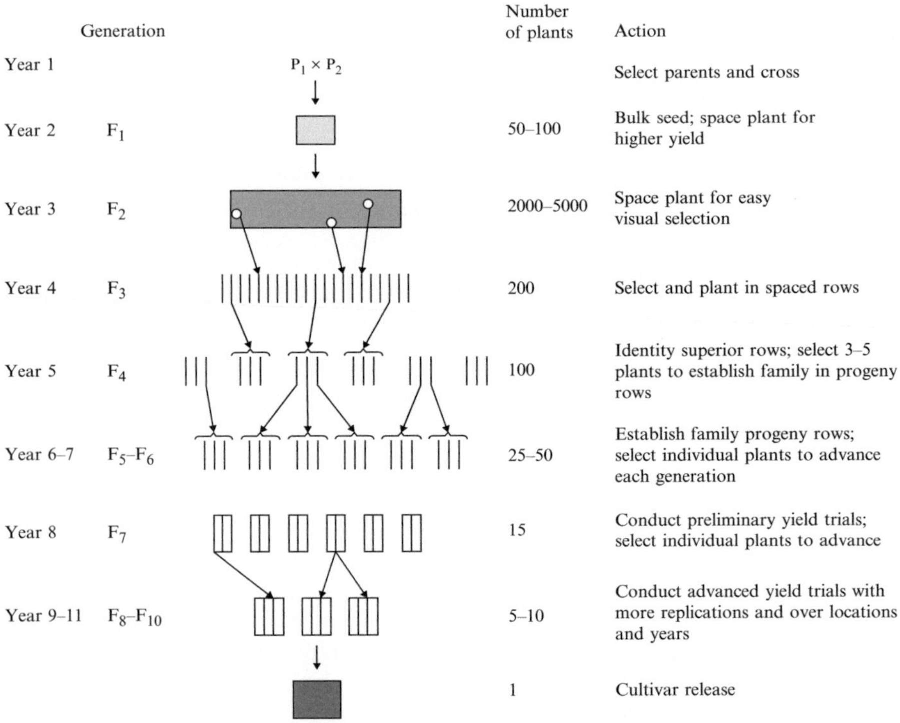

Fig. 3 Steps of pedigree selection. (Source: Acquaah 2012)

Bulk population breeding is a strategy that has been primarily used for wheat, barley and some beans in which after a first cross, all the products are cultivated, primarily under natural conditions for some generations, and artificial selection of the bulk population is delayed and performed after some years (Ortiz et al. 2007). The single seed descent method has been proposed to speed up the process and overcome the limitations of the previous methods. In Fig. 4, a general scheme of plant breeding activity is reported, including the search or arrangement of a new trait, its selection, and the release of improved varieties for their use after their multiplication.

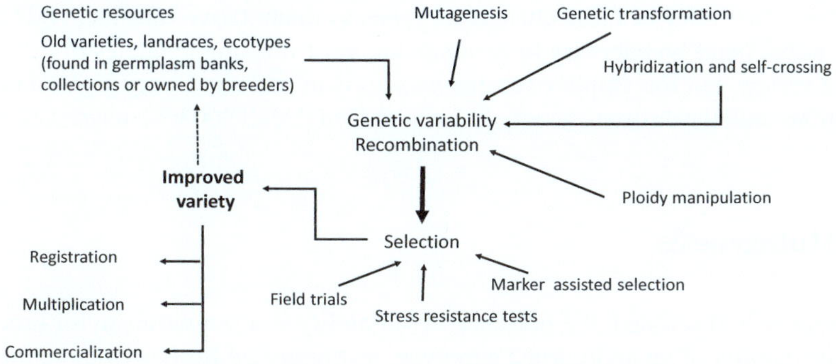

Fig. 4 Exemplificative diagram of genetic improvement programs for plant breeding

For cross-pollinated species, breeding mostly focuses on the improvement of a population rather than of single plants. In these species, hybridization could be difficult due to the necessity of avoiding extraneous pollen "contamination" different from that of the species or variety that has been chosen as the pollen donor. Cross-pollinated species are subjected to inbreeding depression. The genetic structure of the varieties of these plants corresponds to a number of heterozygous plants, often released as a population (Posselt 2010).

4 Role of Hybridization

F1 hybrids can be obtained either from cross- or self-pollinated species. F1 hybrids offer several advantages, and their diffusion represents one the most important milestones in the breeding history of some important crops. Maize represents the first species for which the importance of heterosis with increased yields than the better parents was shown at the beginning of 1900 and was soon deeply exploited. The release of hybrid varieties implies the possibility for breeding companies to protect their seeds or improve their parental lines (that are not sold to farmers). In fact, hybrids highly uniform at the F1 stage undergo segregation and a reduction of their potential at the F2 stage, discouraging growers from re-planting (Rouphael et al. 2010).

In addition to obtaining F1 hybrid varieties, crossing methods can be used to introduce a particular trait to a specific variety, such as resistance to a disease or a morphological trait, such as the gene for shortness in wheat. If breeders find a resistance trait that is under the control of one or few genes, this trait can be transferred in a sexual compatible genotype via backcrossing. Backcrossing is a method in which a recurrent parent is used as a donor to add the specific trait (Semagn et al. 2006).

Of course, many other characters are not easily transferred via backcrossing, since they are controlled by several genes and in a complex manner (quantitative traits). In addition, the expression of complex traits is influenced by non-genetic factors. Because of the large number of genes involved in complex trait expression, breeders often refer to these clusters of genes as quantitative trait loci (QTL). For these traits, plant breeders try to combine the most favourable alleles of each gene in one variety, but this implies substantial efforts to perform a very high number of selections until the desired phenotypes are obtained (Udall 2003; Semagn et al. 2010).

5 Mutagenesis

As previously discussed, the presence of variability is a pre-requisite for breeding. The availability of an appropriate genotype, distinguished for one or more interesting traits, is necessary to start a breeding programme. In nature, mutations arise spontaneously and are important for natural evolution. If this is not the case, variability can be induced by mutagenesis in particular. Mutagenesis in breeding can be used directly for new variety development, or indirectly, since the mutants can be used as the parent in breeding programmes (Broertjes and van Harten 1988; Ahloowalia et al. 2004).

Several techniques can be used to induce mutations. Physical and chemical agents can be applied to seeds or plant tissues from which it is possible to regenerate plants to be evaluated for the presence of new traits. In the case of vegetatively propagated crops, and primarily in fruit tree species, most of the results have been obtained using physical agents, while chemical agents, such as colchicine, often use to obtain tetraploids, has been used to a lesser extent (Ahloowalia and Maluszynski 2001).

Among the physical mutant agents, different sources of radiation must be described. In particular, gamma rays, X-rays and cobalt isotopes were among the first mutagens used. These agents are to be used on buds, scions, or even on *in vitro* multiplied shoots to be propagated after treatment. Among the most interesting results so far achieved with mutation breeding, at least for woody plants, we report the selection of pink grapefruit varieties, seedless mandarins or clementines (Tango mandarin, Monreal apireno clementine) and the mutation determined in sweet cherry at the locus S, responsible for the self-incompatibility character to obtain self-fertile varieties (Bado et al. 2015). In this case, the inducing treatment was performed on pollen (Table 2).

A particular aspect to be considered in the evaluation of mutations concerns their stability. In fact, there are rare cases of the regression of mutations and the subsequent reappearance of the original character, i.e., from the spur phenotype to the normal phenotype in the apple tree. This could depend on the layers of the bud apex involved in the mutagenic event. In particular, mericlinal chimaeras, in which only a few portions of the outer layer are mutated, are unstable, and the regression towards the original phenotype are frequent. Alternatively, the sectoral or periclinal chimaeras are more stable, and the solid or total mutations are stable (Fig. 5). For the reasons described above, obtaining new varieties through this technique involves a long selection phase with successive multiplication cycles during which the single buds must be evaluated individually and possibly further propagated, which lengthens the whole process (Oladosu et al. 2016).

Table 2 Origin of some sweet cherry varieties and their allelic profiles at locus S. S_4'allele is responsible for self-compatibility

	Origin	Allelic status at locus S
JI2420	Self-compatible selection obtained by mutagenesis from cultivar Napoleon. It contains S_4' mutated allele	S_4S_4'
Stella	Lambert (S_3S_4) × JI2420 (S_1S_4')	S_3S_4'
Lapins	Stella (S_3S_4') × Van (S_1S_3)	S_1S_4'
Blaze star	Lapins (S_1S_4') × Durone Compatto di Vignola (?)	$S_4'S_6$
Sunburst	Van (S_1S_3) × Stella (S_3S_4')	S_3S_4'
Early star	Burlat (S_3S_9) × Stella compact (S_3S_4')	$S_4'S_9$
Sweetheart	Van (S_1S_3) × Newstar ($S_4'S_1$)	S_3S_4'

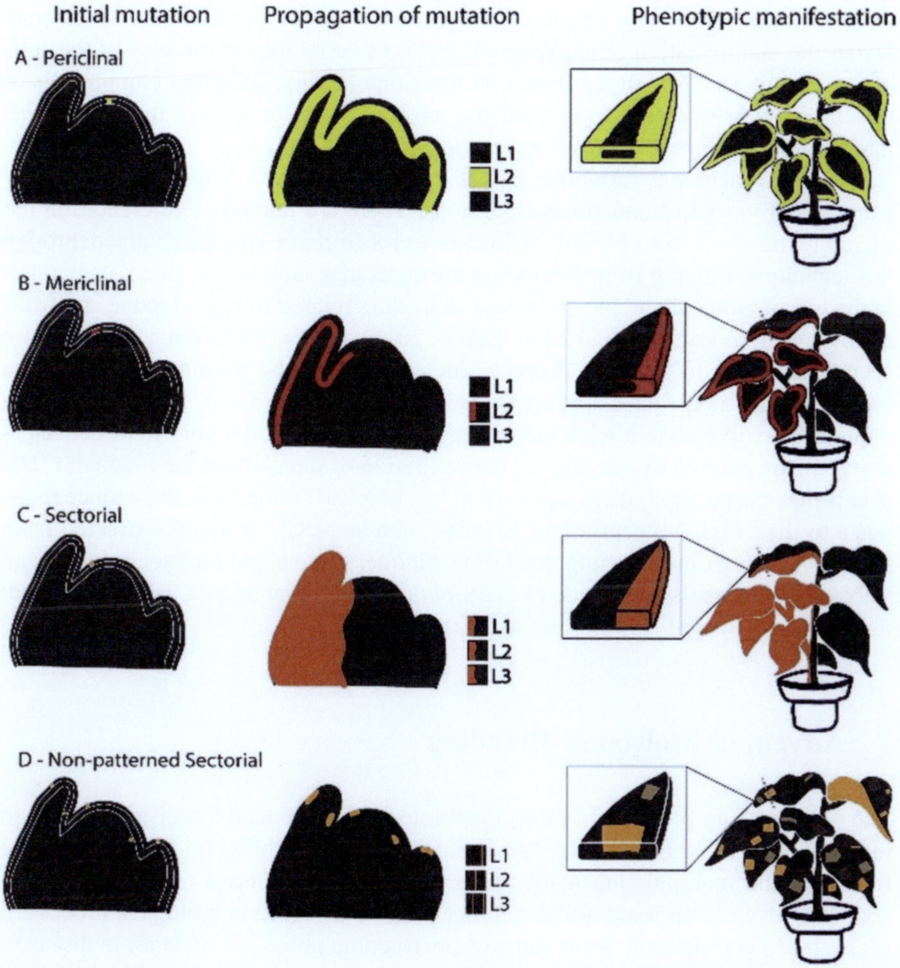

Fig. 5 Origin and phenotype in different kind of induced bud mutation. (**a**) periclinal, (**b**) mericlinal, (**c–d**) sectorial. (Source: Frank and Chitwood 2016)

6 Somaclonal Variability and *in vitro* Selection

Another interesting strategy that could be adopted for woody species refers to the possibility of selecting somaclonal mutants that can be obtained by imposing selective conditions to specifically prepared *in vitro* cultures (Krishna et al. 2016). Somaclonal variability represents a portion of the variability that can arise in cells or tissues cultured *in vitro*. This variability may involve different characters and can be "fixed" in selected cellular lines. The origin of this variability involves different types of mutations at the gene, chromosomal or whole genome level, including ploidy variation (Sattler et al. 2016). A fundamental prerequisite of this technique is that the selected cellular lines are able to produce whole plants once the selection cycles are complete. The origin of somaclonal variability may be very different, since the different variations may arise spontaneously during in vitro culture or manifest themselves as a result of the particular culture conditions, which may favour the manifestation of unexpressed levels of variability in the pool of the cells placed in culture. Growth regulators, in particular auxins, and other components of the culture substrate can also favour the onset of variations, since they can determine accelerated growth rates (Larkin and Scowcroft 1981).

Finally, somaclonal variability can be increased, even in a targeted manner, by using selective agents and stress conditions. There are numerous references in the literature that show examples of modifications for fruit tree species obtained through this technique, ranging from changes in the vegetative habit of the plant, to changes in the morphological and physiological characters related to reproductive and fruiting biology (Anis and Ahmad 2016; Lestari 2016). *In vitro* selection protocols more specifically refer to selective processes determined by the presence of a selective pressure of a stress factor applied to the in vitro culture (Smith 2000). This technique can be applied to a wide range of agents allowing better control of the selective pressure exerted by varying the concentration or intensity of the treatment. The agents that can be applied are represented by chemical compounds that induce resistance to their high concentrations in soils, such as NaCl, or toxins extracted from pathogens. For example, using this last technique, some apple rootstocks have been selected that possess resistance to *Phytophthora* and lemons that can resist fungal disease (Gentile et al. 1992; Predieri 2001).

7 Advent of Molecular Breeding

The extraordinary progress of molecular biology has provided a decisive impulse to research in agriculture since the 1980s, enabling the possibility of deciphering gene functions and manipulating many genes of agronomic interest in the genome. In addition, several important molecular mechanisms involved in biological processes, such as seed germination, the fruiting cycle, ripening process, resistance to diseases, and the adaptability of cultures to abiotic stress have been well studied and fully characterized (Giovannoni 2001; Seymour et al. 1993). Other important tools for modern breeding stem from the manipulation techniques using the *in vitro* culture of different organs, such as cells, protoplasts, and anthers that enable the use of different breeding techniques, such as somatic hybridization, *in vitro* selection, genetic transformation, or the obtainment of useful materials for subsequent applications, such as haploids and double haploids. Certainly, a first tangible result of the development of knowledge in the field of molecular genetics concerned the constitution and the progressive "saturation" of genetic maps. In many cultivated species, these maps contain useful information for geneticists, such as the position of the genes responsible for traits of agronomic interest and their association with the available markers (Barabaschi et al. 2016). In addition, this information is often shared with related species, such as in the genus Prunus, and enables the easy extension of this information from one species to another. Interesting examples include information on the frequency of recombination, assumed in turn as the measurement of the distance between two genes (or between a gene and a marker) (Fig. 6). This information can also allow the "cloning by position" approach of the genes of interest, as was the case for the scab resistance gene in apple. Additional approaches have been based on the substantial sequencing of ESTs from different tissues, organs or experimental conditions. Currently, collections of expressed sequences, often organized in microchips (arrays) with thousands of oligonucleotides used for the analysis of expression, are available for exhaustive and cheap analyses for all the most important cultivated species (Ganal et al. 2012; Voss-Fels and Snowdon 2016).

More recently, the new generation sequencing (NGS) approaches offer a significant impact to genetics and breeding studies. The progressive decrease of sequencing costs enabled the whole genome sequencing of an increasing number of the most important cultivated plant species (Barabaschi et al. 2011; Wendel et al. 2016), and for many other species, genome sequencing is in progress. For recent updates visit: http://www.ncbi.nlm.nih.gov/genomes/leuks.cgi; http://phytozome.jgi.doe.gov/pz/portal.html.

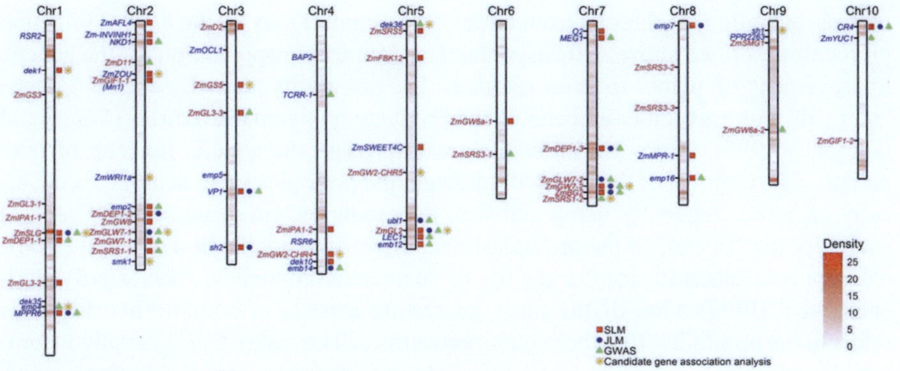

Fig. 6 Comparative analysis of QTLs and genes identified from maize mutant studies or based on rice seed size or weight genes. (Source: Liu et al. 2017)

Such tools enable the resequencing of the different varieties, and this has simplified the identification of genes of interest and QTL. The latter are of particular importance given the polygenic nature of many characters, including those most involved in yield determination and qualitative aspects of the fruit, such as size, shape, consistency, important biosynthetic pathways of synthesis and the degradation of carbohydrates and organic acids (Collard et al. 2005).

The investigation of the relationship between phenotype and genotype will be increasingly affordable in the future with molecular based screening protocols and with their interaction with all the technologies able to accurately measure many parameters related to different characters and allow time course evaluations. Among the characters to be taken into consideration are many that concern qualitative parameters and organoleptic characteristics with many genes involved in complex biosynthetic pathways. NGS also includes, in addition to the re-sequencing of varieties (Whole Genome Re-sequencing, WGRS) applications for genotyping by sequencing (GBS) and the resequencing of the transcriptome or RNA-Seq with useful outputs for complex analyses, such as chips (Illumina technology) and arrays (Affymetrix technology) that enable the analysis of thousands of well-distributed markers in the entire genome of a species as already occurred in grapevine, apple, peach and many other herbaceous and woody crops (Edwards and Batley 2010; Thomson 2014; Kim et al. 2016).

Resequencing also allows the identification of genes that are present only in some accessions and enables researchers to decipher the whole pan-genome of a species (Morgante et al. 2007).

8 Somatic Hybridization and Ploidy Manipulation

One of the most reliable applications of *in vitro* culture techniques, especially for herbaceous crops, refers to the possibility of obtaining protoplasts, i.e., cells deprived of the cell wall. Such cells can be used for different biotechnological applications, including the direct introduction of exogenous DNA, or the *in vitro* selection protocols described above. However, the most interesting application for the genetic improvement of plants to date refers to the possibility that these cells can be "merged" with each other and enable the production of somatic hybrids (Waara and Glimelius 1995). These are individuals who possess the genetic makeup of both parents. The removal of the cell wall to obtain the protoplasts can be done mechanically or, more frequently using cell wall degrading enzymes, while the protocols useful for the "fusion" of the protoplasts can refer to the application of small electrical impulses (electrofusion) or the use of chemical compounds, such as polyethylene glycol (PEG). One of the most interesting aspects of somatic hybridization refers to the possibility that the process concerns cells isolated from sexually incompatible individuals, whether or not they belong to the same species. In these cases, if the fusion process is symmetrical, the individual obtained contains the genetic information of both protoplasts.

Alternatively, the "addition" of the genetic material to one of the two cells can only concern the cytoplasmic material, and therefore, the extranuclear DNA harboured in the chloroplasts and mitochondria. In this case, the hybridization product is called a "cybrid". Somatic hybridization has been successfully applied to citrus, and particularly for rootstocks, since efficient regeneration protocols are available starting from protoplasts for many of these species. This approach has also been successfully used in the *Solanaceae* and other horticultural species (Cardi and Earle 1997; Cardi 2001; Orczyk et al. 2003; Prabhu Shankar et al. 2013; Rotino et al. 2014; Singh et al. 2015).

An additional application of *in vitro* culture techniques refers to the possibility of obtaining cells with alterations in the ploidy level (Sattler et al. 2016). Currently the most interesting application concerns of tetraploids, i.e., individuals with a 4n chromosome kit. These individuals, used in the breeding programmes of mandarins and mandarin-like, and hybridized with diploid individuals have led to the obtainment of triploid individuals, constitutionally sterile, and which, in particular parthenocarpic species, such as mandarins, allow the production of seedless fruits (Ollitrault et al. 2008). An additional application refers to the possibility of obtaining haploid individuals from the *in vitro* cultures of ovules and anthers, and, from these, double haploids (fertile and homozygous). This type of material is especially important for basic genetic studies, the identification of genes for specific characters and genome sequencing (Jiao and Schneeberger 2017).

9 GMO Parabola

Heritable variation can arise through natural phenomena or can be artificially induced via mutagenesis. Another method for artificial variation induction is the direct modification of the genetic material. Both hybridization and mutagenesis require a long time before the expected results can be achieved. Modern breeding tools have been developed in the last 40 years and are available for molecular breeding and genetic modifications. The most powerful method is the recombinant DNA (rDNA) technology (Gosal and Wani 2018). Through this technique, it is possible to transfer and integrate genes coming from the most diverse sources stably inside the genome, bypassing the sexual process and thus, expanding the transfer possibility to individual species that are very distant phylogenetically. For example, a gene from a bacterium can be transferred and integrated into the genome of a cultivated plant species (Broothaerts et al. 2005). In other terms, breeders have the opportunity to enlarge the gene pool, which can provide a source of variation and useful traits. In fact, DNA and its structure, as well as the mechanisms of coding, transcription, duplication and transmission to the descendants, represent a common heritage shared between all living beings. Of course, such a powerful technique implies different consequences and the need of new and accurate selection schemes. The products of this method of genetic improvement are known as Genetically Modified Organism (GMO, or GMP in the case of plants) and, according to the European directive 2001/18 can be defined as "an organism, with the exception of human beings, in which the genetic material has been altered in a way that does not occur naturally by mating and/or natural recombination". The first genetically modified (transgenic) plants were obtained in the early 1980s, and since then, there have been many applications regarding different traits and an increasing number of species and varieties. In addition, the potentials of the technique allow the hypothesis of uses that go beyond the traditional use of plants for food or energy and open new frontiers related to the use of plants for other purposes, such as the production of vaccines or other substances with pharmacological action, detoxification and remediation from pollutants. Genetically modified plants, and in particular, soybeans, cotton and corn, have been the subject of over 20 years of increasing diffusion in some areas of the globe but with strong limitations in most of the countries of the European Union (Clive 2015; Gould et al. 2016).

The techniques used for the constitution of genetically modified varieties refer to two different methods, one indirect based on the use of the agrobacterium (*Agrobacterium tumefaciens*) and a direct one, which involves the insertion of DNA using the biolistic method (Fig. 7).

In both cases, the prerequisite for a successful establishment of transgenic plants is the availability of efficient *in vitro* culture protocols of cells for each specific genotype to be transformed. These must be suitable both to be transformed and to give rise to new plants. However, only those cells that can both receive the T-DNA (transfer-DNA) from the agrobacterium and regenerate whole plants will allow a transgenic plant to be obtained (Chrispeels and Sadava 2003). In fact, for some varieties recalcitrant to *in vitro* culture, such as some Prunus species or some cereals, the absence of suitable protocols represents the primary obstacle to the application of transgenesis (Flachowsky et al. 2009). The research aimed at obtaining tissues with a high regenerative capacity *in vitro* in the various species and varieties is important to achieve a high number of cells, somatic, competent for transformation and regeneration. Reduced rates of regeneration render it impossible to obtain transformed clones, and this is the primary limitation for many tree species. However, highly efficient meristematic platforms for genetic transformation have been developed for many species (Sabbadini et al. 2015; Limera et al. 2017). Another aspect to be considered is that the use of young tissues, more easily manipulated *in vitro*, or the rejuvenation to which explants undergo during the cultivation and regeneration phases, prolongs the time needed to select the transformed clones for woody plants in particular. To overcome this problem, the use of tissues from adult plants for the transformation (for example in sweet orange, with subsequent grafting on vigorous rootstocks) or the transformation with genes able to induce fruit set in plants independently from their juvenile stage, have been proposed (Cervera et al. 1998; Wu et al. 2015). Most of the protocols refer to regeneration from somatic tissues, such as internodes and leaf portions. These must possess high regeneration efficiency, especially for organogenesis (the formation of new organs starting from somatic tissue). Less frequent and more recent is the use of protocols that provide for regeneration using somatic embryogenesis (Kumar and Van Staden 2017).

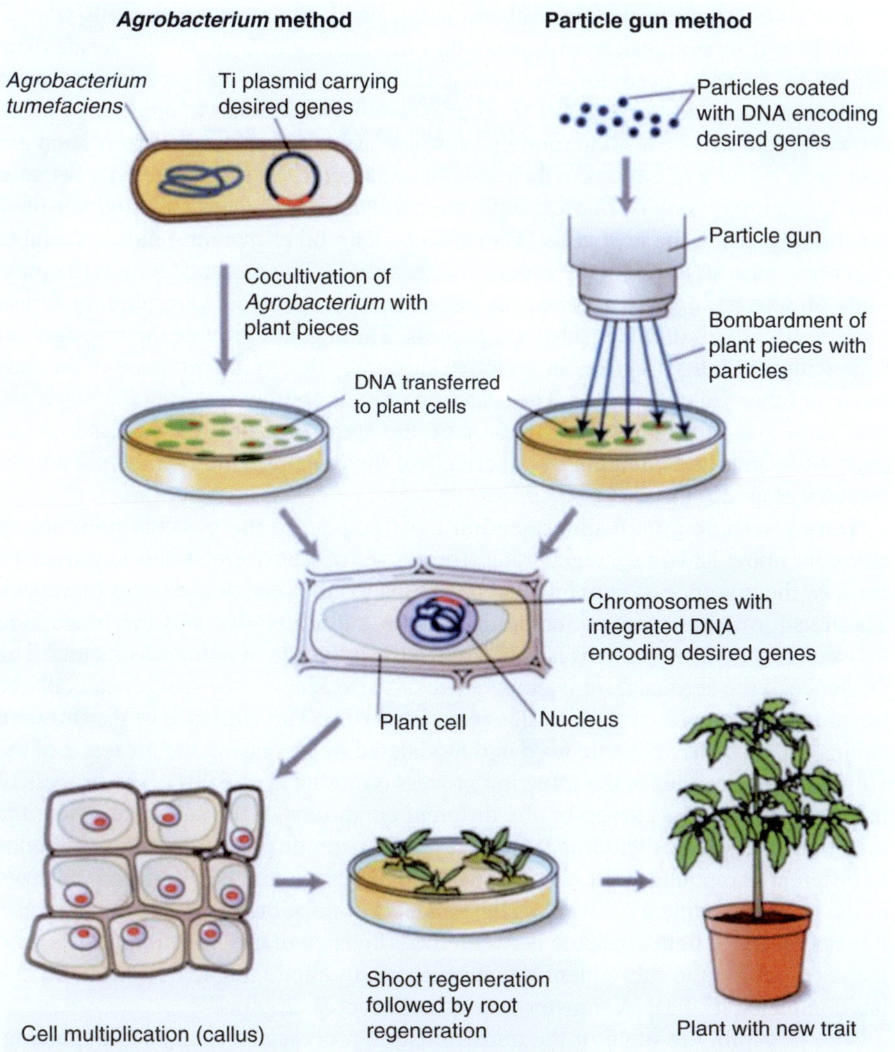

Fig. 7 Two different ways to create transgenic plants. (Source: Mirkov 2003)

The mechanism by which agrobacterium can transfer portions of its genetic material to other living organisms represents the molecular genetic assumption that allowed the establishment of the first genetically modified plants. The identification of this mechanism represents an exciting page of the history of the sciences of the last century. *Agrobacterium tumefaciens* is naturally responsible for the crown gall disease that affects many cultivated species and whose pathogenesis is determined by the transfer within the infected cells of a portion of the bacterial DNA (T-DNA) contained inside the circular nucleic acid molecule (plasmid) of the bacterial cell (Chilton 1979). Through the introduction of T-DNA into the plant cell, the bacterium is able to modify the biosynthetic pathways of the plant whose cells start to multiply and to synthesize substances that the bacterium uses to proliferate. The strains of bacteria used for the first genetic transformation experiments were deprived of the genes responsible for the disease. Instead, genes of agronomic interest were included. Today, the modern plasmids used for genetic transformation are said to be "disarmed", because they are not pathogenic, and "engineered", because they contain the gene of interest with the regulatory sequences essential for their functioning inside the host cells. They also contain other marker genes, selectable or reporter, useful to direct the process or monitor it. The presence within the plasmids of selectable marker genes, in particular the neomycin phosphotransferase gene, nptII, allows efficient selection process. These genes provide the transformed cells with the ability to grow on selective substrates due to the presence of an antibiotic or other selection factor. The selection process continues during the regeneration until the molecular confirmation of the transformation to the analysis of expression and the evaluation of the effects of the transformation on the phenotype (Stewart et al. 2011).

During selection, particular attention must be paid to the possible presence of chimeric individuals, i.e., regenerated from a set of transformed and unprocessed cells, or the presence of explants that have escaped the selection process (escapes). The transformation process, including *in vitro* culture phases and the interaction between two living organisms is subjected to the influence of numerous factors. The virulence of the agrobacterium strain is one of these, since some species can only be transformed by more virulent strains (e.g., EHA105). The virulence of the different strains can be somewhat enhanced and modulated by increasing the presence of the *vir* genes responsible for the infection process (Ghorbel et al. 2001). The process of infection can also be altered by the different conditions of the substrate where the infection takes place, including temperature, pH, osmotic pressure, and the presence of different compounds (Hu et al. 2016). Also important are the physiological stage of the plant and explants, as well as the residence time of the bacterium. In this case, it is necessary to find a balance between the efficiency of the infection process and the need, during the subsequent selection phase, to eliminate the bacterium once it has completed its T-DNA transfer action (Leifert et al. 2000).

The possibility to monitor the transformation process and to adjust the parameters that reduce its efficiency is crucial. To this end, and to be able to eliminate the use of selectable marker genes, considered as potentially harmful to human health, it is possible to use reporter marker genes and, among these, especially the vital marker genes, such as the green fluorescent protein (GFP) that allows the monitoring of the transformation process permitting the recognition in a set of cells or tissues in which the transformation event successfully occurred. The transformed cells re-emit a green colour following exposure to illumination with particular light sources (ultraviolet). This gene reporter ultimately makes it possible to monitor the transformation process until the complete transgenic plant is obtained (Stewart 2001).

When selectable marker genes are used, it should be possible to remove them through gene cleaning techniques once the selection process has been completed.

This approach allows one of the most discussed security aspects to be resolved with regard to the acceptability of GMPs even if, to be honest, the *npt*II gene was considered by the European Food Safety Authority (EFSA) to be "safe for use as selectable marker" (Delaney et al. 2017).

10 New Breeding Techniques

The development of new techniques for genetic improvement is necessary to increase the efficiency of the use of plant genetic resources speeding up the release of new crop varieties. The latter goal is very important to promptly respond to present and future agricultural challenges in a frame of climate changes that are becoming increasingly dramatic.

During the last decades, the improvement of sequencing technologies has made it possible to decipher the genomes of many important crops (from grapes to the last, in order of release, wheat) (Galbiati et al. 2017). Functional genomic, that enable the establishment of a strict relationship between gene and phenotype, has allowed the development of so-called second-generation biotechnologies and their application in breeding (new breeding techniques or NBTs), aiming to isolate genes underlying the characters of interest and their precise modification or transfer into targeted varieties (Gruskin 2012; He et al. 2014; Lusser et al. 2012; Varshney et al. 2014).

Among the NBTs, the most promising methods for the genetic improvement of crops are the cisgenesis and genome editing approaches that, exploiting the accuracy of the biotechnological approaches, maximize the similarity with the traditional methods of genetic improvement, and specifically, gene transferring by crossing and mutagenesis, respectively.

The first one is based on the transfer of only genes and regulatory sequences from one species that acts as donor to another that is the recipient without sexual reproduction. The crucial point is that the method is applied between genotypes that are of the same or sexually compatible species, and the result of this method will be a genotype that could have been obtained by sexual crossing and further selection. However, even if gene transfer within the same or evolutionary close species can also achieved by conventional breeding, the cisgenic approaches considerably reduce both the time for selection, and this is particularly important for fruit tree species, and the linkage drag (Tanuja and Kumar 2017).

The insertion of the cisgene can mediated by *Agrobacterium* or by the particle gun system similar to what is used for the production of GMO, but, in contrast to that procedure, the plant that will be regenerated after cigenesis will only contain the genes that encode the characters to be modified and no other DNA fragment. For this reason, a number of technologies can be adopted to avoid the presence of a selectable marker, generally used for the *in vitro* selection of transformed cells (Holme et al. 2013).

Cisgenesis has already been applied to improve pathogen resistance and quality traits in several crops. Among those, poplar, durum wheat, apple, and grape some of the first cisgenetic plants have been already subjected to additional manipulations to eliminate microbial regulatory sequences or selectable markers. Recently, cisgenetic lines of apple resistant to *Venturia inaequalis* (Wurdig et al. 2015) and potato expressing resistance to late blight have been obtained (Jo et al. 2014), both marker free, using an alternative recombinase system and PCR for the selection of transformed plants.

To enable a wider application of cisgenesis to genetically improve plants, some drawbacks have to be overcome. These include the random insertion of the cisgene in the host genome and the number of gene copies integrated. In addition, an efficient protocol for *in vitro* regeneration is required.

In Europe, the regulation of the product obtained by cigenesis is at the moment under the same regulation as the GMO plants, even if the EFSA concluded that cisgenic plants pose the same risk of plants obtained by conventional breeding (EFSA Panel on Genetically Modified Organism (GMO), 2012), and several reports confirmed a greater acceptance of cisgenic products compared to transgenic ones by consumers (Delwaide et al. 2015).

The other promising NBT is genome editing, more recent than cisgenesis but very powerful. The term genome editing refers to a number of techniques that share the possibility to specifically modify the genomic sequences of interest. They are based on the induction of cuts in the double-strand DNA, which are then repaired with two different process: the non homologous end-joining (NHEJ) and the homology-directed repair (HDR). The breaks in the double strand DNA can be induced by four systems based on specific enzymes: meganucleases, zinc finger nucleases (ZFN), trascription activator-like effector nucleases (TALEN) and cluster regular interspaced short palindormic repeats/CRISPR-associated nucleases (CRISPR/Cas). The latter has been developed in 2013 and represents the most suitable method, due to its high ease of use and flexibility, to determine a precise useful modification without the insertion of further sequences in a cultivated variety.

Essentially, genome editing can be considered an upgrade of conventional mutagenesis, utilised without chemicals or physical mutagens, because the induction of the mutations is not random in the genome but is limited to the genes of interest. Because of the conventional random mutagenesis, unwanted mutations can occur throughout the entire genome, and large-scale screens of mutagenized populations are needed to identify those plants with the mutations of interest.

The delivery of various components for genome editing into plant cells has been accomplished by transformation methods, such as Agrobacterium-mediated procedures, biolistic methods and protoplast systems. The choice will depend on different conditions, including the possibility of protoplast isolation and plant regeneration. During the last 5 years, many studies have been published that report the modification of traits of agronomic interest in different crops using genome editing approaches (Cardi 2016). Since genome-edited plants are free from sequences encoding nucleases and other components, they are indistinguishable from similar plants obtained by natural or induced mutations.

The NBTs also represent a useful tool to exploit plant genetic resources to identify genes and superior alleles and to use this knowledge to edit gene sequences in elite crop varieties, targeting specific sites and inducing relevant mutations leaving the genetic background and other traits untouched. To this extent, genome-editing technology has been recently used to establish a *de novo* domestication of a tomato wild species. In particular, Zsögön and colleagues (2018) edited six loci underlying agronomical traits of interest through the use of a CRISPR-Cas9 genome editing strategy. By doing so, it was possible to combine those desirable traits with useful traits of the wild species *Solanum pimpinellifolium*. The engineered *S. pimpinellifolium* variety showed a three-fold increase in fruit size and a ten-fold increase in the number of fruits compared to the wild parent, while the lycopene accumulation increased by 500%. This approach can be readily used in other crops as well, and can represent a valuable strategy to stem genetic erosion.

Of course, the application of both cisgenesis and genome editing methods requires the knowledge of the gene that encodes the specific agronomic traits and the availability of an efficient system of *in vitro* culture for the genotype that is going to be modified. Unfortunately, the latter represents a severe bottleneck for many species and genotypes that reduces the possibility to apply the NBTs to many important crops and to specific cultivars. For this reason, many laboratories are developing strategies to significantly increase the *in vitro* regeneration of different crops and make the regeneration procedure independent of the genotype used. Some interesting results have been already obtained for monocots (Lowe et al. 2016).

A vigorous debate is underway in Europe regarding the opportunity to include plants obtained by genome editing into the GMP regulations. A very recent (July 2018) sentence of the European Court of Justice states that the plants obtained by mutagenesis have to be considered GMOs except those obtained by means of certain mutagenesis techniques that have a long safety record. This indicates that the NBTs, such as genome editing, will fall into the GMO regulations.

References

Acquaah G (2007) Principles of plant genetics and breeding. Blackwell, Oxford, p 385

Acquaah G (2012) Principles of plant genetics and breeding. Wiley, Oxford

Ahloowalia BS, Maluszynski M (2001) Induced mutations – a new paradigm in plant breeding. Euphytica 118:167–173

Ahloowalia B, Maluszynski M, Nichterlein K (2004) Global impact of mutation-derived varieties. Euphytica 135:187

Al-Khayri JM, Jain SM, Johnson DV (eds) (2015) Advances in plant breeding strategies: breeding, biotechnology and molecular tools. Springer, Cham

Alvarez JB, Guzmán C (2018) Interspecific and intergeneric hybridization as a source of variation for wheat grain quality improvement. Theor Appl Genet 131:1–27

Anis M, Ahmad N (eds) (2016) Plant tissue culture: propagation, conservation and crop improvement. Springer, Singapore

Annicchiarico P (2002) Genotype × environment interactions: challenges and opportunities for plant breeding and cultivar recommendations, FAO plant production and protection paper no. 174. FAO, Rome
Anwar R, Mattoo AK, Handa AK (2018) Ripening and senescence of fleshy fruits. In: Paliyath G, Subramanian J, Lim L-T, Subramanian K-S, Handa AK, Mattoo AK (eds) Postharvest biology and nanotechnology. John Wiley & Sons, Inc., pp 15–51
Arzani A, Ashraf M (2016) Smart engineering of genetic resources for enhanced salinity tolerance in crop plants. Crit Rev Plant Sci 35(3):146–189
Bado S, Forster BP, Nielen S, Ali AM, Lagoda PJ, Till BJ, Laimer M (2015) Plant mutation breeding: current progress and future assessment. Plant Breed Rev 39:23–88
Barabaschi D, Guerra D, Lacrima K, Laino P, Michelotti V, Urso S, Valè G, Cattivelli L (2011) Emerging knowledge from genome sequencing of crop species. Mol Biotechnol 50:250–266
Barabaschi D, Tondell i A, Desiderio F, Volante A, Vaccino P, Valè G, Cattivelli L (2016) Next generation breeding. Plant Sci 242:3–13
Basile B, DeJong TM (2018) Control of fruit tree vigor induced by dwarfing rootstocks. Hortic Rev 46:39–97
Battenfield SD, Guzmán C, Gaynor RC, Singh RP, Peña RJ, Dreisigacker S, Fritz AK, Poland JA (2016) Genomic selection for processing and end-use quality traits in the cimmyt spring bread wheat breeding program. Plant Genome 9(2):1–12
Bebber DP, Holmes T, Gurr SJ (2014) The global spread of crop pests and pathogens. Glob Ecol Biogeogr 23(12):1398–1407
Blum A (1985) Breeding crop varieties for stress environments. CRC Crit Rev Plant Sci 2(3):199–238
Blum A (2018) Plant breeding for stress environments. CRC Press, Boca Raton
Broertjes C, van Harten AM (1988) Applied mutation breeding for vegetatively propagated crops. Elsevier, Amsterdam
Broothaerts W, Mitchell HJ, Weir B, Kaines S, Smith LM, Yang W, Mayer JE, Roa-Rodriguez C, Jefferson RA (2005) Gene transfer to plants by diverse species of bacteria. Nature 433:629–633
Brummell DA, Harpster MH (2001) Cell wall metabolism in fruit softening and quality and its manipulation in transgenic plants. Plant Mol Biol 47:311–339
Byrne DH, Sherman WB, Bacon TA (2000) Stone fruit genetic pool and its exploitation for growing under warm climatic conditions. In: Erez A (ed) Temperate fruit crops in warm climates. Kluwer Academic Publishers, Dordrecht, pp 157–230
Cadic A, Widehem C (2001) Breeding goals for new ornamentals. Acta Hortic 2001(552):75–83
Cardi T (2001) Somatic hybridization between Solanum commersonii and S. tuberosum. In: Nagata T, Bajaj YPS (eds) Biotechnology in agriculture and forestry, Somatic hybridization in crop improvement, Part II, vol 49. Springer, Heidelberg, pp 245–263
Cardi T (2016) Cisgenesis and genome editing: combining concepts and efforts for a smarter use of genetic resources in crop breeding. Plant Breed 135:139–147
Cardi T, Earle ED (1997) Production of new CMS Brassica oleracea by transfer of 'Anand' cytoplasm from B.-rapa through protoplast fusion. Theor Appl Genet 94:204–212
Cattivelli L, Valè G (2017) Next Generation Breeding: le conoscenze genomiche rivoluzionano il miglioramento genetico. In: Galbiati M, Gentile A, La Malfa S, Tonelli C (eds) Biotecnologie sostenibili. Edagricole - Edizioni Agricole di New Business Media srl, Milano, pp 1–13
Cervera M, Juarez J, Navarro A, Pina JA, Duran-Vila N, Navarro L, Pena L (1998) Genetic transformation and regeneration of mature tissues of woody fruit plants bypassing the juvenile stage. Transgenic Res 7(1):51–59
Chen J, Glémin S, Lascoux M (2017) Genetic diversity and the efficacy of purifying selection across plant and animal species. Mol Biol Evol 34(6):1417–1428
Chilton MD (1979) Agrobacterium Ti plasmids as a tool for genetic engineering in plants. Basic Life Sci 14:23–31
Chrispeels MJ, Sadava DE (2003) Plants, genes, and crop biotechnology, 2nd edn. Jones and Bartlett Publishers, Boston

Clive J (2015) 20th anniversary (1996 to 2015) of the global commercialization of biotech crops and biotech crop highlights in 2015, ISAAA brief no. 51. ISAAA, Ithaca

Collard BCY, Jahufer MZZ, Brouwer JB, Pang ECK (2005) An introduction to markers, quantitative trait loci (QTL) mapping and marker-assisted selection for crop improvement: the basic concepts. Euphytica 142:169–196

Company RSI, Kodad O, Martí AFI, Alonso JM (2015) Mutations conferring self-compatibility in *Prunus* species: from deletions and insertions to epigenetic alterations. Sci Hortic 192:125–131

Condon AG, Richards RA, Rebetzke GJ, Farquhar GD (2004) Breeding for high water-use efficiency. J Exp Bot 55:2447–2460

Cruz-Hernández A, Paredes-López O (2012) Fruit quality: new insights for biotechnology. Crit Rev Food Sci Nutr 52:272–289

Cuenca J, Garcia-Lor A, Navarro L, Aleza P (2018) Citrus genetics and breeding. In: Advances in plant breeding strategies: fruits. Springer, Cham, pp 403–436

Darnell RL (2000) Blueberries. In: Erez A (ed) Temperate fruit crops in warm climates. Kluwer Academic Publishers, Dordrecht, pp 429–444

Delaney B, Goodman RE, Ladics GS (2017) Food and feed safety of genetically engineered food crops. Toxicol Sci 162:361–371

Delwaide A-C, Nalley LL, Dixon BL, Danforth DM, Nayga RM Jr, van Loo EJ, Verbeke W (2015) Revisiting GMOs: are there differences in European consumers' acceptance and valuation for cisgenically vs. transgenically bred Rice? PLoS One 10:e0126060

Duc G, Agrama H, Bao S, Berger J, Bourion V, De Ron AM et al (2015) Breeding annual grain legumes for sustainable agriculture: new methods to approach complex traits and target new cultivar ideotypes. Crit Rev Plant Sci 34(1–3):381–411

Edgerton MD (2009) Increasing crop productivity to meet global needs for feed, food, and fiber. Plant Phys 149:7–13

Edwards D, Batley J (2010) Plant genome sequencing: applications for crop improvement. Plant Biotechnol J 8:2–9

Fehr WR (2007) Breeding for modified fatty acid composition in soybean. Crop Sci 47(Suppl 3):S72–S87

Flachowsky H, Hanke MV, Peil A, Strauss SH, Fladung M (2009) A review on transgenic approaches to accelerate breeding of woody plants. Plant Breed 128:217–226

Frank MH, Chitwood DH (2016) Plant chimeras: the good, the bad, and the 'Bizzaria. Dev Biol 419:41–53

Fuchs M (2017) Pyramiding resistance-conferring gene sequences in crops. Curr Opin Virol 26:36–42

Galbiati M, Gentile A, La Malfa S, Tonelli C (2017) Biotecnologie sostenibili. Edagricole-New Business Media. Università & formazione, Collana

Ganal MW, Poll ey A, Graner E-M, Plieske J, Wieseke R, Luerssen H, Durstewitz G (2012) Large SNP arrays for genotyping in crop plants. J Biosci 37:821–828

Gentile A, Tribulato E, Deng ZN, Vardi A (1992) In vitro selection of nucellar lemon callus and regeneration of plantstolerant to Phoma tracheiphia toxin. Adv Hortic Sci 6:151–154

Gepts P (2000) A phylogenetic and genomic analysis of crop germplasm: a necessary condition for its rational conservation and use. In: Gustafson J (ed) Proceedings of the stadler symposium. Plenum, New York, pp 163–181

Gepts P (2004) Crop domestication as a long-term selection experiment. Plant Breed Rev 24(2):1–44

Ghorbel R, La-Malfa S, López MM, Petit A, Navarro L, Peña L (2001) Additional copies of virG from pTiBo542 provide a super-transformation ability to agrobacterium tumefaciens in citrus. Physiol Mol Plant Pathol 58(3):103–110

Giovannoni J (2001) Molecular regulation of fruit ripening. Annu Rev Plant Physiol Plant Mol Biol 52:725–749

Gomes HI (2012) Phytoremediation for bioenergy: challenges and opportunities. Environ Technol Rev 1:59–66

Gosal SS, Wani SH (2018) Plant genetic transformation and transgenic crops: methods and applications. In: Gosal S, Wani S (eds) Biotechnologies of crop improvement, vol 2. Springer, Cham
Gould F, Amasino RM, Brossard D, Buell CR, Dixon RA, Falck-Zepeda JB et al (2016) Genetically engineered crops: experiences and prospects, National academies of sciences. National Academy of Sciences, Washington, DC, p 388
Grabowski GA, Golembo M, Shaal tiel Y (2014) Taliglucerase alfa: an enzyme replacement therapy using plant cell expression technology. Mol Genet Metab 112:1–8
Gruskin D (2012) Agbiotech 2.0. Nat Biotechnol 30:211–214
Hall A (2018) Crop responses to environment. CRC Press, Boca Raton
Hasanuzzaman M, Hossain MA, Da Silva JAT et al (2012) Plant response and tolerance to abiotic oxidative stress: antioxidant defense is a key factor. Crop stress and its management: perspectives and strategies. Springer, Berlin, pp 261–315
Hay RKM (1995) Harvest index – a review of its use in plant-breeding and crop physiology. Ann Appl Biol 126:197–216
He J, Zhao X, Laroche A, Lu ZX, Liu H, Li Z (2014) Genotyping-by-sequencing (GBS), an ultimate marker-assisted selection (MAS) tool to accelerate plant breeding. Front Plant Sci 5:484
Ho DP, Ngo HH, Guo W (2014) A mini review on renewable sources for biofuel. Bioresour Technol 169:742–749
Holme IB, Wendt T, Holm PB (2013) Intragenesis and cisgenesis as alternatives to transgenic crop development. Plant Biotechnol J 11:395–407
Hu W, Li W, Xie S, Fagundez S, McAvoy R, Deng Z, Li Y (2016) Kn1 gene overexpression drastically improves genetic transformation efficiencies of citrus cultivars. Plant Cell Tissue Organ Cult 125(1):81–91
Hussain B (2015) Modernization in plant breeding approaches for improvingbiotic stress resistance in crop plants. Turk J Agric For 39:515–530
Jansen G, Flamme W (2006) Coloured potatoes (Solanum tuberosum L.)–anthocyanin content and tuber quality. Genet Resour Crop Evol 53(7):1321–1331
Jiao W-B, Schneeberger K (2017) The impact of third generation genomic technologies on plant genome assembly. Curr Opin Plant Biol 36:64–70
Jo KR, Kim CJ, Kim SJ, Kim TY, Bergervoet M, Jongsma MA, Visser RG, Jacobsen E, Vossen JH (2014) Development of late blight resistant potatoes by cisgene stacking. BMC Biotechnol 14:50
Johnson R (2000) Classical plant breeding for durable resistance to diseases. J Plant Pathol 82:3–7
Khush GS (2001) Green revolution: the way forward. Nat Rev Genet 2:815–822
Kim C, Guo H, Kong W, Chandnani R, Shuang LS, Paterson AH (2016) Application of genotyping by sequencing technology to a variety of crop breeding programs. Plant Sci 242:14–22
Krishna H, Alizadeh M, Singh D, Singh U, Chauhan N, Eftekhari M, Sadh RK (2016) Somaclonal variations and their applications in horticultural crops improvement. 3 Biotech 6(1):54
Kumar V, Van Staden J (2017) New insights into plant somatic embryogenesis: an epigenetic view. Acta Physiol Plant 39:194
Larkin PJ, Scowcroft WR (1981) Somaclonal variation: a novel source of variability from tell cultures for plant improvement. Theor Appl Genet 60:197–214
Laurens F, Aranzana MJ, Arus P et al (2018) An integrated approach for increasing breeding efficiency in apple and peach in Europe. Hortic Res 5:11
Lee M (1998) Genome projects and gene pools: new germplasm for plant breeding? PNAS 95:2001–2004
Leifert C, Cassels AC, Doyle BM, Curry RF (2000) Quality assurance systems for plant cell and tissue culture: the problem of latent persistence of bacterial pathogens and agrobacterium-based transformation vector systems. Acta Hortic 530:87–91
Lestari EG (2016) Combination of somaclonal variation and mutagenesis for crop improvement. Jurnal Agro Biogen 8(1):38–44
Limera C, Sabbadini S, Sweet JB, Mezzetti B (2017) New biotechnological tools for the genetic improvement of major woody fruit species. Front Plant Sci 8:1418

Liu J, Huang J, Guo H et al (2017) The conserved and unique genetic architecture of kernel size and weight in maize and rice. Plant Physiol 175:774

Lopes MS, El-Basyoni I, Baenziger PS et al (2015) Exploiting genetic diversity from landraces in wheat breeding for adaptation to climate change. J Exp Bot 66:3477–3486

Lowe K, Wu E, Wang N, Hoerster G, Hastings C, Cho M-J, Scelonge C, Lenderts B, Chamberlin M, Cushatt J et al (2016) Morphogenic regulators baby boom and Wuschel improve monocot transformation. Plant Cell 28:1998–2015

Lusser M, Parisi C, Plan D, Rodriguez-Cerzo E (2012) Deployment of new biotechnologies in plant breeding. Nat Biotechnol 30:231–239

Lyrene PM (2005) Breeding low-chill blueberries and peaches for subtropical areas. HortScience 40:1947–1949

Marini RP, Fazio G (2018) Apple rootstocks: history, physiology, management, and breeding. Hortic Rev 45:197–312

Mirkov (2003) The molecular basis of genetic modification and improvement of crops. In: Chrispeels MJ, Sadava DE (eds) Plants, genes and crop biotechnology, 2nd edn. Jones and Bartlett Publishers, Boston

Morgante M, De Paoli E, Radovic S (2007) Transposable elements and the plant pan-genomes. Curr Opin Plant Biol 10:149–155

Murphy DJ (2007) In: Murphy J (ed) People, plants, and genes: the story of crops and humanity. Oxford University Press, Oxford

Oladosu Y, Rafii MY, Abdullah N, Hussin G, Ramli A, Rahim HA, Miah G, Usman M (2016) Principle and application of plant mutagenesis in crop improvement: a review. Biotechnol Biotechnol Equip 30:1–16

Ollitrault P, Dambier D, Luro F, Froelicher Y (2008) Ploidy manipulation for breeding seedless triploid Citrus. Plant Breed Rev 20:323–354

Orczyk W, Przetakiewicz J, Nadolska-Orczyk A (2003) Somatic hybrids of *Solanum tuberosum* – application to genetics and breeding. Plant Cell Tiss Org 74:1–13

Ortiz R, Trethowan R, Ferrara GO, Iwanaga M, Dodds JH, Crouch JH, Crossa J, Braun HJ (2007) High yield potential, shuttle breeding, genetic diversity, and a new international wheat improvement strategy. Euphytica 157:365–384

Paine JA, Shipton CA, Chagg ar S, Howells RM, Kennedy MJ et al (2005) Improving the nutritional value of Golden Rice through increased pro-vitamin a content. Nat Biotechnol 23:482–487

Pereira HM, Navarro LM, Martins IS (2012) Global biodiversity change: the bad, the good, and the unknown. Annu Rev Environ Resour 37:25–50

Perl A, Sahar N, Spiegel-Roy P, Gavish S, Elyasi R, Orr E, Bazak H (2000) Conventional and biotechnological approaches in breeding seedless table grapes. Acta Hortic 528:613–618

Pertot I, Caffi T, Rossi V, Mugnai L et al (2017) A critical review of plant protection tools for reducing pesticide use on grapevine and new perspectives for the implementation of IPM in viticulture. Crop Prot 97:70–84

Posselt UK (2010) Breeding methods in cross-pollinated species. In: Boller B (ed) Fodder crops and amenity grasses, handbook of plant breeding, vol 5. Springer, New York

Prabhu Shankar L, Tom E, Dieter D, Erik VB, Johan VH (2013) Asymmetric somatic plant hybridization: status and applications. Am J Plant Sci 04:1–10

Predieri S (2001) Mutation induction and tissue culture in improving fruits. Plant Cell Tissue Organ Cult 64:185–210

Rotino GL, Sala T, Toppino L (2014) Eggplant. In: Pratap A, Kumar J (eds) Alien Gene transfer in crop plants, Achievements and impacts, vol 2. Springer, New York, pp 381–409

Rouphael Y, Schwarz D, Krumbein A, Colla G (2010) Impact of grafting on product quality of fruit vegetables. Sci Hortic 127:172–179

Rutkoski JE, Crain J, Poland J, Sorrells ME (2017) Genomic selection for small grain improvement. In: Varshney R, Roorkiwal M, Sorrells M (eds) Genomic selection for crop improvement. Springer, Cham

Sabbadini S, Pandolfini T, Girolomini L, Molesini B, Navacchi O (2015) Peach (Prunus persica L.). In: Wang K (ed) Agrobacterium protocols. Springer, New York, pp 205–215
Sattler MC, Carvalho CR, Clarindo WR (2016) The polyploidy and its key role in plant breeding. Planta 243:281–296
Schwarz D, Rouphael Y, Colla G, Venema JH (2010) Grafting as a tool to improve tolerance of vegetables to abiotic stresses thermal stress, water stress and organic pollutants. Sci Hortic 127:162–171
Semagn K, Bjornstad A, Ndjiondjop MN (2006) Progress and prospects of marker assisted backcrossing as a tool in crop breeding programs. Afr J Biotechnol 5:2588–2603
Semagn K, Bjornstad A, Xu Y (2010) The genetic dissection of quantitative traits in crops. EJB 13:14
Seymour GB, Taylor JE, Tucker GA (1993) Biochemistry of fruit ripening. Chapman & Hall, London. ISBN 0-412-40830-9
Singh AK, Mishra NK, Kumar S, Pandey S (2015) Applications of somatic hybridization for the improvement of horticultural crops. Hortic Biotech Res 1:39–45
Smith RH (2000) Plant tissue culture: techniques and experiments, 2nd edn. Academic, San Diego, p 231
Smulders MJM, Arens P (2018) New developments in molecular techniques for breeding in ornamentals. In: Van Huylenbroeck J (ed) Ornamental crops. Handbook of plant breeding, vol 11. Springer, Cham
Soltis PS, Soltis DE (2009) The role of hybridization in plant speciation. Annu Rev Plant Biol 60:561–588
Stewart C (2001) The utility of green fluorescent protein in transgenic plants. Plant Cell Rep 20:376–382
Stewart CN, Touraev A, Citovsky V, Tzfira T (2011) Plant transformation technologies. Wiley-Blackwell Scientific Publishing, Ames, p 296
Tanuja P, Kumar AL (2017) Transgenic fruit crops – a review. Int J Curr Microbiol App Sci 6(8):2030–2037
Tester M, Langridge P (2010) Breeding technologies to increase crop production in a changing world. Science 327:818–822
Thomson MJ (2014) High-throughput SNP genotyping to accelerate crop improvement. Plant Breed Biotech 2:195–212
Udall JA (2003) Breeding for quantitative traits in plants. Crop Sci Madison 43:1578
van de Wouw M, Kik C, van Hintum T, van Treuren R, Visser B (2009) Genetic erosion in crops: concept, research results and challenges. Plant Genet Res 8:1–15
Varshney RK, Terauchi R, McCouch SR (2014) Harvesting the promising fruits of genomics: applying genome sequencing technologies to crop breeding. PLoS Biol 12:e1001883
Voss-Fels K, Snowdon RJ (2016) Understanding and utilizing crop genome diversity via high-resolution genotyping. Plant Biotech J 14:1086–1094
Waara S, Glimelius K (1995) The potential of somatic hybridization in crop breeding. Euphytica 85:217–233
Wallace DH, Baudoin JP, Beaver J, Coyne DP, Halseth DE, Masaya PN, Munger HM, Myers JR, Silbernagel M, Yourstone KS, Zobel RW (1993) Improving efficiency of breeding for higher crop yield. Theor Appl Genet 86:27–40
Wendel JF, Jackson SA, Meyers BC, Wing RA (2016) Evolution of plant genome architecture. Genome Biol 17:37
White PJ (2002) Recent advances in fruit development and ripening: an overview. J Exp Bot 53:1995–2000
Wu H, Acanda Y, Shankar A, Peeples M, Hubbard C, Zale J (2015) Genetic transformation of commercially important mature citrus scions. Crop Sci 55(6):2786–2797
Würdig J, Flachowsky H, Saß A, Peil A, Hanke M-V (2015) Improving resistance of different apple cultivars using the Rvi6 scab resistance gene in a cisgenic approach based on the Flp/FRT recombinase system. Mol Breed 35:95

Zhu XG, Long SP, Ort DR (2010) Improving photosynthetic efficiency for greater yield. Annu Rev Plant Biol 61:235–261

Zsögön A, Čermák T, Naves ER, Notini MM, Edel KH, Weinl S, Freschi L, Voytas DF, Kudla J, Peres LEP (2018) De novo domestication of wild tomato using genome editing. Nat Biotechnol 36:1211–1216

Part VI
Agricultural Sustainability in Changing Climate

Sustainable Agriculture and Climate Change

Liyong Xie and Hongliang Zhao

Abstract Climate change, a change in the statistical distribution of weather patterns for an extended period, is caused by natural factors such as variations in solar radiation, and human activities. Climate change is expected to affect the frequency, distribution, intensity, and location of extreme events, and thus will affect the sustainability of agriculture. In this regard, management and adaptation strategies to mitigate the climate change effects are direly needed to ensure stable yields in the world and food security. Sustainable agriculture provides a potential solution to enable agricultural systems to feed a growing population while successfully operating within the changing environmental conditions. International cooperation in research and actions is very important and may yield great benefits to cope with the climate change challenges. This chapter introduce the issues of climate change, describe its characteristics, discusses its impacts on sustainable agriculture and highlight the potential risks involved. Strategies for sustainable agriculture under climate change have also been proposed.

Keywords Adaptation strategies · International cooperation · Natural factors · Stable yields

1 Introduction and Context

1.1 Basic Issue

Over 200 years of modern agriculture with great achievement to feed the increasing population in the latest history, more and more people rethink the effects of modern agriculture on resource and environment. More new models of agriculture develop-

L. Xie (✉) · H. Zhao
College of Agronomy, Shenyang Agricultural University,
Shenyang, People's Republic of China
e-mail: xly0910@syau.edu.cn

M. Farooq, M. Pisante (eds.), *Innovations in Sustainable Agriculture*,
https://doi.org/10.1007/978-3-030-23169-9_14

ment were discussed heated, including sustainable agriculture, which was accepted widely around the world nowadays. Sustainable agriculture has been described in many international reports or literatures, which could understand that an integrated system of plant and animal production practices having a site-specific application that will last over the long term, to satisfy human food and fiber needs, to enhance environmental quality and the natural resource base upon which the agricultural economy depends, to make the most efficient use of non-renewable and on-farm resources and integrate natural biological cycles and controls, to sustain the economic viability of farm operations, and to enhance the quality of life for farmers and society as a whole (Gold 2009). The concept emphasized on two points, namely resources-saving and environmental-friendly, aiming to resolve problem of agriculture and natural resource and environment through the cooperation and collaboration with different skills.

In meanwhile, climate change, another environmental issue, also draw more attention in the world. Climate change is a change in the statistical distribution of weather patterns when that change lasts for an extended period, refer to a change in average weather conditions, or in the time variation of weather around longer-term average conditions (IPCC 2014a). Climate change is caused by natural factors such as variations in solar radiation, and human activities which have been identified recent climate change. Climate change will affect the frequency, intensity, and location of extreme events. Climate-related disasters are among the main drivers of sustainable agriculture development, both in the aftermath of a disaster and in the long run. Drought is a major driver of crop production and contributes to a negative impact on nutrition. Floods and tropical storms also affect agriculture by destroying livelihood assets. The world population reached 7.0 billion in 2011, and expected to reach 8.0 billion in the midst of 2020. Global food production must increase by 50% to meet the projected demand of the world's population by 2050 (see Table 1). The relationship between climate change and food production depends to a large degree on when and which adaptation actions are taken.

The progress towards food security was uneven across regions (Table 1). The MDG hunger target was achieved in Latin America, the eastern and south-eastern regions of Asia, the Caucasus and Central Asia and the northern and western regions of Africa. In total, 73 developing countries out of the 129 monitored reached the target. In sub-Saharan Africa, Southern Asia, the Caribbean and Oceania progress was too slow, and the prevalence of undernourishment is still relatively high (over 14% of the total population) (Table 1).

Sustainable agriculture is facing not only population boost, water resources decrease, and soil environment degradation, but also encountering climate change. Furthermore, the latter is more complicated and uncertain (Reyes et al. 2018) (Table 2). World and regional figures include other regions and countries not reported separately. Aggregate food production is an index, by weight, of cereals, meats, fruits and vegetables, oilseeds and pulses. Per capita food consumption is a projection of daily dietary energy supply. Estimates of the number of people at risk of hunger are based on a quadratic specification of the relationship between national-level calorie supply and the share of population that is undernourished as defined by

Table 1 Estimated prevalence of undernourishment in the world over the last 25 years

Regions/sub-regions	1990–1992	2000–2002	2005–2007	2010–2012	2014–2016
World	**18.6**	**14.9**	**14.3**	**11.8**	**10.8**
Developing regions	23.3	18.2	17.3	14.1	12.9
Northern Africa	<5.0	<5.0	<5.0	<5.0	<5.0
Sub-Saharan Africa	33.2	30.0	26.5	24.2	22.9
Latin America and the Caribbean	14.7	11.4	8.4	6.4	5.5
Caribbean	27.0	24.4	23.5	19.8	19.8
Latin America	13.9	10.5	7.3	5.5	<5.0
Eastern Asia	23.2	16.0	15.2	11.8	9.6
Eastern Asia excluding China	9.6	14.6	13.9	15.1	14.6
Southern Asia	23.9	18.5	20.1	16.1	15.7
Southern Asia excluding India	24.5	21.0	19.0	17.5	17.0
South-Eastern Asia	30.6	22.3	18.3	12.1	9.6
Western Asia	6.4	8.6	9.3	8.9	8.4
Oceania	15.7	16.5	15.4	13.5	14.2
Caucasus and Central Asia	14.1	15.3	11.3	8.9	7.0
Developed regions	<5.0	<5.0	<5.0	<5.0	<5.0
Least Developed Countries (LDCs)	40.0	36.5	31.4	28.0	26.5
Landlocked Developing Countries (LLDCs)	35.6	33.6	28.1	24.5	22.7
Small Island Developing States (SIDS)	24.5	22.5	21.3	18.2	18.0

Source: FAO (2016)

the FAO (FAO 2008). Values reported for 2010 are calibrated model results. Projections for 2030 and 2050 assume changes in population and income as reflected in the Intergovernmental Panel on Climate Change's (IPCC) Shared Socioeconomic Pathway 2 (IPCC 2014b). Climate change impacts are simulated using the IPCC's Representative Concentration Pathway 8.5 and the HadGEM general circulation model (Table 2).

It is important to excellent management of agriculture to adapt to climate change so that to ensure stable yields in the world and food security, at least taking some potential adaptation co-benefits of food system mitigation strategies and explore the potential implications of such strategies on food systems (Niles et al. 2018). At the same time, there are several countries have been pioneers in developing ways to simultaneously increase agricultural yields and quality, protect water and soil environment. Studies showed that a warming climate has a negative effect on crop production and generally reduces yields of staple cereals such as wheat, rice, and maize. Elevated carbon dioxide (CO_2) could benefit crops yields in the short term by increasing photosynthesis rates; however, there is big uncertainty in the magnitude of the CO_2 effect and the significance of interactions with other factors, especially in a long term (Hovenden and Newton 2018).

Table 2 Impact projections of food production, consumption and hunger to 2050 with and without climate change

	Aggregate food production (index, 2010–1.00)					Per capita food consumption (KCAL per capita per day)					Hunger (millions of people at risk)				
		Without climate change		With climate change			Without climate change		With climate change			Without climate change		With climate change	
	2010	2030	2050	2030	2050	2010	2030	2050	2030	2050	2010	2030	2050	2030	2050
World	**1.00**	**1.37**	**1.69**	**1.33**	**1.60**	**2795**	**3032**	**3191**	**2982**	**3079**	**838.1**	**528.2**	**405.8**	**592.3**	**476.9**
Developing countries	1.00	1.42	1.76	1.39	1.71	2683	2961	3137	2909	3020	823.3	513.3	392.2	576.7	461.1
Developed countries	1.00	1.24	1.47	1.15	1.29	3384	3439	3513	3406	3435	14.8	14.9	13.6	15.7	15.8
Asia and Pacific	**1.00**	**1.37**	**1.64**	**1.36**	**1.63**	**2656**	**3003**	**3185**	**2954**	**3072**	**539.8**	**249.8**	**181.8**	**280.9**	**204.6**
East Asia	**1.00**	**1.23**	**1.35**	**1.26**	**1.41**	**3009**	**3509**	**3628**	**3459**	**3516**	**187.2**	**59.2**	**54.7**	**60.3**	**56.8**
China	1.00	1.23	1.34	1.26	1.40	3044	3604	3733	3552	3616	173.4	44.8	41.0	44.7	41.0
Japan	1.00	1.24	1.52	1.31	1.69	2770	2787	2842	2757	2773	2.3	2.0	1.2	2.3	1.9
Korea, Rep.	1.00	1.25	1.43	1.26	1.44	3139	3347	3429	3310	3347	0.6	0.4	0.4	0.4	0.4
South Asia	**1.00**	**1.57**	**2.05**	**1.50**	**1.91**	**2361**	**2669**	**2959**	**2623**	**2848**	**268.5**	**138.3**	**87.7**	**161.6**	**97.0**
Afghanistan	1.00	1.33	1.73	1.35	1.77	2149	2239	2452	2206	2349	7.0	9.4	7.9	10.1	10.4
Bangladesh	1.00	1.41	1.63	1.33	1.46	2426	2714	2911	2653	2781	26.0	11.3	6.9	14.8	8.7
India	1.00	1.63	2.16	1.56.	2.01	2354	2697	2998	2651	2883	189.7	73.9	45.0	90.5	44.9
Nepal	1.00	1.33	1.60	1.37	1.71	2425	2695	3186	2625	3028	2.7	2.0	0.8	2.4	1.5
Pakistan	1.00	1.33	1.63	1.27	1.50	2379	2540	2862	2514	2787	37.6	38.0	24.4	39.9	28.0
Southeast Asia and Pacific	**1.00**	**1.48**	**1.89**	**1.46**	**1.84**	**2551**	**2852**	**3051**	**2796**	**2931**	**84.1**	**52.3**	**39.4**	**58.9**	**50.8**
Indonesia	1.00	1.62	2.02	1.63	2.05	2540	2990	3281	2910	3110	32.4	12.9	7.2	15.3	11.1
Malaysia	1.00	1.83	2.95	1.79	2.84	2838	3173	3462	3143	3384	0.9	0.8	0.9	0.8	0.9
Myanmar	1.00	1.35	1.55	1.34	1.53	2169	2473	2592	2420	2487	10.5	6.5	4.8	7.2	6.0
Philippines	1.00	1.33	1.68	1.31	1.65	2503	2641	2777	2602	2691	12.1	12.2	11.0	13.2	13.1
Thailand	1.00	1.18	1.26	1.12	1.14	2742	3012	3183	2975	3103	6.2	3.1	1.8	3.5	2.3
Viet Nam	1.00	1.25	1.36	1.20	1.24	2512	2710	2828	2654	2712	12.9	9.5	7.2	10.8	9.7

Africa and Middle East	**1.00**	**1.60**	**2.24**	**1.55**	**2.11**	**2623**	**2795**	**3002**	**2735**	**2873**	**238.7**	**229.8**	**185.0**	**258.7**	**227.1**
Africa south of the Sahara	**1.00**	**1.65**	**2.37**	**1.57**	**2.17**	**2358**	**2587**	**2853**	**2518**	**2713**	**209.5**	**195.7**	**150.5**	**223.0**	**188.7**
Congo, Dem. Rep.	1.00	1.72	2.49	1.67	2.38	1943	2392	2998	2325	2848	37.6	20.3	6.6	25.2	6.6
Ethiopia	1.00	1.65	2.45	1.66	2.48	2066	2307	2614	2266	2533	32.7	32.3	22.5	34.7	26.5
Kenya	1.00	1.76	3.12	1.79	3.14	2133	2395	2708	2300	2524	10.2	8.9	5.0	10.8	8.2
Nigeria	1.00	1.62	2.31	1.56	2.16	2751	2943	3136	2866	2984	9.7	8.5	11.6	10.6	11.6
South Africa	1.00	1.50	1.87	1.49	1.80	2962	3229	3397	3157	3258	1.9	1.5	1.6	1.5	1.6
Sudan	1.00	1.74	2.47	1.44	1.76	2329	2465	2714	2431	2635	11.4	12.7	9.0	13.7	10.9
Tanzania, United Rep.	1.00	1.64	2.42	1.56	2.22	2178	2396	2602	2309	2439	15.6	17.8	17.8	20.4	23.0
Uganda	1.00	1.89	3.05	1.77	3.71	3391	2505	2794	2520	2647	8.5	10.4	11.3	11.8	13.8
Middle East and North Africa	**1.00**	**1.51**	**2.01**	**1.50**	**2.00**	**3125**	**3250**	**3377**	**3208**	**3275**	**29.3**	**34.2**	**34.5**	**35.7**	**38.4**
Algeria	1.00	1.54	2.02	1.42	1.71	2977	3098	3163	3061	3071	1.9	1.9	1.9	2.0	2.2
Egypt	1.00	1.47	1.96	1.43	1.91	3395	3580	3783	3520	3645	1.6	2.2	2.5	2.2	2.5
Iran	1.00	1.48	1.96	1.52	2.06	3079	3109	3228	3067	3126	4.7	5.2	4.4	5.7	5.3
Iraq	1.00	1.77	3.16	1.75	3.09	2342	2651	2773	2618	2685	7.8	7.5	8.5	7.9	9.6
Morocco	1.00	1.61	2.27	1.42	1.82	3287	3592	3856	3553	3755	1.7	1.9	2.0	1.9	2.0
Saudi Arabia	1.00	1.76	2.74	1.76	2.71	2936	3055	3128	3020	3046	1.3	1.4	1.5	1.5	1.8
Turkey	1.00	1.40	1.60	1.44	1.70	3596	3661	3698	3620	3597	1.8	2.2	2.4	2.2	2.4
The Americas	**1.00**	**1.37**	**1.69**	**1.27**	**1.48**	**3188**	**3290**	**3392**	**3244**	**3297**	**42.5**	**35.7**	**27.7**	**39.3**	**32.7**
Latin America and the Caribbean	**1.00**	**1.46**	**1.83**	**1.42**	**1.72**	**2878**	**3036**	**3184**	**2985**	**3081**	**39.5**	**32.1**	**24.0**	**35.8**	**28.7**
Argentina	1.00	1.42	1.75	1.42	1.74	3171	3327	3426	3297	3354	0.7	0.6	0.7	0.6	0.7
Brazil	1.00	1.52	1.95	1.41	1.66	3142	3336	3492	3292	3398	3.7	3.2	3.1	3.2	3.1
Colombia	1.00	1.44	1.75	1.52	1.96	2645	2804	2957	2759	2868	5.0	3.9	2.7	4.5	3.6

(continued)

Table 2 (continued)

	Aggregate food production (index, 2010–1.00)					Per capita food consumption (KCAL per capita per day)					Hunger (millions of people at risk)				
		Without climate change		With climate change			Without climate change		With climate change			Without climate change		With climate change	
	2010	2030	2050	2030	2050	2010	2030	2050	2030	2050	2010	2030	2050	2030	2050
Mexico	1.00	1.35	1.62	1.31	1.54	3040	3134	3240	3054	3096	5.3	5.4	5.3	6.1	6.1
Peru	1.00	1.46	1.78	1.71	2.44	2472	2752	2886	2700	2782	3.6	2.0	1.4	2.3	1.8
Venezuela	1.00	1.41	1.76	1.30	1.50	2536	2626	2763	2579	2669	1.4	1.3	0.7	1.6	1.2
North America	**1.00**	**1.29**	**1.58**	**1.15**	**1.29**	**3714**	**3725**	**3735**	**3689**	**3654**	**3.0**	**3.6**	**3.7**	**3.6**	**4.0**
Europe and Former Soviet Union	**1.00**	**1.18**	**1.33**	**1.14**	**1.26**	**3275**	**3390**	**3491**	**3359**	**3414**	**17.1**	**13.0**	**11.4**	**13.4**	**12.5**
Former Soviet Union	**1.00**	**1.26**	**1.42**	**1.20**	**1.36**	**3092**	**3321**	**3423**	**3288**	**3338**	**9.7**	**5.9**	**5.2**	**6.2**	**5.5**
Russia	1.00	1.26	1.44	1.23	1.44	3227	3450	3532	3417	3452	1.8	1.2	1.1	1.2	1.2
Ukraine	1.00	1.21	1.31	1.11	1.18	3201	3434	3581	3400	3499	0.6	0.4	0.3	0.4	0.3
Uzbekistan	1.00	1.28	1.49	1.27	1.45	2563	2849	3024	2820	2935	2.4	0.8	0.8	0.9	0.8
Europe	**1.00**	**1.15**	**1.28**	**1.11**	**1.21**	**3370**	**3424**	**3523**	**3395**	**3450**	**7.4**	**7.0**	**6.2**	**7.3**	**6.9**

Source: CGIAR (2018)

1.2 Similar Expectations

It looks that there is no correlation between climate change and sustainable agriculture, however, there are so many similar core and object both for sustainable agriculture and climate change. They are crossing and containing each other, all of improvement and achievement for each of them is benefit to another. Agriculture has an enormous environmental footprint, and is simultaneously leading to huge amounts of environmental changes globally and being hugely impacted by these climate changes. The human population is continuing to grow rapidly at a rate which will require an increase in food production globally. This is complicated by the fact that the Earth is undergoing rising amounts of environmental risks. Sustainable agriculture provides a potential solution to enable agricultural systems to feed a growing population while successfully operating within the changing environmental conditions (Rockstrom et al. 2016). In sustainable agriculture, changes in lower rates of soil and nutrient loss, improved soil structure, and higher levels of beneficial microorganisms are not quick. The changes could helpful to reduce greenhouses gases emission from soil land and mitigate global warming, even benefits to controlling weeds, pests, etc. (Carolan 2006).

Climate change is one of the most important environmental issues facing human beings in the world today, and it affected directly and indirectly on the sustainable agriculture development in a widely scope and levels. Global warmer and drier severely exacerbate the water shortage and soil erosion in most arid and semi-arid regions, adverse impacted on crops yield harvest and grain quality and diseases and pest worsen as well. Climate change is also a developmental issue facing all countries, it brings burden on the balance of economic development and resources utilization and environment protection (Editorial board 2015). In agricultural fields, disasters such as flood and drought, showed an extensive tendency, which limited the climate resources utilization and production potential capacity, exacerbate stability of agriculture (Xie and Feng 2009).

1.3 Common Concerns

Climate change and sustainable agriculture are two greatest challenges facing human beings and they are highly interlinked. On the one hand, climate change puts pressure on food production and security, on the other hand, agriculture activities contribute to anthropogenic greenhouse gas emissions. Climate change impacts on agricultural supply chains and on food systems are not linear and involve multiple interactions between social, political and biophysical systems (Lin and Xie 2014). Identifying the uncertainties and the benefits expected from climate smart strategies is required. In addition, improved means for communicating risks and uncertainties to policy-makers and stakeholders need to be investigated. It is well known that although actions with zero or negative cost may exist, their adoption will require overcoming multiple barriers, e.g., through better information, decision support and capacity building. The analysis of barriers for action has therefore to be completed by the design of decision-support tools, integrating different aspects of a given regional and sub-regional and national and local context.

Climate smart agriculture which was first articulated in 2009 in an FAO publication, is a way to achieve short-and-long-term agricultural development priorities as facing climate change and serves as a bridge to other development priorities (Mann et al. 2009). It emerged as a way to square the goals of climate change mitigation and adaptation with the need to increase productivity in the agricultural sector through the promise of a triple-win solution. It seeks to support countries and other actors in securing the necessary policy, technical and financial conditions to enable them to sustainably increase agricultural productivity and incomes in order to meet national food security and development goals, build resilience and the capacity of agricultural and food systems to adapt to climate change, seek opportunities to mitigate emissions of greenhouse gases and increase carbon sequestration (Saj et al. 2017; Xie et al. 2014). Metrics to address jointly agricultural sustainability and climate change dimensions are still draw more attention both from scientific field and practice reality. Developing a common understanding of these metrics is required for guiding sustainable agriculture. Such metrics are enabling firstly to identify sustainable agriculture practices and then to measure their impacts in a consistent way on sustainability.

2 The Main Characteristics of Climate Change

2.1 The Average Situation

According to the Intergovernmental Panel on Climate Change (IPCC) Assessment Report 5 (AR5) (IPCC 2014a). The globally averaged temperature showed a warming of 0.85 °C based on that of the period 1880–2012, The total increase between the average of the 1850–1900 period and the 2003–2012 period is 0.78 °C and the total increase between the average of the 1850–1900 period and the reference period for projections, 1986–2005 is 0.61 °C, based on the single longest dataset available. For the longest period when calculation of regional (1901–2012), almost the entire globe has experienced surface warming. Climate change has already affected and will continue to affect the natural world, human well-being, and the global economy. Relative to 1986–2005, the global mean surface temperature by the end of the twenty-first century will increase by 0.3–4.8 °C. Based on World Meteorological Organization (WMO) Statement on the State of the Global Climate in 2017 (WMO 2018), global mean temperatures the year 2017 were 0.46 ± 0.1 °C above the 1981–2010 average, and about 1.1 ± 0.1 °C above pre-industrial levels. By this measure, 2017 and 2015 were effectively indistinguishable as the world's second and third warmest years on record, ranking only behind 2016, which was 0.56 °C above the 1981–2010 average (Fig. 1).

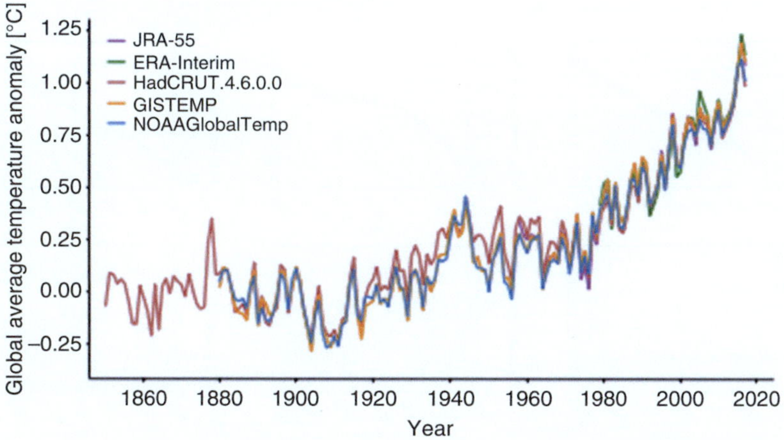

Fig. 1 Global mean temperature anomalies, with respect to the 1850–1900 baseline, for the five global datasets. (Source: WMO 2017)

2.2 Atmospheric Greenhouse Gasses Concentration

Observations of CO_2 concentrations are globally averaged temperature are generally well within the range of the extent of the earlier IPCC projections. The atmospheric abundance of CO_2 was 390.5 ppm (390.3–390.7) in 2011; this is 40% greater than in 1750. Based on WMO Statement on the State of the Global Climate in 2017 (WMO 2018), Greenhouse Gases (GHGs) concentrations in the end of 2016 reached new highs with CO_2 at 403.3 ± 0.1 parts per million (ppm), methane (CH_4) at 1853 ± 2 parts per billion (ppb) and nitrous (N_2O) at 328.9 ± 0.1 ppb. These values constitute, respectively, 145%, 257% and 122% of pre-industrial (before 1750) levels. The increase in CO_2 from 2015 to 2016 was larger than the increase observed from 2014 to 2015 and the average over the last decade, and it was the largest annual increase observed in the post-1984 period (Fig. 2).

2.3 Extreme Climate and Weather Events

According to the IPCC special report, Managing the Risks of Extreme Events and Disasters to Advance Climate Change Adaptation (SREX), there has been an overall decrease in the number of cold days and nights, and an overall increase in the number of warm days and nights. These changes have also occurred at the continental scale in North America, Europe, and Australia. There is a warming trend in daily temperature extremes in much of Asia. Daily temperature extremes in Africa and South America generally varied depending on the region (see Fig. 3). In many regions over the globe that the length or number of warm spells, or heat waves, has increased (IPCC 2012).

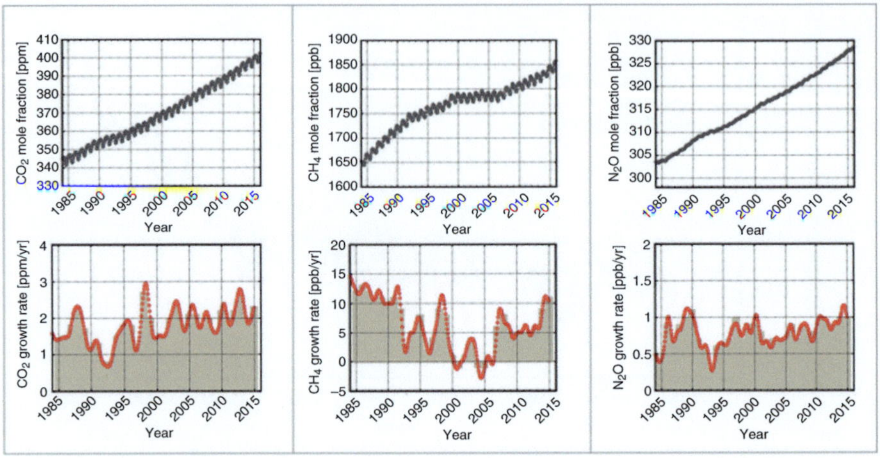

Fig. 2 The trend of atmospheric greenhouse gases in recent years. Top row: Globally averaged mole fraction (measure of concentration), from 1984 to 2016, of CO_2 in parts per million (left), CH_4 in parts per billion (middle) and N_2O in parts per billion (right). The red line is the monthly mean mole fraction with the seasonal variations removed; the blue dots and line depict the monthly averages. Bottom row: The growth rates representing increases in successive annual means of mole fractions for CO_2 in parts per million per year (left), CH_4 in parts per billion per year (middle) and N_2O in part per billion per year (right). (Source: WMO 2017)

The number of heavy precipitation events over land has increased in more regions than it has decreased. Some regions of the world have experienced more intense and longer droughts, in particular in southern Europe and West Africa, but in some regions, droughts have become less frequent, less intense, or shorter, e.g., in central North America and North-western Australia. There has been a poleward shift in the main Northern and Southern Hemisphere extra-tropical storm tracks.

3 Impacts of Climate Change on Sustainable Agriculture and Potential Risk

3.1 Agricultural Production and Food Quality

Generally, climate change will negatively impact four the major crops (wheat, rice, and maize) production for local temperature increases of 2 °C or more above late-twentieth-century levels. Projected impacts vary across crops and regions and adaptation scenarios, with about 10% of projections for the period 2030–2049 showing yield gains of more than 10% and about 10% of projections showing yield losses of more than 25%, compared to the late twentieth century. After 2050, the risk of more severe impacts increases. Besides, climate change will increase progressively the inter-annual variability of crop yields in many regions (see Fig. 4).

Exposure and vulnerability at scale of risk management in the example	Information on Climate Extreme Across Spatial Scales			Options for risk management and adaptation in the example
	GLOBAL Observed (since 1950) and projected (to 2100) global changes	**REGIONAL** Observed (since 1950) and projected (to 2100) changes in the example	**Scale of risk management: available information for the example**	
Droughts in the context of food security in West Africa				
Less advanced agricultural practices render region vulnerable to increasing variability in seasonal rainfall, drought, and weather extremes. Vulnerability is exacerbated by population growth, degradation of ecosystems, and overuse of natural resources, as well as poor standards for health, education, and governance.	Observed: *Medium confidence* that some regions of the world have experienced more intense and longer droughts, but in some regions droughts have become less frequent, less intense, or shorter. Projected: *Medium confidence* in projected intensification of drought in some seasons and areas. Elsewhere there is overall *low confidence* because of inconsistent projections. [Table 3.1, 3.5.1]	Observed: *Medium confidence* of an increase in dryness. Recent years characterized by greater interannual variability than previous 40 years, with the western Sahel remaining dry and the eastern Sahel returning to wetter conditions. Projected: *Low confidence* due to inconsistent signal in model projections.	Sub-seasonal, seasonal, and interannual forecasts with increasing uncertainty over longer timescales. Improved monitoring, instrumentation, and data associated with early warning systems, but with limited participation and dissemination to at-risk populations.	Low-regrets options that reduce exposure and vulnerability across a range of hazard trends: • Traditional rain and groundwater harvesting and storage systems • Water demand management and improved irrigation efficiency measures • Conservation agriculture, crop rotation, and livelihood diversification • Increasing use of drought-resistant crop varieties • Early warning systems integrating seasonal forecasts with drought projections, with improved communication involving extension services • Risk pooling at the regional or national level [2.5.4; 5.3.1, 5.3.3, 6.5; Table 6-3, 9.2.3, 9.2.11]

Fig. 3 Droughts in the context of food security in West Africa. Global-scale trends in a specific extreme may be either more reliable (e.g., for temperature extremes) or less reliable (e.g., for droughts) than some regional-scale trends, depending on the geographical uniformity of the trends in the specific extreme. During the period from 1970 to 2008, over 95% of deaths from natural disasters occurred in developing countries. Middle income countries with rapidly expanding asset bases have borne the largest burden. During the period from 2001 to 2006, losses amounted to about 1% of GDP for middle income countries, while this ratio has been about 0.3% of GDP for low income countries and less than 0.1% of GDP for high income countries, based on limited evidence. In small exposed countries, particularly Small Island Developing States, losses expressed as a percentage of GDP have been particularly high, exceeding 1% in many cases and 8% in the most extreme cases, averaged over both disaster and non-disaster years for the period from 1970 to 2010. (Source: IPCC 2012)

Studies had demonstrated that lots of negative sensitivity of crop yields to, both from several crops and regions, which existed throughout the growing season, extreme daytime temperatures around 30 °C. Temperature trends were important for determining both past and future impacts of climate change on crop yields in the global, and precipitation projections remained important factors for assessing future impacts for a regional scales. Stimulatory effects of CO_2 in most cases and the damaging effects of elevated tropospheric ozone (O_3) on crop yields (Porter et al. 2014).

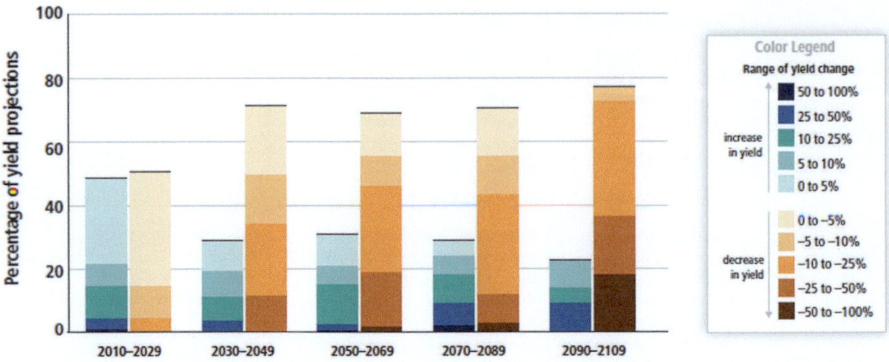

Fig. 4 Summary of projected changes in crop yields, due to climate change over the twenty-first century. The figure includes projections for different emission scenarios, for tropical and temperate regions, and for adaptation and no-adaptation cases combined. Relatively few studies have considered impacts on cropping systems for scenarios where global mean temperatures increase by 4 °C or more. For five timeframes in the near-term and long-term, data (n = 1090) are plotted in the 20-year period on the horizontal axis that includes the midpoint of each future projection period. Changes in crop yields are relative to late-twentieth-century levels. Data for each timeframe sum to 100%. (Source: IPCC 2014a)

Climate change could result in many crops growth period go ahead, the total growth season shortened, Study showed the growth season of maize could shortened by 4.3–13%, 10.8–22.5%, and 12.3–30.3% respectively (Tao and Zhang 2013) if temperature increased 1 °C, 2 °C and 3 °C, whereas wheat could shortened by 3.94%, 6.90% and 9.67% (Liu and Tao 2013). If temperature increased in night 2.5 °C, wheat could bring forward 5 days for maturity and 4 days for milk ripe stage and shorten 5 days for grain filling process. When annual mean temperature increased 1.0 °C, the growth period of rice could be shortened 4.1–4.4 days (Zhang et al. 2013). The growth period of winter wheat, about 30% total planting area in the whole country, was shorten, whereas the period of heading to maturity of winter wheat about 60% total planting area increased (Tao et al. 2012; Xiao et al. 2013).

The different methods consistently showed negative temperature impacts on crop yield at the global scale showed that, without CO_2 fertilization, effective adaptation, and genetic improvement, each degree-Celsius increase in global mean temperature would, on average, reduce global yields of wheat by 6.0%, rice by 3.2%, maize by 7.4%, and soybean by 3.1% (Zhao et al. 2017) (See Fig. 5). Case study on China showed that significant adverse impacts of climate change on crops yield has been occurred at least for the three main crops, rice, maize and wheat. The situation should be still keeping in the future (Tao et al. 2016; DaMatta et al. 2010). Besides, an increased heat extreme temperature stress or a decreased cold extreme temperature stress could be expected a large spatial variability of yield loss in the future (Zhang et al. 2017c).

Climate change affects crop quality by altering carbon and nutrient uptake and biochemical processes that produce secondary compounds or redistribute and store compounds during grain development. Cereals grown in elevated CO_2 show a

a

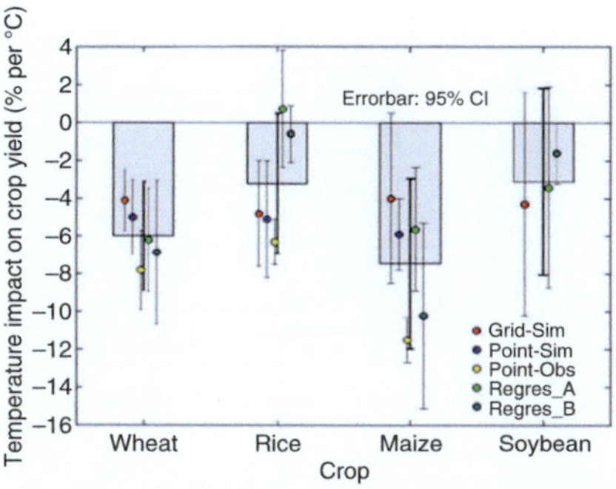

b

Scenario	Wheat	Rice	Maize	Soybean	Mean
	Yield changes (%) due to temperature changes by the end of century				
RCP2.6	−6.9	−3.3	−8.6	−3.6	−5.6
	[−15.0, −1.4]	[−9.2, 0.8]	[−18.6, −1.8]	[−11.2,1.7]	[−14.4,−0.1]
RCP4.5	−11.4	−5.5	−14.2	−5.9	−9.2
	[−21.7, −3.9]	[−13.8, 1.0]	[−27.9, −4.9]	[−17.0, 3.1]	[−21.2, −0.3]
RCP6.0	−14.0	−6.8	−17.4	−7.2	−11.3
	[−25.7, −5.1]	[−16.8, 1.3]	[−33.1, −5.8]	[−20.2, 3.6]	[−25.6, 0.1]
RCP8.5	−22.4	−10.8	−27.8	−11.6	−18.2
	[−40.2, −8.5]	[−125.3, 2.4]	[−50.4, −9.7]	[−31.0, 6.0]	[−38.6, −0.7]

Fig. 5 Global crop yield changes in response to temperature increase. (**a**) Impacts on crop yields of a 1 °C increase in global temperature in grid-based simulations (Grid-Sim), point-based simulations (Point-Sim), field-warming experiments (Point-Obs), and statistical regressions at the country level (Regres_A) (9) and the global level (Regres_B) (8). Circles, means of estimates from each method or medians for Grid- and Point-Sim. Filled bars, means of the multimethod ensemble. Error bars show 95% CIs for individual methods (gray lines) and the ensemble of methods (black lines). The loss in yield for each degree Celsius increase in global mean temperature is largest for maize (with multimethod average ±2 SE) of −7.4 ± 4.5% per degree Celsius. For wheat, the average estimate is a 6.0 ± 2.9% loss in global yield with each degree-Celsius increase in temperature. Results agree more closely on the impact on wheat (−7.8 to −4.1% per degree Celsius) than on maize yields. Global increase in temperature of 1 °C will reduce global rice yield by an average of 3.2 ± 3.7%, much less than for maize and wheat. The global average reduction in soybean yield is 3.1% per degree-Celsius rise. (**b**) Projected changes in yield due to temperature changes by the end of the twenty-first century. CIs of 95% are given in square brackets. (Source: Zhao et al. 2017)

decrease in protein (Porter et al. 2014). Extreme temperatures and elevated CO_2 concentrations reduce milling quality of rice by increasing chalkiness but can improve taste through reduced amylase concentration (Xie et al. 2016; Yang et al. 2007). For example, high temperature could speed filling rate of rice, shorten filling period, as to adverse effect on grains filling degree, brown rice recovery, and milled rice percentage. Moreover, grain protein content of rice decreased, the amylose content increased, and some of trace element benefit to human health, such as Fe and Zn could decreased (Xie and Feng 2009). Double CO_2 concentration could reduce the nitrogen content in grains by 9–16% for C3 crops, and by 7% for C4 crops. Increasing of both temperature and CO_2 concentration could reduce protein content, lysine content and fat content in maize grain, generally, the quality of maize was decreased under the circumstance of climate change (Porter et al. 2014). Meta-analysis finds decreases between 10% and 14% in edible portions of wheat, rice, barley, and potato, but only 1.5% in soybeans, a nitrogen-fixing legume, when grown in elevated CO_2, and decreases in zinc, sulphur, phosphorus, magnesium, and iron in wheat and barley grain; increases in copper, molybdenum, and lead; and mixed results for calcium and potassium (Porter et al. 2014) (see Fig. 6).

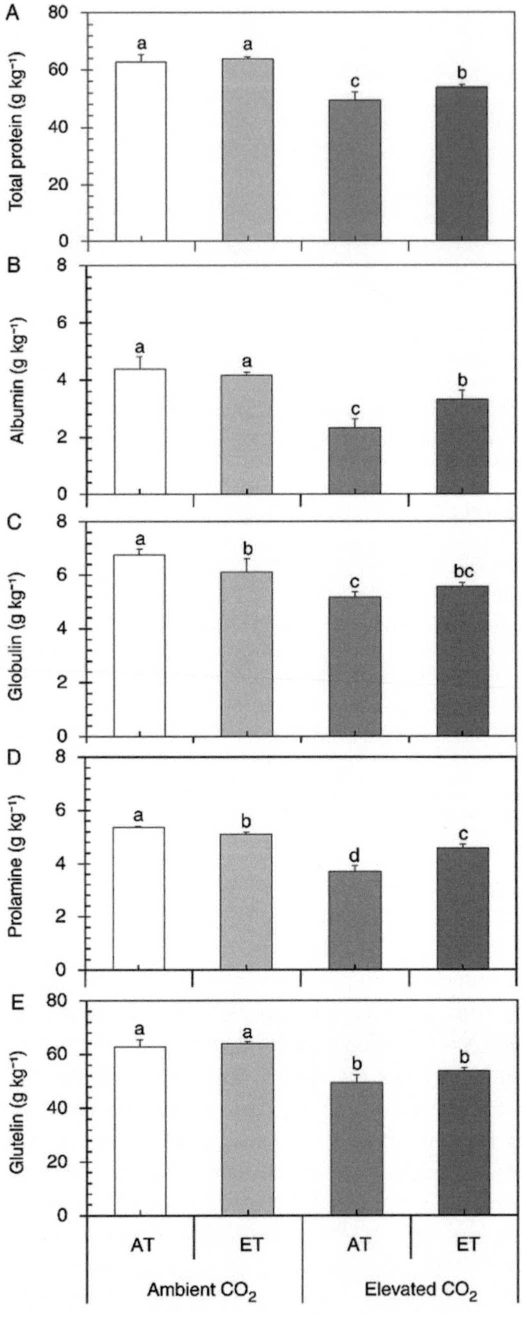

Fig. 6 The impact of elevated CO_2 and temperature on grain quality of rice grown under open-air field conditions. Concentrations of (**a**) total protein, (**b**) albumin, (**c**) globulin, (**d**) prolamine and (**e**) glutelin in rice grains. Rice plants were exposed to ambient or elevated (200 μmol mol^{-1} above ambient) CO_2 in combination with either ambient or elevated (1 °C above ambient) temperature

3.2 Diseases Pest and Weed Damage

Climate change affects the geographic range of specific species of insects and diseases for a given crop growing region. Migratory insects could colonize crops over a larger range in response to temperature increases, with subsequent reductions in yield. Climate change is also a factor in extending the northward migration of agronomic and invasive weeds in North America (Ziska et al. 2011). Milder winters may enable crop pests better able to over-winter, and their active scope could be enlarged to the region with high latitude and damage more, besides, the pest generations could become more densely, and their reproductive capacity become more stronger (Huo et al. 2012). Because of that, the frequency of pest epidemics would increase, and the impact could be even more serious when combined with the effects of drought and warm weather.

Taking northeast of China as example, the most important diseases were rice blast (*Pyricularia grisea*) during last century, but rhizoctonia (*Rhizoctonia solani*) and flax leaf spot (*Xanthomonas axonopodis*) have become regular diseases and cause heavy damage now. Only Crioceridae (*Oulema oryzae*) and rice leaf miner (Hydrellia griseola) occurred during crop seedling stage 50 years ago, but stem borer (*Chilo suppressalis*) has become the main pest now. The complex interaction of temperature, water supply, higher CO_2 concentration and changing growth conditions impact on the host-pathogen interaction, and the details of these interactions will require careful research to assess the potential for crop loss in the future (Chakraborty and Newton 2011).

Meanwhile, climate change and CO_2 concentration might enhance the distribution and increase the competitiveness of agronomical important and invasive weeds. Rising CO_2 could reduce the effectiveness of some herbicides. The effects of climate change on disease pressure on food crops also showed that could change geographical ranges of pests and diseases. In most of regions, diseases and pest occurrence intensified and damages worsen.

Fig. 6 (continued) from tillering to maturity. Bars indicate ± SD. Bars not sharing the same letter differ significantly at $P < 0.05$. *AT* ambient temperature, *ET* elevated temperature. The nutritional quality of rice is mainly determined by the protein concentration of rice grains. Elevated CO_2 and temperature had opposite effects on total protein concentration of rice grains: elevated CO_2 decreased total protein concentration by 21.1% at ambient temperature and 15.6% at elevated temperature; in contrast, elevated temperature increased total protein concentration, with the increment more pronounced at elevated CO_2 than at ambient CO_2. Elevated CO_2 also reduced the concentration of each protein component, i.e. albumin, globulin, prolamine and glutelin. The impact of temperature elevation varied with atmospheric CO_2 level: a trend of decrease by elevated temperature at ambient CO_2 but an increase at elevated CO_2 on the concentrations of albumin, globulin and prolamine was observed, which was also indicated by the significant $CO_2 \times$ temperature interactions for the three components. Temperature elevation had no clear effect on glutelin concentration. (Sources: Jing et al. 2016)

For example, in northeast of China raw rice has never been a serious weed before but is now present in many regions and causes significant damage (Xie et al. 2014). The differential effects of climate and elevated CO_2 will change the competitive ability of crops and weeds so that changes in weed communities can be expected. A Free Air CO_2 Enrichment (FACE) study elevated CO_2 enhanced the competitive ability of rice (C3) was enhanced relatively, while that of the common weed barnyard grass (C4) weakened (Zeng et al. 2011). However, similar the host-pathogen interactions, the effects of climate change on weed-crop competition will require a detailed understanding of the response to several factors, including water supply and temperature, although the actual interactions and impacts will be highly specific (Ziska and Goins 2006; Ziska 2003). Besides, the competitive ability of long term between C3 and C4 under elevated CO_2 still need clarify for long term experiment (Reich et al. 2018).

With respect to control, a number of studies have, to date, indicated a decline in herbicide efficacy in response to elevated CO_2 and/or temperature for some weed species, both C3 and C4 (Manea et al. 2011). The decomposition of microorganism in soil organic matter is accelerated with global warming, which required more fertilizer application for meeting crop growth and development. Studies indicated that with 1 °C increasing, available nitrogen release could increase 4%, and release duration shortened 3.6 days. That means, if you want to maintain the same fertilizer effect, you must apply to extra 4% fertilizer at the same time (Editorial 2015).

3.3 Soil and Water Environment Pollution

Climate change not only result in extreme weather events and diseases and pests increasing, but also led to soil resources and water resources decreasing, soil erosion and desertification, and rural and agricultural environment worsen more (Lin et al. 2018; Mohawesh et al. 2015). Global grain yield depends on chemical input, including fertilizer and pesticide in some certain. Studies estimated that varieties and field management taken half roles in the increasing yield to most crops. However, many studies showed fertilizer utilization has increased 30% yield increment for most crops (Mueller et al. 2012). At the same time, overuse of chemical maters brings lots of new environmental issues, soil and water pollution, restrict the soil productivity as desertification and erosion, even affected the air as greenhouses gas emission, harmful to agriculture sustainable development. So, it is important to control the chemical materials overuse, and protect the agricultural environment. Climate change also brought adverse effects on forest, grassland, wetland and biodiversity those important environmental system closed to agriculture (Phalan et al. 2011; Balmford et al. 2012).

Under the circumstance of climate change, Potential soil evaporation capacity enlarged, soil moisture decreased, and soil organic matter content decreased (Zhang et al. 2017a; Piao et al. 2010). As a result, soil agriculture quality decreased, soil environmental quality and soil health quality decreased, salinization of soil tend to be serious, Nitrogen fertilizer application need to increase (Basu et al. 2018; Lal 2018). Case study on the northeast of China showed that, in recent years, the crop land organic matter content decreased dramatically, though some regions soil quality increased a little. The same situation also occurrence in the north west of China (Lin et al. 2018).

Sustainable agriculture is helpful to adapt to climate change, protect agricultural environment and maintain biodiversity (Carberry et al. 2013; Zhang et al. 2015). Some new ideas on agriculture, such as ecology-agriculture, cycle-agriculture, green-agriculture, concern about both from resources utilization and environment protection, could be as supplement to sustainable agriculture (Sayer et al. 2013; Sayera and Cassman 2013).

3.4 *Food Access and Food Security*

All aspects of food security are potentially affected by climate change, including food access, utilization, and price stability (Fig. 7). The food supply stability could be affected through the changes in precipitation induced by the climate change where increases in the prevalence and severity of droughts and floods give rise to swings in food availability and prices, increased prices during drought years outweigh the decline in prices during the wetter years, resulting in small increases in prices under climate variability assumptions (Gohar and Cashman 2016). However, there remains limited quantitative understanding of how non-production elements of food security will be affected, and of the adaptation possibilities in these domains.

In recent decades, there have been several periods of rapid food and cereal price increases following climate extremes in key producing regions, indicating a sensitivity of current markets to climate extremes, among other factors (Wossen et al. 2018; Porter et al. 2014). Even small-scale farmers in developing countries because of certification and market barriers, also could influence food security primarily (Jouzi et al. 2017). Certainly, a range of potential adaptation options exist across all food system activities, not just in food production, but benefits from potential innovations in food processing, packaging, transport, storage, and trade are insufficiently researched.

4 Suggestions for Sustainable Agriculture under Climate Change

Climate change and its consequences have affected sustainable development, especially in agriculture. When and how does take adaptive actions to climate change not only involved in achievements, but also related to closely budgets input, including financial, materials and human resources. The basic work to climate change is improving the recognition to vulnerability and resilience of climate change, evaluating scientific adaptive capacity, and then strengthening integrated capacity to adapt

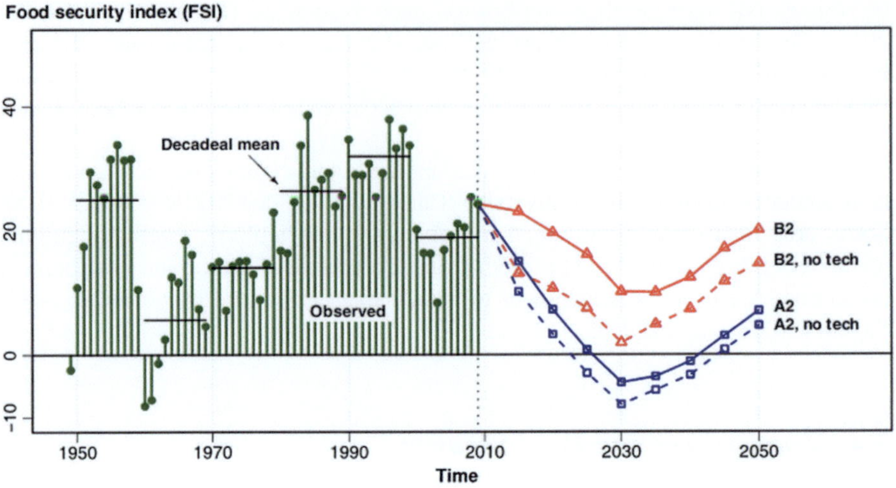

Fig. 7 Relative food surplus (%) in per capita terms, or the food security index (FSI), as evaluated using census data during 1949–2009 and as projected in 2030 and 2050. Horizontal bars represent the average FSI levels at the decadal time scale. The food security index (FSI) values in China, evaluated from a food supply–demand point of view using census and estimated data for the pre-2009 and model simulated crop yields under socio-economic and agronomic scenarios in association with the SRES A2 and B2 climate change scenarios. China's food security status was significantly improved soon after the long-lasting wars that ended in the late 1940s. At the end of the first 5-year plan, the FSI increased from −2.4 in 1949 to 31.4 in 1957, showing that the supply–demand relationship turned from a 2.4% deficit to a 31.4% surplus. The peak FSI value of 38.5 appeared in 1984, coinciding with the record harvest of 390 MT in the same year. Although higher productions (~500 MT) were achieved consecutively during 1996–1999, the FSI values in the same period were not higher than that of 1984, reflecting the combined effects of a larger population and a higher standard of living. Extreme climatic events and natural hazards, which caused notable production losses during 2000–2003, were responsible for the second largest drop in the FSI values after the period of the Great Leap Forward (1957–1961). China has achieved record harvests for six consecutive years since 2004, reaching the level of 530 million tons in 2009. However, the average FSI level of 18.8 during the first decade of the twenty-first century is considerably lower than the average level of 31.9 in 1990s or 26.3 in 1980s, showing the *dragging* effect of steadily increasing consumption levels on FSI. This suggests that food utilization (safe, balanced and nutritious food, etc.) is gaining momentum and attention is needed on how to integrate it into existing food security assessment frameworks which took food availability as the primary indicator. (Source: Ye et al. 2013)

to climate change in all levels, including rural region in particularly. Though, it is not depending on rural region itself only to adapt to climate change, the government, the administration apartment and research institute and all the stakeholder's relatives should be collaborative and unitive to the challenges.

4.1 *Management in Agriculture Production (Technology-Oriented)*

In general, the basic and important purpose of agriculture adaptation is ensuing sustainable development of agriculture, so keeping the crop yields enough and stable in certain region is the basic requirement. Adaptation to climate change requires that appropriate adjustments to crop management are undertaken (Martin-Guay et al. 2018; Pittelkow et al. 2015). For example, adjusting agriculture layout and distribution, aiming at consequences of recognize of radiation, heat and precipitation from global warming, developing new crop planting zone with new planting technology, including field management technology, crop cultivars selecting technology, and new rotation system (Bai and Tao 2017; Dong et al. 2016). Ensure the stress insistence and restorability of agricultural production system. Ensure that high yield and stability in key crop production regions.

The process of adapting to climate change has seen significant changes in the types of major crops grown, the cultivars selected and the crop rotations used in most of regions and countries, for example, proportion of rice grown has increased remarkably and the wheat area has significantly reduced in the Northeast of China, as a result, the rotations employed have changed into maize-rice systems from wheat-maize systems (Xie et al. 2011). The current rotations have developed to take full advantage of more favorable thermal conditions and the relatively abundant water resources. This has resulted in a substantial increase in the total rice output specifically and food production in general.

Over the past few decades, more mid to late season maturity species and cultivars with drought tolerance, water logging, and salinity tolerance have been grown in high latitude, including semi-arid region. These changes have extended the growing season and achieved greater production levels in China. In response to changes in the recent climate, crop breeding has paid more attention to develop high yielding cultivars better adapted to high temperature and long growth duration, as well as with resistance to cold damage. Despite the progress to date, there is still a need to broaden the range of species and varieties available to ensure that adaptation to climate warming continues. Researchers found that, during 1981–2009, the growth period of rice in most region of China shortened gradually, but some of them in Northeast of China has prolonged a little. This is just because rice variety changed under climate change, which was got more heat resources than previous varieties (Tao and Zhang 2013). Due to changes in temperature and heat conditions, crop growth and development process will change so that new agronomic packages will be needed. As an example of the importance of integrating cultivar, management and environment, simulation studies have shown that maize yields could increase when maize sowing times are adjusted and current maize cultivars are replaced with middle-late maturing (Zhang and Huang 2013). From this, important strategies to adapt to future climates are to adjust the sowing date in order to meet the better soil moisture and temperature, even miss the peat occurrence date, change to late-maturing high-yielding crop cultivars, and change to conservation tillage systems.

Which could improve the capacity to prevent extreme climate events and disease, pest and weed in the field.

With the temperature increasing and heat resources enhancement, the northern boundary of multiple cropping in China is expected to move towards the region of high latitude and high altitude. Several studies showed that the north boundary of double cropping in China during 2011–2040 and 2041–2050 probably expanded to 130 km and 160 km, respectively, comparing to that of 1951–1980, and the north boundary of triple cropping could move 40–200 km and 70–300 km (Yang et al. 2010). Besides, maize planting area is expected to expand among the semi-arid region in northwest of China. Adaptation will need to continue incorporating refinements that build on the past adaptations, as well as the inclusion of strategies only just being used. The adoption of automation and precision farming technology could promote sustainable agriculture, reduce the cost of production, and improve land utilization efficiency and output efficiency (Editorial board 2015).

4.2 Public Perceptions and Capacity Building (Public-Oriented)

The key strategy, for developing countries, to overcome climate change ensuing sustainable agriculture is promoting adaptability and reducing vulnerability through strengthening infrastructure and capacity construction and increasing public awareness. In the case of the agricultural sector, reactive adaptation is particularly important rather than pro-active response, especially in developing countries since this sector is a substantial source of national income for the developing countries and it is to be effective within a very short gestation gap. The benefit of adaptation is huge compared to costs of adaptation, and the early action costs are little for the early periods, whereas if the adaptation is to be taken late, the cost would increase (Al-Amin and Ahmed 2016; Wijaya et al. 2018).

Sustainable agriculture is a process of agriculture development with plenty of contents, including adaptation to climate change. Due to climate change, the extreme climate and weather events become more and more frequently and brought more and more damages and losses both from safety of life and property. So, strengthening the capacity construction should pay more attention (Xie et al. 2014). Firstly, improving the predictive accuracy on meteorological disasters and the timelines of disasters early warning. With global warming, some weather disasters, such as hot damage, chilly damage, and rainy sparse sunlight, come to more serious in some regions. Besides, as there exit uncertainties on the impacts scale from spatial and temporal of climate change on agriculture, so adaptive actions on sustainable agriculture must have more foreseeability. Secondly, developing a flexible and adjustable risk management mechanism, in all levels of national, regional and community, is benefit to minimum the vulnerability and risk in sustainable agriculture both from present and future (Lin and Xie 2014). Thirdly, apart from adjusting field management technologies, agricultural extension system should be established and strengthened, as to enhance the conversion rate of scientific research achievements. Through the comprehensive agricultural extension system, speeding new varieties extension, implementing effective utilization for water and fertilizer, and integrated managing diseases and pest. Besides, new media and the internet should be included for agriculture extension.

In concert with the large hydrological projects, improved small scale water storages to store rainfall will help reduce surface runoff and improve soil water drainage, increased local infiltration of precipitation, and improve water and soil conservation and fertilizer responses (Pradhan et al. 2018). New effective water-saving technology for agriculture, including drip irrigation and trickle irrigation, should be developed, to improve water use efficiency, which will assist the development and optimization of the existing water resources therefore improving the adaptive capacity of the water resources system to deal with climate change impacts. Irrigation reduces crop susceptibility by mitigating the impacts on productivity of the crops' water shortage during drought periods. Adopting irrigation technologies not only benefits food producers but also provides some benefits for consumers as well (Gohar and Cashman 2016).

4.3 *Effective Governance and Government Roles (Governance-Oriented)*

It requires effective governance during dealing with the two issues, climate change and sustainable agriculture. Adapting to climate change is more than just implementing a new technology but is also about enhancing the broader resilience of the community in ways that will ensure its long-term viability. To achieve this, it will be necessary for different components of government and other institutional actors to work together to improve the adaptation capacity of farmers in the future (Brown et al. 2016). Besides farmer's autonomous adaptation by empirical agronomic measures, the governments and scientists should provide systemic programs and corresponding effective and economical techniques, to reduce climate risk and vulnerability, make farming more productive and environmentally sustainable (Luo et al. 2017).

Among them, the governmental function of strong organization and leadership is not replaced by any part, because it requires lots of inputs including human input, material input, and financial input, any individual, farmer and enterprise cannot bear this responsibility (Castellano and Moroney 2018; Kerr et al. 2018; Valipour et al. 2015). In order to address climate change, large-scale collective (e.g. policy) decisions on infrastructure, land-use patterns, built environment, and social mobilization are required. Institutional barriers need to be addressed, and partnerships created between the research, private, and practitioner communities for faster adoption of sustainable technologies and greater knowledge diffusion. Strategic coordination between levels of government and other relevant actors is necessary for the effective implementation of policies and strategies that generate social learning and policy innovation that can influence behavioral change at the macro-, meso-, and micro-levels (Burch et al. 2014).

For example, over-exploitation of groundwater has affected the normal water resources cycle, meanwhile relatively low efficiency of water utilization in agriculture, and consequently there has been a decline in water resource security, which climate change and increased drought will only exacerbate. Improved management of the current water resources is the core measure to adapt to future climate change (Xia et al. 2016). The key strategy is to strengthen water infrastructure construction, develop improved mechanisms and structures for farmland water conservation and in making those plans consider the impact of drying climates on storage capacity (Editorial board 2015). This should be done by linking specific water requirements to construction projects. As well as providing improved water security for irrigation, new water storages linked to long-term flood prevention and control will reduce production losses due to flood, drought and other climate related disasters. It is no doubt that only effective governance organized by government could be implemented the whole thing. Based on definitions for the integration of climate change adaptation policy into disaster risk reduction policy, taking Zambia as an example, addressed the importance of its horizontal (inter-ministerial) and vertical (intra-ministerial) dimensions, pilli-Sihvola suggested that increasing potential inefficiencies in governance and policy implementation (Pilli-Sihvola and Vaatainen-Chimpuku 2016).

Studies supported the point that assets at both household and community levels have significant effects on farmer's decision on taking physical adaptation to climate change. Higher level of household assets in terms of education, social capital and wealth facilitates farmers to take their adaptation decision (Hou et al. 2015). The community assets such as village's access to government's technical service and easiness of communication or information flow also can play an important role in facilitating farmers to make their decision to take adaptation measures. So, a crucial area to improve adaptation capacity both from farmers and community is to improve their education and social capital, and enhancing the adaptive capabilities of the poor in vulnerable regions in responding to climate risks should be one of prioritized areas for policy interventions. Furthermore, the government's technical service related to resistance of disaster should be enhanced since it is of particular importance in facilitating farmers to take adaptation measures for reducing crop loss from drought, there is a great room to play in providing information and service on resistance of extreme weather events to local farmers (Wang et al. 2014; Kurukulasuriya and Mendelsohn 2017).

4.4 *International Cooperation in Research and Actions (Institution-Oriented)*

The international climate regime is dealing on public goods and requires scarce strategic resources allocation. The evolution of the climate regime combines efforts from scientific studies, international political and economic development, stakeholders, and other aspects (Richard 2017). Global climate governance is in the right direction of seeking on-confrontational and win–win cooperation. Governments, companies, and civil societies are gathering together and sharing the benefits of green transformation (Zhang et al. 2017b). The international level cooperation in climate change have yielded varying levels of success, leading to an increasing focus on the multi-level governance, focusing on the recognition that GHG emissions and vulnerability emerge out of a complex web (Burch et al. 2014).

Despite the presence of institutional arrangements for the integration of climate change and development strategies, the capacity to move towards CCD in development planning at sub-national levels is limited due to a number of barriers (Pilato et al. 2018). The climate change strategy and policy of adaptation and mitigation requires consideration of complex interaction of economy, environment as well as social needs of a nation and it is important to align efforts of both developed and developing nations (Bosomworth 2015). The efficiency of the global climate regime could be evaluated from the aspects of environmental performance, economic performance, distribution impacts, and institutional capability (Nordhaus 2007). So sustainable agriculture requires to launch long term adaptive policy and institution to adapt to climate change and ensure to avoid and mitigate agricultural damage (Schaafsma et al. 2018). In some extend, global mechanism and institution on the research field and practice field should be established. Because there is no significant indirect effects of a global pathway where past climate experiences affect global climate change concerns and then the adoption of adaptation practices. Instead, the adoption of climate adaptation practices is influenced mostly by a local pathway where past experiences influence local concerns about future climate change (Niles et al. 2015). Therefore, the international cooperation and collaboration both from research and practice should be encouraged. Sharing the advanced knowledge on climate change is benefitted to adjusting global cognition and actions, and local and regional successful practice to climate change could enlighten other regions and nations. So, institution construction included multiple levels, such as regional level, national level and international level.

Countries differ in terms of climate change impacts and progress made in adaptation strategies and actions. This creates both challenges and opportunities to international cooperation and learning. Given the urgent need for developing response strategies to global change and the emphasis on learning processes, an improved understanding of learning through international cooperation is needed (Vinke-de Kruijf and Pahl-Wostl 2016; Hazard et al. 2018). In various regions, specific knowledge transfer opportunities were created due to an extreme event or policy development. The governance system was supportive in some regions due to the presence of cooperation structures and leadership and restrictive in other regions due to lack of support, leadership or political will. Study confirmed that combinations of partner-specific, process-specific and process-external conditions influence learning at different levels. This implies that participants and partners can make a difference, especially when it comes to transferring lessons learned (Vinke-de Kruijf and Pahl-Wostl 2016).

5 Conclusion

The impacts of climate change on sustainable agriculture and food production are evident in several regions of the world. Negative impacts of climate trends have been more common than positive ones. These are expected to continue with negative impacts on nutrition and food security in coming days. The future agriculture will bear more and more responsibility, however, as the primary industry, agriculture still facing more potential risk to feed the large population under the circumstance of climate change, particularly in some developing countries. Agriculture is one of the most vulnerable industries which was affected by climate change, with a big vulnerability and a small resilience. There are a very common conscious that climate change had already adverse impact on sustainable agriculture, and the adverse impacts has large uncertainties, which means how extend adverse impacts on sustainable agriculture could be rather than whether the impact was occurred. It was convinced that taking promptly actions would be more helpful to adapt to climate change in agriculture development.

With concerned actions, we strongly encouraged that to develop the adaptive measures and technologies based on local situation, only the local adaptation measures could be international adaptation measures. Meanwhile, promoting international cooperation in research and actions also was necessary and benefit. Global warming is becoming one of the obstacles for agriculture sustainable development, which should have been overcome, on one hand, a suitable climate in helpful to maintain good environmental and ecological conditions to promote agriculture sustainable development, on the other hand. Both climate change and sustainable agriculture are main development issues facing human beings in the world today, there was countless ties between climate change and sustainable agriculture. The world should try to work together, enhance communication and cooperation, as to get the win-win situation.

References

Al-Amin AQ, Ahmed F (2016) Food security challenge of climate change: an analysis for policy selection. Futures 83:50–63

Bai HZ, Tao FL (2017) Sustainable intensification options to improve yield potential and ecoefficiency for rice-wheat rotation system in China. Field Crop Res 211:89–105

Balmford A, Green R, Phalan B (2012) What conservationists need to know about farming? Proc R Soc B Biol Sci 279(1739):2714–2724

Basu S, Rabara RC, Negi S (2018) AMF: the future prospect for sustainable agriculture [J]. Physiol Mol Plant Pathol 102:36–45

Bosomworth K (2015) Climate change adaptation in public policy: frames, fire management, and frame reflection. Environ Plan C Gov Policy 33(6):1450–1466

Brown PR, Bridle KL, Crimp SJ (2016) Assessing the capacity of Australian broadacre mixed farmers to adapt to climate change: identifying constraints and opportunities. Agric Syst 146:129–141

Burch S, Shaw A, Dale A et al (2014) Triggering transformative change: a development path approach to climate change response in communities. Clim Pol 14(4):467–487

Carberry PS, Liang WL, Twomlow T et al (2013) Scope for improved eco-efficiency varies among diverse cropping systems. Proc Natl Acad Sci U S A 110:8381–8386

Carolan M (2006) Do you see what I see? Examining the epistemic barriers to sustainable agriculture. Academic search complete

Castellano RLS, Moroney J (2018) Farming adaptations in the face of climate change. Renew Agric Food Syst 33(3):206–211

CGIAR: 2018 Global Food Policy Report (2018)

Chakraborty S, Newton AC (2011) Climate change, plant diseases and food security: an overview. Plant Pathol 60:2–14

DaMatta FM, Grandis A, Arenque BC et al (2010) Impacts of climate changes on crop physiology and food quality [J]. Food Res Int 43(7):1814–1823

Dong ZQ, Pan ZH, Wang S et al (2016) Effective crop structure adjustment under climate change[J]. Ecol Indic 69:571–577

Editorial board (2015) The 3rd China's national assessment of report on climate change. Science Press, Beijing

FAO (2008) World agriculture towards 2015/2030. Food and Agriculture Organization

FAO (2016) Monitoring food security and nutrition in support of the 2030 agenda for sustainable development: taking stock and looking ahead

Gohar AA, Cashman A (2016) A methodology to assess the impact of climate variability and change on water resources, food security and economic welfare. Agric Syst 147:51–64

Gold M (2009) What is sustainable agriculture? United States Department of Agriculture, Alternative Farming Systems Information Center

Hazard L, Steyaert P, Martin G et al (2018) Mutual learning between researchers and farmers during implementation of scientific principles for sustainable development: the case of biodiversity-based agriculture [J]. Sustain Sci 13(2):517–530

Hou LL, Huang JK, Wang JX (2015) Farmers' perceptions of climate change in China: the influence of social networks and farm assets [J]. Clim Res 63(3):191–201

Hovenden M, Newton P (2018) Plant responses to CO_2 are a question of time. Science 360(6386):263–264

Huo ZG, Li MS, Wang L et al (2012) Impacts of climate warming on crop diseases and pests in China. China Agric Sci 45(10):1926–1934

IPCC (2012) Managing the risks of extreme events and disasters to advance climate change adaptation. Cambridge University Press

IPCC (2014a) Climate change 2014 synthesis report summary for policymakers. Cambridge University Press, Cambridge/New York

IPCC (2014b) Climate change 2014: mitigation of climate change. Contribution of Working Group III to the fifth assessment report of the Intergovernmental Panel on Climate Change. Cambridge University Press, Cambridge/New York

Jing LQ, Wang J, Shen SB et al (2016) The impact of elevated CO_2 and temperature on grain quality of rice grown under open-air field conditions. J Sci Food Agric 96(11):3658–3667

Jouzi Z, Azadi H, Taheri F et al (2017) Organic farming and small-scale farmers: main opportunities and challenges [J]. Ecol Econ 132:144–154

Kerr RB, Nyantakyi-Frimpong H, Dakishoni L et al (2018) Knowledge politics in participatory climate change adaptation research on agroecology in Malawi [J]. Renew Agric Food Syst 33(3):238–251

Kurukulasuriya P, Mendelsohn R (2017) Impact and adaptation of South-East Asian farmers to climate change: conclusions and policy recommendations. Clim Chang Econ 8(3):1–6

Lal R (2018) Sustainable intensification of China's agroecosystems by conservation agriculture. Int Soil Water Conserv Res 6(1):1–12

Lin ED, Xie LY (2014) Revelation of agro meteorology learning from climate change 2014: impact, adaptation, and vulnerability. Chin J Agro Meteorol 35(4):359–364

Lin ED, Guo LP, Ju H (2018) Challenges to increasing the soil carbon pool of agro-ecosystems in China. J Integr Agric 17(4):723–725

Liu YJ, Tao FL (2013) Probabilistic change of wheat productivity and water use in China for global mean temperature changes of 1°C, 2°C, and 3°C. J Appl Meteorol Climatol 52:114–129

Luo XS, Muleta D, Hu ZH et al (2017) Inclusive development and agricultural adaptation to climate change [J]. Curr Opin Environ Sustain 24:78–83

Manea A, Leishman MR, Downey PO (2011) Exotic C4 grasses have increased tolerance to glyphosate under elevated carbon dioxide. Weed Sci 59:28–36

Mann W, Leslie L et al (2009) Food security and agricultural mitigation in developing countries: options for capturing synergies. FAO, Rome

Martin-Guay MO, Paquette A, Dupras J et al (2018) The new Green Revolution: sustainable intensification of agriculture by intercropping [J]. Sci Total Environ 615:767–772

Mohawesh Y, Taimeh A, Ziadat F (2015) Effects of land use changes and soil conservation intervention on soil properties as indicators for land degradation under a Mediterranean climate. Solid Earth 6(3):857–868

Mueller ND, Gerber JS, Johnston M et al (2012) Closing yield gaps through nutrient and water management. Nature 490(7419):254–257

Niles MT, Lubell M, Brown M (2015) How limiting factors drive agricultural adaptation to climate change. Agric Syst 200:178–185

Niles MT, Ahuja R, Barker T et al (2018) Climate change mitigation beyond agriculture: a review of food system opportunities and implications [J]. Renew Agric Food Syst 33(3):297–308

Nordhaus W (2007) Critical assumptions in the stern review on climate change. Science 317(5845):1682–1682

Phalan B, Onial M, Balmford A (2011) Reconciling food production and biodiversity conservation: land sharing and land sparing compared. Science 333(6047):1289–1291

Piao SL, Ciais P, Huang Y et al (2010) The impacts of climate change on water resources and agriculture in China. Nature 467(7311):43–51

Pilato G, Sallu SM, Gaworek-Michalczenia M (2018) Assessing the integration of climate change and development strategies at local levels: insights from Muheza District, Tanzania. Sustainability 10:1–11

Pilli-Sihvola K, Vaatainen-Chimpuku S (2016) Defining climate change adaptation and disaster risk reduction policy integration: evidence and recommendations from Zambia. Int J Disaster Risk Reduct 19:461–473

Pittelkow CM, Liang XQ, Linquist BA et al (2015) Productivity limits and potentials of the principles of conservation agriculture [J]. Nature 517(7534):365–482

Porter JR, Xie L, Challinor AJ et al (2014) Food security and food production systems. In: Field CB, Barros VR et al (eds) Climate change 2014: impacts, adaptation, and vulnerability. Part A:

Global and sectoral aspects. Contribution of Working Group II to the fifth assessment report of the Intergovernmental Panel on Climate Change. Cambridge University Press, Cambridge/New York, pp 485–533

Pradhan A, Chan C, Roul PK et al (2018) Potential of conservation agriculture (CA) for climate change adaptation and food security under rainfed uplands of India: a transdisciplinary approach [J]. Agric Syst 163:27–35

Reich PB, Hobbie SE, Lee TD et al (2018) Unexpected reversal of C_3 versus C_4 grass response to elevated CO_2 during a 20-year field experiment. Science 360(6386):317–320

Reyes JJ, Wiener JD, Doan-Crider D et al (2018) Building collaborative capacity: supporting tribal agriculture and natural resources in a changing climate. Renew Agric Food Syst 33(3):222–224

Richard SJ (2017) The structure of the climate debate. Energy Policy 104:431–438

Rockstrom J, Williams J, Daily G et al (2016) Sustainable intensification of agriculture for human prosperity and global sustainability. Ambio 46(1):4–17

Saj S, Torquebiau E, Hainzelin E et al (2017) The way forward: an agroecological perspective for Climate-Smart Agriculture [J]. Agric Ecosyst Environ 250:20–24

Sayer J, Sunderland T, Ghazoul J et al (2013) Ten principles for a landscape approach to reconciling agriculture, conservation, and other competing land uses. Proc Natl Acad Sci U S A 110:8349–8356

Sayera J, Cassman KG (2013) Agricultural innovation to protect the environment. Proc Natl Acad Sci U S A 110(21):8345–8348

Schaafsma M, Utila H, Hirons MA (2018) Understanding trade-offs in upscaling and integrating climate- smart agriculture and sustainable river basin management in Malawi [J]. Environ Sci Policy 80:117–124

Tao FL, Zhang Z (2013) Climate change, high-temperature stress, rice productivity and water use in Eastern China: a new super ensemble-based probabilistic projection. J Appl Meteorol Climatol 52(3):531–551

Tao FL, Zhang Z, Zhang S et al (2012) Response of crop yields to climate trends since 1980 in China. Clim Res 54(3):233–247

Tao FL, Zhang Z, Zhang S et al (2016) Variability in crop yields associated with climate anomalies in China over the past three decades. Reg Environ Chang 16(6):1715–1723

Valipour M, Ahmadi MZ, Raeini-Sarjaz M et al (2015) Agricultural water management in the world during past half century [J]. Arch Agron Soil Sci 61(5):657–678

Vinke-de Kruijf J, Pahl-Wostl C (2016) A multi-level perspective on learning about climate change adaptation through international cooperation. Environ Sci Policy 66:242–249

Wang YJ, Huang JK, Wang JX (2014) Household and community assets and farmers' adaptation to extreme weather event: the case of drought in China. J Integr Agric 13(4):687–697

Wijaya A, Glasbergen P, Leroy P et al (2018) Governance challenges of cocoa partnership projects in Indonesia: seeking synergy in multi-stakeholder arrangements for sustainable agriculture [J]. Environ Dev Sustain 20(1):129–153

WMO (2017) Global atmosphere watch

WMO (2018) Statement on the state of the global climate in 2017

Wossen T, Berger T, Haile MG et al (2018) Impacts of climate variability and food price volatility on household income and food security of farm households in East and West Africa[J]. Agric Syst 163:7–15

Xia J, Mo XG, Wang JX et al (2016) Impacts of climate change and adaptation in agricultural water management in North China [J]. Clim Change Agric Water Manag Dev Countries 8:63–77

Xiao D, Tao F, Liu Y et al (2013) Observed changes in winter wheat phenology in the North China Plain for 1981–2009. Int J Biometeorol 57(2):275–285

Xie L, Feng Y (2009) Climatic resources utilization to rice production in North China. China Agricultural Science and Technology Press, Beijing

Xie LY, Li Y, Lin M (2011) Response and adaptation to climate change of agriculture and environment in Northeast China [J]. Chin J Eco-Agric 19(1):197–201

Xie LY, Li Y, Qian FK et al (2014) Responses of food production systems to climate change: sensitivity and vulnerability. Chin J Popul Resour Environ 12(3):227–232

Xie LY, Lin ED, Zhao HL et al (2016) Changes in the activities of starch metabolism enzymes in rice grains in response to elevated CO_2 concentration. Int J Biometeorol 60(5):727–736

Yang LX, Wang YL, Dong GC et al (2007) The impact of free-air CO_2 enrichment (FACE) and nitrogen supply on grain quality of rice. Field Crop Res 102(2):128–140

Yang XG, Liu ZJ, Chen F (2010) The possible effects of global warming on cropping systems in China I. the possible effects of climate warming on northern limits of cropping systems and crop yields in China. China Agric Sci 43(2):329–336

Ye LM, Xiong W, Li ZG et al (2013) Climate change impact on China food security in 2050. Agron Sustain Dev 33(2):363–374

Zeng Q, Liu BA, Gilna B et al (2011) Elevated CO_2 effects on nutrient competition between a C-3 crop (*Oryza sativa* L.) and a C-4 weed (*Echinochloa crusgalli* L.). Nutr Cycl Agroecosyst 89(1):93–104

Zhang TY, Huang Y (2013) Estimating the impacts of warming trends on wheat and maize in China from 1980 to 2008 based on county level data. Int J Climatol 33(3):699–708

Zhang TY, Huang Y, Yang XG (2013) Climate warming over the past three decades has shortened rice growth duration in China and cultivar shifts have further accelerated the process for late rice. Glob Chang Biol 19(2):563–570

Zhang X, Davidson EA, Mauzerall DL et al (2015) Managing nitrogen for sustainable development [J]. Nature 528(7580):51–59

Zhang MX, Zhang J, Liu MG et al (2017a) Simulation on the future change of soil organic carbon from phaeozems under different management practices in Northeast China. Sustainability 9(7):1129

Zhang YX, Chao QC, Zheng QH et al (2017b) The withdrawal of the U.S. from the Paris Agreement and its impact on global climate change governance. Adv Clim Chang Res 8(4):213–219

Zhang Z, Chen Y, Wang CZ et al (2017c) Future extreme temperature and its impact on rice yield in China [J]. Int J Climatol 37(14):4814–4827

Zhao C, Liu B, Piao SL et al (2017) Temperature increase reduces global yields of major crops in four independent estimates [J]. Proc Natl Acad Sci U S A 114(35):9326–9331

Ziska LH (2003) Evaluation of the growth response of six invasive species to past, present and future atmospheric carbon dioxide. J Exp Bot 54:395–404

Ziska LH, Goins EW (2006) Elevated atmospheric carbon dioxide and weed populations in glyphosate treated soybean. Crop Sci 46:1354–1359

Ziska LH, Blumenthal DM, Runion GB et al (2011) Invasive species and climate change: an agronomic perspective [J]. Clim Chang 105(1–2):13–42

Carbon Sequestration for Sustainable Agriculture

Muhammad Sanaullah, Tahseen Afzal, Tanvir Shahzad, and Abdul Wakeel

Abstract The climate of earth has been experiencing an unprecedented change possibly due to the rapidly increasing amount of greenhouse gases (GHGs) in the atmosphere. If the release of GHGs into atmosphere continued at current rate, global warming will make the earth's atmosphere uninhabitable for living beings in near future. There is an urgency to discreetly devise multiple strategies to offset the current release of GHGs into atmosphere. The CO_2 has a prominent share in global warming amongst all GHGs in atmosphere. Soil carbon sequestration is a promising approach to offset the raising amount of CO_2 in the atmosphere. Both partially degraded and agricultural soils have a considerable potential to minimize the elevated CO_2 levels in the atmosphere. On a global scale, the soils can retain twofold more C than that present in the atmosphere or captured in vegetation. The temperature, soil moisture and elevated CO_2 levels are the dominant climatic factors affecting the soil C sequestration. Soil C sequestration is also strongly influenced by various edaphic factors i.e. soil texture, soil structure, soil porosity, soil compaction, soil mineralogy, and soil microbial community composition etc. Additionally, agricultural practices like land-use changes, plant residue management, agro-chemicals etc. influence soil organic carbon (SOC) stocks, either directly e.g. by altering the amount of C being added in the soil, or indirectly e.g. influencing soil aggregation and thereby accelerating microbial decomposition processes. Besides offsetting the rapidly increasing atmospheric GHGs, soil C sequestration may potentially improve the soil quality and advances the food security. It may play a crucial role in sustainable agriculture (SA) because it is highly sustainable and environment-friendly approach. It can enhance the soil quality by improving soil health parameters followed by improved crop production on sustainable basis.

M. Sanaullah (✉) · T. Afzal · A. Wakeel
Institute of Soil and Environmental Sciences, University of Agriculture, Faisalabad, Pakistan
e-mail: msanaullah@uaf.edu.pk

T. Shahzad
Department of Environmental Sciences and Engineering, Government College University Faisalabad, Faisalabad, Pakistan

M. Farooq, M. Pisante (eds.), *Innovations in Sustainable Agriculture*,
https://doi.org/10.1007/978-3-030-23169-9_15

Keywords Greenhouse gases · Carbon sequestration · Global warming · Soil texture · Soil structure · Soil porosity · Soil compaction · Soil mineralogy · Soil organic carbon · Sustainable agriculture

1 Introduction

Climate change is a continuous phenomenon of nature and the climate of earth has experienced it throughout history. The phenomenon of climate change may proceed at an unprecedented rate from decades to millennia IPCC (2007), but the current warming trend is crucial because >95% of its probability is the result of anthropogenic activities since the mid-twentieth century. Undoubtedly, specific ranges of GHGs induce warming in the Earth's atmosphere and make it livable for living-beings, however the current warming is happening approximately ten times faster than the average rate of ice-age-recovery warming (NRC 2007). The increasing concentration of GHGs in the atmosphere as a result of anthropogenic activities is the major driver behind the global warming (Santer et al. 1996). Carbon dioxide (CO_2) is the most prevalent and long-lived component of the atmosphere's gaseous composition. Current projections of climate models suggest that the rise in average surface temperature of planet is about 1 °C since the late nineteenth century and is largely attributed to higher atmospheric concentration of CO_2 (IPCC 2007). Globally, almost 50% of the warming effect is attributed to the sole CO_2 out of all GHGs (IPCC 2014). The CO_2 is delivered to atmosphere through natural processes such as soil respiration and volcano eruptions as well as through anthropogenic activities such as combustion of fossil fuels, deforestation and land-use changes. The anthropogenic CO_2 emission through humans has increased about threefold since the inception of industrial revolution. The industrial sector being a major contributor in human-emitted CO_2, has elevated the level of atmospheric CO_2 from 280 parts per million (ppm) to 400 ppm during the previous 150 years (IPCC 2014).

The sustainability in agriculture specifically lies on the principle that we must fulfill our present needs without making compromise on the ability of future generations to meet their demands. Sustainable agriculture is an ecosystem-based approach to agriculture while ensuring three basic intents i.e. healthy environment, securing economic profitability and maintaining social and economic equity. Sustainable management of agriculture may protect the biodiversity of ecosystem as well as fostering their development and maintenance. The sustainable approach to agriculture discourages the use of resource-intensive inputs i.e. toxic pesticides, synthetic fertilizers, genetically-modified seeds, and different malpractices that make soil, water, or other natural resources liable to degradation. It emphasizes on the utilization of recent resource-conservative techniques i.e. crop rotation, conservation tillage, and pasture-based livestock husbandry to ensure the efficient utilization of non-renewable resources. It is an integration of practices that are used for the acquirement of plant as well as animal production over long-term basis. Detrimental practices such as land degradation, excessive tilling of the soil (leading to erosion) and inadequate drainage (leading to salinization) pose potential risks to environment. The aim of sustainable agriculture is to replenish the soil while minimizing the reliance on the use or demand of non-renewable resources (Gold 2009).

Technically-sound and economically-feasible strategies are being demanded to lessen the effect and amount of global atmospheric CO_2 as well as improve the soil health and quality. Carbon sequestration is a promising approach to decrease the concentration of CO_2 in the atmosphere with the aim of offsetting global climate change along with improving the soil health for better plant growth. In broader sense, carbon sequestration describes both natural and deliberate processes by which CO_2 is either directly eliminated from the atmosphere or redirected from emission sources and stored in the C sinks i.e. ocean and terrestrial environments (vegetation, soils, and sediments). Carbon sequestration is typically related to the capture and storage of C in the terrestrial ecosystems as humic substances or carbonates particularly in subsoil for extended period of time, approximately 15–50 years (Eswaran et al. 1993). The net amount of C sequestered by a soil is the equilibrium between C uptake and C release on long-term basis.

The fulfillment of food demand of rapidly increasing population without any significant increase in the agricultural area and environmental degradation is a great challenge for agriculture sector (Stevenson et al. 2013). The C sequestration is one of the most stunning approaches to ensure the food security and agricultural sustainability along with the reduction of global atmospheric CO_2 level. Agricultural management practices i.e. minimum- or zero-tillage systems, application of synthetic fertilizers along with organic amendments, crop rotation, and crop residue incorporation can potentially escalate the soil C storage and ultimately agricultural sustainability (Six et al. 2002a).

This chapter briefly explores the impacts of different climatic and edaphic factors on soil C sequestration. The magnitude of C sequestration in soil and agricultural practices are closely interlinked. The impacts of different agricultural practices on soil C sequestration i.e. land-use changes, crop rotation, intensive tillage, plant residue management, organic amendments, agro-chemicals, and irrigation have been discussed in detail. In addition, the role of C sequestration in sustainable agriculture and its possible effects on soil health i.e. physical, chemical, and biological health have been described in detail. The scope of soil C sequestration towards SA in response to cropping intensity and patterns has been concisely discussed. Furthermore, the potential impacts of soil C sequestration on crop productivity and global environment have also been described.

2 Factors Affecting Carbon Sequestration

The quantity of carbon sequestered into soil can be regulated by various factors such as climate, soil parent material, topography, vegetation type, and soil microbial community composition (Jenny 1941). These factors can be categorized into climatic and edaphic factors as well.

2.1 Climatic Factors

The rising temperature, changing soil moisture regimes, and elevated CO_2 levels have drastic direct or indirect impact on soil microbial activities which can potentially influence the extent of soil C sequestration. Among climatic factors, temperature and precipitation are the major covariates that control soil organic carbon (SOC) accumulation under various climatic conditions and at different elevations. Heterotrophic microorganisms are the dominant agents for the decomposition of organic matter. This microbes-mediated decomposition process is highly influenced by climatic conditions i.e. temperature, moisture, elevated atmospheric CO_2 levels, and prevailing soil conditions which consequently lead to the release of plant nutrients particularly nitrogen (N), phosphorus (P), and sulfur (Murphy et al. 2007).

2.1.1 Precipitation

Soil moisture within the optimum range (field capacity to permanent wilting point) induces positive impacts on soil carbon sequestration because it enhances the vegetation on soil and favors plant growth ultimately increase the plant residue input in soil. Any fluctuation in soil moisture from the optimum range has negative impacts on soil C sequestration probably due to reduced addition of plant inputs. Limited moisture (<permanent wilting point) in the soil reduces the vegetation and likely negatively impacts the soil C sequestration. Drought influences the plant-microbes interactions and ecosystem functioning in terrestrial conditions (Sanaullah et al. 2011, 2012), it may also perturb the C partitioning between atmosphere and terrestrial ecosystems. For instance, longer period drought in North America in 2002 resulted in a decrease of C uptake from 650 to 330 Mt by soil and vegetation (Peters et al. 2007). The amount of C sequestered into soil is continuously decreasing since the inception of frequent drought and heat periods (Mearns et al. 1984).

The unsaturated/anaerobic conditions in soil also have reducing impacts on C sequestration owing to enhanced activity of anaerobes in the soil which respire C in the form of methane gas. Numerous studies documented that flooding condition in rice fields drastically influence the SOC dynamics by inducing anaerobic conditions through depletion of O_2 (Bronson et al. 1997). The change in SOC dynamics in flooded soil is governed through alteration in redox potential of soil, soil pH and by reduction of C, N, and sulfur (Fageria et al. 2011). The SOC is depleted from flooded soils as CO_2 and CH_4 via sequential oxidation-reduction reactions mediated by a variety of microbial species (Faulkner 2004).

The nitrous oxide (N_2O) emissions from nitrification and de-nitrification processes relatively occur at higher redox levels than the redox level of CH_4 production (Masscheleyn et al. 1993). This may induce a higher SOC loss and causes additional impact on global warming through the emission of N_2O emission in comparison with the continuously flooded systems. Similarly, alternate wetting and drying increases the labile SOC portion which is easily prone to loss followed by soil disturbance (Shrestha et al. 2002). Accordingly, intermittent drainage of flooded soil induces more rapid and larger loss of SOC, compared with continuously flooded soils due to high aerobic activities (Cassman et al. 1996).

2.1.2 Temperature

The soil C cycling processes are highly sensitive to temperature and even a small rise of temperature may have the potential to release a substantial amount of C from soil (Classen et al. 2015). Higher temperature enhances the decomposition of organic matter by accelerating the activities of soil microorganisms. The shifts in the activities, processing, and community composition of microbes are highly influenced by rising temperature (Bradford et al. 2008). The increase in soil microbial activities induced by high temperature could have positive feedback to global warming, whereas this increase can be negated in case of lower moisture conditions in soil.

The process of photosynthesis in all plants is highly sensitive to temperature. Increasing temperature has a negative impact on soil C sequestration possibly by adversely affecting the plant growth and through accelerating the microorganism activities in the soil. Higher SOC content in cooler climatic condition is attributed to moderate or low temperature and higher precipitation rates that favor the extensive vegetative growth due to lower soil organic matter (Ransom et al.) decomposition (Gami et al. 2001). Comparatively lower SOC in tropical regions than temperate ones is owing to its higher oxidation and mineralization rates under high ambient temperature (Ghimire et al. 2017).

In a 10-year soil warming study, the warming induced an increase in CO_2 flux by 28% in initial 6 years, while the impact of warming on CO_2 release was found non-significant in the subsequent years (Melillo et al. 2002). Dan et al. (2016) investigated the impact of temperature on the decomposition rate of SOM and they concluded that the changes in SOM decomposition were more sensitive to increasing temperature rather than changing soil moisture, and the thermal sensitivity of SOM (which expresses a change in SOM decomposition rate in response to 10 °C rise in soil temperature) was observed significant among different altitudes. In addition, the interaction of both moisture and temperature significantly accelerated the SOM decomposition.

2.1.3 Elevated CO_2

Elevated atmospheric CO_2 levels have both direct and indirect impacts on soil quality and plant growth. It is well-understood that elevated CO_2 enhances the plant growth and photosynthesis, particularly in case of highly fertile soils (Curtis and Wang 1998). It may increase the C flux to roots via rhizo-exudations of labile sugars, proteins, and organic acids (Zak et al. 1999). The increase in rhizo-depositions may indirectly influence the soil processes by promoting plant growth and returning atmospheric CO_2 to soil through biological disintegration of litter, roots, and rhizo-deposits (Kuikman et al. 1991; Van de Geijn and Veen 1993). Under elevated CO_2, alterations in the rates of addition and composition of these rhizo-deposits can strongly influence the composition, biomass, and activity of rhizosphere microorganisms. Elevation in atmospheric CO_2 levels may induce the alterations in the entire ecosystem. The microbe-soil-plant root system can also be modified by elevated atmospheric CO_2 level possibly by changing the soil moisture content and enhancing the root growth as well as rhizodeposition.

Elevated CO_2 has resulted in a relative increase in plant root to shoot ratios (Rogers et al. 1992) but this increase may vary appreciably among plant species, possibly owing to differences among plant species and environmental conditions. Patterson et al. (1996) demonstrated that an established pasture ecosystem had higher below-ground net primary productivity by 25% at 525 ppm CO_2 concentration, and 33% at 700 ppm CO_2 concentration, compared with a 350 ppm CO_2 concentration control. Elevated level of atmospheric CO_2 enhances the water-use efficiency (WUE) of plants *viz.* C fixed/water transpired. Consequently, higher root development induced by elevated CO_2 can improve the nutrient acquisition of plants in relatively lower amount of water (Rogers et al. 1992).

Marhan et al. (2010) investigated the influence of elevated atmospheric CO_2 for 5 years on net soil C sequestration in a temperate spring wheat agro-ecosystem. The observations included that elevated CO_2 levels pragmatically enhanced the plant C inputs into soil and resulted in accumulation of additional C (40–50 g C m^{-2}) relative to soil receiving only ambient CO_2.

2.2 *Edaphic Factors*

Compared to atmosphere, almost 2.5 times more C can be retained by the combination of soil and vegetation (Singh et al. 2010). The capacity of soil to entrap a large amount of C can potentially help to cushion the anthropogenic climate change by reducing the GHGs emissions. The C storage capacity of soils is nearly double than atmosphere which renders the soil a largest terrestrial sinks for carbon globally (Willey et al. 2009) which is largely dependent on various soil attributes, mentioned below:

2.2.1 Soil Texture

The clayey particles are negatively charged, and clayey soils have more micro-pores than macro-pores. The SOM can be more protected in clayey soils, as compared to coarse-textured ones. This higher protection of organic matter (OM) within clayey particles enhances the scope of soil C sequestration in these soils. The physical protection and stabilization of SOC in the mineral soils is largely associated with their cation exchange capacity (CEC) or clay contents (Hassink et al. 1997; Baldock and Skjemstad 2000). Similarly, Masscheleyn et al. (1993) proved the positive correlation between clay particles and protected SOC. The interactions of clay mineral and SOC are largely influenced by types of clay minerals and organic C present, and their respective specific surface areas (Krull et al. 2001). The specific surface area (SSA) of fine-textured soils is also higher which enhances the OM adsorption on clay particles due to their higher SSA than coarse-textured soils (Keil et al. 1994; Ransom et al. 1998).

2.2.2 Soil Structure

Soil structure renders biological stability to SOM via its impacts on the availability of water and oxygen (O_2), physical protection from microbial communities, and affecting the dynamics of soil aggregation (Krull et al. 2001). Free particulate organic matter (f-POM) is more prevalent in less-structured soils and therefore, higher OM degradation occurs in these soils. While occluded particulate organic matter (o-POM) occurs predominantly in well-structured soils which eventually results in reduced SOM decomposition. In this way, well-structured soils have positive impact on soil C sequestration, whereas poorly-structured soils may reduce C storage in the soil.

The encapsulation of organic C within aggregates renders it more preservation and stability against biological attack due to higher physical stability and lower turnover of aggregates (Amelung and Zech 1996). More protection of SOC has been achieved via development of macro-aggregates (>250 μm) in zero-tilled soils as compared with conventionally-tilled soils. Stable macro-aggregates in cultivated soils have been shown to contain more and relatively younger C than the C present in micro-aggregates (Six et al. 2002b). In case of higher aggregate stability and their lower turnover, physical protection of SOC within stable macro-aggregates is higher (Amelung and Zech 1996). The greater protection and aggregation were attributable to the intra-aggregate particulate organic matter (i-POM) which was relatively more in forested ecosystems than agricultural ones. This greater protection rendered to SOC within soil aggregates can potentially enhance the extent of soil C sequestration and aggregation of soil is imperative for long-term storage of C in soil.

2.2.3 Soil Porosity

The continuum of pores varying in size from micro-pores (<0.1 mm) to the macro ones (>20 mm) in soil constitutes the soil porosity. The processes of OM decomposition and mineralization are optimized by the enough availability of water and O_2 within soil pores. Clayey soil having higher proportion of micro-pores has pragmatic impacts on soil C sequestration possibly by providing protection against soil micro-biota. The alterations in the pore size distribution through progressive conversion of clayey soil to sandy soil having higher proportion of macro-pores which can enhance the mineralization of SOC possibly due to an increase in air-filled porosity (Franzleubbers 1999).

The pore size distribution of soils has strong impact on the access of microorganisms towards organic substances available within soil pores. Kilbertus (1980) implicated that soil bacteria can reach to the SOC present in pores having >3 mm size. For the decomposition of SOC present within pores having <3 mm size, microbes-secreted enzymes travels towards it via diffusion and the product of enzyme reaction also gets back towards the microbes through diffusion (Krull et al. 2001).

2.2.4 Soil Compaction

Soil compaction is the compression of soil particles into a smaller volume and reduced pore spaces for air and water. It can possibly reduce the SOC by declining the addition of plant litter and pooling of C in above soil layers. The extent of soil C sequestration is largely reduced in case of compacted soils possibly due to increase in soil bulk density and reduction in plant inputs to the soil. Brevika et al. (2002) investigated the impact of soil compaction for 150 years and found that bulk density was relatively higher in compacted soil from 5 to 20 cm soil depth than the non-compacted one. While, SOC was higher in non-compacted soil. On the other hand, C mineralization in compacted soil severely reduced at a bulk density of 1.6 Mg m^{-3} via reduction of microbial activity and the decrease of C mineralization in such soils helped in increased soil C sequestration (Tan and Chang 2007).

2.2.5 Soil Mineralogy

Soil C storage capacity is greatly affected by the soil mineralogical characteristics. The adsorption and protection capacity of clay minerals for organic C is potentially determined by the mineralogy, surface charge distribution as well as precipitation of Fe and Al oxides on clay minerals (Mizota and Reeuwijk 1989; Batjes 1996). The interactions of organic substances with clay minerals are caused by following two major mechanisms i.e. complexes formation through physical and chemical retention and stabilization via sorption of organic particles onto the surfaces of clay particles (cation and anion exchange, H-bonding, and polyvalent cation bridging) and physical stabilization of organic matter by its piercing into interlayer spaces of expanding clay minerals (Stevenson 1994). The stabilization of organic matter via these mechanisms induces its encapsulation and protection resulting in reduction of its accessibility to decomposing soil microbial community or their secreting enzymes. The role of aluminum (Al), iron (Fe), and allophane minerals increases the stability of OM by suppressing the microbial activities through their toxic impacts on microbes especially in Andosols (Varadachari et al. 1994; Parfitt et al. 2002).

2.2.6 Soil Microbial Community Composition

Soil microbes are the predominant drivers of cycling of C and nutrients in ecosystem and their activities are largely influenced by abiotic and biotic factors i.e. temperature, moisture, amount and quality of plant residue inputs etc. The ratio of fungi to bacteria can be altered due to changes in precipitation and soil moisture content (Williams 2007). A wide range of soil microbes i.e. fungi, bacteria, actinomycetes, and microalgae play a fundamental role in degradation of SOM, nutrient cycling, and various chemical transformations in soil (Murphy et al. 2007). The amount of C in soil becomes lower in case of an increase in the activity of soil communities possibly due to higher decomposition rates relative to inputs from plant litter, which may escalate the positive C feedback to atmosphere (Wieder et al. 2013). A large proportion of terrestrial C is stored by wetlands, peatlands, and permafrost likely due to limited microbial communities prevailing there (Castro et al. 2010). Any structural changes in microbial community composition can influence the SOM decomposition as well as their interactions with plants (Bardgett et al. 2008). Six et al. (2006) provided a convincing suggestion that more stable C can be formed in soil in case of higher prevalence of fungal biomass rather than bacterial ones. The fungi can form a higher biomass per unit of C assimilated due to higher C-use efficiency. The storage of C is considered more persistent in case of fungal dominance and more labile in case of bacterial dominance (Bailey et al. 2002).

3 Carbon Sequestration and Agriculture Practices

The carbon budget of a soil is the net result of the C entering as plant litter and rhizodeposition and leaving it in the form of CO_2 and CH_4. All the agricultural practices influence C sequestration in soil, either directly e.g. by modifying the amount of C which enters in soil, or indirectly e.g. modifying the aggregation thereby the protection afforded to SOC from microbial decomposition (Fig. 1). Moreover, the direction and the quantitative effect of an agricultural practice on carbon sequestration may vary across soil types. The significance and mechanisms by which different agricultural practices modify C sequestration are given below:

3.1 Land-Use Changes

Land-use changes determine the amount of time for which the land will remain covered with vegetation round the year. Moreover, the disturbance or absence thereof in the soil structure as a result of a particular land use determine the turnover rate of SOC.

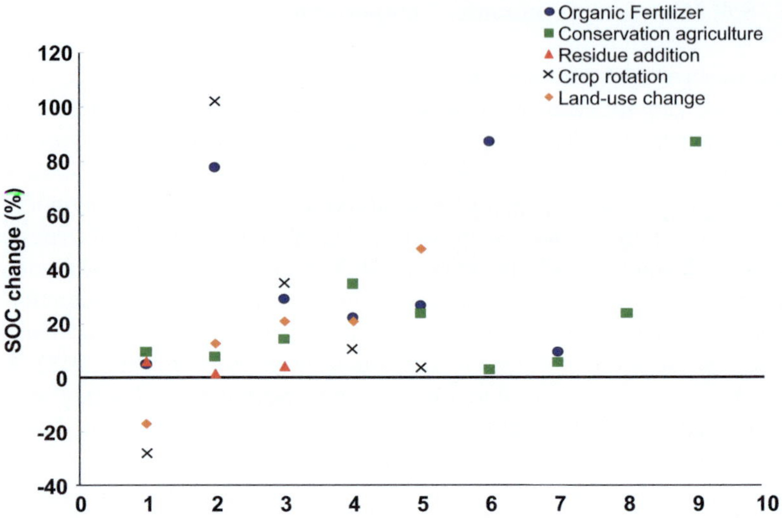

Fig. 1 Soil organic carbon changes (%) in response to different agricultural practices. (Sources: Lal 2004a, b, Sainju et al. 2008, Álvaro-Fuentes et al. 2009, Don et al. 2011, Ghimire et al. 2012, Parihar et al. 2017, Datta et al. 2018, Gai et al. 2018, Li et al. 2018, Liu et al. 2018, Swanepoel et al. 2018, Walia and Dick 2018, Wegner et al. 2018, and Zhang et al. 2018)

Deforestation or land clearing for agriculture decreases SOC stocks by halting or significantly reducing the amount of plant derived C input in soil and by increasing the decomposition of the SOC. For instance, a meta-analysis of 385 studies on land-use changes in tropics concluded that the 25%, 30% and 12% of SOC is lost when primary forest is converted into croplands, perennial croplands and pastures, respectively (Don et al. 2011). Moreover, intensive cropping to raise farm profits and ensure food supply has further depleted the SOC stocks around the globe (Lal 2004a, b). The conversion of degraded croplands into permanent pastures, shrub lands, or forests builds up part of the SOC being lost due to intensive agriculture and land clearing in the long run. Afforestation increases the SOC stocks with time since afforestation being the major determinant of the size of the increase (Laganière et al. 2010). The other factors which determine the rate of SOC sequestration are aggregation of soil is extremely indispensable for long-term storage of C in soil, soil type, previous land use, tree species planted (broadleaved *vs.* pine plantations) and soil clay content.

Converting a cropland to forest results in highest increase in SOC stocks among the land use change types i.e. 26–53% (Guo and Gifford 2002; Laganière et al. 2010). Moreover, the rate of SOC accumulation is higher in broadleaved (i.e. hardwood) than pine (i.e. softwood) and in evergreen than deciduous forests (Laganière et al. 2010; Deng et al. 2014).

In a grand 'Grain-for-green' program in China, research projects were carried out to evaluate the SOC stocks in degraded croplands converted into forests, shrublands or grasslands. In a meta-analysis of 135 publications on the subject in the said program, all the land-use changes from croplands to permanent vegetation cover significantly increased SOC stocks (Deng et al. 2014). Lands converted to forests showed a significant decrease in SOC in the first 5 years. However, the SOC stocks increased afterwards at varying rate for over 40 years of chrono-sequence and in surface (0–20 cm) and subsoil (20–100 cm) layers. The rates of soil C change were −0.93, 0.89, 1.30, 0.05 and 0.13 Mg ha^{-1} $year^{-1}$ for 0–20 cm soil during the periods 0–5, 6–10, 11–30, 31–40 and >40 years, respectively. The same were −3.15, 0.83, 3.59, 1.15 and 0.02 Mg ha^{-1} $year^{-1}$ for 20–100 cm during 0–5, 6–10, 11–30, 31–40 and >40 years, respectively. Similarly, the conversion from cropland to shrubland increases SOC stocks in the first 10 years of restoration. However, thereafter, gradual decrease was recorded till 30 years after restoration and the SOC was significantly higher after 30 years. The rates of change in SOC were again higher in 0–20 cm than 20–100 cm soil layer. Like shrublands, the grasslands established on degraded croplands showed increase in SOC in the first 30 years of restoration with the highest SOC values achieved in the 11–30 years' duration and afterwards SOC decreased significantly.

The effect of deforestation and subsequent land-use changes on soil organic C has been explored by Eleftheriadis et al. (2018) and a comparison between two lands; the first type of land had undergone deforestation since three dates (25, 34, and 72 years) and the other one was adjacent land which was still under forest, was made for the soil organic C status. The amount of coarse particulate organic matter (cPOM) was most significantly higher in forest plots (14.8 g kg^{-1}) against that of in cultivated plots (2.8 g kg^{-1}), compared with others soil health attributes. While the SOC in fine sand fraction in agricultural soil (8.6 g kg^{-1} soil) was markedly lesser than the same in forested soil (40.6 g kg^{-1} soil). In case of land-use changes, the shift from forest to cultivated soil caused a considerable decrease in SOC, N and cPOM contents of soil by 70%, 65% and 80% respectively.

A meta-analysis of 385 land-use change studies conducted in tropics showed that the conversion of croplands into any form of permanent vegetation cover i.e. forests, shrublands or grasslands significantly increased the SOC stocks (Don et al. 2011). Conversely, converting any land-use type into cropland resulted in loss of SOC stocks; where the SOC loss varied according to initial land-use type. For example, the highest SOC loss occurred when primary forests were converted into croplands (−25%), whereas the SOC loss was equal to −12% when the same are converted to grasslands. Afforestation on the same degraded croplands achieved SOC stocks even higher than the loss occurred due to deforestation in the first place (i.e. +50%). Conversion of forest initially into grasslands, and then after some years of grazing, converting that grassland into a cropland is a typical scenario in tropics. This cascade of land-use changes results into the loss of SOC at each conversion step. However, this loss is mainly restricted in the uppermost layer (~0 to 20 cm), and the subsoil usually maintains its SOC stocks due to higher inputs of crop residues in case conventional tillage is practiced (Don et al. 2011). The reconversion of croplands into grasslands, as in case of reconversion to forests, builds the lost SOC.

Hinge et al. (2018) investigated a change in regional carbon fluxes in response to different land-use changes in Northeast India. They experienced a varied trend between different regions in terms of effect of LUCs on SOC status. In some states, a net accrual of SOC by up to 3.91, 0.22, 0.13 Tg C was experienced. In contrast, net reduction in carbon biomass was observed by nearly 0.43, 1.51, 0.31, and 0.49 Tg C in some of the states. Importantly, the conversion of forest to grassland led to a marked increase in carbon emissions. The similar trend was observed upon shift from grassland to cultivated land. So, the importance of proper land management also arises to combat the climate change.

3.2 Crop Rotation

Growing of crops in succession on a piece of land to avoid exhausting the soil and suppress weeds, pests and diseases is known as crop rotation. The aboveground biodiversity that is observed in natural ecosystems on spatial scales is, therefore, absent in cropping systems and can only be found on temporal scale because of crop rotation. Greater biodiversity in prairie grasslands has been shown to increase primary productivity, resource use efficiency, nutrient availability and to a large extent ecosystem stability (Tilman et al. 1997; Lambers et al. 2004; Waldrop et al. 2006). In cropping systems, the suppression of weeds and diseases by adopting suitable rotation is known since long. Recently, several belowground benefits of adopting crop rotation e.g. enhanced nutrient cycling, SOC content, microbial biomass and activity, improvement in soil structure through aggregation etc. have been reported, and these co-benefits of crop rotation become more emphasized in combination with no-till practices (Nunes et al. 2018).

The legumes have the capability to become a potential alternative for mineral N fertilization. Inclusivity of legumes in cropping sequence has been reported to minimize the losses of C and N from different agricultural ecosystems (Drinkwater et al. 1998) along with the net accumulation of SOM, as SOC is incorporated into the soil because of higher rate of addition and turnover of biomass (Gregorich et al. 2001). Minimal application of mineral nitrogen is one of the most suggested strategies for climate-smart soil management (Paustian et al. 2016).

The combined influencing impact of inclusion of leguminous (clovers and beans) and non-leguminous (corn) in cropping sequence in combination with different mineral fertilization treatments (zero fertilization, PK fertilization and NPK fertilization) on the buildup of SOC and N was studied by Hobley et al. (2018). The results demonstrated that addition of legumes into crop rotation enhanced the C density with higher available K (>152.4 mg kg^{-1}) but in the absence of N fertilizer. During the 34 years of the study, the increase in profile SOC in the PK fertilized leguminous system demonstrated an average annual return of 4.1‰, compared with the corn-based system. In this way, the leguminous system in the absence of N fertilization was found well-aligned with the objectives of the 4 per 1000 imagination of the Lima-Paris Action Agenda (Chabbi et al. 2017). Parihar et al. (2017) found that inclusion of chickpea and mustard-mung bean into the cropping sequence enhanced the SOC storage by 102% and 34%, respectively (Fig. 1).

In a recent meta-analysis of 122 studies on crop rotation, adding one or more crops in a monoculture was found to increase SOC content by 3.6% and total N content by 5.3% (McDaniel et al. 2014). Another meta-analysis showed that enhancing the rotational diversity of crops can sequester 20 + 12 g C m^{-2} $year^{-1}$, significantly higher than monocropping or less diverse crop rotations (West and Post 2002). The quantity and quality of the plant derived organic matter entering into soils improves with increasing complexity of crop rotation thereby resulting into enhanced microbial processing and stabilization of the plant derived C as SOC (Dijkstra et al. 2006; Tiemann et al. 2015). The benefits of a crop rotation in terms of SOC and nutrient stocks and cycling depend on the type of the crops present in the rotation. For instance, adding a cover crop into a rotation increased the total SOC content by 8.5% and total N content by 12.8% in the foresaid meta-analysis (i.e. McDaniel et al. 2014), however the benefits are multiplied when legumes are used as cover crop. On the contrary, adding soybean which is a legume, in corn monocropping did not increase SOC than the corn monocropping (McDaniel et al. 2014), which underlines that the positive changes in soil functions brought by legumes cannot be generalized and that they depend on the residue chemistry, productivity and specific physiology of a legume crop.

Crop rotational diversity also improves the fast cycling of soil-carbon pools. For instance, it may enhance soil microbial biomass by 21%, an estimate based on meta-analysis of 122 studies (McDaniel et al. 2014). The improvement in microbial biomass and a change from bacteria-dominated to fungal-dominated microbial community structure considerably enhances the soil aggregation which in turn stabilizes and improves the sequestration of organic matter (Tiemann et al. 2015). Moreover, increasing the complexity of a crop rotation ensures the inputs of higher quality organic matter which has been shown to be major contributor to SOC (Cotrufo et al. 2015).

3.3 Intensive Tillage

Minimal to zero tillage operation to ensure minimum working of the soil are known as conservation tillage. The cultivation of forests and grasslands i.e. replacement of permanent vegetation cover with annual crops, has resulted into an estimated loss of 20% of the initial C stock or about 1500 g m^{-2} C has been lost from the top 30 cm of the soils (Mann 1986). Another estimate reported 30% loss of the initial C after 20 years of cultivation of a forest or grassland and most of this loss occurs during first 5 years of cultivation (Davidson and Ackerman 1993). Most of the loss of C is due to the intensive working on soils which exposes the soil-protected organic matter and improves aeration fostering microbial decomposition. Consequently, conservation tillage i.e. reduced till (RT) or no-till (NT) is proposed as a strategy to sequester C.

In 67 long term agricultural experiments, West and Post (2002) compared the rates of C sequestration among treatments pairs of conventional till vs. RT vs.

NT. They found that the C sequestration did not differ between conventional till and RT. However, shifting from conventional tillage to NT resulted into a significant increase in C sequestration representing 48 ± 13 g C m^{-2} $year^{-1}$. Moreover, this shift of tillage in wheat-fallow crop rotations did not change the C sequestration rates. Therefore, when the authors excluded this cropping system from their calculations, the C sequestration enhanced up to 57 + 14 g C m^{-2} $year^{-1}$ by shifting to NT. The authors also concluded that the change in tillage system for various crop rotations also differed in terms of C sequestration. Overall, the C sequestration benefits were significantly higher for all crop rotations compared to monocultures when the shift to NT was adopted. In the light of this study based on 93 paired treatments of conventional tillage versus NT from 67 experiments, it was concluded that the no till was indeed an excellent management practice for sequestering C in agro-ecosystems.

There was a caution in these conclusions which was revealed when the soil was sampled deeper than 30 cm in some long-term conservation tillage experiments, it was shown that the rooting densities are higher in upper 0–5 cm for no-till systems, whereas they are significantly higher in deep layers in the plowed soils (Qin et al. 2004). It is, therefore, intuitive to expect higher deposition of root-derived C in deep layers in plowed systems in addition to mixing of the surface residues in deep layers due to plowing. This indicated the importance of assessing the C sequestration in the surface and subsurface layers while evaluating the benefits of no-till. In a synthesis of studies comparing conventional till with no-till systems in Canada, accounting in the deep soil layers for C sequestration produced surprising revelations. When the sampling depth was 30 cm or less, 37 studies out of 45 reported higher SOC in NT than that of conventionally tilled soils. However, when the sampling depths was more than 30 cm, 35 of the 51 NT treatments showed less SOC than the conventional tilled plots. The similar SOC content in no-till and conventional tillage treatments were confirmed in a recent meta-analysis of 69 paired treatments of conventional versus no-till where soil was sampled deeper than 40 cm (Luo et al. 2010). This study found that adopting no-till significantly increased SOC by 3.15 ± 2.42 t ha^{-1} in the top 10 cm soil whereas it decreased the SOC in the 20–40 cm by 3.30 ± 1.61 t ha^{-1}. Overall, the adoption of no-till did not enhance SOC when compared to conventional tillage; it just re-distributed the SOC along the soil profile. Zhang et al. (2018) estimated the effects of no-tillage and conservation tillage practices on SOC dynamics and GHGs i.e. CO_2, N_2O emissions through modelling approach. They found that SOC stocks significantly increased (3755 ± 942 kg C ha^{-1} $year^{-1}$) in case of no-tillage practices. However, soils receiving conventional tillage also showed increased SOC content (3443 ± 1078 kg C ha^{-1} $year^{-1}$). The reduced SOC stocks under no-tillage system was attributed to lower rates of soil respiration. They concluded that no-tillage system with the inclusivity of crop residues can potentially enhance the SOC storage, as compared with traditional tillage practices. The adoption of no-tillage significantly enhanced the SOC stocks up to 23%, as compared with conventional tillage (Álvaro-Fuentes et al. 2009). Swanepoel et al. (2018) reported the impact of reduced tillage practices on SOC stocks and found that SOC contents increased from 54.9 to 57.9 Mg C ha^{-1} in response to reduced till practices. Ghimire et al. (2012) found that no tillage practices increased the SOC contents from 32.4 to 35.4 Mg C ha^{-1} in collation with traditional practices.

In the light of the above results, it can be concluded that the jury is still out on the effect of conservation tillage on overall C sequestration in a soil profile. While the no-till management safeguards SOC present in surface from microbial decomposition as well as enhances rhizodeposition, it gradually results in loss of the subsurface SOC. On the other hand, conventional tillage, while exposing the surface soil to aeration and temperature and facilitating accessibility of SOC for microbial decomposition, also enhances the root densities in the subsurface layers resulting in higher rhizodeposition. Moreover, it also mixes the surface residues into the subsurface layers thereby enhance the SOC in subsurface layers. While the other benefits of no-till or reduced till i.e. control of erosion, weed suppression etc., cannot be denied (Baker et al. 2007).

3.4 Plant Residue Management

The SOC is primarily result of plant derived carbon stabilized through microbial processing with its stability depending upon various edaphic, climatic and environmental factors. The quality and quantity of the input carbon also determines the SOC stocks. Extractive agricultural management practices -the removal of all or a major part of aboveground biomass from the field- are a norm in agriculture, particularly in developing countries (Lal 2004a, b). It is considered that these practices are making the world soils depleted in SOC and reversing these practices i.e. leaving or incorporating major portion of aboveground biomass of a crop in the field can restore and increase the SOC stocks (Lal 2004a, b, 2006). Increasing the crop residue inputs and static or reduced losses of C from soils should result in increased SOC until a new dynamic equilibrium is achieved after which increasing the inputs cause little SOC increase (Aune and Lal 1998). However, the increase in SOC in response to a certain amount of plant residue input depends on its conversion-efficiency into SOC which is modulated by many factors i.e. residue quality in terms of chemical complexity, tillage practices, climate, soil type and all of the management practices (Duiker and Lal 1999). Among these, the factors that can be controlled to a certain extent on a farm are tillage and other management practices if the aim is to sequester SOC.

In a long term field experiment of wheat straw application combined with different tillage methods, the SOC was found to be increased linearly for first 7 years of experiment in all tillage practices (Duiker and Lal 1999). However, in the 8th year of experiment, these trends continued for plow and no-till but not for the ridge till. Similarly, it is suggested that the SOC content are linearly related with the amount of residue inputs (Christopher and Lal 2007). However, in many studies it has been shown to be true for few early years; then on, there is little or no increase in SOC in response to increased crop residue inputs. For example, soil C levels were similar between residue-removed and residue-incorporated treatments after 10 years with the overall SOC decreasing than the beginning (Soon 1998). In an experiment spanning over 30 years, varying fertilizer treatments and rotations were compared with different treatments varying up to 50% in terms of residue return to the soil. Despite the large difference in the amount of residue return, the SOC levels at the end of the experiment did not differ (Campbell et al. 1991). Rumpel (2008) compared the SOC stocks and composition in residue-burned and residue-incorporated plots and found that the SOC stocks were similar between two residue management strategies even after 30 years. These results signify that the residue retention/incorporation does not always translate into increase soil C sequestration. Rather, sometimes the C input may also lead to a decrease in existing SOC– a phenomenon defined as the priming effect whereby addition of external organic matter accelerates the mineralization of existing SOC (Kuzyakov 2002; Fontaine et al. 2004; Shahzad et al. 2012). It appears that most of the SOC is the microbial-processed organic matter which needs a certain amount of nitrogen, phosphorus, and sulfur atoms against assimilation of a certain amount of C, H and O atoms (Richardson et al. 2014). These elements have fixed ratios (more or less) in different microbial groups and to maintain these ratios, which are known as stoichiometric ratios, soil microorganisms need them in certain proportion. If soil microorganisms have C source abundantly available but are limited by inorganic nutrients, e.g. nitrogen, they would mine/decompose the extant soil organic matter, which is usually rich than the plant residues, to obtain that nutrient in order to assimilate the externally added plant residues (Fontaine et al. 2011; Kaneez-e-Batool et al. 2016). In doing so, they destabilize the extant and stable SOC thereby resulting in neutral or negative SOC budget even after addition of plant residues. Similar to soil microorganisms, the C:N:P:S ratios of soil organic matter of a large dataset of world soils were found almost constant indicating that the microorganisms process them in a similar way in different soils according to their own stoichiometry (Kirkby et al. 2011). Moreover, these elements were found correlated with C content of the relevant soils indicating that the need of the nutrients of a particular soil to stabilize externally added plant residue can be estimated replenished (Kirkby et al. 2011). Similarly, Walia and Dick (2018) determined the impact of crop residue addition on SOC buildup and found that addition of crop residues along with mineral fertilizers increased the SOC storage from 4.38% to 4.44%.

Hence, it is evident that increasing the plant residue input to a soil could enhance soil C sequestration – if the mineral nutrients to stabilize the input C are sufficiently present in that soil.

3.5 *Organic Amendments*

Organic amendments cover the addition of animal manure (livestock, pig or poultry etc.), biosolids and sludge application in agricultural soils. Consistent amendment with animal manure over long term have been shown to markedly increase the soil

C stocks when compared to reference sites receiving or not receiving mineral fertilizers. For example, 10 years of poultry litter addition increased the surface SOC stocks by 3.2 Mg C ha^{-1} in comparison to mineral fertilized plots in southeastern USA (Sainju et al. 2008). Similarly, 25 years of annual cattle manure applications raised the SOC stocks by 19.1 Mg C ha^{-1} in surface soil (0–30 cm) compared to unfertilized plots in Nepal (Gami et al. 2009). In China, more than 3.8 Mg C ha^{-1} SOC stocks were recorded after 25 years of pig manure application than the mineral fertilizer alone treatments (Huang et al. 2010). Maillard and Angers (2014) conducted a meta-analysis of 42 articles reporting 130 observations from 49 sites around the world where animal manure (cattle, pig, and poultry) was applied for at least 3 years (average duration of the studies was 18 years). They compared the SOC stocks in manure applied treatments with mineral fertilized and unfertilized reference treatments. The average SOC stocks were higher by 9.4 Mg C ha^{-1} when compared to unfertilized treatments whereas they were higher by 5.6 Mg C ha^{-1} when compared to mineral fertilized treatments. Overall, the cumulative manure-C input was found in a strong linear relationship with the SOC stocks explaining 53% of variability in SOC when compared to mineral fertilized or unfertilized treatments. Li et al. (2018) conducted a long-term study for 26 years to evaluate the effects of organic manure and mineral fertilizers on SOC accumulation and reported a 86% increase in SOC stock through applying the organic manure in comparison with only mineral fertilizers. Gai et al. (2018) also found that applying organic manure caused 27% increase in SOC stocks for 22 years. Datta et al. (2018) conducted the study to compare the impact of mineral fertilizers and FYM on SOC sequestration rate. Compared with mineral fertilizers, integrative application of FYM and inorganic fertilizers significantly enhanced the total organic C of soil by 6.0%, and higher SOC sequestration rate (0.15 Mg C ha^{-1} $year^{-1}$) as well.

There are two ways whereby an organic amendment could increase SOC concentration and/or stocks: (i) direct contribution after microbial processing of the manure-C, (ii) enhanced organic C input from plants after the manure application raises the plant productivity thereby leading to higher plant derived C inputs in soil (Maillard and Angers 2014). Consequently, it may be argued that organic amendment may be more effective in enhancing SOC stocks than the plant-derived organic inputs i.e. straw application, residue incorporation etc. The conversion of part of the applied C into SOC is measured as manure/straw/residue-C retention coefficient (%), where higher retention coefficient means that higher percentage of the applied C is becoming part of the SOC.

Based on seven individual studies, Bhogal et al. (2009) calculated the C-retention coefficient for manure as 23% after 9 years of application whereas the same for crop residues was 22% after 23 years of application. This study, however, could not be a true comparison between these organic amendments possibly due to a large difference in the number of years following application of these amendments. In a relatively short-term study, where the application duration of sheep feces and crop residues was similar (9 years), C-retention coefficient was 30% for sheep feces and 19% for crop residues. Consequently, it can be argued that animal manure is more efficient for enhancing SOC stocks than crop residues (Thomsen and Christensen

2010). However, the meta-analysis of Maillard and Angers (2014) concluded that the C-retention coefficient for manure application studies was 12% for an average duration of 18 years of study. This low coefficient for animal manure C based on 42 long term studies indicates that it is incorrect to conclude that organic amendment in the form of manure is more efficient for enhancing the SOC stocks. However, in the short run, the animal manure is indeed more efficient than crop residues for enhancing SOC stocks.

3.6 Agro-chemicals

There are two major classes of agro-chemicals that are used in agriculture i.e. pesticides and fertilizers. The former does not affect the C cycling and storage directly, although they may affect it indirectly through influencing plant productivity and soil biodiversity which in turn determine the quality and quantity of plant C going into soils and the C cycling processes in soil, respectively (Pelosi et al. 2014).

Among chemical fertilizer nutrients, the interaction of nitrogen with C cycling has been extensively studied since both these cycles are tightly coupled in soils (Stockmann et al. 2015). External addition of organic matter stimulates the mineralization of existing organic C, a process known as priming effect (Kuzyakov 2010; Shahzad et al. 2012). Under limited N conditions, priming of existing organic matter upon C inputs decreases soil C stocks instead of increasing it. Whereas, excess N availability reduces priming effect in addition to stabilizing the externally added C in soil (Fontaine et al. 2004; Dijkstra et al. 2013).

Although, phosphorus (P) is the second most abundant mineral form of fertilizer that is applied to agricultural soils, its interaction with C-cycle in soil has been little studied. From the few studies that have been carried out on the topic, it emerges that the P deficient soils have lower basal respiration (Jing et al. 2017). Moreover, P deficiency leads to higher mineralization of the externally added C thereby reducing its retention as soil C (Jing et al. 2017). On the other hand, the additional availability of labile C empowers soil microbes to stimulate the release of P from unavailable forms into available ones (Spohn and Kuzyakov 2013; Guo et al. 2017). Sainju et al. (2008) evaluated the effects of nitrogen fertilizer sources and long-term tillage practices on SOC accrual. They reported that application of poultry litter enhanced the SOC stocks by 9% and C sequestration rate of 510 kg C ha^{-1} $year^{-1}$ compared with mineral N fertilization. They found that applying poultry litter can potentially increase the SOC stocks in the long-run while enhancing the soil quality (Fig. 1).

Another nutrient whose role in C retention and stabilization in soils has been studied to some extent is sulfur (S). The stoichiometric ratios of C:N:P:S across the world soils and particularly, in Australian soils have been found fairly consistent indicating two things: (i) the soil C is the microbially-processed C and the stabilization of plant-derived C as such does not seem plausible, (ii) the nutrient demand for C stabilization in a particular ratio is real (Kirkby et al. 2011).

Summing up, many studies showed that consistent inputs of C may not necessarily lead to increase the soil C stocks (Baker et al. 2007). However, the increased stabilization of soil C stocks is mediated by the increased availability of inorganic nutrients i.e. N, P and S (Richardson et al. 2014). Therefore, to build soil C from rich C inputs, management of inorganic nutrients is an important consideration.

3.7 Irrigation

Soil microbes need certain amount of moisture to perform their activities at optimum level. Their extracellular enzymatic activity is facilitated by presence of suitable amount of moisture (Schimel et al. 2011). Briefly, extracellular enzymes released by microbes diffuse towards soil C as well as the monomers formed after decomposition travel to microbes through diffusion (Xiang et al. 2008; Dungait et al. 2012). In agroecosystems, the moisture conditions do not remain same throughout a crop's cycle. There are drying-rewetting (DRW) cycles that are more acute in arid and semi-arid regions. These DRW cycles have varying effects on soil C mineralization. A spike of soil respiration occurs immediately following rewetting of a dry soil – a phenomenon is known since long as "Birch effect" (Birch 1958). This is an important process that occurs across all soil types although in varying amounts (Butterly et al. 2010). The extra mineralization may come from two sources; extra availability of soil C and microbial stress. Briefly, the rapid rewetting ensures the disruption of aggregates thereby releasing soil C for microbial decomposition. Moreover, gradual rewetting may also cause cracking thereby releasing extra substrates (Fierer and Schimel 2002; Schimel et al. 2011). According to microbial stress mechanism, the microbes accumulate solutes (osmolytes) inside their cells to retain the limited amount of moisture available in time of drought. The rapid release of C after rewetting has been attributed to these microbial solutes released from cells. Initially, it was thought that the cell death occurs in this process thereby resulting extra availability of soil C. However, the later investigations have shown that the cells survive this release of osmolytes following rewetting (Fierer and Schimel 2002; Schimel et al. 2007).

4 Role of Carbon Sequestration in Sustainable Agriculture

The complex interactions among climate and soil processes (physical, chemical, and biological) determine the level of organic C in soil at any time. Soil C sequestration may play a crucial role in sustainable agriculture because it is highly sustainable and eco-friendly approach. It does not only offset the anthropogenic rise of atmospheric CO_2 but also improves the soil quality. Improved soil quality is highly indispensable determinant of global food security.

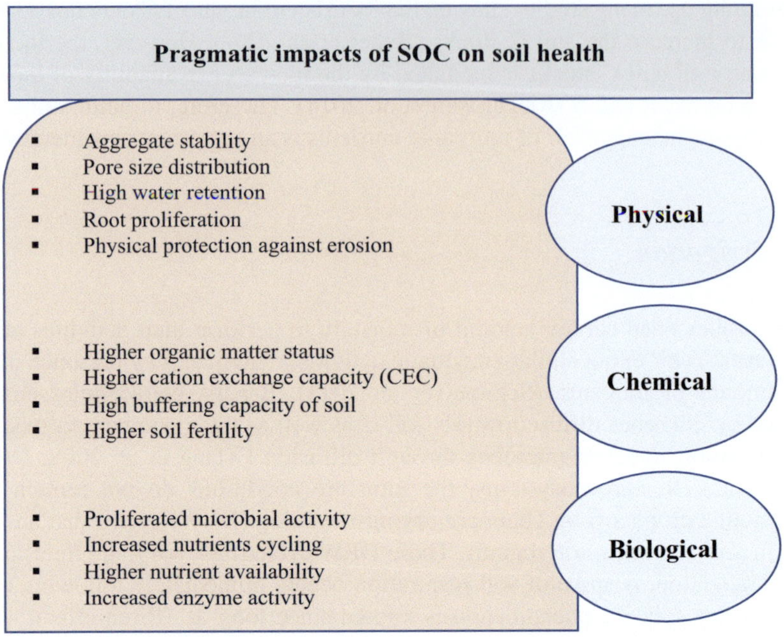

Fig. 2 The pragmatic impact of carbon sequestration on soil health parameters

4.1 *Effect on Soil Health*

Soil organic C has not only multifarious benefits on physical, chemical, and biological health of soil but also buffers the polluting substances present in the soil (Fig. 2). Soil carbon sequestration has two co-benefits i.e. mitigating the global warming by entrapping the atmospheric CO_2 as well as improving the soil quality and health. The balance between C inputs (plant residues and rhizo-deposition) and C outputs from soil via microbial degradation of SOM as well as plant respiration determines the absolute amount of available SOC. Under any climate and soil type, the rate of C incorporation into soil is a good indicator in determining the extent of C sequestration in soil (Paustian et al. 1997). The SOC is an indispensable component of the natural C cycle and soils can retain nearly twice as much C as in the atmosphere as well as vegetation on a global scale. Miller and Miller (2000) reported that application of organic matter to cropland can potentially impact the soil properties but the effects may not be temporally detectable.

4.1.1 Soil Physical Health

Aggregate stability is the driving factor for better soil fertility and can be escalated through appropriate management of organic amendments and improving the soil C content, which consequently improve the soil structure. The improvement in soil structure could enhance the pore size distribution of soil which is highly suitable for gaseous exchange, water retention, better root proliferation, and microbial activity (Van-Camp et al. 2004). Soils with greater proportion of organic matter are less vulnerable to water erosion than soils with lower proportion i.e. in arid and semi-arid regions (Durán and Pleguezuelo 2008). The SOM with labile components (easily decomposable) has transient and large impacts on aggregate stability, whereas the impacts of recalcitrant organic material (lignin and cellulose) are relatively lower but may persist for longer time periods.

The soil structure is possibly stabilized by these two mechanisms with concurrent increase of organic matter: (1) by improved cohesive forces among particles within aggregates and (2) increase in their hydrophobicity. While, Van-Camp et al. (2004) reported that the subsequent proliferation in microbial activity after the addition of organic matter may also enhance the stability of soil structure. Abiven et al. (2008) found that numerous biological agents have been observed responsible for the aggregation as well as aggregate stability. Microbially-synthesized polysaccharides especially at the inception of OM degradation may enhance the inter-particle cohesive forces and adsorption of mineral particles on organic residues. Thus, enhancing soil C content may improve the aggregation and structure of soil.

The soil aggregate stability was found higher (28.3%) in case of non-leguminous organic matter followed by their combined treatment (22.4%) at the end of experiment probably due to higher amount of humic acids which directly improved the formation of complexes between clay and organic matter (Tejada et al. 2008). Another study is evident that after ten cycles of rice-wheat cropping, the amount of hydrophobic aggregates was substantially higher in plots receiving rice-straw compost amendment in comparison with the inorganic amendments (Sodhi et al. 2009), the greater aggregate stability was assigned to intermittent incorporation of organic matter into soil which consequently enhance the microbial activity and formation of microbial products, leading to greater binding of soil particles.

4.1.2 Soil Chemical Health

Chemically, SOM has a strong influence on the CEC and the buffering capacity of soil for pH (Walsh and McDonnell 2012). Crop residue is a prime substrate for the accumulation of SOC and enhancement of soil productivity. The plant residual material upon decomposition and mineralization yield essential plant nutrients, and advances soil fertility.

The increase in soil C storage through the addition of organic amendments caused a considerable increase in the CEC of soil possibly due to the negative charge of soil organic matter (Kaur et al. 2008). This increase in CEC of soil by organic amendments improves the availability of nutrients to plants. The humic substances rank among chemically most active substances in the soil having cation and anion exchange capacities much larger than those of clay particles and being the long-lived decisive chemical component may persist in the soil from decades to millennia (Mayhew 2004). Numerous researches provide the evidences that consistent addition of organic matter into soil for a longer period caused an increase in both soil C and N stock, and enhanced the accumulation of C and N, rendering it higher stability likely due to aggregation (Sodhi et al. 2009).

4.1.3 Soil Biological Health

The SOM serves as the primary source of food and energy for soil microbial and plant biomass. A chemically and biologically fertile soil without the ability to physically support the plant growth is unable to attain required agronomic potential. Soil productivity is evaluated through the influence of SOM on the physical, chemical, and biological characteristics of the soil.

The decomposer microorganisms utilize the organic C which is subsequently either incorporated into microbial cells or released as carbon dioxide via soil respiration. Consequently, the macronutrients i.e. N, P, and S, present in the soil organic matter are mineralized. Later on, these inorganic nutrients are either immobilized or released into soil available nutrient pool (Baldock and Skjemstad 2000). The use of organic amendments i.e. composts and farm yard manures, can sustain the microbial biomass population for a relatively longer period of time possibly due to slow release of nutrients following their decomposition, as compared with inorganic fertilizers (Murphy et al. 2007). Another pragmatic impact of composts on the microbial activity was observed which tended to improve the higher availability of plants nutrients. Ginting et al. (2003) found the residual effects of composts and manures on soil microbial biomass after 4 years of application and documented that these organic amendments promote the microbial biomass from 20% to 40%, as compared with inorganic fertilizers. Research showed that single addition of municipal waste for 17 years caused an increment of 70% in the soil organic matter in semi-arid Mediterranean conditions (Bastida et al. 2008).

Several researches evidenced that humic substances positively influence the activities of enzymes i.e. urease, β-glucosidase, alkaline phosphatase, and o-diphenyloxidase, and thereby improve the soil health and biochemistry as well (Kaur et al. 2008). Both the quality and quantity of organic matter are important drivers for controlling the profusion and activities of microbial communities implicated in nutrient cycling. For instance, addition of composted material enhanced the organic C and enzyme activities more and at an accelerated rate than the addition of fresh paper mill residuals (Leon et al. 2006).

4.2 Cropping Pattern and Intensity

The impact of cropping intensity and fallowing of land on the rate of C sequestration was determined by Maysoon et al. (2010). The different crops *viz.* winter wheat, corn, pearl millet, dry pea, and fallows were included in the cropping intensities. The results showed that crop rotations with higher intensity caused a considerable increase in the level of soil C sequestration than those ones with fallows. The presence of dry pea (legume) in crop rotation induced non-significant impact on C sequestration when observed with rotations with summer fallows. The higher cropping intensity induced the formation of macro-aggregates with grasses in no-tilled soil, while aggregate development was recorded minimum in less intensive rotations. Higher macro-aggregates formation under higher cropping intensity favored the preservation of particulate organic matter and soil organic matter in most intensive crop rotations. Therefore, most intensive crop rotations posed pragmatic impacts on soil C storage as well as its protection and stabilization within soil aggregates.

Shukla et al. (2017) evaluated the level of soil C sequestration in two cropping systems *viz.* rice-wheat and sugarcane-ratoon-wheat. The observations showed that the bulk density was higher in rice-wheat system (1.145 Mg m^{-3}) than the sugarcane-ratoon-wheat system (1.10 Mg m^{-3}). Soil porosity was observed (58.68%) higher in sugar-ratoon-wheat system, as compared with rice-wheat system. The amount of C sequestered was found (19.51%) higher in sugarcane-ratoon-wheat system than the other one up to 0–15 cm soil depth. In addition, nutrient status and availability was also higher in the soil with sugarcane-ratoon-wheat system than the soil with other system. The management practices enhancing the soil C storage also improve the soil health and crop productivity. Different regions have variable potential to sequester the C probably due to variations in their climate, soil conditions, and cropping systems.

4.3 *Effect on Crop Yield and Productivity*

Behind offsetting the amount of atmospheric GHGs, both restoration and improvement of soil C pool are vital to increase the agronomic production as well as food security. Increasing C sequestration is indispensable to improve the soil quality and health as well as for greater efficacy of applied agricultural inputs (agro-chemicals and water). The lucrative impacts of precious farm inputs i.e. improved varieties, advanced agro-chemicals, and improved management practices can only be realized through improving the soil C status.

Soil C storage for long time poses beneficial impacts on plants, soil health and productivity and mitigating the climate change (Goh 2004). The pragmatic effects of SOM on crop productivity are attributed to its higher nutrient status, suitability for soil physical condition, favorable conditions for soil health and involvement in nutrient cycling processes. The grain yield and soil productivity improved with increasing SOC levels (Edmeades 2003). The benefits of increased soil C sequestration on soil health and productivity are congruent with their role in mitigation of climate change. Beneficial effects include SOM acting as a source of major plant nutrients, a promoter of favorable soil physical conditions, soil biota population, and nutrient cycling processes. Tittarelli et al. (2007) highlighted that approximation of organic matter and nutrient supply is the simplistic method to determine the agronomic efficiency of stabilized fraction of organic matter (humus). The agricultural management practices that enhance the soil organic matter content are considered sustainable practices because they simultaneously improve the environmental quality through preservation of farming output (Lal 2004b).

The fertility level of soil is highly based on the soil organic C which may yield huge amount of nutrients i.e. N, P, Ca, S, and Zn upon mineralization. The SOM as a highly reactive and omnipresent soil quality parameter, influences the physical stability as well as productivity of soil (Lal 2006). Regular addition of organic materials enhances the soil fertility with respect to its strong influence on the aggregate stability of soil and this rise in soil fertility possibly leads to potential higher soil productivity and cropping yield. Higher organic matter as a result of C sequestration improves the fertility status of soils by enhancing the supply and availability of nutrients through its decomposition. The increased C stock in soil has a pragmatic impact on agricultural soil and crop productivity and can potently ensure the food security.

4.4 Effect on Global Environment

Increasing SOM contents in soil has a great potential to cushion the climate change through potentially reducing the amounts of GHGs in the atmosphere. Current numerous studies provide the evidence that partially degraded and most of the agricultural soils can substantially minimize the raising global CO_2 levels. Currently, soil organic matter has been emerging as a critical component of soil for potentially acting as a source or sink of atmospheric GHGs responsible for global warming (Lal 2002). Globally, soils retain about twofold more C than that present in atmosphere or captured in vegetation (Trumbore et al. 1996). The extent of C sequestration varies from 100 to 1000 kg ha^{-1} $year^{-1}$ for soil organic C and 5–15 kg ha^{-1} $year^{-1}$ for soil inorganic C, reliant on the soil properties, climatic conditions, land-use changes, and cropping patterns (Lal 2009). Climate change prediction models project that the annual decrease of 3.5–4 Gt in CO_2 emission might have the potential to control the increase in temperature by 1.5–2 °C till 2050. But this annual reduction in atmospheric CO_2 concentration can be substantiated only by enhancing the C storage in the deeper layers of soils (Minasny et al. 2017).

5 Conclusion

Carbon sequestration is a promising approach to decrease the concentration of CO_2 in atmosphere with the aim of offsetting the global climate change and improving the soil health for better plant growth ensuring the food security. Amongst climatic factors, SOM decomposition is more sensitive to increasing temperature than precipitation and elevated CO_2 level. While, soil texture is the predominant edaphic factor that may affect SOC accumulation. The impact of agricultural practices on C sequestration largely differs among different soil types. Soil physical and biological characteristics are the major drivers that could potentially control the amount of SOC in response to different agricultural practices during a short time period. The loss of C through intensive tillage can be halted by adopting the conservation tillage or no-till practices; however, its benefits in increasing the organic fraction of soil can be enhanced substantially in combination with proper cropping sequence and pattern. The addition of legumes into the cropping schemes may be a suitable option to enhance the SOC storage in the soil. The strategy of C sequestration may be an advantageous tool in achieving the goals of sustainable agriculture because it not only improves the physical, chemical, and biological health of soil but also buffers the polluting substances present in soil.

References

Abiven S, Menasseri S, Chenu C (2008) The effects of organic inputs over time on soil aggregate stability – a literature analysis. Soil Biol Biochem 41:1–12

Álvaro-Fuentes J, López MV, Arrúe JL, Moret D, Paustian K (2009) Tillage and cropping effects on soil organic carbon in Mediterranean semiarid agroecosystems: testing the century model. Agric Ecosyst Environ 134:211–217

Amelung W, Zech W (1996) Organic species in ped surface and core fractions along a climosequence in the prairie, North America. Geoderma 74:193–206

Aune J, Lal R (1998) Agricultural productivity in he tropics and critical limits of properties of oxisols, ultisols and alfisols. Tropical Agriculture (Trinidad) 74:96–103

Bailey VL, Smith JL, Bolton H (2002) Fungal-to-bacterial ratios in soils investigated for enhanced C sequestration. Soil Biol Biochem 34:997–1007

Baker JM, Ochsner TE, Venterea RT, Griffis TJ (2007) Tillage and soil carbon sequestration-what do we really know? Agric Ecosyst Environ 118:1–5

Baldock JA, Skjemstad JO (2000) Role of the soil matrix and minerals in protecting natural organic materials. Org Geochem 31:697–710

Bardgett RD, Freeman C, Ostle NJ (2008) Microbial contributions to climate change through carbon cycle feedbacks. ISME J 2:805–814

Bastida F, Kandeler E, Hernández T, García C (2008) Long-term effect of municipal solid waste amendment on microbial abundance and humus-associated enzyme activities under semiarid conditions. Microb Ecol 55:651–661

Batjes NH (1996) Total carbon and nitrogen in the soils of the world. Eur J Soil Sci 47:151–163

Bhogal A, Nicholson FA, Chambers BJ (2009) Organic carbon additions: effects on soil bio-physical and physico-chemical properties. Eur J Soil Sci 60:276–286

Birch HF (1958) The effect of soil drying on humus decomposition and nitrogen availability. Plant Soil 10:9–31

Bradford MA, Davies CA, Frey SD, Maddox TR, Melillo J, Mohan JE, Reynolds JF, Treseder KK, Wallenstein M (2008) Thermal adaptation of soil microbial respiration to elevated temperature. Ecol Lett 11:1316–1327

Brevika E, Fentonb T, Moranb L (2002) Effect of soil compaction on organic carbon amounts and distribution, South-Central Lowa. Environ Pollut 116:137–141

Bronson KF, Neue HU, Singh U, Abao EB (1997) Automated chamber measurements of methane and nitrous oxide flux in a flooded rice soil: in residue, nitrogen, and water management. Soil Sci Soc Am J 61:981–987

Butterly CR, Marschner P, McNeill AM, Baldock JA (2010) Rewetting CO_2 pulses in Australian agricultural soils and the influence of soil properties. Biol Fertil Soils 46:739–753

Campbell CA, Zentner RP, Bowren KE, Townley-Smith L, Schnitzer M (1991) Effect of crop rotations and fertilization on soil organic matter and some biochemical properties of a thick Black Chernozem. Can J Soil Sci 71:377–387

Cassman KG, DeDatta SK, Amarante ST, Liboon SP, Samson MI, Dizon MA (1996) Long-term comparison of the agronomic efficiency and residual benefits of organic and inorganic nitrogen sources for tropical lowland rice. Exp Agric 32:427–444

Castro HF, Classen AT, Austin EE, Norby RH, Schadt CW (2010) Soil microbial community responses to multiple experimental climate change drivers. Appl Environ Microbiol 76:999–1007

Chabbi A, Lehmann J, Ciais P, Loescher HW, Cotrufo MF, Don A, SanClements M, Schipper L, Six J, Smith P, Rumpel C (2017) Aligning agriculture and climate policy. Nat Clim Chang 7:307

Christopher SF, Lal R (2007) Nitrogen management affects carbon sequestration in North American cropland soils. Crit Rev Plant Sci 26:45–64

Classen AT, Sundqvist MK, Henning JA, Newman GS, Moore JAM, Cregger MA, Moorhead LC, Patterson CM (2015) Direct and indirect effects of climate change on soil microbial and soil microbial-plant interactions: what lies ahead? Ecosphere 6:1–21

Cotrufo MF, Soong JL, Horton AJ, Campbell EE, Haddix ML, Wall DH, Parton WJ (2015) Formation of soil organic matter via biochemical and physical pathways of litter mass loss. Nat Geosci 8:776–779

Curtis PS, Wang X (1998) A meta-analysis of elevated CO_2 effects on woody plant mass, form, and physiology. Oecolegia 113:299–313

Dan W, Nianpeng H, Qing W, Yuliang L, Qiufeng W, Zhiwei X, Jianxing Z (2016) Effects of temperature and moisture on soil organic matter decomposition along elevation gradients on the Changbai Mountains, Northeast China. Pedosphere 26:399–407

Datta A, Mandal B, Badole S, A KC, Majumder SP, Padhan D, Basak N, Barman A, Kundu R, Narkhede WN (2018) Interrelationship of biomass yield, carbon input, aggregation, carbon pools and its sequestration in vertisols under long-term sorghum-wheat cropping system in semi-arid tropics. Soil Tillage Res 184:164–175

Davidson EA, Ackerman IL (1993) Changes in soil carbon inventories following cultivation of previously untilled soils. Biogeochemistry 20:161–193

Deng L, Liu Gb, Shangguan Zp (2014) Land-use conversion and changing soil carbon stocks in china's 'grain-for-green' program: a synthesis. Glob Chang Biol 20:3544–3556

Dijkstra FA, Cheng W, Johnson DW (2006) Plant biomass influences rhizosphere priming effects on soil organic matter decomposition in two differently managed soils. Soil Biol Biochem 38:2519–2526

Dijkstra FA, Carrillo Y, Pendall E, Morgan JA (2013) Rhizosphere priming: a nutrient perspective. Front Microbiol 4:1–8

Don A, Schumacher J, Freibauer A (2011) Impact of tropical land-use change on soil organic carbon stocks – a meta-analysis. Glob Chang Biol 17:1658–1670

Drinkwater LE, Wagoner P, Sarrantonio M (1998) Legume-based cropping systems have reduced carbon and nitrogen losses. Nature 396:262

Duiker SW, Lal R (1999) Crop residue and tillage effects on carbon sequestration in a luvisol in central ohio. Soil Tillage Res 52:73–81

Dungait JAJ, Hopkins DW, Gregory AS, Whitmore AP (2012) Soil organic matter turnover is governed by accessibility not recalcitrance. Glob Chang Biol 18:1781–1796

Durán ZVH, Pleguezuelo CR (2008) Soil-erosion and runoff prevention by plant covers. A review. Agron Sustain Dev 28:65–86

Edmeades DC (2003) The long-term effects of manures and fertilisers on soil productivity and quality: a review. Nutr Cycl Agroecosyst 66:165–180

Eleftheriadis A, Lafuente F, Turrión M-B (2018) Effect of land use, time since deforestation and management on organic C and N in soil textural fractions. Soil Tillage Res 183:1–7
Eswaran H, Berg EVD, Reich P (1993) Organic carbon in soils of the world. Nature
Fageria NK, Carvalho GD, Santos AB, Ferreira EPB, Knupp AM (2011) Chemistry of lowland rice soils and nutrient availability. Comun Soil Sci Plant Anal 42:1913–1933
Faulkner SP (2004) Soils and sediment: understanding wetland biogeochemistry. In: Spray SL, McGothlin KL (eds) Wetlands. Rowman & Littlefield Publishers, Lanham, pp 30–54
Fierer N, Schimel JP (2002) Effects of drying – rewetting frequency on soil carbon and nitrogen transformations. Soil Biol Biochem 34:777–787
Fontaine S, Bardoux G, Abbadie L, Mariotti A (2004) Carbon input to soil may decrease soil carbon content. Ecol Lett 7:314–320
Fontaine S, Henault C, Aamor A, Bdioui N, Bloor JMG, Maire V, Mary B, Revaillot S, Maron PA (2011) Fungi mediate long term sequestration of carbon and nitrogen in soil through their priming effect. Soil Biol Biochem 43:86–96
Franzleubbers AJ (1999) Microbial activity in response to water-filled pore space of variably eroded southern piedmont soils. Appl Soil Ecol 11:91–101
Gai X, Liu H, Liu J, Zhai L, Yang B, Wu S, Ren T, Lei Q, Wang H (2018) Long-term benefits of combining chemical fertilizer and manure applications on crop yields and soil carbon and nitrogen stocks in north china plain. Agric Water Manag 208:384–392
Gami SK, Ladha JK, Pathak H, Shah MP, Pasquin E, Pandey SP, Hobbs PR, Joshy D (2001) Long term changes in yield and soil fertility in a twenty year rice-wheat experiment in nepal. Biol Fertil Soils 34:73–78
Gami SK, Lauren JG, Duxbury JM (2009) Soil organic carbon and nitrogen stocks in nepal long-term soil fertility experiments. Soil Tillage Res 106:95–103
Ghimire R, Adhikari KR, Chen Z-S, Shah SC, Dahal KR (2012) Soil organic carbon sequestration as affected by tillage, crop residue, and nitrogen application in rice–wheat rotation system. Paddy Water Environ 10:95–102
Ghimire R, Lamichhane S, Acharya BS, Bista P, Sainju UM (2017) Tillage, crop residue, and nutrient management effects on soil organic carbon in rice-based cropping systems: a review. J Integr Agric 16:1–15
Ginting D, Kessavalou A, Eghball B, Doran JW (2003) Greenhouse gas emissions and soil indicators four years after manure and compost applications. J Environ Qual 32:23–32
Goh KM (2004) Carbon sequestration and stabilization in soils: Implications for soil productivity and climate change. Soil Sci Plant Nutr 50(4):467–476
Gold M (2009) What is sustainable agriculture? United States Department of Agriculture. Alternative Farming Systems Information Center
Gregorich EG, Drury CF, Baldock JA (2001) Changes in soil carbon under long-term maize in monoculture and legume-based rotation. Can J Soil Sci 81:21–31
Guo LB, Gifford RM (2002) Soil carbon stocks and land use change: a meta analysis. Glob Chang Biol 8:345–360
Guo H, Ye C, Zhang H, Pan S, Ji Y, Li Z, Liu M, Zhou X, Du G, Hu F, Hu S (2017) Long-term nitrogen and phosphorus additions reduce soil microbial respiration but increase its temperature sensitivity in a Tibetan Alpine Meadow. Soil Biol Biochem 113:26–34
Hassink J, Whitmore AP, Kubát J (1997) Size and density fractionation of soil organic matter and the physical capacity of soils to protect organic matter. Eur J Agron 7:189–199
Hinge G, Surampalli RY, Goyal MK (2018) Regional carbon fluxes from land-use conversion and land-use management in Northeast India. J Hazard Toxic Radioact Waste 22:8
Hobley EU, Honermeier B, Don A, Gocke MI, Amelung W, Kogel-Knabner I (2018) Decoupling of subsoil carbon and nitrogen dynamics after long-term crop rotation and fertilization. Agric Ecosyst Environ 265:363–373
Huang S, Peng X, Huang Q, Zhang W (2010) Soil aggregation and organic carbon fractions affected by long-term fertilization in a red soil of subtropical China. Geoderma 154:364–369

IPCC (2007) Solomon S, Qin D, Manning M, Chen Z, Marquis M, Averyt KB, Tignor M, Miller HL (eds). Climate change 2007: the physical science basis. Contribution of Working Group I to the fourth assessment report of the Intergovernmental Panel on Climate Change. Cambridge University.

IPCC (2014) Climate change 2014: mitigation of climate change, vol 3. Cambridge University Press

Jenny H (1941) Factors of soil formation. McGraw-Hill, New York

Jing Z, Chen R, Wei S, Feng Y, Zhang J, Lin X (2017) Response and feedback of C mineralization to P availability driven by soil microorganisms. Soil Biol Biochem 105:111–120

Kaneez-e-Batool N, Shahzad T, Mahmood F, Hussain S, Riaz M, Maqbool Z, Anwar F, Rehman K, Rashid MI (2016) Carbon mineralization in response to nitrogen and litter addition in surface and subsoils in an agroecosystem. Arch Agron Soil Sci 62:1–9

Kaur T, Brar BS, Dhillon NS (2008) Soil organic matter dynamics as affected by long-term use of organic and inorganic fertilizers under maize–wheat cropping system. Nutr Cycl Agroecosyst 81:59–69

Keil RG, Tsamakis E, Fuh CB, Giddings C, Hedges JI (1994) Mineralogical and textural controls on the organic composition of coastal marine sediments: hydrodynamic separation using splitt-fractionation. Geochemica et Cosmochimica Acta 58:879–893

Kilbertus G (1980) Microhabitats in soil aggregates. Their relationship with bacterial biomass and size of procaryotes present. Revue d'Ecologie et de Biologie du Sol 17:543–557

Kirkby CA, Kirkegaard JA, Richardson AE, Wade LJ, Blanchard C, Batten G (2011) Stable soil organic matter: a comparison of C:N:P:S ratios in australian and other world soils. Geoderma 163:197–208

Krull E, Baldock J, Skjemstad J (2001) Soil texture effects on decomposition and soil carbon storage. Paper read at Net Ecosystem Exchange CRC workshop proceedings

Kuikman PJ, Lekkerkerk LJA, Veen JAV (1991) Carbon dynamics of a soil planted with wheat under elevated atmospheric CO_2 concentration. In: Wilson WS (ed) Advances in soil organic matter research. The impact on agriculture and the environment. Redwood Press, Melksham, pp 267–274

Kuzyakov Y (2002) Review: factors affecting rhizosphere priming effects. J Plant Nutr Soil Sci 165:382–396

Kuzyakov Y (2010) Priming effects: interactions between living and dead organic matter. Soil Biol Biochem 42:1363–1371

Laganière J, Angers DA, Paré D (2010) Carbon accumulation in agricultural soils after afforestation: a meta-analysis. Glob Chang Biol 16:439–453

Lal R (2002) The potential of soils of the tropics to sequester carbon and mitigatc the greenhouse effects. Adv Agron 76:1–30

Lal R (2004a) Soil carbon sequestration impacts on global climate change and food security. Science 304:1623–1627

Lal R (2004b) Soil carbon sequestration to mitigate climate change. Geoderma 123:1–22

Lal R (2006) Enhancing crop yields in the developing countries through restoration of the soil organic carbon pool in agricultural lands. Land Degrad Dev 17:197–209

Lal R (2009) Soil degradation as a reason for inadequate human nutrition. Food Sec 1:45–57

Lambers JHR, Harpole WS, Tilman D, Knops J, Reich PB (2004) Mechanisms responsible for the positive diversity-productivity relationship in minnesota grasslands. Ecol Lett 7:661–668

Leon MCC, Stone A, Dick RP (2006) Organic soil amendments: impacts on snap bean common root rot (Aphanomyes euteiches) and soil quality. Appl Soil Ecol 31:199–210

Li J, Wen YC, Li XH, Li YT, Yang XD, Lin Z, Song ZZ, Cooper JM, Zhao BQ (2018) Soil labile organic carbon fractions and soil organic carbon stocks as affected by long-term organic and mineral fertilization regimes in the North China Plain. Soil Tillage Res 175:281–290

Liu W, Qiao C, Yang S, Bai W, Liu L (2018) Microbial carbon use efficiency and priming effect regulate soil carbon storage under nitrogen deposition by slowing soil organic matter decomposition. Geoderma 332:37–44

Luo Z, Wang E, Sun OJ (2010) Can no-tillage stimulate carbon sequestration in agricultural soils? A meta-analysis of paired experiments. Agric Ecosys Environ 139:224–231

Maillard É, Angers DA (2014) Animal manure application and soil organic carbon stocks: a meta-analysis. Glob Chang Biol 20:666–679

Mann LK (1986) Changes in soil carbon storage after cultivation. Soil Sci 142:279–288

Marhan S, Kandeler E, Rein S, Fangmeier A, Niklaus PA (2010) Indirect effects of soil moisture reverse soil C sequestration responses of a spring wheat agro-ecosystem to elevated CO2. Glob Chang Biol 16:469–483

Masscheleyn PH, DeLaune RD, Patrick WH (1993) Methane and nitrous oxide emissions from laboratory measurements from laboratory measurements of rice soil suspension: effects of oxidation-reduction status. Chemosphere 26:251–260

Mayhew L (2004) Humic substances in biological agriculture. Acres USA 34

Maysoon MM, Benjamin JG, Vigil MF, Nielson DC (2010) Cropping intensity impacts on soil aggregation and carbon sequestration in the Central Great Plains. Soil Sci Soc Am J 74:1712–1719

McDaniel MD, Tiemann LK, Grandy AS (2014) Does agricultural crop diversity enhance soil microbial biomass and organic matter dynamics? A meta-analysis. Ecol Appl 24:560–570

Mearns LO, Katz RW, Schneider SH (1984) Extreme high-temperature events: changes in their probabilities with changes in mean temperature. J Clim Appl Meteorol 23:1601–1613

Melillo JM, Steudler PA, Aber JD, Newkirk K, Lux H, Bowles FP, Catricala C, Magill A, Ahrens T, Morrisseau S (2002) Soil warming and carbon-cycle feedbacks to the climate system. Science 298:2173–2176

Miller DM, Miller WP (2000) Chapter 9: land application of wastes. In: Sumner M (ed) Handbook of soil science. CRC Press

Minasny B, Malone BP, McBratney AB, Angers DA, Arrouays D, Chambers A, Chaplot V, Chen Z-S, Cheng K, Das BS, Field DJ, Gimona A, Hedley CB, Hong SY, Mandal B, Marchant BP, Martin M, McConkey BG, Mulder VL, O'Rourke S, Richer-de-Forges AC, Odeh I, Padarian J, Paustian K, Pan G, Poggio L, Savin I, Stolbovoy V, Stockmann U, Sulaeman Y, Tsui C-C, Vågen T-G, van Wesemael B, Winowiecki L (2017) Soil carbon 4 per mille. Geoderma 292:59–86

Mizota C, Reeuwijk LPV (1989) Clay mineralogy and chemistry of soils formed under volcanic material in diverse climatic regions, Soil monograph 2. ISRIC, Wageningen

Murphy DV, Tockdale EA, Brookes PC, Goulding KWT (2007) Impact of microorganisms on chemical transformation in soil. In: Abbott LK, Murphy DV (eds) Soil biological fertility – a key to sustainable land use in agriculture. Springer, Dordrecht, pp 37–59

NRC (2007) Surface temperature reconstructions for the last 2,000 years. National Academies Press, Washington, DC

Nunes MR, van Es HM, Schindelbeck R, Ristow AJ, Ryan M (2018) No-till and cropping system diversification improve soil health and crop yield. Geoderma 328:30–43

Parfitt RL, Parshotam A, Salt GI (2002) Carbon turnover in two soils with contrasting mineralogy under long-term maize and pasture. Aust 1 Soil Res 40:127–136

Parihar CM, Yadav MR, Jat SL, Singh AK, Kumar B, Pooniya V, Pradhan S, Verma RK, Jat ML, Jat RK, Saharawat YS (2017) Long term conservation agriculture and intensified cropping systems: effect on growth, yield, water and energy-use efficiency of maize in North-Western India. Pedosphere

Patterson E, Rattray EAS, Kellham K (1996) Effect of elevated atmospheric CO_2 concentration on c-partitioning and rhizosphere C for three plant species. Soil Biol Biochem 28:195–201

Paustian K, Collins HP, Paul EA (1997) Management controls on soil carbon. In: Paul EA, Paustian K, Elliot ET, Cole CV (eds) Organic matter in temperate agroecosystems. CRC Press, Boca Raton, pp 15–49

Paustian K, Lehmann J, Ogle S, Reay D, Robertson GP, Smith P (2016) Climate-smart soils. Nature 532:49

Pelosi C, Barot S, Capowiez Y, Hedde M, Vandenbulcke F (2014) Pesticides and earthworms. A review. Agron Sustain Dev 34:199–228

Peters W, Jacobson AR, Sweeney C, Andrews AE, Conway TJ, Masarie K, Miller JB, Bruhwiler LMP, Pétron G, Hirsch AI, Worthy DEJ, van der Werf GR, Randerson JT, Wennberg PO, Krol MC, Tans PP (2007) An atmospheric perspective on north american carbon dioxide exchange: Carbontracker. Proc Natl Acad Sci 104:18925–18930

Qin R, Stamp P, Richner W (2004) Impact of tillage on root systems of winter wheat. Agron J 96:1523–1530

Ransom B, Kim D, Kastner M, Wainwright S (1998) Organic matter preservation on continental slopes: importance of mineralogy and surface area. Geochim Cosmochim Acta 62:1329–1345

Richardson AE, Kirkby CA, Banerjee S, Kirkegaard JA (2014) The inorganic nutrient cost of building soil carbon. Carbon Manag 5:265–268

Rogers HH, Peterson CM, McCrimmon JN, Cure JD (1992) Response of plant roots to elevated carbon dioxide. Plant Cell Environ 15:749–752

Rumpel C (2008) Does burning of harvesting residues increase soil carbon storage. J Soil Sci Plant Nutr 8:44–51

Sainju UM, Senwo ZN, Nyakatawa EZ, Tazisong IA, Reddy KC (2008) Soil carbon and nitrogen sequestration as affected by long-term tillage, cropping systems, and nitrogen fertilizer sources. Agric Ecosyst Environ 127:234–240

Sanaullah M, Blagodatskaya E, Chabbi A, Rumpel C, Kuzyakov Y (2011) Drought effects on microbial biomass and enzyme activities in the rhizosphere of grasses depend on plant community composition. Appl Soil Ecol 48:38–44

Sanaullah M, Chabbi A, Rumpel C, Kuzyakov Y (2012) Carbon allocation in grassland communities under drought stress followed by 14C pulse labeling. Soil Biol Biochem 55:132–139

Santer BD, Wigley TML, Barnett TP, Anyamba E (1996) A search for human influences on the thermal structure of the atmosphere. Nature 382:39–46

Schimel J, Balser TC, Wallenstein M (2007) Microbial stress-response physiology and its implications for ecosystem function. Ecology 88:1386–1394

Schimel JP, Wetterstedt JÅM, Holden Pa, Trumbore SE (2011) Drying/rewetting cycles mobilize old C from deep soils from a California annual grassland. Soil Biol Biochem 43:1101–1103

Shahzad T, Chenu C, Repinçay C, Mougin C, Ollier J-L, Fontaine S (2012) Plant clipping decelerates the mineralization of recalcitrant soil organic matter under multiple grassland species. Soil Biol Biochem 51:73–80

Shrestha RK, Ladha JK, Lefroy RDB (2002) Carbon management for sustainability of an intensively managed rice-based cropping system. Biol Fertil Soils 36:215–223

Shukla SK, Swaha S, Maity SK, Solomon S, Awasthi SK, Gaur A, Pathak AD, Jaiswal VP (2017) Soil carbon sequestration and crop yields in rice–wheat and sugarcane–ratoon–wheat cropping systems through crop residue management and inoculation of Trichoderma viride in subtropical India. Sugar Tech 17:347–358

Singh HB, Anderson BE, Brune WH, Cai C, Cohen RC, Crawford JH, Cubison MJ, Czech EP, Emmons L, Fuelberg H, Huey G, Jacob DJ, Jimenez JL, Kaduwela A (2010) Pollution influences on atmospheric composition and chemistry at high northern latitudes: Boreal and California forest fire emissions. Atmos Environ 44:4553–4564

Six J, Feller C, Denef K, Ogle SM, Sa JCD, Albrecht A (2002a) Soil organic matter, biota and aggregation in temperate and tropical soils – effects of no-tillage. Agronomie 22:755–775

Six J, Callewaert P, Lenders S, de Gryze S, Morris SJ, Gregorich EG, Paul EA, Paustian K (2002b) Measuring and understanding carbon storage in afforested soils by physical fractionation. Soil Sci Soc Am J 66:1981–1987

Six J, Frey SD, K TR, Batten KM (2006) Bacterial and fungal contributions to carbon sequestration in agroecosystems. Soil Sci Soc Am J 70:555–569

Sodhi GPS, Beri V, Benbi DK (2009) Soil aggregation and distribution of carbon and nitrogen in different fractions under long-term application of compost in rice–wheat system. Soil Tillage Res 103:412–418

Soon YK (1998) Crop residue and fertilizer management effects on some biological and chemical properties of a Dark Grey Solod. Can J Soil Sci 78:707–713
Spohn M, Kuzyakov Y (2013) Phosphorus mineralization can be driven by microbial need for carbon. Soil Biol Biochem 61:69–75
Stevenson FJ (1994) Humus chemistry, genesis, composition reactions. Wiley, New York, pp 429–471
Stevenson JR, Villoria N, Byerlee D, Kelley T, Maredia M (2013) Green revolution research saved an estimated 18 to 27 million hectares from being brought into agricultural production. Proc Natl Acad Sci USA 110:8363–8368
Stockmann U, Adams MA, Crawford JW, Field DJ, Henakaarchchi N, Jenkins M, Minasny B, McBratney AB, Remy VD, Courcelles D, Singh K, Wheeler I, Abbott L, Angers DA, Baldock J, Bird M, Brookes PC, Chenu C, Jastrow JD, Lal R, Lehmann J, Donnell AGO, Parton WJ, Whitehead D, Zimmermann M (2015) The knowns, known unknowns and unknowns of sequestration of soil organic carbon. Agric Ecosyst Environ 164:80–99
Swanepoel CM, Rötter RP, van der Laan M, Annandale JG, Beukes DJ, du Preez CC, Swanepoel LH, van der Merwe A, Hoffmann MP (2018) The benefits of conservation agriculture on soil organic carbon and yield in Southern Africa are site-specific. Soil Tillage Res 183:72–82
Tan X, Chang SX (2007) Soil compaction and forest litter amendment affect carbon and net nitrogen mineralization in a boreal forest soil. Soil Tillage Res 93:77–86
Tejada M, Gonzalez JL, García-Martínez AM, Parrado J (2008) Application of a green manure and green manure composted with beet vinasse on soil restoration: effects on soil properties. Bioresour Technol 99:4949–4957
Thomsen IK, Christensen BT (2010) Carbon sequestration in soils with annual inputs of maize biomass and maize-derived animal manure: evidence from 13C abundance. Soil Biol Biochem 42:1643–1646
Tiemann LK, Grandy AS, Atkinson EE, Marin-Spiotta E, McDaniel MD (2015) Crop rotational diversity enhances belowground communities and functions in an agroecosystem. Ecol Lett. n/a–n/a
Tilman D, Knops J, Wedin D, Reich P, Ritchie M, Siemann E (1997) The influence of functional diversity and composition on ecosystem processes. Science 277:1300–1302
Tittarelli F, Petruzzelli G, Pezzarossa B, Civilini M, Benedetti A, Sequi P (2007) Quality and agronomic use of compost. In: Diaz LF, de bertoldi M, Bidlingmaier W, Stentiford E (eds) Compost science and technology. Waste management series, vol 8. Elsevier, pp 119–145
Trumbore SE, Chadwick OA, Amundson R (1996) Rapid exchange between soil carbon and atmospheric carbon dioxide driven by temperature change. Science 272:383–396
Van de Geijn SC, Veen JAV (1993) Implications of increased carbon dioxide levels for carbon input and turnover in soils. Vegetatio 104/105:283–292
Van-Camp L, Bujarrabal B, Gentile AR, Jones RJA, Montanarella L, Olazabal C, Selvaradjou SK (2004) Reports of the Technical Working Groups established under the thematic strategy for soil protection. Office for Official Publications of the European Communities, Luxembourg, p 872
Varadachari CH, Mondal AH, Nayak DC, Ghosh K (1994) Clay-humus complexation: effect of ph and the nature of bonding. Soil Biol Biochem 26:1145
Waldrop MP, Zak DR, Blackwood CB, Curtis CD, Tilman D (2006) Resource availability controls fungal diversity across a plant diversity gradient. Ecol Lett 9:1127–1135
Walia MK, Dick WA (2018) Selected soil physical properties and aggregate-associated carbon and nitrogen as influenced by gypsum, crop residue, and glucose. Geoderma 320:67–73
Walsh E, McDonnell K (2012) The influence of measurement methodology on soil infiltration rate. Int J Soil Sci 7:168–176
Wegner BR, Osborne SL, Lehman RM, Kumar S (2018) Seven-year impact of cover crops on soil health when corn residue is removed. BioEnergy Res 11:239–248
West TO, Post WM (2002) Soil organic carbon sequestration rates by tillage and crop rotation: a global data analysis. Soil Sci Soc Am J 66:1930–1946

Wieder WR, Bonan GB, Allison SD (2013) Global soil carbon projections are improved by modelling microbial processes. Nat Clim Chang 3:909–912
Willey JM, Sherwood LM, Woolverton CJ (2009) Prescott's principles of microbiology. McGraw-Hill, New York
Williams MA (2007) Response of microbial communities to water stress in irrigated and drought-prone tallgrass prairie soils. Soil Biol Biochem 39:39
Xiang S-R, Doyle A, Holden PA, Schimel JP (2008) Drying and rewetting effects on C and N mineralization and microbial activity in surface and subsurface California grassland soils. Soil Biol Biochem 40:2281–2289
Zak DR, Holmes WE, MacDonald NW, Pregitzer KS (1999) Soil temperature, matrix potential, and the kinetics of microbial respiration and nitrogen mineralization. Soil Sci Soc Am J 63:575–584
Zhang JY, Sun CL, Liu GB, Xue S (2018) Effects of long-term fertilisation on aggregates and dynamics of soil organic carbon in a semi-arid agro-ecosystem in China. PeerJ 6:20

Use of Biochar in Sustainable Agriculture

Nirmali Gogoi, Banashree Sarma, Subham C. Mondal, Rupam Kataki, and Ankit Garg

Abstract Biochar is an emerging soil amendment in agriculture with many facets contributing towards agricultural sustainability. The increased population burden had shrunken the global cultivable area, putting tremendous pressure in agricultural productivity. This has led to an increased use of chemicals in the form of pesticides, herbicides, insecticides or inorganic fertilizers polluting the whole environment. Increased use of inorganic nitrogenous fertilizer sources lead to leaching of nitrogen that contaminate the water bodies and deteriorate soil heath. It also increases emission of greenhouse gas (GHG) nitrous oxide from the agricultural fields contributing towards global warming. Use of biochar in agriculture has shown encouraging results in mitigating soil pollution and decreasing soil acidity. Reduced greenhouse gas emission and improved soil fertility is obtained under biochar application due to its physico-chemical properties such as higher porosity, alkalinity and nutrient contents. Thus, the role of biochar in soil fertility, pollution remediation, greenhouse gas emissions, abiotic stress and disease management makes it an important tool of sustainable agriculture.

Keywords Biochar · Carbon sequestration · Agriculture · GHG · Pollutants · Soil fertility

N. Gogoi (✉) · S. C. Mondal
Department of Environmental Science, Tezpur University, Tezpur, Assam, India
e-mail: nirmali@tezu.ernet.in

B. Sarma (✉)
Nature Learning Center- Assam, Assam State Biodiversity Board, Guwahati, Assam, India

R. Kataki
Department of Energy, Tezpur University, Tezpur, Assam, India

A. Garg
Department of Civil and Environmental Engineering, Shantou University, Shantou, China

M. Farooq, M. Pisante (eds.), *Innovations in Sustainable Agriculture*,
https://doi.org/10.1007/978-3-030-23169-9_16

1 Introduction

Excessive use of inorganic fertilizers in agriculture for yield improvement has contributed immensely to greenhouse gas emission leading to global warming, climate change as well as soil and water pollution (Zheng et al. 2007; Annabi et al. 2011). Moreover, long-term cultivation builds soil acidity, deplete soil organic matter stock, and erode the top soil layer leading to formation of degraded land (De Meyer et al. 2011). Thus, the rising concerns about these global environmental issues necessitate the use of alternative sources of fertilizers. In addition, the requirement of a sustainable agricultural system along with economic improvement of the farmers demanded key changes in agricultural crop management practices. Manures and composts have been used as alternative sources of inorganic fertilizers from long back. But, they contain pathogens and have the potentiality to release the greenhouse gases i.e. methane, nitrous oxide, etc. (Lehmann et al. 2011) and increase the mineralization of SOC stock (Sarma et al. 2017a). In recent studies, biochar is identified as a promising resource for soil's fertility management (Lone et al. 2015). Biochar also gained attention for its higher adsorption capacity and extensively studied to reduce the pesticide bioavailability (Ahmad et al. 2014; Khorram et al. 2016). Being a renewable source of amendment as well as its agronomic, environmental and economic benefits (Fig. 1), the importance of biochar is expanding globally (Liu et al. 2013; Stavi and Lal 2013).

The production of biochar is done through the process of pyrolysis in an oxygen-deficient environment where the thermal conversion of biomass and organic material is completed to a charred compound at varying temperatures (Joseph et al. 2010). The produced biochar is highly porous, rich in carbon (C) and also contains other plant nutrients like nitrogen (N), phosphorous (P), sulfur (S) along with ash, hydrogen (H), oxygen (O), (Duku et al. 2011; Lehmann and Joseph 2015). The

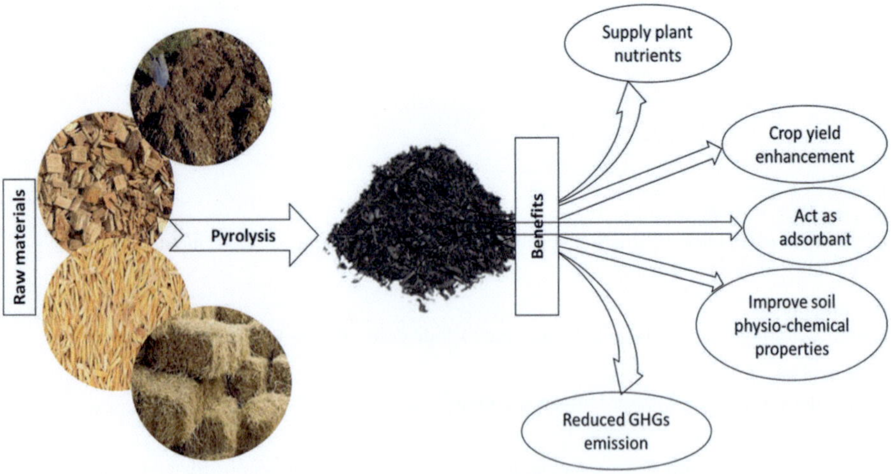

Fig. 1 Potential applications of biochar application in sustainable agriculture

porous surface of biochar also contains numerous functional groups (Lehmann and Joseph 2009) and extractable fulvic- and humic-like elements (Lin et al. 2012). Furthermore, due to the presence of more aromatic C, the molecular structure of biochar possesses higher degree of stability both in terms of microbially and chemically (Cheng et al. 2008).

The production conditions like duration of pyrolysis, temperature during pyrolysis, type of raw materials used are the dominant parameters governing the characteristics of biochar (Joseph et al. 2010; Bruun et al. 2011). A variety of raw materials such as hard- and soft-wood chips, agricultural waste, sewage sludge, organic waste, industrial waste, manure especially poultry manure, etc. are being used as feed stock for biochar production (Sohi et al. 2010; Hossain et al. 2010; Yuan et al. 2011; Khan et al. 2017; Yang et al. 2017).

The analysis of biochar has exhibited the presence of H, N and micro-nutrients like Mg, Ca, Na, Zn, Fe (Zhang et al. 2015; Sarma et al. 2017b). Studies reported an increase in ash and C content of biochar and surface area with increasing pyrolysis temperature from 300 to 600 °C (Tan et al. 2017). Biochar also has a number of polar and non-polar groups, which help in adsorbing heavy metal ions and nitrates (Schmidt et al. 2015; O'Connor et al. 2018).

1.1 Effects on Soil Nutrient Status/Soil Fertility

Biochar affects soil fertility in numerous ways. It can add nutrients by itself or make them more available for plant uptake or reduce decomposition rates of other organic material and thereby increasing soil C concentration in long run. Moreover, the large surface area of biochar facilitates increased cation exchange capacity (CEC), which in turn restricts nutrient leaching (Lehmann and Joseph 2009). Lehmann et al. (2003) also reported a significant decrease in leaching of applied fertilizers after addition of biochar with an increased plant uptake of nutrients like P, K and Ca. The improved CEC of soil after biochar application help in absorbance of added fertilizers to the biochar surfaces and thereby ensure nutrient sufficiency to plants (Steinbeiss et al. 2009). The availability of nutrients and its uptake are affected due to alteration of soil pH because of biochar addition (Lehmann and Joseph 2009). The basic ions in biochar also have a significant effect on soil pH. The base ions in biochar exchange with the H^+ and Al^{3+} ions, that decreases the acidity of soil and increase soil alkalinity (Van Zwieten et al. 2010a). The nutrient availability in biochar amended soil is a function of many factors including the type of feedstock used for the production of biochar (Lehmann et al. 2003). For example, the total nutrient concentration in biochar can be high, however the proportion of plant available nutrients vary among different biochars. Nutrients like N and S in organic compounds, are tightly bound and therefore less available to plants, which has been demonstrated in previous studies (Lehmann and Joseph 2009; Nelissen et al. 2015; Nguyen et al. 2017; Sarma et al. 2018). Wolf (2008) found an average C concentration of 47.6% in biochar when produced from agricultural wastes. However, Gaskin et al. (2010) showed that C concentration in biochar produced from poultry manure and pine chips can range between 40% and 78%. In general, biochar has a high C/N ratio (mean value of 67%) which indicates that immobilization of N can occur when applied to soil. Thus, the application of biochar contributes towards higher N retention (Güereña et al. 2013) by decreasing leaching losses of NO_3^-. This reduction of NO_3^- leaching under biochar addition is owing to its sorption to biochar, immobilization by microbes (Zheng et al. 2013) along with eventual uptake by the plant (Steiner et al. 2008). Rondon et al. (2007) reported that addition of *Eucalyptus deglupta* biochar produced at low temperature (350 °C) improve legume growth and yield in alkaline soil through better fixation of biological nitrogen due to lowering of soil pH. This increased biological N fixation due to biochar application in alkaline soil under legume cultivation increases nutrient availability, particularly P and N that contributes to the increased crop yield. The increased biological N fixing due to biochar application can be attributed to biochar-induced availability of P (Tagoe et al. 2008), K (Mia et al. 2014), and micronutrients like boron and molybdenum (Rondon et al. 2007). Because of the stability of C in biochar, it cannot easily be digested by microbes and therefore, it acts as source of slow release of nutrients especially N (Lone et al. 2015). The surface area can be colonized and small pores act as refugee site for microbes to avoid grazers (Thies and Rillig 2009). The porous nature of biochar shows a positive influence on mycorrhizal fungi i.e. biochar provides the physical environment required for its growth (Solaiman et al. 2010). Biochar increases the association between mycorrhizal fungi and plant roots that promoted higher P availability (Garcia-Montiel et al. 2000). However, the better microbial population straightaway after application of biochar can be accredited to the labile components of the added biochar (Smith et al. 2010). Thus, by providing a positive impact on the availability of both micro- and macronutrients in soil, biochar helps in soil fertility improvement.

1.2 Effects of Biochar on Soil Physico-chemical Properties

Biochar incorporation into soil has an impact on soil aggregate stability, porosity and other hydrological functions such as water holding capacity. Application of biochar influences soil physico-chemical characteristics through several mechanisms. For example, the improvement of soil pores is due to its higher porous surfaces, which also augments spaces for microbes between the pores and surrounding soil aggregates, and thus significantly alters soil characteristics.

The persistence of soil pore is also enhanced under biochar application due to better aggregate stability (Hardie et al. 2014). Many studies have specified the positive effects of biochar on soil water retention capacity (Githinji 2014; Sarma et al. 2017b). Biochar, due to its lighter texture, decreases soil bulk density, which in turn increases the porosity and water retention (Abel et al. 2013; Karhu et al. 2011). However, these changes are largely dependent on biochar properties such as rate of application, particle size and feedstock (Li et al. 2018; Narzari et al. 2015). The biochar-induced responses are also dependent on soil type. Jeffery et al. (2015) found no significant improvement in hydrological functions of sandy soil due to addition of biochar. Likewise, no significant difference was obtained between biochar treated loamy sand soil and soil without biochar with regard to its physical properties and water availability (Hardie et al. 2014). Haefele et al. (2011) reported variable effect of biochar application in different soil types with decreased bulk density in the top layer of clay soil but found no effects on loamy sand soil. In addition to soil physical and water retention properties, biochar potentially alters soil pH (Glaser et al. 2002). Unger (2008) and Chintala et al. (2014) reported increased pH of acidic soil while in alkaline soil pH decreased. Column experiments reported by Bruun et al. (2014) showed increased pH of sub-soil from 6.8 to 9.2 with the addition of slow pyrolysed straw biochar but found no effect when the same quantity of wood biochar produced at pyrolysis temperature of 450–480 °C was added. In an incubation experiment, Laird et al. (2010) reported increased pH by one unit after application of mixed biochar at a rate of 20 g kg^{-1} soil. Thus, the alteration of bulk density, porosity, hydraulic characteristics and soil pH under biochar application is not consistent and vary with soil type and biochar properties.

Biochar application is also associated with the modification of soil nutrients availability and soil CEC (Table 1). When added to soil, biochar is exposed to water and oxygen that causes spontaneous surface oxidation leading to increased anion contents and hence a higher CEC value (Agegnehu et al. 2017). A recent study by

Table 1 Influence of biochar application on soil nutrients/pollutants

Feed materials	Parameters studied	Results obtained +/−	References
Switchgrass, water oak and biosolids	Sorption of nitrogen (as NH_4^+, NO_2^- and NO_3^-)	Switchgrass: +NH_4^+; −NO_2^- and NO_3^-	Li et al. (2018)
		Biosolids: decrease in NH_4^+ retention with increasing pyrolysis temperature	
Rice straw	Nitrification and nitrate leaching	Increase in nitrification and decrease in leaching	Zhao et al. (2013)
Peanut hull, pine chip	N, P, K, S, Ca, Mg	Peanut hull: + N P K Ca Mg	Gaskin et al. (2010)
		Pine Chip: −Ca, K	
Sugarcane bagasse, peanut hull, Brazilian pepperwood and bamboo	Sorption of nitrate, ammonium, and phosphate	Reduced nitrate, ammonium, and phosphate leaching	Yao et al. (2012)
Mixed wood	Ca, Mg, K, P, Sr, Al	+K, Sr, P; decrease in Ca and Mg loss; −Al	Major et al. (2010a)
Papermill waste	Exchangeable Al, Ca	−Exchangeable Al +Exchangeable Ca	Van Zwieten et al. (2010a)
Miscanthus (*Miscanthus x giganteus*) straw, willow (*Salix* sp.) wood chip	NO_3 –N; NH_4^+–N	NO_3 –N >NH_4^+–N	Prendergast-Miller et al. (2014)
Sewage sludge	NH_4^+, NO_3^-, PO_4^{3-}, K^+ leaching	Reduced NH_4^+, NO_3^-, PO_4^{3-}, K^+ leaching	Yuan et al. (2016)
Chicken manure	Ca, Fe, K, P, Cu, Mn, Ni, Zn	+ Ca, Fe, K, P, Cu, Mn, Ni, Zn	Hass et al. (2012)

Gautam et al. (2017) reported increased K^+ availability upon biochar addition (5 t ha^{-1}) in silty loam Nepalese soil. Biochar also affect the soil C/N ratio, which is a key function for alteration of many other soil parameters (Wang et al. 2015). However, the total N content decreased with an increase in pyrolytic temperature during biochar production (Feng et al. 2012). Nutrient contents in biochar are widely studied and found to be the function of feedstock used for biochar production. Biochar derived from manure is rich in soil nutrients such as N, P, Ca, Mg, and K (Cantrell et al. 2012) while that of sludge- derived biochar has higher N and P concentrations (Hossain et al. 2011). Thus, application of biochar with various nutrient contents results in alteration of the soil chemical properties (Table 1).

1.3 Biochar and Its Impacts on Crop Growth and Yield

Biochar is being promoted as a potential soil amendment, which illustrates differential responses to crop growth and yield (Table 2). These biochar-mediated positive, negative or no effects on crop growth can be attributed to the factors such as types of crop and soil, soil mineralogy, biochar properties such as rate and age of biochar, raw materials used and pyrolysis conditions during biochar production (Sarma et al. 2017b). In biochar amended Ferrosol, increased biomass was noted in radish while

Table 2 Impact of biochar application on crop growth and yield

Biochar sources	Crop types	Effect on soil properties/soil quality changes	References
Teak and rose wood biochars (4–16 t ha^{-1})	Rice and sorghum	Improved plant growth and 2–3 times yield increment	Asai et al. (2009) and Steiner et al. (2007)
Oil palm fruit bunch biochar (10, 20 and 40 t ha^{-1}) green waste compost	Rice	Increase in grain yield under organic system of rice intensification by 141–472%	Bakar et al. (2015)
Mango wood (8, 16 t ha^{-1}), corn stover (2.6–91 t ha^{-1})	Maize	Increase in plant biomass form 30–43% and grain yield by 22%	Rajkovich et al. (2012)
Maize straw (20,30 and 40 t ha^{-1})	Choy sum and amaranth	Increase in grain yield by 28–48%	Jia et al. (2012)
Citrus wood biochar (1, 3 and 5% by volume in pots)	Pepper and tomato	Increase in leaf area, canopy dry weight, number of nodes and yield	Graber et al. (2010)
Biochar from whole tree green waste of acacia green fowl manure (10 t ha^{-1})	Apple	Increase in tree trunk girth without any effect on yield or quality	Eyles et al. (2015)
Rice husk char (25, 50, and 150 g kg^{-1})	Lettuce and cabbage	Increase in biomass by 903%	Carter et al. (2013)

no significant effect was found for wheat and soybean. While in Calcarosol, biochar application increased soybean biomass, but reduced wheat and radish biomass (Van Zwieten et al. 2010b). The responses of biochar to plant growth is directly due to supply of nutrients or indirectly due to amelioration of soil physico-chemical properties (Agegnehu et al. 2016a) and increased microbial activities (Wang et al. 2016). Vaccari et al. (2011) reported higher wheat yield due to addition of hardwood biochar. Similar results were also reported by Sarma et al. (2017b). Major et al. (2010b) reported no significant increase in maize yield under biochar addition during 1st year of experimentation. However, 28%, 30% and 140% increased maize yield over control was recorded in subsequent 3 years of experimentation with the addition of 20 t ha^{-1} biochar. This signifies the long term benefits of biochar in terms of crop yield which might be due to slow release of nutrients along with stabilization of higher organic matter and retention of cations (Lehmann 2007). Martinsen et al. (2014) stated that addition of K due to biochar application aid an increased maize biomass under nutrient stress condition. The growth rate of sorghum was also increased (15–32 times) in a nutrient- poor soil after amended with algal biochar (Bird et al. 2012). Biochar application (25 t ha^{-1}) along with farmyard manure (5 t ha^{-1}) results in improved maize growth with lower weed population till 30–60 days of sowing (Arif et al. 2012). Application of biochar also significantly increased N uptake in wheat grown in fertilizer-amended ferrosol, resulting in a concomitant increase of 250% biomass production compared to the control attributable to improved fertilizer use efficiency. Nevertheless, a number of studies on biochar-mediated impacts on crop growth and yield reported negligible or insignificant responses in temperate regions (Schmidt et al. 2014). Thus, the impact of biochar on crop growth and development is a function of numerous factors including the soil properties, climatic conditions, crops and characteristics of the applied biochar.

2 Biochar Application and Emission of GHG from Agricultural Soil

As soil amendment, biochar has shown promising results in increasing soil fertility and altering soil physio-chemical properties that enhance crop production (Table 2). Due to its recalcitrant nature, biochar helps in C sequestration (Harvey et al. 2012). However, the presence of some amount of labile C in biochar is mineralized over time and leads to CO_2 emissions from the applied soils (Hilscher et al. 2009; Kuzyakov et al. 2009; Zimmerman 2010). Many studies conducted to observe the CO_2 emission rate from the soil under biochar application. These studies delineate that though the emissions are significantly high at initial stages of application, but become sable after the labile C has mineralized (Ameloot et al. 2013). Besides CO_2, biochars are also known to interfere with emission of other greenhouse gases such as N_2O and CH_4 from soil (Wang et al. 2011; Saarnio et al. 2013; Scheer et al. 2011).

2.1 Biochar Application and Soil N_2O Emission

Agriculture is the main source of the global anthropogenic N_2O emission (Smith et al. 2008), mostly due to the extensive use of synthetic nitrogenous (N) fertilizers. Though some studies described that N_2O is produced via. abiotic redox reactions (Samarkin et al. 2010; Rubasinghege et al. 2011), but predominantly it is generated by the microbial transformations of reactive N in soil (Baggs 2011; Thomson et al. 2012; Butterbach-Bahl et al. 2013). Moreover, soil can act not only as a source but also as a sink for N_2O (Chapuis-Lardy et al. 2007; Thomson et al. 2012). Nitrous oxide can be used by some heterotrophic bacteria, as well as by some autotrophic bacteria for respiration and is further transformed to N_2 (Chapuis-Lardy et al. 2007). Nitrification and denitrification are the two processes that drive soil N_2O production (Davidson et al. 1986) and enhancing crop available N is an effective strategy to reduce N losses through direct and indirect emissions (Mosier et al. 1998). Singh et al. (2010) and Van Zwieten et al. (2010b) suggested that biochar application could affect N transformation and N fate in soils. Furthermore, when biochar applied to soybean crop and grass covered fields Rondon et al. (2005) reported 50–80% reduction in soil N_2O emissions. Stimulation or inhibition of N_2O emission by biochar depends upon the initial soil moisture during the soil rewetting period (Yanai et al. 2007). Soil bioavailable or reactive N in soil constitutes both organic as well as mineral N species, including nitrate and ammonium, which are readily utilized by microorganisms and plants (Huygens et al. 2008). Biochar application in soils reduces the N stock that are available for denitrification as soil ammonium retention is much enhanced under biochar application (Singh et al. 2010; Steiner et al. 2010). Biochar addition in soil decreases the soil NH_4^+ -N and NO_3-N concentration that reduces the soil N_2O emission potential (Wang et al. 2011). Zwieten et al. (2009) also reported that the degree of reduction in N_2O emission under biochar application also depends on the feedstock used to produce biochar, soil type and rate of application along with soil moisture content. Biochar properties such as pH and C/N ratio affect the soil N dynamics (Cavigelli and Robertson 2001; Yanai et al. 2007; Rondon et al. 2007; Warnock et al. 2007; Zwieten et al. 2009). Alkaline nature of biochar increases the soil pH which favors the activity of enzyme such as N_2O reductase in denitrifying bacteria (Yanai et al. 2007), while inhibiting the reductases activity involved in the conversion of nitrite to nitrate and ultimately to N_2O (Zwieten et al. 2009). Biochar alters the microbial abundance and community composition (Lehmann et al. 2011) including ammonium oxidizer community composition (Dempster et al. 2012) in soil. Theoretically, the ratio of nitrifiers to denitrifiers within the soil affected under biochar application. The effect of biochar on soil N_2O emission is not always positive. For example, increased biological nitrogen fixation in legumes under biochar application (Nishio 1996; Rondon et al. 2007) can likely increase N within the soil system and may ultimately rise the potential for N_2O emissions. Studies done so far suggest that though biochar has the ability to reduce soil N_2O emission, but the degree of reduction depends upon the soil type, soil moisture content and biochar properties, that vary with the feedstock material and pyrolysis conditions.

2.2 Biochar Application and Soil CH_4 Emission

Methane is produced naturally under waterlogged anaerobic conditions by methanogenic archaea by the process of methanogenesis (Conrad 2007). Methanotrophy on the other hand takes place in soils under aerobic condition and is considered the sink of methane. Methanotrophic bacteria such as groups of α- and γ-proteobacteria are the primary drivers of this process that completely feed on methane along with some bacteria of genera *Methylocapsa* and *Methylocella* that are facultative methanotrophs (Pratscher et al. 2011; Knief 2015). These two processes undergoing in soil decide whether the soil act as a source or sink of CH_4. Biochar is known to alter soil physical and chemical properties such as increase pH, soil aeration and decrease bulk density. Higher soil porosity and potential to decrease bulk density under biochar application facilitates soil aeration. Theoretically, this situation should promote methanotrophy over methanogenesis. But, the studies done so far do not provide ample evidences about the processes driven by soil biochar addition with regards to CH_4 emission. For example, Rondon et al. (2005) found complete suppression of CH_4 emissions when soil was amended with biochar at the rate of 20 g kg^{-1}. Whereas, Knoblauch et al. (2008), Feng et al. (2012), Dong et al. (2013) and Reddy et al. (2014a), Zhang et al. (2010) and Spokas and Bogner (2011) observed no significant changes in soil CH_4 emissions when amended with biochar with respect to non-amended soils. However, quite a good number of studies were conducted on soil CH4 emission under biochar application, but the mechanisms proposed are only assumed, hypothesized or remain unclear. The mechanism which affects CH_4 fluxes due to biochar include sorption of CH4 to its surface (Yaghoubio et al. 2014) and enhancement of soil aeration that may upsurge diffusive CH_4 uptake (Van Zwieten et al. 2010a; Karhu et al. 2011), as microbial CH_4 oxidation in upland soils is mostly substrate-limited (Castro et al. 1994). Moreover, under anoxic conditions, the labile C pool of biochar may act as methanogenic substrate and thus promote CH_4 production (Wang et al. 2012). Biochar has also shown to encourage methanotrophic CH_4 consumption under aerobic/anaerobic interfaces in anaerobic conditions and reduce CH_4 emissions via the "biofilter" function of CH_4 consumption (Feng et al. 2012; Reddy et al. 2014b). The studies done so far reveal that application of biochar has positive, negative and no significant effect on CH_4 emission and further studies needed to confirm the mechanisms involved there in.

2.3 Biochar Application and Soil CO_2 Emission

Converting biomass into biochar can stabilize organic C and thus has the potential to reduce CO_2 emissions (Lehmann et al. 2006; Steiner et al. 2007; Major et al. 2010b), though biochar can be decomposed by microbes to some extent (Czimczik and Masiello 2007). Abiven and Andreoli (2011) and Jones et al. (2011) found that biochar application does not enhance the mineralization of soil organic matter, but the decomposition of biochar itself increases the soil respiration rate. However, degradable portion of biochar is very small and decomposed very quickly as compared to the time it required to sequester the non-mineralized portion (Jones et al. 2011; Bruun et al. 2012). Kuzyakov et al. (2009) found that the decomposition of biochar was relatively quick during the first 3 months following its addition to soil and slow, partial decomposition occurred during the following 3.2 years of the experiment. Pyrolysis temperature plays a vital role in mineralization of the biochar and thus CO_2 emission from the soil. Zimmerman et al. (2011) concluded that biochar produced at temperatures below 400 °C stimulate C mineralization which decreases with increasing pyrolysis temperature. The differences in pyrolysis temperature induces significant changes in physiochemical structure as well as composition of the biochar (Asadullah et al. 2010; Li et al. 2006; Scott and Glasspool 2007) responsible for CO_2 evolution as observed in several studies (Baldock and Smernik 2002; Nguyen and Lehmann 2009). Furthermore, the decomposition rate of biochar and ultimately CO_2 emission vary under varying soil environments such as water regime (Nguyen and Lehmann 2009) and native soil organic carbon concentrations (Kimetu and Lehmann 2010). The pyrolysis temperature and the feedstock material determine the degree of polarity (i.e. O/C ratio) as well as the aromaticity (i.e. H/C ratio) of the produced biochar. Low aromaticity and high polarity indicate the presence of higher amount of labile C and thus more CO_2 emission form the soil after its application (Khodadad et al. 2011). Similarly, high aromaticity and low polarity of biochar designate lower amount of labile and higher amount of stable C that leads to high sequestration potential of the biochar and ultimately lower amount of CO_2 emission from the soil (Chun et al. 2004).

Thus, biochar can be used to extract more CO_2 from atmosphere into soils and its use can be an effective approach to combat global warming in the coming future.

3 Pollutant Bioavailability in Agricultural Soil Under Biochar Application

To feed the immensely growing population of the world, the need of higher agricultural production from the limited cultivable area is become a matter of concern in present time. This has led to the dependence of the farmers to the inorganic fertilizers and pesticides and non-judicious use of these chemicals cause pollution of the agricultural soils. The global use of pesticides, insecticides and herbicides is in huge amount and the accurate data for which is not available (Benbrook 2016). In the 1990s, about \$13,280 million worth of pesticides, herbicides and insecticides were sold globally (Khan 2016). This indicates the magnitude of pesticides, herbicides and insecticides used globally. Soil remediation methods such as soil washing, flushing, vapor extraction and bioremediation are proposed by many researchers to remediate the contaminated soil (Kong et al. 2014). However, the feasibility of all these methods in agricultural soil is debatable because of the difficulties arising in their application such as high maintenance cost, nutrient leaching, soil erosion or simply the unavailability of soil due to lack of fallow period (Kumpiene et al. 2008; Powlson et al. 2011; Kong et al. 2014). Hence, cost effective, in situ method by application of soil amendments to remediate the polluted agricultural soil is of immense importance (Lehmann and Joseph 2009).

During initial period, as soil amendment biochar gained importance in agriculture and other qualities of biochar such as C sequestration, GHG emission reduction (Spokas et al. 2009; Zhang et al. 2010; Gomez-Eyles et al. 2011) has also come into light. The role of biochar in reducing bioavailability of pesticides and heavy metals has also not gone unnoticed (Cabrera et al. 2011; Barrow 2012; Chen et al. 2012; Ahmad et al. 2014). Properties of biochar like high porosity, presence of various functional groups, higher surface area and cation exchange capacity CEC make it potent soil amendment for correction of agricultural pollution (Park et al. 2011; Jiang et al. 2012; Zhang et al. 2014).

Martin et al. (2012) reported that biochar is one of the most efficient sorbents of several groups of pesticides. The physio-chemical properties of biochar such as organic carbon content, specific surface area and porosity determine the pesticide adsorption capacity of biochar (Spokas et al. 2009; Dechene et al. 2014; García-Jaramillo et al. 2014; Sopena et al. 2012; Wang et al. 2010; Cabrera et al. 2014; Xu et al. 2008). Later on it is documented that biochar not only improves the pesticide sorption capacity of the soil but also affects the sorption mechanism and the bioavailability of pesticide residue for organisms (Yang and Sheng 2003; Yu et al. 2006; Kookana et al. 2011). Pesticide mobility in biochar amended soils have been studied previously by various researchers (Wang et al. 2010; Sopena et al. 2012; Spokas et al. 2009; Cabrera et al. 2009). Wang et al. (2010) reported a higher adsorption of terbuthylazine in soil when amended with pine wood biochar. Sopena et al. (2012) observed that biochar obtained from *Eucalyptus dunni* had greater adsorption capacity for herbicide like isoproturon. Similarly, Spokas et al. (2009) found higher adsorption capacity of sawdust biochar to atrazine and acetochlor, whereas Cabrera et al. (2009) documented almost a complete sorption of herbicides bentazone and aminocyclopyrachlor by silt loam soil when amended with high specific surface area biochar obtained from wood pellets. These studies specify that the addition of C rich amendments to soil generally decrease pesticide leaching in soil due to an increased adsorption by the process of its entrapment into the micro pore network and/or pore deformation of the biochar (Larsbo et al. 2013; Li et al. 2013; Marin-Benito et al. 2013). Studies also documented reversible results on pesticide adsorption in biochar-amended soils. The process of this reversible adsorption can take place via different mechanisms such as the swelling of sorbent during adsorption process that results in deformation of macropore network (Sopena et al. 2012; Khorram et al. 2015) and weak bonding between biochar and the tested particles of pesticides (Tatarkova et al. 2013; Khorram et al. 2015). Adsorption and entrapment are not the only processes that are responsible for the reduced bioavailability of the pollutants in the soil. Studies have also revealed that microbial stimulation by biochar is also responsible for enhanced pesticide biodegradation leading to reduced bioavailability of the pesticides (Lopez-Pineiro et al. 2013; Zhang et al. 2005, 2006).

For example, Qiu et al. (2009) found the degradation of atrazine increased with respect to the rate of application of wheat biochar. This increased degradation may be attributed to the better nutrient content, particularly P responsible for the enhanced microbial activity. Biochar has also documented promising results in enhancing sorption of polycyclic aromatic hydrocarbons into the soil (Chen and Yuan 2011). However, Sorption to biochar is determined by the presence of relative carbonized and non-carbonized fractions and their surface and bulk properties (Cheng et al. 2008) of the applied biochar.

The application of biochar is not only useful to mitigate the effects of pesticide pollution, as mentioned above but also to manage the heavy metal pollution caused by the over and/or non- judicious use of inorganic fertilizers. Studies done at both laboratory and field have demonstrated the efficiency of biochar in decreasing the mobility and bioavailability of heavy metals in soil (Jiang et al. 2012; Abdelhafez et al. 2014; Lu et al. 2014). Heavy metals having variable charges present in soil are available for plant uptake and can easily move into the plant because of the low pH and cation exchange capacity of the agricultural soils (Guo et al. 2006). The application of biochar modifies the soil chemical properties such as electrical conductivity, pH, soil organic matter, CEC, dissolved organic C, and macro- and micronutrients promoting a suitable environment for reducing the bioavailability of heavy metals to the plants (Beesley et al. 2010; Park et al. 2011; Khan et al. 2017).

Mukherjee et al. (2011) reported that the presence of significant negative charge on biochar surface is responsible for attraction of positively charged metals and organic compounds from soil solution. During pyrolysis at low temperature (200–400 °C), the surface of the biochar encompasses oxygen containing functional groups, which enables the creation of surface complexes between cations like Cu^{2+}, Ni^{2+}, Cd^{2+}, Pb^{2+}, and Zn^{2+} at the biochar surface (Beesley and Marmiroli 2011; Uchimiya et al. 2011). This negative charge so developed, leads to increase soil pH after biochar application in polluted soils as negative surfaced biochar attracts H^+ ions from the soil solution. A soil with increased pH further increases the metal sorption from the soil solution due to the deprotonation of the pH dependent cation exchange sites on the soil surface, especially in soils with lower pH (Rees et al. 2014). Thus, it reduces the concentration of heavy metals or organic contaminants in the soil solution (Beesley et al. 2010), as well as their availability for uptake by organisms (Semple et al. 2004).

Biochar application in soil also enhances the soil microbial activity that are responsible for the immobilization of heavy metals in soils. The soil microbes associated with plant roots (rhizospheric microbes) can significantly influence the metal uptake by the plant (Shilev et al. 2001). The enzymes such as urease, catalase and acid phosphatase produced by the microorganisms, are capable of inducing chelation with heavy metal ions and subsequently decrease their availability to the plants (Lehmann et al. 2011; Xu et al. 2014).

The use of biochar in remediation of agricultural pollution (organic as well as inorganic) has tremendous potential. The primary mechanisms include adsorption and desorption of the pollutants to make it less bioavailable for plants or immobilization in the soil.

4 Role of Biochar in Abiotic Stress Amelioration

Increased anthropogenic interferences in the global biogeochemical cycles have led to more exposure of agricultural ecosystem to abiotic stresses. These includes climate change induced water and soil pollution causing soil quality deterioration and higher frequencies of adverse climatic events such as drought and heat stress.

The ability to biomagnify make heavy metals more risky to both human health and the environment (Roy and McDonald 2015). Therefore, an efficient and affordable method is of immense important to address this issue. Use of biochar as soil amendment is gaining popularity because of its ability to mitigate soil against heavy metal toxicity. The higher surface area of the biochar implies greater capacity for complexing heavy metals. Several previous studies with scanning electron microscopy has demonstrated surface sorption of heavy metals on biochars (Beesley and Marmiroli 2011; Lu et al. 2012). This sorption can be due to complexation of heavy metal with different functional groups present in the biochar or exchange of heavy metals such as Ca^{2+} and Mg^{2+}; K^{+}, Na^{+} and S with cations linked with biochar (or due to physical adsorption (Lu et al. 2012; Uchimiya et al. 2011). Moreover, the functional groups associated with oxygen are recognized as heavy metal stabilizer (particularly for weaker acids like Pb^{2+} and Cu^{2}) (Uchimiya et al. 2011). Méndez et al. (2009) observed that sorption of Cu^{2+} was associated to the increased oxygenated surface groups along with higher average pore diameter, elevated superficial charge density and Ca^{2+} and Mg^{2+} exchange content of biochar. This indicates the higher dependence of sorption mechanism on soil type and the cations present in both biochar and soil. Some other compounds present in the ash, such as carbonates, phosphates or sulfates (Cao et al. 2009; Karami et al. 2011; Park et al. 2013) also help in stabilization of heavy metals by precepitating these compounds with the pollutants. Enhancement of soil microorganisms by application of biochar also plays an important role in heavy metal immobilization (Jones et al. 2012; Park et al. 2011). Biochar can alter the microbial diversity by the following mechanisms (Warnock et al. 2010):

- Providing more nutrients to improve plant growth and attracting beneficial microbes that stimulate fungal growth.
- Modifying the rhizosphere environment to improve soil physico-chemical parameters.
- Providing a safe sanctuary for beneficial fungi and bacteria.
- Stopping or decreasing the toxic effect of allelochemicals, soil-persistent agrochemicals, and various types of soil containments.
- Improving signaling during plant–fungal interactions.

The use of biochar can also be beneficial in addressing soil salinity that is a result of interference in biogeochemical cycles and is another major concern of reduced crop productivity. Under saline environment, plant has to survive with two major stresses, that is osmotic stress and ionic stress. The rising salt levels in the soil solution surrounding the roots cause osmotic stress inhibiting water uptake, cell expansion and lateral bud development (Munns and Tester 2008). Whereas, ionic stress develops gradually with the excess accumulation of toxic ions (e.g. Na^+) to levels beyond plant specific thresholds. This leads to higher leaf mortality, chlorosis, necrosis and decreased activity of cellular metabolism including photosynthesis (Yeo and Flowers 1986; Glenn et al. 1999; Zahir et al. 2012; Panuccio et al. 2014).

The application of biochar as soil amendment has been revealed as effective in reducing salinity stress by improving soil physico-chemical and biological properties. Improvement on various Na removal processes such as Na leaching, Na adsorption ratio, and electrical conductivity were documented under application of biochar (Chaganti et al. 2015; Diacono and Montemurro 2015; Oo et al. 2015; Drake et al. 2016; Sun et al. 2016). Addition of biochar in agricultural soil is found to decrease the concentration of Na^+ in the xylem sap of potato, while increasing K^+ concentrations and Na^+/K^+ ratio in the xylem sap as compared to the control (Lashari et al. 2015; Akhtar et al. 2015). Similarly, biochar decreased Na uptake in lettuce and maize crop grown under salt stress and increased K uptake (Kim et al. 2016; Hammer et al. 2015). Soil salinity is also responsible for production of reactive oxygen species in plants (Parihar et al. 2015; Fazal and Bano 2016; Farhangi-Abriz and Torabian 2017). Application of biochar under saline conditions reduced antioxidant enzyme (ascorbate peroxidase and glutathione reductase) activities and oxidative stress (Kim et al. 2016; Farhangi-Abriz and Torabian 2017) in both maize and bean crop.

The role of biochar in mitigating drought stress has also been widely studied. The properties of biochar such as higher water holding capacity, porosity and surface area makes it a suitable amendment to use in mitigating drought stress in agriculture. The use of biochar as soil amendment increases the soil water holding capacity, which is a key aspect in improving plant growth and yield under drought stress (Artiola et al. 2012; Pereira et al. 2012; Akhtar et al. 2014; Bruun et al. 2014; Lu et al. 2015; Agegnehu et al. 2016b; Foster et al. 2016). This increased soil water holding capacity under biochar addition might be due to higher CEC and porous structure of biochar (Artiola et al. 2012). Increased soil aggregate stability was also recorded under biochar application, which might be effective in enhancing soil water retention under limited water conditions (Baiamonte et al. 2015). For example, application of biochar reported to increase wilting resistance in tomato seedlings grown in sandy soil with lower moisture content as compared to control (Mulcahy et al. 2013; Vaccari et al. 2015). The effects of biochar on plant morpho-physiological characters under drought have also been reported. For example, better plant height and leaf area of okra (Batool et al. 2015) and maize (Haider et al. 2015), increased biomass in field-grown wheat (Olmo et al. 2014) was documented under biochar application. Biochar application as soil amendment improves the plant photosynthesis under drought-stress (Akhtar et al. 2014; Lyu et al. 2016; Paneque et al. 2016; Xiao et al. 2016) due to significant improvement of chlorophyll content, stomatal conductance, photosynthetic rate, water use efficiency, relative water contents, and stomatal density of drought-stressed tomato leaves.

Another major aspect of biochar in abiotic stress management is mitigation of heat stress in crop plants. An increased surface air temperature by 4–5.8 °C over the next few decades has been predicted by global circulation models (IPCC 2007). Heat stress in crucial stages of plant development can cause deleterious effects in crops. For example, Jagadish et al. (2010) found that temperatures greater than 35 °C lasting more than 1 h during anthesis can lead to high levels of sterility in rice. Matsui and Omasa (2002) reported high temperatures stress on the day before flowering resulted in poor anther dehiscence during subsequent anthesis. Fahad et al. (2015) has found that application of biochar helped in maintaining higher relative water content in leaves and inside panicle under high temperature stress. They also documented higher grain yield along with better grain quality with the addition of biochar. However, not much study has been done in this regard and further studies are required for better understanding of the interaction between heat stress and soil biochar application for improved crop growth.

Thus, application of biochar as soil amendment can be a useful tool in ameliorating crop abiotic stress. However, prior study on characteristics of both applied biochar and soil type is needed to receive maximum benefit.

5 Role of Biochar in Disease Suppression

Intensive agriculture largely contributes towards degradation of soil organic matter and microbial diversity with the increased incidence of soil borne plant pathogens (Campbell 2006). Conventionally used chemicals for controlling these disease pathogens have some drawbacks apart from triggering soil pollution if not used judiciously. Biochar has shown promising results in suppressing plant diseases caused by soil borne as well as foliar pathogens (Elad 2010; Elmer and Pignatello 2011; Jaiswal et al. 2014). Nutrients provided by biochar (Silber et al. 2010) or increased availability of soil nutrients in presence of biochar (Graber et al. 2014) enhances plant vigor and thus reduce the ability of the pathogen to infect the plant. By improving morphological, histological and functional characters of plant tissues and maintenance of higher level of inhibitory compounds in tissues biochar application makes plant enable for a quick response to pathogen attack (Datnoff et al. 2007). This enhanced nutrition under biochar application may result due alteration of root architecture (Prendergast-Miller et al. 2014) which can have concerns for host susceptibility since finer roots always provide a larger surface area to attack by soil-borne pathogens (Newsham et al. 2005). Biochar also alters the soil pH or together bring changes in Eh–pH system (Husson 2012; Yuan and Xu 2012). These biochar-induced changes at rhizospere could strongly modify the pathogen viability as many soil pathogens grow well under narrow Eh–pH ranges (Husson 2012). Thus, soil Eh-pH system plays an important role in microbial community development, diversity, structure and pathogen virulence (Husson 2012). The redox activity of biochar may interfere the chemical and biological electron transfer reactions in the rhizosphere (Graber et al. 2014) including microbial processes reliant upon electron transfer (Lovley 2012). For example microbial processes involving N cycling (Cayuela et al. 2013) and chemical processes linking reduction and solubilization of nutrients like Fe and Mn (Graber et al. 2014). Rhizospheric resistance towards invasion by soil pathogens is directly related to the diversity of the microbiota present therein (Matos et al. 2005; Irikiin et al. 2006). The microbes promoted due to application of biochar can compete with pathogens for resources. Moreover, compounds produced by those microbes may inhibit or parasitize the pathogens (Kolton et al. 2011). Kolton et al. (2011) observed a significant enhancement of the bacteroides-affiliated *Flavobacterium* genus in biochar-applied soil. *Flavobacterium* are highly antagonistic for the fungal pathogens *Sclerotium rolfsii*, *Lasiodiplodia theobromae*, *Colletotrichum musae*, and *Phytophthora cactorum* (Alexander and Stewart 2001; Hebbar et al. 1991). Biochar also shows fungi toxic effect in soil that is responsible for disease suppression in plants. During pyrolysis a wide range of organic compounds, potentially fungi toxic are also produced (Spokas and Bogner 2011). Other chemical changes occurred during pyrolysis include degradation of *O*-alkyl carbons associated with carbohydrate, and a contemporary neo-formation of aliphatic and aromatic C compounds (Spokas and Bogner 2011).A study conducted by Jaiswal et al. documented that damping-off caused by *R. solani* on *Cucumis sativus* (2014) and *Phaseolus vulgaris* (2015) was suppressed by different biochars. This attracts special attention because it has been reported earlier that effective control of *R. solani* with organic amendments, including composts, is extremely difficult (Krause et al. 2001; Scheuerell et al. 2005; Termorshuizen et al. 2006). Thus, it can be concluded that biochar application can be an effective strategy to suppress diseases caused by soil borne and foliar pathogens.

6 Conclusion and Future Perspective

Considering all the factors, it is concluded that biochar can be a very useful tool in agriculture for improvisation of soil quality in terms of water retention, soil porosity, and nutrient content as well as soil microbial and enzymatic activity. Furthermore, it is also helpful in reducing nutrient losses and greenhouse gas *viz.* N_2O and CH_4 emissions. Though biochar may induce CO_2 emission for a short period of time, but in long run it sequester carbon. Studies related to application of biochar in polluted soil indicate its potentiality in lowering bioavailability. However, the wide biochar properties, rate of biochar application, the effect of biochar on crop productivity, and biochar longevity leads to the different results while using as soil amendment .Although, considerable studies have done in context to biochar behavior in agricultural soils, but proper conditions for biochar production using different feedstocks should be optimized for obtaining maximum benefit from the produced biochar. Moreover, site specific comparative studies among biochar from different feedstocks and their impact on overall crop production and soil quality needs to be studied.

References

Abdelhafez AA, Li J, Abbas MH (2014) Feasibility of biochar manufactured from organic wastes on the stabilization of heavy metals in a metal smelter contaminated soil. Chemosphere 117:66–71

Abel S, Peters A, Trinks S, Schonsky H, Facklam M, Wessolek G (2013) Impact of biochar and hydrochar addition on water retention and water repellency of sandy soil. Geoderma 202:183–191

Abiven S, Andreoli R (2011) Charcoal does not change the decomposition rate of mixed litters in a mineral cambisol: a controlled conditions study. Biol Fertil Soil 47(1):111–114

Agegnehu G, Nelson PN, Bird MI (2016a) Crop yield, plant nutrient uptake and soil physicochemical properties under organic soil amendments and nitrogen fertilization on Nitisols. Soil Tillage Res 160:1–13

Agegnehu G, Bass AM, Nelson PN, Bird MI (2016b) Benefits of biochar, compost and biochar–compost for soil quality, maize yield and greenhouse gas emissions in a tropical agricultural soil. Sci Total Environ 543:295–306

Agegnehu G, Srivastava AK, Bird MI (2017) The role of biochar and biochar-compost in improving soil quality and crop performance: a review. Appl Soil Ecol 119:156–170

Ahmad M, Rajapaksha AU, Lim JE, Zhang M, Bolan N, Mohan D et al (2014) Biochar as a sorbent for contaminant management in soil and water: a review. Chemosphere 99:19–33

Akhtar SS, Li G, Andersen MN, Liu F (2014) Biochar enhances yield and quality of tomato under reduced irrigation. Agric Water Manag 138:37–44

Akhtar SS, Andersen MN, Liu F (2015) Biochar mitigates salinity stress in potato. J Agron Crop Sci 201:368–378

Alexander BJR, Stewart A (2001) Glasshouse screening for biological control agents of *Phytophthora cactorum* on apple (*Malus domestica*). N Z J Crop Hort Sci 29(3):159–169

Ameloot N, Graber ER, Verheijen FGA, De Neve S (2013) Interactions between biochar stability and soil organisms: review and research needs. Eur J Soil Sci 64(4):379–390

Annabi M, Bahri H, Chibani R, Angar H, Bahri B, Ben Hadj Salah H (2011) Soil carbon storage under no-tillage practice in Northern Tunisia. In: Proceedings of 5th world congress on conservation agriculture

Arif M, Ali K, Munsif F, Ahmad W, Ahmad A, Naveed K (2012) Effect of biochar, FYM and nitrogen on weeds and maize phenology. Pak J Weed Sci Res 18:475–484

Artiola JF, Rasmussen C, Freitas R (2012) Effects of a biochar-amended alkaline soil on the growth of romaine lettuce and bermudagrass. Soil Sci 177:561–570

Asadullah M, Zhang S, Min Z, Yimsiri P, Li CZ (2010) Effects of biomass char structure on its gasification reactivity. Bioresour Technol 101(20):7935–7943

Asai H, Samson BK, Stephan HM, Songyikhangsuthor K, Homma K, Kiyono Y, Inoue Y, Shiraiwa T, Horie T (2009) Biochar amendment techniques for upland rice production in Northern Laos: 1. Soil physical properties, leaf SPAD and grain yield. Field Crops Res 111(1–2):81–84

Baggs EM (2011) Soil microbial sources of nitrous oxide: recent advances in knowledge, emerging challenges and future direction. Curr Opin Environ Sustain 3:321–327

Baiamonte G, De Pasquale C, Marsala V, Cimò G, Alonzo G, Crescimanno G, Conte P (2015) Structure alteration of a sandyclay soil by biochar amendments. J Soils Sediment 15:816–824

Bakar RA, Razak ZA, Ahmad SH, Seh-Bardan BJ, Tsong LC, Meng CP (2015) Influence of oil palm empty fruit bunch biochar on floodwater pH and yield components of rice cultivated on acid sulphate soil under rice intensification practices. Plant Prod Sci 18(4):491–500

Baldock JA, Smernik RJ (2002) Chemical composition and bioavailability of thermally altered *Pinus resinosa* (Red pine) wood. Org Geochem 33:1093–1109

Barrow CJ (2012) Biochar: potential for countering land degradation and for improving agriculture. Appl Geogr 34:21–28

Batool A, Taj S, Rashid A, Khalid A, Qadeer S, Saleem AR, Ghufran MA (2015) Potential of soil amendments (biochar and gypsum) in increasing water use efficiency of *Abelmoschus esculentus* L. Moench. Front Plant Sci 6:1–13

Beesley L, Marmiroli M (2011) The immobilisation and retention of soluble arsenic, cadmium and zinc by biochar. Environ Pollut 159(2):474–480

Beesley L, Moreno-Jiménez E, Gomez-Eyles JL (2010) Effects of biochar and greenwaste compost amendments on mobility, bioavailability and toxicity of inorganic and organic contaminants in a multi-element polluted soil. Environ Pollut 158(6):2282–2287

Benbrook CM (2016) Trends in glyphosate herbicide use in the United States and globally. Environ Sci Eur 28(1)

Bird MI, Wurster CM, de Paula Silva PH, Paul NA, de Nys R (2012) Algal biochar: effects and applications. GCB Bioenergy 4:61–69

Bruun EW, Müller-Stöver D, Ambus P, Hauggaard-Nielsen H (2011) Application of biochar to soil and N_2O emissions: potential effects of blending fast-pyrolysis biochar with anaerobically digested slurry. Eur J Soil Sci 62(4):581–589

Bruun EW, Ambus P, Egsgaard H, Hauggaard-Nielsen H (2012) Effects of slow and fast pyrolysis biochar on soil C and N turnover dynamics. Soil Biol Biochem 46:73–79

Bruun EW, Petersen CT, Hansen E, Holm JK, Hauggaard-Nielsen H (2014) Biochar amendment to coarse sandy subsoil improves root growth and increases water retention. Soil Use Manag 30(1):109–118

Butterbach-Bahl K, Baggs EM, Dannenmann M, Kiese R, Zechmeister-Boltenstern S (2013) Nitrous oxide emissions from soils: how well do we understand the processes and their controls? Philos Trans R Soc Lond B: Biol Sci 368:20130122. https://doi.org/10.1098/rstb.2013.0122

Cabrera A, Cox L, Fernández-Hernández A, Civantos CGO, Cornejo J (2009) Field appraisement of olive mills solid waste application in olive crops: Effect on herbicide retention. Agric Ecosyst Environ 132(3–4):260–266

Cabrera C, Cox L, Spokas KA, Celis R, Hermosin MC, Cornejo J (2011) Comparative sorption and leaching study of the herbicides fluometuron and 4-chloro-2-methylphenoxyacetic acid (MCPA) in a soil amended with biochars and other sorbents. J Agric Food Chem 59(23):12550–12560

Cabrera A, Cox L, Spokas L, Hermosin MC, Cornejo J, Koskinen WC (2014) Influence of biochar amendments on the sorption–desorption of aminocyclopyrachlor, bentazone and pyraclostrobin pesticides to an agricultural soil. Sci Total Environ 470–471:438–443

Campbell A (2006) Compost use for pest and disease suppression in NSW. edn. Recycled Organics Unit, The University of New South Wales

Cantrell KB, Hunt PG, Uchimiya M, Novak JM, Ro KS (2012) Impact of pyrolysis temperature and manure source on physicochemical characteristics of biochar. Bioresour Technol 107:419–428

Cao X, Ma L, Gao B, Harris W (2009) Dairy-manure derived biochar effectively sorbs lead and atrazine. Environ Sci Technol 43(9):3285–3291

Carter S, Shackley S, Sohi S, Suy TB, Haefele S (2013) The impact of biochar application on soil properties and plant growth of pot grown lettuce (*Lactuca sativa*) and cabbage (*Brassica chinensis*). Agronomy 3(2):404–418

Castro MS, Mellilo JM, Steudler PA, Chapman JW (1994) Soil moisture as a predictor of methane uptake by temperate forest soils. Can J For Res 24:1805–1810. https://doi.org/10.1139/x94-233

Cavigelli MA, Robertson GP (2001) Role of denitrifier diversity in rates of nitrous oxide consumption in a terrestrial ecosystem. Soil Biol Biochem 33:297–310

Cayuela ML, Sánchez-Monedero MA, Roig A, Hanley K, Enders A, Lehmann J (2013) Biochar and denitrification in soils: when, how much and why does biochar reduce N_2O emissions? Sci Rep 3:1732

Chaganti VN, Crohn DM, Šimůnek J (2015) Leaching and reclamation of a biochar and compost amended saline–sodic soil with moderate SAR reclaimed water. Agric Water Manag 158:255–265

Chapuis-Lardy L, Wrage N, Metay A, Chotte JL, Bernoux M (2007) Soils, a sink for N2O? A review. Global Chang Biol 13:1–17

Chen B, Yuan M (2011) Enhanced sorption of polycyclic aromatic hydrocarbons by soil amended with biochar. J Soils Sediments 11(1):62–71

Chen B, Yuan M, Qian L (2012) Enhanced bioremediation of PAH contaminated soil by immobilized bacteria with plant residue and biochar as carriers. J Soils Sediments 12(9):1350–1359
Cheng CH, Lehmann J, Engelhard MH (2008) Natural oxidation of black carbon in soils: changes in molecular form and surface charge along a climosequence. Geochim Cosmochim Acta 72(6):1598–1610
Chintala R, Schumacher TE, McDonald LM, Clay DE, Malo DD, Papiernik SK, Clay SA, Julson JL (2014) Phosphorus sorption and availability from biochars and soil/biochar mixtures. CLEAN–Soil, Air, Water 42(5):626–634
Chun Y, Sheng G, Chiou CT, Xing B (2004) Compositions and sorptive properties of crop residue-derived chars. Environ Sci Technol 38(17):4649–4655
Conrad R (2007) Microbial ecology of methanogens and methanotrophs. Adv Agron 96:1–63
Czimczik CI, Masiello CA (2007) Controls on black carbon storage in soils. Glob Biogeochem Cycle 21:GB3005
Datnoff LE, Elmer WH, Huber DM (2007) Mineral nutrition and plant disease. American Phytopathological Society Press, St. Paul
Davidson EA, Swank WT, Perry TO (1986) Distinguishing between nitrification and denitrification as sources of gaseous nitrogen production in soil. Appl Environ Microbiol 52:1280–1286
De Meyer A, Poesen J, Isabirye M, Deckers J, Raes D (2011) Soil erosion rates in tropical villages: a case study from Lake Victoria Basin, Uganda. Catena 84(3):89–98
Dechene A, Rosendahl I, Laabs V, Amelung W (2014) Sorption of polar herbicides and herbicide metabolites by biochar-amended soil. Chemosphere 109:180–186
Dempster DN, Jones DL, Murphy DV (2012) Clay and biochar amendments decreased inorganic but not dissolved organic nitrogen leaching in soil. Aust J Soil Res 50:216–221
Diacono M, Montemurro F (2015) Effectiveness of organic wastes as fertilizers and amendments in salt-affected soils. Agriculture 5(2):221–230
Dong D, Yang M, Wang C, Wang H, Li Y, Luo J, Wu W (2013) Responses of methane emissions and rice yield to applications of biochar and straw in a paddy field. J Soils Sediments 13(8):1450–1460
Drake JA, Cavagnaro TR, Cunningham SC, Jackson WR, Patti AF (2016) Does biochar improve establishment of tree seedlings in saline sodic soils? Land Degrad Dev 27:52–59
Duku MH, Gu S, Hagan EB (2011) Biochar production potential in Ghana—a review. Renew Sustain Energy Rev 15(8):3539–3551
Elad Y, David DR, Harel YM, Borenshtein M, Kalifa HB, Silber A, Graber ER (2010) Induction of systemic resistance in plants by biochar, a soil-applied carbon sequestering agent. Phytopathology 100(9):913–921
Elmer WH, Pignatello JJ (2011) Effect of biochar amendments on mycorrhizal associations and Fusarium crown and root rot of asparagus in replant soils. Plant Dis 95(8):960–966
Eyles A, Bound SA, Oliver G, Corkrey R, Hardie M, Green S, Close DC (2015) Impact of biochar amendment on the growth, physiology and fruit of a young commercial apple orchard. Trees 29(6):1817–1826
Fahad S, Hussain S, Chauhan BS, Saud S, Wu C, Hassan S, Tanveer M, Jan A, Huang J (2015) Weed growth and crop yield loss in wheat as influenced by row spacing and weed emergence times. Crop Protect 71:101–108
Farhangi-Abriz S, Torabian S (2017) Antioxidant enzyme and osmotic adjustment changes in bean seedlings as affected by biochar under salt stress. Ecotoxicol Environ Saf 137:64–70
Fazal A, Bano A (2016) Role of plant growth-promoting rhizobacteria (pgpr), biochar, and chemical fertilizer under salinity stress. Commun Soil Sci Plant Anal 47:1985–1993
Feng YZ, Xu YP, Yu YC, Xie ZB, Lin XG (2012) Mechanisms of biochar decreasing methane emission from Chinese paddy soils. Soil Biol Biochem 46:80–88
Foster EJ, Hansen N, Wallenstein M, Cotrufo MF (2016) Biochar and manure amendments impact soil nutrients and microbial enzymatic activities in a semi-arid irrigated maize cropping system. Agric Ecosyst Environ 233:404–414

García-Jaramillo M, Cox L, Cornejo J, Hermosín MC (2014) Effect of soil organic amendments on the behavior of bentazone and tricyclazole. Sci Total Environ 466:906–913

Garcia-Montiel DC, Neill C, Melillo J, Thomas S, Steudler PA, Cerri CC (2000) Soil phosphorus transformations following forest clearing for pasture in the Brazilian Amazon. Soil Sci Soc Am J 64:1792–1804. https://doi.org/10.2136/sssaj2000.6451792x

Gaskin JW, Speir RA, Harris K, Das KC, Lee RD, Morris LA, Fisher DS (2010) Effect of peanut hull and pine chip biochar on soil nutrients, corn nutrient status, and yield. Agron J 102(2):623–633

Gautam DK, Bajracharya RM, Sitaula BK (2017) Effects of biochar and farmyard manure on soil properties and crop growth in an agroforestry system in the Himalaya. Sustain Agric Res 6(4):74

Githinji L (2014) Effect of biochar application rate on soil physical and hydraulic properties of a sandy loam. Arch Agron Soil Sci 60(4):457–470

Glaser B, Lehmann J, Steiner C, Nehls T, Yousaf M, Zech W (2002) Potential of pyrolyzed organic matter in soil amelioration. In: 12th ISCO conference'. Beijing, pp 421–427

Glenn EP, Brown JJ, Blumwald E (1999) Salt tolerance and crop potential of halophytes. Crit Rev Plant Sci 18(2):227–255

Gomez-Eyles JL, Sizmur T, Collins CD, Hodson ME (2011) Effects of biochar and the earthworm Eisenia fetida on the bioavailability of polycyclic aromatic hydrocarbons and potentially toxic elements. Environ Pollut 159(2):616–622

Graber ER, Harel YM, Kolton M, Cytryn E, Silber A, David DR, Tsechansky L, Borenshtein M, Elad Y (2010) Biochar impact on development and productivity of pepper and tomato grown in fertigated soilless media. Plant Soil 337(1–2):481–496

Graber ER, Frenkel O, Jaiswal AK, Elad Y (2014) How may biochar influence severity of diseases caused by soilborne pathogens? Carbon Manag 5(2):169–183

Güereña D, Lehmann J, Hanley K, Enders A, Hyland C, Riha S (2013) Nitrogen dynamics following field application of biochar in a temperate North American maize-based production system. Plant Soil 365(1–2):239–254

Guo G, Zhou Q, Ma LQ (2006) Availability and assessment of fixing additives for the in situ remediation of heavy metal contaminated soils: a review. Environ Monit Assess 116(1–3):513–528

Haefele SM, Konboon Y, Wongboon W, Amarante S, Maarifat AA, Pfeiffer EM, Knoblauch C (2011) Effects and fate of biochar from rice residues in rice-based systems. Field Crops Res 121(3):430–440

Haider G, Koyro HW, Azam F, Steffens D, Müller C, Kammann C (2015) Biochar but not humic acid productamendment affectedmaize yields via improving plant-soil moisture relations. Plant Soil 395:141–157

Hammer EC, Forstreuter M, Rillig MC, Kohler J (2015) Biochar increases arbuscular mycorrhizal plant growth enhancement and ameliorates salinity stress. Appl Soil Ecol 96:114–121

Hardie M, Clothier B, Bound S, Oliver G, Close D (2014) Does biochar influence soil physical properties and soil water availability? Plant Soil 376(1–2):347–361

Harvey OR, Kuo LJ, Zimmerman AR, Louchouarn P, Amonette JE, Herbert BE (2012) An index-based approach to assessing recalcitrance and soil carbon sequestration potential of engineered black carbons (biochars). Environ Sci Technol 46(3):1415–1421

Hass A, Gonzalez JM, Lima IM, Godwin HW, Halvorson JJ, Boyer DG (2012) Chicken manure biochar as liming and nutrient source for acid Appalachian soil. J Environ Qual 41(4):1096–1106

Hebbar P, Berge O, Heulin T, Singh SP (1991) Bacterial antagonists of sunflower (*Helianthus annuus* l.) fungal pathogens. Plant Soil 133(1):131–140

Hilscher A, Heister K, Siewert C, Knicker H (2009) Mineralisation and structural changes during the initial phase of microbial degradation of pyrogenic plant residues in soil. Org Geochem 40(3):332–342

Hossain MK, Strezov V, Chan KY, Nelson PF (2010) Agronomic properties of wastewater sludge biochar and bioavailability of metals in production of cherry tomato (*Lycopersicon esculentum*). Chemosphere 78(9):1167–1171

Hossain MK, Strezov V, Chan KY, Ziolkowski A, Nelson PF (2011) Influence of pyrolysis temperature on production and nutrient properties of wastewater sludge biochar. J Environ Manag 92(1):223–228

Husson O (2012) Redox potential (Eh) and pH as drivers of soil/plant/microorganism systems: a transdisciplinary overview pointing to integrative opportunities for agronomy. Plant Soil 362(1–2):389–417

Huygens D, Boeckx P, Templer P, Paulino L, Van Cleemput O, Oyarzun C, Muller C, Godoy R (2008) Mechanisms for retention of bioavailable nitrogen in volcanic rainforest soils. Nat Geosci 1:543–548

IPCC Intergovernmental Panel on Climate Change (2007) Climate change 2007: impacts, adaptation and vulnerability. In: Parry ML, Canziani OF, Palutikof JP, van der Linden PJ, Hanson CE (eds) Contribution of working group II to fourth assessment report of the Intergovernmental Panel on Climate Change. Cambridge University Press, Cambridge, p 1000

Irikiin Y, Nishiyama M, Otsuka S, Senoo K (2006) Rhizobacterial community-level, sole carbon source utilization pattern affects the delay in the bacterial wilt of tomato grown in rhizobacterial community model system. Appl Soil Ecol 34(1):27–32

Jagadish SVK, Cairns J, Lafitte R, Wheeler TR, Price AH, Craufurd PQ (2010) Genetic analysis of heat tolerance at Anthesis in Rice. Crop Sci 50(5):1633

Jaiswal AK, Elad Y, Graber ER, Frenkel O (2014) Rhizoctonia solani suppression and plant growth promotion in cucumber as affected by biochar pyrolysis temperature, feedstock and concentration. Soil Biol Biochem 69:110–118

Jaiswal AK, Frenkel O, Elad Y, Lew B, Graber ER (2015) Non-monotonic influence of biochar dose on bean seedling growth and susceptibility to Rhizoctonia solani: the "Shifted R max-Effect". Plant Soil 395(1–2):125–140

Jeffery S, Meinders MB, Stoof CR, Bezemer TM, van de Voorde TF, Mommer L, van Groenigen JW (2015) Biochar application does not improve the soil hydrological function of a sandy soil. Geoderma 251:47–54

Jia J, Li B, Chen Z, Xie Z, Xiong Z (2012) Effects of biochar application on vegetable production and emissions of N_2O and CH_4. Soil Sci Plant Nutr 58(4):503–509

Jiang J, Xu RK, Jiang TY, Li Z (2012) Immobilization of Cu (II), Pb (II) and Cd (II) by the addition of rice straw derived biochar to a simulated polluted Ultisol. J Hazard Mater 229:145–150

Jones DL, Murphy DV, Khalid M, Ahmad W, Edwards-Jones G, DeLuca TH (2011) Short-term biochar-induced increase in soil CO_2 release is both biotically and abiotically mediated. Soil Biol Biochem 43(8):1723–1731

Jones DL, Rousk J, Edwards-Jones G, DeLuca TH, Murphy DV (2012) Biochar-mediated changes in soil quality and plant growth in a three year field trial. Soil Biol Biochem 45:113–124

Joseph SD, Camps-Arbestain M, Lin Y, Munroe P, Chia CH, Hook J, Van Zwieten L, Kimber S, Cowie A, Singh BP, Lehmann J (2010) An investigation into the reactions of biochar in soil. Soil Res 48(7):501–515

Karami N, Clemente R, Moreno-Jiménez E, Lepp NW, Beesley L (2011) Efficiency of green waste compost and biochar soil amendments for reducing lead and copper mobility and uptake to ryegrass. J Hazard Mater 191(1–3):41–48

Karhu K, Mattila T, Bergström I, Regina K (2011) Biochar addition to agricultural soil increased CH4 uptake and water holding capacity–Results from a short-term pilot field study. Agric Ecosyst Environ 140(1–2):309–313

Khan SU (2016) Pesticides in the soil environment. Elsevier, Amsterdam

Khan KY, Ali B, Cui XQ, Feng Y, Yang XE, Stoffella PJ (2017) Impact of different feedstocks derived biochar amendment with cadmium low uptake affinity cultivar of pak choi (*Brassica rapa ssb. chinensis* L.) on phytoavoidation of Cd to reduce potential dietary toxicity. Ecotoxicol Environ Saf 141:129–138

Khodadad CLM, Zimmerman AR, Uthandi S, Green SJJ, Foster JS (2011) Taxa-specific changes in soil microbial composition induced by pyrogenic carbon amendments. Soil Biol Biochem 43:385–392

Khorram MS, Wang Y, Jin X, Fang H, Yu Y (2015) Reduced mobility of fomesafen through enhanced adsorption in biochar-amended soil. Environ Toxicol Chem 34(6):1258–1266
Khorram MS, Zheng Y, Lin D, Zhang Q, Fang H, Yu Y (2016) Dissipation of fomesafen in biochar-amended soil and its availability to corn (*Zea mays* L.) and earthworm (Eisenia fetida). J Soils Sediments 16(10):2439–2448
Kim HS, Kim KR, Yang JE, Ok YS, Owens G, Nehls T, Wessolek G, Kim KH (2016) Effect of biochar on reclaimed tidal land soil properties and maize (*Zea mays* L.) response. Chemosphere 142:153–159
Kimetu JM, Lehmann J (2010) Stability and stabilisation of biochar and green manure in soil with different organic carbon contents. Aust J Soil Res 48:577–585
Knief C (2015) Diversity and habitat preferences of cultivated and uncultivated aerobic methanotrophic bacteria evaluated based on pmoA as molecular marker. Front Microbiol 6:1346
Knoblauch C, Zimmermann U, Blumenberg M, Michaelis W, Pfeiffer EM (2008) Methane turnover and temperature response of methane-oxidizing bacteria in permafrost-affected soils of northeast Siberia. Soil Biol Biochem 40(12):3004–3013
Kolton M, Harel YM, Pasternak Z, Graber ER, Elad Y, Cytryn E (2011) Impact of biochar application to soil on the root-associated bacterial community structure of fully developed greenhouse pepper plants. Appl Environ Microbiol 77(14):4924–4930
Kong LL, Liu WT, Zhou QX (2014) Biochar: an effective amendment for remediating contaminated soil. In: Reviews of environmental contamination and toxicology, vol 228. Springer, Cham, pp 83–99
Kookana RS, Sarmah AK, Van Zwieten L, Krull E, Singh B (2011) Chapter three—biochar application to soil: agronomic and environmental benefits and unintended consequences. In: Donald LS (ed) Advances in agronomy, vol 112. Academic, New York, pp 103–143
Krause MS, Madden LV, Hoitink HA (2001) Effect of potting mix microbial carrying capacity on biological control of Rhizoctonia damping-off of radish and Rhizoctonia crown and root rot of poinsettia. Phytopathology 91(11):1116–1123
Kumpiene J, Lagerkvist A, Maurice C (2008) Stabilization of As, Cr, Cu, Pb and Zn in soil using amendments–a review. Waste Manag 28(1):215–225
Kuzyakov Y, Subbotina I, Chen H, Bogomolova I, Xu X (2009) Black carbon decomposition and incorporation into soil microbial biomass estimated by ^{14}C labeling. Soil Biol Biochem 41:210–219. https://doi.org/10.1016/j.soilbio.2008.10.016
Laird D, Fleming P, Wang B, Horton R, Karlen D (2010) Biochar impact on nutrient leaching from a Midwestern agricultural soil. Geoderma 158(3–4):436–442
Larsbo M, Löfstrand E, de Veer DV, Ulén B (2013) Pesticide leaching from two Swedish topsoils of contrasting texture amended with biochar. J Contam Hydrol 147:73–81
Lashari MS, Ye Y, Ji H, Li L, Kibue GW, Lu H, Zheng J, Pan G (2015) Biochar–manure compost in conjunction with pyroligneous solution alleviated salt stress and improved leaf bioactivity of maize in a saline soil from central China: a 2-year field experiment. J Sci Food Agric 95:1321–1327
Lehmann J (2007) Concepts and questions: bioenergy in the black. Front Ecol Environ 5:381–387
Lehmann J, Joseph S (2009) Biochar for environmental management: an introduction. In: Lehmann J, Joseph S (eds) (2015) Biochar for environmental management: science, technology and implementation. Routledge, London
Lehmann J, Joseph S (2015) Biochar for environmental management: science, technology and implementation. Routledge, New York
Lehmann J, da Silva JP, Steiner C, Nehls T, Zech W, Glaser B (2003) Nutrient availability and leaching in an archaeological Anthrosol and a Ferralsol of the Central Amazon basin: fertilizer, manure and charcoal amendments. Plant Soil 249(2):343–357
Lehmann J, Gaunt J, Rondon M (2006) Bio-char sequestration in terrestrial ecosystems—a review. Mitig Adapt Strateg Glob Chang 11:395–419
Lehmann J, Rillig MC, Thies J, Masiello CA, Hockaday WC, Crowley D (2011) Biochar effects on soil biota – a review. Soil Biol Biochem 43:1812–1836

Li X, Hayashi J, Li CZ (2006) FT-Raman spectroscopic study of the evolution of char structure during the pyrolysis of a Victorian brown coal. Fuel 85:1700–1707

Li J, Li Y, Wu M, Zhang Z, Lu J (2013) Effectiveness of low-temperature biochar in controlling the release and leaching of herbicides in soil. Plant Soil 370(1):333–344

Li S, Barreto V, Li R, Chen G, Hsieh YP (2018) Nitrogen retention of biochar derived from different feedstocks at variable pyrolysis temperatures. J Anal Appl Pyrolysis 133:136–146

Lin Y, Munroe P, Joseph S, Henderson R, Ziolkowski A (2012) Water extractable organic carbon in untreated and chemical treated biochars. Chemosphere 87(2):151–157

Liu Z, Quek A, Hoekman SK, Balasubramanian R (2013) Production of solid biochar fuel from waste biomass by hydrothermal carbonization. Fuel 103:943–949

Lone AH, Najar GR, Ganie MA, Sofi JA, Tahir Ali T (2015) Biochar for sustainable soil health: a review of prospects and concerns. Pedosphere 25(5):639–653

Lopez-Pineiro A, Pena D, Albarran A, Sánchez-Llerena J, Becerra D (2013) Behavior of MCPA in four intensive cropping soils amended with fresh, composted, and aged olive mill waste. J Contam Hydrol 152:137–146

Lovley DR (2012) Electromicrobiology. Ann Rev Microbiol 66:391–409

Lu H, Zhang W, Yang Y, Huang X, Wang S, Qiu R (2012) Relative distribution of Pb^{2+} sorption mechanisms by sludge-derived biochar. Water Res 46(3):854–862

Lu K, Yang X, Shen J, Robinson B, Huang H, Liu D, Bolan N, Pei J, Wang H (2014) Effect of bamboo and rice straw biochars on the bioavailability of Cd, Cu, Pb and Zn to *Sedum plumbizincicola*. Agric Ecosyst Environ 191:124–132

Lu W, Kang C, Wang Y, Xie Z (2015) Influence of biochar on the moisture of dark brown soil and yield of maize in Northern China. Int J Agric Biol 17(5):1007–1012

Lyu S, Du G, Liu Z, Zhao L, Lyu D (2016) Effects of biochar on photosystem function and activities of protective enzymes in *Pyrus ussuriensis* Maxim. under drought stress. Acta Physiol Plant 38:1–10

Major J, Lehmann J, Rondon M, Goodale C (2010a) Fate of soil applied black carbon: downward migration, leaching and soil respiration. Glob Chang Biol 16:1366–1379

Major J, Rondon M, Molina D, Riha S, Lehmann J (2010b) Maize yield and nutrition during 4 years after biochar application to a Colombian savanna oxisol. Plant Soil 333:117–128

Marin-Benito JM, Brown CD, Herrero-Hernández E, Arienzo M, Sánchez-Martín MJ, Rodríguez-Cruz MS (2013) Use of raw or incubated organic wastes as amendments in reducing pesticide leaching through soil columns. Sci Total Environ 463–464:589–599

Martin SM, Kookana RS, Van Zwieten L, Krull E (2012) Marked changes in herbicide sorption–desorption upon ageing of biochars in soil. J Hazard Mater 231–232:70–78

Martinsen V, Mulder J, Shitumbanuma V, Sparrevik M, Børresen T, Cornelissen G (2014) Farmer-led maize biochar trials: effect on crop yield and soil nutrients under conservation farming. J Plant Nutr Soil Sci 177:681–695

Matos A, Kerkhof L, Garland JL (2005) Effects of microbial community diversity on the survival of *Pseudomonas aeruginosa* in the wheat rhizosphere. Microb Ecol 49(2):257–264

Matsui T, Omasa K (2002) Rice (*Oryza sativa* L.) cultivars tolerant to high temperature at flowering: anther characteristics. Ann Bot 89(6):683–687

Méndez A, Barriga S, Fidalgo JM, Gascó G (2009) Adsorbent materials from paper industry waste materials and their use in Cu (II) removal from water. J Hazard Mater 165(1–3):736–743

Mia S, Van Groenigen JW, Van de Voorde TF, Oram NJ, Bezemer TM, Mommer L, Jeffery S (2014) Biochar application rate affects biological nitrogen fixation in red clover conditional on potassium availability. Agric Ecosyst Environ 191:83–91

Mosier AR, Duxbury JM, Freney JR, Heinemeyer O, Minami K (1998) Assessing and mitigating N_2O emissions from agricultural soils. Clim Chang 40:7–38

Mukherjee A, Zimmerman AR, Harris W (2011) Surface chemistry variations among a series of laboratory-produced biochars. Geoderma 163:247–255

Mulcahy DN, Mulcahy DL, Dietz D (2013) Biochar soil amendment increases tomato seedling resistance to drought in sandy soils. J Arid Environ 88:222–225

Munns R, Tester M (2008) Mechanisms of salinity tolerance. Annu Rev Plant Biol 59:651–681
Narzari R, Bordoloi N, Chutia RS, Borkotoki B, Gogoi N, Bora A, Kataki R (2015) Biochar: an overview on its production, properties and potential benefits. In: Biology, biotechnology and sustainable development, pp 13–40. https://doi.org/10.13140/RG.2.1.3966.2560
Nelissen V, Ruysschaert G, Manka'Abusi D, D'Hose T, De Beuf K, Al-Barri B, Cornelis W, Boeckx P (2015) Impact of a woody biochar on properties of a sandy loam soil and spring barley during a two-year field experiment. Eur J Agron 62:65–78
Newsham KK, Fitter AH, Watkinson AR (2005) Multi-functionality and biodiversity in arbuscular mycorrhizas. Trends Ecol Evol 10:407–411
Nguyen BT, Lehmann J (2009) Black carbon decomposition under varying water regimes. Org Geochem 40:846–853
Nguyen TT, Xu CY, Tahmasbian I, Che R, Xu Z, Zhou X, Wallace HM, Bai SH (2017) Effects of biochar on soil available inorganic nitrogen: a review and meta-analysis. Geoderma 288:79–96
Nishio M (1996) Microbial fertilizers in Japan, FFTC extension bulletin. Food and Fertilizer Technology Center, Taipei
O'Connor D, Peng T, Zhang J, Tsang DC, Alessi DS, Shen Z, Bolan NS, Hou D (2018) Biochar application for the remediation of heavy metal polluted land: a review of in situ field trials. Sci Total Environ 619:815–826
Olmo M, Alburquerque JA, Barrón V, del Campillo MC, Gallardo A, Fuentes M, Villar R (2014) Wheat growth and yield responses to biochar addition under Mediterranean climate conditions. Biol Fert Soil 50:1177–1187
Oo AN, Iwai CB, Saenjan P (2015) Soil properties and maize growth in saline and nonsaline soils using cassava-industrial waste compost and vermicompost with or without earthworms. Land Degrad Develop 26:300–310
Paneque M, José M, Franco-Navarro JD, Colmenero-Flores JM, Knicker H (2016) Effect of biochar amendment on morphology, productivity and water relations of sunflower plants under non-irrigation conditions. Catena 147:280–287
Panuccio MR, Jacobsen SE, Akhtar SS, Muscolo A (2014) Effect of saline water on seed germination and early seedling growth of the halophyte quinoa. AoB Plants 6. https://doi.org/10.1093/aobpla/plu047
Parihar P, Singh S, Singh R, Singh VP, Prasad SM (2015) Effect of salinity stress on plants and its tolerance strategies: a review. Environ Sci Pollut Res 22:4056–4075
Park JH, Choppala GK, Bolan NS, Chung JW, Chuasavathi T (2011) Biochar reduces the bioavailability and phytotoxicity of heavy metals. Plant Soil 348:439–451
Park JH, Choppala G, Lee SJ, Bolan N, Chung JW, Edraki M (2013) Comparative sorption of Pb and Cd by biochars and its implication for metal immobilization in soils. Water Air Soil Pollut 224(12):1711
Pereira RG, Heinemann AB, Madari BE, Carvalho MTDM, Kliemann HJ, Santos APD (2012) Transpiration response of upland rice to water deficit changed by different levels of eucalyptus biochar. Pesq Agropec Bras 47:716–721
Powlson DS, Whitmore AP, Goulding KWT (2011) Soil carbon sequestration to mitigate climate change: a critical re-examination to identify the true and the false. Eur J Soil Sci 62(1):42–55
Pratscher J, Dumont MG, Conrad R (2011) Ammonia oxidation coupled to CO_2 fixation by archaea and bacteria in an agricultural soil. Proc Natl Acad Sci 108(10):4170–4175
Prendergast-Miller MT, Duvall M, Sohi SP (2014) Biochar–root interactions are mediated by biochar nutrient content and impacts on soil nutrient availability. Eur J Soil Sci 65(1):173–185
Qiu Y, Pang H, Zhou Z, Zhang P, Feng Y, Sheng DG (2009) Competitive biodegradation of dichlobenil and atrazine coexisting in soil amended with a char and citrate. Environ Pollut 157(11):2964–2969
Rajkovich S, Enders A, Hanley K, Hyland C, Zimmerman AR, Lehmann J (2012) Corn growth and nitrogen nutrition after additions of biochars with varying properties to a temperate soil. Biol Fertil Soils 48(3):271–284

Reddy KR, Yargicoglu EN, Yue D, Yaghoubi P (2014a) Enhanced microbial methane oxidation in landfill cover soil amended with biochar. J Geotech Geoenviron 140(9):04014047

Reddy KR, Xie T, Dastgheibi S (2014b) Evaluation of biochar as a potential filter media for the removal of mixed contaminants from urban storm water runoff. J Environ Eng 140(12):04014043

Rees F, Simonnot MO, Morel JL (2014) Short-term effects of biochar on soil heavy metal mobility are controlled by intra-particle diffusion and soil pH increase. Eur J Soil Sci 65:149–161. https://doi.org/10.1111/ejss.12107

Rondon M, Ramirez JA, Lehmann J (2005) Charcoal additions reduce net emissions of greenhouse gases to the atmosphere. In: Proceedings of the 3rd USDA Symposium on Greenhouse Gases and Carbon Sequestration in Agriculture and Forestry, Baltimore, MD. 21–24 Mar, University of Delaware, USA, p 208

Rondon MA, Lehmann J, Ramirez J, Hurtado M (2007) Biological nitrogen fixation by common beans (*Phaseolus vulgaris* L.) increases with biochar additions. Biol Fertil Soils 43:699–708

Roy M, McDonald LM (2015) Metal uptake in plants and health risk assessments in metal-contaminated smelter soils. Land Degrad Dev 26(8):785–792

Rubasinghege G, Spak SN, Stanier CO, Carmichael GR, Grassian VH (2011) Abiotic mechanism for the formation of atmospheric nitrous oxide from ammonium nitrate. Environ Sci Technol 45:2691–2697

Saarnio S, Heimonen K, Kettunen R (2013) Biochar addition indirectly affects N2O emissions via soil moisture and plant N uptake. Soil Biol Biochem 58:99–106

Samarkin VA, Madigan MT, Bowles MW, Casciotti KL, Priscu JC, McKay CP, Joye SB (2010) Abiotic nitrous oxide emission from the hypersaline Don Juan Pond in Antarctica. Nat Geosci 3:341–344

Sarma B, Borkotoki B, Gogoi N, Kataki R (2017a) Responses of soil enzymes and carbon mineralization to applied organic amendments: a short-term study in acidic sandy loam soil. J Indian Soc Soil Sci 65(3):283–289

Sarma B, Borkotoki B, Narzari R, Kataki R, Gogoi N (2017b) Organic amendments: effect on carbon mineralization and crop productivity in acidic soil. J Clean Prod 152:157–166

Sarma B, Gogoi N, Bharali M, Mali P (2018) Field evaluation of soil and wheat responses to combined application of hardwood biochar and inorganic fertilizers in acidic sandy loam soil. Exp Agric 54(4):507–519

Scheer C, Grace P, Rowlings D, Kimber S, Van Zwieten L (2011) Effect of biochar amendment on the soil-atmosphere exchange of greenhouse gases from an intensive subtropical pasture in northern New South Wales, Australia. Plant Soil 1–2:47–58

Scheuerell SJ, Sullivan DM, Mahaffee WF (2005) Suppression of seedling damping-off caused by *Pythium ultimum*, P. irregulare, and Rhizoctonia solani in container media amended with a diverse range of Pacific Northwest compost sources. Phytopathology 95(3):306–315

Schmidt HP, Kammann C, Niggli C, Evangelou MW, Mackie KA, Abiven S (2014) Biochar and biochar-compost as soil amendments to a vineyard soil: influences on plant growth nutrient uptake, plant health and grape quality. Agric Ecosyst Environ 191:117–123

Schmidt HP, Pandit BH, Martinsen V, Cornelissen G, Conte P, Kammann CI (2015) Fourfold increase in pumpkin yield in response to low-dosage root zone application of urine-enhanced biochar to a fertile tropical soil. Agriculture 5(3):723–741

Scott AC, Glasspool IJ (2007) Observations and experiments on the origin and formation of inertinite group macerals. Int J Coal Geol 70:53–66

Semple KT, Doick KJ, Jones KC, Burauel P, Craven A, Harms H (2004) Defining bioavailability and bioaccessibility of contaminated soil and sediment is complicated. Environ Sci Technol 38:228A–231A

Shilev SI, Ruso J, Puig A, Benlloch M, Jorrin J, Sancho E (2001) Rhizospheric bacteria promote sunflower (*Helianthus annuus* L.) plant growth and tolerance to heavy metals. Minerva Biotecnol 13:37–39

Silber A, Levkovitch I, Graber ER (2010) pH-dependent mineral release and surface properties of cornstraw biochar: agronomic implications. Environ Sci Technol 44(24):9318–9323

Singh BP, Hatton BJ, Singh B, Cowie AL, Kathuria A (2010) Influence of biochars on nitrous oxide emission and nitrogen leaching from two contrasting soils. J Environ Qual 39:1224–1235
Smith LC, deKlein CA, Catto WD (2008) Effect of dicyandiamide applied in a granular form on nitrous oxide emissions from a grazed dairy pasture in Southland, New Zealand. N Z J Agric Res 51(4):387–396
Smith JL, Collins HP, Bailey VL (2010) The effect of young biochar on soil respiration. Soil Biol Biochem 42(12):2345–2347
Sohi SP, Krull E, Lopez-Capel E, Bol R (2010) A review of biochar and its use and function in soil. In: Advances in agronomy, vol 105. Academic, Burlington, pp 47–82
Solaiman ZM, Blackwell P, Abbott LK, Storer P (2010) Direct and residual effect of biochar application on mycorrhizal root colonisation, growth and nutrition of wheat. Soil Res 48(7):546–554
Sopena F, Semple K, Sohi S, Bending G (2012) Assessing the chemical and biological accessibility of the herbicide isoproturon in soil amended with biochar. Chemosphere 88(1):77–83
Spokas KA, Bogner JE (2011) Limits and dynamics of methane oxidation in landfill cover soils. Waste Manag 31(5):823–832
Spokas KA, Koskinen WC, Baker JM, Reicosky DC (2009) Impacts of woodchip biochar additions on greenhouse gas production and sorption/degradation of two herbicides in a Minnesota soil. Chemosphere 77(4):574–581
Stavi I, Lal R (2013) Agroforestry and biochar to offset climate change: a review. Agron Sustain Dev 33(1):81–96
Steinbeiss S, Gleixner G, Antonietti M (2009) Effect of biochar amendment on soil carbon balance and soil microbial activity. Soil Biol Biochem 41(6):1301–1310
Steiner C, Teixeira W, Lehmann J, Nehls T, de Macêdo J, Blum W, Zech W (2007) Long term effects of manure, charcoal and mineral fertilization on crop production and fertility on a highly weathered Central Amazonian upland soil. Plant Soil 291:275–290
Steiner C, Glaser B, Geraldes Teixeira W, Lehmann J, Blum WE, Zech W (2008) Nitrogen retention and plant uptake on a highly weathered central Amazonian Ferralsol amended with compost and charcoal. J Plant Nutr Soil Sci 171(6):893–899
Steiner C, Das CK, Melear N, Lakly D (2010) Reducing nitrogen loss during poultry litter composting using biochar. J Environ Qual 39:1236–1242. https://doi.org/10.2134/jeq2009.0337
Sun J, He F, Shao H, Zhang Z, Xu G (2016) Effects of biochar application on Suaeda salsa growth and saline soil properties. Environ Earth Sci 75:1–6
Tagoe SO, Horiuchi T, Matsui T (2008) Effects of carbonized and dried chicken manures on the growth, yield, and N content of soybean. Plant Soil 306(1–2):211–220
Tan XF, Liu SB, Liu YG, Gu YL, Zeng GM, Hu XJ, Wang X, Liu SH, Jiang LH (2017) Biochar as potential sustainable precursors for activated carbon production: Multiple applications in environmental protection and energy storage. Biores Technol 227:359–372
Tatarkova V, Hiller E, Vaculik M (2013) Impact of wheat straw biochar addition to soil on the sorption, leaching, dissipation of the herbicide (4-chloro-2-methylphenoxy) acetic acid and the growth of sunflower (*Helianthus annuus* L.). Ecotoxicol Environ Saf 92:215–221
Termorshuizen AJ, Van Rijn E, Van Der Gaag DJ, Alabouvette C, Chen Y, Lagerlöf J, Malandrakis AA, Paplomatas EJ, Rämert B, Ryckeboer J, Steinberg C (2006) Suppressiveness of 18 composts against 7 pathosystems: variability in pathogen response. Soil Biol Biochem 38(8):2461–2477
Thies JE, Rillig MC (2009) Characteristics of biochar: biological properties. In: Lehmann J, Joseph S (eds) Biochar for environmental management: science and technology. Earthscan Press, UK, pp 85–105
Thomson AJ, Giannopoulos G, Pretty J, Baggs EM, Richardson DJ (2012) Biological sources and sinks of nitrous oxide and strategies to mitigate emissions. Philos Trans R Soc Lond B: Biol Sci 367:1157–1168
Uchimiya M, Wartelle LH, Klasson KT, Fortier CA, Lima IM (2011) Influence of pyrolysis temperature on biochar property and function as a heavy metal sorbent in soil. J Agric Food Chem 59:2501–2510. https://doi.org/10.1021/jf104206c

Unger RC (2008) The effect of bio-char on soil properties and corn grain yields in Iowa. Graduate Theses and Dissertations. 11719. https://lib.dr.iastate.edu/etd/11719

Vaccari FP, Baronti S, Lugato E, Genesio L, Castaldi S, Fornasier F, Miglietta F (2011) Biochar as a strategy to sequester carbon and increase yield in durum wheat. Eur J Agron 34:231–238

Vaccari FP, Maienza A, Miglietta F, Baronti S, Di Lonardo S, Giagnoni L, Lagomarsino A, Pozzi A, Pusceddu E, Ranieri R, Valboa G (2015) Biochar stimulates plant growth but not fruit yield of processing tomato in a fertile soil. Agric Ecosyst Environ 207:163–170

Van Zwieten L, Kimber S, Morris S, Chan K, Downie A, Rust J, Joseph S, Cowie A (2010a) Effects of biochar from slow pyrolysis of papermill waste on agronomic performance and soil fertility. Plant Soil 327:235–246

Van Zwieten L, Kimber S, Morris S, Downie A, Berger E, Rust J, Scheer C (2010b) Influence of biochars on flux of N2O and CO2 from ferrosol. Aust J Soil Res 48:555–568

Wang HL, Lin KD, Hou ZN, Richardson B, Gan J (2010) Sorption of the herbicide terbuthylazine in two New Zealand forest soils amended with biosolids and biochars. J Soils Sediments 10(2):283–289

Wang J, Zhang M, Xiong Z, Liu P, Pan G (2011) Effects of biochar addition on N_2O and CO_2 emissions from two paddy soils. Biol Fert Soils 47:887–896

Wang D, Maughan MW, Sun J, Feng X, Miguez F, Lee D, Dietze MC (2012) Impact of nitrogen allocation on growth and photosynthesis of Miscanthus (Miscanthus × giganteus). GCB Bioenergy 4:688–697

Wang X, Cui H, Shi J, Zhao X, Zhao Y, Wei Z (2015) Relationship between bacterial diversity and environmental parameters during composting of different raw materials. Biores Technol 198:395–402

Wang J, Xiong Z, Kuzyakov Y (2016) Biochar stability in soil: meta-analysis of decomposition and priming effects. GCB Bioenergy 8:512–523

Warnock DD, Lehmann J, Kuyper TW, Rillig MC (2007) Mycorrhizal responses to biochar in soil concepts and mechanisms. Plant Soil 300:9–20

Warnock DD, Mummey DL, McBride B, Major J, Lehmann J, Rillig MC (2010) Influences of non-herbaceous biochar on arbuscular mycorrhizal fungal abundances in roots and soils: results from growth chamber and field experiments. Appl Soil Ecol 46:450–456

Wolf D (2008) Biochar as a soil amendment a review of the environmental implications. Available via http://orgprints.org/13268/1/Biochar_as_a_soil_amendment_a_review.pdf. Accessed 8 July 2010

Xiao Q, Zhu LX, Shen YF, Li SQ (2016) Sensitivity of soil water retention and availability to biochar addition in rainfed semi-arid farmland during a three-year field experiment. Field Crops Res 196:284–293

Xu C, Liu W, Sheng GD (2008) Burned rice straw reduces the availability of clomazone to barnyardgrass. Sci Total Environ 392(2–3):284–289

Xu NN, Lin DS, Xu YM, Xie ZL, Liang XF, Guo WJ (2014) Adsorption of aquatic Cd^{2+} by biochar obtained from corn stover. J Agro-Environ Sci. (in Chinese 33:958–964

Yaghoubi P, Yargicoglu E, Reddy K (2014) Effects of biochar-amendment to landfill cover soil on microbial methane oxidation: initial results. Geo-Congress 2014 Technical papers. American Society of Civil Engineers, pp 1849–1858

Yanai Y, Toyota K, Okazaki M (2007) Effects of charcoal addition on N_2O emissions from soil resulting from rewetting air-dried soil in short-term laboratory experiments. Soil Sci Plant Nutr 53:181–188

Yang YN, Sheng GY (2003) Enhanced pesticide sorption by soils containing particulate matter from crop residue burns. Environ Sci Technol 37(16):3635–3639

Yang X, Lu KP, McGrouther K, Che L, Hu GT, Wang QY, Liu XY, Shen LL, Huang HG, Ye ZQ, Wang HL (2017) Bioavailability of Cd and Zn in soils treated with biochars derived from tobacco stalk and dead pigs. J Soils Sediment 17:751–762

Yao Y, Gao B, Zhang M, Inyang M, Zimmerman AR (2012) Effect of biochar amendment on sorption and leaching of nitrate, ammonium, and phosphate in a sandy soil. Chemosphere 89(11):1467–1471

Yeo AR, Flowers TJ (1986) Salinity resistance in rice (*Oryza sativa* L.) and a pyramiding approach to breeding varieties for saline soils. Funct Plant Biol 13(1):161–173

Yu XY, Ying GG, Kookana RS (2006) Sorption and desorption behaviors of diuron in soils amended with charcoal. J Agric Food Chem 54(22):8545–8550

Yuan JH, Xu RK (2012) Effects of biochars generated from crop residues on chemical properties of acid soils from tropical and subtropical China. Soil Res 50(7):570–578

Yuan X, Huang H, Zeng G, Li H, Wang J, Zhou C, Zhu H, Pei X, Liu Z, Liu Z (2011) Total concentrations and chemical speciation of heavy metals in liquefaction residues of sewage sludge. Biores Technol 102(5):4104–4110

Yuan H, Lu T, Wang Y, Chen Y, Lei T (2016) Sewage sludge biochar: nutrient composition and its effect on the leaching of soil nutrients. Geoderma 267:17–23

Zahir ZA, Akhtar SS, Ahmad M, Nadeem SM, Nadeem SM (2012) Comparative effectiveness of *Enterobacter aerogenes* and *Pseudomonas fluorescens* for mitigating the depressing effect of brackish water on maize. Int J Agric Biol 14(3):337–344

Zhang P, Sheng G, Feng Y, Miller DM (2005) Role of wheat-residue-derived char in the biodegradation of benzonitrile in soil: nutritional stimulation versus adsorptive inhibition. Environ Sci Technol 39(14):5442–5448

Zhang P, Sheng G, Feng Y, Miller DM (2006) Predominance of char sorption over substrate concentration and soil pH in influencing biodegradation of benzonitrile. Biodegradation 17(1):1–8

Zhang H, Lin K, Wang H, Gan J (2010) Effect of *Pinus radiata* derived biochars on soil sorption and desorption of phenanthrene. Environ Pollut 158(9):2821–2825

Zhang X, He L, Sarmah A, Lin K, Liu Y, Li J, Wang H (2014) Retention and release of diethyl phthalate in biochar-amended vegetable garden soils. J Soils Sediments 14:1790–1799

Zhang J, Liu J, Liu R (2015) Effects of pyrolysis temperature and heating time on biochar obtained from the pyrolysis of straw and lignosulfonate. Biores Technol 176:288–291

Zhao L, Cao X, Mašek O, Zimmerman A (2013) Heterogeneity of biochar properties as a function of feedstock sources and production temperatures. J Hazard Mater 256:1–9

Zheng J, Zhang X, Li L, Zhang P, Pan G (2007) Effect of long-term fertilization on C mineralization and production of CH4 and CO2 under anaerobic incubation from bulk samples and particle size fractions of a typical paddy soil. Agric Ecosyst Environ 120:129–138

Zheng R, Chen Z, Cai C, Wang X, Huang Y, Xiao B, Sun G (2013) Effect of biochars from rice husk, bran, and straw on heavy metal uptake by pot-grown wheat seedling in a historically contaminated soil. BioResources 8(4):5965–5982

Zimmerman AR (2010) Abiotic and microbial oxidation of laboratory-produced black carbon (biochar). Environ Sci Technol 44(4):1295–1301

Zimmerman AR, Gao B, Ahn M-Y (2011) Positive and negative carbon mineralization priming effects among a variety of biochar-amended soils. Soil Biol Biochem 43:1169–1179

Zwieten VL, Singh B, Joseph S, Kimber S, Cowie A, Chan YK (2009) Biochar and emissions of non-CO_2 greenhouse gases from soil. In: Lehmann J, Joseph S (eds) Biochar for environmental management science and technology. Earthscan Press, UK, pp 227–249

Managing Drylands for Sustainable Agriculture

J. L. Arrúe, J. Álvaro-Fuentes, D. Plaza-Bonilla, D. Villegas, and C. Cantero-Martínez

Abstract Major constraints to rainfed production systems in the world's drylands include low and highly variable rainfall, nutrient deficiencies and land degradation by wind and water erosion. Although the same principles to cope with these limitations could be in theory applied to all dryland situations, there is no a universal recipe for sustainable dryland agriculture. In this chapter, the authors recall some of the challenges that have been identified for semiarid rainfed farming systems, namely soil conservation, water use efficiency, nutrient use efficiency and climate change mitigation, as well as some sustainable cropping and management strategies that have been formulated and recommended to address them appropriately. To this end, the authors provide examples supporting those practices mainly from semiarid Mediterranean agroecosystems. Among all the strategies discussed in this Chapter, and despite their limitations, the maintenance of a protective crop residue cover and the reduction of tillage operations appear to be the simplest technological options not only to control soil erosion but also to improve water and nutrient use efficiency and mitigate greenhouse gas emissions. The authors conclude that sustainable agricultural management in drylands should be primarily based on conservation agriculture practices and associated local-based crop residue management systems.

Keywords Conservation tillage · Semiarid lands · Dryland farming · Wind erosion · Water use efficiency · Nutrient use efficiency · Greenhouse gas emission · Crop residue management · Mediterranean agroecosystems

---

J. L. Arrúe (✉) · J. Álvaro-Fuentes · D. Plaza-Bonilla
Soil and Water Department, Estación Experimental de Aula Dei (EEAD), Spanish National Research Council (CSIC), Zaragoza, Spain
e-mail: arrue@eead.csic.es; jorgeaf@eead.csic.es; dplaza@eead.csic.es

D. Villegas
Sustainable Field Crops Program, Institute of Agriculture and Food Research and Technology (IRTA), Lleida, Spain
e-mail: dolors.villegas@irta.cat

C. Cantero-Martínez
Crop and Forest Sciences Department, EEAD-CSIC Associated Unit, Agrotecnio, University of Lleida (UdL), Lleida, Spain
e-mail: carlos.cantero@pvcf.udl.cat

M. Farooq, M. Pisante (eds.), *Innovations in Sustainable Agriculture*,
https://doi.org/10.1007/978-3-030-23169-9_17

1 Introduction

According to the classification of the United Nations Convention to Combat Desertification (UNCCD 2000), dryland areas are characterized by low values of the mean annual precipitation to potential evapotranspiration ratio (ranging from 0.05 to 0.65) and account for about 41% of the surface of the Earth (Middleton and Thomas 1997). Semi-arid zones, which are the more extensive, occur in all the continents and cover up to 18% of the land surface. Dryland ecosystems are extremely diverse and include the Mediterranean systems, the cold deserts of Chile and Mongolia, the Sahel and Sahara of Africa, the Arctic Circle, and the high altitude drylands of Afghanistan and Iran (Farooq and Siddique 2016). More than two and a half billion people live in dryland areas, which represents about 37% of the total world population (Koohafkan and Stewart 2008). Furthermore, drylands support 50% of the world's livestock, grow 44% of the world's cultivated ecosystems, store 46% of the planet's carbon inventory and lose 23 ha per minute due to drought and desertification, which implies a loss of 20 million tons of potential grain production every year (CGIAR 2017).

In addition to lack of water, the constraints to sustainable agriculture in drylands include erratic rainfall, land degradation by wind and water erosion, and limited access to agricultural technologies. Dryland soils are typically characterized by moisture deficit, low levels of soil organic matter (SOM) and biological activity and poor fertility. When inappropriately used for agriculture, these soils are prone to fertility loss, erosion, desertification, salinization, and other land degradation processes. On the other hand, drylands are highly vulnerable to climate and land use changes. Globally, drylands expanded over the twentieth century by 4–8% and are expected to increase in extent and aridity in coming decades (Schlaepfer et al. 2017). In summary, rainfed production systems in the world's drylands are constrained by land degradation, nutrient deficiencies, and increasing water scarcity.

The importance of drylands to key emerging issues on the global agenda, including climate change and food security, has been recognized over the last two decades in several reports (e.g. Safriel and Adeel 2005; UN 2011; FAO 2016). These reports address dryland management, taking into account environmental concerns and the well-being of dryland communities, as well as the underlying causes of dryland degradation. Consequently, the subject of sustainable agricultural systems for drylands has been addressed in different international conferences and research projects. Some examples are the International Symposium for Sustainable Dryland Agriculture Systems (Omanya and Pasternak 2005) or the MEDRATE project on the evaluation of agricultural technologies to improve efficiency and environment conservation in Mediterranean arid and semi-arid production systems (Cantero-Martínez and Gabiña 2004) or the SUMAMAD project on sustainable management of marginal drylands of Northern Africa and Asia (Lee and Schaaf 2008). The CGIAR Research Program on Dryland Systems, led by the International Center for Agricultural Research in the Dry Areas (ICARDA), to improve agricultural productivity and income in the world's dry areas is another example worth to be mentioned (CGIAR 2013).

The World Overview of Conservation Approaches and Technologies (WOCAT) programme considers conservation agriculture (CA) as an option for sustainable land management in drylands (Schwilch et al. 2012). Conservation agriculture (CA) is characterized by three linked principles: tillage reduction, soil cover maintenance and crop rotations. Compared to conventional agriculture, CA induces beneficial changes in soil properties and processes which favor the delivery of multiple soil ecosystem services (Palm et al. 2014). While current crop production systems have often resulted in soil degradation and in extreme cases desertification, the adoption of CA technologies has led to a reversion of these processes (Friedrich et al. 2012). Although CA principles can be universally applicable to all agricultural landscapes and land uses, the benefits of implementing CA successfully are perhaps the least in dryland farming areas (Stewart and Thapa 2016). In dryland farming areas, even when no-tillage is used, there are often insufficient crop residues remaining after planting the next crop to cover 30% of the soil surface, which is the threshold for an agricultural system to be considered under CA. This is particularly true in extensive cereal–fallow rotation agroecosystems, when only one crop is produced every 2 years, that results in long fallow periods between crops. Moreover, in many dryland farming areas, especially in developing countries of Asia and Africa, many crop residues are removed as fodder for cattle or for cooking and heating fuel, which can lead to lower SOM concentrations and lower soil quality (Plaza-Bonilla et al. 2015). However, the adoption of CA should be recommended for all dryland farming systems (Stewart and Thapa 2016).

Dryland farming, dryland agriculture, rainfed agriculture and rainfed farming are terms often used in an interchangeable way. In this chapter we shall consider rainfed agriculture to be synonymous of dryland agriculture, term that stresses soil and water conservation and sustainable crop production under the constraints characteristic of conditions in semi-arid drylands. Currently, the target is to increase resilience and to achieve and maintain the sustainability of dryland agriculture. A review of literature concerning rainfed agriculture and sustainable land management technologies is outside the scope of this Chapter. Instead, this Chapter discusses some of the challenges that have been identified for semiarid non-irrigated dryland agriculture, namely soil conservation, efficient use of both water and nutrients, and climate change mitigation, with emphasis on some sustainable management practices and measures that have been formulated and recommended to address those issues appropriately.

2 Integrated Challenges for Sustainable Agriculture in Drylands

2.1 Soil Conservation

Globally, soil erosion is a leading cause of land degradation in drylands (Farooq and Siddique 2016) and still remains a major issue. Dryland environments are usually prone to soil erosion by water due to the lack of a significant soil cover, which is usually aggravated by the high intensity of rainstorms (typical in some dryland areas such as the Mediterranean basin), a reduced soil structural stability, which is generally associated to a limited amount of soil organic carbon (SOC), and a high human pressure (Plaza-Bonilla et al. 2015). In some areas, the presence of steep slopes also enhances the susceptibility to soil erosion in drylands.

Soil nutrient losses due to wind erosion is also another major characteristic of drylands. According to the Global Assessment of Soil Degradation (GLASOD) map (Oldeman 1994), wind erosion is the most important soil degradation process on drylands, followed by water erosion, chemical degradation (i.e. soil salinization) and physical degradation (i.e. soil compaction and crusting) (Sterk et al. 2001). In general, wind erosion can be successfully controlled by reducing the wind velocity at the soil surface to a non-erosive level or by increasing soil roughness to a level creating sufficient resistance of the soil to wind forces (Sterk 2003; Unger et al. 2006). In other words, wind speed at the soil-air interface must be reduced to below the threshold value needed for the initiation of soil particle movement (Unger et al. 2006). In dryland agroecosystems, wind erosion can be offset through the adoption of appropriate farming practices. Some the strategies that have been proposed to control wind erosion include: maintain a vegetative cover (e.g., cover crops, crop residues); establish windbreaks; reduce intensive grazing; minimize or eliminate tillage (e.g. conservation tillage practices); reduce tillage speed and do not bury residues; implement strip cropping and mulch tillage; apply soil stabilizers or conditioners; and roughen the soil surface and reduce field length (Blanco and Lal 2008). The protection of the soil surface with a cover of growing plants or crop residues, as well as the use of conservation tillage systems, have also proven to be suitable practices for water erosion control in drylands.

2.1.1 Crop Residue Management

Since most of drylands used for agriculture are under cereal cultivation (Koohafkan and Stewart 2008), it appears that mulching with post-harvest crop residues, either standing or flat, is without doubt the most widely applied soil conservation measure to control wind erosion in drylands. It not only prevents wind erosion, but also protects the soil from crust formation and water erosion (Sterk et al. 2001). Consequently, sustainable soil management in those areas must be based on agricultural practices leaving crop residues on the soil surface. Therefore, CA practices, like conservation tillage (reduced or no-tillage) systems, are highly recommended. In many semiarid regions, however, low biomass production, as well as inadequate agricultural practices, leads to limited amounts of crop residues and, thereby, insufficient soil protection. This is the case of semiarid areas where the main cropping system is the traditional cereal–fallow rotation (one crop in 2 years), which implies a long-fallow period of 16–18 months. In these areas, where strong and dry winds are also frequent, the risk of wind erosion increases during fallowing due to insufficient crop residue cover and highly pulverized soils by repeated traditional tillage (López et al. 2001).

The effectiveness of crop residues depends on their persistence in time and the amount of surface residues can be reduced considerably by tillage and decomposition. Therefore, information about tillage effects on residue incorporation into the soil is essential to evaluate effective crop residue management strategies for soil conservation purposes in semiarid drylands. On an equal weight basis, standing residues are more effective than flat residues for controlling wind erosion. Likewise, flat residues in contact with the soil are also decomposed more rapidly than standing residues (López et al. 2003). Results reported by López et al. (2005) on the evolution of surface barley residues during the long-fallow period (18 months) in the traditional cereal-fallow rotation in semiarid Aragon indicate that the lack of residue-disturbing operations in no-tillage makes this practice the best strategy for fallow management in that region. With no-tillage, soil surface is protected by sufficient amount of standing and flat residues in the most critical period of wind erosion during long-fallow. Moreover, with this tillage system, the soil surface still maintains a residue cover of 10–15% after long-fallowing and percentages of standing residues ranging from 20% to 40% of the total mass after 11–12 months of fallow (López et al. 2005).

Likewise, tillage has a considerable effect on the placement and distribution of crop residues at the soil surface. For instance, a single pass of a mouldboard plough can reduce the percentage of residue remaining on the soil surface by 90–100% (López et al. 2001); after a pass of chisel plough, the reduction may vary between 40% and 70% (López et al. 2001). In any case, choosing tillage operations creating ridges at the surface which are perpendicular to the prevailing erosive wind direction and a reduction in the number of tillage operations should also be recommended (López et al. 2000; Sterk et al. 2001).

2.1.2 Tillage Reduction

Results from tillage comparison experiments have indicated that reduced tillage, and especially no-tillage, can be considered as a viable alternative for wind erosion control during the fallow period in semiarid dryland areas (Moreno et al. 2011). Thus, reduced tillage, based on the use of light vertical implements, with chiselling as primary operation, provides higher soil protection than conventional tillage through a lower wind-erodible fraction (aggregates <0.84 mm in diameter) and a significantly higher percentage of soil cover with crop residues and clods (aggregates >38 mm in diameter). The frontal area of this nonerodible material and soil roughness was reduced four times after mouldboard ploughing (López et al. 1998). Compared with traditional tillage no significant dust emission and saltation transport has been observed under reduced tillage (Sterk et al. 1999; Gomes et al. 2003).

Soil structure controls a significant number of soil processes and properties such as water and nutrient availability for plant growth, crop residue decomposition dynamics and soil erosion susceptibility. Soils in the semiarid Mediterranean agroecosystems are characterized by a low SOC content and a weak soil structure due to the continuous use of intensive tillage and long-fallowing. Álvaro-Fuentes et al.

(2007, 2008) demonstrated that the use of no-tillage and the suppression of long-fallowing may lead to an increase in soil aggregation and structural stability in these agroecosystems. A greater soil aggregate size and stability observed under no-tillage management implies a lower soil susceptibility to degradation processes. In semiarid regions, where rainfall and consequently soil water content is the most limiting factor for crop production, an improvement in aggregate stability and soil structure may lead to better water infiltration and retention in the soil profile and ultimately to greater crop yields. In semiarid Mediterranean agroecosystems, the adoption of no-tillage and the intensification of cropping systems are two management strategies that can be beneficial also to enhance SOC sequestration by improving macroaggregate stability and the consequent stabilization of SOC as particulate organic matter occluded inside microaggregates formed within stable macroaggregates (Álvaro-Fuentes et al. 2009a). More recently, Plaza-Bonilla et al. (2013b) in a no-tillage chronosequence under Mediterranean conditions found that the proportion of water-stable macroaggregates in the soil surface increased according to the increase of the years under no-tillage.

Soil erosion by water continues to be a serious threat to crop production in drylands. Where dryland agriculture is practiced, farming methods that increase the amount of water stored in the soil profile are also beneficial for soil conservation. Those methods reduce the volume and flow rate of water across the soil surface, which is one of the basic principles for controlling water erosion. Numerous practices have been developed for water erosion control, but the most cost-effective method for reducing runoff and soil erosion from agricultural soils is, undoubtedly, to maintain a good vegetative cover on the land. This can be done either through the use of conservation tillage systems, which maintain crop residues on the soil surface, or by establishing cover crops. Using crop residues to protect the soil surface from rainfall by dissipating the energy of the raindrops can reduce water erosion by up to 90% compared to an unprotected, intensively tilled field. Crop residue management provides a way for limiting both soil particle detachment and removal from the field. Soil management in vineyards and olive orchards using cover crops in lanes has shown its effectiveness in reducing water erosion in Southern Spain (Gómez et al. 2009, 2011).

However, the benefits of conservation tillage, especially no-tillage, as a practice for soil erosion control have not been tested in all the dryland agricultural areas of the world. Moreover, the adoption of CA technologies by dryland farmers in developing countries is being constrained by different socio-economic and biophysical limitations. In the semiarid and arid regions of West and Southeastern Africa, for instance, biomass production is low and crop residues are also needed for other uses like fuel, fodder and construction (Sterk et al. 2001; Lahmar et al. 2012). In Central Asia, a major limitation is the lack of appropriate manufactured machinery (Thomas 2008).

Although the principles of soil conservation (erosion control) are apparently simple, achieving soil protection and conservation in sustainable dryland agricultural systems continues to be a challenging task because of the many situations and factors involved, including soil, climate, resource and technology availability, and farmer's knowledge and capabilities. Although the same principles may apply, not all control methods can be appropriate to all situations (Unger et al. 2006). Despite the above mentioned limitations, the maintenance of a protective vegetative cover appears to be the most straightforward strategy to control soil erosion in dryland agroecosystems. Therefore, agricultural activity in those areas should be based on CA practices and associated local-based crop residue management systems.

2.2 *Cropping Strategies for Efficient Water Use*

2.2.1 Factors Limiting Water Use Efficiency in Dryland Farming Systems

Drylands are characterized by low and erratic precipitation and in many cases by a high evaporative demand. Another important limitation that affects the global efficiency of water use by crops in drylands is the difficulty to store water in the soil profile. Indeed, the volume stored depends on soil depth and structure. Soil depth may be increased by technical means, but they are not economically viable. Soil structure, however, may be improved by using certain management techniques. Slope, infiltration, salinity or stoniness may also further limit the soil water holding capacity, therefore affecting the efficient use of water by the crop (Stanhill 1986).

In dryland conditions, the main yield constraint is the shortage of water; therefore, an efficient use of this limiting resource is of paramount importance (Connor et al. 2011). Crop productivity can be then analyzed in terms of water-use efficiency (WUE), which is defined as the quotient between dry biomass (DB) and evapotranspiration (ET). The transpiration efficiency (TE), physiological term directly related with the evaporative demand, is only referred to the efficient circulation of water through the plant, given that transpiration (T) is the component of ET related to crop growth and production (Tanner and Sinclair 1983). In turn, WUE and TE are related by through the equation (Cooper et al. 1983): $WUE = TE * (1 + E/T)$. While TE is considered at crop level, WUE is used to compare among management practices.

2.2.2 Technological Options for Improving Water Use Efficiency in Dryland Conditions

In drylands, an efficient use of water is based on a few principles, with the main objective of closing the gap between crop ET and water availability. Since there are many interactions between water availability, water use and crop growth, the general management strategies to improve WUE should focus on: (i) maximizing available water for the crop; (ii) maximizing transpiration (T) by diminishing direct evaporation from the soil (E); and (iii) improving transpiration efficiency (TE) (Connor et al. 2011). There are many technological options to maximize available water, all of them based on increasing water storage, for instance by reducing runoff, improving infiltration, reducing water consumption by weeds, choosing soils with high water holding capacity, promoting deep root systems, and using long cycle cultivars. To maximize T and minimize E farmers can use mulching, adjust plant density, and sow earlier to promote both a rapid crop establishment and a shallow root growth. To improve TE, high evaporative demand periods must be avoided through phenological adjustment, sowing at the right moment, controlling pests and diseases, and in some cases by adjusting fertilization (Cooper et al. 1987). Some strategies for improving WUE and success cases are discussed next.

Water harvesting implies the principle of concentration of a resource non-uniformly distributed and spatially limited. Water harvesting technologies such as terracing, dams or contouring may contribute to improve WUE (Oweis 1997). Conservation tillage is another technological option of utmost importance for WUE improvement in drylands (Cantero-Martínez et al. 2007; Lampurlanés et al. 2016). This type of soil management, which increases soil water-holding capacity through soil structure improvement (Alvaro-Fuentes et al. 2009a), enhances water infiltration and transport throughout the soil profile. Water accumulation in the deeper soil layers represents a valuable asset, since it stimulates the development of a deep root system for greater water extraction from the soil, therefore reducing evaporation losses (Lampurlanés et al. 2001; Morell et al. 2011a). In general, CA systems are based on the maintenance of crop residues on the soil surface and alternatively on the sowing of a cover crop to be killed or maintained by mowing and left to form a mulch that improves water infiltration and reduces direct soil evaporation. Although globally CA practices are suitable for dryland systems, they are sometimes associated with certain problems, such as soil compaction, and high incidence of weeds, pests and diseases. Consequently, CA systems need to be locally adjusted (Fig. 1). Whenever mulch is insufficient, the benefits of CA decrease, and when mulch is excessive it may lead to a number of problems such as allelopathies, lack of proper germination or poor crop establishment. Therefore, a crop residue management tailored for each individual farm is advised. As previously indicated, in some regions crop residues are valued as animal feed, which is a limitation that will be solved only when farmers find alternative feeding sources with a higher nutritive value and at a reasonable price.

Fallowing has been practiced for millennia in water-limited environments, but this practice has not always proven to be beneficial in improving precipitation storage efficiency (Passioura and Angus 2010). The purpose of fallow is to gather up water and hold it within the soil during the non-cropping period. Even when this is fulfilled, the main reason for the low efficiency of fallow lies in the high direct evaporation that occurs soon after water is harvested. This is especially important in intensively ploughed fallows, where soil structure is damaged by tillage (Alvaro-Fuentes et al. 2009a), and immediate erosion, lower infiltration and direct evaporation take place during the tillage operation. To efficiently use fallow, it must be performed in deep soils with high water holding capacity in the lower layers, where direct evaporation does not take place, and capillarity is interrupted (French 1978). Weed control must also be effective, preferably avoiding an intensive cropping (Cantero-Martínez et al. 1995a; Lampurlanés et al. 2002), and done by using herbicides or by cutting the plant cover but maintaining the stubble. In shallow soils, with

Fig. 1 Direct drill planter into adequate crop residue cover. (Photo by C. Cantero-Martínez)

a low water holding capacity, water is easily lost by evaporation and most of the times due to tillage operations carried out during the fallow period.

Crop rotations may also contribute to improve WUE. Improvements may come from a better weed, pest and disease control, but also from a suitable distribution of water availability for the different crops in the sequence. Peterson and Westfall (2004) showed that in water-limited environments, an intensified cropping system, with a sequence of winter and summer crops, improved yields when compared with a wheat-fallow system. The improvement was mainly due to a better synchrony between water availability and crop growth timing of the rotation. It is also possible to improve both water management and WUE by considering different water use patterns (in amount and time) for the different species in the rotation (Fereres et al. 1993). Another tool available is that of choosing among determinate or indeterminate crops. Determinate crops have a narrower window of flowering time so that a water stress during flowering may lead to a substantial yield loss. Indeterminate crops, with staggered flowering and ripening and longer periods of risk, are more unlikely to suffer from severe water stress over the entire flowering time. Also, certain crop species may be cultivated for either grain or forage. Forage is harvested earlier and thus may avoid terminal stress (Alvaro-Fuentes et al. 2009b).

Plant breeding is another option to improve WUE in dryland agriculture. Under water-limiting conditions, crop production should be based on escape mechanisms and crop tolerance to water shortage episodes, thus adjusting phenology to the periods of water availability. The most suitable cultivars for these conditions are those able to efficiently use all the water that rain and management may provide. Crop growth and development are regulated by three genetic systems: vernalization, photoperiod and earliness per se. In the most economically important crops, vernalization and photoperiod genes are known and molecular markers are available or are currently being developed, thus facilitating the selection of the most favourable alleles for each environment. A different breeding goal is the improvement of genetic WUE, defined in crop physiology as the net assimilation rate versus stomata conductance. Since net assimilation rate is quite constant within a given species, genetic differences are due to variation in stomata conductance (Blum 2005). For this reason, genetic WUE and yield are antagonists, since WUE is increased when water is conserved by stomata closure, therefore reducing photosynthesis. An increased capture of water resources through a more efficient root system has been pointed out as a better target for breeding traits (Blum 2009). Carbon isotope discrimination, mainly in grains, has also been considered as a WUE indicator, but its use in breeding is limited by the cost of the analysis and by its effectiveness depending on the target environment (Araus et al. 2003; Rebetzke et al. 2002). Because most of the underlying physiological processes responsible for WUE are not well understood, or studies do not clearly link physiology and productivity (Flexas et al. 2013; Tardieu 2012), some of the proposed tools for breeding are based on high-throughput methodologies. These methods allow us to evaluate many possible genetic combinations with remote sensing technologies, and try to detect the best genotypes (with lower canopy temperature, higher biomass or indirect indicators or WUE) with a semi-empirical method (Salekdeh et al. 2009).

Strategies to maximize ET, improving T and reducing E, include a crop cycle as long as possible. Nevertheless, this may be a disadvantage if it implies a higher incidence of weeds, pests and diseases. In a study under Mediterranean dryland conditions, a delay in sowing time was associated with a better control of weeds (García et al. 2014), suggesting an escape to some pests and diseases, thus improving WUE and yields (Plaza-Bonilla et al. 2017a). Wherever terminal drought is an important limitation, which may compromise production, a shorter crop growth cycle, implying moving the synthesis of harvestable biomass to a moment when water availability, and usually temperature, is more favorable, may also be beneficial. High plant density and narrow row spacing allow a lower E and higher crop competitiveness against weeds. On the other hand, low plant density and wide row spacing may improve crop performance under very water-limiting conditions, where it is important to match water availability period with the maximum growth of forage or grain production. This strategy also implies weed, pest and disease control. In summary, optimum sowing time and plant density must be determined locally, depending on the site-specific water availability, but always keeping in mind other factors that might be involved, especially those of biotic nature.

Water and nutrient use are strongly interrelated in many agricultural systems (discussed further in next Section). Under humid conditions without water limitations, nutrient availability usually becomes the limiting factor for crop performance, and an effective fertilization management is of utmost importance. On the other hand, under water limiting conditions, water determines the use of the nutrients by the crop. In rainfed areas, the role of fertilization has a low impact in improving WUE. For the case of N, Sadras (2004) described the biomass production as a co-limitation between water and N. For example, N application with enough available water at the initial or at the mid-growth stage may improve biomass production, but it also may lead to excessive transpiration associated with a higher leaf area and, thus, to premature depletion of soil water and, ultimately, to a failure in setting harvestable organs if more water is not available on time (Cantero-Martínez et al. 1995b). In a long-term study under Mediterranean conditions, reducing soil tillage and N fertilization doses increased both water and N use efficiencies (Angás et al. 2006; Morell et al. 2011b). In that study, overall N fertilizer applications could be reduced to about half of the traditional rate applied by the farmers without a loss of yield (Cantero-Martínez et al. 2016). In those semiarid conditions, achieving a suitable balance between applied fertilization and available water remains a challenge. Moreover, management practices should be oriented to optimize water and N use simultaneously (Quemada and Gabriel 2016). For this reason, fertilization must be integrated with other techniques such as soil tillage, which may help to hold water in the soil and make it available later for crop use.

2.3 *Efficient Nutrient Use*

Given the increasing food demand, drylands become more important for satisfying global population needs and achieving food security. Adequate plant nutrition, based on soil fertility and the application of fertilizers, plays a crucial role in food security (Roy et al. 2006). An efficient nutrient use requires first the comprehension of the limitations posed by the intrinsic characteristics of the agroecosystems, focusing on climatic, edaphic and socioeconomic aspects. Crop nutrient management in dryland areas is highly diverse. Nutrient availability is usually lower than the optimal in the developing countries (Nawaz and Farooq 2016). On the contrary, some nutrients such as nitrogen (N) and phosphorus (P) have been applied in excess in other dryland areas (Ju et al. 2009). In this section the main constraints for an efficient nutrient use in dryland areas are described. Being the concept contextualized, a range of technological options to make nutrient use more efficient in dryland areas are discussed, with emphasis on pragmatic alternatives toward a holistic approach at the farm level.

2.3.1 Factors Limiting Nutrient Use Efficiency in Dryland Farming Systems

When no irrigated, dryland systems are characterized by crop yield limitations due to low soil water availability. This is a consequence of low rainfall and high evapotranspiration during the crop growing period and/or inadequate soil water storage during periods of high rainfall and low evaporative demand (Lampurlanés et al. 2016). Nutrient transport in the soil and absorption by roots are dependent on soil water availability. Therefore, an efficient use of nutrients in dryland areas must reduce water limitation by adopting adequate crop management practices.

Soil nutrient losses through water and wind erosion represents a serious threat for crop nutrient availability in dryland areas. For instance, Ramos and Martínez-Casasnovas (2004) studied the impact of an extreme rainfall event on water erosion and nutrient losses in a Mediterranean rainfed vineyard under continuous tillage with bare soil for most of the growing season. The authors reported a loss of 109 kg N ha^{-1}, 109 kg P ha^{-1} and 36 kg K ha^{-1} in the soil sediments. Wind erosion preferentially removes finer soil particles, which present greater surface to hold plant nutrients. For instance, Sterk et al. (1996) measured wind-blown nutrient transport in an experimental plot cropped with pearl millet (*Pennisetum glaucum* L.) in Niger and estimated the nutrient losses as 57 kg K ha^{-1}, 18 kg N ha^{-1} and 6 kg P ha^{-1}.

Organic matter confers a range of physical, biological and chemical functions to soils crucial for crop growth and represents an important source of nutrients for crops. In dryland areas, inadequate soil management such as intensive tillage, overgrazing and elimination of vegetation cover has exacerbated the loss of SOM (Plaza-Bonilla et al. 2015). Although crops in rainfed drylands have low N needs due to their low productivity, in many circumstances the low levels of soil mineralizable N can limit crop yield.

Soil characteristics can also affect nutrient availability for crops. Given their important extent, dryland areas present a significant soil diversity. However, some generalizations can be made; among them, dryland soils of arid and semiarid areas tend to have high pH values resulting from salts content, commonly in the form of calcium and magnesium carbonates. The calcareous nature of most of these soils results in the formation of apatite compounds with calcium phosphates, reducing severely the availability of P for crops, which can then restrict the uptake of other nutrients by the crops. Potassium tends to be relatively high due to the K-rich parent materials from which most of dryland soils were developed (Li et al. 2009). Furthermore, in soils with pH values from neutral to calcium saturation the solubility of micronutrients such as Fe, Zn, Mn and Cu is reduced (Hagin and Tucker 1982).

2.3.2 Technological Options for an Efficient Nutrient Use in Dryland Areas

In dryland farming systems an efficient use of nutrients relies on adequate crop management practices that increase the amount of water available to the crop and make water use more efficient. The implementation of conservation tillage is key to reach the last objectives. Research has highlighted the influence of the interaction between tillage systems and N fertilization on crop yield and agronomic nitrogen use efficiency (NUE) in different rainfed dryland areas. For instance, in a rainfed area of Spain, the use of no-tillage (NT) allowed for higher mineral N rates to barley (*Hordeum vulgare* L.) compared with conventional tillage (CT) due to greater water availability in NT (Cantero-Martínez et al. 2016). The last authors reported no response to N fertilizer application when using CT. In the same area and under the same crop, Plaza-Bonilla et al. (2017b) found greater NUE under NT than CT in a 4-year field experiment, as a result of greater soil water available for the crop. Similarly, Halvorson et al. (2000) reported greater spring wheat (*Triticum aestivum* L.) grain yield under NT compared with CT when applying 101 kg N ha^{-1} in annual cropping systems in the Central Great Plains (USA).

Besides its effect on NUE, the use of conservation tillage helps to maintain or increase the SOM pool, enhancing nutrient cycling. López-Garrido et al. (2011) observed greater amount of SOC, total N, and extractable P and K in the rooting zone (0–20 or 0–30 cm depth, depending on the variable) under conservation tillage than intensive tillage in a long-term (16 years) experiment. Moreover, in a 4-year experiment, greater amount of extractable K was observed when using conservation tillage, compared with intensive tillage, as a result of the maintenance of crop residues in the soil surface. The increase in SOM also leads to increasing P solubility (Salinas-Garcia et al. 1997), a process of great importance in calcareous soils to increase P availability to crops.

Crop fertilization best management practices rely on the approach of the "4R" Nutrient Stewardship Initiative, based on the use of the right nutrient source, at the right rate, right time and in the right place (IFA 2009). Within this framework, nutrient diagnosis by chemical analysis of soils and plants is the first step in prescriptive management. Soil analysis makes fertilization more efficient by accounting residual nutrients and mineralization of nutrients, particularly N. However, many dryland farmers are still reluctant to use it as a basis for their nutrient management decisions. Therefore, the usefulness of soil analysis must be emphasized by highlighting the potential economic savings associated to its use, and its capacity to detect soil nutrient imbalances, which can severely compromise crop yields (Li et al. 2009). Unfortunately, some dryland areas lack reliable analytical facilities, as is the case of the West Asia-North Africa region (Ryan et al. 2012).

Being a highly dynamic nutrient, N requires routine soil tests of its availability in the form of nitrate to supplement soil N supply according to crop demand. For field crops, soil samples must be taken to a depth of at least 60 cm, right before the period of rapid growth of the crop. Under rainfed dryland conditions the potential crop yield and right nutrient rate for a given season are difficult to estimate given the unpredictability of rainfall. Therefore, the development of improved forecasts and low-cost decision-making tools based on available soil water is of paramount importance. A more precise estimation of crop yield potential for a given site-year combination would ease farmers' decisions for a more efficient use of fertilizers (Wienhold et al. 2000), particularly N. In the current situation, this information is not usually available in dryland areas. Consequently, producers should establish the N rate in accordance to the average year, while maintaining flexibility to respond to higher yielding years (Westfall et al. 1996). The traditional practice based on the application of high N rates to secure crop yields based on risk aversion should be avoided (Plaza-Bonilla et al. 2017b). Corrective N management can be carried out with the use of diagnostic tools, such as leaf color charts, chlorophyll meters or canopy reflectance sensors (Kirkegaard and Robertson 2013). In some situations, precision agriculture based options such as the variable rate technologies are becoming popular. Robertson et al. (2009) studied the economic impact of implementing variable rate technologies according to soil characteristics and routine soil analysis concluding that they are economically profitable. However, it must be taken into account that the last authors performed their work in an Australian context of large farms, intensive agriculture and high capital investments. Therefore, the need for a detailed study of soil characteristics and routine soil tests makes this technology unaffordable for farmers in developing countries.

In the current context of high prices of fertilizers, organic sub-products are becoming more valued by the farmers as sources of nutrients. Organic fertilizers represent a very interesting source of nutrients which can be used as an alternative or in combination with mineral fertilizers to balance the disagreement between crop nutrient needs and organic fertilizer composition. This is due to the fact that compared to plant needs, P in organic fertilizers is usually at a greater proportion than N (Li et al. 2009). The application of organic fertilizers in dryland soils of low SOM content improves soil aggregate stability (Abiven et al. 2009; Plaza-Bonilla et al. 2013a), preventing soil erosion and enhancing the nutrient status of the soil. Traditionally, farmers apply organic products at a fixed rate, without regard to the actual concentration of nutrients, which can vary according to a range of factors related to livestock characteristics and farm management. Low-cost devices based on the close relationship between electrical conductivity and ammonium content allow a precise application of slurries in the field (Scotford et al. 1998; Yagüe and Quílez 2012). Currently, these devices can be installed in slurry tankers offering the operator real-time information of the actual ammonium N concentration of the product that is being applied (Fig. 2). The tractor speed is then manually or automatically regulated to reach the desired N rate to be applied.

Right time of application is of great importance for enhancing efficiency of highly mobile nutrient forms such as mineral N, minimizing losses. Nitrogen availability must be in synchrony with crop needs in order to increase NUE. For instance, in rainfed Mediterranean areas winter season crops sown in autumn are preponderant in annual cropping systems. Crop N requirement is limited during the autumn-

Fig. 2 A slurry tanker equipped with disc injectors (*left*) and detail of the nitrogen rate control system based on electrical conductivity (*right*). (Photos by C. Cantero-Martínez)

winter period due to low temperatures, while N losses can occur, since an important fraction of annual rainfall takes place in autumn. Traditional N fertilization practices involved the application of mineral or organic N before sowing, which leads to low NUE (Angás et al. 2006; Cantero-Martínez et al. 1995b). Recent research performed in the area has demonstrated that pre-sowing application of N leads to lower NUE of winter cereal compared with top-dressing applications at tillering, independently of the source of N, mineral N or organic fertilizer as pig slurry (Plaza-Bonilla et al. 2017b). If too much N is available early in the season large biomass growth and excessive water use takes place in the vegetative stage, compromising water availability for the reproductive stages, a process known as "haying off" (Rose and Bowden 2013). Another important aspect to be taken into account in dryland areas is the interaction between the time of application and the source of N. For instance, in calcareous soils of high pH important volatilization N losses can occur if urea or ammonium forms are applied under high air temperatures and/or low soil moisture levels (Li et al. 2009). Therefore, farmers must carry out fertilizer applications under adequate environmental conditions. Technological strategies are available to reduce N volatilization losses. Among them, a shallow incorporation of fertilizers in the soil is one of the most common and of a low cost. However, in rainfed areas managed under conservation tillage that strategy can compromise the benefits of maintaining the soil surface covered by residues. When possible, the application of some irrigation (10–15 mm) is a useful strategy to reduce N volatilization significantly, reduction that can be of similar magnitude than the obtained using urease inhibitors (Sanz-Cobena et al. 2011). The use of controlled release N fertilizers (coatings, urease, nitrification inhibitors) is another technology available aiming at improving NUE. However, these materials are not commonly used in rainfed dryland areas due to their high cost (Ryan et al. 2012), which ranges from 1.3 to 12 times the price of standard fertilizers (Lammel 2005). Therefore, currently controlled release N fertilizers represent a small proportion of the global market of fertilizers and are mainly devoted to high added value crops (Chien et al. 2009).

The use of break crops and the diversification of crop rotations are seen as one of the simplest and most effective means of increasing NUE. For instance, Australian researchers observed a greater amount of water and mineral N extracted from the soil by wheat after a break crop (Angus et al. 1998). The authors hypothesized that a reduction in root disease allowing greater suction by roots could explain this result. Combining different crop rooting patterns can increase the nutrient use efficiency at the crop and rotation scales (Thorup-Kristensen 2006). The introduction of legumes is a specific case of diversification where N fixation is used to improve the N budget of the farms. Atmospheric N fixation by annual and perennial legume crops represents a sustainable input of N which is available for the subsequent crops, reducing farming systems dependence on external inputs. However, the benefits on N cycling depend on the type of legume, since high N harvest index grain legumes remove most of the N fixed with the grain harvest. Moreover, it must be taken into account that the symbiosis between the legumes and *Rhizobium* is highly sensitive to drought stress, reducing legume productivity and the amount of N fixed for the subsequent crop (Serraj et al. 1999).

Phosphorus application can have a profound effect on crops yield in some dryland areas. For instance, Shangyou et al. (1997) reported a 34–44% increase in wheat grain yield when applying increasing P_2O_5 rates from 75 to 300 kg ha^{-1} in a dryland area of Northwest China. Moreover, the last authors observed a 10% increase in NUE when applying N and P in combination compared with the sole application of N. Adapted placement methods can also improve greatly the efficiency in the use of nutrients, particularly phosphorus. As it has been explained above, soluble P fixation or immobilization in calcareous soils is a common process. Phosphorus fixation can be partially overcome by fertilizer banding, allowing for lower application rates than broadcasting (Ryan et al. 2012) and increasing P use efficiency (Mallarino and Borges 2006). The use of CA is useful to reduce P losses by erosion. Many manufacturers design direct drilling machines which allow the combined use of seeds and fertilizer at sowing, by placing fertilizer granules some centimeters beside and below the seeds, without a significant disturbance of the soil cover. Differently to calcareous soils, in acid soils, the efficiency in the use of P can be compromised by the aluminium (Al) toxicity to crops. In this line, in acid soils of Southern Australia liming and the use of Al-tolerant cultivars increased P use efficiency of crops and pastures (Simpson et al. 2011). Crop rotations can also impact positively P cycling by increasing the excretion of P-solubilizing compounds, as it occurs when legumes and rapeseed are included in the rotation, and/or by maintaining mycorrhizal activity (Grant et al. 2002; McNeill and Penfold 2009).

2.4 Climate Change Mitigation

Agriculture is responsible for the 10–14% of the total anthropogenic greenhouse gas (GHG) emissions (Paustian et al. 2016). Since 1990, GHG emissions in the agricultural sector have increased annually at 1% reaching 5.4 Gt CO_2 eq year^{-1} in 2012 (Tubiello et al. 2015). Agricultural soils contribute to approximately 35% of the total GHG emissions emitted by agriculture (Smith et al. 2014). However, concurrently, soils have the potential to withdraw significant amounts of atmospheric carbon dioxide (CO_2) through the buildup and stabilization of photosynthetically plant-derived carbon into the soil. This double role of agricultural soils in relation to climate change (sink and source of GHG) determines the two main options to mitigate GHG emissions in agricultural soils: (i) enhancing removals, and (ii) reducing emissions (Smith et al. 2008).

2.4.1 Enhancing Carbon Removals

Soils store large amounts of carbon (C) in the forms of organic and inorganic C. The soil organic C (SOC) pool, up to 1 m depth, is two times greater than the C in the atmosphere as CO_2 (1500 Pg C vs. 830 Pg C, respectively). Furthermore, the inorganic C pool, formed mainly by carbonates, stores approximately 695–748 Pg C (up to 1 m depth), not far to the amount of C located in the atmosphere as CO_2 (Batjes 1996). In drylands soils with high carbonate content, the withdrawal of atmospheric CO_2 may result from the sequestration of organic and/or inorganic C.

In croplands, historical cultivation has resulted in the loss of significant amounts of SOC. Land use change from grassland or forest to cropland decreases SOC levels (Guo and Gifford 2002). Tillage breakdowns stable soil macroaggregates favouring microbial decomposition of the organic compounds that were physically occluded inside these aggregates (Cambardella and Elliot 1993). Likewise, the reduction in C inputs associated to agriculture conversion also contributes to the observed losses of SOC after the change in land use. Nevertheless, C depleted soils have the potential to increase SOC levels and, in turn, to sequester atmospheric CO_2 (FAO 2017).

A main feature of dryland soils is the low SOC concentration with values normally below 0.5% by weight (Lal 2004). Compared to soils located in humid areas, SOC buildup in dryland areas is restricted by limited water availability affecting net primary productivity and, hence, crop residue inputs. In dryland areas, limited crop growth is also driven by the inherent low soil fertility and the low use of fertilizers since the excessive use of N fertilizers in water-limited conditions could be an economical loss with associated environmental side effects (Morell et al. 2011b). Dryland agriculture, despite the inherent harsh conditions, should be regarded by the C sequestration potential of these areas. Drylands occupy large extensions on the earth surface. Consequently, the total C stored in dryland soils is large. Moreover, the depletion in SOC make dryland soils to be far from their C saturation level and, hence, with a concomitant high capacity to restore SOC if proper management is adopted. According to estimates provided by Lal (2004), the potential of dryland soils for SOC sequestration could be between 12 and 20 Pg C over a 50-year period.

Several management options have been recognized for the increase of SOC contents in dryland soils (Lal 2002, 2004; Plaza-Bonilla et al. 2015). Since the levels of SOC depend on the balance between C inputs and outputs, strategies oriented to the increase of SOC should focus on maximizing C inputs and/or minimizing C losses. Accordingly, management strategies can be divided on the basis of this premise. Increases in C inputs can be achieved by maximising the amount of crop residues returned into the soil per surface unit. In drylands, water scarcity limits crop growth and in turn the production of plant biomass and crop residues (Cantero-Martínez et al. 2007). Therefore, the increase in crop residue returned to the soil is a challenging commitment in drylands and, basically, management should be oriented to avoid the removal or the burning of crop residues after harvest. Other strategies could be oriented to raise crop residue production by increasing WUE and/or reducing the fallow period. The reduction in tillage intensity has been identified as a successful strategy to increase WUE, as indicated in a previous section, and, in turn, for crop residue production in drylands (Lampurlanés et al. 2016). According to a recent meta-analysis, no-tillage compared to conventional tillage showed greater potential for increasing yield in drylands compared with more humid climates (Pittelkow et al. 2015). The greater crop production in no-tillage systems is translated into higher amount of crop residues returned to the soil compared with intensive tillage systems, thus favouring the sequestration of SOC. For Mediterranean drylands,

Álvaro-Fuentes and Cantero-Martínez (2010) estimated a mean SOC sequestration rate of 0.23 Mg C ha^{-1} year^{-1}. This value was estimated according to the results obtained from eight long-term experiments located throughout Spain. Indeed, these same authors estimated that if the entire agricultural dryland area in Spain was converted to no-tillage, the total SOC sequestered would offset about 17% of the annual total GHG emissions generated from agricultural activities in Spain (Álvaro-Fuentes and Cantero-Martínez 2010).

The positive impact of no-tillage on SOC sequestration, however, is not only explained by an increase in crop growth and residue production but also by the significant effect of no-tillage on SOC decomposition. No-tillage increases physical protection of SOC by the formation of stable macroaggregates that protect SOC from microbial attack (Six et al. 1999). These authors, in dryland conditions of central United States, observed that no-tillage resulted in lower soil aggregate turnover favouring the stabilization of SOC in microaggregates formed within macroaggregates (Six et al. 2000). However, it is important to highlight that SOC sequestration through the adoption of no-tillage has several limitations. Organic C accumulation is a finite process with different sequestration duration depending on the climate, historical land use and management (West et al. 2004). In a tillage chronosequence located in a Mediterranean dryland agroecosystem, no-tillage accrued SOC during 20 years. However, almost 75% of this total C was accumulated during the first 11 years after the adoption of no-tillage (Álvaro-Fuentes et al. 2014). Concurrently, in this same study, it was observed that the maximum sequestration rate occurred 5 years after the adoption of no-tillage (Álvaro-Fuentes et al. 2014).

Intensification of cropping systems through the reduction of the fallow period may be also an interesting option to sequester SOC in drylands. In rainfed Great Plains, after 12 years of different cropping systems, the shift from a wheat-fallow system to a more intensified wheat-wheat system resulted in the buildup of 2.8 Mg C ha^{-1} in the 0–30 soil depth (Peterson and Westfall 2004). Similarly, in rainfed Mediterreanan conditions, after 20 years, no-tillage wheat-wheat sequestered about 70% more SOC in the first 90 cm soil depth than the wheat-fallow system (López-Bellido et al. 2010). Long-term rotation experiments established by ICARDA in Syria have also shown an increase in SOC levels when different wheat crop rotations were compared with the conventional wheat-fallow rotation despite only the wheat-medic rotation was significantly greater (Jenkinson et al. 1999).

As commented before, in drylands available water limits crop residue production and, hence, SOC accrual. The conversion from rainfed to irrigated land could be a viable strategy to increase net primary productivity and the return of crop residues into the soil. However, irrigation not always leads to an increase in SOC stocks compared to rainfed conditions (Trost et al. 2013). Denef et al. (2008), in two sites located in semiarid Great Plains, observed positive effect of irrigation on SOC buildup in the 0–20 cm in both sites but only in one site when soil was considered to 75 cm depth. In more humid conditions, the positive influence of irrigation on SOC buildup tends to vanish (Trost et al. 2013).

In drylands, C removal may also come from the sequestration of inorganic C when carbonates are present in the soil. The C sequestered through the formation of secondary carbonates was estimated in 0.1–0.2 Mg C ha^{-1} $year^{-1}$ (Lal 2004). The formation and precipitation of secondary carbonates occur when CO_2 partial pressure increases (due to the decomposition of SOM or root respiration) and this CO_2 reacts with water forming bicarbonates (HCO_3^-). The leaching of these bicarbonates and Ca^{+2} cation to deep soil horizons, where lower partial pressure of CO_2 prevails, facilitates precipitation of $CaCO_3$ (Lal and Kimble 2000). The impact of agricultural management on the sequestration of soil inorganic C is not clear. Despite these mixed effects, the presence of carbonates in arid soils may have a positive impact on SOC stabilization within soil aggregates (Bronick and Lal 2005). In soils with carbonates, the presence of exchangeable Ca^{+2} facilitates cation bridging and, hence, physical SOC protection (Virto et al. 2011). Moreover, the precipitation of secondary carbonates on organic compounds also facilitates protection of SOC against microbial decomposition (Virto et al. 2011). However, despite the potential for soil inorganic C sequestration in arid and semiarid drylands, there is still an important lack of knowledge regarding the most suitable options and strategies for enhancing C removal through the formation of secondary carbonates.

2.4.2 Reducing GHG Emissions

Agricultural soils are emitters of the three main greenhouse gas gases (i.e., carbon dioxide, CO_2; nitrous oxide, N_2O; and methane, CH_4). Although soil CO_2 is produced by microbial respiration during SOM decomposition and by root respiration, only the CO_2 derived from microbial activity contributes to SOC levels. However, since the increase in SOC is a viable strategy to withdrawal atmospheric CO_2 (as commented in the previous section), the reduction in emissions as a mitigation strategy is primarily related to the other two non-CO_2 gases (i.e., N_2O and CH_4). In soils, N_2O is mainly produced by microorganisms during nitrification and denitrification processes. Denitrification occurs mainly under high soil water conditions when oxygen levels are limited and, hence, NO_3^- and NO_2^- act as alternative electron acceptors. In contrast, nitrification occurs in aerated soils with maximum nitrification rates at 60% water-filled pore space (WFPS) (Linn and Doran 1984). In drylands, in which water deficit is a main feature, soil N_2O is mainly produced during nitrification (Galbally et al. 2008). Nitrification is the process in which NH_4^+ is converted to NO_3^-. Consequently, high NH_4^+ levels in the soil will result in higher nitrification rates and, in turn, in greater N_2O production. Accordingly, N fertilization is the agricultural practice with the highest impact on N_2O emissions which results in a positive relationship between N rates and N_2O emissions (Bouwman et al. 2002). In southern Great Plains, results from a 3-year N fertilization experiment in rainfed wheat showed that N fertilization rate was positively related with cumulative N_2O emissions (Wilson et al. 2015). In this last experiment, the authors also observed that the highest N_2O emissions were obtained during the driest year since low crop yields and low N uptake by the crop resulted in more N available in the soil (Wilson et al. 2015). Consequently, the adjustment of N rate to crop needs and the application of the fertilizer in the right timing are recommendations to effectively reduce N_2O emissions in croplands (Venterea et al. 2012). The application of N fertilizers could also be reduced with the introduction of legume crops into the rotations. In drylands, crop diversification, through the use of rotations, has been confirmed as an interesting strategy to reach sustainable yields preserving soil and water conservation (Díaz-Ambrona and Mínguez 2001; Lenssen et al. 2014). The inclusion of legumes reduces N_2O emissions compared to N-fertilized systems (Jensen et al. 2012). According to this last study, comparing mean values from several sites and years, they found that soils under legumes emitted about 60% less N_2O compared to N-fertilized crops (1.29 vs. 3.22 kg N_2O-N ha^{-1}, respectively) (Jensen et al. 2012). In dryland Australia, data from lupin and bare fallow plots revealed that the contribution of biological N fixation to N_2O emission was minimal (Barton et al. 2011).

The selection of certain tillage systems could also contribute to decrease N_2O emissions in drylands. In a meta-analysis with 239 comparisons, it was found that in dryland condition no-tillage and reduced tillage decreased soil N_2O emissions in long-term experiments (>10 years) (van Kessel et al. 2013). Similarly, Six et al. (2004) for dryland conditions found greater N_2O emitted in no-tillage and reduced tillage in the short-term compared to conventional tillage and the opposite trend in the long-term. In dryland Spain, two experiments with different years since establishment (3 and 15 years) were selected to evaluate the effect of N fertilization and tillage systems on soil N_2O emissions in barley cropping (Plaza-Bonilla et al. 2014a). After 2 years of soil gas emission measurements, differences in mean N_2O emissions between no-tillage and conventional tillage differed in both experiments (Fig. 3). In the long-term experiment, after 13 years, similar N_2O emissions were observed between both tillage systems. But, in the short-term, after 3 years, greater N_2O emissions were observed in no-tillage compared with conventional tillage (Fig. 3).

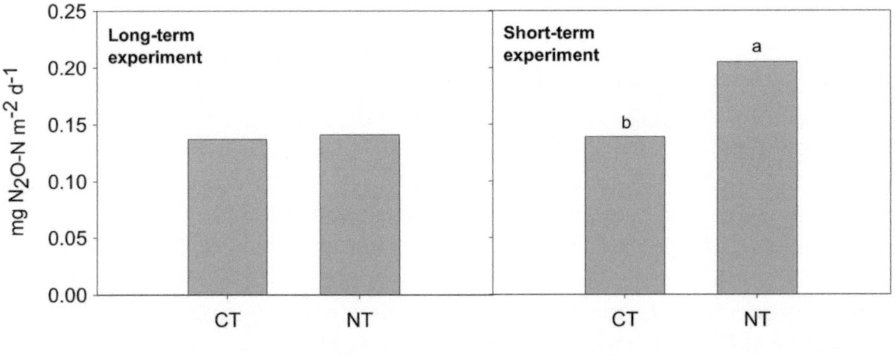

Fig. 3 Mean soil N_2O emissions during 2 years in two tillage (*CT* conventional tillage, *NT* no-tillage) experiments located in semiarid NE Spain: a long-term experiment located in Lleida province (13 years since establishment) and a short-term experiment located in Huesca province (3 years since establishment). Within the same experiment, different lowercase letters indicate significant differences between tillage systems ($P < 0.05$). (Plaza-Bonilla et al. 2014a)

Soils in dryland areas are more susceptible to uptake CH_4 rather than to emit it. Net CH_4 emissions are common in anaerobic environments in which methanogenic microoganisms reduce carbon compounds to produce CH_4. In contrast, under aerobic conditions typical of dryland soils, CH_4 oxidation occurs through the activity of methanotrophic bacteria (Hütsch 2001). In croplands, tillage, N fertilization and grazing may reduce CH_4 oxidation potential of dryland soils (Galbally et al. 2008). In dryland Spain, the adoption of long-term no-tillage resulted in twofold higher CH_4 oxidation compared to conventional tillage system (Plaza-Bonilla et al. 2014b). Accordingly, with an optimal selection of management practices, dryland agroecosystems could uptake significant amounts of atmospheric CH_4 contributing to offset CH_4 emissions from other sectors.

3 Conclusions

The protection of the soil surface with a cover of growing plants or crop residues and the reduction or elimination of tillage are suitable practices for wind and water erosion control in drylands. Mulching with post-harvest crop residues, either standing or flat, is without doubt the most widely applied soil conservation measure to control wind erosion in these areas. Thus, the adoption of CA practices is highly recommended for sustainable management of rainfed farming systems.

Regarding the available technologies to improve WUE in dryland agriculture, water harvesting, fallowing, conservation tillage, crop rotations and plant breeding are feasible options. Sowing time and plant density can also affect WUE but they must be adjusted locally. The first step in prescriptive management to increase nutrient use efficiency in dryland agriculture is nutrient diagnosis by chemical analysis of soils and plants. However, under rainfed dryland conditions the right nutrient rate for a given season is difficult to estimate and adoption of precision agriculture based options such as the variable rate technologies are expensive and unaffordable for farmers in many regions. On the contrary, simplest and affordable technological options to improve NUE include the implementation of conservation tillage systems, the use of break crops and the diversification of crop rotations.

The two main options to mitigate GHG emissions in dryland soils are C sequestration and reduction of GHG emissions. Strategies recognized for C sequestration in dryland soils include maximising the amount of crop residues returned to the soil, which implies avoiding the removal or the burning of crop residues after harvest, the reduction of tillage intensity, the intensification of cropping systems and the reduction of the fallow period. To reduce N_2O emissions in drylands the adjustment of N rate to crop needs and the application of N fertilizer at the right timing are highly recommended. Likewise, the adoption of long-term no-tillage for reducing N_2O emissions and increasing CH_4 uptake in dryland soils are also advised.

Among all the strategies analyzed, the maintenance of a protective crop residue cover and the reduction of tillage operations appear to be the simplest technological options not only to control soil erosion but also to improve WUE and NUE and mitigate GHG emissions. However, the benefits of conservation tillage, especially no-tillage, have not been tested in all the dryland agricultural areas of the world. Moreover, the adoption of CA technologies by dryland farmers in developing countries can be constrained by socio-economic and biophysical limitations. Despite these limitations, sustainable agricultural management in drylands should be primarily based on CA practices and associated local-based crop residue management systems.

References

Abiven S, Menasseri S, Chenu C (2009) The effects of organic inputs over time on soil aggregate stability-A literature analysis. Soil Biol Biochem 41:1–12

Álvaro-Fuentes J, Cantero-Martínez C (2010) Short communication. Potential to mitigate anthropogenic CO_2 emissions by tillage reduction in dryland soils of Spain. Span J Agric Res 8:1271–1276

Álvaro-Fuentes J, Arrúe JL, Gracia R, López MV (2007) Soil management effects on aggregate dynamics in semiarid Aragon (NE Spain). Sci Total Environ 378:179–182

Álvaro-Fuentes J, Arrúe JL, Gracia R, López MV (2008) Tillage and cropping intensification effects on soil aggregation: temporal dynamics and controlling factors under semiarid conditions. Geoderma 145:390–396

Álvaro-Fuentes J, Cantero-Martínez C, López MV, Paustian K, Denef K, Stewart CE, Arrúe JL (2009a) Soil aggregation and soil organic carbon stabilization: effects of management in semiarid Mediterranean agroecosystems. Soil Sci Soc Am J 73:1519–1529

Álvaro-Fuentes J, Lampurlanés J, Cantero-Martínez C (2009b) Alternate crop rotations under no-tillage rotations under Mediterranean rainfed conditions: I. Biomass, grain yield and water-use efficiency. Agron J 101:1227–1234

Álvaro-Fuentes J, Plaza-Bonilla D, Arrúe JL, Lampurlanés J, Cantero-Martínez C (2014) Soil organic carbon storage in a no-tillage chronosequence under Mediterranean conditions. Plant Soil 376:31–41

Angás P, Lampurlanés J, Cantero-Martínez C (2006) Tillage and N fertilization effects on N dynamics and barley yield under semiarid Mediterranean conditions. Soil Tillage Res 87:59–71

Angus FE, Peoples MB, van Herwaarden AF (1998) Water and nitrogen in crop and pasture systems in southern Australia. In: Management of nutrients and water in rainfed arid and semi-arid areas, Proceedings of a consultants meeting organized by the Joint FAO/IAEA Division of Nuclear Techniques in Food and Agriculture, Vienna, 26–29 May 1997, IAEA-TECDOC No.1026, Vienna

Araus JL, Villegas D, Aparicio N, García del Moral LF, El Hani S, Rharrabti Y, Ferrio JP, Royo C (2003) Environmental factors determining carbon isotope discrimination and yield in durum wheat under Mediterranean conditions. Crop Sci 43:170–180

Barton L, Butterbach-Bahl K, Kiese R, Murphy DV (2011) Nitrous oxide fluxes from a grain-legume crop (narrow-leafed lupin) grown in a semiarid climate. Glob Change Biol 17:1153–1166

Batjes NH (1996) Total C and N in the soils of the world. Eur J Soil Sci 47:151–163

Blanco H, Lal R (2008) Principles of soil conservation and management. Springer, Heidelberg

Blum A (2005) Drought resistance, water-use efficiency, and yield potential – are they compatible, dissonant, or mutually exclusive? Aust J Agric Res 56:1159–1168

Blum A (2009) Effective use of water (EUW) and not water-use efficiency (WUE) is the target of crop yield improvement under drought stress. Field Crops Res 112:119–123

Bouwman AF, Boumans LJM, Batjes NH (2002) Emissions of N_2O and NO from fertilized fields: summary of available measurement data. Glob Biogeochem Cycles 16(4):1058

Bronick CJ, Lal R (2005) Soil structure and management: a review. Geoderma 124:3–22
Cambardella CA, Elliott ET (1993) Carbon and nitrogen distribution in aggregates from cultivated and native grassland soils. Soil Sci Soc Am J 57:1071–1076
Cantero-Martínez C, Gabiña D (2004) Evaluation of agricultural practices to improve efficiency and environment conservation in Mediterranean arid and semi-arid production systems: MEDRATE project. In: Cantero-Martínez C, Gabiña D (eds) Mediterranean rainfed agriculture: Strategies for sustainability, Options Méditerranéennes: Série A. Séminaires Méditerranéennes no. 60. CIHEAM, Zaragoza
Cantero-Martínez C, O'Leary G, Connor DJ (1995a) Stubble retention and nitrogen fertilization in a fallow-wheat rainfed cropping system. I. Soil water conservation, growth and yield. Soil Tillage Res 34:79–94
Cantero-Martínez C, Villar JM, Romagosa I, Fereres E (1995b) Nitrogen fertilization of barley under semi-arid conditions. Eur J Agron 4:309–316
Cantero-Martínez C, Angás P, Lampurlanés J (2007) Long-term yield and water use efficiency under various tillage systems in Mediterranean rainfed conditions. Ann Appl Biol 150:293–305
Cantero-Martínez C, Plaza-Bonilla D, Angás P, Álvaro-Fuentes J (2016) Best management practices and nitrogen fertilization in Mediterranean rainfed conditions: combining field and modelling approaches. Eur J Agron 79:119–130
CGIAR (2013) New research approaches to improve dryland agriculture to deliver more prosperous future. CGIAR Research Program on Dryland Agricultural Production Systems. Available at: http://www.icarda.org/publications-resources/drylans-systems
CGIAR (2017) The world's dry areas. CGIAR Research Program on Dryland Systems. Available at: http://drylandsystems.cgiar.org/content/worlds-dry-areas
Chien SH, Prochnow LI, Cantarella H (2009) Recent developments of fertilizer production and use to improve nutrient efficiency and minimize environmental impacts. Adv Agron 102:267–322
Connor DJ, Loomis RS, Cassman KG (2011) Crop ecology: productivity and management in agricultural systems. Cambridge University Press, New York
Cooper PJM, Keatinge J, Hughes G (1983) Crop evapotranspiration—A technique for calculation of its components by field measurements. Field Crops Res 7:299–312
Cooper PJM, Gregory PJ, Tully D, Harris HC (1987) Improving water use efficiency of annual crops in rainfed farming systems of west Asia and North Africa. Exp Agric 23:113–158
Denef K, Stewart CE, Brenner J, Paustian K (2008) Does long-term center-pivot irrigation increase soil carbon stocks in semi-arid agro-ecosystems? Geoderma 145:121–129
Díaz-Ambrona CH, Mínguez MI (2001) Cereal-legume rotations in a Mediterranean environment: biomass and yield production. Field Crops Res 70:139–151
FAO (2016) Trees, forests and land use in drylands. The first global assessment. Food and Agriculture Organization of the United Nations, Rome. Available at: http://www.fao.org/3/a-i5905e.pdf
FAO (2017) Soil organic carbon: the hidden potential. Food and Agriculture Organization of the United Nations, Rome
Farooq M, Siddique KHM (2016) Research and developmental issues in dryland agriculture. In: Farooq M, Siddique KHM (eds) Innovations in dryland agriculture. Springer International Publishing AG, Cham, pp 31–46
Fereres E, Orgaz F, Villalobos F (1993) Water use efficiency in sustainable agricultural systems. In: Buxton DR, Shibles R, Forsferg RA, Blad BL, Asay KH, Paulsen GM, Wilson RF (eds) International crop science I. CSSA, Madison, pp 83–94
Flexas J, Niinemets Ü, Gallé A, Barbour MM, Centritto M, Díaz-Espejo A, Douthe C, Galmés J, Ribas-Carbo M, Rodriguez PL, Rosselló F, Soolanayakanahally R, Tomas M, Wright IJ, Farquhar GD, Medrano H (2013) Diffusional conductances to CO_2 as a target for increasing photosynthesis and photosynthetic water-use efficiency. Photosynth Res 117:45–59
French RJ (1978) The effect of fallowing on the yield of wheat. II The effect on grain yield. Aust J Agric Res 29:669–684

Friedrich T, Derpsch R, Kassam A (2012) Overview of the global spread of conservation agriculture. Field Actions Science Reports. Special Issue 6, Reconciling poverty eradication and protection of the environment. Available at: http://factsreports.revues.org/1941
Galbally IE, Kirstine WV, Meyer CP, Wang YP (2008) Soil-atmosphere trace gas exchange in semiarid and arid zones. J Environ Qual 37:599–607
García AL, Royo-Esnal A, Torra J, Cantero-Martínez C, Recasens J (2014) Integrated management of *Bromus Diandrus* in dryland cereal fields under no-till. Weed Res 54:408–417
Gomes L, Arrúe JL, López MV, Sterk G, Richard D, Gracia R, Sabre M, Gaudichet A, Frangi JP (2003) Wind erosion in a semi-arid agricultural area of Spain: the WELSONS project. Catena 52:235–256
Gómez JA, Giráldez JV, Fereres E (2009) The influence of cover crops and tillage on water and sediment yield, and on nutrient, and organic matter losses in an olive orchard on a sandy loam soil. Soil Tillage Res 106:137–144
Gómez JA, Llewellyn C, Basch G, Sutton PB, Dyson JS, Jones CA (2011) The effects of cover crops and conventional tillage on soil and runoff loss in vineyards and olive groves in several Mediterranean countries. Soil Use Manage 27:502–514
Grant CA, Peterson GA, Campbell CA (2002) Nutrient considerations for diversified cropping systems in the Northern Great Plains. Agron J 94:186–198
Guo LB, Gifford RM (2002) Soil carbon stocks and land use change: a meta-analysis. Glob Change Biol 8:345–360
Hagin J, Tucker B (1982) Fertilization of dryland and irrigated soils. Springer, Berlin
Halvorson AD, Black AL, Krupinsky JM, Merrill SD, Wienhold BJ, Tanaka DL (2000) Spring wheat response to tillage and nitrogen fertilization in rotation with sunflower and winter wheat. Agron J 92:136–144
Hütsch BW (2001) Methane oxidation in non-flooded soils as affected by crop production-invited paper. Eur J Agron 14:237–260
IFA (2009) The global "4R" nutrient stewardship framework for developing and delivering fertilizer best management practices. IFA task force on fertilizer best management practices, AgCom/09/32, A/09/68. International Fertilizer Industry Association, Paris
Jenkinson DS, Harris HC, Ryan J, McNeill AM (1999) Organic matter turnover in a calcareous clay soil from Syria under a two-course cereal rotation. Soil Biol Biochem 31:687–693
Jensen ES, Peoples MB, Boddey RM, Gresshoff PM, Hauggaard-Nielsen H, Alves BJR, Morrison MJ (2012) Legumes for mitigation of climate change and the provision of feedstock for biofuels and biorefineries. A review. Agron Sustain Dev 32:329–364
Ju XT, Xing GX, Chen XP, Zhang SL, Zhang LJ, Liu XJ, Cui ZL, Yin B, Christie P, Zhu ZL, Zhang FS (2009) Reducing environmental risk by improving N management in intensive Chinese agricultural systems. Proc Natl Acad Sci USA 106:3041–3046
van Kessel C, Venterea R, Six J, Adviento-Borbe MA, Linquist B, Van Groenigen KJ (2013) Climate, duration, and N placement determine N_2O emissions in reduced tillage systems: a meta-analysis. Glob Change Biol 19:33–44
Kirkegaard JA, Robertson MJ (2013) Agronomic principles of water- and nutrient-use efficiency: case studies in dryland grain production in Australia. In: Rengel Z (ed) Improving water and nitrogen-use efficiency in food production systems. Wiley, Hoboken, pp 211–233
Koohafkan P, Stewart BA (2008) Water and cereals in drylands. The Food and Agriculture Organization of the United Nations and Earthscan, London
Lahmar R, Bationo BA, Lamso ND, Guéro Y, Tittonell P (2012) Tailoring conservation agriculture technologies to West Africa semi-arid zones: building on traditional local practices for soil restoration. Field Crops Res 132:158–167
Lal R (2002) Carbon sequestration in dryland ecosystems of West Asia and North Africa. Land Degrad Dev 13:45–59
Lal R (2004) Carbon sequestration in dryland ecosystems. Environ Manag 33:528–544

Lal R, Kimble JM (2000) Pedogenic carbonates and the global carbon cycle. In: Lal R, Kimble JM, Eswaran H, Stewart BA (eds) Global climate change and pedogenic carbonates. CRC Press, Boca Raton, pp 1–14

Lammel J (2005) Cost of the different options available to the farmers: current status and prospects. In: IFA International workshop on enhanced-efficiency fertilizers, Frankfurt. International Fertilizer Industry Association, Paris

Lampurlanés J, Angás P, Cantero-Martínez C (2001) Root growth, soil water content and yield of barley under different tillage systems on two soils in semiarid conditions. Field Crops Res 69:27–40

Lampurlanés J, Angás P, Cantero-Martínez C (2002) Tillage effect on water storage efficiency during fallow, and soil water content, root growth and yield of the following barley crop on two different soils in semiarid conditions. Soil Tillage Res 65:207–220

Lampurlanés J, Plaza-Bonilla D, Álvaro-Fuentes J, Cantero-Martínez C (2016) Long-term analysis of soil water conservation and crop yield under different tillage systems in Mediterranean conditions. Field Crops Res 189:59–67

Lee C, Schaaf T (eds) (2008) Sustainable management of marginal drylands. UNESCO-MAB, Division of Ecological and Earth Sciences, Paris

Lenssen AW, Sainju UM, Jabro JD, Iversen WM, Allen BL, Evans RG (2014) Crop diversification, tillage and management system influence spring wheat yield and water use. Agron J 106:1445–1454

Li SX, Wang ZH, Hu TT, Gao YJ, Stewart BA (2009) Nitrogen in dryland soils of China and its management. Adv Agron 101:123–181

Linn DM, Doran JW (1984) Effect of water-filled pore-space on carbon-dioxide and nitrous-oxide production in tilled and nontilled soils. Soil Sci Soc Am J 48:1267–1272

López MV, Sabre M, Gracia R, Arrúe JL, Gomes L (1998) Tillage effects on soil surface conditions and dust emission by wind erosion in semiarid Aragón (NE Spain). Soil Tillage Res 45:91–105

López MV, Gracia R, Arrúe JL (2000) Effects of reduced tillage on soil surface properties affecting wind erosion in semiarid fallow lands of Central Aragón. Eur J Agron 12:191–199

López MV, Gracia R, Arrúe JL (2001) An evaluation of wind erosion hazard in fallow lands of semi-arid Aragón (NE Spain). J Soil Water Conserv 56:212–219

López MV, Moret D, Gracia R, Arrúe JL (2003) Tillage effects on barley residue cover during fallow in semiarid Aragon. Soil Tillage Res 72:53–64

López MV, Arrúe JL, Álvaro-Fuentes J, Moret D (2005) Dynamics of surface barley residues during fallow as affected by tillage and decomposition in semiarid Aragon (NE Spain). Eur J Agron 23:26–36

López-Bellido RJ, Fontán JM, López-Bellido FJ, López-Bellido L (2010) Carbon sequestration by tillage, rotation, and nitrogen fertilization in a Mediterranean vertisol. Agron J 102:310–318

López-Garrido R, Madejón E, Murillo JM, Moreno F (2011) Short and long-term distribution with depth of soil organic carbon and nutrients under traditional and conservation tillage in a Mediterranean environment (southwest Spain). Soil Use Manage 27:177–185

Mallarino AP, Borges R (2006) Phosphorus and potassium distribution in soil following long-term deep-band fertilization in different tillage systems. Soil Sci Soc Am J 70:702–707

McNeill AM, Penfold CM (2009) Agronomic management options for phosphorus in Australian dryland organic and low-input cropping systems. Crop Pasture Sci 60:163–182

Middleton N, Thomas D (1997) World atlas of desertification, 2nd edn. Arnold, London

Morell FJ, Cantero-Martínez C, Álvaro-Fuentes J, Lampurlanés J (2011a) Root growth of barley as affected by tillage systems and N fertilization in a semiarid Mediterranean agroecosystem. Agron J 103:1270–1275

Morell FJ, Lampurlanés J, Álvaro-Fuentes J, Cantero-Martínez C (2011b) Yield and water use efficiency of barley in a semiarid Mediterranean agroecosystem: long-term effects of tillage and N fertilization. Soil Tillage Res 117:76–84

Moreno F, Arrúe JL, Cantero-Martínez C, López MV, Murillo JM, Sombrero A, López-Garrido R, Madejón E, Moret D, Álvaro-Fuentes J (2011) Conservation agriculture under mediterranean conditions in Spain. In: Lichtfouse E (ed) Biodiversity, biofuels, agroforestry and conservation agriculture. Sustainable Agriculture Reviews No. 5, Springer, Dordrecht, pp 175–192

Nawaz A, Farooq M (2016) Nutrient management in dryland agriculture systems. In: Farooq M, Siddique KHM (eds) Innovations in dryland agriculture. Springer International Publishing AG, Cham, pp 115–142

Oldeman LR (1994) The global extent of soil degradation. In: Greenland DJ, Szabolcs I (eds) Soil resilience and sustainable land use. CAB International, Wallingford

Omanya GO, Pasternak D (eds) (2005) Sustainable agriculture systems for the drylands. Proceedings of the International Symposium for Sustainable Dryland Agriculture Systems, 2–5 December 2003, Niamey, Niger. International Crops Research Institute for the Semi-Arid Tropics (ICRISAT)

Oweis T (1997) Supplemental irrigation: a highly water-efficient practice. ICARDA, Aleppo

Palm C, Blanco-Canqui H, De Clerck F, Gatere L, Grace P (2014) Conservation agriculture and ecosystem services: an overview. Agric Ecosyst Environ 187:87–105

Passioura JB, Angus JF (2010) Improving productivity of crops in water-limited environments. Adv Agron 106:38–67

Paustian K, Lehmann J, Ogle S, Reay D, Robertson GP, Smith P (2016) Climate-smart soils. Nature 532:49–57

Peterson GA, Westfall DG (2004) Managing precipitation use in sustainable dryland agroecosystems. Ann Appl Biol 144:127–138

Pittelkow CM, Linquist BA, Lundy ME, Liang X, van Groenigen KJ, Lee J, van Gestel N, Six J, Venterea RT, van Kessel C (2015) When does no-till yield more? A global meta-analysis. Field Crops Res 183:156–168

Plaza-Bonilla D, Álvaro-Fuentes J, Cantero-Martínez C (2013a) Soil aggregate stability as affected by fertilization type under semiarid no-tillage conditions. Soil Sci Soc Am J 77:284–292

Plaza-Bonilla D, Cantero-Martínez C, Viñas P, Álvaro-Fuentes J (2013b) Soil aggregation and organic carbon protection in a no-tillage chronosequence under Mediterranean conditions. Geoderma 193:76–82

Plaza-Bonilla D, Álvaro-Fuentes J, Arrúe JL, Cantero-Martínez C (2014a) Tillage and nitrogen fertilization effects on nitrous oxide yield-scaled emissions in a rainfed Mediterranean area. Agric Ecosyst Environ 189:43–52

Plaza-Bonilla D, Cantero-Martínez C, Bareche J, Arrúe JL, Álvaro-Fuentes J (2014b) Soil carbon dioxide and methane fluxes as affected by tillage and N fertilization in dryland conditions. Plant Soil 381:111–130

Plaza-Bonilla D, Arrúe JL, Cantero-Martínez C, Fanlo R, Iglesias A, Álvaro-Fuentes J (2015) Carbon management in dryland agricultural systems. A review. Agron Sustain Dev 35:1319–1334

Plaza-Bonilla D, Álvaro-Fuentes J, Bareche J, Masgoret A, Cantero-Martínez C (2017a) Delayed sowing improved barley yield in a no-till rainfed Mediterranean agroecosystem. Agron J 109:1249–1260

Plaza-Bonilla D, Cantero-Martínez C, Bareche J, Arrúe JL, Lampurlanés J, Álvaro-Fuentes J (2017b) Do no-till and pig slurry application improve barley yield and water and nitrogen use efficiencies in rainfed Mediterranean conditions? Field Crops Res 203:74–85

Quemada M, Gabriel JL (2016) Approaches for increasing nitrogen and water use efficiency simultaneously. Glob Food Sec 9:29–35

Ramos MC, Martínez-Casasnovas JA (2004) Nutrient losses from a vineyard soil in Northeastern Spain caused by an extraordinary rainfall event. Catena 55:79–90

Rebetzke GJ, Condon AG, Richards RA, Farquhar GD (2002) Selection for reduced carbon isotope discrimination increases aerial biomass and grain yield of rainfed bread wheat. Crop Sci 42:739–745

Robertson M, Carberry P, Brennan L (2009) Economic benefits of variable rate technology: case studies from Australian grain farms. Crop Pasture Sci 60:799–807

Rose T, Bowden B (2013) Matching soil nutrient supply and crop demand in the growing season. In: Rengel Z (ed) Improving water and nitrogen-use efficiency in food production systems. Wiley, Hoboken, pp 93–103

Roy RN, Finck A, Blair GJ, Tandon HS (2006) Plant nutrition for food security: a guide to integrated nutrient management, FAO fertilizer and plant nutrition Bull. No. 16. Food and Agriculture Organization of the United Nations, Rome

Ryan J, Sommer R, Ibrikci H (2012) Fertilizer best management practices: a perspective from the dryland West Asia-North Africa region. J Agron Crop Sci 198:57–67

Sadras VO (2004) Yield and water-use efficiency of water-and nitrogen-stressed wheat crops increase with degree of co-limitation. Eur J Agron 21:455–464

Safriel U, Adeel Z (2005) Dryland systems. Chapter 22. In: Hassan R, Scholes R, Ash N (eds) Ecosystems and human well-being: current state and trends, vol 1. Island Press, Washington, DC

Salekdeh GH, Reynolds M, Bennett J, Boyer J (2009) Conceptual framework for drought phenotyping during molecular breeding. Trends Plant Sci 14:488–496

Salinas-Garcia JR, Matocha JE, Hons FM (1997) Long-term tillage and nitrogen fertilization effects on soil properties of an Alfisol under dryland corn/cotton production. Soil Tillage Res 42:79–93

Sanz-Cobena A, Misselbrook T, Camp V, Vallejo A (2011) Effect of water addition and the urease inhibitor NBPT on the abatement of ammonia emission from surface applied urea. Atmos Environ 45:1517–1524

Schlaepfer DR, Bradford JB, Lauenroth WK, Munson SM, Tietjen B, Hall SA, Wilson SD, Duniway MC, Jia G, Pyke DA, Lkhagva A, Jamiyansharav K (2017) Climate change reduces extent of temperate drylands and intensifies drought in deep soils. Nat Commun 8:14196. https://doi.org/10.1038/ncomms14196

Schwilch G, Hessel R, Verzandvoort S (2012) Desire for greener land. Options for sustainable land management in drylands. University of Bern-CDE, Alterra-Wageningen UR, ISRIC-World Soil Information and CTA-Technical Centre for Agricultural and Rural Cooperation, Bern

Scotford IM, Cumby TR, White RP, Carton OT, Lorenz F, Hatterman U, Provolo G (1998) Estimation of the nutrient value of agricultural slurries by measurement of physical and chemical properties. J Agric Eng Res 71:291–305

Serraj R, Sinclair TR, Purcell LC (1999) Symbiotic N_2 fixation response to drought. J Exp Bot 50:143–155

Shangyou S, Tinglu F, Yong W (1997) Comprehensive sustainable development of dryland agriculture in Northwest China. J Sustain Agr 9:67–84

Simpson RJ, Oberson A, Culvenor RA, Ryan MH, Veneklaas EJ, Lambers H, Lynch JP, Ryan PR, Delhaize E, Smith FA, Smith SE, Harvey PR, Richardson AE (2011) Strategies and agronomic interventions to improve phosphorus-use efficiency of farming systems. Plant Soil 349:89–120

Six J, Elliott ET, Paustian K (1999) Aggregate and soil organic matter dynamics under conventional and no-tillage systems. Soil Sci Soc Am J 63:1350–1358

Six J, Elliott ET, Paustian K (2000) Soil macroaggregate turnover and microaggregate formation: a mechanism for C sequestration under no-tillage agriculture. Soil Biol Biochem 32:2099–2103

Six J, Ogle SM, Breidt FJ, Conant RT, Mosier AR, Paustian K (2004) The potential to mitigate global warming no-tillage management is only realized when practiced in the long run. Glob Change Biol 10:155–160

Smith P, Martino D, Cai Z, Gwary D, Janzen H, Kumar P, McCarl B, Ogle S, O'Mara F, Rice C, Scholes B, Sirotenko O, Howden M, McAllister T, Pan G, Romanenkov V, Schneider U, Towprayoon S, Wattenbach M, Smith J (2008) Greenhouse gas mitigation in agriculture. Philos Trans R Soc Lond B Biol Sci 363:789–813

Smith P, Bustamante M, Ahammad H, Clark H, Dong H, Elsiddig EA, Haberl H, Harper R, House J, Jafari M, Masera O, Mbow C, Ravindranath NH, Rice CW, Robledo Abad C, Romanovskaya A, Sperling F, Tubiello F (2014) Agriculture, Forestry and Other Land Use (AFOLU). In: Edenhofer O, Pichs-Madruga R, Sokona Y, Farahani E, Kadner S, Seyboth K, Adler A, Baum

I, Brunner S, Eickemeier P, Kriemann B, Savolainen J, Schlömer S, von Stechow C, Zwickel T, Minx JC (eds) Climate change 2014: mitigation of climate change, Contribution of Working Group III to the Fifth Assessment Report of the Intergovernmental Panel on Climate Change. Cambridge University Press, Cambridge, UK

Stanhill G (1986) Water use efficiency. Adv Agron 39:53–85

Sterk G (2003) Causes, consequences and control of wind erosion in Sahelian Africa: a review. Land Degrad Dev 14:95–108

Sterk G, Herrmann L, Bationo A (1996) Wind-blown nutrient transport and soil productivity changes in southwest Niger. Land Degrad Dev 7:325–335

Sterk G, López MV, Arrúe JL (1999) Saltation transport on a silt loam soil in north-east Spain. Land Degrad Dev 10:545–554

Sterk G, Riksen M, Goossens D (2001) Dryland degradation by wind erosion and its control. Ann Arid Zone 40:351–367

Stewart BA, Thapa S (2016) Dryland farming: concept, origin and brief history. In: Farooq M, Siddique KHM (eds) Innovations in dryland agriculture. Springer International Publishing AG, Cham, pp 3–29

Tanner CB, Sinclair TR (1983) Efficient water use in crop production: research or research? In: Taylor HM, Jordan WR, Sinclair TR (eds) Limitations to efficient water use in crop production, Am. Soc. Agron, Madison, pp 1–27

Tardieu F (2012) Any trait or trait-related allele can confer drought tolerance: just design the right drought scenario. J Exper Bot 63:25–31

Thomas RJ (2008) Opportunities to reduce the vulnerability of dryland farmers in Central and West Asia and North Africa to climate change. Agric Ecosyst Environ 126:36–45

Thorup-Kristensen K (2006) Effect of deep and shallow root systems on the dynamics of soil inorganic N during 3-year crop rotations. Plant Soil 288:233–248

Trost B, Prochnow A, Drasting K, Meyer-Aurich A, Ellmer F, Baumecker M (2013) Irrigation, soil organic carbon and N_2O emissions. A review. Agron Sustain Dev 33:733–749

Tubiello FN, Salvatore M, Ferrara AF, House J, Federici S, Rossi S, Biancalani R, Condor Golec RD, Jacobs H, Flammini A, Prosperi P, Cardenas-Galindo P, Schmidhuber J, Sanz Sanchez MJ, Srivastava N, Smith P (2015) The contribution of agriculture, forestry and other land use activities to global warming, 1990–2012. Glob Change Biol 21:2655–2660

UN (2011) Global drylands: a UN system-wide response. United Nations Environment Management Group. Available at: http://www.unccd.int/Lists/SiteDocumentLibrary/Publications/Global_Drylands_Full_Report.pdf

UNCCD (2000) An introduction to the United Nations Convention to Combat Desertification. Available at http://www.unccd.int

Unger PW, Fryrear DW, Lindstrom MJ (2006) Soil conservation. In: Peterson GA, Unger PW, Payne WA (eds) Dryland agriculture, Agron. Monogr. 23, 2nd edn. ASA, CSSA, and SSSA, Madison, pp 87–112

Venterea RT, Halvorson AD, Kitchen N, Liebig MA, Cavigelli MA, Del Grosso SJ, Motavalli PP, Nelson KA, Spokas KA, Singh BP, Stewart CE, Ranaivoson A, Strock J, Collins H (2012) Challenges and opportunities for mitigating nitrous oxide emissions from fertilized cropping systems. Front Ecol Environ 10:562–570

Virto I, Gartzia-Bengoetxea N, Fernández-Ugalde O (2011) Role of organic matter and carbonates in soil aggregation estimated using laser diffractometry. Pedosphere 21:566–572

West TO, Marland G, King AW, Post WM, Jain AK, Andrasko K (2004) Carbon management response curves: estimates of temporal soil carbon dynamics. Environ Manage 33:507–518

Westfall DG, Havlin JL, Hergert GW, Raun WR (1996) Nitrogen management in dryland cropping systems. J Prod Agric 9:192–199

Wienhold BJ, Luchiari A, Zhang R (2000) Challenges confronting soil management for dryland agriculture. Ann Arid Zone 39:333–346

Wilson TM, McGowen B, Mullock J, Arnall DB, Warren JG (2015) Nitrous oxide emissions from continuous winter wheat in the Southern Great Plains. Agron J 107:1878–1884

Yagüe MR, Quílez D (2012) On-farm measurement of electrical conductivity for the estimation of ammonium nitrogen concentration in pig slurry. J Environ Qual 41:893–900

Crop-Livestock Interaction for Sustainable Agriculture

Paul Kayode Baiyeri, Harriet Ndofor Foleng, Ndubuisi Samuel Machebe, and Cynthia Ebere Nwobodo

Abstract Crop-livestock is a form of land use system whereby livestock husbandry and cropping are practiced in association. The production system safeguards ecosystem balance and assures sustainable production; besides, the production system does not impact on the environment negatively. The burgeoning human population globally has increased the spate of food insecurity and malnutrition especially in the developing countries. There is therefore an urgent need for a better, efficient and sustainable production system capable of providing both crop and animal products for the teeming human population. Crop-livestock integrated production system seems to provide opportunity for the production and supply of food of both crop and animal origin without detrimental impact on the fragile environment. In this integrated system, crop and livestock interact to create synergy that allows ecosystem balance and sustainable production intensification. In this chapter, comprehensive documentation of the concept, principles, and practices of crop – livestock production system have been made. Also, the importance of this integrated production system on ecosystem balance and sustainable crop and livestock production have been presented.

Keywords Integrated farming · Sustainable system intensification · Crop-livestock interaction

---

P. K. Baiyeri (✉)
Department of Crop Science, University of Nigeria, Nsukka, Nigeria
e-mail: paul.baiyeri@unn.edu.ng

H. N. Foleng · N. S. Machebe
Department of Animal Science, University of Nigeria, Nsukka, Nigeria

C. E. Nwobodo
Department of Agricultural Extension, University of Nigeria, Nsukka, Nigeria

M. Farooq, M. Pisante (eds.), *Innovations in Sustainable Agriculture*,
https://doi.org/10.1007/978-3-030-23169-9_18

1 Introduction

Agriculture and food systems are facing major challenges of not only ensuring food and nutrition security of the ever-growing human population but also the need to ensure that the environment is judiciously exploited. Population growth together with changes in consumer demand preferences, associated with rising incomes, drive greater demand for food while food systems are increasingly threatened by land degradation, climate change, and other stressors (SDSN 2013). Agriculture systems must change to meet the rising demand, to contribute more effectively to the reduction of poverty and malnutrition, and to become more ecologically sustainable (SDSN 2013). Stewardship in the use of natural resources is therefore sine-quo-non-to an agricultural system that boasts to carter for the present and future population of the world. A sustainable agriculture and food system should provide nutrition for all, minimize negative impact on ecosystems and human health, improve livelihoods and resilience whilst being economically viable (CNS-FAO 2016). There is a need to not just emphasize increase in per unit agricultural yield but to accommodate productivity in agricultural landscapes that encompass valued ecosystem services such as water infiltration, carbon sequestration, and conservation of biodiversity (Giovannucci et al. 2012). According to Pretty (2008), new approaches are needed to integrate biological and ecological processes into food production, minimize the use of non-renewable inputs that cause harm to the environment or to the health of farmers and consumers. Crop-livestock is an integrated production system that portends ecosystem balance together with sustainable productivity. The system ensures production and supply of food of both crop and animal origin without unfavorable impact on the environment. Crop-livestock is a form of land use system whereby livestock husbandry and cropping are practiced in association (Jabbar and Cobbina 1990). Sustainable agriculture is the ability of farmland to produce food and other agricultural products to satisfy human needs indefinitely as well as having sustainable impacts on the broader environment (UNEP 2010). Climate smart agriculture (CSA) on the other hand is a continuous process of improving agriculture innovations in policies, technologies, management and financing that aim to: sustainably increase agricultural productivity and incomes; strengthen the adaptive capacity and resilience of people, food production systems and ecosystems in agricultural landscapes; and reduce and/or remove greenhouse gases emissions, where possible.

2 Concept and Principles of Sustainable Agriculture

Sustainable agriculture is a multidimensional concept. Sustainable agriculture does not refer to the use of labor-intensive, low cost input and low technology methods. It comprises site-specific ranching and farming practices designed to meet current and future needs of food, fiber, energy, and ecosystem services including but not limited to soil conservation, clean water and biodiversity (Menalled et al. 2008). Pretty (2008) argues that sustainability in agricultural systems center on the need to the develop technologies and practices that are accessible and effective to farmers, lead to improvements in food productivity and that do not have adverse effects on environmental goods and services. Sustainable agriculture development represents a new approach of agricultural development that ensure the economic and social benefits for the actual generation without compromising the capacity of the future generation to fulfill the proper agriculture demands and without injure to the fundamental ecological process (Asachi 2010). Sustainable Agriculture is knowledge intensive and context specific, requiring significant capacity building for smallholders in developing countries. There is a greater need for institutional support if farmers are to engage into significant sustainable agricultural production capable of providing food, feed and other materials for the teeming population and in ensuring environmental sustainability. Agricultural extension needs to be intensified in order to bring about the most desired move into meaningful sustainable agricultural production system.

2.1 Sustainable Crop Production Methods

Sustainable crop production includes adhering to food production that do not harm the environment, that provides fair treatment to workers, and that supports and sustains local communities (Food Program 2018) Sustainable crop farmers focus on ensuring that farming practices can be sustained over time and do not cause undue damage to the environment. Sustainable crop production intensification (SCPI) aims to increase production per unit area, taking into consideration all relevant factors affecting productivity and sustainability, including social, political, economic and environmental impacts. With focus on the environment, SCPI aims to maximize options for crop production intensification through the management of biodiversity and ecosystem services (FAO 2018). Sustainable crop production methodologies include such methods as: multi-cropping involving crop rotation, intercropping; minimal to no pesticide use methods such as integrated pest management, intercropping and companion planting, mulching, groundcover, and manual control, release of beneficial insects and organisms; soil health management including organic matter use, crop rotation, intercropping, and mulching, less or no tillage and reduction of heavy machinery use; as well as socio-economic dimension, which involves ensuring fair treatment of farm workers, supporting farm communities, and sustaining local food systems (Food Program 2018).

2.2 Sustainable Livestock Production Systems Methods

The current livestock production practices place severe pressure on the environment, through emissions to air, water and soil and through utilization of natural resources, including land, water and fossil energy (Royal Netherlands Society for Agricultural Sciences 2013). At present, livestock contribute 14.5% to global greenhouse emissions (FAO 2017). It uses about 70% of total available agricultural land and represents about 8% of global water withdrawals (Royal Netherlands Society for Agricultural Sciences 2013). The Royal Netherlands Society of Agricultural (2013) highlighted four major concerns for a sustainable livestock production system. First, ensuring health and safety of animals and humans: this includes eradication of infectious animal and zoonotic diseases emerging from livestock farming and transport; mitigation of consequences of use of antibiotics, such as the development of microbe resistance to antibiotics. Use of bio-security measures such as isolation of new animals before adding to the main flock, vaccination, applying strict sanitation and disinfectant measures are important strategies for promoting health of animals and humans (Alberta SPCA 2014). Second is the customized care: ensuring robustness, dignity and integrity of the production of animal including compliance to human standards for animal welfare aspects as well as respecting the specific biological traits and requirements of individual animals in the group. Close confinement of animals, individual rearing of social animals such as pigs and cattle, and other systems of housing and managing animals that do not meet the needs of animals constitute unsustainable animal production practices (Maurice 2017). The third concern is 'no nuisance', meaning that a sustainable animal production system should avoid exposure of the environment to critical emissions of waste materials i.e. dust, noise, malodor or pathogens originating from livestock production systems. This also includes sustainable management of natural resources, biodiversity and prevention of land degradation. Pollution prevention strategies such as use of animal waste as organic manure, energy source can be adopted to ensure environmental sustainability. The fourth concern is credit performance (from socio-economic standpoint): sustainable animal production system should guarantee a responsible and trustworthy livestock production sector with sound perspectives for farmers in local, national, regional or global food production chains. Livestock enterprise should be appropriate for the resources available to the farmer. Efficient use of feed increases sustainability. Use of quality genetic breeding stock, combined with knowledge of healthy reproductive processes reduce health problems and feed costs thereby making production more sustainable (Alberta SPCA 2014). There is no single solution for achieving a sustainable supply of animal-source food. A trade-off between the four major concerns for a sustainable animal production requiring a more comprehensive and synergistic approach is recommended (Royal Netherlands Society for Agricultural Sciences 2013).

3 Concept of Crop-Livestock Production

Growth in demand for food products all over the world is rising at a fast rate due to increasing population, urbanization and sustained rise in per capita income (Powell et al. 2004; Kumar et al. 2014). Thus, the world is under enormous strain to reduce food insecurity, soaring food prices and deepening poverty (Nalubwama et al. 2014) because of the projected increase in human population of about 8.3 billion by 2020 as reported by the United Nations World population prospects (UNPP 2008). According to FAO report (FAO 1993), currently 800 million people worldwide are suffering from malnutrition and hunger especially in developing country, due mainly to insufficient production and inadequate distribution and lack of income to acquire food of adequate quantity and quality to satisfy their needs. This calls for a better, efficient and sustainable agricultural production system capable of providing crop and animal products for the teeming human population. A system capable of providing succour for the production and supply of food of both crop and animal origin without detrimental impact on the environment is crop-livestock integrated production system. Herrero et al. (2009) indicated that almost one billion people globally depend on crop livestock system as their primary source of livelihood. Statistically, mixed crop livestock system is reported to produce 50% of global cereals, 34% of beef and 30% of milk (Herrero et al. 2009; Duncan et al. 2013).

Jabbar and Cobbina (1990) defined crop-livestock systems as a form of land use systems in which livestock husbandry and cropping are practiced in association. The system may include (i) sole livestock farming practiced in proximity to and functional association with crop farming, and (ii) crop livestock mixed farming under the same management with various degrees of combinations (Jabbar and Cobbina 1990). Other specific examples of such system are rice-wheat-livestock system (buffalo, cattle, and goats) as practiced in North-West India, crop-pasture system, agropastoralist (produces crop and raises cattle principally by grazing natural pastures) (Okoruwa et al. 1996; Kumar et al. 2014), cereal based-livestock system also practiced in many regions of West Africa (Ezeaku et al. 2015), integrated tree crops-ruminant systems as witnessed in Malaysia (Devendra 2011), etc. It is as an integrated system of agricultural production whereby crop and livestock interact to create synergy with recycling allowing the use of available resources. In this system crop residues are used for animal feed, while livestock and livestock by-products production and processing are used to enhance agricultural productivity by intensifying nutrients that improve soil fertility and at the same time reduces the use of chemical fertilizers (Singh et al. 2011). In general, productivities of livestock and croplands are inextricably linked. Reports available indicates (Jabbar and Cobbina 1990; Powell et al. 2004) that crop-livestock integrated system is a key solution for enhancing livestock production and safeguarding the environment when its resources are prudently and efficiently managed.

3.1 Principles and Practices of Crop-Livestock Production System

Crop–livestock production systems is a classification of livestock systems that combine crop (mainly millet, cowpea, sorghum, cotton and groundnut) and livestock activities (cattle, sheep, goats and camels) in different proportions to enhance productivity and food supply. The system is practiced mostly in arid and semi-arid areas, moist savannah zones, and may include rain fed mixed crop livestock production system or irrigated mixed crop-livestock system. Improved grain yield in alley farming and fodder bank technology are typical examples epitomizing the benefits inherent in crop livestock production system of farming. For instance, Jabbar and Cobbina (1990) noted that the higher maize grain output witnessed in alley cropping following a short-term grazed fallow is due to soil fertility build-up through decomposition and mineralization of dead plant biomass and a fallow factor. In the system, ruminants especially sheep and goats can graze on the fallow land and their activities enhance nutrient recycling through the dung that is eventually returned to the soil in the course of grazing (Jabbar and Cobbina 1990; Powell et al. 2004). In fodder bank system, forage legumes are established and managed at very high density to provide dry season supplementation for ruminant livestock. Forage legumes like *Stylosanthes guianensis* (Stylo), *Centrosema pubescens* (butterfly pea), etc. are cropped and managed specifically for dry season feeding of ruminants to improve growth rate especially at a time when pasture quantity and quality has greatly declined due to drought. Eventually, fodder bank plots are returned to cropping after establishment and grazing. Report available shows that maize crop yield on fodder bank plots returned to cropping is one-and-half to twofolds better compared to adjacent control plots (Jabbar and Cobbina 1990).

In regions where crop-livestock production is practiced, the system has been known to significantly improve livelihoods especially in rural populations in most part of the world especially in developing economies. Farmers produce food grains and manage livestock to boost their income. Thus, farmer's income, agriculture and rural economy are heavily dependent on livestock. In an integrated system, livestock and crops are produced within a coordinated framework such that the waste products of one component serve as a resource for the other. For instance, animals produce power for agricultural operations and pulling cart and use of dung as organic fertilizers (Singh et al. 2011). Manure from animals is used as organic fertilizers to enhance crop production whereas crop residues and by-products are used to complement animal feed that are often inadequate especially in the dry season leading to improved animal nutrition and productivity. The farmer also benefits from the sale of animal products like meat, milk and eggs as these products are a major source of food for humans. A major characteristic of the system is that crop (mainly millet, cowpea, sorghum, rice, wheat, cotton and groundnut) and livestock activities (cattle, buffalo, sheep, goats and camels) are combined in different proportions to maximize both crop and livestock productivity. This is an important system both for agriculture and livestock production, and in terms of livelihoods in places where it is practiced.

Adamawa State is 1 of the 36 states in Nigeria located in the Northeastern part of the country. The state has a total land area of approximately 38,741 km^2, out of which about 22,604 km^2 is known to be arable (Adebayo 1999). Southern Guinea Savannah, Northern Guinea Savannah and the Sudan Savannah happens to be the major vegetation found in the state with an interspersion of thickest tree savannah, open grass savannah and finging forests in the river-valley (Lawal 2012). In 2006, the population of the State was 3.2 million according to NPC (2006) report and majority of the people are farmers, cultivating different variety of crops including maize, millet, sorghum, rice, yam, cowpea and groundnut and rearing of animals mainly cattle sheep and goat. Agro-pastoralism, a type of mixed farming is a common practice in the area even though mono-cropping is also found (Lawal 2012).

There is an obvious shift from mono-cropping to agricultural intensification and development of crop livestock system. Studies show a rise in the number of livestock owned by crop farmers. Most crop farmers invest their surplus revenue from crop sales in animal production when crop prices are high. Again, in time of drought when prices of livestock are low, crop farmers in the state acquire animals from poor farmers or pastoralist. As found in other parts of Nigeria, agro-pastoralist in the state are aware of the benefits in crop-livestock integration farming. Farmers use the hoof action of livestock like cattle to prepare land for growing small cereals. In a like manner, goats are sometimes used to weed crops after being allowed to satisfy their initial appetite on natural pastures. The animals selectively eat the herbs without destroying the cereal crops in the field. Altogether, the crop-livestock interaction that exist in the state include (i) the use of draught animal mainly bulls as source of power for carrying farm equipment and implements for land preparation, weeding and other activities in crop field. The animals are also used for the transport of farm inputs, crop residues, manure, crop produce etc. to and from the farm as the case maybe; (ii) Manures from animals are readily available for crop production to increase soil fertility and nutrient cycling. In farms where fallow crop field was grazed by animals, crop productivity was found to increase; (iii) Crop residues though not enough are used as feed materials for animals; (iv) farmers rely on the sale of meat and milk from animals as additional source of income for the family.

Despite the numerous advantages inherent in agro-pastoral system, there was a technical inefficiency among the agro-pastoralist in Adamawa State and this impacts negatively on the output levels of the farmer. As shown in Table 1 (Lawal et al. 2015), among the challenges faced by farmers in the state are inadequate and high cost of fertilizers, high cost of paid labour, high cost of supplementary inputs, pest and diseases, lack of improved seeds, etc. From Table 1, it is glaring that the most serious challenges faced by farmers in the state are those associated with crop production.

Table 1 Distribution of respondents based on constraints associated with livestock- crop farming system

Constraints	Frequency	%a	Ranking[b]
Inadequate and high cost of fertilizer	98	49.00	1
High cost of paid labour	96	48.00	2
Low fertility status of farm land	93	46.50	3
High cost of transportation	81	40.50	4
Lack of adequate grazing land	69	34.50	5
Clashes with pastoralists	55	27.50	6
Lack or inadequate water during the dry season	39	19.50	7
High cost of agrochemicals (herbicide and pesticide)	37	18.50	8
Drought/flooding	26	13.00	9
Insecurity for life and property	20	10.00	10
Lack of improved seeds	15	7.50	11
Attack by pests and diseases	5	2.50	12
High cost of supplementary feeds	3	1.50	13

Source: Adapted from Lawal et al. (2015)
[a]Multiple responses existed hence >100%
[b]Ranks are in descending order of severity

3.2 *Features of Crop-Livestock Interaction*

Crop and livestock production have been linked for years as epitomized by exchange of grain, crop residues and water for manure between sedentary crop farmers and migratory pastoralists (McCown et al. 1979; Mortimore 1991). For some reasons like increasing human and livestock population and changing climate, cropping system anchored on shifting cultivation and livestock production system hitherto based on transhumant and commercial grazing of rangelands was transformed to more sedentary, mixed farming system where crop and livestock are part of the same farm system (Winrock International 1992; Wolfe 2011). Crop and livestock interact within the same production system and their interaction often offers increased efficiencies and productivity to farmers (Fig. 1). Crop-livestock integration could represent a model of sustainable agriculture according to principles of nutrient recycling and efficient use of land and resources (Moraine et al. 2014). Invariably, the system involves nutrient cycling to increase soil fertility, processing of crop residues as forage for livestock, among others. To the rural farmer, the benefits of investing in crop-livestock integration system include increased income and income stability.

Within an integrated crop-livestock production system, there are both positive and negative interactions. Positive interactions occur from synergies and complementarities whereas negative interactions come from antagonistic effects and competition between enterprises. As regard positive interactions, Wolfe (2011) enumerated the mutually beneficial interaction between livestock and crop management to include;

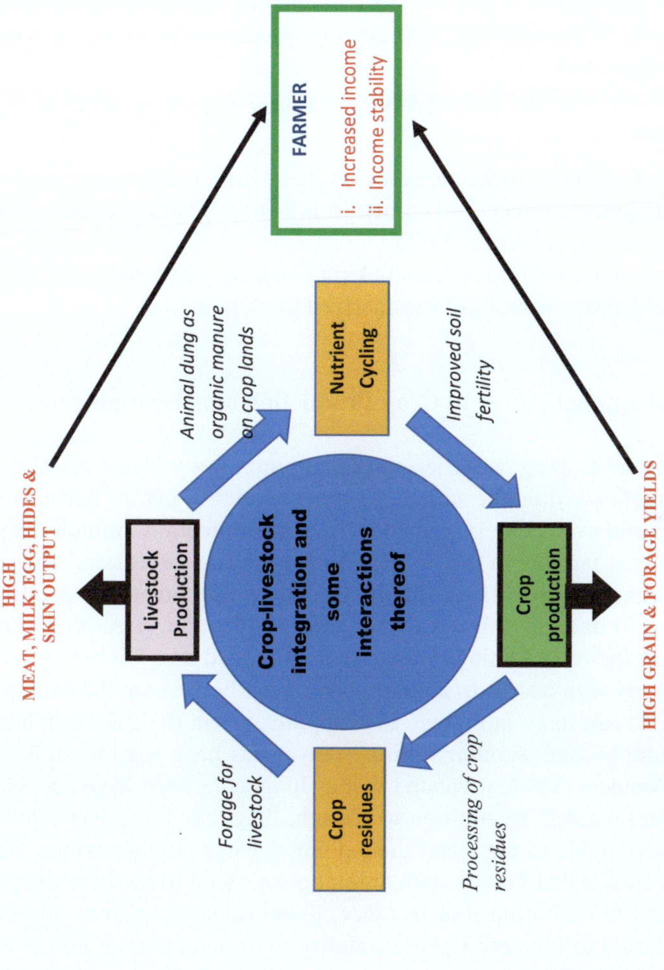

Fig. 1 Example of crop-livestock interaction in integrated crop livestock farm

(i) The benefits of nitrogen-fixing legumes that improve both forage quality and soil fertility
(ii) The return of dung and urine to the soil
(iii) The use of deep-rooted pasture species that extract more soil water for production and environmental benefits; and
(iv) The use of pasture-crop rotation to break disease and pest cycles.

Also, examples of positive complementary effects include;

(i) The consumption of weeds by livestock thereby reducing weed populations
(ii) The benefit to animals and crop sowing operations from the consumption of crop residues and
(iii) The utilization of the grassy understory by cattle in developing rubber tree plantations.

Examples of negative interaction include the distribution of some crop weeds by livestock (antagonistic effect) and competition between both enterprises for labour, resources or investment (Wolfe 2011).

According to the role crop or livestock play in a crop-livestock production enterprise, the interactions thereof are summarized to include:

3.2.1 Use of Animal Power in Crop Production and Transport

In most of the developing countries and in communities where crop-livestock production is practiced, draught animals, most commonly cattle, but also donkeys, horses, mules and even camels contribute to crop production through the provision of power to assist farmers in the production, harvesting, processing and marketing of crops. Equipment for primary tillage, weeding, and sometimes planting, etc. is often attached to draught animals during crop production. A positive relationship exists between the use of animal power and cultivated area (Sumberg and Gilbert 1992) as farmers with access to animal power usually increase the area, they cultivate to make its use more attractive. In cropping system that relies on hand-tillage in land preparation and weeding, animal power has been used to alleviate labour bottleneck. Wanders (1994) indicated that as long as farmers have access to weeders, ridgers and planters in addition to plough, they take full advantage of animal power that is available to them and in so doing achieve labour savings. Substantial evidence has shown that famers with animal power tend to increase the proportion of land planted to cash crop. For instance, planting of groundnuts and other cash crops are reported to increase the profitability of draught animal power more than staple food crops, such as millet, maize and sorghum (Delgado and McIntire 1982; Panin 1987). In a similar report, Shumba (1984) concluded that the use of animal power can improve the timeliness of planting and thus increase crop yields in areas where growing seasons are short and time of planting is crucial.

Similarly, animal power is used also for transportation of equipment, manure, and other crop materials, harvest or residues, to points where they are readily accessible for use in the farm or for marketing. To achieve this purpose, farmers usually purchase carts for transport. The use of carts for transport drastically reduces labor constraints and workloads in cropping activities, and helps step up production (Zenebe and Fekade 1998). The use of carts enhances the timely removal of crops from fields and thus contributes significantly to reduction in post-harvest losses from pest (Anderson and Dennis 1994). During planting, carts are used to move manures and fertilizers to the field and farmers are better able to market their produce at the end of harvest.

3.2.2 Nutrient Cycling

Livestock play pivotal role in regulating nutrient cycling in mixed crop-livestock enterprise. In addition to other products like meat, milk and eggs, animals also provide manure and other types of animal waste that play crucial role in the overall sustainability of crop-livestock system. Basically, manure contains several nutrients including nitrogen, phosphorus and potassium and organic matters (Singh et al. 2011). Soil structure and fertility is maintained when these materials are ploughed back into the soil. Accordingly, crop production and yield are improved and the risk of soil degradation is reduced. In an integrated crop-livestock production system where a short-term graze fallow system is practiced, nutrient recycling is enhanced and crop yield is improved due to soil fertility build-up through decomposition and mineralization of dead plant biomass and the dung that was returned to the soil in the course of grazing by animals (Jabbar and Cobbina 1990). On the other hand, the integration of livestock into cropping systems converts some crop residues into meat and milk. In general, both crop and livestock production are significantly improved if soils, crops, fertilizer, and manure are managed intensively in a manner that enhances nutrient recycling (Powell et al. 2004).

3.2.3 Use of Crop Residues as Feed Materials for Animals

The importance of crop residues in an integrated crop livestock production has been well document (Singh et al. 2011; Powell et al. 2004; McIntire et al. 1992). In respect to crop livestock interaction, crop residues are used as ready source of feed in form of forage for livestock especially in dry season. During the same time of the year, the quality of crop residue in terms of digestibility, crude protein and phosphorus content, is far better than those of natural rangeland. Among other uses include as fuel, construction material and source of income to the farmer when sold to buyers (McIntire et al. 1992). According to Powell et al. (2004), farmers may allow animal free access to whole residues on harvested fields or in case of availability of labour, farmers harvest, transport and store crop residues for feed or sale (McIntire et al. 1992). In situations where all crop residues are not used for livestock feed, they are often traded and sometime their monetary value approaches that of grain (Singh et al. 2011). In drier part of the world especially in West Africa, Bayer (1986) report indicate that cattle spend 80% of grazing time on harvested fields. Allowing crop residue in the field improves soil structure, and soil chemical properties because crop residue provides the physical barrier in the field that enhanced the accumulation of soil organic matter (Buerkert et al. 1996; Powell et al. 2004).

3.2.4 Manure Availability for Crop Production

Manure remains the most important soil fertility amendment and an important link in areas where crop livestock mixed farming is emerging. Manure is very critical for sustaining soil productivity and so are often used as crop fertilizers. Manure returns organic matter to the soil, helping to maintain its structure as well as its water retention and drainage capacities (Singh et al. 2011). Manure availability for cropping is limited by livestock types, numbers and manure output per animal, the location of livestock during the time when manure is needed, feed supply from range and crop land and the efficiency of manure collection (Powell et al. 2004; Schlecht et al. 1995; Singh et al. 2011). Manure quality can also be enhanced via improved feeding, such as using urea-treated straw (Singh et al. 2011). There are also indications that the efficiency of manure use can be increased by joint application of manure and fertilizers and manipulation of the relative amounts and times of application (Singh et al. 2011; Brouwer and Powell 1998) because soil acidification linked with repeated applications of Nitrogen fertilizer is reduced when fertilizer and manure are applied together (de Ridder and van Keulen 1990). Factors like rainfall, temperature, soil type, manure nutrient content, and farmer management affect manure application rates and crop response (Powell et al. 2004). In wet-dry tropical environments the daily manure output of grazing cattle during the wet season can be twice the manure output during the dry season because manure output is affected by wide fluctuation in feed availability and quality across wet and dry seasons (Omaliko 1980). However, manure application improves soil conditions and increases crop yields. In addition to the use of livestock manure as organic fertilizers in a crop livestock integrated farm, it also serves as source of biofuel for burning to replace scarce firewood resources. Dung are seasonally collected and used to produce "dungcakes", dried and stored for burning (Österle et al. 2012).

3.3 Challenges Faced by Crop-Livestock Production System

Crop-livestock production system of agriculture, like all other production systems of agriculture is faced with some challenges irrespective of its numerous advantages to the farmer in ensuring food security.

3.3.1 Soil Compaction

Livestock production requires proper management and care of the livestock. In crop livestock integrated system, lack of proper management impact negatively on the soil and its properties (Schiere et al. 2000). Improper management of livestock can exacerbate surface soil compaction, and this may lead to water runoff, loss of nutrients and cultivation/sowing difficulties. Consequently, crops eventually planted in such soils are less likely to produce well.

3.3.2 Huge Economic Trust on Labour and Infrastructure

Crop and livestock agriculture depend largely on availability of labour in all aspects of it production. In an integrated system of crop livestock production, although animal serves as source of farm power, labour is also required in the control and management of livestock for proper efficiency of production. Equipment for ploughing, ridging, weeding, fertilizer application, etc. used in crop livestock integrated system are often under the control of human labour when in use. Also, livestock requires proper attention and availability of infrastructures like fences, yards/corrals, water and feeding points and other facilities (Entz et al. 2005). Pastures and livestock require proper management in farms that practices crop livestock integration.

3.3.3 Lack of Incentives to Farmers

Crop livestock integrated system requires that farmers have access to knowledge, assets (e.g. land) and inputs (e.g. seeds, fertilizers, etc.) to manage the system in a way that is economically and environmentally sustainable in the long run (Kumar et al. 2014). Unfortunately, farmers are short of sufficient access to adequate incentives to promote rural based crop livestock mixed farming. An integrated crop livestock system is a form of organic agriculture and farmers need to be encouraged to sustain the practice because of its inherent benefits and should be paid adequately to promote rural based crop livestock mixed farming system.

3.3.4 Complexity in Management

It is a common knowledge that addition of livestock to cropping system results in greater complexity in management, increasing the number and difficulty of decision that must be made by the manager (Wolfe 2011; Moraine et al. 2014). Unfortunately, many farmers engaged in crop livestock system are usually generalists having limited time or specialist knowledge to manage each enterprise according to the combined expectations of an agronomist and animal scientist. In some instance where some crop livestock farmers have large size of farms, reports (Pannell et al. 2006; Wolfe 2011) show that in most cases such farms are run by a single farm family with occasional outside labour making it complex and difficult for farmers to adapt to current trends in farm management that enhance productivity. In effect, farmers find it difficult to handle the increased complexity of managing towards multiple goals including agricultural, environmental, economic, social and political goals (Wolfe 2011). In essence, managing the potential interactions between crop production and livestock production requires farmers/managers that are capable to appreciate and run these interactions in a way to release synergy and counterbalance resentment.

3.3.5 Imbalance in Resource Distribution Between Crop and Livestock

There is evidence that livestock are often neglected in mixed crop livestock enterprise compared to crop (Duncan et al. 2013). In support, FAO (2013) report emphasized that research, development and extension efforts are geared towards intensification of crop production. That notwithstanding, and according to FAO (2013) reports, several evidence available shows that four out of five of the highest value commodities in an integrated crop livestock system are livestock products. For instance, compared with livestock products, policy initiatives encourage heavy subsidies on crop inputs like seeds, fertilizer, and irrigation even though high valued products in the system come from livestock (Erenstein and Thorpe 2010; Duncan et al. 2013).

3.3.6 Pest and Disease Constraint

Incidences of pest and disease is a major constraint hindering crop and animal productivity. In a crop livestock system, this have negative effect on crop production (both grain and fodder) and causes poor growth, high mortality and reduced productivity in animals. The unfortunate scenario is that inputs (pesticides, herbicide, drugs, etc.) to counteract these negative forces are generally scarce or priced well above the means of smallholder farmers (Ezeaku et al. 2015). Consequently, the reciprocal benefits from crop and livestock in the system (crop residues as fodder for livestock, manure for crops and soil fertility improvement, etc.) are drastically reduced.

3.3.7 Economic Constraints

The sustainability of crop livestock system is threatened by decline in productivity and profitability (Moraine et al. 2014), due mainly to some economic factors including increasing prices of farm inputs for crop (seed/seedlings, pesticides, irrigation water, farm machineries, lands, etc.), and livestock (drugs, feed, electricity, fuel, etc.) productions. Altogether, the lack of incentives and source of funding to farmers in the system culminate in low productivity and supply of crop and animal products from the system. In the long run, some farmers are forced out of business since the system is no more sustainable.

3.3.8 Impact of Climate Change on Crop and Livestock

The negative effect of climate change on crop and animal production is well known (Aydinalp and Cresser 2008; Nardone et al. 2010). Variability in seasons, irregular rainfall patterns, flooding, drought, high temperature, etc. causes a lot of harm to crops and livestock. These damages impose high economic losses to the farmer. Animal growth rate, meat and milk yield and quality, egg production and quality, reproductive performance, metabolic and health status, and immune response are negatively affected by hot ambient temperature (Nardone et al. 2010). The process of desertification reduces the carrying capacity of rangelands and the buffering ability of agro-pastoral and pastoral systems (Nardone et al. 2010). It is pertinent to state that in crop livestock production system, the livestock component may be more affected by global warming because global warming causes soil infertility, water scarcity, reduced grain and pasture yield and quality. Moreover, the survival and productivity of livestock and the interaction thereof in a crop livestock system depends on increased yield of crop, and availability of good quality forages and crop residues as feed material for livestock. Lower rainfalls, increased drought effects on crops and on pastures coupled with the effect of high temperature and solar radiation impairs animal productivity. In crop production, temperature increases can have both positive and negative effects on crop yields, however, numerous evidences show that temperature increases lead to reduce yields and quality of many crops, especially cereal and feed grains. In addition, just as in livestock production, rising temperatures cause changes in the incidence and distribution of pests and pathogens, increased rates of soil erosion and degradation, increased tropospheric ozone levels, high runoff and groundwater recharge rates, etc. (Adams 1986; Adams et al. 1998). In general, availability of water for crop and pasture production and irrigation agriculture is hindered in the process leading to low productivity of crops and decline in yield and quality of available forages and crop residues for livestock.

3.4 How Does Crop-Livestock Interaction Enhance Sustainable Agricultural Production?

Sustainability was defined by Conway (1987) as the ability of a system to maintain productivity irrespective of major disturbances such as is caused by intensive stress e.g. the effect of soil erosion and degradation or farmers' indebtedness or by a large awful situation like incidence of a drought or a food or a new pest or pathogens. Therefore, sustainable agricultural system may be defined as an agricultural system capable of maintaining increased production and supply of crops and livestock products in the midst of numerous environmental, physical, economic and social stressors. As mentioned earlier, even though crop livestock interaction in a crop livestock production system may be complementary or competitive, there are huge benefits and contributions of the system towards the enhancement of sustainable agricultural production if well managed. In general, crop livestock interaction may enhance sustainable agricultural production in the following ways:

- There is higher aggregate output of crop and animal products for a larger population in crop livestock system considering the fact that a given amount of land can support more people in the system than under either crop or livestock system. Crop livestock interaction leads to diverse and effective utilization of farm resources and as such contributes to stable increase in both food crop and livestock production to meet the increases in population growth and food security.
- Crop livestock interaction creates room for diversification of production, consumption and investment, and contributes to steadiness of the system by minimizing risk, employing and allocating profits to more people (Jabbar and Cobbina 1990).
- Crop livestock interaction enhances agricultural productivity by decreasing the overall cost of production. For instance, poor farmers practising crop livestock system are opportune to make good use of available resources from the farm like the use of manures from livestock operations as replacement for chemical fertilizers, which are often unaffordable. In places where crop livestock system is well practised, it provides income growth prospects for many poor rural people involved in the livestock sector (Delgado et al. 1999).
- The pitfalls of high-tech commercial agriculture may be avoided or minimized by encouraging crop-livestock interaction in the early stage of production. A major pitfall of high-tech commercial agriculture is the continued reliance on the use of agrochemicals. Agrochemicals are a major source of environmental pollution and a significant risk for human, animals and wildlife. Crop livestock interaction contributes to a reduction in major environmental treats emanating from agriculture like soil erosion, leaching and associated effects on water quality (Jabbar and Cobbina 1990). It helps to improve and conserve the productive capacities of soils, with physical, chemical and biological soil recuperation (Delgado et al. 1999) thereby making agricultural production effective and sustainable.

- The system creates room for better agronomic and economic resilience to external influences such as climate change and input price volatility thus making agricultural production more sustainable (Moraine et al. 2014).

4 Sustainable Livestock Production Systems

By 2050 the world's population is expected to reach 9.6 billion, with nearly all of the population increase in the developing countries (United Nations 2013). Alexandratos and Bruinsma (2012) estimated that global food demand would increase by 1.1% per year from 2005/2007 to 2050. They also expect that in this period, global demand for meat will grow by 1.3% per year and for milk and dairy products by 1.1% per year. Searchinger (2013) indicated three categories of solutions to sustainably feed this larger, more urban and richer world population in 2050:

- Solutions that reduce the growth in food consumption, by reducing amongst others waste, obesity and excessive consumption;
- Solutions that increase food production on existing agricultural land, by e.g. increasing yield; and
- Solutions that reduce the environmental impact of food production, by e.g. efficient use of inputs.

4.1 Types of Livestock Production Systems

The term "livestock" refers to any breed or population of animal kept by humans for a useful, commercial purpose. They are domesticated animals raised in an agricultural setting to produce labor and commodities such as meat, egg, milk, fur, wool and leather. Livestock production plays a major role in the life of farmers in developing countries. It provides food, income, employment and many other contributions to rural development. They provide a flow of essential food products throughout the year, are a major source of government revenue and export earnings, sustain the employment and income of millions of people in rural areas, contribute draught energy and manure for crop production and are the only food available to many Africans (Michio et al. 2003). Livestock production systems have been classified by Steinfeld and Mäki-Hokkonen (1995) to include:

1. Solely livestock production systems. Livestock systems in which more than 90% of dry matter fed to animals comes from rangelands, pastures, annual forages and purchased feeds and less than 10% of the total value of production comes from non-livestock farming activities.
2. Landless livestock production systems. Subset of the solely livestock production systems in which less than 10% of the dry matter fed to animals is farm-produced

and in which annual average stocking rates are above ten livestock units per hectare of agricultural land. The developed countries dominate the picture of landless intensive production with more than half of total meat production.
3. Landless monogastric production system. This system is defined using monogastric species, mainly chickens and pigs, where feed is introduced from outside the farm, thus separating decisions concerning feed use from those of feed production, and particularly of manure utilization on fields to produce feed and/or cash crops. The system is typically competing with traditional land-based production systems for shares in the urban markets.
4. Landless ruminant production system: This production system is defined using ruminant species, principally cattle, where feed is mainly introduced from outside the farm system. Landless ruminant production systems are highly concentrated in only a few regions of the world. The system is based almost exclusively on high-producing, specialized breeds and their crosses, which, nevertheless, have not been bred specifically for performance under "landless" conditions. The system is highly capital-intensive, leading to substantial economies of scale. It is also feed-intensive and labour-extensive.
5. Temperate zones and tropical highlands grassland-based system: In these areas, the grazing system is constrained by low temperatures. In the temperate zones, there are 1–2 months of mean temperatures, corrected to sea level, to below 5 °C, whereas in the tropical highlands daily mean temperatures during the growing period are in the range of 5–20 °C.
6. Mixed-farming systems: Livestock systems in which more than 10% of the dry matter fed to animals comes from crop by-products or stubble or more than 10% of the total value of production comes from non-livestock farming activities.
7. Temperate zones and tropical highlands rain-fed system: This system is defined as a combination of rain-fed crop and livestock farming in temperate or tropical highland areas, in which crops contribute at least 10% of the value of total farm output. The main common feature of these two regions is that low temperatures during all or part of the year limit and determine vegetation that is quite distinct from that found in tropical environments (Steinfeld and Mäki-Hokkonen 1995).

4.2 *Classification of Sustainable Livestock Production Systems*

One common and more studied classification is that between organic and conventional systems. Sustainable conventional production systems are typified by the production systems used by most farms, which use technologies for increased productivity, such as high-yielding breeds, modern breeding and feeding techniques, modern medication, machines and equipment, and (artificial) fertilizers and pesticides. Conventional livestock production systems comply with local legal requirements in force for all livestock producers irrespective of their production system. Therefore, it should be noted that farming practices could differ substantially among conventional livestock producers.

Organic production systems were typified as a holistic production management system that promotes and enhances agroecosystems health including biodiversity, biological cycles, and soil biological activity (Codex Alimentarius Commision 2007). It emphasizes the use of management practices, taking into account that regional conditions require locally adapted systems. This is accomplished by using, where possible, cultural, biological and mechanical methods to fulfill any specific function within the system, as opposed to using synthetic materials. In addition to the legal requirements for conventional livestock production systems, the organic production system has to comply with legal requirements and clearly defined standards on organic livestock production.

Comparing organic versus conventional dairy farms, studies with dairy cattle and pigs reported a price premium for organic milk and organic pork above the conventional farm gate price. It should be noted that, organic animal production compared to conventional production used more labour per animal, have lower building costs per animal, higher levels of price risk, yield risk and income risk (van Wagenberg et al. 2016). van Wagenberg et al. (2016) enumerated the advantages of organic livestock production compared to conventional as follows:

Lower building costs per animal; higher income per animal or full time employee; lower eutrophication and acidification potential per unit of land; lower impact on local biodiversity and fossil phosphorus depletion per unit of product and higher activity levels and better leg health, a similar or lower likelihood of antibiotic resistance in bacteria isolated from the farms' environment, animals or animal products.

Conventional livestock production compared to organic had:

- a lower labour need to produce an animal,
- a lower level of income risk per animal,
- a higher output (in kg product) per animal per time unit,
- a better reproductive performance,
- a lower feed conversion ratio,
- a lower land use,
- a generally lower acidification and eutrophication potential per unit of product,
- a lower risk of parasitic infections, and
- a similar or lower microbiological contamination in animal products (van Wagenberg et al. 2016).

The following lessons for sustainable livestock production systems have been identified:

1. Best practices and technologies that enhance productivity could help to achieve food security while contributing to environmental sustainability and economic farmer sustainability (Capper and Bauman 2013; van Wagenberg et al. 2016).
2. Selection of feed products could be based on a balance between a low environmental impact and high quality of the feed ration that ensures a high animal productivity and feed efficiency. To enhance future food security, food-feed competition is an important aspect to consider.

3. Livestock grazing on marginal land, which is less suitable to produce human edible protein through e.g. crop production, could also play an important role in achieving food security (Fraser 2014).
4. Benefits of increasing feed production productivity per area of land such as lower land use should be balanced against increased local environmental impact.
5. Antibiotics should be used prudently, based on a balance of the risk of development of antimicrobial resistance with animal welfare related to treatment of diseases.
6. The use of high yielding and robust breeds adapted to their environment should be balanced with animal welfare and environmental performance.
7. Technological applications to reduce nutrient losses from manure management or application of manure processing techniques can contribute to reducing waste and improving nutrient cycling and environmental performance of livestock systems.
8. Improvements in housing design offer opportunities to reduce welfare issues.
9. Sustainable livestock production should be approached as a multi-criteria optimization problem in which a balanced combination of indicators should be optimized.

4.3 Key Role of Crop-Livestock Production in the Developing World

Dalibard (1995) reported some of the key role crop-livestock production systems play in the developing countries:

1. Draught animal power: Livestock provide renewable energy for agriculture, saving a considerable amount of fossil energy that otherwise would be used mainly for manufacturing and operating heavy agricultural machinery, as well as for producing fertilizers. Draught animal power can also play a key role in large-scale production systems, as is the case on a 30,000 ha sugar estate (La Romana) in the Dominican Republic, where some 18,000 oxen haul sugar cane from the fields to the railway system leading to the sugar mill (Preston and Murgueitio 1992). Sustainability is guaranteed in such a system, since the oxen's main source of energy comes from the sugar-cane tops and, in the end, the manure goes back to the sugar cane. In many parts of the world, draught animal power is often the best alternative leading to increased yields of crop production
2. Fertilizer production: The value of manure as fertilizer could be increased through better animal feeding. In mixed-farming systems, livestock are fed on crop residues and by-products as well as pasture. In turn, some nutrients and organic matter are returned to the soil through livestock manure, ensuring the maintenance of soil fertility resulting to improved cation exchange, better absorption of water and the prevention of runoff and soil surface crusting. It is

therefore clear that manure plays a key role in sustainable crop-livestock production systems.

3. Weed control: Livestock can be successfully used for weed control and therefore contribute to a decrease in water pollution by herbicides. The fossil energy that would otherwise be used for making and spreading herbicides is also saved. While the livestock is grazing in these plantations, most commonly sheep also provide organic fertilizer.
4. Source of fuel: The use of cow dung as fuel in India and other countries results in an enormous number of trees being saved. When water is not a limiting factor, however, dung can be utilized much more efficiently for fuel production through biodigesters.

5 Ethical Challenges Facing Sustainable Livestock Production

Increased demand is due, in part, to a predicted increase in world population from 7.2 billion to between 9 and 10 billion people in 2050 (United Nations 2013). The increase in population puts additional pressure on the availability of land, water, and energy needed for animal and crop agriculture. Global environmental challenges, including global climate change and the growing threat of disease transmission to and from agricultural animals add further challenges to sustainably meeting the demand for animal agriculture in 2050.

The ethical challenges facing sustainable livestock production are:

Climate-related risks: Climate variability is one of the major characteristic risk in crop-livestock systems. Changes in temperature influence sowing and harvesting dates – thus, grazing and mixed rain-fed systems of livestock production will be the most damaged by climate change (Nardone et al. 2010).

Stocking rate: When more animals are stocked than the stipulated stocking rate

- Decrease in grazing cattle number but increase in sedentary draught and fattened cattle and in small ruminants
- Decrease in rangeland area versus cropped area
- More herd mobility in dry season because of less rangeland availability
- Change in livestock species contribution to herd composition.
- Change in mobility pattern with longer distances, longer durations
- Herd routes moving southwards to more humid areas to find forage resources
- More conflicts with sedentary people in the south owing to competition for water and pasture resources

Growth in demand for animal protein due to:

- Population growth
- Increasing global affluence

- Increase in per capita animal protein intake
- Impact of global environmental change on:
- Climate (Steinfield and Gerber 2010)
- Water and land scarcity

Jarvis (1990) reported that despite the generally favorable effect, which the use of livestock has on agricultural resources, livestock are also a component of several unsustainable production systems. Among the most widely cited include:

- The overgrazing of arid and semi-arid lands, leading to range degradation and productivity decline,
- The destruction of humid rainforests for the establishment of pastures which degrade soils and quickly cease production,
- The pollution of watercourses from animal wastes resulting from intensive dairy and meat production, leading to reduced production and consumer welfare elsewhere,
- The production of methane gas as a result of ruminant livestock production contributing to the threat of a global "greenhouse" effect, and
- And the pollution of soils, watercourses and subterranean water supplies from the application of fertilizers, herbicides, and pesticides in the production of livestock feed grains.

6 Future Perspectives

The current system of livestock production system in Africa may not respond to the future demands since too many animals are producing at low levels when compared to natural resource base. Food security can be achieved by closing 'yield gaps,' increasing crop and livestock production efficiency, reducing waste in the food supply chain; crop/livestock diversification and integration; conserving crop wild relatives and agro biodiversity and by adopting greenhouse gas abatement and production enhancement technologies in agriculture and animal husbandry (Tiwari et al. 2014).

According to Tiwari et al. (2014), application of these measures together could double the food production with available resources without increasing environmental impacts. Smallholder's intensification and linking them with corporate bodies and modern retail food supply chains needs urgent attention since they hold majority of livestock in the country and play a major role in food security and environmental stability (Tiwari et al. 2014). To avoid harmful effects of global warming, small changes in our day to day life style is a crucial turning point which need due attention. The recent advances in science and novel technologies/concepts need to be fully explored for their optimum potentials like genetic engineering, disease resistant varieties, embryo transfer technology, artificial insemination, superior genetics and breeding practices, cloning, nutrigenomics, immunomodulatory among others (Tiwari et al. 2014). These altogether may help increase and boost both agricultural and animal produces including crops, cereals, foods, milk, meat and other products and ultimately lead to production of healthier, safer and high quality food apart from boosting production and safeguarding food security for everyone.

References

Adams RM (1986) Agriculture, forestry and related benefits of air pollution control. Am J Agric Econ 68:885–894

Adams RM, Hurd BH, Lenhart S, Leary N (1998) Effects of global climate change on agriculture: an interpretative review. Clim Res 11:19–30

Adebayo AA (1999) Climate I and II. In: Adebayo AA, Tukur AL (eds) Adamawa State in map. Paraclete Publishers, Yola, pp 20–26

Alberta SPCA (2014) Need to know: sustainable livestock production systems. Accessed from: www.ctsanimals.ca/va3040/resources/toKnow/pdfBC+SH_3040_6-2.pdf on 12th July 2018

Alexandratos N, Bruinsma J (2012) World agriculture towards 2030/2050: the 2012 revision. EAS working paper 12-03. FAO, Rome. Online. Available at. http://www.fao.org/docrep/016/ap106e/ap106e.pdf. Accessed 16 June 2018

Anderson M, Dennis R (1994) Improving animal-based transport: options, approaches, issues and impact. In: Starkey P, Mwenya E, Stares J (eds) Improving animal traction technology, Proceedings of the workshop of ATNESA, 1st, Lusaka, Zambia. 18–23 January 1992. Technical Center for Agricultural and Rural Co-operation (CTA), Wageningen, pp 378–395

Asachi G (2010) Sustainable agriculture development- concepts, principles, eco-efficiency, eco-equity, eco-conditioning. Cercet Agron Moldova XLIII 3(143):91–100

Aydinalp C, Cresser MS (2008) The effects of global climate change on agriculture. Am Eurasian J Agric Environ Sci 3(5):672–676

Bayer W (1986) Agropastoral herding practices and grazing behaviour of cattle in the subhumid zone of Nigeria. ILCA Bull 24:8–13

Brouwer J, Powell JM (1998) Increasing nutrient use efficiency in West-African Agriculture: the impact of micro-topography on nutrient leaching from cattle and sheep manure. Agric Ecosyst Environ 71:229–239

Buerkert B, Allison BE, von Oppe M (eds) (1996) Wind erosion in Niger. Implications and control measures in millet-base farming system. Kluwer Academic Publishers, Stuttgart

Capper JL, Bauman DE (2013) The role of productivity in improving the environmental sustainability of ruminant production systems. Annu Rev Anim Biosci 1:468–489

CNS-FAO (2016) Working towards sustainable agriculture and food systems, a discussion paper. 19p. http://www.blw.adminoch/btw/en/home/international/institutionen/multistakeholder-partnerschaften/cns-fao.html

Codex Alimentarius Commission (2007) Organically produced foods. 3rd Edition. Publication of Joint WHO/FAO of the United Nations, Food Standard Programme Codex Alimentarius Commission. Electronic Publishing and Support Branch Communication Division, FAO. Viale delle Terme di Caracalla 00153 Rome, Italy

Conway G (1987) The properties of agro-ecosystems. Agric Syst 24(2):95–117

Dalibard C (1995) Livestock's contribution to the protection of the environment. www.fao.org/docrep/V8180T/v8180T14.htm

Delgado CL, McIntire J (1982) Constraints on oxen cultivation in the Sahel. Am J Agric Econ 64:188–196

Delgado C, Rosegrant M, Steinfeld H, Ehui S, Courbois C (1999) Livestock to 2020: the next food revolution, Food, agriculture and the environment discussion paper 28. International Food Policy Research Institute, Washington, DC

deRidder N, van Keulen H (1990) Some aspects of the role of organic matter in sustainable intensified arable farming systems in the West African semi-arid tropics. Fert Res 26:299–310
Devendra C (2011) Integrated tree crops-ruminants systems in South East Asia: advances in productivity enhancement and environmental sustainability. Asian Aust J Anim Sci 24(5):587–602
Duncan AJ, Tarawali SA, Thorne PJ, Valbuena D, Descheemaeker K, Homann-Kee Tui S (2013) Integrated crop-livestock systems- a key to sustainable intensification in Africa. Trop Grassl Forrajes Trop 1:202–206
Entz MH, Bellotti WD, Powell JM, Angadi SV, Chen W, Ominski KH, Boelt B (2005) Evolution of integrated crop–livestock production systems. In: McGilloway DA (ed) Grass-land: a global resource. Wageningen Academic Publ, Wageningen, pp 137–148
Erenstein O, Thorpe W (2010) Crop-livestock interactions along agro-ecological gradients: a meso-level analysis in the Indo-Gangetic Plains, India. Environ Dev Sustain 12:669–689
Ezeaku IE, Mbah BN, Baiyeri KP, Okechukwu EC (2015) Integrated crop livestock farming system for sustainable agricultural production in Nigeria. Afr J Agric Res 10(47):4268–4274
FAO (1993) Livestock and improvement of pasture, feed and forage, Committee on Agricukture, Twelfth Session, Item 7, 26 April–4 May 1993. Food and Agriculture Organization, Rome
FAO (2013) FAOSTAT statistical database. faostat.fao.org
FAO (2017) Africa sustainable livestock 2050. Technical Meeting and Regional Launch. Addis Ababa, Ethiopia 21–23 February
FAO (2018) Compendium for sustainable crop production intensification. Accessed from: http://www.fao.org/agriculture/crops/thematic-sitemap/theme/spi/scpi-home/framework/en/ on 12th July, 2018
Food Program (2018) Sustainable crop production. Accessed from http://www.sustainabletable.org/249/sustainable-crop-production on 12th July, 2018
Fraser N (2014) Sustainable livestock for people, for the planet. Paper presented at the International Meat Secretariat (IMS) World Meat Congress, June 16, 2014, Beijing. Online: at http://www.meat-ims.org/wp-content/uploads/2014/06/Plenary-IMS-Beijing-2014-Fraser.editsJD-1-2.pdf
Giovannucci D, Scherr S, Nierenberg D, Hebebrand C, Shapiro J, Milder J, Wheeler K (2012) Food and agriculture: the future of sustainability. Retrieved from: http://www.un.org/esa/dsd/dsd_sd21st/21_pdf/agriculture%20and%20food%20report.pdf on the 7th of Jul 2018
Herrero M, Thornton PK, Notenbaert A, Msangi S, Wood S, Kruska R, Dixon J, Bossio D, van de Steeg J, Freeman HA, Li X, Parthasarathy Rao P (2009) Drivers of change in crop-livestock systems and their impacts on agro-ecosystems services and human well-being to 2030. ILRI (International Livestock Research Institute), Nairobi. www.goo.gl/KbZxDa
Jabbar MA, Cobbina J (1990) Crop-livestock interaction for sustainable food crop production in tropical Africa. Summary of a lecture delivered at the training course on sustainable food production systems in Tropical Africa, held at IITA, Ibadan, 18 June–6 July 1990
Jarvis LS (1990) Strategies for sustainable animal agriculture in developing countries. http://www.fao.org/docrep/004/T0582E/T0582E06.htm
Kumar S, Ansari MQ, Naresh RN, Kumar V (2014) Integrated crop and livestock management for enhanced productivity, profitability and sustainability of the rice-wheat system in North West India. Int J Life Sci Biotechnol Pharma Res 3(2):74–84
Lawal H (2012) Resource use efficiency in livestock–crop faming systems in Adamawa State, Nigeria. Unpublished PhD thesis, Department of Agricultural Economics University of Maiduguri. Borno State, Nigeria, 138pp
Lawal H, Iheanacho AC, Amaza PS (2015) Analysis of technical efficiency of sedentary pastoralists in livestock-crop farming in selected local government areas of Adamawa State, Nigeria. J Agric Econ Environ Soc Sci 1(1):1–8
Maurice D (2017) Components of sustainable animal production and the use of silvopastoral system. Bras R Zootec 46(8):683–688. https://doi.org/10.1590/s1806-92902017000800009. vs
McCown RL, Haaland G, de Haan C (1979) The interaction between cultivation and livestock production in semi-arid Africa. In: Hall AE et al (eds) Agriculture in semi-arid environments. Springer, Berlin, pp 297–332

McIntire J, Bourzat D, Pingali P, (1992) Crop-livestock interaction in sub-Saharan Africa (English) World Bank regional and sectoral studies. The World Bank, Washington, DC. https://documents.worldbank.org/curated/en/505681468768678913/crop-livestock-interaction-in-sub-Saharan-Africa

Menalled F, Bass T, Buschena D, Cash D, Malone M, Maxwell B, McVay K, Miller P, Soto R, Weaver D (2008) An introduction to the principles and practices of sustainable farming. Accessed from: http://msuextension.org/publications/AgandNaturalResources/MT200813AG.pdf on 9th July, 2018

Michio S, Arunasiri I, Katsuyuki O, Nobuaki K (2003) Development of livestock sector in Asia: an analysis of present situation of livestock sector and its importance for future development

Moraine M, Duru M, Nicholas P, Leterme P, Therond O (2014) Farming system design for innovative crop livestock integration in Europe. Animal 8(8):1204–1217

Mortimore M (1991) A review of mixed farming systems in the semi-arid zone of sub-Saharan Africa, Livestock economics division working document no. 17. ILCA International Livestock Centre for Africa (ILCA), Addis Ababa

Nalubwama S, Vaarst M, Kabi F, Kiggundu M, Bagamba F, Odhong C, Mugisha A, Halberg N (2014) Challenges and prospects of integrating livestock into smallholder organic pineapple production in Uganda. Livest Res Rural Dev 26(6). http://www.lrrd.org/lrrd26/6/nalu26113.htm

Nardone A, Ronchi B, Lacetera N, Ranieri MS, Bernabucci U (2010) Effects of climate changes on animal production and sustainability of livestock systems. Livest Sci 130(1–3):57–69

National Population Commission (NPC) (2006) Population and housing census, Population distribution by Sex, State, LGA, and Senatorial district. http://www.population.gov.ng/images/Vol%2003%20Table%20DSx%20LGAPop%20by%20SDistrict-PDF.pdf

Okoruwa V, Jabbar MA, Akinwumi JA (1996) Crop-livestock competition in the West African derived Savanna: application of a multi=objective programming model. Agric Syst 52:439–453

Omaliko CPE (1980) Dung deposition, breakdown and grazing behavior of beef cattle at two seasons in a tropical grassland ecosystem. J Range Manag 34:360–362

Österle N, Angassa A, Tadesse A, Ebro A, Sauerborn J, Treydte AC (2012) Crop-livestock farming systems varying with different altitudes in Southern Ethiopia. Sci Technol Arts Res J 1(4):1–13

Panin A (1987) The use of bullock traction for crop cultivation in Northern Ghana. An empirical economic analysis, ILCA Bulletin. International Livestock Centre for Africa (ILCA), Addis Ababa

Pannell DJ, Marshall GR, Barr N, Curtis A, Vanclay F, Wilkinson R (2006) Understanding and promoting adoption of conservation practices by rural landholders. Aust J Exp Agric 46:1407–1424

Powell MJ, Pearson RA, Hiernaux H (2004) Crop-livestock interactions in the West African Drylands. Agron J 96:469–483

Preston TR, Murgueitio E (1992) Strategy for sustainable livestock production in the tropics. CIPAV-SAREC, Cali. 89 pp

Pretty J (2008) Agricultural sustainability: concepts, principles and evidence. Philos Trans R Soc B 363:447–465

Royal Netherlands Society for Agricultural Sciences (2013) Livestock farming with care: towards sustainable production of animal-source food. NJAS Wagening J Life Sci 66:3–5

Schiere JB, Baumhardt RL, Van Keulen H, Whitbread AM, Bruinsma AS, Goodchild AV, Gregorini P, Slingerland MA, Wiedemann-Hartwell B (2000) Mixed crop-livestock systems in semi-arid regions. In: Peterson GA, Unger PW, Payne WA (eds) Dryland agriculture. American Society of Agronomy/Crop Science Society of America/Soils Science Society of America, Madison, pp 227–291

Schlecht E, Mahler F, Sangar M, Susenbeth A, Becker (1995) Quantitative and qualitative estimation of nutrient intake and faecal excretion of Zebu cattle grazing natural pasture in semi-arid Mali. In: Powell JM et al (eds) Livestock and sustainable nutrient cycles in mixed-farming systems of Sub-Sahara Africa. Volume II: technical papers, Proceeding of an International

Conference held in Addis Ababa, Ethiopia, 22–26 November, 1993. ILCA International Livestock Centre for Africa (ILCA), Addis Ababa, pp 85–98

Searchinger T (2013) Creating a sustainable food future: a menu of solutions to feed 9 billion plus people in 2050 (WRI, World Bank, UNDP, UNEP, Washington DC). This book length report evaluates the challenge and solutions for meeting food needs while holding down land use change and agricultural greenhouse gas emissions, and can be found at www.wri.org/wrr (click to download "interim findings")

Shumba EM (1984) Animals and the cropping system in the communal areas of Zimbabwe. Zimb Sci News 18:99–102

Singh YP, Naresh RK, Singh PK, Singh HL, Rathi RC, Singh S (2011) Crop-livestock interaction to food security and sustainable development of agriculture for small and marginal farmers in North West India. J Rural Agric Res 11(2):6–13

Steinfield H, Gerber P (2010) Livestock production and the global environment: Consume less or produce better? Proceedings of the National Academy of Science of the United States of America107 (43):18237–18238

Steinfeld H, Mäki-Hokkonen JH (1995) Steinfeld is Senior Officer (Programme, Policy & Planning) and J. Mäki-Hokkonen is Senior Officer (Livestock Production Systems), Animal Production and Health Division, FAO, Rome, Italy

Sumberg J, Gilbert E (1992) Agricultural mechanization in the Gambia: drought, donkeys and minimum tillage. Afr Livest Res 1:1–10

Sustainable Development Network of the United Nations (SDSN) (2013) Solutions for sustainable agriculture and food systems. A technical report for the Post-2015 Development Agenda, 18 September

Tiwari R, Dileep Kumar H, Triveni D, Singh BP, Pachaiyappan K, Dhama K (2014) Future challenges of food security and sustainable livestock production in India in the changing climatic scenario. Asian J Anim Vet Adv 9:367–384. https://doi.org/10.3923/ajava.2014.367.384

United Nations (2013) World population prospects: the 2012 revision, key findings and advance tables. United Nations, New York. Online. Available at http://esa.un.org/wpp/documentation/pdf/WPP2012_%20KEY%20FINDINGS.pdf

United Nations Environmental Protection Agency (UNEP) (2010) Sustainable agriculture and the sustainable use of agricultural biodiversity: concepts, trends and challenges. An information note submitted by Bioversity International for the Fourteenth Meeting of the Subsidiary Body on Scientific, Technical and Technological Advice of the Convention on Biological Diversity, 10–21 May 2010, Nairobi, Kenya

UNPP World Population Prospect (2008) The 2008 Revision Population Database. www.un.org/esa/population/publications/wpp2014_highlights.pdf

van Wagenberg CPA, de Haas Y, Hogeveen H, van Krimpen MM, Meuwissen MPM, van Middelaar CE, Rodenburg TB (2016) Sustainability of livestock production. Systems comparing conventional and organic livestock husbandry free at https://doi.org/10.18174/374992 or at www.wur.eu/economic-research (under Wageningen Economic Research publications)

Wanders A (1994) Supply and distribution of implements for animal traction: an overview with region-specific scenarios. In: Starkey P et al (eds) Improving animal traction technology, Proceedings of the 1st workshop of ATNESA, 18–23 January, 1992, Lusaka, Zambia. Technical Centre for Agricultural and Rural Co-operation (CTA), Wageningen, pp 226–243

Winrock International (1992) Assessment of animal agriculture in Sub-Saharan Africa. Winrock International Institute for Agricultural Development, Morrilton

Wolfe EC (2011) Interactions between crop and livestock activities in rainfed farming systems. In: Tow P, Cooper I, Partridge I, Birch C (eds) Rainfed farming systems. Springer, Dordrecht, pp 271–298

Zenebe S, Fekade T (1998) The role of pack donkeys in the major grain market (Yehil Berenda) of Addis Ababa, Ethiopia. In: Improving donkey utilisation and management. Proceeding of an ATNESA workshop, 5–7 May 1997, DebreZeit, Ethiopia

Part VII
Use of IT Tools and Modelling for Sustainable Agriculture

Information Technology for Sustainable Agriculture

Muhammad Jehanzeb Masud Cheema and Muhammad Azeem Khan

Abstract Feeding global population by 2050 requires 60% increase in the agricultural production. Agricultural transformation has a role to play for food security, poverty reduction and economic growth. However, sustainability and climate risk management are the challenges. The recent advancements in information technology (I.T) delivered smart devices, computing and sensor technologies. Application of those smart technologies have the potential to enable agricultural industry meets its productivity and sustainability challenge as well as solving indigenous agricultural problems of the developing nations. The geospatial data archives and real-time data from satellites, UAVs, RFIDs in combination with weather data, digitized soil data, and other real-time data streams coming from in-situ smart sensors can now give us a better understanding of the interaction between crops, weather and soils than ever before. Further the analytics of that big data assisted by machine learning can provide decision support in this regard. The customized I.T packages are required where, e-farm production system based on precision agriculture techniques, crops and livestock management, precision irrigation applications, crop water and pest/disease management, wireless moisture sensing networks, wireless communication in UAVs used for vegetation health detection, rainfall monitoring system based on mobile communication data, cloud services for knowledgebase on soils, nutrients, yields by making soil, nutrient and yield maps and disseminating through mobile networks and variable rate application based on GPS and GIS systems. Every passing day, the use of internet and smartphones is enhancing rapidly. The cloud-based services for big data analytics in agriculture and data sharing apps with linkages to integrated platforms and models are the future of farming in both modern and developing world.

M. J. M. Cheema (✉) · M. A. Khan
Center for Advanced Studies in Agriculture and Food Security, University of Agriculture, Faisalabad, Pakistan

Faculty of Agricultural Engineering and Technology, University of Agriculture, Faisalabad, Pakistan
e-mail: mjm.cheema@uaf.edu.pk

M. Farooq, M. Pisante (eds.), *Innovations in Sustainable Agriculture*,
https://doi.org/10.1007/978-3-030-23169-9_19

Keywords Unmanned aerial vehicles · Big data · Real-time data analytics · GPS · GIS · Plant phenomics · Internet of everything · Internet-of-things

1 Introduction

The world's population is expected to increase by one-third until 2050. Majority of those additional two billion people will be living in the developing countries. It is expected that more people will be living in large population centers and big cities. If the current trends continue then the agricultural production will have to increase by 60% to meet the demands for food and feed by 2050 (Faurès et al. 2013). Therefore, farming industry has become ever more important than before. In this scenario, agricultural transformation from conventional to technology based, can play a pivotal role to feed growing global population and provide the foundation for economic growth and poverty reduction. Challenge is to sustainably increase the food production to provide food as well as economic opportunities in both rural and urban communities (Ahmad and Farooq 2010).

The progress in modern agricultural industry transformation can be traced back to decade of 1960s. The decadal progress in agriculture sector is provided in Table 1. In mid-1960s, the breakthrough in wheat and rice production in Asia brought up the famous Green Revolution that symbolized the process of using agricultural science to develop modern techniques for the developing world. It started from Mexico with the "quiet" wheat revolution and in 1960s and 1970s India, Pakistan and the Philippines received world attention for their agricultural progress. Over the past four decades in Developing Asia, the irrigated area has more than doubled – to 176 million hectares (mha). Fertilizer consumption has increased more than 30-fold and tractor in use has increased from 0.2 to 4.6 million that can be considered as a positive. What the world would have been like without the technological advances that have occurred, had the global grain yields of 1950s still prevailed in twenty-first Century we would have needed more than 1.8 billion ha additional land to equal the current global harvest (Borlaug 2002).

An unprecedented systemic transformation is needed at a speed and scale to meet the current and future challenges of food security. The future agricultural systems

Table 1 Decade wise agricultural technology innovation and advancement during the past 60 years

Year	Event
1960–1970s	Green revolution
1980–1990s	Farm chemicals, mineral fertilizers, mechanization
1990–2000	Biotechnology, knowledge (quality standards, traceability, SPS, etc)
2000–2010	Precision agriculture and related technologies
2010-to date	Big data, I.o.T, greater automation

need to be more productive, use inputs efficiently, have less variability and greater stability in their outputs. Moreover, the changing climate also needs the agricultural systems to be more resilient to, risks, shocks and long-term climate variability. The objective of reducing inputs and production costs as well as increasing yield and profitability can be addressed by changing the farming methods from conventional to conservation. However, more productive and more resilient agriculture can be practiced through a major shift in the way land, water, soil nutrients and genetic resources are managed for enhanced resource use efficiency. This transformation can also improve producers' access to markets and a transformed agricultural system can reduce greenhouse gas emissions per unit of production and increase carbon sinks.

The real transformation has been experienced by the advent of Information technology (I.T). It has brought up a revolution; the world is becoming ever more connected through it. The building blocks needed to make I.T based Agriculture a reality, has started to come together in the past few years. The estimated sale of I.T based agricultural gadgets has been increase manifolds as estimated by BI Intelligence (2015) and depicted in Fig. 1. We now have the ability to run complex computations at huge scales: a 100 times increase in computing power. The development of cloud computing resources and technologies can allow us to run calculations across hundreds of machines and turn that back into one simple actionable solution. A shift in progress of technology delivered smart devices those are smaller, faster and cheaper. While computing and sensor technologies are in use in the developed world for the last two decades, their application in agriculture has not been fully utilized. Smart digital services have the potential to help agricultural industry meet its productivity and sustainability challenges. Overall, technologies can be utilized effectively to solve indigenous agricultural problems of the developing nations.

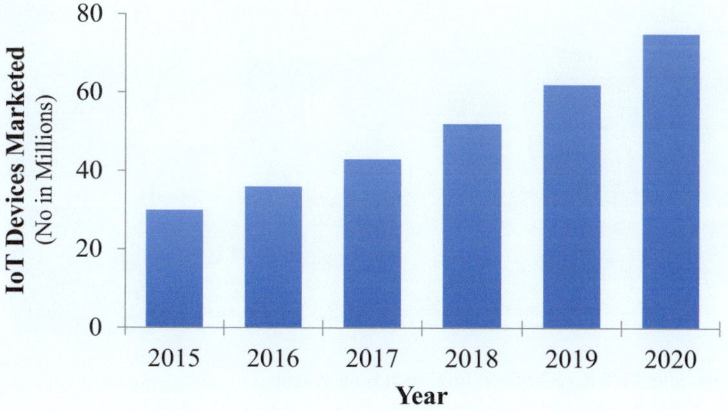

Fig. 1 Agricultural IoT devices marketed in various countries. (Estimates based on BI Intelligence, 2015)

2 Big Data in Agriculture

The world is becoming digitized. We now can know more about the land that we farm than ever before. The availability of geospatial data archives and real-time data sets from satellites and Unmanned Aerial Vehicles (UAVs) in combination with weather data, digitized soil data, and other real-time data streams coming from in-situ smart sensors can give us a better understanding of the interaction between crops, weather and soil that we would never had before, in ways that are cheaper and faster (Fig. 2). All that bulk of data (Big Data) along with computational power is worthless unless we can turn it into a simple actionable solution that we can put in the hands of the farmer.

Over the past years an increasing trend towards data-driven agriculture has been identified among farmers. According to an estimate provided by BI Intelligence (2016), It is estimated that 0.65 million data points are being generated every day from the agricultural farms that will reach up to 0.65 million. It is expected that more than 2.3 million data points will be generated by 2030 as many growers are pushing big data solutions to obtain qualitative and quantitative data to improve decisions regarding agricultural management and to increase their crop yields.

Data in agriculture can be big in many ways. For example, in case of determining spatiotemporal variation of actual crop water use in irrigated areas at regional scales, a large number of Landsat scenes will cover the area every 16 days, where each Landsat scene occupy a storage space of approximately 1000 MB, and if approximately 20 images were usable for each year of study and for every scene. It makes

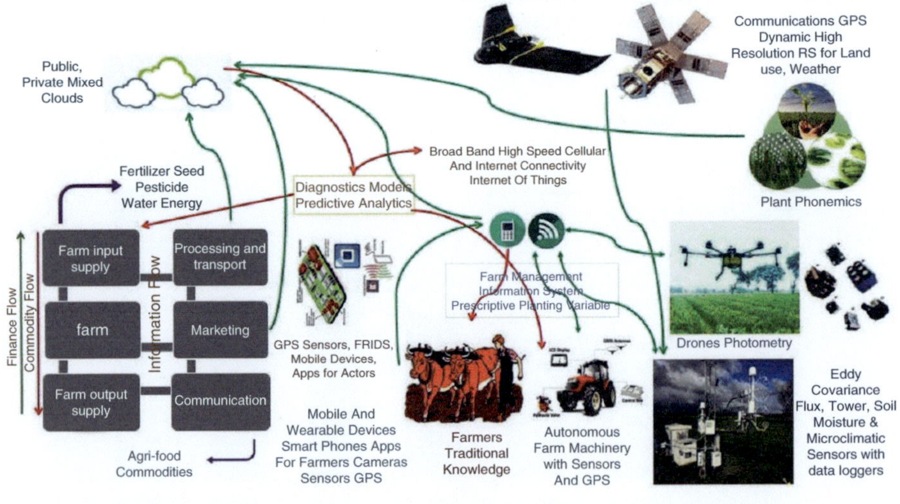

Fig. 2 Conceptual I.T based agricultural data flow system

a total more than 500 images per year (for a typical regional scale study) and more than 1500 images in a 3-year study. For that case, satellite-imagery (one of the input-information) only amounts to more than 15 terabytes, whereas the products generated during the course of scientific work (evapotranspiration in this case) are 10 times more than the original input data. Similarly, a flight of unmanned aerial vehicle covering agriculture area of one village (approx. 60 squares) for multispectral and thermal sensing generated 1.25 TB of raw data, that will be multiplied accordingly for additional flights during the growing season for pattern recognition, change detection and crop yield estimation through image processing and analytics. These projects are two examples of the work being carried out in digital and precision/climate-smart agriculture that is challenging the state of the art in computation, data storage and sharing. The climate-smart agriculture's reliance on I.T. and digital data extends far beyond these examples.

Data can also be big because of its lasting significance agro-ecological zones, soil survey maps, or the observation of other unique events. Also, data can be big because of descriptive challenges that may require contextual explanation and metadata. Because digital data can easily be shared and replicated and so re-combinable, they present tremendous reuse opportunities, accelerating investigations already under way and taking advantage of past investments in science. To encourage and enable reuse, data must be well preserved in workable formats. In some cases, data loss is economic loss (e.g. experiment have to be re-run), or it may also be an opportunity lost forever. Data are assets, and to achieve the greatest pay-off from these investments, researchers and institutions should document and implement data-management and data-sharing plans that address the full life cycle of data. The life cycle management of scientific data presents many challenges and opportunities. The challenges are great, but they can be solved by focused efforts and collaboration between funders, institutions and scientists. To facilitate data reuse, data disciplinary standards should be encouraged that will ultimately open the doors for development of disciplinary repositories for specific classes of data accessible and manageable through specialized software tools. The US National Institutes of Health (NIH) genetic sequence database (GenBank), the US Department of Agriculture (USDA) cropland database (CropScape), the US Geological Survey (USGS) land-cover and water database (GLOVIS) and the US National Virtual Observatory (USVO) are good examples of what is possible here.

The repository of big data help to optimize inputs and increase crop yield thus helping farmers to get better agricultural produce. The historic trends of particular crop at every field can helpful in better predicting harvesting potentials. The evidence-based analysis of this big data can assist agricultural experts to predict and monetize crop yields while the farmers can get information regarding their crops and marketing places.

3 Data Analytics Management and Forecasting

3.1 Data Analytics

The analytics of the big data can add value to lives of farmers and sustainability of the farms. Farmers cannot make sustainable farming choices based exclusively on large amounts of data collected at farm level. In addition to that, they also have limited time to digest large amount of data to find trends and/or irregularities. Therefore, it is difficult for them to gain actionable insights from all the crop sensors, drones and automated tractors. A farm management tool or software with the right blend of data and machine learning can help a farmer quickly decide which crops (down to the level of the plant) need insecticide, and then apply the chemicals spraying in a way that limits environmental impact and increase resource use efficiency. Moreover, it can also provide an opportunity to the banks and other monetary institutions to better assess the agricultural loans and risks associated with it. For example, in Pakistan, a pilot has been successfully tested to analyze the farmer's ability to pay back loans to the banks. The data on land records; crop and soil health gathered by the multi spectral cameras on board UAV provided sufficient information on farmers' pay back ability. The analytics of the data helped in improving farmers' income and productivity through provision of financial & technical services and adoption of good practices.

Data analysis will be complex for more challenging problems of the agricultural system as well as at the individual farm level. However, with more involvement of deep learning and artificial intelligence in agriculture, it will gradually become possible to find the exact causes of low productivity and making real time decisions to take actions.

3.1.1 Real Time Data Analytics

An alternative term Operational Intelligence (OI) is also used for Real-Time Data Analytics (RDA), a method that tracks the behavior of live systems, integrating streaming data with customer needs. RDA is implemented in situations requiring insight on a daily, frequent basis and enables farmers to analyze what's happening on the farm and take immediate actions and see the results of their actions. RDA complements intelligent decision making by providing insight into new unstructured and semi-structured data in real-time. It handles Big Data in ways intelligent decision-making cannot.

The RDA assisted streaming analytics can help agribusinesses make functioning decisions, adapt to the business environment and serve customers better when it really matters. RDA refers to a class of analytics that uses real-time data processing and data visualization into the processes and events while they occur. It is well equipped to address the integral challenges of big data by continuously monitoring and running query analysis against a range of high velocity and high volume big data sources. It associates and analyzes large volumes of (real time) streaming data arriving from diverse sources in order to link related events, discover problems, and reveal opportunities intelligently, that need immediate attention.

3.1.2 Intelligent Decision Making

The intelligent decision making in agriculture is based on structured and static data collected from different sources to identify long-term trends in historical data of days, weeks, or even months long. Data generated by smart sensors collected at farms, on the field or during transportation offer a wealth of information about soil, seeds, livestock, crops, costs, farm equipment or the use of water and fertilizer. Data analytics in agriculture, can help farmers analyze real time data of weather, soil moisture, market trend or prices, GPS signals from farm machines and provide insights on how to optimize the resources to increase yield. It can further improve farm planning to make smarter decisions about the level of resources needed and their distribution strategy in order to prevent waste. Daily operations can be made more efficient with improved system performance. Additionally, the risk of natural mistakes can be reduced as management has a clear picture of what aspects are working successfully and those which can be classified as waste.

The intelligent decision making has variety of uses in agricultural industry in particular as forecasted data, environmental impact, competitive advantage and waste reduction. The Agri-business Intelligence organizations have deployed many such dedicated analytic platforms including data accelerators, data appliances, and cloud based solutions to speed up the performance. These technologies may still not deliver true real-time data for informed decision-making, but they are enabling the update of data far more frequently and deliver answers to queries sooner.

Taking everything into consideration, the intelligent decision making in smart farming leads to decreased expenditure and increased profits. The improved production through precision agriculture along with tailored marketing for sales using informed intelligence through smart farming to identify trends could lead to an extremely efficient agricultural business.

3.1.3 Tools for Big Agricultural Data Analysis

As discussed earlier, the real-time data analytics depends on advanced technologies. All models used to process big data need to be fault-tolerant, scale with data, and have flexibility in the use of resources. High performance computing platforms that can provide real-time operation in seconds and milliseconds are of critical importance. The events need to be handled as they arrive for real-time insights, so high performance should characterize the chosen solution. Fresh streaming data can be pooled with archived data to support decision-making effectively and accurately. For example, the parallelism of Hadoop is excellent in batch processing of collected data but it has high latency. However, integration of a streaming data platform for continuous data analytics and streaming integration, real-time and interactive processing requirements presents a challenge that can be solved with further research in this domain.

Data needs to be collected, processed, stored, and then finally operated for analytics including machine learning. In a typical streaming architecture data is gathered from several smart sensors that can be spread across a geographical area in an agricultural system. Then a distributed funnel mechanism is needed to put data into a set of available servers. Chen and Zhang (2014), reviewed technological solutions for data streams processing and open-source real-time processing systems, including Hadoop Online, S4, Storm, Flume, Spark streaming, Kafka, Scribe, HStreaming, Impala, and relevant messaging technologies. Despite the diversity the real-time systems are very similar. However, to come up with an efficient framework for gathering, processing and analyzing Big Data in agriculture in a near real-time or even better in a real-time perspective further research is required.

3.2 Data Management and Forecasting

Every industry has their own experiences and challenges when it comes to dealing with big data and advanced analytics. Agriculture is no different, but progress is being made. An excellent example of using big data analytics to support sustainable farming practices is a research conducted by CIAT (International Center for Tropical Agriculture) in Colombia with assistance from Colombian government and CGIAR Research Program on Climate Change, Agriculture and Food Security. The CIAT research used data from FEDEARROZ (National Federation of Rice Growers in Colombia) to learn the exact causes of shrunken rice crop yields between 2007 and 2012. By analyzing past data and considering climate predictions, they made site-specific recommendations for crop variety and exact planting dates that are projected to increase yield to numbers that surpassed previous rice harvests. These site-specific interventions to reduce inputs and increase yields are referred to the term precision farming or interchangeably Precision Agriculture (PA).

Real-time assessment of operational conditions (e.g. weather or disease alerts) and to carry out agile actions require reconfiguration features. These features usually include intelligent assistance in implementation, maintenance and use of the technology. For that, a PA grower must utilize information gathered from a series of smart devices and systems that communicate via the Internet to study and observe crop fields. These smart devices may be, temperature and soil moisture sensors, GPS farming apps and modules for automated machinery, satellite data, as well as drones used for aerial imaging. Where precision agriculture only takes in-field variability into account, the smart farming enhances PA by analyzing management tasks not only on location but also on data, enhanced by contextual and situational awareness, triggered by real-time events (Wolfert et al. 2014).

4 Precision Agriculture

Lack of information at right time is causing massive losses to the farmers' income and agricultural sustainability. The variability of soil from one field to the other, the variability in weather and crops that are planted, means the difference in optimal input for the field can vary three to four times from one field to the next, from 1 year to the other. So as a farmer, there is no way other than to put together these complex variables and come to a solution. Unfortunately, majority of our farmers stick back to the practices they know. They fall back to what is familiar or conventional and so majority of them barely having enough food to feed their families.

Nowadays, farming practices are being supported by biotechnology, remote sensing, cloud computing and Internet-of-Things (IoT) (Li 2018) or Internet of Everything (IoE) (Dey et al. 2015). Internet of Everything refers to the intelligent connection of farmers/people, processes, data and things. IoE builds on top of the IoT, enhancing the power of the Internet to improve business outcomes. Rapid developments in the IoT and Cloud Computing are making smart farming possible. It is helpful in managing farms and crops database including farm location and size, cultivated area, inputs, time of sowing and harvesting as well as yield. Development of crop information management system can help the farmers to make quality and timely decisions. A system with a common repository that is able to collect, manage, analyze and disseminate the quality decision based on the data streams coming from different sources including, satellites, UAVs, in-situ sensors, and ambient weather conditions. Such big repository needs to be developed associated with high performance computing powers that can handle the flow of big data.

Sustainably producing the right quantity and quality of food to take on food security challenge can be enabled by technologies. Invent of artificial intelligence and I.T have opened new horizons in the field of agriculture. Now work is being carried out to make real time agricultural robots and auto steering technologies to reduce labor requirements. Advanced precision agriculture technologies that deploy machine vision, big data analytics and advanced robotics could allow farmers to apply the optimal amount of inputs for each crop and assist with the management of livestock and aquaculture, thereby boosting yields and reducing water use and greenhouse gas emissions.

To produce the highest yield with the least impact on the environment and use less water, resources and chemicals, while feeding our increasing population is a huge challenge. Now, new innovative technologies are taking on that challenge. These modern tools provide farmers with a collection of data about the status of the soil, insects, crops, livestock, water, and weather, etc. The ability to collect such detailed data has the potential to revolutionize agriculture and move farming toward more efficient and sustainable practices.

The PA concept is well adopted in the various developed countries of the world. For example, it started in USA and Europe in early 1980s and early 1990s

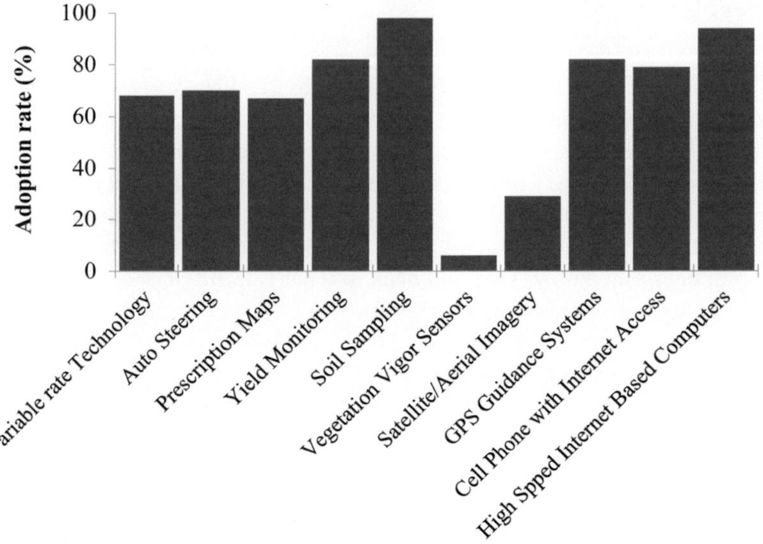

Fig. 3 Precision agriculture technologies adoption in Nebraska, USA. (Source: University of Nebraska-Lincoln)

respectively. Figure 3 provides an insight of adoption rate of various technologies in Nebraska, USA. The spatial variability of soils was mapped and yield monitors were developed to map the spatial yield variability. The researchers further developed precision seed drills and variable rate applicators for spot application of seed, fertilizer and pesticides.

The agricultural systems with on-board sensors were developed for automated fertilizer spreading and agro-chemical spraying. Use of Unmanned Arial Vehicle for agriculture (AgUAV) is also a hot topic of research and these are being used for crop monitoring, yield mapping and agrochemical spraying. Japanese are working extensively on sensor development, digital and hyper spectral image processing on-board UAVs. Information technologies platforms are in use for predictions of optimized local farming practices for the farmers. They provide information to the participant farmers on crop and soil health, weather conditions, socio-economic characteristics, labor and inputs availability and other related variables (Cheema et al. 2018). The flow of information is well described by Gebbers and Adamchuk (2010) and summarized in Fig. 4.

Information technology can be used in providing a precision agriculture package by developing e-farm production system based on Precision Agriculture techniques, Crop and livestock management (RFIDs), precision irrigation applications, Crop water and pest/disease management, Wireless moisture sensing networks, Wireless communication in UAVs used for vegetation health detection, Rainfall monitoring system based on mobile communication data, Cloud services for knowledgebase on soils, nutrients, yields by making soil, nutrient and yield maps and disseminating through mobile networks and variable rate application based on GPS and GIS systems.

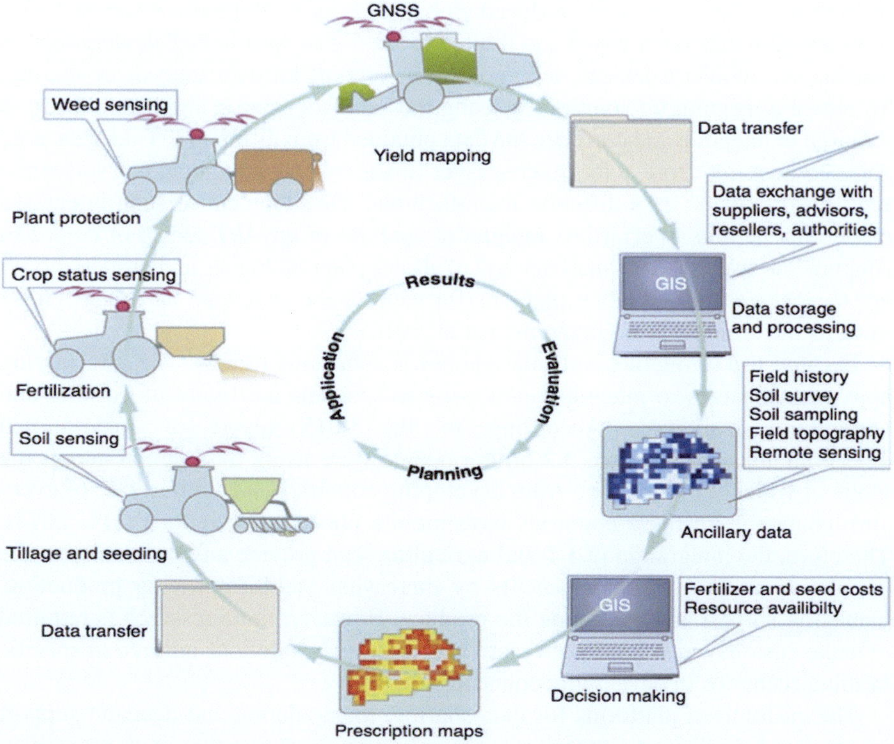

Fig. 4 Precision agriculture information flow for a cropping season. (Source: Gebbers and Adamchuk (2010))

5 Communication and Information Sharing

The innovations in Information Communication Technology (ICT) brought changes in many sectors yet agriculture and food systems have been slow to benefit from these innovations. There exists a great potential to harness the power of digital and I.T. It can promote timely as well as evidence-based decisions to improve the entire agriculture sector of the developing nations. Awareness and capacity building, better planning and community involvement is needed for agricultural breakthroughs.

Since ICT particularly the IoT and related big data analytics provide electronic monitoring of crops, as well as related environmental, soil, fertilization, and irrigation conditions. Such a monitoring data can be used to identify which crop varieties will meet the challenge enhanced productivity at the particular farm anywhere in world. The plant phenomics (an area of biology concerned with the measurement of phenomes—the physical and biochemical traits of organisms—as they change in response to genetic mutation and environmental influences) is used to characterize the crop varieties. Therefore, association of monitoring and related data analysis results with specific crop varieties (i.e., plant genes and phenotypes) will revolutionize the way food is produced globally. Majority of the smart farming solutions are currently point based and limited to the use of specific IoT devices (e.g., a specific soil moisture sensor), with no further support for data analysis or sharing. To obtain a meaningful solution by using these IoT devices, a significant effort is required to integrate and correlate the data obtained from different IoT devices, e.g., data sets acquired from, a fertilizer sprayer (made by one manufacturer), soil moisture sensors (made by a different manufacturer). Development of such integrated model that permits a) effortless integration and use of any IoT sensor or device b) supports the scalable data analytics and c) allows plant biologists and farmers/growers to analyze and visualize plant performance data, will take us a step closer towards the ultimate goal (Jayaraman et al. 2016).

Because use of internet and smartphones is enhancing rapidly. The data sharing apps with linkages to integrated platforms and models are future of farming and agricultural marketing. According to the 2015 report of International Telecommunications Union, 3.2 billion people were using the Internet across the globe of which 2 billion were from developing countries and similarly 92 of every 100 inhabitants in these countries have mobile phone subscriptions (ITU 2015). Therefore, the integration of I.T and agriculture can provide an opportunity to the farmers to attain maximum benefits by increasing yield, improving production, managing harvest and marketing the produce. However, more research is required to make customized solution and platforms to provide solutions tailored to specific farming scenarios in developing countries.

The customized platforms for data sharing, manipulation and decision support are the key here. Where farmers can remotely monitor their equipment, crops livestock, stats about feeding and produce as well as market trends and information. That technology can even assist them to run prediction analysis for the crops, livestock and market. One of the biggest names in farming equipment John Deere is using integrated platform to connect its self-driving farm vehicles to the internet and display data of crop yields (BI 2015). All of these techniques are making data driven farming a reality. More and more data sources either in-situ, on-farm, low altitude or satellites can provide useful information to improve production, minimize inputs and environmental services from agriculture.

6 Conclusion

The world is on the crossover of a technological revolution in the agriculture industry. The future of farming is in collecting and analyzing big data. Advanced analytics and precision agriculture will be the key to harness the technology and convert knowledge to farming practices in order to maximize efficiency. IoT based technologies with the involvement of modern instruments (sensors, computers, UAVs and satellites) have the potential to systematically change the traditional system of farming towards a low input, high efficiency and sustainable farming. These technologies will inevitably prove essential for taking on, what would likely be an insurmountable challenge, sustainably increasing our food supplies.

The application of information technologies in agriculture wills not only help farmers to improve economy but also help the country to regulate overall economy and trade. Current wave of mobile technology proliferation in rural areas could provide opportunity to the farmers to improve agricultural productivity based on decisions made backed by better information based on big data analytics. With just a click on a smartphone, farmers can remove the guessing from their daily work and make the best choices that benefit their business and the sustainability of the land.

References

Ahmad M, Farooq U (2010) The state of food security in Pakistan: future challenges and coping strategies. Pak Dev Rev:903–923

Business Insider (BI) Intelligence (2015) The internet of things: examining how the IoT will affect the world, Technical report. Business Insider, New York. November 2015

BI Intelligence (2016) The internet of everything. Report. Available at http://www.businessinsider.com/intelligence/bi-intelligence-iot-research-bundle

Borlaug NE (2002) The green revolution revisited and the road ahead. Nobelprize. org, Stockholm

Cheema MJM, Mahmood HS, Latif MA, Nasir AK (2018) Chapter 8: Precision agriculture and ICT: future farming. In: Khan IA, Khan MS (eds) Developing sustainable agriculture in Pakistan. CRC Press/Taylor & Francis Group, Boca Raton, pp 33487–32742

Chen CP, Zhang CY (2014) Data-intensive applications, challenges, techniques and technologies: a survey on big data. Inf Sci 275:314–347

Dey S, Saha JK, Karmakar NC (2015) Smart sensing: chipless RFID solutions for the Internet of everything. IEEE Microw Mag 16(10):26–39

Faurès JM, Bartley D, Bazza M, Burke J, Hoogeveen J, Soto D, Steduto P (2013) Climate smart agriculture sourcebook. FAO, Rome, p 557

Gebbers R, Adamchuk VI (2010) Precision agriculture and food security. Science 327(5967):828–831

ITU (2015) Key ICT indicators for developed and developing countries and the world. Available at: www.itu.int/en/ITU-D/Statistics/.../ITU_Key_2005-2013_ICT_data.xls. Accessed 7 Apr 2016

Jayaraman PP, Yavari A, Georgakopoulos D, Morshed A, Zaslavsky A (2016) Internet of things platform for smart farming: experiences and lessons learnt. Sensors 16(11):1884

Li Y (2018) Problems of application of internet of things in agricultural informationization and recommendations for Shandong Province. Asian Agric Res 10(2):51–54

Wolfert J, Sørensen CG, Goense D, (2014) A future internet collaboration platform for safe and healthy food from farm to fork, global conference (SRII), 2014 annual SRII. IEEE, San Jose, pp 266–273

Spatializing Crop Models for Sustainable Agriculture

Fabrizio Ginaldi, Sofia Bajocco, Simone Bregaglio, and Giovanni Cappelli

Abstract Crop models mathematically represent dynamic point-scale interactions between plant, weather, soil and management practices. They have been increasingly applied large scale (i.e. from farm-level to regional and global applications) to understand and quantify the trade-off between productivity, management and the sustainability of cropping systems, in terms of responsible use of resources (e.g. water and nitrogen) and of adaptation to or mitigation of climate change impacts. This contribution reviews the most recent information about spatializing crop models and provides a comprehensive overview of major assumptions and criticalities related to this methodological approach. The first paragraph focuses on the definition of crop models, presenting their historical evolution and main fields of application. A bibliometric analysis was carried out on 1017 scientific papers published between 1990 and 2018 in order to identify the most frequent scientific topics concerning the adoption of crop simulation modelling for sustainable agriculture. The second section describes the main sources of uncertainty in spatializing crop models, addressing two main aspects. Firstly, basic assumptions and validity domains of processes/phenomena represented may still not be valid when applied in a different spatial resolution. Secondly, reference input data needed to characterize the cropping system under study, to run models and test their performance at large scale can often be scarce and/or uncertain due to aggregation/disaggregation issues. The third section defines the minimum amount of data about environment (i.e. site, weather, soil), management (e.g. sowing and harvest date, cultivars and crop operations adopted) and crop type, needed to operate crop models at a given location under current/future climate scenarios. Necessary methodological indications for building a multi-layer georeferenced database facilitating coupling with biophysical models are also provided. Ways of integrating proxy variables (e.g. obtained from

F. Ginaldi (✉) · S. Bregaglio · G. Cappelli
Research Centre for Agriculture and Environment, Council for Agricultural Research and Economics, Bologna, Italy
e-mail: fabrizio.ginaldi@crea.gov.it

S. Bajocco
Research Centre for Agriculture and Environment, Council for Agricultural Research and Economics, Roma, Italy

M. Farooq, M. Pisante (eds.), *Innovations in Sustainable Agriculture*,
https://doi.org/10.1007/978-3-030-23169-9_20

pedo-transfer functions and remote sensing data) and crop models have been reported. The last section presents two case studies dealing with the spatialized application of crop models to promote the sustainability of agriculture. A European case study is centred on the definition of farmer adaptation strategies to alleviate climate change impacts, while a regional case study evaluates the efficiency of water management and water footprint of tomato cultivation in Southern Italy.

Keywords Biophysical models · Crop model uncertainty · Cropping system sustainability · Model data requirement · Remote sensing assimilation · Spatially-explicit modelling · Spatial-temporal scale

1 Crop Simulation Models in a Nutshell

The main challenge of global agriculture is the need to enhance crop productivity to guarantee food security (Ray et al. 2013), whilst at the same time achieving cropping system sustainability (Ramankutti et al. 2018). Crop simulation modelling is being widely used to support this objective because of its ability to quantify the complex, non-linear and mutual interactions between the crop genotype, farmer management and pedo-environmental conditions, thus permitting evaluation of the environmental and economic performance of an agricultural system (Li et al. 2015). The main requirement of a crop simulation model is the capability to reproduce the functioning of the target cropping system – intended here as the nexus of land, atmosphere, and human processes (Malek et al. 2017) – in order to provide a simplified representation of its behaviour and reactions to variations of farmer management and pedo-climatic conditions. Our dissertation focuses on dynamic crop simulation models (Jones et al. 2017a, b), which are software applications embedding algorithms, which are meant to reproduce the functioning of the different domains of the agricultural systems, with model outputs as the values of state variables over time (e.g., soil water content, leaf area index, aboveground biomass). This branch of science was born in the era of the Cold War and space exploration, when new technologies in computer science and knowledge about system analysis were initially employed to analyse and reproduce the interactions of components in complex systems (Sinclair and Seligman 1996). The first crop models provided a simple estimation of crop productivity as a function of light interception and photosynthesis, through the adoption of empirical relationships considering the basic biochemical and biophysical mechanisms converting solar energy into plant biomass (Loomis and Williams 1963; de Wit 1965; Duncan et al. 1967). The complexity of crop models suddenly increased over the following decades, thanks to a better description of carbon assimilation which considers the effect of stomatal conductance in regulating leaf gas exchange (e.g., Cowan 1978), and the description of plant phenology and its influence on the partitioning of assimilates among plant organs. The addition of these various components led to a number of models of daunting complexity, such as GOSSYM (Whisler et al. 1986), CERES (Ritchie and Otter 1984), and SOYGRO (Wilkerson et al. 1983). In more recent years, the focus of crop modelling has shifted from the assessment of crop productivity to the integrated analysis of the system, in order to tackle new challenges such as the mitigation of greenhouse gas emissions, the enhancement of ecosystem services and environmental performance of agricultural systems, loss reduction associated to pest and disease, improvement of the qualitative aspects of crop production, design of improved ideotypes, and the adaptation and mitigation of climate change impact. Examples of dynamic models for cropping systems are those in the DSSAT suite of models (Jones et al. 2003), and APSIM (Keating et al. 2003), CropSyst (Stöckle et al. 2003), and EPIC (Williams et al. 1983, 1989).

A bibliometric analysis of the adoption of crop simulation modelling for sustainable agriculture, from 1990 to 2018 (1017 documents), groups the most frequent scientific topics into three clusters: (i) water resource and its corresponding domain (i.e. irrigation, water balance, precipitation, etc.); (ii) climate change, economic benefit and food security, both at field and global scale; (iii) bioenergy, biomass production, soil organic carbon and greenhouse gases (Fig. 1). Crop models are

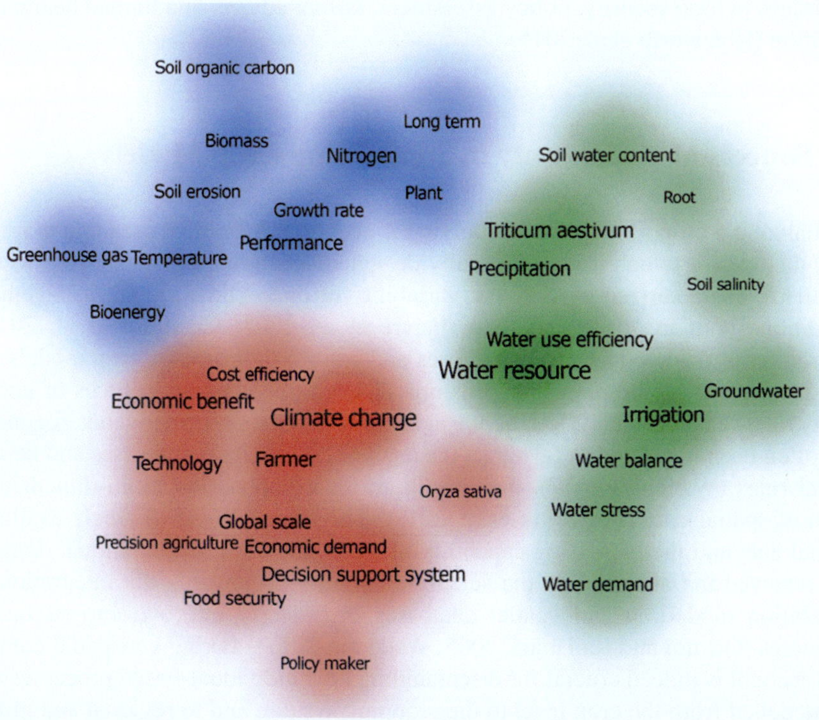

Fig. 1 Density mapping of the terms most used in the crop modelling research for sustainable agriculture (1990–2018). Clustering is determined by topic co-occurrence. The larger the halo, the higher the frequency of the topic. Analysis performed with VOSViewer (http://www.vosviewer.com)

increasingly applied on a large scale, from farm-level applications to regional and global studies, to investigate the influence of global trends such as market dynamics and climate change, on crop productivity (Porwollik et al. 2017). Yield gap analyses with crop models have been performed on different spatial and temporal scales in the context of food security, land use and climate change research (Mueller et al. 2012; Challinor et al. 2014; Asseng et al. 2015). Topics associated with crop model projections have been widely discussed, including those attributed to climate forcing data (Rosenzweig et al. 2014), model structure and parameterization (Rötter et al. 2012), and the effectiveness of CO_2 fertilization (Deryng et al. 2016). Research projects and analyses focused on four main staple food crops: wheat (*Triticum aestivum* L.), rice (*Oryza sativa* L.), maize (*Zea mays* L.), and soybean (*Glycine max* (L.) Merr.). These crops had been listed in the Global Gridded Crop Model Intercomparison (GGCMI) project as Priority 1 crops; they represent key agricultural goods given the global harvest area they cover, production quantities, trade levels, and contribution to human diet (Porwollik et al. 2017).

Nowadays, a large number of crop models are available but little emphasis has been placed on their improvement. As a result, great untapped potential in model development still remains, and filling this gap would contribute to tackling emerging issues in food security, policy assessment, farmer advice, and human health and nutrition (Holzworth et al. 2015).

2 Sources of Uncertainty in Spatializing Crop Models

Spatializing a crop model means applying it over a geographical area that is larger than the one for which it was originally designed (i.e. a homogeneous area within the field), characterized by a higher variability of pedo-climatic and management conditions both in space and time (Faivre et al. 2004; Challinor et al. 2018). Therefore, as basic concepts, hypotheses and validity domains of crop models are usually derived on the plot scale, this upscaling implies various sources of uncertainty (Hansen and Jones 2000; Faivre et al. 2004). In the past, spatial heterogeneity has often been neglected in favour of the analysis of temporal processes and behavioural rules (Wallentin 2017), mainly due to the complex and multi-dimensional form of spatial data (Porwollik et al. 2017). The large amount of freely available spatial data and the increasing capability of managing high computational demand has renewed the attention of the scientific community towards the integration of simulation modelling and spatial databases in order to improve current model resources (Grimm and Railsback 2005; Wallentin 2017). Taking geospatial context into account is indeed crucial for disentangling how individual-based processes can be modelled from the crop level to the cropping system, and to regional and global scales (Manson and O'Sullivan 2006). Goodchild (2001) defined four criteria for a model to be spatially explicit: (i) it has to depict the location of its inputs; (ii) its design has to involve concepts like spatial configuration and neighbourhood; (iii) the outputs vary if the model is run in different locations, and; (iv) the spatial structure of model input and output is different (Wallentin 2017). Spatializing crop models requires making assumptions concerning the selection of the crop model(s) to be applied, the handling of the input data, and the design of the simulation experiments (Donatelli et al. 2012). The resulting decisions set the limits of applicability of the analysis results, and should be considered *a priori*, to avoid introducing conceptual errors.

Once the aim of the study (i.e. what exactly the model should do) and the conditions of applications (e.g. spatial and temporal scale, data availability) are defined, the next step of a simulation study is the choice of model (Donatelli and Confalonieri 2011). The recommended criteria when selecting the crop model are (i) structure (modelling approaches, equations and parameter values); (ii) time step; (iii) feasibility for use in spatially-explicit applications, and; (iv) data handling capacity.

A crop model is a simplified representation of the real system; the suitability for a specific study is subject to its ability to simulate the processes that drive the aspects of cropping systems which are the target for analysis. The relationships coded in the model equations have some level of empiricism, but that empiricism has to be enclosed into one or more levels below the level of the prediction (Acock and Acock 1991) as highlighted in the scheme of the organizational levels in a cropping system (Fig. 2).

The choice of an appropriate time step in dynamic simulations depends on the processes being simulated and its temporal resolution must be short enough to allow capturing variations of the system (e.g. if a crop can wilt irreversibly in a week, a monthly time step cannot be used, Donatelli and Confalonieri 2011). In crop- and cropping-system models the time step is frequently 1 day (i.e. all processes are simulated every day), or 1 h or even less.

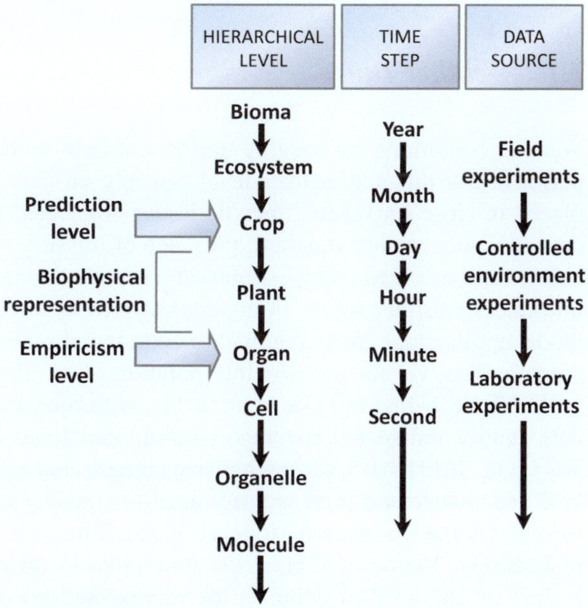

Fig. 2 Level of prediction and level of empiricism in process-based models. (Redrawn from Acock and Acock 1991)

A crop model can be run, independently or dependently from one simulation unit (SU) to another. In the first case, no interactions among SUs exist. Such an assumption is plausible for coarse spatial resolutions, e.g. 25 × 25 km or lower, in which each SU includes and represents a multitude of fields, which markedly differ in soil, nutrient, climatic, and management conditions, and whose inflows and outflows cannot be determined and quantified at simulation runtime. The consideration of local- and regional-scale heterogeneities requires an explicit simulation of the flows within the area, and the spatial interactions between SUs. This can be obtained by interfacing the crop model with a spatial one, which accounts for simulation unit interactions (e.g. a hydrological model handling lateral water flows, a farm system model accounting for constraints due to work organisation).

Besides model structure, the quality of input data (i.e. weather, soil) is one of the major sources of uncertainty in predictions (Confalonieri et al. 2016) as input variables are the main driver of crop model simulations (Grassini et al. 2015).

First of all, analysis of the input data requirement demands definition of the appropriate scale of investigation. In such a context, it is important to distinguish between the two components of scale – resolution and extent – which both significantly influence model outcomes (Wallentin 2017). While the resolution affects the detail of the information modelled, the extent defines the size of the target area: the same input data collected at different resolutions within the same area could provide markedly different model outputs (Wu 2004).

Moreover, there is often a trade-off between data quality and spatial coverage; the quality of measured data is profoundly uneven across global agricultural regions (Donatelli et al. 2012; Grassini et al. 2015; Mourtzinis et al. 2017). Here is a brief overview of the principal issues concerning specific input data:

2.1 Weather Data

Weather conditions are roughly uniform at field scale, therefore, crop simulations performed at this and/or finer level can rely on data collected at weather stations placed in close proximity. Specific issues connected to the unavailability of these data (i.e. poor quality standard, presence of missing values or absence of key variables) can be solved using estimation methods (Donatelli et al. 2004), though this introduces a further source of uncertainty (Rivington et al. 2006). Large-scale crop model applications (from regional to continental scale) require an up-scaling of the climatic data, via site-specific interpolation of weather data on a regular grid (e.g. ECMWF ERA-Interim, Dee et al. 2011), with increasing uncertainty depending on data quality and spatial coverage (Hansen and Jones 2000; Ewert et al. 2011; Van Wart et al. 2013). Also, comprehensive climate change impact assessments targeting land use change and food security need to consider climate projections from contrasting Global Circulation Models (GCM), if they are to take into account plausible realizations. Various GCMs exist and provide different projections, depending mainly on the level of detail in the representations of the global climatic system

(Parker 2010). Furthermore, while the spatial resolution of GCMs (generally hundreds of kilometres in latitude and longitude) is sufficient to simulate the average global climate, their output is often unsuitable when the scale of interest is refined. For instance, while a GCM may estimate monthly precipitation correctly, the daily precipitation may be spread across the month in a very unrealistic way (raining a little every day for example). Such distortions of daily weather variability can seriously bias crop model simulations (Semenov and Porter 1995; Mearns et al. 1996; Hansen and Jones 2000; Baron et al. 2005; van Bussel et al. 2011).

At higher resolution, several factors complicate climate modelling, including local topography, vegetation cover, land use, the presence of atmospheric aerosols and other pollutants. In order to refine the spatial and temporal resolution of GCMs to obtain inputs suitable for crop modelling at a local and regional scale, three different strategies can be pursued: (i) statistical downscaling of GCMs outputs; (ii) the coupling of GCMs with weather generators (WGs) or with; (iii) Regional Climate Models (RCMs). The first method makes use of empirical relationships between fine and coarse scale data, derived from observational data. The main concern is the assumption that current relationships in climatic data will remain unchanged in future scenarios. The second methodology involves perturbing site-specific calibrated parameter sets of the WG by the use of GCM-driven information, such as the mean changes in temperature and precipitation for a given future time slice, under a particular emission scenario. The principal drawback of this approach is that the spatial consistency of generated weather is often unpreserved. Finally, while the spatial resolution of the climate simulations is improved when coupling GCMs with RCMs (in the order of 50 km or less, Christensen and Christensen 2007), the simulation process is slow, computationally expensive, and the temporal distortion of precipitation and (to a lesser extent) temperature are still present in the generated weather series. As is well known, such series need to be bias-corrected (e.g. Christensen et al. 2008) prior to being used for feeding hydrological and crop models otherwise they may lead to unrealistic results (Teutschbein and Seibert 2010). Bias correction in turn requires ground-based observations, and may be limited by their unavailability, poor quality or heterogeneity (Challinor et al. 2003). Furthermore, as pointed out by Duveiller et al. (2015), so-obtained GCM-RCM weather projections are still inadequate for crop simulation and need to be further processed: global solar radiation and wind speed may still have unrealistic distributions when compared to observed data, whereas specific input variables needed for running crop models are still not present in available databases (e.g., evapotranspiration).

2.2 Soil Data

Running crop models with high resolution soil data (e.g., texture, depth, slope) could enable performing more accurate and detailed spatial simulations, albeit requiring large computational and time costs. Water-limited crop model simulations are indeed sensitive to soil parameters derived from soil texture and soil depth, as they determine the basic hydraulic characteristics. However, since the coverage of soil profiles and the quality of information available in public databases is not often uniform over large-scale simulation areas, the predominant soil profile within each SU is often used as a proxy, even though this approximation could lead to a marked underestimation of output variability. In fact, soil-type-related yield variability could outweigh the simulated inter-annual variability in yield due to weather under specific management conditions (i.e. unfertilized cropping systems, Folberth et al. 2016). This concept does not apply when performing crop model applications within a frame of precision agriculture, where the variability in pedological features must be explicitly considered to support the application of spatially variable rates of fertilizers, irrigation water, or chemical treatments within a single field (Sadler et al. 2016).

2.3 Production System

In crop model simulations at large spatial scales, the production systems are often abstracted at "crop" level, ignoring the local farm typologies and cropping system structures. Rather than using crop growth models, it is more appropriate to tackle the issue by modelling cropping systems. When aimed at being used as supporting tools to design adaptation strategies for farmers, these studies must explicitly consider the impact of alternative management strategies. Indeed, farmers are able to timely respond to environmental changes modifying management practices. Basic practices which can be simulated by most of the current crop models are the choice of the variety, which is coded in the model mainly with respect to the crop cycle length, the shifting of the sowing dates based on weather conditions, and the implementation of alternative irrigation and fertilization plans. The capability of handling crop rotations, even if representing a more complex task, is nowadays necessary for a crop model. Crop modelling studies must test different management scenarios in order to anticipate future trends in crop productivity as affected by farmer management in order to identify possible solutions to better adapt to climatic changes.

2.4 Model Calibration

The calibration of the parameters of a crop model is often based on the adjustment of their values within their biophysical ranges, in order to improve the model's performance in reproducing field experimental data. This activity is supported by literature, which makes available reference values of the main parameters used by many crop models in different trials over large areas. Such parameter sets need to be refined through interaction with local experts and stakeholders when spatial simulations are carried out, in order to improve the information on the actual cropping systems simulated as opposed to the idealized crop types. However, the results of spatialized simulations are able primarily to identify main trends and to capture extensive regional signals, and need to be interpreted with caution since they are sensitive to the specific model settings.

3 Data Requirements for Spatializing Crop Models

Spatializing crop models requires information on the heterogeneity of pedo-climatic conditions and management practices within the simulated geographic region; such data need a level of detail that is consistent with model requirements (e.g. model time step versus weather data resolution) and the study's objectives (Donatelli and Confalonieri 2011). While in temperate and flat areas soil can represent the main source of yield variability (regardless of scale; Hoffmann et al. 2016), thermal and pluviometric patterns drive crop production in rainfed agriculture or in morphologically complex hillside areas, where land slope and aspect are substantial in determining yield levels and fruit quality (e.g. vineyards; Esteves and Manso Orgaz 2001).

Basic layers of information for spatialized crop modelling studies are mainly related to pedo-environmental conditions and management practices (Hunt and Boote 1998; Kasampalis et al. 2018). The minimum dataset of information (MD) for running crop models depends on the simulated production level (i.e. potential (P) vs water (WL)/nitrogen (NL)/disease (DL) limited) (Table 1).

The potential level represents the productivity of a crop grown under non-limiting conditions for water, nutrients, weeds and pest/diseases pressure, under prevailing environmental conditions. It is determined by incoming solar radiation, air temperature, atmospheric CO_2, and by genetic traits that modulate e.g., the length of the growing season and light interception (canopy structure) (van Ittersum et al. 2013). In water/nutrient depleted production systems, additional data are necessary in order to set soil conditions (i.e. water and nitrogen content) at model initialization (Müller et al. 2017).

It is to be noticed that, as the simulation scale increases, performing a detailed model calibration/validation with field measurements or gridded datasets becomes more difficult. This is because spatially distributed information related to main phenological variables and growth dynamics (leaf area index, aboveground biomass) is rare (Faivre et al. 2004; Müller et al. 2017). In this context, gridded yields or official sub-regional/national statistics should be approached with extreme caution since such data can include disturbing factors that are yet to be considered by the model or are of unknown origin e.g. technological/time trend, nitrogen shortage/surplus, hailstorms events (Donatelli and Confalonieri 2011).

It is our aim to obtain a representative set of data in areas with significantly different pedo-climatic and management conditions (Coucheney et al. 2015), whilst integrating available experimental data with literature information (e.g. modelling experiments performed in the same or similar environments).

Table 1 Minimum amount of input data for operating crop growth models (Hunt and Boote 1998), according to the production level of interest

Layer	Data	Production level
	Environment	
Site	Latitude, longitude, elevation	P, WL, NL, DL
	Slope and aspect	
	Land use and crop distribution maps	
	Irrigation maps	WL, NL
	Pest/diseases distribution maps	DL
Weather	Daily global solar radiation	P, WL, NL, DL
	Maximum and minimum air temperature	P, WL, NL, DL
	Precipitation, potential/actual evapotranspiration	WL, NL, DL
	Wind speed	WL, NL, DL
	Relative air humidity	WL, NL, DL
	Vapour pressure deficit	WL, NL
	Leaf wetness	DL
Soil	Soil type, soil depth, bulk density, texture	WL, NL
	Organic carbon, pH, soil nitrogen	NL
	Initial water, ammonium and nitrates by soil layer	WL, NL
	Management	
Farming practices	Cultivar, planting date, depth and method, row spacing, plant density	P, WL, NL, DL
	Irrigation and water management (dates, methods and amounts)	WL, NL
	Fertilizer applications (dates, methods and amounts)	NL
	Residue (organic fertilizer) applications (material, depth of incorporation, amount and nutrient concentrations)	NL
	Chemical (e.g., pesticide) applications (material, amount)	DL
	Harvest schedule	P, WL, NL, DL

P potential, *WL* water-limited, *NL* nitrogen-limited, *DL* disease-limited

Model application in climate change impact assessment requires additional data concerning future climate projections (i.e. GCM-RCM based realizations – e.g. CMIP – https://www.wcrp-climate.org/wgcm-cmip, CORDEX – http://www.cordex.org/, projects), atmospheric CO_2 concentration and adaptation strategies (e.g. adoption of improved varieties and/or more efficient irrigation systems, chemical applications) (Donatelli et al. 2015; Challinor et al. 2018).

From an operational point of view, most existing simulation studies rely on the outputs of crop models coupled with databases containing MDs in the areas of interest (Fig. 3). To this end, the study area (e.g. a region, an agrozone, a producing district, a watershed, etc.) is divided into a finite number of smaller areas called simulation units (SUs), characterised by homogeneous pedo-climatic conditions

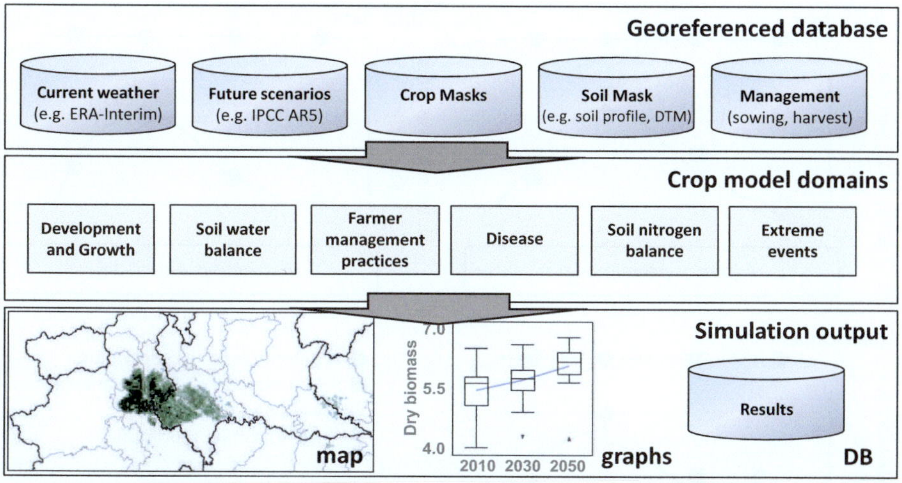

Fig. 3 Schematic representation of the simulation environment

and/or cropping/farming systems (Faivre et al. 2004). Then, MDs are georeferenced in a multi-layer and spatially explicit database and univocally assigned to each SU depending on spatial attributes (e.g. geographic coordinates of centroids), via the use of GIS-based software applications.

Environmental information often relies on data of different resolutions which need therefore to be aggregated/disaggregated before being used via dedicated procedures (e.g. based on weighted average, interpolation, selection of the most representative unit/class of data within the SU) (Faivre et al. 2004; Hoffmann et al. 2016). If unavailable, information can be replaced by proxy variables, as those obtained by pedo-transfer functions (as in the case of hydrological properties computed from soil texture; Donatelli and Confalonieri 2011) or by the use of remote sensing data (Faivre et al. 2004; Kasampalis et al. 2018).

Information related to crop management is highly variable and discontinuous in space and can change from year to year according to farm, consortium, regional, national or supranational decisions (Faivre et al. 2004). Thus, the spatial characterization of such data is rare and the available information does not permit characterization of the system in a detailed way (Müller et al. 2017). The tools available to fill these gaps are literature search, expert opinion or remote sensing data. During the execution of the crop model, management practices can then be implemented through automatic rules triggering the occurrence of agricultural operations based on management decisions (e.g. scheduled events for sowing and irrigation) and/or some states of the system (e.g. irrigation starts when available soil water in the root zone is lower than a critical threshold) (Donatelli et al. 2016).

Remote sensing collects spatial information regularly, with wide coverage and low cost, and therefore has been an advantageous tool for the detection of natural and agricultural resources over the last decades (Kasampalis et al. 2018). Several ways of integrating remote sensing data and crop models have been suggested

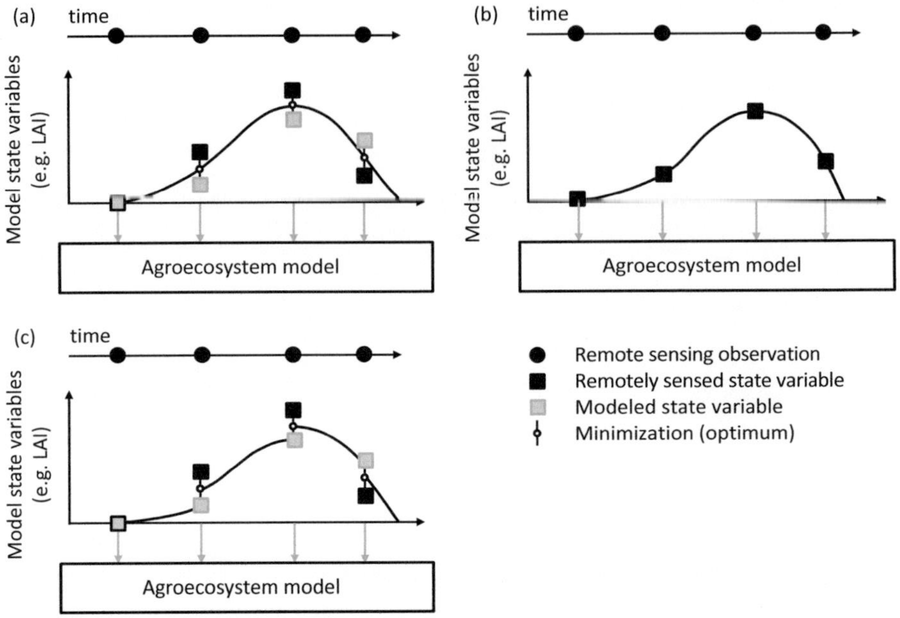

Fig. 4 Schematic representation of different methods for the assimilation of remotely sensed state variables in agroecosystem models: (**a**) calibration, (**b**) forcing, and (**c**) updating (Dorigo et al. 2007)

(Delécolle et al. 1992; Dorigo et al. 2007; Liang 2004), including calibration (i.e. model parameters or initial states are adjusted to obtain an optimal agreement between the simulated and the RS-observed state variables), forcing (i.e. a state variable in the model is replaced by using the observed RS data), and updating method (i.e. model state variables are continuously updated whenever a RS observation is available), as shown in Fig. 4.

Leaf area index (LAI), fractional cover (fCOVER), fraction of photosynthetically active radiation absorbed by the canopy (fAPAR), and plant chlorophyll concentration are among the most relevant canopy state variables which are commonly assimilated in crop simulation models. Remote sensing can also provide plant phenological information (Xin et al. 2002; Karnieli 2003; Dorigo et al. 2007; Bajocco et al. 2019): regular provision of the phenological crop status will markedly improve the spatial outputs of agroecosystem models (Delecolle et al. 1992; Jin et al. 2018).

Remote sensing data are available at multiple spatial scales, from sub-meter (e.g. for precision farming applications), to more than a kilometre (e.g. for regional applications), and at variable temporal resolution, from daily to twice weekly coverage. The choice of the scale and temporal resolution depends on the questions to be answered (Jones et al. 2017a). Newer satellite sensors have been launched to obtain higher spatial and temporal resolution (such as Sentinel-2, Landsat-8, RapidEye, WorldView-2, etc.). The so-called big data revolution is the framework in which the collaboration between remote sensing data and crop models can find new challenges and solutions (Kasampalis et al., 2018).

4 Application Examples of Crop Model Use in Sustainable Agriculture

This section presents two case studies which deal with the spatialized application of crop models to promote the sustainability of agriculture. The main themes of these case studies cover two out of the three clusters mapped in Fig. 1, i.e. climate change impact and the use of water resources. The common goal of these studies is the assessment of the sustainability of cropping systems under contrasting future climate scenarios; they differ both in resolution and extent of the spatial scale, the latter ranging from continental (Europe, Donatelli et al. 2015) to regional (Apulia, Italy, Ventrella et al. 2017). The European case study focused on the definition of farmer adaptation strategies to counteract climate change impacts, whereas the regional case study evaluates the efficiency of water management and water footprint of tomato cultivation in Southern Italy.

4.1 European Scale

The long-term sustainability of agroecosystems and associated livelihoods is unattainable without actively reacting to global climatic and socio-economic changes with feasible and effective adaptation strategies. Climate change adaptation is an adjustment in natural or human systems in response to real or expected climatic *stimuli* or their effects, which moderate harm or exploit beneficial opportunities (IPCC 2001). Potential adaptation policies include the improvement of technology and management practices, agro-environmental climate payments, the design of more sustainable farming systems, Common Agriculture Policy payments for agricultural practices that are beneficial to the climate and the environment ("greening" measures), the introduction of new crop varieties, land use related policies, etc.

In 2015, Donatelli et al. provided an impact assessment of climate change scenarios on agriculture over the EU27 Member States, focusing on three (20-year) time horizons centred on 2000 (baseline), 2020 and 2030. The Authors simulated water-limited yields for three priority crops (wheat, rapeseed and sunflower) and tested some technical adaptation options which could offset climate change impacts. The CropSyst model (Stöckle et al. 2003) was coupled with a georeferenced database including information on (i) land use, (ii) crop distribution, (iii) soil properties, (iv) farming practices (i.e. sowing and harvest dates) and (v) current/climate change scenarios for two contrasting realizations of the same IPCC emission scenario (A1B). Spatially-distributed simulations were executed at 25 × 25 km resolution considering crop responses to different atmospheric CO_2 concentrations, and future yield projections were evaluated as a percentage change compared to the baseline.

Results primarily showed that different realizations of the same emission scenario led to large variations in crop performances in the same time slice. Without adaptation, simulated wheat yield variations in 2030 strictly reflected the spatial

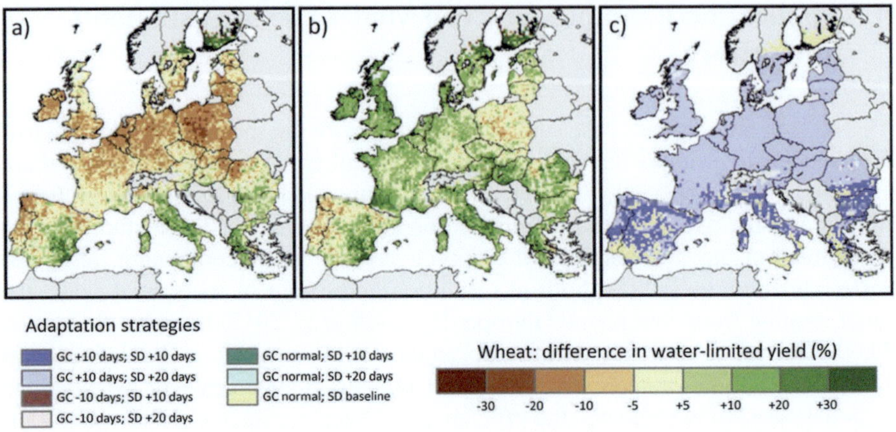

Fig. 5 Percentage change in simulated water-limited yield without (**a**) and with (**b**) adaptation measures for winter wheat in 2030 with respect to the 2000 baseline under the HadCM3 realization (Liu et al. 2013; Semenov et al. 2014) of the A1B scenario. The best adaptation strategies among all tested ones are mapped in (**c**). Adaptation strategies evaluate (i) cultivars with both a shorter and a longer growth cycle (GC) with respect to the current reference variety and (ii) delays of the baseline sowing date (SD) (Donatelli et al. 2015)

pattern of rainfall changes across Europe (Fig. 5a): indeed, projected declines in the amount of rainfall generally resulted in yield reductions and *vice versa*. Further reasons for the increase in yields in Southern Europe were the CO_2 fertilization effect and the shortening of the crop cycle that may have reduced the occurrence of water stress in summer. Simple adaptation techniques such as changes in sowing dates and varieties (in terms of duration of the crop cycle) were effective in alleviating the adverse effects of climate change in most areas (Fig. 5b). In general terms, the best yield was realized by delaying the wheat planting date by 10 days, and using a variety with a longer growth cycle (Fig. 5c; results did not account for a possible greater pressure of plant diseases, for instance wheat rusts).

Figure 5 illustrates large spatial variability in the performance of wheat systems. It enables identifying critical spots for focusing breeding and policy-making efforts, and it highlights opportunities for European wheat agriculture in future time horizons. It should be noted that taking into account the effect of future technological changes and economic consequences (e.g. costs of alternate technologies or levels of fertilizer application in response to changes in prices) would tend to further reduce adverse impacts of climate change. One aspect that requires additional investigation is the impact of extreme events which may lead to crop failure, even in the context of possibly improved weather patterns.

4.2 Regional Scale

Italy is the sixth major tomato producer and supplier of tomato processing worldwide. About a fifth of the national harvested area and productions are concentrated in the Capitanata plain, an area of about 4000 km^2 in the northern part of the Apulia region (Southern Italy). The cultivation of tomatoes in this area plays a key socio-economic role, although the achievement of high quality products is largely driven by the intensive use of chemicals (i.e. fertilizer/pesticides) and irrigation (300–800 mm), with great impact on local natural resources. The major constraint to crop growth is water stress, due to the prevalent semi-arid climate, characterized by precipitation between 25 and 110 mm and temperature peaks over 40 °C in summer. Significant action is thus needed to support tomato growers to enhance production levels while saving irrigation water. Ventrella et al. (2017) applied the DSSAT-CROPGRO model to simulate the growth of industrial tomato and to quantify the green (GW; crop evapotranspiration deriving from rain stored in the soil) and the blue water (BW; crop evapotranspiration deriving from irrigation), the blue water requirement (BWR; ratio between yield and BW) and the water footprint (WFP), under both rainfed and fully-irrigated conditions. Spatially distributed simulations covered the whole tomato area and a period of 30 years for baseline and future climate realizations (IPCC AR4 SRES A2 and A5 scenarios), based on average temperature raises of 2 and 5 °C respectively.

Future scenarios affected all indicators significantly, especially in the drier areas where high thermal and rainfall anomalies are foreseen. In general, the largest increase in BW consumption and BWR were simulated in the northern and south-eastern part of the Capitanata, where available soil water content is already a limiting factor for the crop (Fig. 6). Compared to the baseline, simulated BW under the A5 scenario showed an average increase of about 30%, while yield reductions fluctuated at about −20%. As a consequence, the BWR and WFP are projected to rise steeply to 40 and >65%, respectively. Results confirm that for a global temperature

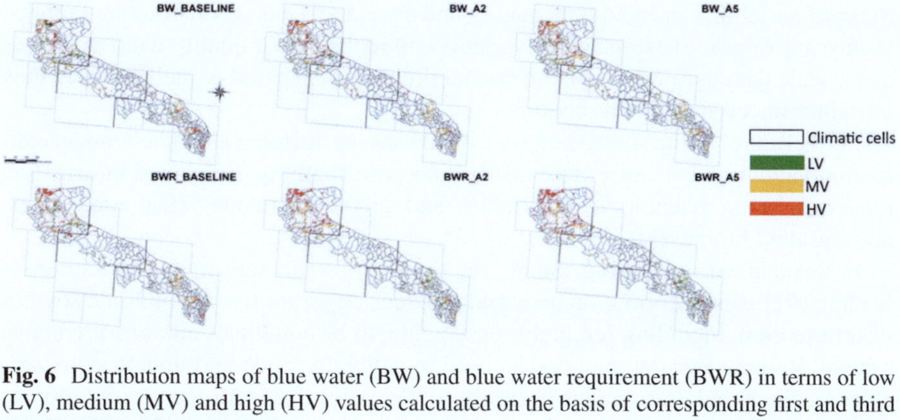

Fig. 6 Distribution maps of blue water (BW) and blue water requirement (BWR) in terms of low (LV), medium (MV) and high (HV) values calculated on the basis of corresponding first and third quartiles (Ventrella et al. 2017)

change of 5 °C potential adaptation measures may not be sufficient to counterbalance the projected negative impact on crop performance in terms of yield and WFP.

These findings could be important to support planning policies for effective allocation of scarce water resources, by concentrating them where water use efficiency is highest (i.e. highest BWR and lowest WFP). Nevertheless, future improvement of WFP simulation under climate change can be obtained by considering the CO_2 effect on stomatal conductance and therefore on crop transpiration.

5 Conclusions

This chapter primarily defines the main research topics involving the application of crop models for promoting sustainable agriculture in the last three decades: water resource management, global climate change and carbon cycle.

Then, focus moved to spatializing crop models with particular attention on defining model data requirements and describing underlying methodological concerns and constraints. In spatial modelling the choice of scale together with input data retrieval and harmonization are two of the most crucial issues to be tackled.

Since spatial simulation output is scale-dependent, the choice of the appropriate scale of analysis is fundamental. Selection of the scale of the system being modelled depends on the goal and the final beneficiary. For example, if the objective refers to the best management practices to adopt, or how to make the land more profitable, the target system should be on a field scale. At farm and larger scales, the goal is understanding how weather, soil, socio-economic factors and crop management practices affect crops and how simulation tools can easily and effectively support policymakers.

The spatialization of crop models needs to link different scales: for example, the scale on which the processes are described by the model, the scale on which input data or information (model parameters and input variables) are be available, or the scale on which output results are expected or sought. In turn, there is also a wide range of variability in input data quality and coverage. Data reconstruction (applied to missing values, to estimate key variables, to replace poor quality data) as well as cross-scale data harmonization are crucial processes in spatial modelling since they introduce uncertainty in predictions.

Furthermore, as the simulation scale increases, performing a detailed model calibration/validation becomes more difficult because spatially distributed information related to main phenological variables and growth dynamics (leaf area index, aboveground biomass) is rare.

In this context, increasing use of "big data" and smart sensors for agriculture is leading to closing information gaps and provides opportunities for multiple sources of information, including remotely sensed data, to be combined into one predictive system. Remote sensing data can be used to calibrate, force and update state variables of the simulation model in runtime. We claim that with the assimilation of model state variables (e.g. leaf area index dynamic over crop cycle, soil water content evolution) via smart sensors, their simulation processes in crop models may be no longer necessary. This practice can be adopted in high-resolution *in-season* simulation studies, provided that output accuracy is preserved.

The final unaddressed issue concerning spatial modelling is a technical one, and regards whether or not current crop/cropping system models are adequate to implement the concepts of spatially explicit simulation modelling discussed so far. The challenge lies in the integration of two complementary toolsets: agent-based models and Geographic Information Systems (GIS). Spatial simulation workflows often make use of GIS in the preparation of spatial input data and in outcome visualization, whereas the model is used to handle spatially-distributed dynamic simulation of biophysical processes.

References

Acock B, Acock MC (1991) Potential for using long-term field research data to develop and validate crop simulators. Agron J 83(1):56–61

Asseng S, Ewert F, Martre P, Rotter RP, Lobell DB, Cammarano D, Kimball BA, Ottman MJ, Wall GW, White JW, Reynolds MP, Alderman PD, Prasad PVV, Aggarwal PK, Anothai J, Basso B, Biernath C, Challinor AJ, De Sanctis G, Doltra J, Fereres E, Garcia-Vila M, Gayler S, Hoogenboom G, Hunt LA, Izaurralde RC, Jabloun M, Jones CD, Kersebaum KC, Koehler AK, Muller C, Naresh Kumar S, Nendel C, O'Leary G, Olesen JE, Palosuo T, Priesack E, Eyshi Rezaei E, Ruane AC, Semenov MA, Shcherbak I, Stockle C, Stratonovitch P, Streck T, Supit I, Tao F, Thorburn PJ, Waha K, Wang E, Wallach D, Wolf J, Zhao Z, Zhu Y (2015) Rising temperatures reduce global wheat production. Nat Clim Change 5(2):143–147

Bajocco S, Ferrara C, Alivernini A, Bascietto M, Ricotta C (2019) Remotely-sensed phenology of Italian forests: going beyond the species. Int J Appl Earth Obs 74:314–321

Baron C, Sultan B, Balme M, Sarr B, Traore S, Lebel T, Janicot S, Dingkuhn M (2005) From GCM grid cell to agricultural plot: scale issues affecting modelling of climate impact. Philos T R Soc B 360:2095

Challinor AJ, Slingo JM, Wheeler TR, Craufurd PQ, Grimes DIF (2003) Toward a combined seasonal weather and crop productivity forecasting system: determination of the working spatial scale. J Appl Meteorol 42:175–192

Challinor AJ, Watson J, Lobell DB, Howden SM, Smith DR, Chhetri N (2014) A meta-analysis of crop yield under climate change and adaptation. Nat Clim Change 4(4):287–291

Challinor AJ, Mueller C, Asseng S, Deva C, Nicklin KJ, Wallach D, Vanuytrecht E, Whitfield S, Ramirez-Villegas J, Koehler A-K (2018) Improving the use of crop models for risk assessment and climate change adaptation. Agric Syst 159:296–306

Christensen J, Christensen O (2007) A summary of the PRUDENCE model projections of changes in European climate by the end of this century. Climatic Change 81:7–30

Christensen JH, Boberg F, Christensen OB, Lucas-Picher P (2008) On the need for bias correction of regional climate change projections of temperature and precipitation. Geophys Res Lett 35:L20709

Confalonieri R, Orlando F, Paleari L, Stella T, Gilardelli C, Movedi E et al (2016) Uncertainty in crop model predictions: what is the role of users? Environ Modell Softw 81:165–173

Coucheney E, Buis S, Launay M, Constantin J, Mary B, Garcia de Cortazar-Atauri I (2015) Accuracy, robustness and behavior of the STICS soil–crop model for plant, water and nitrogen outputs: evaluation over a wide range of agro-environmental conditions in France. Environ Modell Softw 64:177–190

Cowan IR (1978) Stomatal behaviour and environment. Adv Bot Res 4:117–228. Academic Press
de Wit CT (1965) Photosynthesis of leaf canopy. Agr Res Rep (Pudoc, Wageningen, Netherlands) 663:1–57
Dee DP, Uppala SM, Simmons AJ, Berrisford P, Poli P, Kobayashi S et al (2011) The ERA-Interim reanalysis: configuration and performance of the data assimilation system. Q J Roy Meteor Soc 137(656):553–597
Delecolle R, Maas SJ, Guérif M, Baret F (1992) Remote sensing and crop production models—present trends. ISPRS J Photogramm Remote Sens 47:145–161
Deryng D, Elliott J, Folberth C, Müller Pugh CTAM, Boote KJ, Conway D, Ruane AC, Gerten D, Jones JW, Khabarov N, Olin S, Schaphoff S, Schmid E, Yang H, Rosenzweig C (2016) Regional disparities in the beneficial effects of rising CO2 concentrations on crop water productivity. Nat Clim Chang 6(8):786–790
Donatelli M, Confalonieri R (2011) Biophysical models for cropping system simulation. In: Flichman G (ed) Bio-economic models applied to agricultural systems, Springer, p 59–87.
Donatelli M, Carlini L, Bellocchi G, Fila G (2004) CLIMA: a software component for weather generation and estimation. Proc. 8th ESA congress, July 11–15, Copenhagen, Denmark, p 239–240.
Donatelli M, Duveiller G, Fumagalli D, Srivastava A, Zucchini A, Angileri V, Fasbender D, Loudjani P, Kay S, Juskevicius V, Toth T, Haastrup P, M'barek R, Espinosa M, Ciaian P, Niemeyer S (2012) Assessing agriculture vulnerabilities for the design of effective measures for adaptation to climate change (AVEMAC project). EUR – scientific and technical research series
Donatelli M, Srivastava AK, Duveiller G, Niemeyer S, Fumagalli D (2015) Climate change impact and potential adaptation strategies under alternate realizations of climate scenarios for three major crops in Europe. Environ Res Lett 10:075005
Donatelli M, Bregaglio S, Stella T, Fila G (2016) Modelling agricultural management in multi-model simulation systems. In: Ewert F, Boote KJ, Rotter RP, Thorburn P, Nendel C (ed) Crop modelling for agriculture and food security under global change, proceedings of the international crop modelling symposium, 2016, pp 15–17 March 2016, Berlin
Dorigo WA, Zurita-Milla R, de Wit AJ, Brazile J, Singh R, Schaepman ME (2007) A review on reflective remote sensing and data assimilation techniques for enhanced agroecosystem modeling. Int J Appl Earth Obs 9(2):165–193
Duncan WG, Loomis RS, Williams WA, Hanau R (1967) A model for simulating photosynthesis in plant communities. Hilgardia 38:181–205
Duveiller G, Donatelli M, Fumagalli D, Zucchini A, Nelson R, Baruth B (2015) A dataset of future daily weather data for crop modelling over Europe derived from climate change scenarios. Theor Appl Climatol (3–4):573–585
Esteves MA, Manso Orgaz MD (2001) The influence of climatic variability on the quality of wine. Int J Biometeorol 45:13–21
Ewert F, van Ittersum MK, Heckelei T, Therond O, Bezlepkina I, Andersen E (2011) Scale changes and model linking methods for integrated assessment of agri-environmental systems. Agr Ecosyst Environ 142:6–17
Faivre R, Leenhardt D, Voltz M, Bnoit M, Papy F, Dedieu G, Wallach D (2004) Spatialising crop models. Agronomie 24:205–217
Folberth C, Skalský R, Moltchanova E, Balkovič J, Azevedo LB, Obersteiner M, Van Der Velde M (2016) Uncertainty in soil data can outweigh climate impact signals in global crop yield simulations. Nature communications 7:11872
Goodchild M (2001) Issues in spatially explicit modeling. Special workshop on agent-based models of land use/land cover change. Center for Spatially Integrated Social Science, Irvine
Grassini P, van Bussel LGJ, Van Wart J, Wolf J, Claessens L, Yang H, Boogaard H, de Groot H, van Ittersum MK, Cassman KG (2015) How good is good enough? Data requirements for reliable crop yield simulations and yield-gap analysis. Field Crop Res 177:49–63

Grimm V, Railsback SF (2005) Individual-based modeling and ecology. Princeton University Press, Princeton

Hansen JW, Jones JW (2000) Scaling-up crop models for climate variability applications. Agr Syst 65(1):43–72

Hoffmann H, Zhao G, Asseng S, Bindi M, Biernath C, Constantin J et al (2016) Impact of spatial soil and climate input data aggregation on regional yield simulations. PLoS ONE 11(4):e0151782

Holzworth DP, Snow V, Janssen S, Athanasiadis IN, Donatelli M, Hoogenboom G, White JW, Thorburn P (2015) Agricultural production systems modelling and software: current status and future prospects. Environ Modell Softw 72:276–286

Hunt LA, Boote KJ (1998) Data for model operation, calibration, and evaluation. In: Tsuji GY, Hoogenboom G, Thornton PK (eds) Understanding options for agricultural production. Springer, Dordrecht, pp 9–39

IPCC (2001) Climate change 2001: third assessment report, vol I. Cambridge University Press, 2001, Cambridge

Jin X, Kumar L, Li Z, Feng H, Xu X, Yang G, Wang J (2018) A review of data assimilation of remote sensing and crop models. Eur J Agron 92:141–152

Jones JW, Hoogenboom CH, Porter KJ, Boote WD, Batchelor LA, Hunt PWW, Singh U, Gijsman AJ, Ritchie JT (2003) The DSSAT cropping system model. Eur J Agron 18(3–4):235–265

Jones JW, Antle JM, Basso B, Boote KJ, Conant RT, Foster I, Godfray HCJ, Herrero M, Howitt RE, Janssen S, Keating BA, Munoz-Carpena R, Porter CH, Rosenzweig C, Wheeler TR (2017a) Brief history of agricultural systems modeling. Agric Syst 155:240–254

Jones JW, Antle JM, Basso B, Boote KJ, Conant RT, Foster I, Godfray HCJ, Herrero M, Howitt RE, Janssen S et al (2017b) Toward a new generation of agricultural system data, models, and knowledge products: state of agricultural systems science. Agric Syst 155:269–288

Karnieli A (2003) Natural vegetation phenology assessment by ground spectral measurements in two semi-arid environments. Int J Biometeorol 47:179–187

Kasampalis DA, Alexandridis TK, Deva C, Challinor A, Moshou D, Zalidis G (2018) Contribution of remote sensing on crop models: a review. J Imaging 4:52

Keating BA, Carberry PS, Hammer GL, Probert ME, Robertson MJ, Holzworth D, Huth NI, Hargreaves JNG, Meinke H, Hochman Z, McLean G, Verburg K, Snow V, Dimes JP, Silburn M, Wang E, Brown S, Bristow KL, Asseng S, Chapman S, McCown RL, Freebairn DM, Smith CJ (2003) An overview of APSIM, a model designed for farming systems simulation. Eur J Agron 18:267–288

Li T, Hasegawa T, Yin X, Zhu Y, Boote K, Adam M, Bregaglio S, Buis S, Confalonieri R, Fumoto T, Gaydon D, Marcaida M III, Nakagawa H, Oriol P, Ruane AC, Ruget F, Singh B, Singh U, Tang L, Tao F, Wilkens P, Yoshida H, Zhang Z, Bouman B (2015) Uncertainties in predicting rice yield by current crop models under a wide range of climatic conditions. Glob Change Biol 21:1328–1341

Liang NS (2004) Estimation of land surface biophysical variables. In: Kong GA (ed) Qualitative remote sensing of land surfaces, Wiley series in remote sensing. Wiley, New York, pp 246–309

Liu J, Fritz S, van Wesenbeeck CFA, Fuchs M, You L, Obersteiner M, Yang H (2013) A global and spatially explicit assessment of climate change impacts on crop production and consumptive water use. Global Planet Change 64:222–235

Loomis RS, Williams WA (1963) Maximum crop productivity: an estimate. Crop Sci 3:67–72

Malek K, Stockle C, Chinnayakanahalli K, Nelson R, Liu M, Rajagopalan K, Barik M, Adam JC (2017) VIC–CropSyst-v2: a regional-scale modeling platform to simulate the nexus of climate, hydrology, cropping systems, and human decisions. Geosci Model Dev 10:3059–3084

Manson S, O'Sullivan D (2006) Complexity theory in the study of space and place. Environ Plan A 38(4):677–692

Mearns LO, Rosenzweig C, Goldberg R (1996) The effect of changes in daily and interannual climatic variability on CERES-Wheat: a sensitivity study. Climatic Change 32:257–292

Mourtzinis S, Edreira JIR, Conley SP, Grassini P (2017) From grid to field: assessing quality of gridded weather data for agricultural applications. Eur J Agron 82:163–172

Mueller ND, Gerber JS, Johnston M, Ray DK, Ramankutty N, Foley JA (2012) Closing yield gaps through nutrient and water management. Nature 490(7419):254–257

Müller C, Elliott J, Chryssanthacopoulos J, Arneth A, Balkovic J, Ciais P et al (2017) Global gridded crop model evaluation: benchmarking, skills, deficiencies and implications. Geosci Model Dev 10:1403–1422

Parker WS (2010) Predicting weather and climate: uncertainty, ensembles and probability. Special issue: modelling and simulation in the atmospheric and. Clim Sci 41:263–272

Porwollik V, Müller C, Elliott J, Chryssanthacopoulos J, Iizumi T, Ray DK, Ruane AC, Arneth A, Balkovic J, Ciais P, Deryng D, Folberth C, Izaurralde RC, Jones CD, Khabarov N, Lawrence PJ, Liu W, Pugh TAM, Reddy A, Sakurai G, Schmid E, Wang X, de Wit A, Wu X (2017) Spatial and temporal uncertainty of crop yield aggregations. Eur J Agron 88:10–21

Ramankutti N, Mehrabi Z, Waha K, Jarvis L, Kremen C, Herrero M, Rieseberg LH (2018) Trends in global agricultural land use: implications for environmental health and food security. Ann Rev Plant Biol 69:789–815

Ray DK, Mueller ND, West PC, Foley JA (2013) Yield trends are insufficient to double global crop production by 2050. PLoS ONE 8:e66428

Ritchie JT, Otter S (1984) Description and performance of CERES-wheat: a user-oriented wheat yield model. In: Wheat Yield Project, A.R.S (ed) ARS-38. National technical information service. Springfield, Missouri, pp 159–175

Rivington M, Matthews KB, Bellocchi G, Buchan K (2006) Evaluating uncertainty introduced to process-based simulation model estimates by alternative sources of meteorological data. Agr Syst 88:451–471

Rosenzweig C, Elliott J, Deryng D, Ruane AC, Müller C, Arneth A, Boote KJ, Folberth C, Glotter M, Khabarov N, Neumann K, Piontek F, Pugh TAM, Schmid E, Stehfest E, Yang H, Jones JW (2014) Assessing agricultural risks of climate change in the 21 st century in a global gridded crop model intercomparison. Proc Natl Acad Sci U. S A 111(9):3268–3273

Rötter RP, Palosuo T, Kersebaum KC, Angulo C, Bindi M, Ewert F, Ferrise R, Hlavinka P, Moriondo M, Nendel C, Olesen JE, Patil RH, Ruget F, Taká J, Trnka M (2012) Simulation of spring barley yield in different climatic zones of northern and central europe: a comparison of nine crop models. Field Crop Res 133:23–36

Sadler EJ, Barnes EM, Batchelor WD, Paz J, Irmak A (2016) Addressing spatial variability in crop model applications. In Agricultural system models in field research and technology transfer. CRC Press, Boca Raton, pp 258–269

Semenov MA, Porter JR (1995) Climatic variability and the modelling of crop yields. Biospheric Aspects of the Hydrological Cycle 73:265–283

Semenov MA, Stratonovitch P, Alghabari F, Gooding MJ (2014) Adapting wheat in Europe for climate change. J Cereal Sci 59:245–256

Sinclair TR, Seligman NG (1996) Crop modelling: from infancy to maturity. Agron J 88:698–704

Stöckle CO, Donatelli M, Nelson R (2003) CropSyst, a cropping systems simulation model. Eur J Agron 18(3):289–307

Teutschbein C, Seibert J (2010) Regional climate models for hydrological impact studies at the catchment scale: a review of recent modeling strategies. Geogr Compass 4:834–860

van Bussel LG, Müller C, van Keulen H, Ewert F, Leffelaar PA (2011) The effect of temporal aggregation of weather input data on crop growth models' results. Agr Forest Meteorol 151:607–619

van Ittersum MK, Cassman KG, Grassini P, Wolf J, Tittonell P, Hochman Z (2013) Yield gap analysis with local to global relevance – a review. Field Crop Res 143:4–17

van Wart J, Grassini P, Cassman KG (2013) Impact of derived global weather data on simulated crop yields. Global Change Biol 19:3822–3834

Ventrella D, Giglio L, Garofalo P, Dalla Marta A (2017) Regional assessment of green and blue water consumption for tomato cultivated in Southern Italy. J Agric Sci 156:1–13

Wallentin G (2017) Spatial Simulation: a spatial perspective on Individual-based ecology – a review. Ecol Model 35:30–41

Whisler FD, Acock B, Baker DN, Fye RE, Hodges HF, Lambert JR, Lemmon HE, McKinion JM, Reddy VR (1986) Crop simulation models in agronomic systems. Adv Agron 40:142–208
Wilkerson GG, Jones JW, Boote KJ, Ingram KT, Mishoe JW (1983) Modeling soybean growth for crop management. Trans Am Soc Agric Eng 26:63–73
Williams JR, Renard KG, Dyke PT (1983) EPIC: a new method for assessing erosion's effect on soil productivity. J. Soil Water Conserv 38(5):381–383
Williams JR, Jones CA, Kiniry JR, Spanel DA (1989) The EPIC crop growth model. Trans Am Soc Agric Eng 32:497–511
Wu J (2004) Effects of changing scale on landscape pattern analysis: scaling relations. Landscape Ecol 19(2):125–138
Xin J, Yu Z, van Leeuwen L, Driessen PM (2002) Mapping crop key phenological stages in the North China plain using NOAA time series images. Int J Appl Earth Obs 4:109–117

Index

M. Farooq, M. Pisante (eds.), *Innovations in Sustainable Agriculture*,
https://doi.org/10.1007/978-3-030-23169-9

C

D

E

F

G

H

I

T

U

V

W